Computing Archaeology for Understanding the Past

CAA 2000

Computer Applications and Quantitative Methods in Archaeology

Proceedings of the 28th Conference, Ljubljana, April 2000

Edited by

Zoran Stančič
Tatjana Veljanovski

BAR International Series 931
2001

Published in 2016 by
BAR Publishing, Oxford

BAR International Series 931

Computing Archaeology for Understanding the Past - CAA 2000

ISBN 978 1 84171 225 3

BAR Publishing is the trading name of British Archaeological Reports (Oxford) Ltd. British Archaeological Reports was first incorporated in 1974 to publish the BAR Series, International and British. In 1992 Hadrian Books Ltd became part of the BAR group. This volume was originally published by Archaeopress in conjunction with British Archaeological Reports (Oxford) Ltd / Hadrian Books Ltd, the Series principal publisher, in 2001. This present volume is published by BAR Publishing, 2016.

Printed in England

BAR
PUBLISHING

BAR titles are available from:

BAR Publishing
122 Banbury Rd, Oxford, OX2 7BP, UK
EMAIL info@barpublishing.com
PHONE +44 (0)1865 310431
FAX +44 (0)1865 316916
www.barpublishing.com

Contents

Preface

This volume contains the proceedings of the conference titled Computing Archaeology for Understanding the Past which was held in Ljubljana, Slovenia on April 18th-21st 2000. The conference was the first conference ever of two professional organisations: Computer Applications and Quantitative Methods in Archaeology (CAA) and the Union International des Sciences Préhistorique et Protohistorique (UISPP), Commission IV. While CAA started as a European professional association 29 years ago, UISPP was founded on May 28. 1931, as the successor to the Congress Paleoethnologiques International (since 1864) and the Congress International d'Anthropologie et Archeologie Préhistorique (since 1867).

The conference managed to bring together experts and members of two of the largest professional organizations in computer and quantitative methods in archaeology. Dr Lojze Marinček, Minister of Science and Technology, Republic of Slovenia, Prof. Nick Ryan, Chairman CAA and Prof. Robert Whallon President UISPP, Commission IV officially opened the conference. After the official opening, it started with a plenary session in which four distinguished scientists from USA, France, Italy and Belgium presented their views on the advances of methods and theory in quantitative archaeology. Prof. George Cowgill (USA), as one of the most distinguished members of the UISPP and leading scientist in quantitative archaeology in USA compared the history of scientific archaeology in USA and Europe. Prof. Frank Vermeulen (Belgium) and Maurizio Forte (Italy) lectured on current trends and applications of computer methods in archaeological research, analyses and presentation of cultural heritage. Finally Prof. François Djindjian (France) presented basic problems and objectives of computer applications and quantitative methods in archaeology.

After the plenary session the conference split into three parallel sessions and workshops, which were running for the entire period of the conference. Titles of the sessions were: Archaeological Regional Spatial Analyses, Presentation of Archaeological Data, National and Regional Sites and Monuments Records, Predictive Modelling, Museum Presentation of Archaeological Data, Large Areas Spatial Analyses, Documentation and Recording of Sites and Artefacts, Future Trends in Spatial Analyses, Trends in Field Survey Data Analyses, Artefact Analyses and Classification, and Intra Site Spatial Analyses. Internationally recognized experts chaired each session. We are especially proud that the number of oral contributions given by young researchers was very high.

The conference was organized in such a manner that it encouraged interaction between senior leading scientists and young researchers. Special attention was paid to encourage social interaction between conference participants. We hope that those social events were intensive enough that they presented Slovenian culture to international scientists and helped in establishing personal contacts between conference participants.

It is also important to stress that four awards were given at the conference:

- Award for Excellence in the Application of Computer and Quantitative Methods in Archaeology to Prof. Albertuus Voorrips, Faculty of Social and Behavioural Sciences, University of Amsterdam, The Netherlands.
- The Best CAA 2000 Paper Award to Dr Marcos Llobera, Institute of Archaeology, University College London, Centre for Advanced Spatial Analysis, University College London, Pitt-Rivers Museum, University of Oxford, UK, for the paper: More than Meets the Eye.
- The Best CAA 2000 Young Researcher Paper Award to Mr. Kaj Fredsgaard Rasmussen and Mrs. Henriette Günther Sorensen, University of Aarhus, Denmark, for the paper: Publishing on the Internet.
- The Best CAA 2000 Poster to Mrs. Bernarda Županek and Mr. Dimitrij Mlekuž, Ljubljana City Museum, Slovenia, for the poster: Counting the Uncountable: A Quantitative Approach to Religious Differences between the Roman Towns of Emona and Poetovio.

Papers submitted for the conference proceedings were peer reviewed. Despite peer reviewers beeing anonymous we want to acknowledge them here for their work: Jens Andresen, Matthew Bampton, Juan A. Barceló, John Bintliff, Wolfgang Börner, Stephen Bullas, Göran Burenhult, George Cowgill, Stašo Forenbaher, Maurizio Forte, Vince Gaffney, Mark Gillings, Francis Grew, Jeremy Huggett, Hans Kamermans, Keith W. Kintigh, Neil Lang, Peter Leech, Martijn van Leusen, Gary Lock, Torsten Madsen, Mark Mehrer, Umberto Moscatelli, Clive Orton, Gaetano Palumbo, John W.M. Peterson, Julian Richards, David Gilman Romano, Nick Ryan, Dean Snow, Cornelius Steckner, Philip Verhagen, Frank Vermeulen, Albertus Voorrips, Robert Whallon and David Wheatley.

We are very happy that the authors acted in most cases very quickly and corrected their texts according to the comments and advice from the reviewers. Finally, let us thank here all the colleagues and friends who helped us in conference preparation and the finalization of these proceedings: Alenka Koren, Alenka Kregar, Krištof Oštir, Tomaž Herček, Sašo Kuharič, Marko Jevnikar, Tomaž Podobnikar and Uroš Parazajda. We are particularly grateful to Patrick Sunčan Stone, Bojan Miović, Rachel Novšak and Helen Gaffney who proof-read the English texts, improving them significantly. Special thanks go to Krištof Oštir who prepared the layout and photocomposition of the proceedings.

And at the end, nothing would be possible without the financial support of the Scientific Research Centre of the Slovenian Academy of Sciences and Arts, Ministry of Science and Technology and Town of Ljubljana who supported the organization of the conference, the European Commission, Research DG, Human Potential Programme, High-Level Scientific Conferences, HPCF-CT-199-00147 and Computer Applications and Quantitative Methods in Archaeology for giving grants to the young scholars or underprivileged.

Zoran Stančič
Tatjana Veljanovski

Ljubljana, February 2001

1. Documentation and Recording of Sites and Field Survey Data

A New Technique for Recording Archaeological Excavations: Research Progress Report

Geoffrey John Avern

c/o Tamara Lublin, P&G Technical Centre

Rusham Park, Whitehall Lane, Egham, Surrey, TW20 9NW, United Kingdom

e-mail: lublin.ta@pg.com

Abstract

Currently, 3D modelling is used in archaeology for modelling terrains and artefacts, and for virtual reconstructions of buildings and complexes. It is the opinion of the author that the most significant impact of 3D modelling on archaeology is yet to be realised and will be in a different application, that of modelling excavations. This paper is a preliminary report on part of a doctoral research project on the use of high resolution 3D modelling as a means of recording excavations as a quicker, more accurate alternative to drawing.

Key words: 3D, modelling, virtual reality, VR, excavations

1. Introduction

Three-dimensional modelling using computers is not new to archaeology. From early work using terrestrial photogrammetry techniques and Computer Aided Design (CAD) software for recording old buildings, 3D modelling in archaeology has advanced to be used in startling ways. Digital Elevation Models (DEM's) are commonly used to visualise archaeological sites or landscapes. Geographical Information Systems (GIS) are being used for cost surface and viewshed analyses of 3D landscapes. Many examples exist of virtual reconstructions of ancient buildings or ritual complexes created in 3D graphics software (e.g. Forte and Siliotti 1997). The latest Virtual Reality (VR) display environments allow the viewer to immerse themselves in 3D archaeological models (Vote et al., this volume). Somewhat different from the above, another current application is the precise modelling of artefacts in 3D, allowing the creation of virtual collections or museums on the Internet (e.g. Jeffrey Clark, CAA2000 presentation).

The author would argue, however, that the greatest impact of 3D modelling on archaeology has yet to be realised. When 3D modelling becomes the standard technique for recording excavations it will quite literally change the way excavating archaeologists go about their work.

In discussing the use of VR in archaeology, Gillings (in Barcelo et al. 2000:59) states:

> "... the notion of VR models as comprising little more than ingenious "end-products" needs to be challenged. Techniques must be embedded at all levels of archaeological investigation, serving not only as sophisticated visual summaries, but also as primary recording methods, heuristic devices, and display and communication mechanisms."

And,

> "... (there is) the need to develop and adapt a number of routine field and laboratory based methodologies that can be integrated into current archaeological practise."

While we might challenge Gillings by pointing to examples which are much more than high-tech "illustrations", we would be missing his main point, that of the need for greater integration of VR, or 3D modelling, into general archaeological practise. However, we should not be surprised to find ourselves at this juncture when we reflect on how the technology has evolved.

In a market lead by 3D display (especially computer gaming) it is the aspect of display, currently so accessible and widespread, that has been the first to be exploited by archaeologists. It is only now, in the subsequent developmental stages of the general field of 3D imaging, that acquisition devices for 3D data are starting to become widely available, and at which point we can begin to consider their potential impact and their integration into archaeology. This research, then, considers these 3D acquisition devices and how we might use 3D modelling as a standard tool in an integrated archaeological recording system. The author's proposal is that archaeologists should use 3D modelling as a means of making the primary record of an excavation in place of the traditional techniques of drawing and photography. Note that the emphasis here is on high resolution, high accuracy, data-dense models of small excavated areas, complementary to, and for integration with, DEM's and GIS models of entire sites or landscapes.

2. Drawing, photography and 3D modelling

Drawing, as a means of recording an area that has been excavated, can be criticised on a number of levels. Photography also has its disadvantages. Below, we briefly discuss the shortcomings of drawing and photography, and the relative merits of 3D modelling.

2.1. Subjectivity

Drawing is subjective and interpretive. When drawing, consciously or not, we choose what data we feel is relevant and effectively ignore the rest. Yet the excavating archaeologist has the responsibility to record for posterity all the evidence which he/she is in the process of dismantling. To choose to illustrate some aspects and not others are to ignore the possible needs of future archaeologists for the sake of expediency. In an effort to record the missing data we supplement the drawings by photography, which may be fast and simple, but which has its own limitations, *viz* perspective.

The use of high resolution, data-dense modelling effectively combines the best features of drawing and photography; the spatial organisation (conveying the benefits of orthogonal drawings, but with the advantage of being able to choose from an infinite number of projections) and the high detail and colour information of photographs. It becomes possible, for example, to examine the colour information in a 3D model (given that it was colour referenced at the time of capture) and realise, in post-excavation, the ideas of James M. Newhard (CAA2000 presentation) on digital determination of Munsell colour.

2.2. Accuracy

The accuracy of drawings can, or perhaps, should always be viewed with scepticism because of the substantial potential for error. Typically, when drawing a feature we map the coordinates of a number of prominent points of the feature and join them by freehand drawing. The sources of error are many. The thickness of a pencil line on a 1:20 scale drawing (if we allow it to be 1mm) scales up to 2 cm. The freehand parts of a drawing may at least double the error. If we are drawing by quadrates, errors can creep in with each placement of the frame. Further error may creep in when the drawings are collectively re-mapped, re-scaled and inked.

The fact that many of the data points have been gathered by total station with millimetre precision should not delude us as to the accuracy of the entire drawing. We are, in effect, expending considerable effort in fusing data which are poorly compatible in terms of their accuracy.

At this point some may argue that these criticisms are not relevant since we all know the limitations of drawings and work within them. Yet this assumes we share some common, though unquantified, estimate of the reliability of the drawings. It is surely preferable to use a system of recording for our primary record of an excavation which offers greater accuracy with less effort. 3D modelling seems to offer just this.

2.3. Quality

The quality of drawings, essentially their effectiveness in communicating good information, is related to the levels of skill of the person making the drawing. Surely, a system of recording, which produces results of uniformly high quality, must be preferred.

2.4. Speed

Clearly, drawing is slow. In theory we have an obligation to record everything, but in practise we are constrained by limited time and human resources. A means of recording which is quicker, and achievable by a single person, enables us to record much more and/or in much less time. The latter is obviously of great benefit where field time is limited, especially in rescue archaeology, where time can be literally measured in terms of (usually large sums of) money.

Then there are those sites which are particularly difficult to record by drawing. From personal experience at Le Cheslé, an Iron Age fort in southern Belgium, the making of plans and elevations of the steeply-sloped fort wall have been incredibly difficult and slow, requiring constant re-levelling for each successive quadrate as work proceeded down the steep slope (itself, a difficult and slow process under the circumstances). The task would have been much simpler and faster if we had been able to "scan" a 3D acquisition device across the slope.

2.5. Visual interpretation

Archaeologists are practised and adept at interpreting the 3D world from 2D plans and elevations. Yet most will admit that a 3D model is easier to interpret and often gives them a "new perspective". Further, there are situations where plans and elevations are not so illuminating. In the above example of the wall of the Iron Age fort, the choice of point-of-view for the elevation was not simple since two-thirds of the way up the wall, the slope turned to the right as it wrapped around the bastion of the portal. The elevation thus shows much of the upper part of the wall in almost profile view. Another good example is the large group of Neolithic flint mines at Spienne in southern Belgium. A plan of the surface would simply show a large number of circular mine entrances. For any single mine we might make any number of different section views, no single view adequately describing the highly irregular shape of the main chamber. Obviously, viewing and interpreting these two examples would be much easier using a 3D model on computer, especially with the active stereo-viewing technology using Liquid Crystal Shutter (LCS) eye-wear which has been available for at least a decade.

2.6. Utility of data

While speed of recording is a great attraction of 3D modelling, an equally attractive feature is the subsequent utility of the recorded data.

When starting from a 3D model, the considerable time spent in post-excavation in reworking records and diagrams is dramatically reduced and the preparation of reports greatly facilitated. Having criticised drawing above, I will be the first to defend it when publishing on paper. A line drawing to illustrate a particular point, needs display only the data relevant to that point and not other distracting, obfuscating or irrelevant data. It is a relatively simple process to create such line drawings from 3D models. For example, to create a plan view from a 3D model, one selects a vertical viewpoint, changes to an orthogonal projection of the model, and processes the resultant image by thresholding, binarisation and gradient filter for a line-drawn plan of the excavation, in a matter of a couple of minutes.

Photographs, too, have a place in paper publications. Yet a 3D model will supply 2D images too, and with the flexibility of choosing your particular point of view. And further, by simply changing the projection of the model on the computer screen from perspective to orthogonal, an orthophoto has been created.

However, when considering the utility of the data of a 3D model, by far the greatest potential benefit is its ability to be integrated with other digital data. Once we have modelled, for example, the walls and floor of an excavated trench and geo-referenced a number of points, we can accurately place the high resolution, data-dense model into our less data-dense DEM site plan. Such a composite model can be used with GIS, itself, a much simpler and useful means of managing site records than folders of drawings and albums of photographs. Further, GIS allows effective integration of other forms of digitised data by its ability to link to multimedia databases, e.g. to Stratigraphic Unit datasheets, to finds databases

Figure 1: 3D models of a simulated surface of an excavation (approx 70 x 70cm) created by Eyetronics ShapeSnatcher; rendered surface model (left) and textured model in black and white (right).

which include weights and measures, pictures or 3D models of the finds, reports from analytical labs, etc, etc. In effect, by scanning excavations onto the computer at high resolution, we have perhaps the last step required to realise a "total recording system" for excavations.

3. Capturing 3D data

There is an ever-increasing number of devices on the market for recording in 3D. They use a variety of range-finding technologies including stereo-photogrammetry[1], active stereometry (e.g. laser line triangulation), passive stereometry (structured light techniques) and "time-of-flight" methods (e.g. sonar and scanned laser-distance-measuring devices). They have been designed with different applications in mind ranging from metrology and reverse-engineering to a simple means of incorporating complex 3D shapes into computer games. In considering which devices are suitable for our specific application the following five criteria have been applied:

1. *Speed.* Clearly, we want to be able to model the excavation in less time than it would take us to draw it.
2. *Accuracy.* We want a model which is at least as accurate as our drawing would be.
3. *Ease of Use.* A device and its software should be simple and intuitive to use, and not require us to employ a specialist for the recording work.
4. *Suitability for Use in the Field.* The device must be sufficiently portable and robust for use in the typical situations in which we conduct our excavations.
5. *Cost.* If our hope is that the majority of archaeologists will use such a device to record their excavations, they must obviously be able to affordable it.

On applying these criteria to the devices currently on the market we find that almost all, with the exception of stereo-photogrammetry, are quick and accurate. The majority are simple to use, though there are a few whose metrological applications sees them sold with quite complex (though comprehensive) software, making them less user-friendly.

It is their suitability for use in the field which eliminates most devices from consideration for our use. Many systems are large, designed to be fixed in place in the corner of a laboratory or workshop, and certainly not moved around an excavation site. There are also a number of small systems for use on the desktop but which have such small coverage or field of view (e.g. 20 cm) as to render them ineffective for our application.

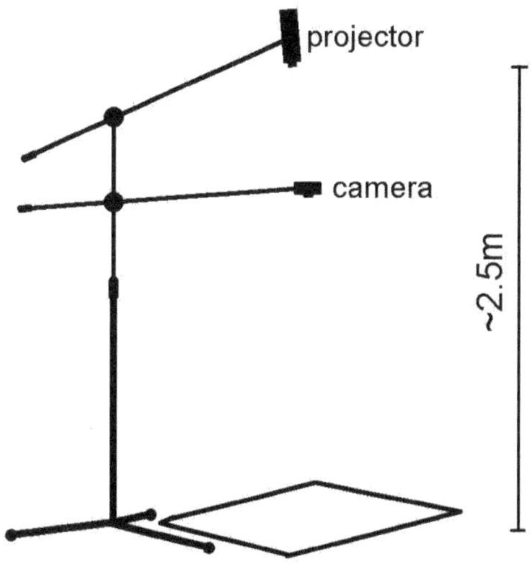

Figure 2: Proposed light-weight portable frame for mounting camera and "projector" for capturing images for 3D modelling by Eyetronics ShapeSnatcher.

4. 3D Acquisition devices

After applying four of the five criteria we are left with perhaps four systems, each deserving at least a little discussion, including that of the final criterion of cost.

4.1. ShapeSnatcher

ShapeSnatcher from Eyetronics is a system using the Structured Light principle. It comes as a program on a CD, packaged with a photographic slide and a small (approximately 20 cm high) calibration box. Other equipments required are a computer, a slide projector and a camera (video or digital still cameras are preferred). The slide is in fact a clear field carrying a very fine, black-lined grid which, using the projector, casts the grid pattern onto the subject. The subject is photographed and then, without moving camera or projector, a second photo is taken of the calibration box placed in front of the subject.

Starting with the calibration image, the software searches for the grid against the known background of the calibration box, and the relevant geometry of the camera-projector system is calculated and saved as a calibration file. The software then addresses the image of the subject, detecting the grid and, with reference to the calibration file, calculating the 3D positions of the grid nodes which become the apices of the resultant wireframe model. Texture data is taken from between the grid lines in the original image and interpolated for those parts obscured by the grid lines. The resulting 3D "patch" can be merged with other patches using an supplementary program, ShapeMatcher. Figure 1 shows a 3D model created by ShapeSnatcher, of some "mocked-up" archaeology (it was done in winter) of some pieces of early sixteenth century Flemish sculpture[2] placed in sand. It was created from three photographs by Eyetronics.

One of the advantages of ShapeSnatcher is that the components required for capturing the unprocessed images are simple and inexpensive. It would be easy to fabricate a more compact "light source and lens" substitute for the slide projector, which might even be battery-powered. Further, a frame such as that in figure 2, based on a microphone stand, not only makes the system portable and easy to relocate but, since it fixes the positions of the "projector" and camera, only one calibration image is required for each group of images taken with the one configuration of the system.

A further advantage of ShapeSnatcher is its versatility, by virtue of the fact that it is scalable. By using different projector lenses and/or grids of different coarseness (available from Eyetronics), the system can be used on a range of subject sizes, including small finds.

The greatest "advantage" or, perhaps, attraction of this system is its cost. At 200,000 Belgian Francs (very roughly US$4500) commercial price, this is by far the cheapest of the systems contemplated here (I understand that there is a special price for educational institutions).

The system is not without drawbacks though. ShapeSnatcher is slower than scanned laser line systems (see below), though faster than manual drawing. The process often requires user input to make refinements to the proposed model. Since it is essentially an optical system, sharp focus and depth of field are issues requiring constant attention when setting up and photographing. Also, since contrast between grid line and subject is fundamental to the detection process, working in anything but the most muted of ambient light gives problems. In practise, I suspect that the best resort is to take photographs at night. Another, perhaps lesser, disadvantage is that the modelling process is performed away from the excavation, so quick reference to the excavation to clarify details is not as simple as if we were creating a model on the spot, as would be the case using ModelMaker (see below).

In the final balance, ShapeSnatcher is quicker than drawing, it gives a good 3D model with colour texture and it is very affordable.

4.2. Metric 3D reconstruction

Eyetronics is also in the process of commercialising a system described by Pollefeys et al. (1998, 2000) whereby a number of photographs of a subject, taken without any record of position or camera parameters, can be used to create a 3D model.

In essence, the process begins by identifying a small number of homologous points in each photograph which are then used to calculate a first approximation of the projective framework (the spatial relationship between subject and points at which the photographs were taken). After a step of further refinement it is then possible to search for correspondences between the images for virtually every point in the images, giving a rigid spatial framework from which a 3D model can be computed using triangulation.

The strong point of this system is that one needs only a digital camera to gather the raw data in the field. Though not of great importance to our application, another interesting feature is that 3D models can be constructed of subjects which no longer exist, as long as there are a few extant photographs.

The drawbacks of this system are the potential for blind spots, the computing power required (I gather it is not possible on today's average desktop computer), the modelling time, and the fact that the modelling is done (probably) off-site. This system is not yet commercially available so we cannot comment on its price.

While I have reservations as to the practicality of this system for modelling excavations, it would seem that this system has enormous potential for modelling entire sites. It would be very interesting indeed to see it applied to aerial photography.

Note: In the two techniques described above the texture map for the rendered model is derived from the original images which have been captured with a point source of illumination. This means that the texture map itself will display highlights and shadows. The following device collects colour information as it scans across the surface of the subject, giving a texture map of (approximately) uniform illumination. In some modelling situations such uniformity is clearly preferable.

4.3. ModelMaker

ModelMaker from 3D Scanners employs laser stripe triangulation as its range-finding technique. Unlike other similar systems which scan the laser stripe across the subject, ModelMaker requires the user to move the unit by hand across and around the subject. The laser unit is mounted on an articulated arm, each joint of which is fitted with sensors measuring flexion and rotation which allow the computer to track the position and orientation of the laser unit. The great advantage of this device is that, while the laser stripe is only some 25 mm or 45 mm wide (depending on the model used), a much larger field can be scanned, limited only by the length of the arm (which comes in a range of lengths). Another great advantage is that, having no fixed point of view, the device can capture data in what would be blind spots to other systems.

There are essentially two disadvantages of this system. As it is sold, it is not really appropriate to take into the field, however, I feel that it may be possible to adapt it to this end. The greater disadvantage is its cost; at approximately UK£ 58,000 it hardly falls within the budget of most archaeologists.

4.4. FastSCAN

FastSCAN from Polhemus also uses laser stripe triangulation for range-finding and, like ModelMaker, the scanning is done by the operator sweeping the unit across and around the subject by hand. The location and orientation of the FastSCAN handpiece is determined by a magnetic tracking system operating between the handset and a small, fixed-position tracking unit.

The advantages are that it seems that the system is portable and robust (I have not yet used it), and well-suited for use on archaeological sites. It claims to be very accurate and very fast, and should deal with the majority of blind spots that we might encounter. Its coverage is 3 m from the fixed position tracking unit, though there is an optional unit available with 10 m range. Multiple handsets can be used simultaneously with the one tracking unit.

The one failure of this system is that it does not capture colour information to enable texturing the model with real colour data. As such, I would not normally have considered it for discussion, except that I feel that it would be a device similar to this, but one which captures colour data, which would be ideal for the excavating archaeologist. An additional shortcoming of the device is that its magnetic tracking system may be affected by proximity to metals, compromising its utility in some situations. The price of this device is approximately US$40,000.

5. Conclusions

The author considers that a high resolution, data-dense 3D model is a better primary record of an excavation than traditional drawings supplemented by photography. The arguments are that 3D modelling is faster and more accurate, and that it includes more data, whose digital format confers greater utility and potential for integration with other data.

Of the many currently available systems, application of the five criteria of speed, accuracy, ease of use, suitability for the field and cost, finds most of them clearly inappropriate for our use. Of the few remaining systems ShapeSnatcher from Eyetronics is currently the best in the light of its "performance for cost". Another system, FastSCAN from Polhemus, while currently lacking colour detection, perhaps points to a future, "ideal" 3D scanning device.

References

FORTE, M. and SILIOTTI, A. (eds.), 1997. *Virtual Archaeology: Great Discoveries Brought to Life Through Virtual Reality* (2nd ed.). Thames and Hudson Ltd, London.

GILLINGS, M., 2000. Plans, Elevations and Virtual Worlds: the Development of Techniques for the Routine Construction of Hyperreal Simulations. In Barceló, J.A., Forte, M. and Sanders, H.D. (eds.), *Virtual Reality in Archaeology,* 59-69. BAR International Series 843.

POLLEFEYS, M., KOCH, R., VERGAUWEN, M. and Van GOOL, L., 1998. Metric 3D Surface Reconstruction from Uncalibrated Image Sequences, 3D structure from Multiple Images of Large-Scale Environments. In Koch, R. and van Gool, L. (eds.), *Lecture Notes in Computer Science*, Vol. 1506, Springer Verlag. (Also on Internet at www.esat.kuleuven.ac.be/~pollefey/reconstruction.html)

POLLEFEYS, M., PROESMANS, M., KOCH, R., VERGAUWEN, M. and Van GOOL, L., 2000. Acquisition of Detailed Models for Virtual Reality. In Barceló, J.A., Forte, M. and Sanders, H.D. (eds.), *Virtual Reality in Archaeology,* 71-77. BAR International Series 843.

Documents available on the World Wide Web:

Eyetronics ShapeSnatcher: www.eyetronics.com

3D Scanners ModelMaker: www.3dscanners.com

Polhemus FastSCAN: www.polhemus.com/fastscan.htm

Notes

1 Stereo-photogrammetry will not be considered in this paper since, in general, it is too slow and the resultant model too simple for the application under consideration. Note that for modelling simple rectilinear features, such as foundations of buildings, whose morphology can be adequately described by a small number of 3D points, stereo-photogrammetry may well be a viable 3D modelling technique.

2 From the Cathedrale de SS. Michael et Gudula in Brussels, Belgium - excavated by Professor P. Bonenfant, Université Libre de Bruxelles.

Integrated Use of DGPS and the Total Station for the Survey of Archaeological Sites: The Case of Colle Breccioso

Francesca Colosi, Roberto Gabrielli

C.N.R.- Istituto per le Tecnologie Applicate ai Beni Culturali

V. Salaria km. 29.300, c.p. 10, 00016- Monterotondo St.- Roma, Italy

e-mail: colosi@mlib.cnr.it, gab@mlib.cnr.it

Dario Rose

V. Tuscolana 1661, Km. 17 – 00044 – Roma, Italy

e-mail: dareios@ciaoweb.it

Abstract

A series of topographical surveys carried out during the past years in the Salto Valley (Rieti - Lazio), have provided much interesting data regarding local archaeological sites, particularly along the southern slopes of the Breccioso Hills which rise between the Corvaro and Spedino plain. An interesting site has recently come to our attention at Colle Breccioso. However, the nature of the site was not clear and its structure is hazy, a consequence of the deterioration of the surrounding ground and increased vegetation coverage.

The situation at the site suggested that a detailed survey of the southern incline and the plateau associated with the site would be valuable. The objective of the survey was to highlight topographic variation and to bring to light any traces of human construction or manipulation.

The survey was carried out using a DGPS Leica SR 510, and a total station. The integration of these two instruments (which had both, differing and complementary capacities) produced satisfactory and innovative results. The processing of the Digital Terrain Model (DTM) of the area highlighted several characteristics of the site and the consequent production of thematic maps from this data could be used to guide future excavations at the site.

Key words: Global Positioning System, Total Station, survey technique, Digital Terrain Model, archaeological site, anthropic presence, Cicolano

1. Archaeological picture

The Cicolano area spreads along the Salto - Imele basin, and contains the watersheds of the Velino chain and the Carseolani Mountains. The region is on the eastern extremity of the Sabina area (Lazio, Italy), and owes its name to the historical, Equicoli people. Today, the Cicolano is contained within the administrative boundaries of Fiamignano, Petrella, Pescorocchiano and Borgorose (this more or less covers the Salto Valley from Caprodosso to S. Anatolia) (Almagià 1909:59, Pietrangeli 1976:75).

Settlement in ancient times is generally associated with an Oppida-Vici model; with fortified sites being placed on the heights (*Oppida*) and a myriad of inhabited settlements and necropolises along the valley floor (*Vici*) (Salmon 1985:85-86, Reggiani 1988:67, Alvino 1993:326, Bonocore and Firpo 1998:286). In ancient times, and as material from some Proto - historic necropolises demonsatrates, the Salto Valley must have had a certain importance in connecting the Fucino basin and the Rieti plain, the Sabina Tiberina and the Faliscan plain and, ultimately, Etruria (Alvino 2000:9).

The Roman conquest of the area occurred at the end of the fourth century BC and after bitter fighting over a period of over one hundred years. The annexation of neighbouring Alba Fucens dates to 303 BC and it may be suggested that it was only in c. 290 BC, the final year of campaigning of Manlio Curio Dentato in Sabina, that the whole of the Salto Valley, a marginal area with regards to the Via Valeria, was brought under Roman domination (Bonocore and Firpo 1998:359).

The real outcome of Roman subjugation was the compulsory abandonment of fortified sites and the destruction of sanctuaries along the valley floor which were in outlying positions to the settlements, but well located as regards the road network (Reggiani 1988:67, Alvino 1995:476).

The ancient N.W. - S.E. route was continued during the Roman period by a road which, on more than one occasion, has been identified as the Via Quinzia. An off-shoot of this road, which diverges from the river valley, wound its way over the Corvaro plain and, crossing the Cartore valley, headed off towards the *Ager Albensis* (Liverani 1985:282, Van Wonterghem 1988:423).

Between these two parallel routes, or to be more precise, between the Spedino and Corvaro plains one can find the Breccioso Hill, which climbs to an elevation of 842 metres. This southern slopes of this prominence contain the site which is the subject of this paper. Here a rectangular structure in *opus caementitum* is set, wedged into the slight slope, and high above a series of low terraces breaking towards the valley and outlined by a few square limestone blocks (figure 1).

In amongst the thick, bushy undergrowth are alignments of stone blocks, surrounded by numerous fragments of tiles and *dolia*. This closed rectangular structure may easily be identified as a cistern through traces of *cocciopesto* found on the site, and as suggested by the name of the site itself (Cisterna Vecchia). Despite this, it is

Figure 1: The Colle Breccioso site. The tank on the top of the hill and the terraces delimited from blocks of limestone are clearly visible.

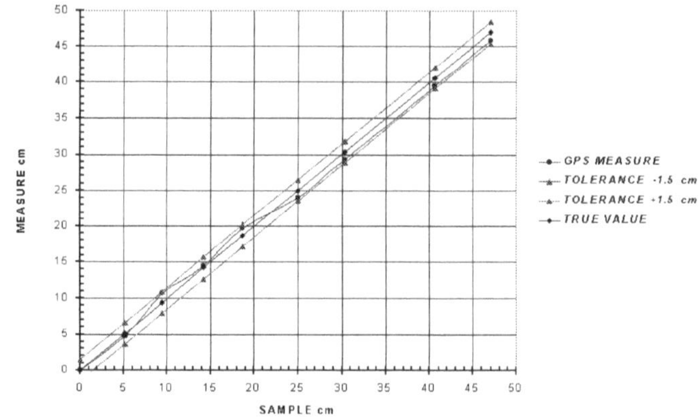

Figure 2: The graph shows the instrument's variable vertical range. The measured value falls into the instrument's tolerance range according to that declared by the producer.

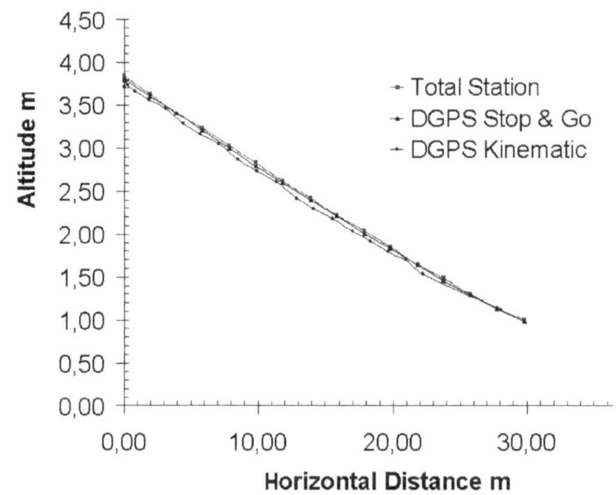

Figure 3: The graph shows the degree of shifting of the Z co-ordinate according to the method used for the acquisition. The values are included in the expected trend of 2-3 cm.

difficult to understand both the structure of the site itself or its larger context.

Up until the end of the 1960s, before the construction of the A24 motorway exit, the Breccioso Hill site was well placed in relation to the neighbouring roads from Corvaro to S. Anatolia and from Corvaro to Torano. As shown by official maps and aerial photographs, these two roads divided right on the lee of the structure. We know that the remains of a roman-imperial building were destroyed at a higher altitude on the hill; perhaps this was a *villa*. The remains of a mosaic in *opus tessellatum* from this site has been incorporated into the war memorial at Spedino. Furthermore, we have some evidence of the existence of the S. Maria de Brizzasecco church, which was already a ruin in the sixteenth century and was probably afterwards substituted by the small chapel of S. Liberatore (Staffa 1987:73). A series of factors together with the probable passing in this zone of a connecting road between the two parts of the ancient road network, its intermediate position between the Corvaro plain, the necropolis of Cau di Cartore and the area of S. Maria del Colle, which is also rich in archaeological remains, creates great interest for study of the settlement.

2. Survey of the site

In order to improve the interpretation of the space layout and to understand the function of the site, we felt that it was essential to investigate the situation of the site, as well as its physical and geomorphological condition. In fact, the nature of the construction itself appears to be clearly linked to the territory within which it is located.

With the aim of highlighting topographic variation of the slope, which was probably subject to terracing in the past, as the limestone blocks show, a detailed survey of the Breccioso Hill was carried out using an experimental Differential Global Positioning System (DGPS) Leica SR510 with a single frequency and differential mode. This instrument guarantees a level of precision to one centimetre when calculating point co-ordinates.

At first, the DGPS was calibrated to measure the level of reliability of the Z coordinate, that is, the value of the height of a topographical point. In fact, as regards research aims, it was important to be able to pick out even minimal height variations in the terrain that are potentially associated with buried structures.

In order to establish the instrument's level of precision in reading relative Z co-ordinates two tests were carried out. First the rod onto which the DGPS aerial was screwed was raised through a series of differing sample heights. Then, a series of measurements were taken using the receiver in the «stop and go» mode, that is, according to a very rapid static procedure where the device is held stationary for 60 seconds to take the reading. In this way, the height of the samples was taken in order to verify the instrument's variable vertical range. The data shown by the chart (figure 2) indicates that the measured values falls within the instrument's claimed tolerance range, i.e. 1 or 2 cm.

The second test allowed us to verify the validity of the height measurements on an incline. A set of equidistant points was fixed with the total station and the distances and relative heights were determined for each point relative to the origin point 0. Using the DGPS to calculate these points, we noticed that in the «stop and go» mode, the reproduction of data in respect to the total station is constant, whilst in the kinematic mode there is a slight divergence. However, each method of measurement was contained within the expected variation of 2-3 cm (figure 3).

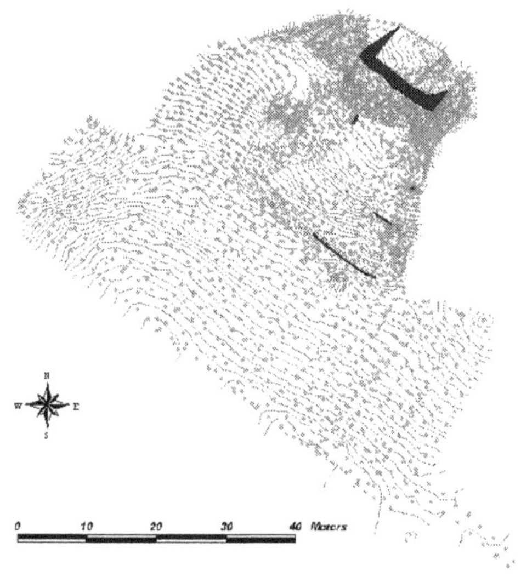

Figure 4: Colle Breccioso. Digital Terrain Model (DTM) of the hill. The surface structures are represented in black.

Having tested the instrumentation the whole of the Colle Breccioso site was then surveyed using the available instruments.

In order to carry out the survey, we chose to use the DGPS instrumentation in full kinematic mode. The operator carried a rucksack to which the aerial was attached on his back, and he then walked slowly over the entire area of coverage, including rugged areas and ensuring coverage of any localised height variation. During this operation the DGPS receiver recorded a point every 2 seconds. In this way it was possible to acquire a large number of points in a relatively short time and with limited energy loss. The operator also has the choice to decide on the density level of the points in relation to the morphology of the ground or any visible anomalies encountered during survey. Obviously, kinematic recording has a level of precision which is lower than that of a static reading, above all in respect of the x and y values, because the recording time of each point is shorter and because the aerial is not fixed but is affected by involuntary movements of the operator.

As the experimental tests showed, the margin of error may vary between 2 and 3 cm: this is quite acceptable for the reconstruction of contour lines or for morphological aspects of the ground. However, the error is too high for specific features, for example the edges of surface structures which may require a pinpoint position.

When the survey required a high level of precision, the aerial was positioned on a rod and the «stop-go» method was used. In these cases, the instrument was fixed upon a point for about 60 seconds thus reducing the possible error to 1.5 - 2 cm.

As part of the terraced zone was not easily accessible, a consequence of vegetation cover, it was necessary to use the total station in areas where vegetation prevented satellite connection. Standing structures including, for example, the tank on the top of the hill, were also surveyed using standard instrumentation (figure 1).

The integration of GPS acquisition methods which are both quick and easy to use, and the co-ordinated use of various instruments such as the DGPS and the total station, made it possible for as few as three people to survey the whole Breccioso Hill in a single day. This result in itself is significant, if one thinks about the time and effort that would usually be necessary for such a survey.

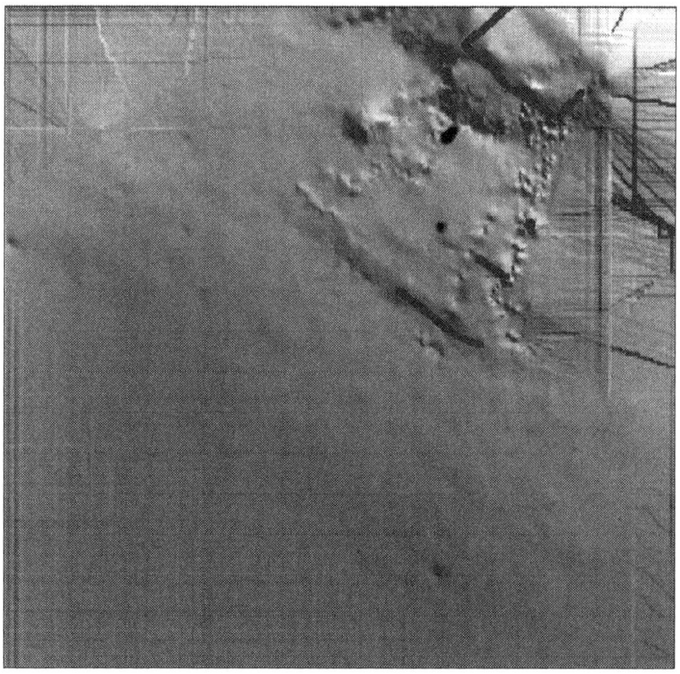

Figure 5: Colle Breccioso. DTM of the studied area overlaid with the shaded elaboration. Four terrace levels are clearly visible.

3. Elaboration of the data

Following survey, the relative system of co-ordinates of the total station was calibrated to the absolute grid of the DGPS, using a simple translation and rotation matrix. In this way, we obtained complete homogeneity within the collected data and could proceed to final interpretation. Data was manipulated within ER Mapper software and, in order to create a digital model of the hill, an interpolation of the height data set was implemented using a Triangular Irregular Network (TIN). In figure 4, the contour lines and all the captured points are given; in those areas where a higher density of measurements is shown, the total station system was used. The surface structures, surveyed using the total station system and then geo-referenced, are represented in black. The cistern on the top of the hill, which is a closed, rectangular form, does not appear as such on the survey, but is open at the northern end where it was impossible to take measurements.

Within the survey area some parallel contour lines appear very close to one another. This confirms the presence of abrupt, non-natural variations in the topography. The southern side of the hill is subject to a succession of low, wide terraces. These structures were demarcated by stone walls located on the ground surface (figure 1).

Figure 5 shows the Digital Elevation Model (DEM) of the area in hill-shaded form. Four levels appear on the map that can be interpreted as artificial terraces. They slope towards the bottom of the valley from which a large plain extends. The succession of minor terraces along the gradual slope suggests an absence of significant ground movement, thus eliminating the need for major buttressing. The whole character of the site is discernible from the map. The terracing was probably associated with a project to ex-

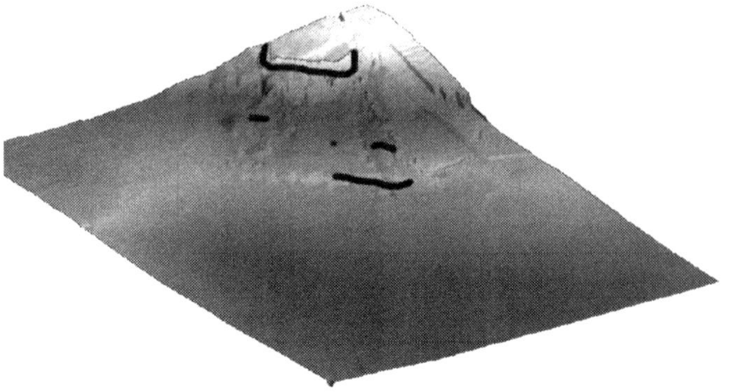

Figure 6: Colle Breccioso. On the map, overlaid with a mesh in pseudo - colour, the Z value was increased in order to highlight even the minimum variations in height.

ploit the hill, and the cistern must have been associated with this endeavour. In figure 6 the Z value was increased in order to highlight minor variations in height. The map emphasises some anomalies that are probably associated with buried structures. On the map one may pick out the position of the complex, carefully placed on the side of the hill and in front of the plain below. Today the site, which has fallen into a state of abandonment, hides the plotted position which is closely connected to the morphology of the slope.

4. Conclusion

In conclusion, the achievement of the micro-survey of the Breccioso Hill was the discovery and analysis of a series of archaeological structures. The maps that were produced from this survey could act as a guide to future excavation and direct the planning of further archaeological investigation.

The integrated use of the DGPS and the total station produced precise and satisfactory results, significantly reduced the registration time of survey points and resolved problems of field data collection. The co-ordinated, integrated survey of the micro-morphology of the terrain and visible structures has provided something extra, not only in respect of the position and nature of the remains, but also as regards its close relationship with the local topography. The use of space on the Breccioso Hill site is now clearer and we are able to provide some suggestions regarding a possible interpretation of the surviving structures. The terraces and the presence of the cistern might perhaps suggest a public or holy building but obviously only further research and excavation at the site will provide such concrete answers.

References

AA.VV., 1989. *Il Fucino e le aree limitrofe nell'antichità*, Atti del Convegno di Archeologia, Avezzano.

ALMAGIA, R., 1909. Il Cicolano. In *Riv. Abruzzese Scienze*, Lettere, Arti, XXIV, 2.

ALVINO, G., 1993. Il tumulo di Corvaro di Borgorose. In *Archeologia Laziale* X.

ALVINO G., 1995. Santuari, culti e paesaggio in un'area italica: il Cicolano. In *Archeologia Laziale* XII.

ALVINO, G., 1996. Alcune riflessioni sulla cultura equicola nella piana di Corvaro. In *Identità e civiltà dei Sabini*, Atti del Convegno di Studi Etruschi ed Italici. Magliano Sabina (RI), Firenze.

ALVINO, G., 2000. Le sepolture a tumulo di età preromana e il tumulo di Corvaro di Borgorose. In *Studi sull'Italia dei Sanniti*, Catalogo della Mostra, Roma.

BONOCORE, M. and FIRPO, G., 1998. *Fonti latine e greche per la storia dell'Abruzzo antico*, L'Aquila.

LIVERANI, P., 1985. Nota sulla via Quinzia. In *Archeologia Classica* 37.

PIETRANGELI, C., 1976. Il Cicolano. In *Rieti e il suo territorio*, Milano.

REGGIANI MASSARINI, A.M., 1988. *Il santuario degli Equicoli a Corvaro*, Roma.

STAFFA, A.R., 1987. *L'assetto territoriale della Valle del Salto fra la tarda antichità e il medioevo*, Xenia 13.

Van WONTERGHEM, F., 1989. La viabilità antica nei territori di Alba Fucens e Carsioli. In Il Fucino e le aree limitrofe nell'antichità. *Atti del Convegno di Archeologia*, Avezzano.

Computerised Techniques for Field Data Acquisition

Enrico Reali

Università degli Studi di Firenze

Via Carlo Liviero 10, 06012 – Città di Castello (PG), Italy

e-mail: realie@technet.it

Tommaso Zoppi

Università degli Studi di Firenze

Via di Coverciano 1, 50135 – Firenze, Italy

e-mail: tzoppi@unifi.it

Abstract

The procedure presented and discussed in this paper was developed in order to reduce time and costs in the process of field data acquisition. By employing equipment which is commonly available and including image rectification software and a CAD programme, it is possible to carry out a photogrammetry analysis of any given structure found within an excavation, accelerating the time of acquisition and thus improving the productivity of the excavation work. Most of the measuring activity is thus transferred from the "field" to the "laboratory", increasing the precision of the drawings and preventing problems derived from difficult field conditions.

The data collected according to this system can improve both the scientific analysis of the information as well as the diffusion of results, since they can be processed into:

- *an archaeological hypertext, which offers access and the possibility to analyse all the information produced by the investigation*
- *a 3D model of the archaeological stratigraphy, which virtually reproduces the original situation and its "dismounting".*

Key words: photogrammetry, image rectification, 3D model, light archaeology, database, building-technique typology, stone-finishing typology

This paper aims to address the problem of significantly reducing the time required in the field for acquiring drawing data. This method is suitable for acquiring spatial data as well as measurements for structures which are almost vertical, and it was also tested on horizontal surfaces.

1. "Light" archaeology

During the last few years, the research team for Medieval Archaeology, directed by Professor Guido Vannini (University of Florence), has adopted a system of territorial analysis, which has been labelled "light archaeology". It is a system based on the integrated use of non-invasive methods which characterise Landscape Archaeology, Environmental Archaeology and the Archaeology of Upstanding Fabrics, and is made possible by information technology. This methodology aims to produce a series of interconnected data (for example, creating dedicated atlases) concerning material sources which were stratigraphically recorded. This approach can also include stratigraphical excavation, especially for clarifying specific aspects which emerged in the course of the research.

Information technology plays an important role in setting up specific research tools that could guarantee standardisation on one side (for example, a specific recording sheet was developed) and flexibility on the other side, in the sense that they can be adapted to different contexts (a site, a topographic unit) and logistic situations (time, resources, competencies available).

2. Field research

Field research carried out by the team of the University of Florence, department for Medieval Archaeology is based on strongly connected procedures of archaeological investigation. As mentioned above, the methodology is based on the Light Archaeology approach and on stratigraphical excavation. Information technology integrates and links all collected data into an articulated system. A photogrammetrical system for recording data in the field has recently been implemented in our group in order to acquire detailed prospects, sections as well as architectural and excavation details.

By employing image rectification software[1] and a digital camera it was possible to experiment in the general use of photogrammetry in a variety of cases, especially where normal acquisition procedures have proven to be difficult, too slow or too costly. Preliminary results which clearly need further processing were recorded using this simple equipment[2] and the results are comparable to those achieved with more traditional methods. Substantial time reduction for acquiring data from excavation layers or upright structures can be noted: this approach allows the optimisation of archaeological field operations and offers great improvement in the computer drawing precision. Furthermore, the advantage of this method is that it can easily be taught to those who generally operate on excavations: after a short introduction to this method and its procedures of numerical data registration, image digitising

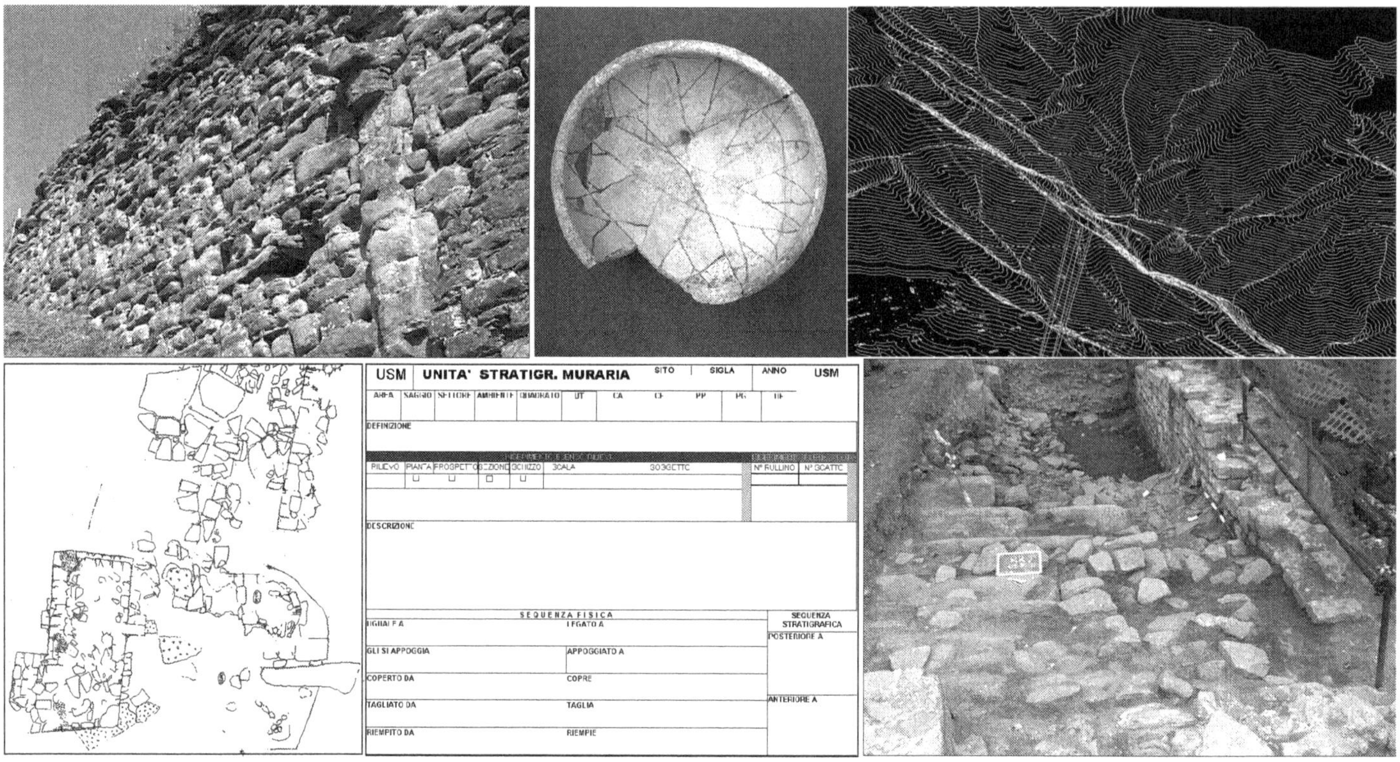

Figure 1: An example of data recorded with the Petradata database.

and rectification, it could be performed by a number of students of Medieval Archaeology.[3]

Furthermore, photogrammetry proved to be fundamental for the acquisition of archaeological graphical data, especially when it would be difficult to operate with traditional methods without high-cost equipment; this is the case, for instance, at physically inaccessible structures as can often be found in Jordan, or unsuitable weather conditions, as found at Rocca Ricciarda, Italy.

However, the method has some limitations; for instance, remarkable difficulties can be encountered in its application to circular structures or to archaeological layers spread on different levels, when it is not possible to provide an accurate graphic restitution without some special device for taking photos perpendicular to the ground.

3. Rocca Ricciarda castle

Because of its limited size, Rocca Ricciarda castle was regarded as an ideal archaeological site for experimenting with the new method of data acquisition.

Rocca Ricciarda is a small medieval castle located near the Arno Valley, Tuscany, in central Italy. The castle has been abandoned since the end of the Early Middle Ages. Written sources are found in abundance and have proven to be very informative, and their study allowed to partially reconstruct the history and the dynamics of the medieval settlement and its historical context in the Arno valley. Research carried out in the archives provided interesting results both in terms of quantity and quality; for example, the sources provide the names of the soldiers in charge of guarding the castle's tower in the years 1260-1285.

In 1997, a team from the University of Florence carried out a campaign of archaeological investigation with the aim to analyse the up-right fabrics of the medieval fort at Rocca Ricciarda. The 1999 campaign was aimed at the total excavation of the settlement in order to obtain a complete stratigraphic record. A restoration project will be launched at the completion of the archaeological investigations.

In this research, we attempted an experimental approach in order to obtain a totally computerised archaeological record, involving data acquisition and document archiving, as well as processing and dissemination of the results, both at a scientific level and on a laymen's level as far as communication and popularisation are concerned.

In an attempt to address various problems which emerge during the course of research, survey and storing archaeological records at the excavation site, computer experiments have paralleled the archaeological investigation. The database management system called *Petradata*[4] integrates all different recording sheets used in the field into a single standardised database. Its main features allow the user to create a database for archive management, as well as consultation and record cataloguing (figure 1). Apart from this it also enables the user to acquire digital images of the excavation, the survey and of the finds, to draw digital maps, prospects, sections, and so on.

The experimental method differs from the traditional method in the intensive use of the computer and it can be described in four phases. The aim of the described operations is that of obtaining a three-dimensional model of a given archaeological layer.

The first phase follows the traditional approach and consists of tracing notable points; usually these are external vertexes of the excavation area and they are established with traditional optic beads and with a theodolite.

In its second phase, the experimental approach differs from the traditional one. The example of application discussed here deals with a stratigraphic unit which consists of a wall of the collapsed tower of the Rocca Ricciarda castle, and which can be best de-

Figure 2: Rocca Ricciarda castle. The rectified photograph of the ruins.

scribed as a deposit formed by a significant number of stones. With the traditional method, all the vertexes of every individual stone should be taken one by one, measuring from the notable points, and at the same time drawing the outline of it on a sheet of paper, which results in a high investment of human resources and a possible loss of accuracy.

With the method tested here, a rectangular grid is set across the area to be recorded, thus providing a horizontal "0 level" for further measurements. The X and Y co-ordinates of the vertexes of this grid are then established: the photographic reading which will be later on computer processed is based on these points. The method proceeds to the graphic restitution of these points with a digital photographic camera; photographs must include at least six vertexes, and they should be as parallel to the ground as possible.

In the third phase, one person downloads the digital photographs and carries out their rectification on the computer by means of specific software; two others carry on with the work in the field. The Z co-ordinate for a number of points on the various stones of the collapsed tower is obtained by measuring from the zero-level created by the grid; these co-ordinates are recorded directly on the rectified photographs of the "zero-level" grid.

In the forth phase, all data is processed onto the computer: rectified photographs are transformed into vectors by the means of a CAD programme and they are eventually imported into a 3D program for three-dimensional representation.

As mentioned above, the case considered here refers to the collapsed wall of a tower of the Rocca Ricciarda castle. The figures presented here illustrate the process of acquisition, analysis and storage of 3D data. The first is a rectified photograph (figure 2) and the second is a detail of the same photograph processed into the CAD system (figure 3). The result is a digitised image of the stones found in the ruins and a 3D shaded model of the same stones. This is obviously a static screenshot but with the CAD system we can rotate and observe the ruins from different points of view.

This system will be of great importance at the reconstruction of the process of archaeological excavation (both for scientific research and for teaching purposes) and at the same time it will put into context the archaeological layers.

Figure 3: Rocca Ricciarda castle. Detail of figure 2 processed into the CAD system.

The final goal of our research is that of creating a 3D model for the entire site of Rocca Ricciarda and place it in the context of its landscape.

4. Shawbak castle

To further illustrate the methods employed by the team, the case of the Shawbak castle (Jordan) will briefly be discussed. Since 1986, an Italian team from the University of Florence (directed by Prof. Guido Vannini, for further details on the research team see the acknowledgements), is carrying out an archaeological research project in Jordan; the research is aimed at analysing the characteristics of a Crusader settlement in the area, and is focused on the Petra system of fortification; the castles of the Petra valley and the castle of al-Shawbak played a fundamental role in the control of the trans-Jordanian territories; moreover, the original features of the Crusader presence in the Holy land are particularly well preserved in the Petra valley because most sites were abandoned after the Crusaders were defeated by Salah ad-Din in the battle of Hattin in 1187.

At present, archaeological data concerning the site of Shawbak is rather limited. In particular, there is, as yet, no detailed map of the site. The earliest map available in literature is a draft drawn by the scholar Mauss in 1871; it shows the perimeter of the fortress, including most of the towers. A more detailed map has been drawn on the basis of an aerial photo survey, sponsored by the Spanish Archaeological Mission in 1978 (Almagro 1980:119). For our preliminary survey of the site we took advantage of the latter map.

In the first stage, the aim of the research at the Shawbak castle was to identify the related building phases and to record the various building techniques in order to gain an understanding of the evolution of the fortified settlements in Trans-Jordan. In a very limited amount of time (approx. 10 days), and with the use of simple equipment, the team carried out the stratigraphical analysis of selected castle structures. A preliminary stratigraphic sequence for a number of structures was established, and this formed the basis for the first attempts to interpret the castle phases; a preliminary atlas illustrating the various stones cutting and finishing was also produced. A number of research topics to be developed in the course of further research have also emerged.

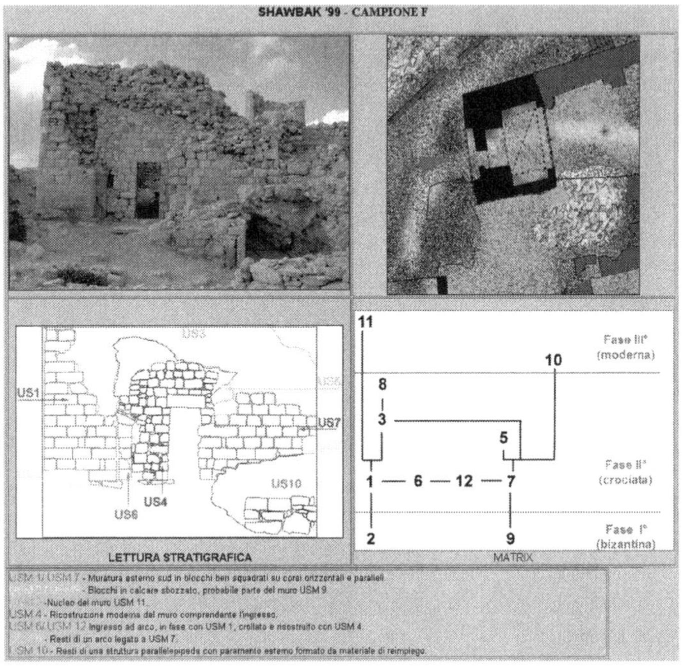

Figure 4: Shawbak castle. The stratigraphic analysis of the inner gate.

After the completion of work, the results will be published on the Internet in a section of the site dedicated to the *Medieval Petra*[5] project.

To illustrate the achieved results, the 1978 map was adopted; on this map, a preliminary interpretation in terms of major building phases is represented by the use of different colours. A series of arrows on this map indicate the points of the photogrammetry shots; the arrows are linked to pages that illustrate the stratigraphical analysis of the structure itself. For example the page related to the inner gate (figure 4), shows a rectified photo, a map, the stratigraphical analysis of the structure and the related matrix (with phases). A detailed analysis of each stratigraphical unit can be seen by clicking on the drawing, and one can also observe the rectified photo, the digitised drawing and the stone surface, classified according to the preliminary typology.

The building-technique typology (figure 5) had been conceived as an Internet page consisting of a short description (including measurements) of the wall structures, rectified photos and the digitised drawings. We are also in the process of preparing a database for this data, which will enable direct access.

The typology of the stone finishing has been conceived as a similar page with a general photo, a detail, and a short description with measurements. A database which will allow direct access and comparison is currently in preparation.

5. Conclusions

The results of the experimental procedure described here have underlined that the developed methodology is applicable to different situations, showing both great flexibility and adaptability.

The experimentation carried out at the castle of Rocca Ricciarda underlined its reduced cost; in the course of the next archaeological campaign we also plan to speed up this process with the support of DGPS equipment.

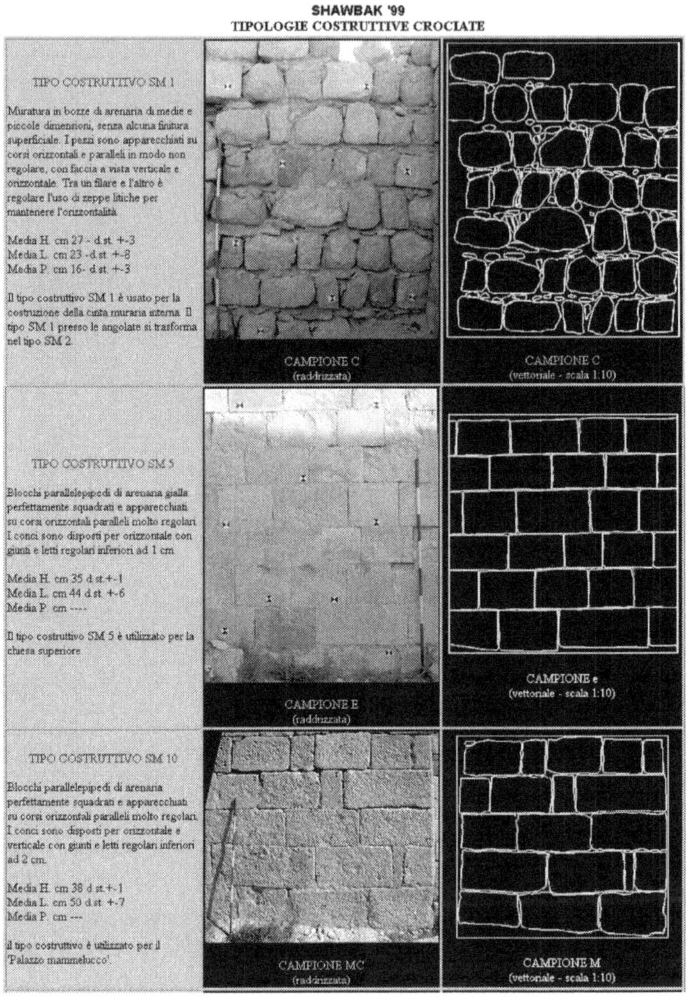

Figure 5: Shawbak castle. An example of the building-technique typology.

The application of this technique to the case of Shawbak showed its speed in the phase of data collection and processing; moreover, the results gained through this method are promptly made available and can be used in the continuation of the field research.

Acknowledgements

The programme of archaeological research which produced a part of the results discussed here is directed by Prof Guido Vannini from the University of Florence; it has been co-directed in the field by Dr Andrea Vanni Desideri (Rocca Ricciarda) and by Dr Cristina Tonghini (Petra); Prof Franco Niccolucci has directed the information technology project. The stratigraphical analysis and the digital elaboration presented here have been carried out in the field by Roberto Gabrielli (CNR), Andrea Magno, Nadia Montevecchi (Rocca Ricciarda); Eugenio Donato (Shawbak); Michele Nucciotti, Enrico Reali, Andrea Vanni Desideri and Tommaso Zoppi (Rocca Ricciarda and Shawbak).

The research project "Medieval Petra" is funded by the Italian Ministry of University, Scientific and Technological Research, by the Italian Ministry for Foreign Affairs and by the University of Florence.

Some maps and sections of the castle of Shawbak are taken from the unpublished thesis of Andrea Iacono and Stefano Mariucci.

References

ALMAGRO, A., 1980. The Photogrammetric Survey of the Citadel of Amman and Other Archaeological site in Jordan. *Annual of the department of antiquities of Jordan* XXIV: 111 - 9 Amman.

BURNS, J.A., 1991. CAD-Photogrammetry: A Powerful Documentation Tool. *CRM*, 14:4-5.

CRESCIOLI, M., NICCOLUCCI, F., 1999. PETRA data: an integrated environment for archaeological data processing. *New Techniques for old times. CAA international Conference, Barcelona 1998, Oxford BAR Series 757*:133-137.

COOPER, M.A.R., ROBSON, S., LITTLEWORTH, R.M., 1992. The Tomb of Christ, Jerusalem; Analytical Photogrammetry and 3D Computer Modelling for Archaeology and Restoration. *International Archives of Photogrammetry and Remote Sensing* 29:778-785.

EITELJORG, H., 1994. AutoCAD Single-Photo Photogrammetry at Pompeii. *CSA Newsletter, A Quarterly Newsletter for Architectural Historians and Archaeologists* 7:3-5.

GISIGER, A., COOPER, E.S., YUAN, Y., LIMP, W.F., 1997. Development and Implementation of a Rapid Low-Cost Photogrammetric Data Archival System for Artefact and Osteological Inventory. In Johnson, I. and North, M. (eds.), *Archaeological Applications of GIS: Proceedings of Colloquium II, UISPP XIIIth Congress*, Forli, Italy, September 1996.

KEMPA, M. and SCHLÜTER, M., 1992. Graphical Representation of an Armenian Castle with AutoCAD. *International Archives of Photogrammetry and Remote Sensing* 29:241-244.

KRAUS, K., 1994. *Fotogrammetria*. Vol.1. Levrotto e Bella, Torino.

NICCOLUCCI, F., CRESCIOLI, M., in press. *Advances in PETRA data management system*. IV Coloquio Internacional de Arqueologia e Informática, Bilbao 1998.

NICCOLUCCI, F., VANNINI, G., TONGHINI, C., CRESCIOLI, M., in press. PETRA: un sistema integrato per la gestione dei dati archeologici. Archeologia computazionale I Workshop Nazionale (Napoli-Firenze, 1999), *Archeologia e Calcolatori* 11/2000.

POTSIOU, C., IOANNIDIS, C., BADEKAS, J., 1992. A Special Information System for the Documentation of Castles. *International Archives of Photogrammetry and Remote Sensing* 29:287-291.

PRINGLE, D., 1997. Secular buildings in the crusader kingdom of Jerusalem. Cambridge.

SELVINI, A., 1994. *Elementi di Fotogrammetria*. CittàStudi, Milano.

VANNINI, G., 1997. *Petra Medievale. Insediamenti di epoca crociata e ayyubide in Transgiordania*. Missioni archeologiche italiane. La ricerca archeologica, antropologica, etnologica, Roma, MAE-"L'Erma" di Bretschneider: 103-106.

VANNINI, G. and TONGHINI, C., 1997. Mediaeval Petra. The stratigraphic evidence from recent archeological excavations at al-Wu'ayra. In *Studies in the History and the Archaeology of Jordan* VI, Amman: 371-384.

VANNINI, G., and VANNI DESIDERI, A., 1995. Archaeological research on Medieval Petra: a preliminary report, *Annual of the Department of Antiquities of Jordan* XXXIX: 509-540.

Documents available on the World Wide Web:

Armenian Architecture - Virtual Ani: http://www.virtualani.freeserve.co.uk/Armenian Architecture - Virtual Ani - Map Of The City.htm

Fields of Application for Photogrammetry and Imaging in Archaeology: http://www.univie.ac.at/Luftbildarchiv/wgv/appl.htm

La fotogrammetria digitale nel restauro e nell'analisi storica di edifici monumentali: alcuni esempi significativi. Atti del primo colloquio internazionale. Bari, 10 - 12 Nov. 1994: http://www.ba.cnr.it/~irisnm08/

Il raddrizzamento fotogrammetrico. seminario di Marco Canciani a.a. 1999-2000: http://www.uniroma3.it/sta/DIS/ricerca/ortho/teoria.htm.

Notes

1 The software for image rectification is Siscam-Archis 32. It allows to turn a central projection frame of a raster image into a orthogonal projection that can easily be transformed into vectors and measured. Two different methods are used for rectifying: geometric omografias and analytical omografias. These two methods allow the performance of image rectification (both knowing the co-ordinates of certain points of the photographed object or lacking any known points).

2 The equipment used consisted of: 2 Personal computers Pentium 266 with 128 Mb Ram; 1 digital camera Nikon Colpix 950; 1 scanner Primax Colorado Direct; Microsoft Access for text data storage; Microsoft Excel for numerical data processing; Adobe Photoshop for image processing; Siscam Archis 32 for image rectification and mosaicing; for final processing the following programmes were used: Autodesk Autocad for vector graphic restitution; Rhinoceros and 3D studio Max for 3 dimensional modelling.

3 The method consists of the acquisition of images with a traditional or digital photographic camera; acquisition of the co-ordinates of an extremely limited number of points on the object; and of image processing by means of PC software.

4 Petradata is an open, flexible and dynamic system of interconnections between recording and storing archaeological data. It integrates the various types of recording sheets used in the field into one standardised database. Data entry is performed using input forms that reproduce the structure of the recording sheets used in the field.

5 The site can be accessed through the server at the Dipartimento di Studi Storici e Geografici, University of Florence, at address: http://www.storia.unifi.it/_PIM/AM/Petra/italian/medievalpetra.htm

Understanding and Using Archaeological Topographic Surveys - The "Error Conspiracy"

Henry Chapman

Wetland Archaeology and Environments Research Centre

Department of Geography, University of Hull

Cottingham Road, HU6 7RY Hull, United Kingdom

e-mail: h.p.chapman@geo.hull.ac.uk

Abstract

Surveys of archaeological sites form a basis for management decisions, and through the displaying of spatial relationships they are an interpretative tool. They are commonly re-used for later work, particularly if damage has occurred in the intermediate period rendering features indistinguishable on the ground. In these cases, the need for a plan to assist in both the management and interpretation of the site is fundamental. Developments in digital archiving and data management have changed the way in which archaeologists work and hold the potential for the re-use of survey data. However, the re-use of data requires an understanding of the accuracy and applicability of earlier material. This paper highlights the inherent dangers in re-using survey data using the example of an Iron Age site near Doncaster, England. The results demonstrate how errors can be perpetuated within the literature and how they continue to be used for the decision-making processes despite this.

Key words: survey, data re-use, error, GPS, aerial photography

1. Introduction

There is a long tradition within archaeology of producing plans and surveys, particularly of sites represented by earthworks. This has been common practice normally aimed at the interpretation and classification of a site by investigating spatial relationships between different archaeological features and the landscape (*cf.* Bowden 1999). It has also been seen as a basis for the management of sites once recognised. For example, legislative protection such as through English Heritage's scheduling procedure requires that an outline is drawn around the site displaying the area to be protected including a buffer zone. In order for such management to be effective, it is important to have an accurate record of the site and its location to inform the decision-making process. Finally, it is important to make surveys of sites for archival purposes to maintain a record in case of future damage or other changes to the site. If a site becomes damaged, an earlier survey may prove invaluable when directing and understanding later work. This highlights the re-usability of survey data, and the importance of maintaining such a record.

The re-usability of archaeological surveys has important implications in relation to the development of digital data management and archiving. Surveys are needed to fulfil certain requirements and are consequently conducted at particular scales. Once placed within a digital environment the potential for re-use expands due to greater ease of access. However, within such an abstract environment, the original consideration of issues such as scale and intention may be forgotten. If a survey is then re-used within an environment that has the ability to create new data from it, such as GIS (*cf.* Savage 1990), the levels of abstraction and inaccuracy are increased.

This paper presents the results from a case study on a site in South Yorkshire, UK. Here, the destruction of earthworks led to subsequent work over two decades being conducted using the earlier survey plans as their basis. The re-use of the earlier survey has proved invaluable in this case with the lack of other alternatives. However, a more recent topographic survey of the site revealed that the earlier survey was inaccurate in a number of ways, and that these inaccuracies have had an effect upon more recent archaeological work. Reflection on these issues has highlighted the dangers inherent in the re-use of previous survey data, and that the unique values of the detail shown in the earlier survey form a paradox. The need to record "metadata" when archiving surveys is demonstrated, and it is noted that previous surveys should be used with extreme caution to avoid the perpetuation of error.

2. Survey data and digital archiving

Developments within digital archiving have enabled datasets to be more freely and usefully accessible (*cf.* Richards 1997). The potential for this has been particularly marked within spatially referenced datasets and the growth of GIS-based excavations and fieldwork (Gillings and Wise 1999). Such an environment enables the results of previous research to be included within the database of ongoing work, making conclusions more easily understood through correlation, and building within a standardised framework.

Surveys are an extremely re-useable form of data. The fact that they are normally spatially referenced means that they are ideal for being included within the framework of digital archaeological datasets. Archaeological surveys of surface features such as earthworks are essentially interpretative drawings. They are representations at a given scale of an interpretation of features seen on the ground, obtained through a combination of the surveyor's knowledge and experience. Surveys therefore provide a different type of data when compared to the distribution of finds, for example, as they cannot be considered as objective. The usefulness of a survey is also determined by the intentions behind it, and this often dictates the representation scale. This determines the level of detail included and also the accuracy at which it may be accepted when re-used. With time such information as the original scale of the survey may be lost, particularly if its record exists as a secondary publication. The issue of maintaining data about data has been focused upon previously (e.g. Wise and Miller 1997, Gillings and Wise 1999), and the issue remains a fundamental theme within digital archiving.

Contemporary digital survey methods have provided a different type of product that offers a wider range of issues concerning the archiving of survey data. Increased efficiency of digital survey techniques such as differential Global Positioning Systems (GPS) have meant that the quantity of data that can be collected in a day has increased providing the opportunity to approach surveys differently (Chapman and Van de Noort in press). The collection of many survey points across an archaeological site creates the possibility of generating a highly accurate Digital Elevation Model (DEM). These can then be manipulated as an interpretative or managerial tool, but can also be incorporated and re-used within other digital datasets. It has previously been stressed that, when collecting data to create a DEM of an archaeological site, the most efficient way of generating a representation of upstanding archaeology is to survey at multiple resolutions (Fletcher and Spicer 1988). This approach meant that visible features could be surveyed at a high resolution reflecting their complexity in the same way as standard interpretative survey methods do. Other areas could be gridded to fill the gaps, providing a level of data similar to that of traditional contour models. Although this is certainly the most appropriate solution to the problem of surveying archaeology that has varying density on the ground, it raises a number of issues relating to the archiving and re-use of digital data that are different to those of previous survey methods. Firstly, it creates a problem of definition. The nature of these types of surveys is a mixture of both subjective interpretative recording and objective gridding of the surface. Secondly, the variable resolution of data points and the digital storage of the data make it very difficult to determine the scale at which the survey was conducted as with earlier approaches to survey. Essentially, at what resolution may the data be considered to still be accurate?

3. Sutton Common - a case study

3.1. Background

Sutton Common lies within an area of low-lying wetland to the northeast of Doncaster, UK (figure 1). Here, the landscape has revealed evidence of human activity from all periods since the Mesolithic, although the focus of the archaeology is on a pair of earthwork enclosures dating to the Iron Age (Whiting 1936, Parker et al. 1997, Van de Noort and Chapman 1999, 2000). These occupy early Holocene sandy "islands" on opposing sides of the relict, peat-filled palaeochannel of the Hampole Beck. The enclosures represent two phases of human activity. The first is characterised by the construction of an oak palisade around the "island" on the eastern side of the channel. The second phase is marked by the construction of earthen ramparts overlying the earlier palisade and enclosing both islands. These are formed by complex multivallation, and elaborate entrances constructed of massive oak posts. Linking the two enclosures across the relict palaeochannel was a sandy causeway defined on either side by an irregular row of stakes.

The area of Sutton Common was enclosed in the 1850s and at this time the positions of the enclosures were mapped by the Ordnance Survey, labelling them as "Crook Hills" (Ordnance Survey

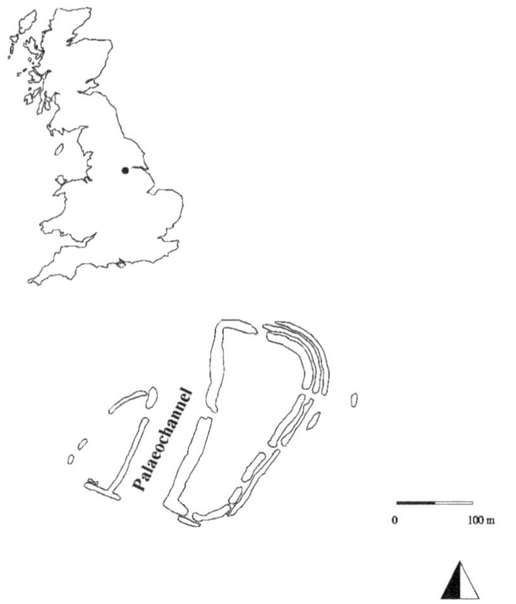

Figure 1: Location of Sutton Common. Banks of the Iron Age enclosures shown after Whiting (1936).

6" scale map (1945-54) sheets 254-301). They were again recorded by Surtees in the 1860s (Surtees 1868) and mentioned in the archaeological literature in the beginning of the 20th century (Allcroft 1908). A series of unrecorded excavations took place throughout the 1920s, but the first scientific archaeological research excavation was not until the 1930s (Whiting 1936). During the work at this time the earthworks of each of the two enclosures were surveyed by Bennett and Hill. The surveys of the two enclosures were published on separate pages within the report, and this secondary record remains their only known source. Consequently, the scale of the original survey is not known, nor are the original intentions. However, annotations accompanying each of the published surveys display text in an extremely small, often illegible handwriting. This indicates that at least some photographic reduction was involved prior to their publication.

The Sutton Common landscape was first subjected to drainage operations at the time of enclosure in the 1830s (Whiting 1936:57), although the effect of this appears to have been extremely limited. Accounts dating from shortly after this event describe the Common as still waterlogged (Surtees 1868, Allcroft 1908). Little concerning the physical landscape appears to have changed until 1979-80 when the area of the large enclosure and part of the small enclosure were bulldozed and transferred to arable farming. In 1982, this was accompanied by the installation of under-field drainage lowering the water levels on the Common by approximately 2 m, accompanied by off site water abstraction (Geomorphological Services Ltd. 1990).

The reduction in ground water levels and the potentially damaging effect of this upon the buried, wet-preserved organic resource prompted a series of excavations and investigations. These were primarily funded by English Heritage and undertaken by the South Yorkshire Archaeology Unit and Sheffield University. This work was aimed at assessing the condition of the buried archaeological and palaeoenvironmental resource and consisted of small trial trenches. For report preparation and for publication, the positions of these trenches were placed in relation to the earthworks using the 1930s surveys by Bennett and Hill as a basis (Parker Pearson and Sydes 1995, 1997). This appears to have been because there

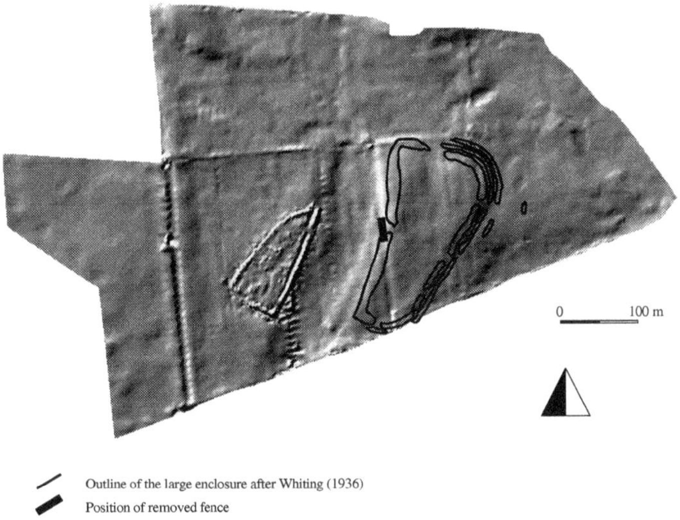

Figure 2: DEM of Sutton Common showing the positions of the georeferenced plan (after Whiting 1936) and the position of the fence used to locate it.

was no real alternative since aerial photographs taken of the Common prior to the bulldozing showed the earthworks as indistinct because of rough vegetation. The resulting maps placed the trenches in relation to the destroyed earthworks, and also brought the two surveys together to form a single plan. This formed the basis for reports on the site throughout the 1980s and 1990s.

3.2. New survey - towards a digital excavation

The ownership of Sutton Common passed to the Carstairs Countryside Trust (CCT) in 1997 for the long-term protection of the upstanding and buried archaeology, and for the wildlife upon it. The new ownership was followed by a topographic survey of the area to create a DEM of the site that could be useful for its future management. The survey was conducted using Spectra Precision's *Geotracer© system 2000 L1* differential GPS (set at an accuracy tolerance of ±0.02 m) at a variable resolution reacting to the variable preservation of earthworks as described above. The survey data were processed within ARC/INFO GIS to create an interpolated DEM at a cell-resolution of 1 m.

Two new phases of excavation of the site were undertaken in order to assess the preservation of the organic archaeological and palaeoenvironmental source material. The locations of the trenches for both were chosen on the basis of the subtleties highlighted within the DEM. The first phase of excavation opened up five small trenches across the site. The second year was more systematic, with eleven trenches positioned at regular intervals across the interior of the larger enclosure. During the second phase of excavation the bases of two posts were discovered that appeared to be parts of a fence that was positioned across the Common during the 1930s and which was included on the plan of the site made at the time (Whiting 1936:60). The alignment shown from these two posts correlated exactly with a surface feature visible on the hillshaded model generated from the DEM. Using this and the scale on the survey drawing, it was possible to georeference the earlier plan and to incorporate it into the GIS (figure 2).

The resulting correlation model revealed that the DEM reflected a high level of survival of features in the area of the larger enclosure despite bulldozing and two decades of arable cultivation.

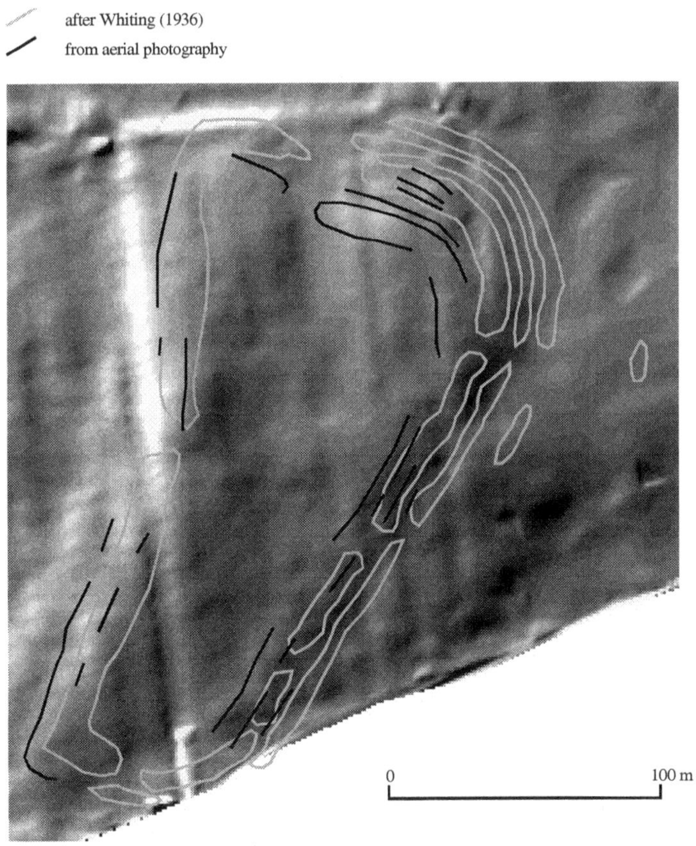

Figure 3: Differences between the positions and dimensions of the larger enclosure.

However, comparison with aerial photographs taken of the site before it was bulldozed and ploughed revealed that the position of the integrated outline was incorrect. The georeferenced outline of the larger enclosure existed further north than it appeared on the aerial photographs, extending onto the modern track. The photographs showed it stopping short of its southern side.

In order to understand the level and areas of error within the incorporated 1930s survey it was important to obtain a more accurate outline of the buried archaeology from the available sources. Aerial photographs taken during the excavation were consulted for their limited cropmark information that may indicate the outlines of the banks and ditches of the larger enclosure. Through a process of colour separation (by J.G.C. Tovey) it was possible to strengthen vegetation differences highlighting the positions of the buried archaeological features. The photographs were rectified and georeferenced using the outlines of the eleven excavation trenches that were visible and that had been surveyed on the ground using the GPS with positions located on the DEM. The outlines of the visible archaeological feature revealed on the colour-separated aerial photographs were digitised and incorporated into the GIS model (figure 3).

3.3. Identification and classification of error

The identification of error within the plans from the 1930s surveys may be classified into three main levels: primary, secondary and tertiary. Primary error relates to errors inherent within the survey itself. A number of errors could be seen in relation to the basic identification of archaeological features. For example, on the western side of the small enclosure, the bank was depicted on the earlier survey as a series of interrupted mounds with breaks

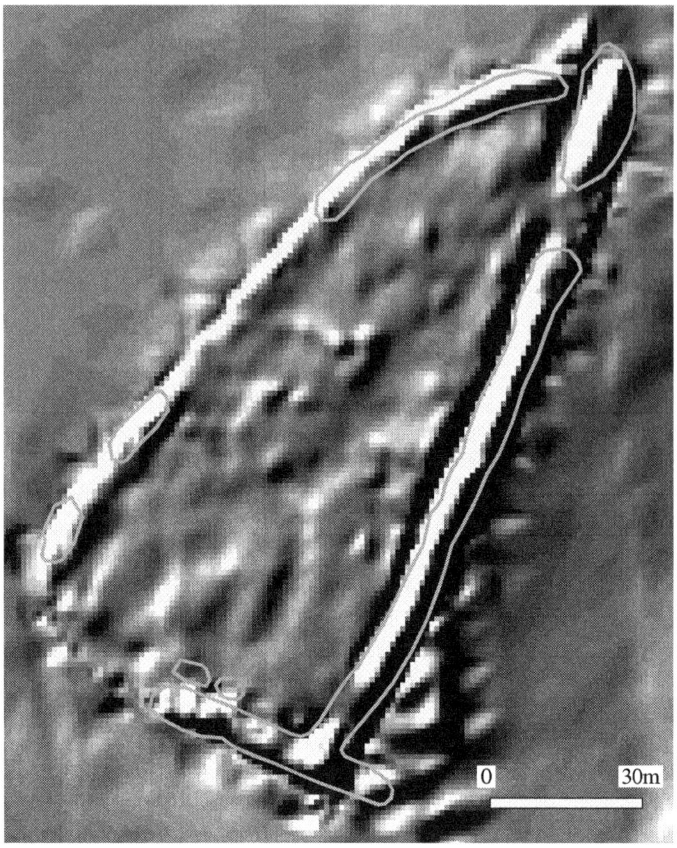

Figure 4: DEM of the smaller enclosure compared with the outline of the banks from the previous survey (after Whiting 1936).

between them (figure 4). The GPS survey, on the other hand, showed a surface that consisted of a near-complete linear bank. It seems that the rough vegetation at the time of survey may have hindered the recognition of the banks on the ground, but the systematic GPS survey was able to pick up these features. A similar result was found on the eastern side of the smaller enclosure where the entrance was formed by a break in the two banks. On the GPS survey this break was less distinct and more spread, indicating a more complex process. This was noted within the excavation of this area that showed evidence of major slumping of material in antiquity (Van de Noort and Chapman 1999).

Secondary error is generated when surveys are depicted according to convention or prepared for publication. At Sutton Common the original scales of the surveys by Bennett and Hill are now known. It appears that they were not redrawn prior to publication as the survey grid lines are visible. However, the size of the writing indicates that they have been considerably reduced, indicating a possible loss of detail and some distortion through the photographic process. Without records of this it is not possible to understand the reliability of the plan in the publication and therefore a second level of potential error is added to the process.

Tertiary error may be generated when a previous survey is re-used and depicted in another way. This level of error relates to the processes that are applied to the plan by the new user, and can include basics such as accuracy of digitisation, chosen points for digitising, data format conversions and its role within further calculations by the GIS. The latter point is particularly important since the reproduction of multiple layers of data and their use for future decision-making processes can abstract it (cf. Savage 1990). Levels of abstraction increase the chances of separating a surface from metadata, meaning that levels of potential error may be forgotten.

3.4. Quantification of error - the larger enclosure

In order to understand the apparent error contained within the 1930s survey compared with the aerial photographic rectification, a quantifiable basis was required. The fragmentary nature of the outlines from the aerial photography meant that measurements were taken to areas that could be identified on these and the other surveys. This was done by constructing a base line along the western edge of the enclosures and measuring off at 90 degrees from percentile positions. These provided actual lengths and width to height ratio lengths of the enclosure.

3.4.1. Size and scale

The length of the two enclosures differed by c. 3%, with the width nearly 13% wider at the northern end and 6.5% towards the southern end for the 1930s survey compared to the later survey. This showed that the scaling of the earlier survey had made the enclosure larger than it was.

3.4.2. Rotation

The earlier survey is rotated by nearly 3° anti-clockwise from that revealed from the aerial photography. The main reason for this appears to have been due to a bend in the western bank of the enclosure north of the western entrance. The earlier survey did not recognise this, perhaps due to the obstructions caused by the fence that separated these two parts of the site at the time. As a result this change in alignment is not reflected in the earlier survey. Rotation has been seen as a source of error in some of the publications that have re-used the earlier survey (cf. Parker Pearson and Sydes 1997:222).

3.4.3. Integrity

In addition to the earlier survey demonstrating inaccurate scaling and some rotation, the internal integrity of the survey appears to have been incorrect. If integrity was apparent then you would expect a similar size increase for all measurements relative to the aerial photographic work. However, while the length was greater by approximately 3%, the width at the northern end was 13% greater. Further, the southern end was only about 6.5% greater. Two conclusions may be drawn from this evidence. Firstly, the width increase was greater than the length increase, suggesting a lateral stretch. Secondly, this lateral stretch was not consistent, with greater expansion displayed in its northern part.

4. Discussion

The discrepancies between the positions of the outlines of the larger enclosure displayed on the 1930s survey and the aerial photographs may have occurred or been influenced by a number of factors. These include surveying errors, but it has also been suggested that lateral movement of sediments is possible through natural processes, particularly on wetland sites (S. Stead pers. comm.). Since Sutton Common is a wetland it may be expected that the natural landscape is dynamic, and that the drainage of the site during the past two decades will have altered the burial conditions. This has been noted by the differential shrinkage of the various sediments across the site, particularly between the biogenic and minerogenic material (Chapman and Van de Noort in press). Although it seems likely that such changes to the natural environment will have altered the surface topography, this will have been to a minor extent. The majority of the site is represented by the reworked minerogenic glacial lake sediments that are less likely to shift compared with the areas of organic material. Also, the basic evidence for the inaccuracy is the false extension of the larger enclosure onto the track to the north. Such excessive error cannot be understood solely through natural processes, although it will have been influential at a lower scale.

Discounting natural processes as having a minor effect on the surface topography of the Common, the reason for the discrepancy in the location of the larger enclosure lies within the accuracy of the original survey. Without knowledge of the intentions and methods of survey it is not possible to understand the relevance of the depiction. Other factors influencing its accuracy have also been mentioned, including potential problems during reproduction for publication, and its conversion into a digital format.

5. Conclusions

The importance of archaeological survey is well established, particularly its potential for re-use, and the integrated advantages that this may have within digital databases. However, without consideration of the many levels of error that may be inherent within the survey, such re-use holds the potential for generating inaccuracies that may not be recognised. This paper has demonstrated how such errors may "conspire" to create results that are far from accurate and that produce an incorrect, or at least misleading, picture of the archaeology on site.

The implications of this for archaeology are far reaching. In terms of interpretation, errors will create different spatial relationships between different monuments, and between monuments and the landscape. Further, such error cannot be understood in terms of fuzzy tolerances due to its non-linear nature. In terms of resource management, the implications are also very high. Sutton Common is protected under English Heritage scheduling policy. Scheduling requires a boundary to be drawn on a map to outline the area of protection. Error in the survey may mean that the wrong area is protected. It is important therefore to critically assess data prior to its use for archaeological requirements.

Acknowledgements

I am grateful to J.G.C. Tovey who provided the expertise and time to correct aerial photographs and to achieve the colour separation work that enabled the archaeological features on them to be recognised and mapped. The work at Sutton Common that formed the background for this research was funded by English Heritage and undertaken by the Centre for Wetland Archaeology at the University of Hull. The site is owned by the Carstairs Countryside Trust (CCT). Current management of the site is part of a partnership including the CCT, the Countryside Commission, English Heritage, English Nature, Grantham Brundell & Farren consulting engineers, Hull University, Wren Environmental and Askern town.

References

ALLCROFT, A.H., 1908. *Earthworks of England.* London: Macmillan.

BANNISTER, A., RAYMOND, S. and BAKER, R., 1998. *Surveying.* Harlow: Longman.

BOWDEN, M. (ed.), 1999. *Unravelling the landscape - an inquisitive approach to archaeology.* Stroud: Tempus Publishing Ltd.

CHAPMAN, H. and Van de NOORT, in press. High resolution wetland prospection, using GPS and GIS: landscape studies at Sutton Common (South Yorkshire), and Meare Villlage East (Somerset). *Journal of Archaeological Science.*

CONDRON, F., RICHARDS, J., ROBINSON, D. and WISE, A., 1999. *Strategies for digital data - findings and recommendations from digital data in archaeology: a survey of user needs.* York: Archaeology Data Service.

Geomorphological Services Ltd., 1990. *Hydrological appraisal of Sutton Common, Askern, South Yorkshire.* Unpublished report for English Heritage.

GILLINGS, M. and WISE, A., 1999. *GIS guide to good practice.* Oxford: Oxbow and the Archaeology Data Service.

PARKER PEARSON, M. and SYDES, R., 1995. Sutton Common: a South Yorkshire wetland. In Cox, M., Straker, V. and Taylor, D. (eds.), *Wetlands - archaeology and nature conservation*: 86-96. London: HMSO.

PARKER PEARSON, M. and SYDES, R., 1997. The Iron Age enclosures and prehistoric landscape of Sutton Common, South Yorkshire. *Proceedings of the Prehistoric Society* 63: 221-59.

RICHARDS, J.D., 1997. Preservation and re-use of digital data: the role of the Archaeology Data Service. *Antiquity* 71: 1057-9.

SAVAGE, S.H., 1990. GIS in archaeological research. In Allen, K.M.S., Green, S.W. and Zubrow, E.B.W. (eds.), *Interpreting space: GIS and archaeology*: 22-32. London: Taylor & Francis.

SURTEES, S.F., 1868. *Footprints of Roman occupation in the southern parts of North Humber Land.* Leeds: Baines.

Van de NOORT, R. and CHAPMAN, H., 1999. *An archaeological assessment in preparation of a management plan at Sutton Common, South Yorkshire.* Hull: Centre for Wetland Archaeology report CWA/RES/EH-Sutton/99-1.

Van de NOORT, R. and CHAPMAN, H., 2000. *Sutton Common, Sutton, South Yorkshire - assessment report.* CWA report CWA/RES/EH/00-1.

WHITING, C.E., 1936. Excavations at Sutton Common 1933, 1934 and 1935. *Yorkshire Archaeological Journal* 33: 57-80.

WISE, A. and MILLER, P., 1997. Why metadata matters in archaeology. *Internet Archaeology* 2 (http://intarch.ac.uk/journal/issue2/wise_toc.html).

3D Visual Information and GIS Technologies for Documentation of Paintings in the M Sepulcher in the Vatican Necropolis

Maurizio Forte

CNR-ITABC, Istituto per le Tecnologie Applicate ai Beni Culturali

Via Salaria, km.29,300, C.P.10, 00016, Monterotondo St., Roma, Italy

e-mail: maurizio.forte@mlib.cnr.it

Stefano Tilia, Angela Bizzarro and Alessandro Tilia

TREERRE sas, Roma, Via Gorgia di Leontini 170, 00124 Roma, Italy

e-mails: stilia@tin.it, bizzarro@tin.it, altilia@tiscalinet.it

Abstract

Digital technologies may be very helpful in the graphic documentation of paintings, not only due to the inherently more detailed and faster method of drawing, but in particular due to the 3D representation of data (structures and materials) in real time, that is, the possibility to have many dynamic views of the same model in order to visualize and interpret features and shapes, as well as the state of conservation of structures and materials. Moreover, it is very important to highlight that paintings or mosaics are intelligible as 3D objects rather than 2D objects, since they ultimately represent solid objects. Thus, using GIS technologies for collecting data, it is possible to process the relative information in detail.

There are many situations one has to face when dealing with graphical documentation of surfaces, whether flat, bossed or curved, with or without decorations. The traditional form of graphic documentation, interpretative by definition, favours different levels of detail, depending on the scale of reduction and the different objectives under study (location of degraded areas as opposed to decorated ones, presence of discomformities and cracks, production of images and thematic maps derived from particular analyses, for example, thermal infrared surface scanning). It is certainly difficult to document a decorated vault. For this reason, we have chosen to consider the walls and the polychrome mosaic vault of the M sepulcher, also known as Cristo Sole (2.-3. cent. AD) which can be found in the Vatican Necropolis. The mosaic presents a small tessera decoration (of which only a fraction is still in place), on the walls and on the highly depressed vault; in places where the tesserae are missing, the mosaic preparatory surface still enables us to interpret the original decoration.

Furthermore, the main goal of our project is a detailed virtual 3D reconstruction of a micro-model of the above mentioned decorations through the use of digital techniques, in order to obtain a final "cognitive" representation. To obtain this, we have implemented new tools for the acquisition of detailed 3D surfaces.

The general positioning (georeferencing) was achieved using a Leica TCR 1103 total station; this instrument also enabled us to obtain vast amounts of surface data for the 3D model thanks to its reflectorless Laser EDM (Electronic Distance Measuring) range finder. Where more detail was necessary, further surface data was gathered using a Microscribe 3D Mechanical Arm Digitiser coupled to a laptop computer and driven by the Rhinoceros® 3D surface modelling software.

The project involves the following main approaches: digital acquisition of surfaces, digital documentation of mosaics and paintings, 3D virtual reconstruction of models and 3D virtual communication of the information by VRML metaphors through the Internet.

Key words: micro-GIS, 3D, M sepulcher, Cristo Sole, virtual reconstruction

1. Introduction

The graphic documentation of wall paintings and of the underlying architectural and/or archaeological structures must aim to facilitate the highest grade of knowledge and communication.

As objective and metrically correct as such documentation may be, it is, however, still subject to the interpretation of the person documenting it, whether obtained through traditional means or through the use of more advanced systems, such as, for example, photogrammetry. This does not mean that we must not search and, through experimentation, choose whichever methodology is best suited to tackle the diverse situations one may encounter.

In every documentation project, the first step is to create a spatial reference system used for proper positioning of all elements under study both in absolute terms and in relation to each other, independent of the scale and complexity. Thanks to recent technological advances, this part of the task has been greatly simplified. New options and equipment are also beginning to aid us in the successive part of the graphic documentation process: the actual "retracing" of the objects of our study, either in high detail (as favoured by historians or archaeologists) or more schematically (to create a base for the conservationist's thematic mappings) has been made much simpler.

Digital technologies may be very helpful in the graphic documentation of wall paintings not only for the greater speed it offers, without compromising any detail, but in particular for the representation of data (structures and materials) in 3D; it offers the possibility to have numerous dynamic views of the same model in

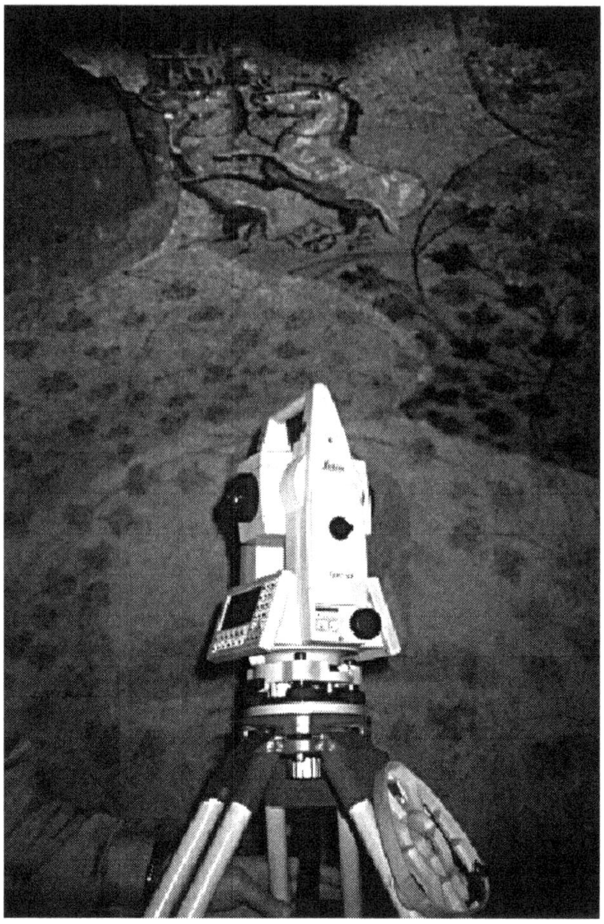

Figure 1: Total laser station for the topographic survey.

order to visualise it and thus to interpret features, shapes, as well as the state of conservation of structures and materials in real time. Moreover, it is very important to highlight that paintings or mosaics are intelligible more as 3D objects rather than 2D objects, because they represent a 3D class. As a result, using GIS technologies, it is possible to process the collected data in detail in order to interpret and analyse the digital content with multidimensional levels of overlay.

For this reason we have planned the following methodological goals:

- Testing new systems for a micro-topological survey in emergency situations (very small spaces, detailed architectonic features, preservation of materials).
- Considering digital graphic representation of mural paintings and mosaics as three-dimensional information (instead of two-dimensional).
- Using 3D GIS techniques for an interactive visualisation of architectonic models and 3D databases (figures 6-9).

On the basis of the above mentioned approaches, we have implemented the following main stages:

- Digital acquisition of all architectonic surfaces with a 3D pantograph (Microscribe, http://www.immerse.com/microscribe/digitizers.html) for a total of 11.000 points (figure 4).
- Rectification and digital photo-mosaic of all walls (3 walls and 1 vault, figures 5-8)).
- Infrared thermal survey (thanks to ENEA's contribution, Ente per Nuove Tecnologie, l'Energia e l'Ambiente).
- 3D interpolation of the micro-relief data (figure 9).
- Digital data-entry in a 3D GIS.
- Texture mapping and draping of all graphic surfaces of the walls on the micro-DEM (figures 6-8).
- Virtual reconstruction of the tomb in VRML (figure 10).

2. The Vatican Necropolis

The work presented here was carried out during the restoration works in the Vatican Necropolis (figure 3). This previously open-air burial ground is now placed under the nave of St. Peter's Basilica. In fact, the Necropolis was covered when the Constantinian Basilica was erected. Furthermore, successive burials have altered the stratigraphy of the site which was also influenced by the structural works for the construction of Bernini's *Baldacchino* and the sequence of constructed altars, the construction of which took place during the centuries gone by. All of this took place on top of the sacred spot considered to be St. Peter's burial place. Today the Necropolis presents itself with walls of considerable size which can be found in a good state of conservation, although the view is at times obstructed by modern pillars needed to sustain the above structures. These proved necessary after the soil removal following the excavations carried out in the early 1940s. The explored area of the Necropolis covers an extension of about 69 x 18 meters and presents itself as a double row of burial edifices separated by a narrow passageway. In particular, the M sepulcher is inserted between two other structures of which it uses the outer walls; this makes it one of the last, chronologically speaking, constructions of the Necropolis (Apolloni Ghetti et al. 1951).

3. The M sepulcher, known as *Cristo Sole* (Christ the "Sun God")

The M sepulcher (figure 2) was discovered in 1574 during an excavation carried out under the pavement of St. Peter's Basilica (Guarducci 1997). Thanks to the discovery of an epigraph (of which only a transcription of its text has reached us) and through the study of the wall typology, it was possible to suppose an originally pagan structure, dated from the mid second century AD, and its successive transformation into a Christian mausoleum at the beginning of the third century. A very rich mosaic decoration originally covered the upper portions of the walls and the whole vault, while the lower portions of the walls were decorated with geometric patterned frescoes. Today the mosaic decoration is partially missing, however, it is possible to recognize the original designs thanks to the traces left by the lost tesserae which indicate M sepulcher as the only entirely Christian burial in the Necropolis. On the preparation layer of the mosaic on the west side we can catch a glimpse of the Good Shepherd; on the frontal (north) scene we can see the Fisherman (Christ or Peter) with a fishing rod; on the eastern most hand side lies the depiction of Jonah being swallowed by the Whale and finally, on the vault lies the image of Christ as the "Sun God" ascending to the sky on a chariot drawn by four horses; the later image is surrounded by an interlacement of vine branches and leaves (Guarducci 1997, Apolloni Gheti et al. 1951). The complexity of the structure, the decorations and the indubitable importance of this sepulcher were decisive factors in

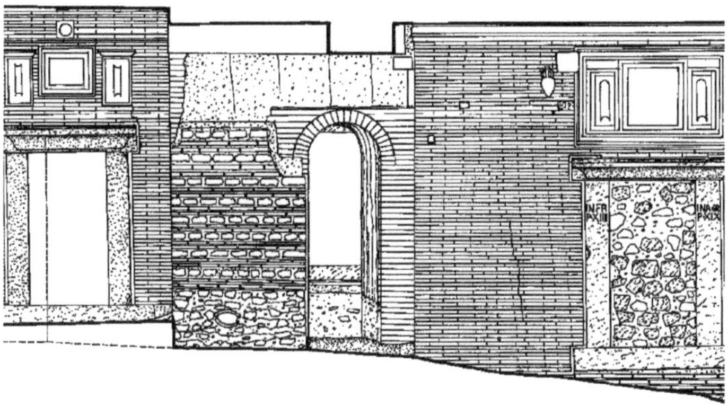

Figure 2: Entrance to the "M" Mausoleum, known as the "Cristo Sole" (Christ the Sun God).

selecting it as an example for testing the integration of different documentation and analysis methodologies.

4. The Micro-topographic survey

The survey and graphic documentation of the Necropolis plan was carried out (thanks to a Sokkia SET 4C total station, figure 1) by creating an accurate surveying base through the establishment of a closed traverse with branches entering into each and every mausoleum; this involved the placement of around 70 instrumental stations. All mausoleums were documented in detail with all documentation referable to this topographical base (figures 2-3).

The surveying instrument also proved very useful at recording the salient points necessary to carry out a metrically correct documentation of wall paintings and architectural decorations even through the subsequent use of traditional means (measuring tape, plumb bob, level etc). When faced with the particular conditions posed by the "M" Mausoleum it was decided that a more unconventional approach might not only simplify the task but also provide the capability of representing the object of study in a radically new way: by virtually recreating the actual surfaces of the walls and vault, in full detail, with the obvious superimposition of the decorations (mosaic or painting) they were adorned with.

Although the concept of a computer generated 3D model was not new, it had always been applied to generate, at the most, schematic exemplifications of structures on which to project the images of any surface texture, obtaining a great general effect, but still hardly useful for the purpose of study. We were encouraged in this by the availability of two new instruments: a total station (Leica TC1103) equipped with a coaxially (along the line of sight) mounted reflectorless EDM laser (this laser does not need reflecting prisms to measure distance) and a 3D digitising arm, that is a sort of 3D pantograph (Microscribe 3DLX).

A few words should be said on this last item: the Microscribe is a 3D digitising (figure 4) precision mechanical arm in which high performance sensors embedded in its articulated joints track the position and orientation of the stylus tip as it is moved in space by the operator. By pressing a button or a foot pedal (depending on the accessory in use), it is possible to record the relative spatial position of the tip in any given moment and, through a serial connection, send such data to a computer. There are many CAD and/or 3D oriented software packages capable of making use of such inputs just as if it were coming from a mouse or digitising tablet (with the fundamental difference of the 3D nature of such data).

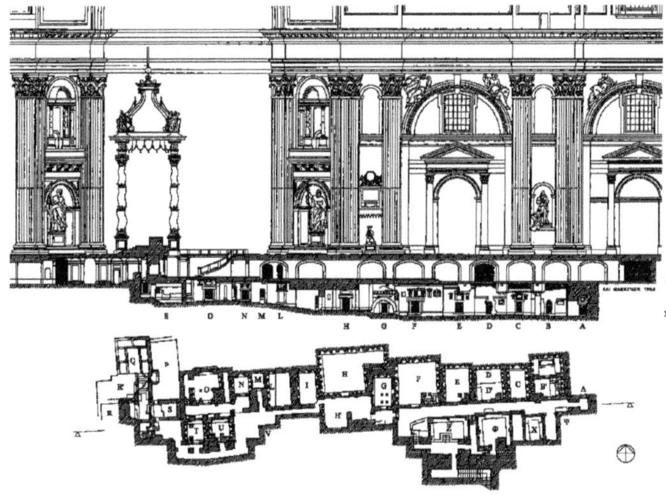

Figure 3: Plan of the Vatican necropolis.

Moreover, the system is highly portable: whether it is set on a table simply sitting on its base or on a tripod it can be placed just about anywhere provided there is a power source and a connection to a computer (a notebook for example). Not only is it possible to record single points but also streams of points by keeping the input button or pedal continuously pressed while the stylus tip is passed over the object we want to digitise; this method is particularly useful in association with software that makes use of "point clouds". Since the first three points recorded during a digitising session are used to establish the reference plane to which all successive points are referenced, it is also possible to fix in a different area, another set of three points during the same session to be used as reference points for a subsequent session. This enables much wider areas to be covered, far beyond the actual reach of the digitiser. As a last consideration we must not underestimate the intuitive and easy operation of this device, with which the operator is always in total control of all activities.

With these instruments, we were able to accurately reproduce a virtual representation of the actual surfaces in extreme detail and accuracy and not only plot points or lines. The "M" Mausoleum presented varying characteristics such as "frescoed" and "once mosaic covered" (though still readable thanks to the traces left by the lost tesserae on the preparation layer) wall surfaces and a vault with a rather undefined shape (a very low cross vault) only partly covered with the polychrome mosaic that once characterised the whole upper portion of the mausoleum.

The first step was to place the laser EDM total station inside the mausoleum onto a predetermined station tied to the main traverse used in the survey of the Necropolis plan. Through the use of this instrument we were able to scan the different wall and ceiling surfaces, carefully distinguishing the brick and plastered surfaces of the left wall, the mosaic preparatory layer, the corbel and frescoed lower portion of the front wall, and the preparatory layer and fresco of the right wall. Furthermore, the total station was used to record points for spatial referencing and geometric rectification of the photographs of various surfaces. Finally, we recorded the points to be used later on in order to reference the 3D digitiser survey sessions (3 points for every session).

The 3D digitiser was used to retrace the outer contour of the only remaining portion of the mosaic placed on the vault, point by point. A series of parallel lines running over the vault's surfaces were

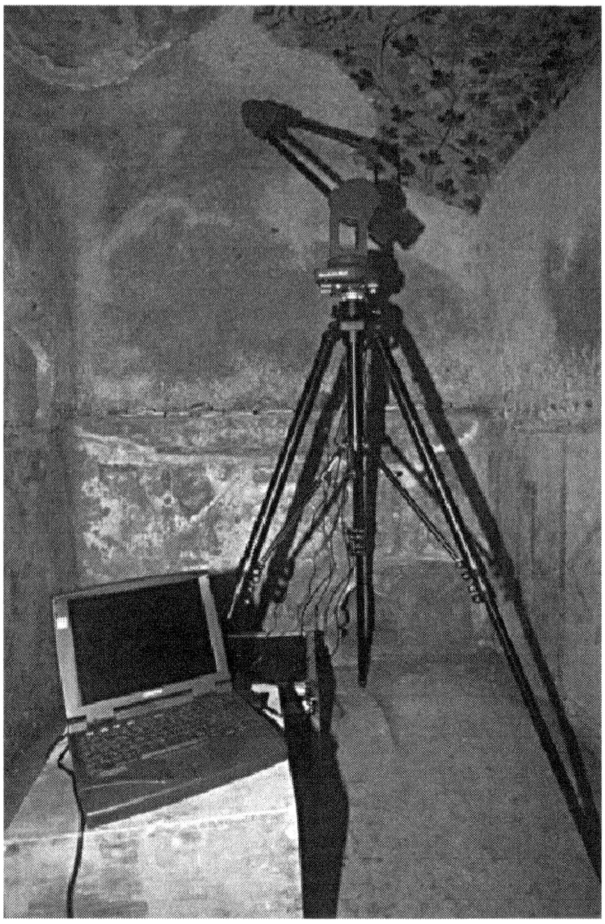

Figure 4: The Microscribe 3D arm.

also digitised to be used as a basis for their reconstruction. This procedure was chosen given the insufficient clearance between the vault and the total station for the laser rangefinder to work properly. All the frescoed designs and the traces of figures left on the lost portions of the mosaics were digitised with the Microscribe Arm and, as a further test, a detail of the face of Christ was digitised tessera by tessera prior to the restoration operations.

At this point we had a clear indication of how helpful these techniques were in obtaining metrically correct graphic documentation directly on site, through the detailed examination of all elements necessary for the interpretation and comprehension of the structures and of the decorative apparatus.

The software used in connection with the digitising arm (Rhinoceros 3D, figures 4-5) not only gave an interface between the instrument and the user, but it also permitted us to digitally recreate shapes, lines, curves and surfaces in their exact spatial position and, most of all, in real time.

This data (through further processing) was instrumental in the construction of our 3D model, to be used either for the extraction of two dimensional representations by projecting it onto whichever plane we desired to observe it on or, by retaining its 3D characteristics, to use it in one of the many applications that made use of spatially defined entities. The model thus created was, for example, used to map the information gathered during a thermal infrared scanning analysis carried out on the frontal (North) wall of the Mausoleum.

To insert the infrared imagery and the result of its processing into our metrically correct representation of the wall, a series of marks,

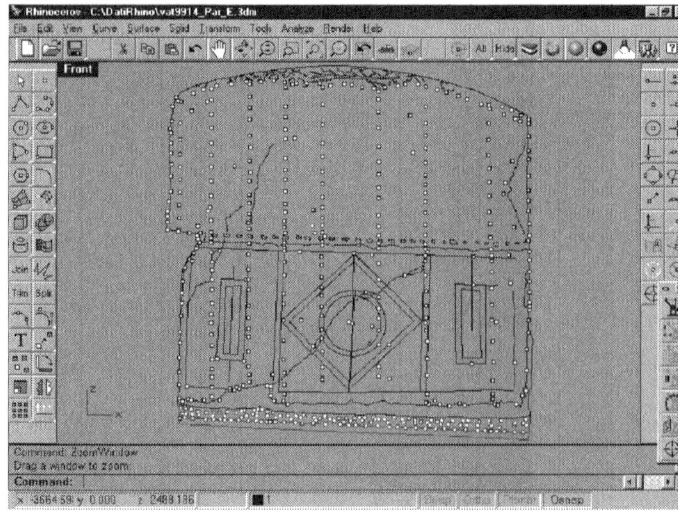

Figure 5: Rhinoceros software interface.

visible both in visible and thermal infrared bands, were placed on the surface and surveyed.

The thermal infrared scanning was carried out with the Marconi TICM II, a high resolution system, operating in the 8-13 micron wavelength range. The images were processed through dedicated software by MEDIA CYBERNETIC, applying the appropriate filters to enhance the readability.

In the example image, captured during the cooling phase, we saw a presence of warmer areas (yellow-red) which correspond to possible slight detachments of the plaster layer from the brick surface. The cooler (blue) areas indicate the presence of a humidity front. By processing the infrared image, we can produce thematic maps and in this particular case we have recorded and outlined the above mentioned areas.

The software used throughout our documentation campaign comprised of AUTOCAD for the acquisition and projection of topographical data, CAD OVERLAY for the processing of orthographic photographs and RHINOCEROS for 3D digitiser data acquisition and subsequent modelling of various surfaces.

RHINOCEROS, in particular, is a tool that enables us to produce mathematically defined (NURBS) surfaces (as opposed to TINs or grid meshes) by combining a series of parallel or criss-crossing curves.

The most complex surface to process was certainly the vault: both because it consisted of two separate surfaces (mosaic and preparatory layer) and also due to its non-regular geometric shape.

Through the use of curves obtained by 3D digitising, the software enabled us to digitally reproduce nearly exact replicas of the surfaces and eventually project images or drawings onto them.

In short, we started with a decorated surface, we surveyed it obtaining a 3D model, and we then digitised its decorative elements in order to obtain a metrically correct base for their successive manual graphic documentation. At this point we were able to reintegrate the surfaces with their now spatially referenced pictorial images in order to be able to visualise the complete model from different points of view.

Figure 6: N wall: GIS overlay of contour levels.

5. GIS and visual information

The main goal of our project was a detailed virtual 3D reconstruction of micro-models of paintings documentation by digital techniques and through different information layers, in order to obtain a final "cognitive" representation (Forte 2000). In the case of a monument such as a tomb, the best cognitive micro-representation is increasing the level of knowledge and of interpretation, practically de-constructing in the first phase the pictorial and architectonic information, finally, recombining every component of the structure into a unique, whole virtual model. The process of de-construction and re-construction is a cognitive dynamic process of knowledge that we think is fundamental for a new approach to digital graphic representation: we disassemble and then reassemble information. Thus a model becomes an articulated set of multi-layered information, where the whole increased model is more significant than each single component.

In this context, 3D GIS (Geographic Information System) and virtual reality technologies have been fundamental in obtaining good results. In fact, even though GIS software is typically used in territorial studies and analyses, it can give very interesting results also in other non-traditional contexts. Furthermore we have to consider GIS as an open planning platform, where the scientific user chooses the directions of research, combining several layers of documentation.

All the work phases have been carried out exchanging all data online (by Internet tools) and creating a network of persons with different roles and functions: topographic survey and acquisition (A. Bizzarro, S. Tilia), data processing (S. Tilia), rectification and topological referencing of images (A. Tilia), GIS implementation and 3D processing (M. Forte), VRML reconstructions and virtual reality applications (A. Tilia, S. Tilia, M. Forte).

In order to process the digital data the following software was used:

- ER Mapper 6.1 (3D visualisation and draping, infrared spectral analysis);

Figure 7: N wall: GIS 3D overlay of different layers (infrared processing, contour levels, drawings, outlines).

- Arcview 3.1 (GIS data-entry and overlay);
- VRML Cosmo Player (3D visualisation of the reconstructed model).

In our case the most important work phase was the superimposition of data, that is the representation of data in a unique referenced co-ordinate system, through the exact use of a GIS implementation.

Starting from the importation of data within the GIS, the main steps of the working project have been the following:

- Data entry of vector (drawings) and raster (images) files.
- Topological referencing of raster (photomosaics and graphics) and vector (decorations, drawings) layers in Arcview (figures 6-9).
- 3D interpolation of data by CAD and GIS software (Arcview 3D Analyst) for the creation of 3D surfaces (interpolation by GRID and by TIN, figure 8).
- 3D Superimposition of all the geometric and graphic layers (figures 6-9).
- 3D Texture mapping of the graphic documentation over the model and draping of all the referenced data over the microDEM of the walls and of the vault (figure 9).
- Real time visualisation of the virtual models (in our case 4 models for the documentation of the tomb) using Er Mapper and the 3D Analyst module of Arcview.
- Merging all types of data systems in order to analyse and to compare the different information layers (raster and vec-

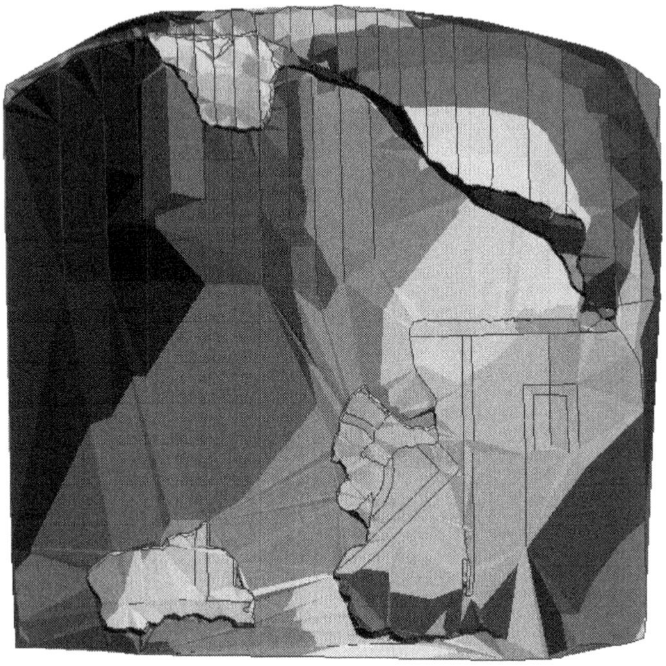

Figure 8: E wall: grid interpolation and vector overlay.

tor, such as images, contour lines, polylines, graphic edges, interpretations and so on).

- Z exaggeration of the interpolated models in order to stress the surfaces for testing the state of conservation of the mural paintings, and the relationships between mosaics, graphic decorations, architectonic elements, lack of materials, etc.
- Model reconstruction in VRML (Virtual Reality Modelling Language, see below) in order to visualise and to interact the data online using the Internet or offline on any PC (figure 10).

One of the most important final steps of GIS applications should be the communication of available data online. The chance to put information (raster, vector, 3D, etc.) on the Internet gives scholars and other interested people the possibility to access remote systems, thus enabling them the comparison and analysis of data and finds. The development of multimedia applications within a GIS environment is now in progress but it will represent one of the most interesting possibilities for communicating information in the future. In particular the Internet or a net of georeferenced 3D databases and graphic libraries could constitute the final goal of GIS, giving the users the possibility to obtain and add information in real time. In our case, using the 3D analyst module of ArcView, the 3D GIS models were reassembled in VRML 2.0, so as to explore them interactively through Internet browsers.

5.1. 3D Interpolation

Interpolation methods allow the creation of 3D surfaces starting from scattered elevation points (or Z values); the choice of the method of interpolation influences the interpretation of the model, thus it is important to know the main techniques in order to obtain good results. In fact topographic, topological or digital terrain data, once "gridded", can be processed and displayed to highlight important physiographic and/or geometric features in territorial or architectural contexts. The 3.0a release of 3D Analyst involves two methods of interpolation, Grid and TIN (Triangulated Irregular Networks).

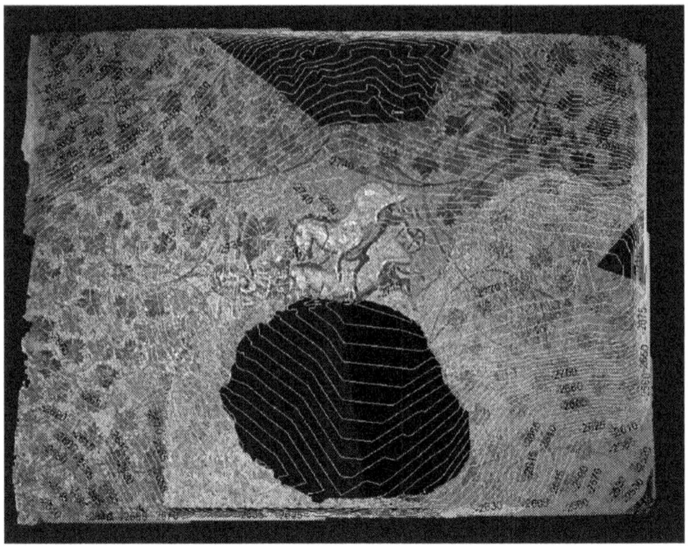

Figure 9: 3D visualization of the vault with contour levels.

- Grid interpolation is a representation of surfaces using a mesh of regularly spaced points. The grid model is simple and processes on them tend to be more efficient than those on other models. Elevation data in the grid format is relatively abundant and inexpensive. However, the mesh structure prevents linear features from being represented sufficiently for large-scale applications.
- On the other hand, Triangulated irregular networks (TIN) represent surfaces using contiguous, non-overlapping triangle facets. One can estimate a surface value anywhere in the triangulation by averaging node values of nearby triangles, giving more weight and influence to those that are closer. The resolution of TINs can vary, because they can be more detailed in areas where the surface is more complex and less detailed in areas where the surface is simpler. The coordinates of the source data are maintained as part of the triangulation, so subsequent analysis, like interpolation, will produce the source data precisely; no information is lost. Unfortunately TINs tend to be expensive to build and process. The cost for having good source data can be high and processing them tends to be less efficient than grids. We can consider this step a detailed "microrelief" of the structures.

6. VRML reconstruction

The Virtual Reality Modelling Language (VRML), is a 3D modelling language for virtual reality applications using 3D scenes. On the basis of the definition of the VRML consortium, VRML is an open standard for 3D virtual and multimedia worlds, connected and linked by the Internet. In December 1997, VRML has been classified as an international standard (ISO/IEC-14772-1: 1997) from ISO (International Standard Organisation) and from the International Electro-technical Commission.

The VRML model was created in order to permit standard access on the web to interactive virtual environments; in fact, over a few years, it has become the most used language for the representation of 3D scenes online. Furthermore the new powerful versions of VRML (2.0 and VRML97) allow specific tools for animations and interactions with the users, very important in the field of virtual reality.

Furthermore, in our project, after the GIS applications we completed the 3D reconstruction of the tomb in VRML (figure10), so that the whole monument could be explored in real time, including textures (photo-mosaics of the walls) and geometry.

The VRML model can be visualised using freeware browsers that can be downloaded from the Internet; to view our reconstruction we use Cosmo Player (created by Silicon Graphics, http://www.cai.com/cosmo/). In this way access to VRML models is available to everybody. For instance the 3D models were created and explored using Cosmo Player 2.1 (Silicon Graphics) as a browser. Using the virtual console of the browser, and visualising the model, it is possible to move forward and back, turn left and right, tilt up and down, seeking, sliding, and so on. The interaction in real time and a user friendly interface allows us to describe the model as a cognitive model: in fact during the virtual navigation we perceive new information and new ways of visualisation, without the need to have GIS software for representing the data.

Figure 10: VRML reconstruction of the whole tomb.

7. Conclusions

The digital project of the "Cristo Sole" tomb demonstrates that a visual information system is very important for a deep knowledge of the monument and of the pictorial content, whether for a scientific analysis, or for technological communication (*digital editing*). In fact the key point of the project was to test the possibility of using GIS systems in "monumental archaeology", mainly integrating 2D and 3D data sources. In this case we can talk of increased reality, because the digital technologies have introduced new levels of information, completely unknown before (Barceló et al. 2000). This approach opens new directions in particular for graphic representation, mainly because it encourages us to visualise multiple dynamic models, not only a simple "drawing". An open platform, such as GIS or another specific visual information system, gives the scientific community the possibility to interact, step by step, with a universe of information, always updateable and constantly increasing. In fact it is important to highlight that a virtual model is an open model, expandable with further documentation and interpretation layers (Barceló et al. 2000).

Moreover the use of a 3D tool (Microscribe arm) for creating the models has shown us another innovative system of micro-survey: at the same time it is possible to draw graphic outlines and to obtain architectonic geometry, verifying, point by point, the shape and dimension of the 3D surface.

Finally, in a few words, we can summarise the main results and the significant perspective that this project opens:

- 3D Data acquisition; new tools of 2D-3D data-acquisition through the use of a digital pantograph (Microscribe arm) have shown great potentialities for micro-topological graphic documentation, mainly in situations where it is difficult to operate with other systems.
- 3D Visual Information by a GIS application and spatial analyses. The creation of a visual information system (in our case GIS) has involved a different approach to the graphic representation of mural paintings. A dynamic visual modelling can open new perspectives in the study, analysis and the scientific communication of a painting's documentation.
- 3D Multi-layered analysis. The combination and the overlay of different sources of data increase the level of knowledge of the site on the basis of data-crossing; a more advanced interpretation depends on the superimposition of referenced data.
- Monitoring of the monument through time. For monuments such as the *Cristo Sole* tomb, it is fundamental to monitor the state of conservation, mainly analysing all the factors of alteration (chemical, physical, architectonic) in a multi-layered visualisation (with the superimposition of graphic information).
- 3D Virtual reconstruction. By assembling every 3D component we have the possibility to visualise and study the model through an interactive and full visualisation, in particular excluding the difficulties existing with real views on site.
- The next stage of possible processing concerns the creation of a complete 3D database of the Vatican Necropolis with further information layers of non-destructive chemical-physical analyses.

Acknowledgements

Special thanks are due to His Eminency Cardinal Virgilio Noè, President of the *Fabbrica di San Pietro*, for having authorised the publication of this paper and to ENEA for the thermal infrared analysis.

References

APOLLONI GHETTI, B.M., FERRUA, A., JOSI, E., KIRSCHBAUM, E., 1951. *Esplorazioni sotto la Confessione di S. Pietro in Vaticano*, Vol. 1, 38-43, Città del Vaticano.

BARCELÓ, J.A., FORTE, M. and SANDERS, D., 2000. *About virtual archaeology: disorders, cognitive interactions and virtuality*. In Barceló, J.A., Forte, M. and Sanders D. (eds.), *Virtual Reality in Archaeology*. ArcheoPress, Oxford (BAR, International Series 843).

GUARDUCCI, M., 1997. *Fondamento della Chiesa. Itinerario nei Sotterranei della Basilica Vaticana*, 20-21, Città del Vaticano, 1997.

FORTE, M., 2000. About virtual archaeology: disorders, cognitive interactions and virtuality. In Barceló, J., Forte, M. and Sanders, D. (eds.), *Virtual reality in archaeology*, Oxford, ArcheoPress (BAR International Series S 843), 247-263.

2. Artefact Analyses and Classification

Past, Present, and Future of Quantitative Methods in United States Archaeology

George L. Cowgill

Department of Anthropology, Arizona State University

Box 872402, Tempe, AZ 85287-2402, USA

e-mail: cowgill@asu.edu

Abstract

I briefly review the history of quantitative methods in American (mainly US) archaeology and assess their present status. I then identify topics that seem most likely to be most active in the near future and express my opinions about the directions that I think will be most productive over a somewhat longer term. What topics, what methods, and what general approaches should be given special attention and encouragement? What roles should quantitative methods play in the archaeology (or archaeologies) of the twenty-first century?

Key words: quantitative methods, United States, typology, seriation, chronology, "resampling", Bayesian inference, electronic knowledge, text-books

1. Introduction

I very much appreciate the invitation to address this plenary session. It gives me an opportunity to update earlier reviews that I published in the 1980s (Cowgill 1986, 1989). In preparing it I have been greatly helped by some more recent reviews, notably that by Aldenderfer (1998), who astutely covers a number of topics in more detail than I can here. I will not attempt to speak for all the Americas. I concentrate on North America, and mainly on the United States. I must also admit that I have reviewed recent literature less thoroughly than is desirable, and I may have missed some important developments. I focus on those of which I am most aware, and which I find most interesting. Non-quantitative computer applications are abundant, but I have little to say about them here.

Quantitative methods in US archaeology most certainly do have a past, a present, and a probably expanding future, but they have never had a really large role in US archaeological theory or practice. I think this is partly because all but the simplest require mathematical skills that are difficult and uncongenial for most archaeologists. It is also because they have never yet achieved results important enough to make most archaeologists feel they had no choice but to overcome their aversion to mastering the concepts and methods. To identify oneself as a "new" or "processual" archaeologist, one didn't really have to learn much. Read a little Hempel, a little Kuhn, a little Binford, maybe a little popular systems literature, acquire a few key words like "technomic", "positive feedback", and "homeostasis", and one could join the bandwagon. More recently, one can read a little Giddens, a little Bourdieu, a little Foucault, a little Hodder, expand one's vocabulary with words like "inscribed", "decentred", "essentialist", "metanarrative", and "agency", and one can proclaim oneself "postprocessual". This is not to say that nothing useful has been done under any of these headings or that none of these terms express useful concepts – it is just that in addition to the good work there has been much that is facile and shallow. In contrast, it has never been that easy to validate one's credentials as a quantitative person – or at least not for long – fakers are too easily identified. At the same time, inept or downright wrong applications of quantitative methods are still easy to find, even in peer-reviewed publications. At first this sounds paradoxical. I think the explanation is that people are cautious about claiming to be qualified to propose innovative methods and the work of those who do claim to innovate is likely to be subject to scrutiny, while the majority who think they are only applying routine and unproblematic methods are unlikely to be reviewed by anyone who knows better about the quantitative aspects of their work. Also, much of the problematic use of quantitative methods is not so much wrong as needlessly limited – available possibilities are not taken advantage of.

I also note that very few people in the US spend most of their time on quantitative methods in archaeology. I judge that I put somewhere around a quarter to a third of my total professional effort on quantitative topics, including research, classroom teaching, and advising graduate student research. It is not so much that I couldn't do more along quantitative lines as that it would have to be at the expense of other things I also want to think about and work on. I think this is fairly common in US archaeology. Often individuals go through a phase where much of their work is quantitative and then, without precisely becoming disenchanted with quantitative methods, they find other research interests increasingly competing for their attention and increasingly rewarding. Also, except for Dwight Read at UCLA, I do not know of any statisticians or mathematicians in the US who have devoted sustained attention to archaeological topics. We generally try to do it by ourselves, or else get some *ad hoc* advice from statisticians on specific projects, which tends to mean that they rummage through their particular tool-kit of techniques they like to see if there is anything they can recommend for our problem, or our problem as they understand it.

At least in the UK, there are a few archaeologists who seem to be spending most of their time on quantitative issues, and a few statisticians and other mathematicians who have spent years seriously addressing quantitative issues in archaeology and developing methods appropriate for these topics. This is probably why most, though not all, of the good English-language textbooks on quantitative methods in archaeology have been written by scholars in the UK.

This is perhaps as good a place as any to deplore another tendency in training archaeologists in the US, which is the overwhelming monolinguality of US education. It is quite possible to obtain a doctorate in many institutions without even a working reading knowledge of any language but English. It is simply unrealistic to expect the majority of our students to be able to read anything in other languages. Often enough this applies to professionals as well. Even publications in French have very slight impact unless they are translated into English. Correspondence Analysis, for example, was virtually unheard of in the US until publications about it began to appear in English (notably those by Torsten Madsen and his group in Denmark).

I am afraid I am beginning on a critical and perhaps even querulous note, and perhaps indulging in a little too much complaining. It is perhaps just as well to get that part done with at the outset, so that I can turn from national self-criticism and look at accomplishments and future possibilities.

The physical sciences are noted for the surprising success of quantitative methods in expressing key concepts and formulating explanatory relationships among the entities involved. This has markedly not been the case in the social sciences, except to a limited extent in some domains of economics, geography, and demography. Some think this is because we haven't tried hard enough and still don't have the right concepts, the right data, or both. Others argue that it is because the so-called social "sciences" are about matters that are intrinsically impossible to express mathematically. I will not try to resolve this debate in this paper. For present purposes it is enough to say that quantitative methods are not without their uses for archaeology.

I turn to a more detailed discussion of quantitative methods among US archaeologists under three headings: data organization and description, use of statistical inference based on probability theory, and mathematicization of theory.

2. Data organization and description

A tremendous amount can be learned simply from well-designed tabulations of counts and proportions of relevant categories. Of course, defining categories in reasonably good ways is a major undertaking in itself – or at least it has been considered such in US archaeology. Nevertheless, in many cases archaeologists have succeeded in establishing categories that are clearly useful for them and meaningful in some sense. Even if they are not the best possible categories and do not employ the best concepts for thinking about categories, they have been quite helpful. At least by the second decade of the twentieth century some US archaeologists were using percentage tables to good effect; especially in deducing chronological sequences (e.g. Kidder and Kidder 1917).

Graphics, especially histograms or bar charts, complement descriptive tables and can be immensely valuable for getting ideas about what is going on. For variables measured on interval scales, simple bivariate scatterplots for pairs of variables are also relatively easy to understand and often highly informative. The importance of well-chosen graphics has been emphasized repeatedly (e.g. Tufte 1983, 1990, Cleveland 1985, 1993) but it bears further repeating.

I emphasize the value of descriptive tabulations and graphics for three reasons. First, there is a strong temptation to underestimate them and to move too quickly to more advanced techniques. This is, of course, all the more the case when demonstrating prowess and seeking prestige is more at stake than actually understanding data as economically and as validly as possible. The ever-growing availability of statistical systems for computers makes it all the easier and more tempting to skip the supposedly pedestrian stuff and leap directly to the more fashionable and mysterious. However, if the data do not reasonably fit the assumptions of more advanced methods, the results of these methods can be very misleading. Simple tabulations and graphics that stay close to the relatively "raw" data are not only instructive in themselves; they are also excellent for revealing departures from the assumptions of more advanced methods, as well as for suggesting which advanced methods are likely, or unlikely, to offer further understanding. I strongly agree with Robert Whallon's (1987) call for more and better use of simple statistics.

Second, when simpler methods are in fact used, even today it is all too easy to find inappropriate or needlessly awkward uses. Bivariate rectangular tables of counts of categories for each of a number of collections or other data sets, for example, may be "standardized" by replacing them with percents of the grand total, rather than with percents of row totals or column totals. The insights gained by replacing counts by percents of the grand total are, at best, very slight, and the much greater insights furnished by percents of row and/or column totals are missed. The prevalence of bad practices such as this is perhaps decreasing in US archaeology, but it is still higher than it should be.

Third, I think we still have a good deal to learn by way of making graphics even more effective than they usually are today. I expect that development of improved graphic practices will be important in the near future. I hope that inappropriate and relatively unhelpful choices will be made less often. There is also too much unthinking reliance on default options offered by computer systems, and often not enough enterprise and imagination in "playing" with ways of showing the data. I don't, of course, suggest that we should fiddle with the display so as to make it seem to mean what we would like it to mean, but rather that if we really want insight we can learn a lot by looking at the data in different ways. Also, the exercise of exploring the extent to which different graphics choices make the data look different is itself instructive.

3. Statistical inference

Unlike quantitative description, statistical inference scarcely existed in US archaeological practice before the 1950s. Archaeologists always, of course, made use of intuitive notions that some observable patterns were too strong and based on too much data to be at all likely to be merely accidental. But very rarely did they make any use of formal statistics grounded in probability theory. Sometimes this resulted in failing to perceive, or at least failing to trust, patterns unlikely to be simply accidental. I suspect that it has more often resulted in treating patterns as "real" even when their empirical support was quite problematic.

In the 1950s interest in methods based on statistical inference was considerably increased through the efforts of Albert C. Spaulding, then at the University of Michigan. Especially notable was his use of what we would now call discrete multivariate analysis for the empirical discovery of patterning in sets of artefacts (e.g. Spaulding 1953). His general approach was to characterize individual objects in terms of a number of variables. Univariate distributions of interval-scale variables were searched for multiple frequency modes, and each such mode was used to set up a nominal category. That is, if a tabulation of measured object lengths revealed three distinct modes, these might be defined as "short", "medium", and "long", and a second "length" variable created that was treated as nominal with three categories (even when, as in this case, ordinal information is preserved). These new variables derived from interval measurements were combined with other variables that were intrinsically nominal, and the set of variables subjected to discrete multivariate analysis (DMA). Spaulding's earlier work generally used only three variables at a time. For example, one could explore the relations and interactions among surface treatment, shape category, and tempering materials for some set of pots; usually those from a single period of a single site. Chi-square was the principal technique used by Spaulding in his earlier papers. Later, as computer-aided (DMA) techniques became available, it was feasible to consider more variables jointly.

In retrospect, I think Spaulding's main contribution was to make US archaeologists far more aware that they might be missing something important if they didn't know a little about statistical methods. On the whole, I do not think that the specific approach used by Spaulding has proven very productive. For one thing, he persisted in reformulating DMA results in terms of "types". The issue here is *not* the ontological status of types, whether they are *discovered* (as Spaulding argued) or *imposed* (as James Ford argued). The answer to that particular debate is that Spaulding and Ford were doing different things with different kinds of data sets (Spaulding working within single periods of single sites and Ford working across multiple sites and periods) and they were both right in describing what they did as discovering or imposing. For present purposes, the relevant point is that it seems DMA results are best and most naturally expressed in terms of the relations and interactions among the variables, and expressing the results in terms of type categories seems forced and less useful.

Also, Spaulding's approach to formulating types has found very little use in US archaeological practice. Nor have other quantitative approaches to typology. For example, I have worked a fair amount with the ceramics of Teotihuacan (an immense prehistoric city in central Mexico), and I have felt fairly comfortable with a typological system that was developed by workers who used methods quite different from Spaulding's. I am interested in improving on this system, but mainly in ways other than through DMA (Cowgill 1990).

There was a surge of interest in quantitative methods in the 1960s, connected with the advent of mainframe computers that could be programmed in languages, such as Fortran, that were not beyond the skills of a few archaeologists, as well as early versions of statistical systems, some of which are still with us in descendant versions. Much of this was connected with the aspirations of the "new archaeology" to make archaeology scientific. A number of multivariate techniques were used or misused, notably multiple regression methods and "factor analysis", which usually turned out, in those cases where the work was well enough documented to allow one to tell, to be some form of principal components analysis. Most of this 1960s work has not stood up well to subsequent criticism, which is probably one of the reasons why the use of quantitative methods has not expanded more rapidly in the US.

Also in the 1960s there were a number of efforts to use quantitative methods of matrix-ordering to seriate archaeological assemblages (Brainerd 1951, for reviews see Cowgill 1972 and Marquardt 1978). Overwhelmingly, the goal was to derive improved chronological orderings. My impression is that this activity took place rather independently of New Archaeology, and was motivated more by the recognition, at least in some quarters, that archaeology of any persuasion suffers greatly when chronologies are uncertain or imprecise. The methods were generally inherently unidimensional and intrinsically dependent on iterative heuristic searches that (a) could not be guaranteed to arrive at optimal arrangements and (b) could not reasonably cope with datasets in which two or more independent axes were required to give a good representation of the patterning. The methods of chronological ordering of the 1960s have been largely replaced by correspondence analysis.

"Numerical Taxonomy" methods also date back to the 1960s and 70s, largely inspired by the influential books of Sokal and Sneath (1963) and Sneath and Sokal (1973). Mainly this meant various forms of agglomerative hierarchical clustering. The tree diagrams generated by these methods may have considerable value for questions of biological phylogeny, but they rarely can capture more than a fraction of the interesting patterning in archaeological data sets. They can have real value as a sort of rather quick and rough exploratory technique, early on in an analysis, but they are too limited (and often misleading) to be respectable end products. It is notable that, nonetheless, they are still treated as end products of analysis by some US anthropologists, though perhaps rarely any longer by US archaeologists.

Of a number of developments in the 80s perhaps most notable was, in a pre-GIS environment, several methods for spatial analysis. Important examples are the use of k-means clustering in the work of Whallon (1984), Kintigh and Ammerman (1982), and Kintigh (1990). Kintigh (1988, 1989) also created programs to control for sample size in comparing collections on the two diversity measures called "richness" and "evenness". These programs avoided some of the problematic assumptions of other methods by using multiple Monte Carlo simulations - what we now call "resampling".

I turn to the present and future, and to methods and topics that particularly interest me. I have little to say about GIS, simply because it hasn't interested me much personally, although I and my students have done a fair amount of spatial analysis, mostly in pre-GIS days (e.g. Cowgill et al. 1984). A number of my colleagues at ASU have used it, mainly as an adjunct to more mathematical methods (Robertson 1999). Kvamme (1998) provides a recent overview of this topic. GIS certainly has a very wide appeal in US archaeology. I think this is mainly because archaeologists tend to be visual more than mathematical, and GIS can produce marvellous pictures that are often very interpretable to the innumerate. Sometimes, perhaps often, such pictures can tell the story so clearly that anything more seems hardly needed. Nevertheless, more mathematically sophisticated methods of spatial analysis have great promise and are still considerably under-utilized in the US. This situation will probably improve, especially if computer systems for these analyses become more available.

I think that many varieties of so-called "postprocessual" archaeology have been too neglectful of chronology building, perhaps because it seems too scientistic and perhaps even ethnocentric – if linear time didn't matter to *them* why should it matter to *us*? But it does and should matter to us. Some of my colleagues who work in parts of the US Southwest where dendrochronology enables them to date phenomena to within a decade or so are able to ask – and answer – important questions that are simply beyond us in many parts of the world. Many of these questions concern matters of agency and context that I think are on the cutting edge, conceptually. So, I both urge and predict that we continue, by every means available, to develop quantitative methods that will enable us to refine our chronologies.

Methods of exploratory data analysis (EDA) are becoming quite popular. This, again, is probably because many of the methods produce such readily comprehensible results. Most anyone can correctly grasp concepts like median and midspread, and box-and-whisker plots are proliferating like rabbits. This is all to the good. I might have mentioned this topic under the heading of data organization and description. It surely has a place there, but the results from EDA also often have a bearing on statistical inference. I hope it is becoming generally recognized that EDA methods belong in our tool kits as partners to so-called "confirmatory" techniques, rather than as replacements for them or somehow in opposition to them.

Multivariate statistics will continue to be used a good deal. I do not know whether any radically new techniques are in the offing.

Resampling methods will surely continue to increase in popularity. They avoid the need for many of the assumptions about shapes of distributions that are often so dubious for archaeological data. Many such methods are already very feasible on antiquated (i.e. last year's model) desktops and laptops, and as computer power per dollar continues to increase exponentially, these methods will become all the more feasible and attractive. An interesting book by the late Julian Simon (1997) expounds this approach on an elementary (perhaps too elementary) level. There are, of course, numerous more advanced books on bootstrapping and other resampling methods (e.g. Efron and Tibshirani 1993).

This leads into a topic that particularly interests me; that of Bayesian concepts and inference. It has long seemed to me that the "classical" (i.e. Neyman-Pearson) approach to statistical inference uses a tortured logic that simply doesn't get across to most students and may be one reason for thinking statistical inference an arcane and perhaps not very applicable topic. Certainly, in practice, most archaeologists who think they are following the classical logic of inference tend to be, in fact, folk Bayesians. Especially, the probability of a sample statistic, conditional on the truth of some hypothesis about the actual value of the corresponding population parameter, is routinely confounded with the probability of the value of the population parameter, conditional on the observed value of the sample statistic.

Shortcomings of a classic outlook on inference can be seen in the otherwise excellent text by Shennan (1997). I feel that he tends to be a little too dubious about formal statistical inference in general, and then, when he does use it, to do so in a rather too formulaic and rigid way that, indeed, warrants some scepticism. In this regard, the text by Drennan (1996) is considerably better in its general outlook on inference, although it does not explain how to carry out any Bayesian methods and covers a more limited range of topics than does Shennan. I have been experimenting with using both books jointly in a first-semester graduate course.

As I see it, Bayesian inference offers both conceptual and practical advantages. With "uninformative" priors, as I understand it, the results are usually the same, or nearly the same, as those obtained by classical inference, so there is little or no practical advantage. The main advantage is probably the conceptual one; that one understands that one can properly speak of the probability of the population parameter, but only because one is knowingly accepting a particular prior probability distribution for the parameter.

When there is a good basis for an informative prior, however, there can also be considerable practical advantages. The impression still seems to be prevalent that Bayesian methods, even if logically preferable, are too difficult computationally to be practical. This may still be true for many statistics, but there are certainly some important exceptions. One such is using proportions of artefact categories in collections to make inferences about the proportions of these categories in the populations represented by the collections. Using beta distributions (Robertson 1999) or Dirichlet distributions (Neiman et al. 2000) the computations are not excessively difficult for modern computers.

In other cases the Bayesian computations are more difficult and require resampling methods, but these have also been put to very good use for important topics, notably by Buck et al. (1996 and numerous papers). An outstanding example is their application to the problem of making optimal chronological inferences from a combination of calibrated radiocarbon dates and other kinds of evidence, such as stratigraphy. The facts that issues of chronology are important and that some of their methods are incorporated in the OxCal program combine to make this Bayesian approach to chronology very attractive in the US, as well as elsewhere. I understand that, in its present version, the OxCal program does not give satisfactory results with some datasets. It seems, however, that this is a problem that could be overcome before long.

The biggest problem I see that impedes wider adoption of Bayesian concepts and methods is that I know of no elementary book in English that explains them. Even the little book by Iversen (1984) is not easy for our students and covers a limited range of topics too briefly. However transparent and elementary the book by Buck et al. may seem to mathematicians and statisticians, the obstinate fact is that it is far beyond the level that can be grasped by any but a tiny fraction of students or professional archaeologists in the US. Coupled with an absurdly high price set by the publisher and essentially no advertising of the book to US archaeologists, I think its very existence has been a rather well-kept secret, except where a few subversives, such as myself, spread the knowledge by word of mouth or through citing it in publications.

By fits and starts I have been working for some years toward a really introductory book that would explain Bayesian concepts, explain useful methods for a few simple cases, and prepare students for books such as that by Buck and her colleagues. I wouldn't like to predict how soon, if ever, the book will be ready to seek a publisher. If any of you know of such books that already exist, or are in preparation, please let me know.

A second area of great interest is the development of electronic knowledge bases. It may seem odd that I do not put this under the heading of data organization and description. Much work of this kind surely belongs there. However, what is emerging as a truly big problem in the US, and doubtless elsewhere, is finding ways to make effective use of databases created over the years within a region, by diverse projects using diverse methods both for collecting and for describing their data. Methods for durably archiving databases and "migrating" them to new systems and new platforms when existing ones become obsolete are receiving considerable attention, with good reason. However, the problem I have in mind here is a different one, essentially that of enabling researchers who didn't collect the data themselves to compare apples and oranges – to acquire the background information needed to search files for relevant data and to understand what it means when it's located. These problems can certainly be considerably alleviated by electronic means. To me, however, the problem of comparing disparate datasets can sometimes be put in the form of an estimation problem: "If project A rather than project B had collected these data, what would they probably have reported?" This is the only way I see to escape the need to reduce comparisons to least common denominators, which necessarily wastes much of the information in the individual datasets. Estimates, of course, come with confidence intervals. The hope is that one can find ways to combine diverse lines of evidence to derive confidence intervals that are not excessively large. So far this is merely a suggestion for a line of research that, if it can be made to work, should become immensely useful. "Fuzzy set" concepts and methods may also prove useful.

4. Mathematization of theory

At present I have relatively little to say under this heading, except to note its scarcity. Most archaeological theory is not expressed in mathematical terms at all, and perhaps there is good reason for this, at least as long as we use the kinds of math that have been developed for the physical sciences. I note with some surprise that at the April 2000 annual meeting of the Society for American Archaeology there was a symposium entitled "The Mathematics of Cultural Evolution: Taking Stock in the Year 2000". For better or worse, I didn't attend this session. However, the session abstract indicates that some of the papers used formal modelling by means of differential equations. Since I didn't hear these papers, I really cannot judge but (speaking as a Bayesian) my prior probability is on the sceptical side. Previous efforts to apply differential equations to sociocultural phenomena have not been very successful, in my opinion. Certainly none have become popular. For one thing, how do you do justice to agency with differential equations? To be sure, economists have long used differential equations, but only at the expense of postulating excessively oversimplified rational actors rather than reasonable approximations of human beings.

I think that computer-intensive modelling probably has more to offer. Such methods can cope with non-linear relationships among variables and can deal with multiple "agents". As yet I don't know enough about these methods to know whether the term "agent" here means anything that could be translated into archaeological concepts of agency. But it sounds promising.

5. Teaching

In his review, Aldenderfer (1998) discusses at some length the teaching of quantitative methods to archaeologists in the US. A number of programs, even at the graduate level, do not require any course in quantitative methods and, of those that do, quite a few send their students to other departments for that training. In my experience, courses taught in other departments, even when they cover techniques applicable in archaeology, simply don't get across to the students very well how to apply them to archaeological situations. And, of course, such courses often spend considerable time on techniques and issues not very relevant for archaeology, and too little time on techniques and issues that *are* relevant for us. I suspect that those departments that do not require any in-house courses in quantitative methods generally have no faculty member qualified to teach such a course.

I think we are still caught in a vicious cycle in which, without more students who are somewhat sophisticated quantitatively, we have not produced quantitative results of such universally recognized significance that they *demand* that the field as whole pay more attention to quantitative training. Of course, as I've said before, it may also be that our subject matter is simply not very amenable to quantitative methods that can yield really big payoffs. This is still a question for the future to decide.

A welcome change since the mid-80s is the appearance of several books that are on an appropriate mathematical level and written by archaeologists or at least by experts with a real feel for archaeological situations and issues. Notable examples on an introductory level are Shennan, first published in 1988, now in a substantially revised second edition (Shennan 1997) and Drennan (1996). Important more advanced books include Baxter (1994) and Buck et al. (1996). These are all extremely useful for anyone trying to teach quantitative methods to students, and hopefully they will have a substantial impact on practice. One thing still missing is an introductory book on Bayesian concepts and methods.

References

ALDENDERFER, M., 1998. Quantitative Methods in Archeology: A Review of Recent Trends and Developments. *Journal of Archaeological Research* 6(2):91-120.

BAXTER, M.J., 1994. *Exploratory Multivariate Analysis in Archaeology*. Edinburgh University Press, Edinburgh.

BRAINERD, G.W., 1951. The Place of Chronological Ordering in Archaeological Analysis. *American Antiquity* 16: 301-313.

BUCK, C.W.C. and LITTON, C., 1996. *Bayesian Approach to Interpreting Archaeological Data.* Wiley, Chichester and New York.

CLEVELAND, W.S., 1985. *The Elements of Graphing Data.* Wadsworth, Monterey, California.

CLEVELAND, W.S., 1993. *Visualizing Data.* AT&T Bell Laboratories, Murray Hill, New Jersey.

COWGILL, G.L., 1972. Models, Methods, and Techniques for Seriation. In Clarke, D.L. (ed.), *Models in Archaeology*: 381-424. Methuen, London.

COWGILL, G.L., 1986. Archaeological Applications of Quantitative and Formal Methods. In Meltzer, D., Fowler, D. and Sabloff, J. (eds.), *American Archaeology: Past and Future*: 369-393. Smithsonian Institution Press, Washington, D.C.

COWGILL, G.L., 1989. Formal Approaches in Archaeology. In Lamberg-Karlovsky, C.C. (ed.), *Archaeological Thought in America*: 74-88. Cambridge University Press, Cambridge.

COWGILL, G.L., 1990. Artifact Classification and Archaeological Purposes. In Voorrips, A. (ed.), *Mathematics and Information Science in Archaeology: A Flexible Framework*: 61-78. Holos Verlag, Bonn.

COWGILL, G.L. ALTSCHUL, J.H. and SLOAD, R.S., 1984. Spatial Analysis of Teotihuacan: a Mesoamerican Metropolis. In Hietala, H.J. (ed.), *Intrasite Spatial Analysis in Archaeology*: 154-195. Cambridge University Press, Cambridge.

DRENNAN, R.D., 1996. *Statistics for Archaeologists: A Commonsense Approach*. Plenum, New York and London.

EFRON, B. and TIBSHIRANI, R.J., 1993. *An Introduction to the Bootstrap*. Chapman and Hall, New York.

IVERSEN, G. 1984. *Bayesian Statistical Inference*. Sage, Beverly Hills, California.

KIDDER, M.A. and KIDDER, A.V., 1917. Notes on the Pottery of Pecos. *American Anthropologist* 19:325-360. Reprinted in: Lyman, R.L., O'Brien, M.J. and Dunnell, R.C. (eds.), *Americanist Culture History*: 93-115. Plenum, New York and London, 1997.

KINTIGH, K.W., 1988. Monte Carlo Evaluation of the Effectiveness of Subsurface Testing for the Identification of Archaeological Sites. *American Antiquity* 53(4):686-707.

KINTIGH, K.W., 1989. Sample Size, Significance, and Measures of Diversity. In Leonard, R. and Jones, G.T. (eds.), *Quantifying Diversity in Archaeology*: 25-36. Cambridge University Press, Cambridge.

KINTIGH, K.W. 1990. Intrasite Spatial Analysis: A Commentary on Major Methods. In Voorrips, A. (ed.), *Mathematics and Information Science in Archaeology: A Flexible Framework*: 165-200. Holos Verlag, Bonn.

KINTIGH, K.W. and AMMERMAN, A., 1982. Heuristic Approaches to Spatial Analysis in Archaeology. *American Antiquity* 47(1):31-63.

KVAMME, K., 1998. Recent Directions and Developments in Geographical Information Systems. *Journal of Archaeological Research* 7(2):153-201.

MARQUARDT, W.H., 1978. Advances in Archaeological Seriation. In Schiffer, M.B. (ed.), *Advances in Archaeological Method and Theory, Volume 1*: 266-314. Academic Press, New York.

NEIMAN, F.D., BON-HARPER, S., McFADEN, L. and WHEELER, D., 2000. Deciphering Plowzone Palimpsests with Bayesian Spatial Smoothing and Correspondence Analysis. Poster presented at the Annual Meeting of the Society for American Archaeology, Philadelphia, 5-9 April 2000.

ROBERTSON, I.G., 1999. Spatial and Multivariate Analysis, Random Sampling Error, and Analytical Noise: Empirical Bayesian Methods at Teotihuacan, Mexico. *American Antiquity* 64 (1):137-152.

SHENNAN, S., 1997. *Quantifying Archaeology* (Second Edition). University of Iowa Press, Iowa City.

SIMON, J.L., 1997. *Resampling: The New Statistics*. Second Edition. Resampling Stats, Inc. Arlington, Virginia.

SNEATH, P.H.A. and SOKAL, R.R., 1973. *Numerical Taxonomy: The Principles and Practice of Numerical Classification*. Freeman, San Francisco.

SOKAL, R.R. and SNEATH, P.H.A., 1963. *Principles of Numerical Taxonomy*. Freeman, San Francisco.

SPAULDING, A.C., 1953. Statistical Techniques for the discovery of Artifact Types. *American Antiquity* 18(4):305-313.

TUFTE, E.R., 1983. *The Visual Display of Quantitative Information*. Graphics Press, Cheshire, Connecticut.

TUFTE, E.R. 1990. *Envisioning Information*. Graphics Press, Cheshire, Connecticut.

WHALLON, R., 1984. Unconstrained Clustering for the Analysis of Spatial Distributions in Archaeology. In Hietala, H.J. (ed.), *Intrasite Spatial Analysis in Archaeology*: 242-277. Cambridge University Press, Cambridge.

WHALLON, R., 1987. Simple Statistics. In Aldenderfer, M. (ed.), *Quantitative Research in Archaeology*: 130-150. Sage Publications, Newbury Park, CA.

Artefact Analysis

François Djindjian

Université de Paris 1-Panthéon-Sorbonne, Institut d'Art et d'Archéologie

3 rue Michelet 75006 Paris, France

e-mail: francois.djindjian@wanadoo.fr

Abstract

Historically, artefact analysis has focused on classifications and typologies. In fact, the typological approach may be considered as a particular cognitive step in the evolution of Archaeology. In this text more ambitious actual approaches of artefact analysis are described, relying on obtaining more cognitive information from artefacts. The applicable quantitative methods, which have been developed and tested to solve classical and modern archaeological problems, are detailed. Finally, a general methodology, founded on recent results in semiology, mathematical modelling and neuroscience, may be considered as a contribution to a renewed role of artefact analysis in archaeological reasoning.

1. Introduction

Artefact analysis is the beginning of archaeology. In order to find an answer to the question: "what is this thing?" a question posed when curious remains or ruins were found, scientists created a new science: archaeology. Artefact analysis is a fundamental platform to nearly all formalised archaeological questions. Thus, theoretical and methodological questions concerning artefact analysis are extremely important for our discipline.

2. Artefact: A definition

"Artefact" is perhaps the most popular word in Archaeology. It means a "thing" manufactured by persons and discovered by archaeologists. Of course the word "artefact" is a concept. From a semiotic point of view, it means a "thing" which is not easy to identify and recognise, archaeologically speaking, a "thing" manufactured by persons in the past. Any experienced archaeologist will of course agree with such a proposition.

Firstly, an "artefact" is a physical find, manufactured by persons. More precisely, I would define an "artefact" as the result of a more or less explicit design and of a more or less controlled manufacturing process: a standard to be identified by archaeologists and a variability to be explained (hazards of manufacturing, quality, reproducibility, personal style, etc).

A surprising issue in such a definition, is that an artefact exists by itself (as a standard), but also through the success of the identification process resulting in the interaction between the archaeologists and the artefacts.

But because an "artefact" is a concept, it would be unfortunate to restrict its definition merely to physical finds. The same definition of an "artefact" may also be applied to "logical" artefacts, i.e. any set of physical finds: for example, spatial associations of artefacts, burning structures, taphonomic assemblages, working areas, burials, settlements, buildings, sites, etc.

Therefore, it is interesting to point out the similarities in the formal definition of "physical" and "logical" artefacts, but it is also interesting to emphasise the re-entering feature of the definition, hierarchically from a physical object to the totally reconstructed civilisation.

Perhaps our proposition is just a formalisation of evidence or common sense which archaeologists know well. But unfortunately, the history of archaeology provides us with thousands of examples of deceptive appearance of artefacts such as natural origin, geological alteration, manufacturing waste, random association with other artefacts, misinterpreted spatial structure, etc. History of artefacts in archaeology is also the history of diagnostic errors.

With a methodological perspective, artefact analysis is a fundamental method in archaeology, with special needs to refer to semiology, taxonomy, systemic analysis, and finally cognitive science. At any level of archaeological reconstitution, artefact analysis is directly connected with the question of artefact knowledge.

3. A brief history of the concept of the artefact

The history of the concept of the artefact in archaeology is strongly associated with the history of theoretical approaches in archaeology since the very beginning of this science.

At first artefacts were merely "unknown things" and for that reason, they were collected and stored world-wide in "curiosity cabinets" of the upper classes.

With the progressive discovery of the remains of great civilisations, artefacts became selected beautiful pieces from burial excavations or famous archaeological sites. In the eighteenth century these pieces were kept in large private collections, and in the beginning of nineteenth century the first museums appeared.

The influence of Natural Science during the second part of the 19th century and the first part of the 20th century has led to a concern for the European prehistory field, a concern accompanied by systematisation and classification of artefacts. Particularly important was the establishment of chronology. The concept of "fossile directeur" has been used and re-used as an apparent projection of the Darwinian theory on human systems, materialising the strong influence of the positivist approach.

The concept of "style" in Art History, whether applied to an individual or to a school, varies with time and region, and is characterised by changes in technological, social or ideological context.

In the 1960s, growing out of Linguistics and Semiology, artificial intelligence has included research on exhaustive representation of artefacts, using standardised vocabularies, indices and thesauri.

At the same time, the quantitative movement in social and human sciences based on applied mathematics, information theory and computers, found itself at the beginning of the development of artefact measuring which involved the use of numerical taxonomy and exploratory data analysis.

Archaeology was also influenced by physics, via Archaeometry and physical/chemical characterisation, used to solve problems concerning the origins of raw material, localisation of manufacturing centres and distribution of artefacts.

In archaeology (viewed as cultural anthropology, revisited in the following general movement of neopositivism and structuralism of New Archaeology after the Second World War), artefacts may be analysed as an individual, functional, social and ecological signature.

In the 1980s, experimental archaeology was concerned with an important development focused on artefact technology, use-wear analysis, paleometallurgy, dwelling reconstitution, etc.

In the 1990s, as a reaction against caricature modelisation of the new archaeology, models are seen more as a projection of the archaeologist himself onto the artefact than as a normal top-down formalisation. The post-modern archaeology, more concerned with the modelling of the modeller, is focusing on the archaeologists' views of the artefact and no more on the artefact itself.

During the same time, under the influence of Neuroscience, cognitive approaches are focusing on the interaction between the archaeologist and artefacts as a learning dynamic process, in a attempt to include both the modelling process and the empirical-inductive approaches.

As a consequence, it is impossible to debate artefact analysis without knowing in each case the exact underlying theoretical context involved. It is nevertheless possible to try to conceptualise on a higher level the formalisation of the artefact analysis in order to replace all those theoretical models as particular cases of a more general approach. This is the aim of the present paper.

4. Theoretical background of the artefact

Certainly some archaeologist would wonder if we are not perversely over complicating a non-complex question: a statue is a statue, a hammer is a hammer, a wall is a wall, and so on. This is generally right in all cases where identifying an artefact is trivial. But, in the more general case, identification is the first question, and that is why it is necessary to consider such a theoretical background.

This is not the place to write a new digest concerning an endless philosophical debate carrying on since Aristotle, Kant or more recently Pierce, amongst others, as regards the reality of the concept of things: Nominalism versus Realism. Do things exist because we name them or is true that we can give them a name because they exist, independently from us?

To avoid such a debate where archaeologists have no better evidence than semiologists, I would dare to put forth a personal proposition, influenced by cognitive approaches, which is a process rather than a definition:

> "Artefact existence and identification is a part of the artefact knowledge which in turn is the result of an uninterrupted interaction between the artefact and archaeologists".

The cognitive process may be formalised by a protocol, with the following steps:

- prior perception,
- identification,
- characterisation,
- naming,
- encyclopaedic knowledge,
- determination (keys to recognition).

I would like to emphasise the fact that my proposition is not only a classical one but also may be seen just as renaming various equivalent approaches, in particular two of them, the semiotic approach and the taxonomic approach.

Following Eco (1999), a semiotic approach to the question of artefact identification may be summarised by the following protocol:

- a perceptive semiosis (phenomenology),
- a cognitive type, obtained via:

  personal and informal perception,

  occurrence accumulation (implicit prototyping),

  naming
- a core content (explicit characterisation),
- a molar content (extended knowledge).

The taxonomic approach may be defined by the following protocol:

- a prior analogical perception,
- a quantitative procedure, involving:

  similarities and dissimilarities,

  description and measures,

  partition searching,

  naming
- characterisation,
- encyclopaedic knowledge.

As it is often the case in numerous sciences, apparently different approaches are very similar.

5. The role of artefact analysis in archaeology

Artefact analysis plays a major role in a number of archaeological constructs. The following list is not exhaustive, but representative of certainly 95 % or more of archaeological studies.

- artefact identification and classification,
- "culture" identification or assemblage of variability studies (time and space systems obtained from artefacts),
- seriation (chronology from artefacts),

- intrasite spatial analysis (artefact spatial distribution studies),
- identifying raw material sources and manufacturing centres (artefact production subsystem studies),
- anthropological studies (artefact as a functional, social, individual, hierarchical, tribal, environmental, symbolic signature),
- identifying distribution networks (artefact exchange and trading subsystem studies),
- intersite spatial analysis (artefact for territory identification, peopling, carrying capacity, demography as well as time and space changes).

It is of course obvious to say that the artefacts are of concern in all archaeological studies, but it is also important to point out that the same artefact may serve many different studies, as if an artefact was a limitless source of intrinsic information.

6. Artefact identification

6.1. The concept of type

Unfortunately the concept of types in archaeology has been neglected over the last twenty years as if the theoretical debate concerning typologies was already definitively closed. The reason for such an attitude is probably due to some Natural Sciences sterile heritage considering that artefacts remain widely defined as an existing type in an existing and definitive typology.

Semiology and cognitive sciences have fortunately enhanced the debate concerning the types and classification, allowing us to introduce some new definitions:

1. A *real type* is a type resulting from a manufacturing design, which must be statistically demonstrated on a representative collection of artefacts. Such a type exists independently of the archaeologist, who nevertheless has discovered it.
2. A *virtual type* is a type resulting from a formalisation realised by the archaeologist for a given archaeological construct. The formalisation is validated by the success of the construct.
3. A classification (or typology) is a virtual organisation of real or virtual types.

6.2. Intrinsic and extrinsic information

By definition, *intrinsic information* is information perceived by an archaeologist about an artefact, formalising a (and not the) representation of this artefact. The richness of intrinsic information is the result of the efficiency of a cognitive interaction between the artefact and the archaeologist. In fact, intrinsic information is not information but knowledge.

On the other hand, *extrinsic information* is information recorded from the context of the artefact: stratigraphy, spatial localisation, environment, etc. The richness of the extrinsic information is the result of the quality and the precision of the survey or excavation. Extrinsic information is archaeological information.

Artefacts, intrinsic and extrinsic information define a system. The correlations between intrinsic and extrinsic information initiate a cognitive process increasing the knowledge of the artefacts system (Djindjian 1980). This may concern both the level of the system structuration, the meaning of the intrinsic information and the explaining role of the extrinsic information.

6.3. The multiple representations of an artefact

Multiple representations of an artefact result from different pieces of intrinsic information the archaeologist is able to perceive as regards the artefact. While such information is knowledge, the richness of this information is the result of an uninterrupted cognitive process, having started at the beginning of the archaeological study and with no end in sight.

A non-exhaustive list is given below:

- size,
- morphology,
- physical and chemical characteristics of the raw material,
- manufacturing technology,
- manufacturing quality,
- decoration,
- function,
- use,
- value,
- social meaning,
- individual signature,
- symbolism,
- etc....

There are as many classifications as there are different sets of intrinsic information selected to describe an artefact.

Until the 1960s, lithic and ceramic typologies in European prehistory were generally built from a non-formalised description implicitly involving morphology (supposed to indicate function) and technology. Such a combination of uncontrolled intrinsic information has often produced a mixture of structurations impossible to interpret (see infra chapter 7).

Among other intrinsic information new archaeology has put the highest emphasis on function, individual signature (named style), and social meanings.

Through various separate studies recent experimental approaches have generally focused on technology and use-wear (supposed to be function), and no longer consider morphology.

But in fact, all sets of intrinsic information are potentially, independently but also correlatively, cognitive.

6.4. Searching for a partition

Artefact identification is achieved through the search for a real partition in a multidimensional representation of space artefacts.

The process of artefact identification was first formalised by Spaulding (1953) through his attribute analysis and by Clarke (1962) with his matrix analysis, using quantitative methods available at that time (respectively Chi-square test of statistical independence, and matrix permutation).

The revolution in multidimensional data analysis between 1965 and 1980 has finally provided algorithmic solutions to numerical taxonomy in general as well as to artefact identification. In archaeology, early methodological references are:

- Cluster analysis (Hodson, Sneath and Doran 1966),
- K-means cluster analysis (Hodson 1971),
- Multidimensional scaling (Hodson 1971, Doran and Hodson 1975),
- Correspondence Analysis with cluster analysis (Djindjian 1976),
- Principal Components Analysis with cluster analysis (Christenson and Read 1977, Whallon 1982).

Correspondence analysis is now used successfully in many countries: Boelicke et al. (1981), Bolviken et al. (1982), Moscati (1986), Slachmuylder (1985), Greenacre (1984), Scollar (1985), Ringrose (1988), Madsen (1988), Baxter (1994).

But several improvements are necessary in order to utilise the efficient algorithms and software and to realise a practical method for artefact identification (Djindjian 1991):

- selection of a single set of intrinsic information for classification,
- unambiguous definition of attributes and variables, vocabulary and measure,
- transformation of a description model in a variable list,
- adapted coding and metrics to handle simultaneous analysis of metric and qualitative variables,
- guided description (by defining an explicit protocol for choosing variables),
- searching for resonance (i.e. the minimal description for the best discrimination of clusters),
- computing the precision of the clustering, in terms of probability.
- comparison of classifications obtained for different sets of intrinsic information, recorded on the same artefacts.

6.5. Ceramic classification

A good example of artefact identification and classification is ceramic classification. The question is as old as archaeology itself, i.e. it has been going on since the 19th century (Montelius, Petrie, Dechelette, etc.) when ceramic classification took a major place in the archaeological corpus. Ceramic morphology, being the direct result of the design of a ceramist, is important intrinsic information with respect to a better understanding of manufacturing standards.

Historically, the quantitative methods for morphological classification of ceramics have used two different types of description: numerous measures between various selected points (for example, Mohen, Whallon) inspired by physical anthropology and profile digitalisation inspired by pattern recognition techniques.

Profile digitalisation, which can be entirely automated by a camera, has been increasingly used, and many authors have contributed to the optimisation of algorithms for a better differentiation of ceramic shapes:

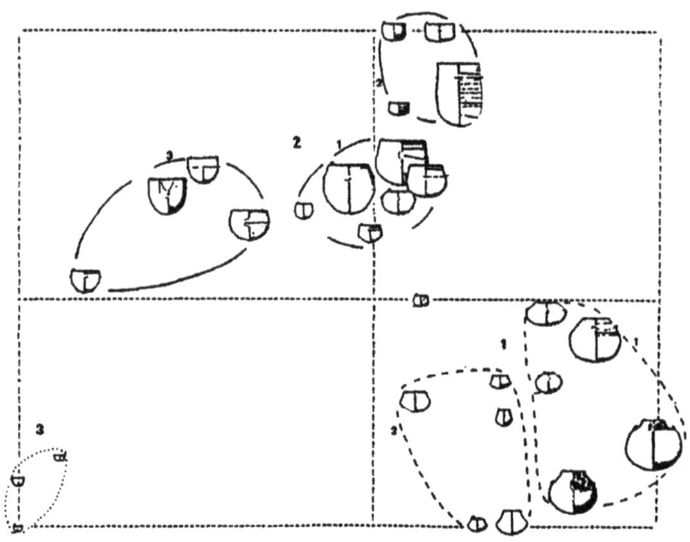

Figure 1: Real Ceramic types obtained by P.C.A. and cluster analysis from the chalcolithic dwelling of Boussargues (Herault, France). (Giligny 1990).

- Sliced method (Wilcock and Shennan 1975),
- Tangent-profile technique (Main, 1986),
- Extended sliced method (Djindjian et al. 1985),
- B-spline curve (Hall and Laflin 1984),
- Fourier series (Gero and Mazzula 1984),
- Centroid and cyclical curve (Tyldesley et al. 1985),
- Two-curves system (Hagstrum and Hildebrand 1990).

For example, the extended sliced method is a coding system specially adapted to separate size from shape and then to decide whether to keep or to eliminate size from the classification.

In the study of an exceptional broken (in situ) ceramic set from the chalcolithic dwelling of Boussargues (Herault, France), Giligny (1990) used a PCA with L_i/L_{max} on eight profile measures. He obtained three real types interpreted as ceramics for storing, cooking and consuming food (figure 1).

In another study on 87 pieces of ceramics from the final Neolithic period at Clairvaux (Ain, France), the same author (Giligny 1990), he classified the shape and size using PCA and profile measures, and extended his classification to other intrinsic pieces of information (technology, decor), furnishing virtual types. Then he obtained by a correspondence analysis a seriation of Neolithic lacustrine sites in Eastern France between 2900 BC and 2200 BC (figure 2).

7. From assemblage variability studies to "Culture" identification

With the question of "Culture" identification, we penetrate the most controversial debate in archaeology, often considered as the greatest gap between European prehistory and cultural anthropology.

In the second half of the 19th century, the discovery of the stratigraphic method provided the only scientific way to establish

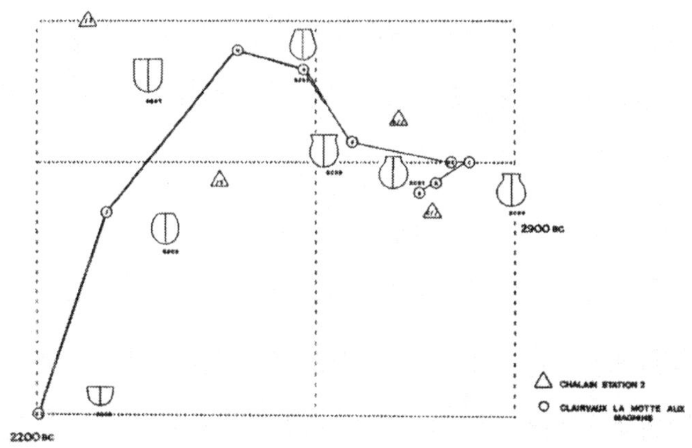

Figure 2: Seriation by C.A. of neolithic lacustrine sites (Eastern France) from virtual ceramic types obtained by separated data analysis (P.C.A., C.A. and cluster analysis) applied on shape and size, technology and decor. (Giligny 1990).

a (relative) chronology. The association of "fossile directeur" chronology and space permitted the definitions of a methodological approach still used today, to structure prehistoric data. But these structures, by some irrepressive semantic drift, have been rapidly interpreted in terms of an in fact never defined word: "culture" (But is it really possible to define it?):

- points like points found in Aurignac,
- Aurignacian points,
- Assemblage "du type d'Aurignac",
- Aurignacian "culture",
- the Aurignacians.

During the first half of the 20th century, the necessity to improve the use of "fossiles directeurs" with a type list, systematising the classification of prehistoric artefacts appeared: Bordes and Bourgon, Sonneville-Bordes and Perrot, Laplace, Tixier, etc.

It would take pages and pages to briefly summarise all the literature written about the complexity of the association between a typology of physical finds (lithics; bone, antler and ivory; ceramics, bronze axes, etc.) and a culture. The controversy between Bordes, Binford and Mellars concerning the Mousterian is famous. Starting from a type list, which is a mix of morphology (supposed to be functional) and technology, the first author concluded there were different subcultures, the second found different functionally specialised sites of the same population, and the third was of the opinion that a chronological evolution existed. Nobody paid adequate attention to an excellent data analysis performed by Callow and Webb (1981) which was applied to the same data used by Bordes and Binford. The study revealed the main variability of data: the first axis is an opposition between side scrapers (Quina retouch) and notches/denticulates; the second axis is an opposition between Levallois tools and others; the third axis is the frequency of bifacial points and upper Palaeolithic tools. The interpretation of this variability is rather easy: the first axis is associated with the distance for supplying good quality of flint raw material to obtain Quina side scrapers. The second axis is a technological variability in chipping flint. The third axis has a chronological significance. All three original authors failed because their methodology was faulty: Bordes because no extrinsic information was able to explain any variability in the data (stratigraphy, space, paleoclimate, etc.); Binford because the typology used was not functional, and the factor analysis sufficiently "manipulated" to obtain the expected result; and Mellars because the structure is only partially chronological.

Nevertheless it is important to point out that the key to understanding the variability in the Mousterian assemblages is the use of exploratory data analysis with repetitive attempts to associate the variability with other intrinsic and extrinsic information.

Methods for studying assemblage variability started with the development of typologies and assemblage quantification in the fifties: Bordes and Bourgon introduced well known cumulative percentage frequency graphs from sedimentology, and Laplace used histograms and the Chi-square test. However, multidimensional data analysis is the only solution.

Doran and Hodson (1966) used multidimensional scaling (as a Q-mode analysis of types), Binford (1968) used a controversial factor analysis (as a R mode analysis of types), Hodson (1969) used a Principal Components Analysis (as a Q-mode analysis of types), Graham and Roe (1970) used a canonical analysis (as a Q mode analysis of attributes), Hodson (1970) used a cluster analysis (as a Q mode analysis of types), Gower (1971) used a constellation analysis (as a Q mode method on types), Djindjian (1976) used a correspondence analysis on the abundance table and cluster analysis on factors (as a R + Q mode on types), Djindjian (1980) used a correspondence analysis on a burt matrix (as a R + Q mode on attributes). In the 1980s, virtually all major assemblages of prehistoric European Palaeolithic data were processed or revised by multidimensional data analysis, allowing major results in the structuration of the early and late prehistoric cultures (for example Djindjian, Kozlowski and Otte 1999).

The failure in understanding the Mousterian nevertheless resulted in a general movement against analytical descriptive typology, with a progressive replacement by experimental technology improving the knowledge of flint chipping process. Therefore, experimental lithic technological studies have confirmed that core reduction technology is more structuring than shaping technology in middle Palaeolithic studies; on the other hand shaping technology is much more structuring than core reduction technology in upper Palaeolithic studies. Nevertheless, some dogmatic attitudes suggest that only core reduction technology is structuring, representing a will to change a paradigm and not to improve Palaeolithic structurations and their explanations. Such a movement is also accompanied by a methodological approach voluntarily limited to only refitting core reduction and opposed to any formalisation of any description or quantification for comparison and statistics.

Such a regression is certainly temporary, when appropriate methodology has already been proposed (Djindjian 1980, 1986). A good example is given by the studies on Aurignacian structuration in La Ferrassie rockshelter (Dordogne, France) by a technological tool description without any reference to typology (3). The technological description is based on different types of blank retouch (lateral retouch, truncation, denticulation, endscraping retouch, endscraping bladelet retouch, etc.) and different types of burin reduction sequences (for a detailed study of burin reduction technology, see Djindjian 1996). The result is a structuration of Aurignacian in Perigord, in five technological (intrinsic) facies revealing a chronological (extrinsic) evolution and strongly correlated with (extrinsic) paleoclimatic variations (I: cold; II = Arcy;

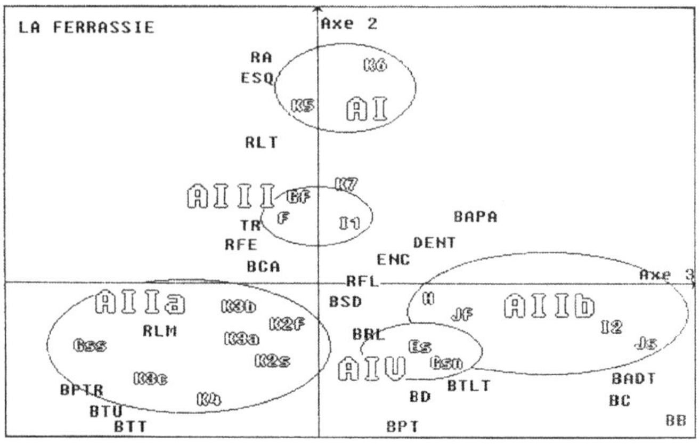

Figure 3: Aurignacian structuration in La Ferrassie rockshelter (Dordogne, France): C.A. and cluster analysis applied on a technological description, in a two-step approach (first step: attribute analysis by C.A. separately on retouched tools and burins produced virtual technological types; second step: analysis of a technological types x layers abundance table by C.A. produced the chronological structuration). (Djindjian 1986, 1996).

III = cold, IV = Maisieres). The result has been validated with seven other Aurignacian rock shelter sequences in Perigord.

In conclusion, the actual trend is oriented to the construction of intrinsic knowledge-based artefact descriptions structuring a real cultural system certified by other intrinsic and extrinsic explanations. Such cognitive approaches could certainly contribute to a convergence between cultural anthropology and European prehistory.

8. Seriation

Seriation is a methodological approach which has played a major role in establishing chronological sequences and period schemes from burials, particularly in late prehistory.

The concept of seriation is easy to understand: artefacts have a limited time of use, and their life curves are supposed to have some Gaussian profile. A burial (or any closed system) is supposed to be an instantaneous sample of artefacts at different steps of their life, and thus provides an indicator of time in a relative chronology.

The formalisation of the concept was more or less finalised in the 19th century (for a well-known example, Petrie for the chronology of pre-dynastic Egypt, 1889). The seriation model is generally represented by the "Petrie" diagonal matrix. A matrix of types x burials may be reorganised by permutation of rows and columns under a diagonal form, if a seriation structure is present.

Rapidly, the question was to reorganise large matrices of any kind of data, using mathematics and computers. The main algorithms are now well known:

- graphs (Ford 1962),
- similarity matrix ordering (Brainerd and Robinson 1951),
- incidence matrix direct ordering (Kendall 1963),
- computerised similarity matrix ordering (various 1963-1968),

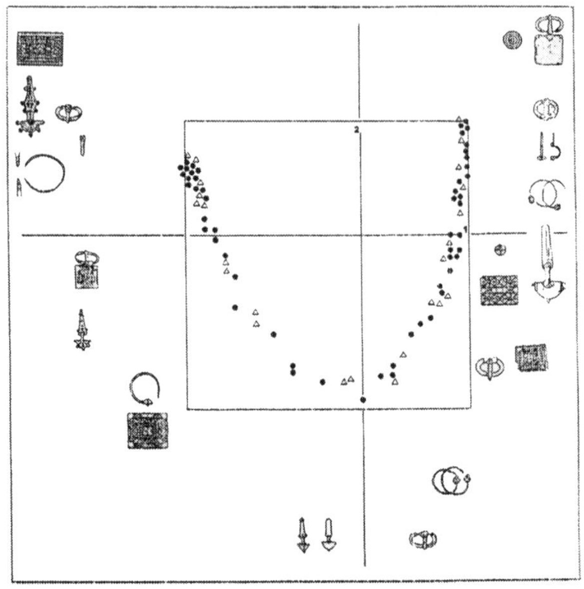

Figure 4: Seriation of the burials of the visigoth cemetery of Duraton (Segovie, Spain). Parabolic effect revealed by a C.A. applied to a 291 burials x 60 types table. (Ciezar 1990).

- rapid methods on similarity matrix (various 1966-71),
- multidimensional scaling (Kendall 1971),
- travelling salesman problem (Wilkinson 1971),
- reciprocal averaging method (various 1972-80),
- correspondence analysis (Djindjian 1976),
- PCA (Marquardt 1976),
- rapid method on incidence data matrix (Ester 1981),
- Toposeriation (Djindjian 1984),
- and so on (Laxton and Restorick 1989, Baxter 1994, etc).

After having succeeded in reorganising a "Petriefiable" matrix, the next question that appeared was how to process a noisy matrix with a supposed serial effect badly described by a maladapted typology and possibly disturbed excavations. This practical constraint is the explanation of the success of the correspondence analysis which reveals a superposed parabolic effect, both on burials and artefact types, furnishing a double chronological order on the first factorial axis, which can be improved by iteration on variables (typology) and units (burials) (Djindjian 1985).

Once the noisy data matrix problem was solved, it appeared obvious that the seriation model was also purely theoretical, and that archaeologists had to process more complex serial structures (Djindjian 1991):

- parasite effects (for example, male-female structure in burials),
- mix of partition and seriation structures,
- existence of non-linear seriation models,
- existence of non chronological seriation models,
- etc.

Finally, multidimensional data analysis (and especially Correspondence Analysis) is certainly able to offer the methodological

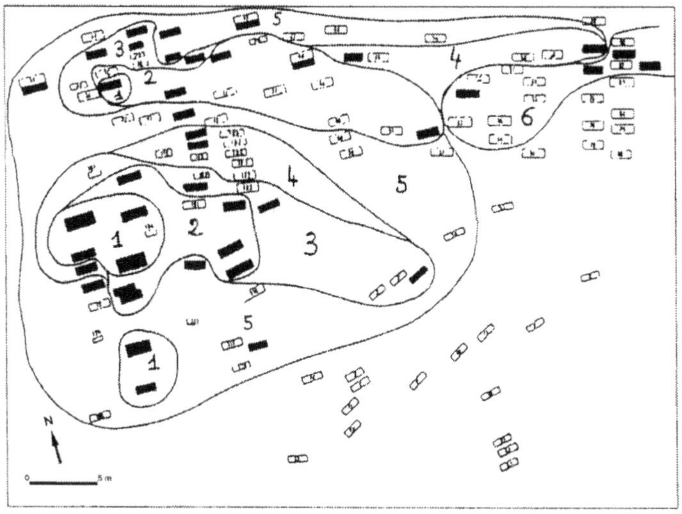

Figure 5: The toposeriation method applied on the Merovingian cemetery of Mezieres (Ardennes, France), revealing the expansion of the cemetery in time and space from three original set of burials. (Djindjian 1985).

tools necessary to manage the interaction between the archaeologist and the artefacts during the process of seriation. Nevertheless, if the seriation structure is exact, and if it is possible to build a periodisation from it, it is (from many archaeological and mathematical reasons) an illusion to believe that the chronological rank of each point (unit and variable) is exact.

Now, the success and the difficulty of a seriation primarily depends on the choice of a description of the artefacts having a strong chronological dependence, in other words a typology a priori correlated with the extrinsic information "chronology".

A good example of seriation by a correspondence analysis is given by the study of the Visigoth cemetery at Duraton (Segovie, Spain) performed by Ciezar (1990). Starting with 291 published burials and a typology of 60 types, the analysed matrix was finally 56 burials x 26 types (figure 4), revealing a good parabolic effect and a periodisation into three phases.

Another example is given by the toposeriation analysis (Djindjian 1985) of the Merovingian cemetery at Mezieres (Ardennes, France). Toposeriation is a method based on an association of seriation by correspondence analysis on a burial x type matrix and of k-means-like classification on a topographic matrix with chronological constraints. The result of the program is both chronological and topographical clusters, which can be figured on the cemetery map (figure 5).

9. Intrasite spatial analysis

Intrasite spatial analysis deals with the study of artefact spatial distributions within an archaeological site. Compared to stratigraphy and typology, intrasite spatial analysis is a relatively new method in excavations, however it certainly plays the same fundamental role in archaeology.

Under the influence of Ecology quantitative methods with single spatial distribution analysis, appeared: Hesse (1971), Dacey (1973), Whallon (1973,1974); later on extended to two spatial distribution association tests: Hodder and Orton (1976), Hietala and Stevens (1977), Clarke (1977), Berry et al. (1980); and finally multidimensional spatial data analysis was actually used: Johnson (1984), Graham (1980), Kintigh and Ammermann (1982), Whallon (1984), Hietala (1984), Djindjian (1988), Djindjian (1999).

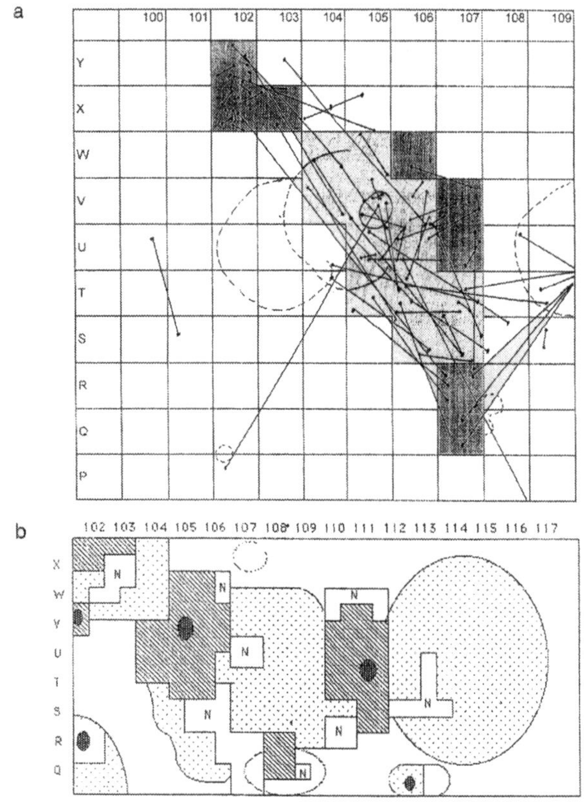

Figure 6: Application of the spatial structuration method and of the spatial structuration method on refitted artefacts to the data of the Magdalenian camp-site of Pincevent (Seine et Marne, France). (Djindjian 1988, 1999).

These multidimensional methods reveal spatial structures that are interpreted in terms of post-depositional effects, garbage areas, working areas or social spatial organisations.

If the quantitative methods can be considered as effective, improvements are now expected in the selection of archaeological data used in spatial distributions. It is evident that an exhaustive choice of artefact types used in other methods (typology, seriation, "culture" identification) is not an efficient strategy to use in order to discover activity areas or social organisations.

Intrasite spatial analysis can certainly be the field where functional data as obtained by use-wear analysis, subsistence data furnished by plant and animal remains, refitted data flints, burnt stones, and bones or social data estimated from ethnographic contemporary models must be used to obtain interpretable spatial structures.

For example, I offer the application of two methods, spatial structuration method (Djindjian 1988) and a complementary method: spatial structuration on refitted artefacts very useful in identifying garbage areas (Djindjian 1999). These methods have beeen applied to the data of the Magdalenian camp-site of Pincevent (Seine-et-Marne, France) published by Leroi-Gourhan and Brezillon (1972). In figure 6, a new interpretation of the spatial structure of the camp-site is given (Djindjian 1999).

10. Identification of raw material sources and manufacturing centres

The question as regards the sources of raw materials and the discovery of manufacturing centres is the most fully formalised question in archaeology, because it mainly involves the use of methods from physics and mathematics. The procedure may be summarised in three steps:

- data acquisition by physical and chemical characterisation of flint, ceramic, metal, stone samples,
- structuration by discriminant data analysis, which results in distinguishing and characterising samples from known raw material sources or manufacturing centres,
- decision by determination of the origin of samples coming from various archaeological sites.

Nevertheless, the method, apparently rigorous, has strong constraints:

- the localisation of sources or centres (all of them),
- a strongly discriminant characterisation,
- the control of post extraction raw material modifications by transformation, mix or pollution.

Barrandon and Irrigoin (1979) have published a beautiful study on paper manufacturing in Europe during 17th and 18th centuries. After analysing (by neutronic activation) a collection of papers, they found two clusters, discriminated by As and Co (intrinsic structuration). They discovered the difference is due to the fact that the paper was manufactured in Angouleme (France) and the Netherlands (extrinsic regional explanation). The difference is explained by the know-how of Dutch manufacturers who add smaltite (CO, Ni) AS_3 in the process in order to obtain a paper whiter than that of their competitors. The presence of smaltite is proven by traces of Ni in the data (discovery of more cognitive intrinsic information). Finally, they discovered that such a technical tradition has only been used after 1748, as shown by the analysis of Dutch papers from 1650 to 1810 (extrinsic chronological information).

11. Artefacts and anthropological models

The use of anthropological models is the only way to exploit intrinsic information, particularly difficult to estimate, such as functional, individual, hierarchical, social, symbolic, etc. information.

The anthropological models are based on intrinsic analogical knowledge deriving from only three sources:

- modern world (considered as the evolution of ancient worlds),
- experimentation,
- contemporaneous ethnographic models.

Unfortunately, such knowledge only has a local and modern value, obtained by analogy and cannot be used without being cautious as regards its scope of application.

Simulation methods are particularly well adapted to this kind of hypothetical data, and a large number of simulation methods approaching these particularly difficult questions have been tested:

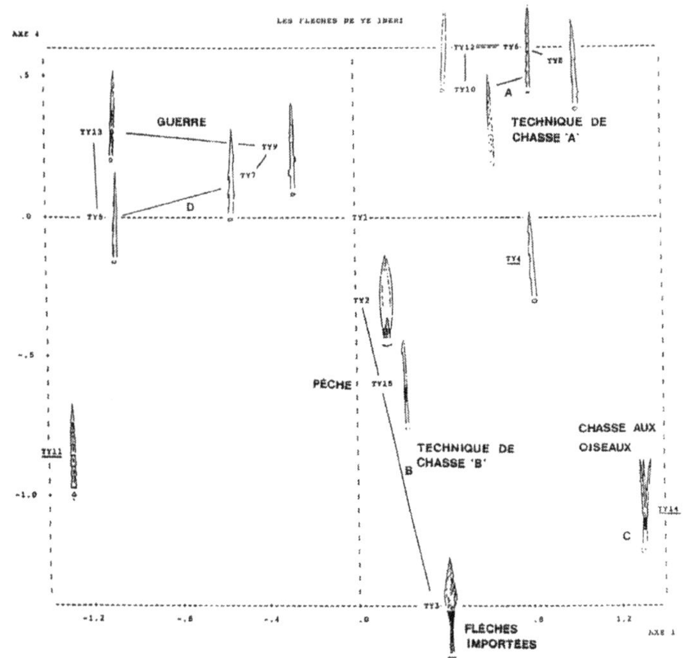

Figure 7: Correspondence between morphological types obtained by artefact analysis and functional types known by ethnographic studies of the same arrow set from contemporary Dani tribes (Irian Jaya, Indonesia). Factorial axis 1-4 of a C.A. (Giligny and Sidi-Maamar 1990).

- Regression by data analysis,
- Operational research models,
- Monte-Carlo simulations,
- Expert systems,
- System Dynamics,
- Decision making.

Examples of simulation methods in archaeology are unfortunately limited but are of great interest: Doran (1970), Thomas (1972), Ammermann and Cavalli-Sforza (1973), Wobst (1974), Zubrow (1975), Hassan (1977), Jochim (1976), Hodder ed. (1978), Renfrew and Cooke (1979), Sabloff (1981), Keene (1979), Belovsky (1987), Doran (1990), etc.

An elementary example of the difficulties that arise due to the differences between available and necessary data for a simulation is given by a study realised by Giligny and Sidi-Maamar (1990) on neolithic arrows from contemporary Dani tribes (Irian Jaya, Indonesia), with data collected by Petrequin and Petrequin (1990). Arrows are found in quivers found in the huts of a single village. Arrows are first analysed by an artefact analysis. Owners of quivers have offered explanations concerning the variability of the arrows. The results of the two approaches are compared. Explanations are numerous, socially complex and often individual. Factor analysis axis 1 and 4 show some correspondence between the morphological typology and the functions of the arrows (figure 7).

12. Identification of artefact distribution networks (exchange and trading systems)

After having identified raw material sources and manufacturing centres (cf. chapter 9), it is possible to identify the distribution network for any artefact.

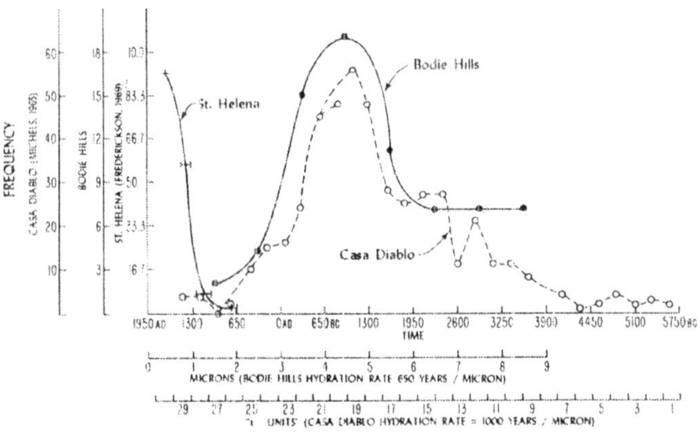

Figure 8: Evolution of the operation of obsidian quarries on the California peninsula over a period of 7000 years (Ericson 1981).

The procedure may be defined in three steps:

- data acquisition by sampling artefacts in archaeological sites and physical or chemical characterisation,
- network structuration by various quantitative methods like:

  distance decay diagrams,

  isodensity decay diagrams,

  gravity models,

  distance decay diagrams through time,

- exchange and trading system reconstitution by simulation methods.

A good example for the evolution of the operation of obsidian quarries in the California peninsula over a period of 7000 years is given by Ericson (1981). After the characterisation of obsidian sources and obsidian dating, Ericson can reconstitute curves of obsidian quarry activity (figure 8).

13. Intersite spatial analysis

Intersite spatial analysis is concerned with the studies of landscape, territory, boundaries, settlement spatial organisation and peopling dynamic models.

In contrast to previous studies, in this analysis, artefacts are archaeological sites themselves.

The procedure may be organised in the following three steps:

- data acquisition by a GIS implementation,
- identification and structuration of the territory by various quantitative methods:

  boundaries:

    artefact assemblages,

    territorial artefacts such as coins, defence sites, etc.

    geographical boundary models,

  spatial organisation:

    central place model,

    rank size model,

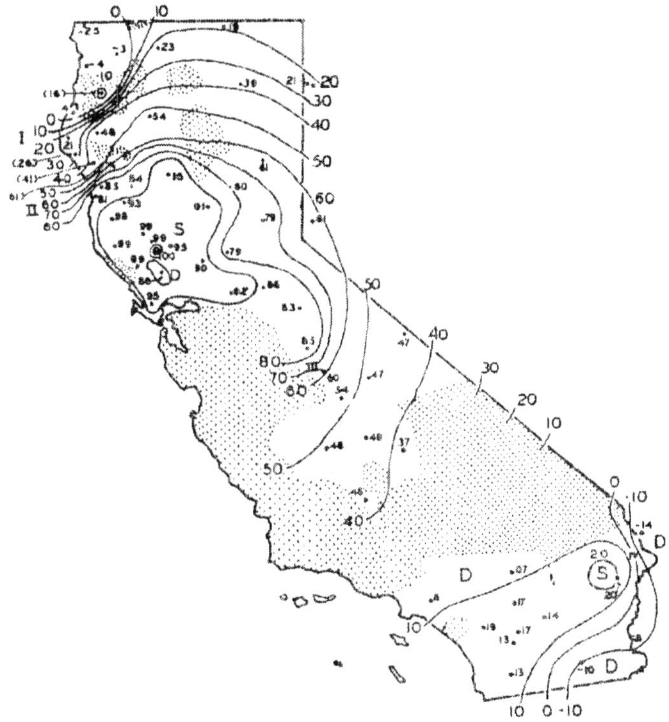

Figure 9: Paleo-indian cultures on the California peninsula (Mielke 1949).

    viewsheds,

    etc.

- reconstitution of the peopling system:

  carrying capacity models,

  localisation models,

  system dynamics.

Nevertheless, intersite spatial analysis has to support numerous constraints such as contemporaneousness and representative spatial sampling of archaeological sites.

A historical example is the result of the study performed by Mielke (1949) concerning Paleoindian cultures on the Californian peninsula. Cultural regional facies are revealed on a map by level curves drawn from a cultural similarity measure computed from artefact analysis (figure 9).

14. Conclusions: A three step cognitive model: Acquisition, structuration and reconstitution

For each archaeological problem, which has been analysed in this paper, I have demonstrated the fundamental role of artefacts and their intrinsic information in the success of the construction of the archaeological interpretation.

For each archaeological problem, I have also demonstrated the efficiency of a three-step model: acquisition, structuration, reconstitution, following a framework already known in numerous social and human sciences (amongst others: Pierce 1992).

This model may be applied at each level where the re-entering concept of "logical" artefact is used.

From the level of the physical find

- acquisition:

  perception, description, recording, coding intrinsic information,

- structuration:

  partition,

- object reconstitution:

  intrinsic and extrinsic added explanations.

To the site level

- acquisition:

  survey, excavations, extrinsic data recording, intrinsic artefact analysis,

- structuration:

  artefact classification, intrasite spatial analysis,

  seriation, raw material origins, subsistence resources, distribution network,

- system reconstitution:

  technology, settlement, craft production, exchange and trading, subsistence, defence…

In the formalisation of any archaeological problem with a three-step framework: acquisition, structuration and reconstitution (modelling), the cognitive approach has reconciled the empirical-inductive approach (structuration) and the hypothetical-deductive approach (modelling). In emphasising the role of acquisition, the cognitive approach has changed the role of the artefacts in archaeological constructs, from the concept of the artefact as a whole, to the concept of artefact as an element of a system. In that way, artefact analysis is a modern concept in the present archaeological method and theory.

References

AMMERMANN, A.J. and SFORZA, C., 1973. A Population model for the diffusion of early farming in Europe. In Renfrew, C. (ed.), *The explanation of culture change*. London: Duckworth: 343-358.

BARRANDON, J.N. and IRRIGOIN, J., 1979. Papiers de Hollande et papiers d'Angoumois de 1650 à 1810. *Archaeometry,* n°10: 101-106.

BERRY, K.J., KVAMME, K.L., MIELKE, P.W., 1980. A permutation technique for the spatial analysis of the distribution of artefacts into classes. *American Antiquity*, n°45: 55-59.

BAXTER, M.S., 1994. *Exploratory Multivariate Analysis in Archaeology*. Edinburgh: Edinburgh University Press.

BELOVSKY, G.E., 1987. Hunter-gatherer foraging: a linear programming approach. *Journal of Anthropological Archaeology*, 7: 163-202.

BINFORD, L.R. and BINFORD, S.R., 1966. A Preliminary analysis of functional variability in the Mousterian of Levallois facies. *American Anthropologist*, n°68: 238-295.

BORDES, F., 1953. Essai de classification des industries moustériennes. *B.S.P.F.*, t.50: 457-466.

BOELICKE, V., STEHLI, P., ZIMMERMANN, A. and ANIOL, R.W., 1981. The definition of phases in Neolithic settlements from classification and seriation of pit inventories. In Cowgill, G.L., Whallon, R. and Ottaway, B.S. (eds.), *Colloque informatique et mathématiques appliquées en Archéologie*, X° congrès UISPP, Mexico: 157.

BOLVIKEN, E. et al., 1982. Correspondence analysis: an alternative to principal components. In Quantitative methods in Archaeology. *World Archaeology*, vol. 14, fasc.1: 41-60.

BRAINERD, G.W. and ROBINSON, N.S., 1951. The place of chronological ordering in archaeological analysis. *American Antiquity*, n°15: 293-301 and 301-313.

CALLOW, P. and WEBB, E., 1981. The application of multivariate statistical techniques to middle Palaeolithic assemblages from south-western France. *Revue d'Archéométrie*, t.5: 330-338.

CHRISTENSON, A.L. and READ, D.W., 1977. Numerical taxonomy, R-mode factor analysis and archaeological classification. *American Antiquity*, n°42: 163-179.

CIEZAR, P.G., 1990. Sériation de la nécropole de Duraton (Ségovie, Espagne). *Histoire et Mesure*, vol.5, n°12: 107-144.

CLARKE, D.L., 1962. Matrix analysis and archaeology with particular reference to British Beaker Pottery. *Proceedings of the Prehistorical Society*, n°28: 371-382.

CLARKE, D.L., 1977. *Spatial Archaeology*. New York: Academic Press, 1977.

DACEY, M.F., 1973. Statistical tests of spatial association in the location of tool types. *American Antiquity*, n°38: 320-328.

DJINDJIAN, F., 1976. *Contributions de l'Analyse des Données à l'étude de l'outillage de pierre taillée.* Mémoire de Maîtrise, Université de Paris 1: 1976.

DJINDJIAN, F., 1980. *Constructions de systèmes d'aides à la connaissance en archéologie préhistorique. Structuration et Affectation: méthodes et algorithmes.* 2 vol. Thèse 3°cycle: Archéologie préhistorique: Université de Paris I, 1980.

DJINDJIAN, F., 1985. Seriation and toposeriation by correspondence analysis. In Voorrips, A. and Loving, S.H. (ed.), *To Pattern the past.* PACT, n°11:119-136

DJINDJIAN, F. (ed.), 1985. Rapport d'activités du séminaire "Informatique et mathématiques appliquées en Archéologie", 1984-1985. Paris: G.R.A.Q.

DJINDJIAN, F., 1986. Recherches sur l'Aurignacien du Périgord à partir des données nouvelles de La Ferrassie. *L'Anthropologie*, t.90, n°1: 89-106.

DJINDJIAN, F., 1988. Improvements in intrasite spatial analysis techniques. In Ruggles, C.L.N and Rahtz, S.P.Q. (eds.), *Computer and quantitative methods in Archaeology 1988.* Oxford: BAR Internnational Series 446: 95-106.

DJINDJIAN, F., 1989. Fifteen years of contributions of the French school of Data Analysis to quantitative Archaeology. In Rahtz, S.P.Q. and Richards, J. (ed.), *Computer and Quantitative Methods in Archaeology 1989*, Oxford: BAR. International Series 446(i): 95-106.

DJINDJIAN, F., (ed.), 1990. *Histoire et Mesure*, vol.5, n°1-2,1990.

DJINDJIAN, F., 1991. *Méthodes pour l'Archéologie.* Paris: Armand Colin, 1991.

DJINDJIAN, F., 1996. Histoires de burins. *Bulletin du Centre Genevois d'Anthropologie*, n°3-4, Genève: Université de Genève, 1993-4: 3-21

DJINDJIAN, F., 1996. L'apport des Sciences cognitives à l'Archéologie. XIII° Congrès UISPP. Forli (Italie) 8-14/09/1996. *Colloque n°I.*, Vol 1: 17-28

DJINDJIAN, F., 1998. GIS Usage in Worldwide Archaeology. *Archeologia e Calcolatori*, n°9: 19-29.

DJINDJIAN, F., 1999. L'analyse spatiale de l'habitat: un état de l'art. *Archeologia e Calcolatori,* n°10, 1999: 17-32.

DJINDJIAN, F., KOZLOWSKI, J.K., OTTE, M., 1999. *Le paléolithique supérieur en Europe.* Paris: Armand Colin, 1999.

DORAN, J., 1970. Systems theory, computer simulations and Archaeology. *World Archaeology*, vol.1: 289-298.

DORAN, J., 1990. Computer-based simulation and formal modelling in Archaeology. In Voorips, A. (ed.), *Mathematics and Information Science in Archaeology: a flexible framework,* Studies in Modern Archaeology, vol.3. Bonn: Holos: 93-114.

DORAN, J.E. and HODSON, F.R., 1966. A Digital computer analysis of Palaeolithic flint assemblages. *Nature,* n°210: 688-689.

DORAN, J.E. and HODSON, F.R., 1975. *Mathematics and Computers in Archaeology.* Edinburgh: Edinburgh University Press.

ECO, U., 1999. *Kant et l'ornithorynque*. Paris: Grasset, 1999.

ERICSON, J.E., 1981. *Exchange and production systems in Californian Prehistory.* Oxford: BAR. International Series 110, 1981.

ESTER, M., 1981. A column-wise approach to seriation. In X° Congrès UISPP, 1981, Mexico. *Colloquium 5 on data management and mathematical methods in Archaeology*: 125-156 (preprints).

FORD, J.A., 1962. *A Quantitative method for deriving cultural chronology*. Washington: Department of social affairs. Pan-American Union, 1962. (Technical manual, 1).

GERO, J. and MAZZULLO, J., 1984. Analysis of artefact shapes using Fourier series in closed form. *Journal of field archaeology*, 11: 315-322.

GILIGNY, F., 1990. La reconnaissance des formes céramiques: une approche typologique formalisée. *Histoire et Mesure*, vol.5, n°12: 89-107.

GILIGNY, F. and SIDI-MAAMAR, H., 1990. Simulation archéologique à partir de l'étude ethno-archéologique des flèches de Ye Ineri (Irian Jaya, Indonésie). *Histoire and Mesure,* vol.5, n°1-2: 145-162.

GRAHAM, J.M. and ROE, D., 1970. Discrimination of British lower and middle Palaeolithic handaxe groups using cannonical variates. *World Archaeology*, n°1: 321-337.

GRAHAM, I.D., 1980. Spectral analysis and distance methods in the study of archaeological distributions. *Journal of Archaeological Science*, 7: 105-130.

GREENACRE, M.J., 1984. *Theory and applications of correspondence analysis*. London: Academic Press, 1984.

HAGSTRUM, M.B., and HILDEBRAND, J.A., 1990. The two-curve method for reconstructing ceramic morphology. *American Antiquity*, 55(2): 388-403.

HALL, N.S. and LAFLIN, S., 1984. A computer aided design technique for pottery profiles, database. In Laflin, S. (ed.), *Computer Applications in Archaeology 1984*. Birmingham: University of Birmingham Computer Centre: 177-188.

HASSAN, F.A., 1977. The dynamic of agricultural origins in Palestine: a theoretical model. In Reed, C. (ed.), *Agricultural origins*, La Hague: Mouton, 1977: 589-610.

HESSE, A., 1971. Les Tarteręts II, site paléolithique de plein air à Corbeil-Essonnes (Essonne): comparaison par le calcul des distributions horizontales des vestiges lithiques. *Gallia-Préhistoire*, n°14: 41-46.

HIETALA, H.J. (ed.), 1984. *Intrasite Spatial Analysis in Archaeology*. Cambridge: Cambridge University Press.

HIETALA, H.J. and STEVENS, D.S., 1977. Spatial analysis: multiple procedures in pattern recognition studies. *American Antiquity*, n°42: 539-559.

HODDER, I. (ed.), 1978. *Simulation studies in Archaeology*. Cambridge, London: University Press. (New directions in Archaeology).

HODDER, I. and ORTON, C., 1976. *Spatial analysis in archaeology*. Cambridge: Cambridge University Press. (New studies in Archaeology: 1).

HODSON, F.R., 1970. Cluster analysis and archaeology: some new developments and applications. *World Archaeology*, vol.1, n°3: 299-320.

HODSON, F.R., 1971. Numerical taxonomy and prehistoric archaeology. In Hodson, F.R., Kendall, D.G. and Tautu, P. (eds.), *Mathematics in the archaeological and historical sciences.* Proceedings of the anglo-romanian conference: Mamaia, 1970. Edinburgh: Edinburgh University Press: 30-45.

HODSON, F.R., SNEATH, P.H.A. and DORAN, J.E., 1966. Some experiments in the numerical analysis of archaeological data. *Biometrika*, vol.53: 311-324.

JOCHIM, M.A., 1976. *Hunter-gatherer subsistence and settlement: a predictive model*. New York: Academic Press.

JOHNSON, T., 1984. Cell frequency recording and analysis of artefact distributions. In Hietala H.J. (ed.), *Intrasite Spatial Analysis in Archaeology*. Cambridge: Cambridge University Press: 75-96.

KEENE, A.S., 1979. Economic optimisation models and the study of hunter-gatherer subsistence settlement systems. In Renfrew, C. and Cooke, K.L. (eds.), *Transformations: mathematical approaches to culture change.* New York: Academic Press: 369-404.

KENDALL, D.G., 1963. A statistical approach to Flinder's Petrie sequence dating. *Bull. Int. stat.*, n°40: 657-680.

KENDALL, D.G., 1971. Seriation from abundance matrices. In Hodson, F.R., Kendall, D.G. and Tautu, P. (eds.), *Mathematics in the archaeological and historical sciences*: Proceedings of the anglo-romanian conference: Mamaia, 1970. Edinburgh: Edinburgh University Press: 215-252.

KINTIGH, K.W. and AMMERMANN, A.J., 1982. Heuristic approaches to spatial analysis in archaeology. *American Antiquity*, 47: 31-63.

LAXTON, R.R. and RESTORICK, J., 1989. Seriation by similarity and consistency. In Rahtz, S. and Richards, J. (eds.), *Computer Applications in Archaeology1989*, BAR International Series 548: 215-224.

LEROI-GOURHAN, A. and BREZILLON, M., 1972. Fouilles de Pincevent: essai d'analyse d'un habitat magdalénien: la section 36. Paris: CNRS, *Gallia Préhistoire*, Supplément n°7.

MADSEN, T., 1988. Multivariate statistics and Archaeology. In Madsen, T. (ed.), Multivariate archaeology. Numerical approaches in Scandinavian archaeology, *Jutland archaeological Publication* 21: 7-27.

MAIN, P.L., 1986. Accessing outline shape information efficiently within a large database. In Laflin, S. (ed.), *Computer Applications in Archaeology 1986,* Birmingham: University of Birmingham Computer centre, 1986: 73-82.

MARQUARDT, W.H., 1976. Advances in archaeological seriation. In Schiffer, M. (ed.), *Advances in archaeological method and theory*. London, New York: Academic Press, Vol.1: 257-314.

MELLARS, P., 1969. The chronology of mousterian industries in the Perigord region of south-west France. *Proceedings of the Prehistoric Society*, 35: 134-171.

MONTELIUS, O., 1885. Sur la chronologie de l'Âge du Bronze, spécialement dans la Scandinavie. *Matériaux pour l'Histoire primitive de l'Homme.* 19° année, 3° série, t.2: 3-8.

MOSCATI, P., 1986. Analisi statistiche multivariate sugli specchi Etrusci. Roma: Academia Nationale dei Lincei. (Contributi del Centro Lineo Interdisciplinare di scienze matematiche e loro aplicazioni, 74).

ORTON, C., 1980. *Mathematics in Archaeology*. Cambridge: Cambridge University Press, 1980.

PETREQUIN, A.M. and PETREQUIN, P., 1990. Flèches de guerre, flèches de chasse, la cas des Dani d'Irian Jaya (Indonésie). *B.S.P.F.*, 87 (10-12): 485-511.

PETRIE, W.M.F., 1889. Sequences in prehistoric remains. *Journal of Anthropological Institute*, n°29, p.295-301.

PEIRCE, C.S., 1992. *Reasoning and The Logic of Things.* The Cambridge conferences. Lectures of 1898. Cambridge: Harvard University Press.

RENFREW, C. and COOKE, K.L. (eds.), 1979. *Transformations: mathematical approaches to culture change.* New York: Academic Press, 1979.

RINGROSE, T., 1988. Correspondence analysis for stratigraphic abundance data. In Ruggles, C.L.N. and Rahtz, S.P.Q. (eds.), *Computer and quantitative methods in Archaeology 1987*, Oxford: BAR International Series 393: 3-14.

SABLOFF, J.A. (ed.), 1981. *Simulations in archaeology*. Albuquerque: University of New Mexico Press.

SLACHMUYLDER, J.L., 1985. Seriation by correspondence analysis for Mesolithic assemblages. In PACT n°11: 137-148.

SCOLLAR, I., WEIDNER, B., and HERZOG, I., 1985. A portable seriation package with dynamic memory allocation in Pascal. In Voorrips, A. and Loving, S.H. (eds.), *To pattern the past.* PACT n°11: 149-160.

SPAULDING, A.C., 1953. Statistical techniques for the discovery of artefact types. *American Antiquity*, n°18: 305-313.

THOMAS, D.H., 1972. A computer simulation model of Great Basin Shoshonean subsistence and settlement patterns. In Clarke, D.L. (ed.), *Models in Archaeology*. London: Methuen: 671-704.

TYLDESLEY, J.A., JOHNSON, J.G., SNAPE, S.R., 1985. Shape in archaeological artefacts: two case studies using a new analytic method. *Oxford Journal of Archaeology*, 4(1): 19-30.

WHALLON, R., 1973. Spatial Analysis of occupation floors: application of dimentional analysis of variance. *American Antiquity,* 38: 266-278.

WHALLON, R., 1974. Spatial analysis of occupation floors: application of nearest neighbour analysis. *American Antiquity*, 39: 16-34.

WHALLON, R., 1982. Variables and dimensions: the critical step in quantitative archaeology. In Whallon, R. and Brown, J.A. (eds.), *Essays on archaeological typology*. Center for American Archaeology Press: 127-161.

WHALLON, R., 1984. Unconstrained clustering for the analysis of spatial distributions in archaeology. In Hietala H.J. (ed.), *Intrasite Spatial Analysis in Archaeology*, Cambridge: Cambridge University Press: 242-277.

WILCOCK, J.D. and SHENNAN, S.J., 1975. Shape and style variation in central German Bell Beaker: a computer assisted study. *Science and Archaeology*, n°15: 17-31.

WILKINSON, E.M., 1971. Archaeological seriation and the travelling salesman problem. In Hodson, F.R., Kendall, D.G. and Tautu, P. (eds.), *Mathematics in the archaeological and historical sciences:* Proceedings of the anglo-romanian conference: Mamaia, 1970. Edinburgh: Edinburgh University Press: 276-283.

WOBST, H.M., 1974. Boundary conditions for Palaeolithic social systems: a simulation approach. *American Antiquity*, n°39, fasc. 2: 147-170.

ZUBROW, E.B.W., 1975. *Prehistoric carrying capacity: a model.* Menlo Park, Cal.: Cummings.

Grouping Ceramic Compositional Data: An S-Plus Implementation

Christian C. Beardah and Mike J. Baxter

Department of Mathematics, Statistics and Operational Research, The Nottingham Trent University

Clifton, Nottingham NG11 8NS, United Kingdom

e-mails: christian.beardah@ntu.ac.uk, michael.baxter@ntu.ac.uk

Abstract

The scientific analysis of the chemical composition of ceramics naturally leads to multivariate data that is often explored using techniques such as principal components analysis and cluster analysis. The general aim is to identify groups of chemically similar artefacts. Separate groupings could be assumed to indicate, for example, distinct origins or processing recipes of the raw materials (clay) used to make the artefacts.

In this paper we aim to show how, with the aid of powerful statistical software such as S-Plus, traditional exploratory multivariate analysis can easily be used alongside, or in combination with, a technique designed specifically for grouping ceramics by chemical composition (Beier and Mommsen 1994). This latter technique involves grouping together artefacts whose chemical compositions are "close" with respect to a mathematical measure of dissimilarity. The measure used takes into account uncertainty of measurement and the possibility of constant shifts in the data due, for example, to dilution of the clay or to instrumental variation.

We shall demonstrate the ease of implementation in S-Plus of this methodology and show how graphical exploratory techniques can also be used to create an integrated approach to the grouping of ceramic chemical compositional data. The methodology will be applied to real data of this type.

Key words: ceramics, compositional data, multivariate analysis, cluster analysis, S-Plus

1. Introduction: archaeological and statistical background

This paper presents on-going work by members of the EU funded GEOPRO TMR Network. The aims of the project are to "design, develop and apply a novel approach which integrates fully geochemical and mineralogical analytical techniques, to further the study of the provenance and technology of archaeological ceramics and their raw materials".

From a statistical point of view the early stages of our work has involved an investigation into "state of the art" techniques for the analysis of data resulting from (a) geochemical and (b) mineralogical analysis of ceramics (Baxter et al. in press). Our ultimate aim is to produce a software package, using commercial tools, for the analysis and presentation of mixed-mode data. By *mixed-mode* data we mean that containing *both* geochemical and mineralogical information.

Mineralogical data can consist of variables of a mixture of types, including continuous, categorical and presence/absence. By contrast, geochemical data is typically continuous in nature. In addition, modern analytical methods such as Neutron Activation Analysis (NAA) can provide measurements for the concentration of over 20 elements in a ceramic. For this reason, geochemical data is usually high-dimensional. Such data, represented mathematically as an n by p data matrix, is often explored using techniques such as principal components analysis and cluster analysis (Baxter 1994). The general aim of such analyses is to identify groups of chemically similar artefacts. Separate groupings could be assumed to indicate, for example, distinct origins or processing recipes of the raw materials (clay) used to make the artefacts.

Here we concentrate solely on the implementation of statistical methods for geochemical data. In particular, we discuss an implementation of a technique designed specifically for grouping ceramics by chemical composition (Beier and Mommsen 1994). This technique involves grouping together artefacts whose chemical compositions are "close" with respect to a mathematical measure of dissimilarity (see section 2.2 for details). The measure used takes into account uncertainty of measurement and the possibility of constant shifts in the data due, for example, to dilution of the clay or instrumental variation.

Although this work is at a relatively early stage, a number of decisions with long-term implications have been taken. One of these is to use the S-Plus package as the basis for the development and distribution of our software. This package has many powerful "off-the-shelf" statistical tools of a kind commonly used in archaeometric study. Its most attractive feature, however, is that it is associated with a powerful programming language, called "S" (Venables and Ripley 2000) which allows many non-standard methods to be programmed with relative ease. The intention is to program non-standard methods of choice and to provide the end user with a user-friendly and accessible "front end", through the S-Plus user interface, which can also be manipulated via the S programming language.

More background detail on statistical issues is given in the next section. This includes a brief description of Beier and Mommsen's method of grouping, based upon Mahalanobis distance. Section 3 of the paper presents an archaeometric case study and is followed in section 4 by a discussion of some software-related issues. Our conclusions are contained in section 5.

2. Statistical issues

2.1. Traditional exploratory techniques: principal components analysis and cluster analysis

When analysing multivariate statistical data of the kind that often arises in archaeometry, it is almost always useful to apply a bat-

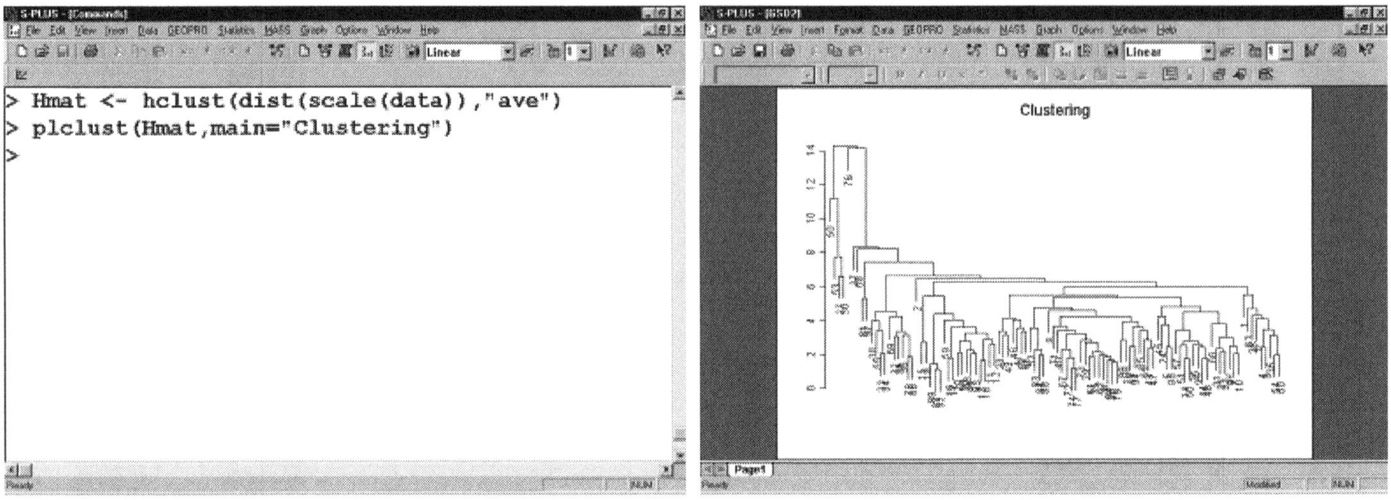

Figure 1: (a) S-Plus commands and (b) resulting output when carrying out a PCA. These data consist of 27 elemental concentrations for each of 100 ceramic samples from Nichoria in the Peloponnese and are part of the Perlman-Asaro databank of Mycenaean samples (Mommsen et al. in press).

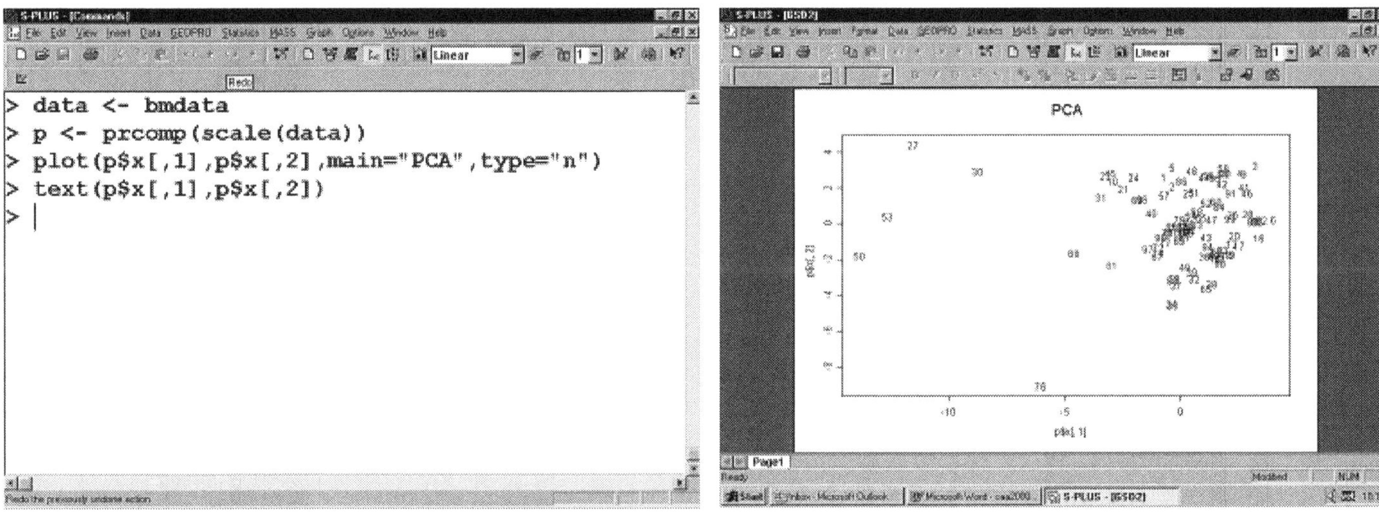

Figure 2: For the same data set as in figure 1, (a) S-Plus commands and (b) resulting output when carrying out an average linkage CA.

tery of techniques, rather than one method in isolation. Indeed the long-term aim of our research is not to recommend a particular method of statistical analysis, but to make a variety of methods available in widely accessible and user-friendly form. We hope that this will enable informed users to develop their own views about the advantages and drawbacks of different approaches.

As previously mentioned, analysis of the chemical composition of ceramics naturally leads to multivariate data that is often explored using techniques such as principal components analysis (PCA) and cluster analysis (CA). These methods are easy to carry out using standard, built-in S-Plus functions and can be called from menus and associated dialogue boxes, or from the command line. For example, figures 1 and 2 show (a) S-Plus commands and (b) the resulting output when carrying out respectively a PCA and an average linkage CA on the same data set. The S programming language may be thought of as a collection of commands, like those shown in figure 1(a) and figure 2(a), that implement powerful statistical and mathematical operations, together with traditional programming constructs such as looping, conditional (if … else) branching and facilities for graphical and textural output.

In the next section we outline a grouping method (Beier and Mommsen 1994) based upon the Mahalanobis distance between a point, and a group of points, in p-dimensional space. Here groups are iteratively "grown" from an initial group that can be as small as just one object. Although the idea behind this methodology is simple, the existing implementation is less so. Our aim is to make the procedure more accessible by implementing it within a commercially accessible package, namely S-Plus.

One way in which this can be achieved is by utilising the "traditional" exploratory techniques discussed here to provide some of the many methods that could be used to obtain an initial grouping. Other possible methods include scanning the data set for pairs of objects that are close together in the Euclidean sense, or using less well known exploratory methods, such as kernel density estimation in combination with contouring (Beardah et al. 2000), to identify potential groups. Several of these possible starting procedures have been provided as options in our S-Plus implementation of Beier and Mommsen's grouping procedure.

2.2. Statistical concepts: grouping using Mahalanobis distance

The grouping method introduced by Beier and Mommsen (1994) is based largely upon the Mahalanobis distance between a point, $\underline{x}$, in p-dimensional space, and a group with mean $\underline{y}$ and p by p covariance matrix S. This is given by

$$d_{xy}^2 = (\underline{x}-\underline{y})^T \Sigma^{-1} (\underline{x}-\underline{y}) \qquad (1)$$

where $\underline{x}$ and $\underline{y}$ are represented by p by 1 vectors.

In simple terms, new objects are added to an existing group if they are "close" to the centre of the group. Groups are iteratively "grown" from an initial group that can be as small as just one object. As discussed in section 2.1, "traditional" exploratory methods could be used to provide this initial grouping. Once a group has ceased to grow, the procedure is re-started from a different initial group. This process is continued until the whole data set has been classified as belonging to a group or not. The original presentation of this methodology (Beier and Mommsen 1994) extends the basic idea discussed above to take into account (a) uncertainty of measurement and (b), the possibility of constant shifts in the data due, for example, to dilution of the clay or instrumental variation. These additional properties are achieved by using modifications to the definition of Mahalanobis distance given in equation (1). The most general modified Mahalanobis "distance" used by Beier and Mommsen is given by

$$d_{xy}^2 = (f_0\underline{x}-\underline{y})^T (f_0^2 S_x + S_y)^{-1} (f_0\underline{x}-\underline{y}).$$

Here S_y denotes the covariance matrix of the group (previously denoted by S), S_x denotes a diagonal matrix, the kth diagonal of which is the squared "uncertainty" associated with the kth element of $\underline{x}$ and f_0 is the "best-relative fit" factor, calculated as the solution of

$$\frac{\partial}{\partial f}\left[(f\underline{x}-\underline{y})^T (f^2 S_x + S_y)^{-1} (f\underline{x}-\underline{y})\right]=0.$$

In other words f_0 is chosen to minimise d^2_{xy}. The calculation of f_0 can be achieved using standard numerical minimisation methods.

It should be noted that, in order for the Mahalanobis distance to be used, the size of the current group, n_g, needs to be greater than the number of variables, p, indeed $n_g >> p$ is preferable. If the size of the current group is too small for the Mahalanobis distance to be utilised, weighted Euclidean distance is used instead. As the analysis of chemical composition of artefacts in archaeology typically involves "large" numbers of variables and "small" sample sizes, this restriction can be problematic for our main application. Fortunately, this potential drawback can be alleviated in several ways. For example, variable selection methods can be used to reduce the number of variables considered, without adversely affecting the outcome of the statistical analysis (Baxter and Jackson in press). Alternatively, we could apply a dimension reducing technique such as PCA to the data before subjecting the transformed data to the grouping procedure, using a subset of components.

Of course the outcome of this method depends upon what we mean by new objects being "close" to the centre of the current group. If we desire the grouping procedure to act in a completely automatic manner, without user-intervention, we may quantify "closeness"

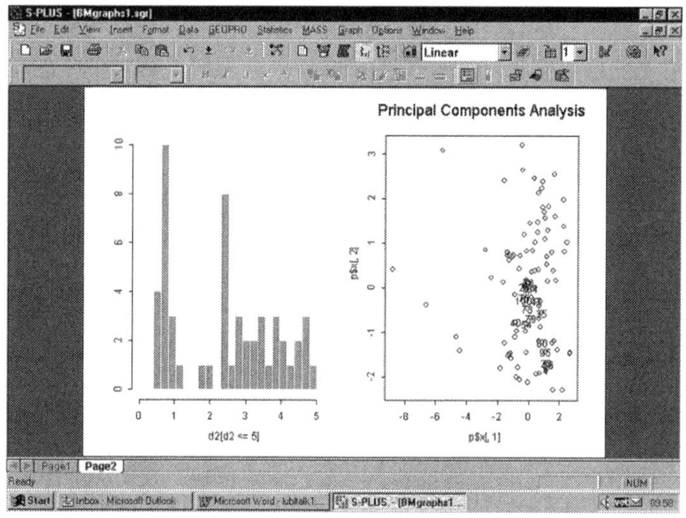

Figure 3: The left-hand plot shows a histogram of squared distance values. A plot of the first two PC scores is shown on the right. Labels have been used to highlight the objects in the current group. These data consist of a subset of 11 elemental concentrations for the same 100 ceramic samples from Nichoria in the Peloponnese considered throughout the paper (Mommsen et al. in press).

by the definition of two threshold values for the weighted Euclidean and Mahalanobis distance, D_E and D_M respectively. If the size of the current group is big enough to measure distances in the Mahalanobis sense then we add a new object $\underline{x}$ to the current group if $d^2_{xy} < D_M$. On the other hand, if the size of the current group is small then we add a new object to the current group if $d^2_{xy} < D_E$. Based on properties of d^2_{xy} under the assumption that groups have a multivariate normal distribution Beier and Mommsen (1994) suggest using D_M=1.6.

Applying the methodology as a "black box" as described above may be appropriate in some circumstances, for example if the data set is very large (Beardah et al. in press). In other situations more routinely encountered in the analysis of ceramics, Beier and Mommsen's grouping procedure is best used interactively, with the user making decisions regarding what is meant by an object being "close" to the centre of a group. To illustrate, suppose a group is in the process of being "grown". The group will have a mean $\underline{y}$ and will consist of n_g objects that are "close" to $\underline{y}$ as measured by either weighted Euclidean or Mahalanobis distance depending upon the current size of the group. To assess whether the group can be grown further, we measure the distance of all objects in the data set from $\underline{y}$. This results in n squared distance values $d_1^2, d_2^2, ..., d_n^2$. As a visual aid to group growth, a histogram of the values d_i^2 for $i=1, ..., n$ can be plotted. For a well-defined group of objects separated from the remainder of the data set a "valley" or "edge" should occur at relatively small values of squared distance. The range of values of squared distance where this occurs will depend upon the number of variables under consideration (Beier and Mommsen 1994). For example, in the case of the $p=27$ elements considered in our example in section 3, we would expect to see the "valley" or "edge" at values of squared distance in the range of 1.5 to 2.0.

The histogram on the left of figure 3 shows an example of the idea discussed above, as featured in our S-Plus implementation of Beier and Mommsen's grouping procedure. As the next step in this particular grouping process we would have to decide whether to use

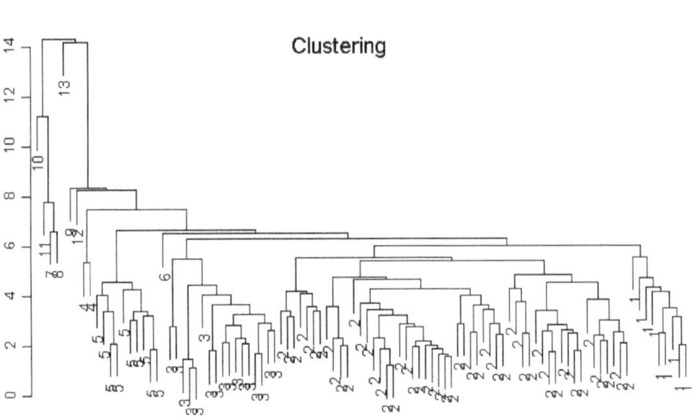

Figure 4: A grouping of the data provided by cutting the dendrogram resulting from an average linkage CA in S-Plus. (In this case the dendrogram was cut at a value of just less than 6.) The user is prompted to enter a group number that will define the initial group for Beier and Mommsen's grouping procedure.

a cut-off value for squared distance of approximately 1.5, or approximately 2.1. As can be seen from the histogram, the latter value would result in the addition to the group of two objects that the former value would leave out. In fact, in this case a cut-off value of approximately 1.5 would result in the group ceasing to grow at this point whereas the higher cut-off value would result in the group growing further.

As an additional aid to making a decision regarding an appropriate cut-off value, the plot of the first two PC scores on the right of figure 3 labels the objects currently in the group. By clicking the mouse on the histogram, the user can experiment with the effect of choosing different cut-off values. After each choice, the PC scores plot on the right is updated and user is asked to accept the current group or to choose an alternative cut-off value. Once a group is accepted, the next iteration of the grouping procedure can take place. This consists of the calculation of (a) the current group mean and (b), the squared distances from each object in the data set to the current group mean. At this stage a new histogram is plotted and the group may or may not grow further, subject to the choice of cut-off value. (Note that as the choice of cut-off value is unrestricted, the user also has the option of *shrinking* the current group.)

Interaction with the graphical display of the type reported above is easily implemented within S-Plus and can also be used in other ways. For example, figure 2(b) of section 2.1 shows the graphical output resulting from carrying out an average linkage CA in S-Plus. As mentioned in section 2.1, such traditional exploratory techniques could be used as a method of choosing initial groups for Beier and Mommsen's procedure. In the case of output from a cluster analysis, the user may achieve this by "cutting" the tree with the mouse at an appropriate point on the dendrogram. The result is a grouping of the data set defined by the point at which the tree was cut. Figure 4 shows such a grouping, labelled by group number. The user is then prompted to enter a group number that will define the initial group for Beier and Mommsen's procedure. PCA output can also be used to identify an initial grouping by clicking on points in a plot of the first two PC scores.

3. Case study

Our implementation of Beier and Mommsen's grouping procedure is now illustrated using a data set consisting of 27 elemental concentrations (measured by Neutron Activation Analysis (NAA) in Berkeley) for each of 100 ceramic samples from Nichoria in the Peloponnese. These data form part of the Perlman-Asaro databank of Mycenaean samples. More information can be found in Mommsen et al. (in press). Using the original implementation of the methodology described here, it is claimed that the majority of these data can be divided into 6 well-separated groups. (Several chemical loners, that is objects not belonging to any group, are also identified.) Rather than presenting a full grouping of the data set, we discuss the growth of three example sub-groups that, for illustration, utilise a variety of methods for providing an initial grouping.

Before describing the grouping process in more detail, it is worth pointing out that presently ours is not a full implementation of Beier and Mommsen's grouping procedure. Their original implementation consists of several thousand lines of FORTRAN code and has evolved over many years to include subtleties that are not yet reflected in our S-Plus implementation. Furthermore, the published grouping of these data (Mommsen et al. in press) is influenced by scientific and statistical decisions not replicated here. For example, we perform our grouping based upon the full data matrix of 100 objects by 27 variables, whereas Mommsen et al. use 25 variables, omitting Ca and Na. It would therefore be surprising if the procedure outlined below gave the same grouping, however we do hope to get a similar grouping.

3.1. Group 1

A preliminary scan through the data set, searching for pairs of objects that are close with respect to weighted Euclidean distance reveals that, for example, objects 74 and 77 are very close together. Using these two objects as our initial grouping the group mean is calculated and, with the aid of a histogram of (squared Euclidean) distance values, we look for objects that are close to our current group. Using a cut-off value of approximately 3.5 adds the four nearest objects to the group, which now consists of objects 26, 61, 64, 70, 74 and 77. The group mean is re-calculated and we again look for objects that are close to our current group. Using a cut-off value of approximately 1.6 adds four more objects to the group, which now consists of objects 26, 61, 62, 64, 70, 73, 74, 75, 77 and 79. The group mean is again re-calculated and the resulting histogram of squared distance values is shown in figure 5. This reveals (a) that the ten members of our current group are all close to the group mean and (b), shows a pronounced "edge" at a distance value of less than 1, meaning that the group as a whole is distant from other objects. The PCA plot on the right also illustrates that the final grouping is quite compact (compare with the full PCA plot in figure 1(b)). Our current group of ten objects is therefore classified as a genuine sub-group within these data. All the members of this sub-group are contained within a sub-group of 13 objects identified by Mommsen et al. (in press). The objects missing from our sub-group are 47, 67 and 72. It is worth noting that the closest object to our sub-group of ten is object 67, but at a distance that makes inclusion difficult to justify. Object 72 is also close to the group, being the fourth nearest object not included.

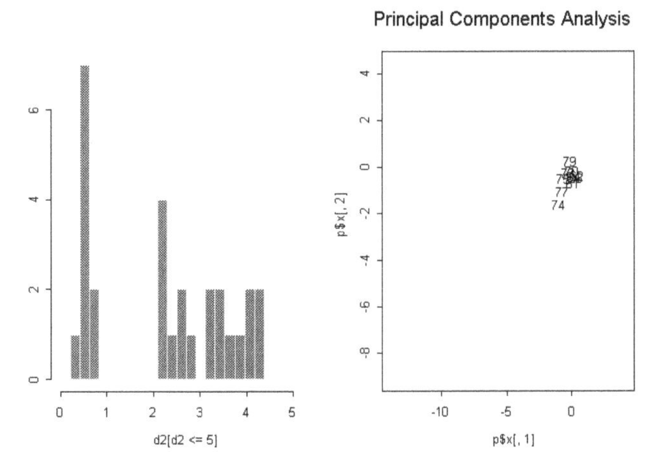

Figure 5: The result of growing the initial grouping consisting of objects 74 and 77. The histogram of squared distances (left) reveals a pronounced "edge" at a value of less than 1 and the labelled PCA plot on the right reveals that the final grouping is quite compact (compare with the full PCA plot in figure 1(b)).

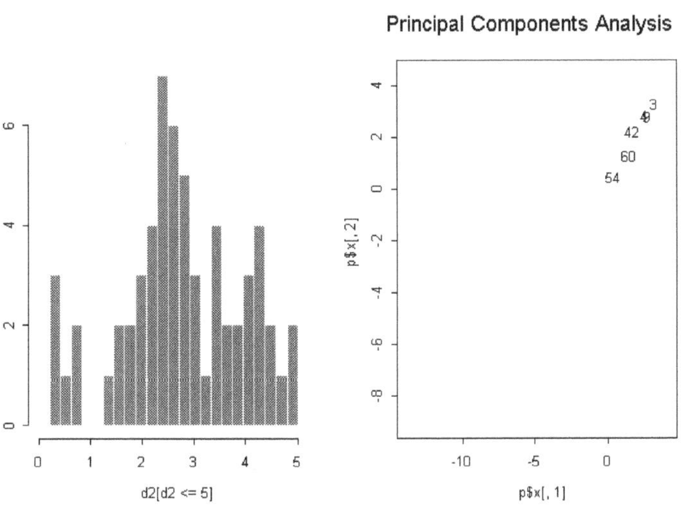

Figure 6: The result of growing the initial grouping consisting of objects 3, 4 and 9.

3.2. Group 2

Having found one sub-group, we now continue, trying to find others. An average linkage CA reveals that objects 33, 34, 38 and 65 are close together. Using these objects as our initial grouping the stages of Beier and Mommsen's grouping procedure give the results presented in table 1. The process stabilises after stage three yielding a sub-group of nine objects. This sub-group is almost identical to a sub-group of ten objects identified in Mommsen et al. (in press). The only difference is that object 37 is missing from our sub-group. Indeed this object lies some distance from the centre of our sub-group.

3.3. Group 3

As a final example, we note that a PCA plot of the first two PC scores reveals that objects 3, 4 and 9 are close together in the extreme top-right of the plot. Selecting these objects directly from the PCA plot (by clicking on them with the mouse) and growing the group yields the results presented in table 2. The process stabilises after stage two yielding a sub-group of six objects. Note

Stage	Group	Group size	Approximate cut-off value
0	33 34 38 65	4	
1	32 33 34 36 38 65 78 80	8	1.8
2	32 33 34 36 38 39 65 78 80	9	1.5
3	32 33 34 36 38 39 65 78 80	9	1.5

Table 1: Growth of the initial grouping of objects 33, 34, 38 and 65 (identified using an average linkage CA).

Stage	Group	Group size	Approximate cut-off value
0	3 4 9	3	
1	3 4 9 42 54 60	6	2.1
2	3 4 9 42 54 60	6	1.3

Table 2: Growth of the initial grouping of objects 3, 4 and 9 (identified using a plot of the first two PC scores).

that this method of selecting an initial grouping is presented mainly to illustrate the kind of interaction that is possible within S-Plus. Though it works well in this case, using the first two PC scores in this way is less preferable than the methods of initial group selection discussed in sections 3.1 and 3.2.

The final histogram of squared distance values is shown in figure 6. Again note the pronounced "valley" around distance values of approximately 1. It is possible to further grow this group as six objects occur at distances of between 1.4 and 2.0 from our sub-group of six objects. However, if this is attempted the group quickly spreads out and the resulting histograms of squared distance fail to reveal the "valley" or "edge" that signifies a coherent sub-group. We therefore classify our six objects as a genuine sub-group within these data. This sub-group is identical to one identified by Mommsen et al. (in press).

4. Software issues

4.1. Extending the graphical user interface

Using S-Plus enables user-friendly calling of user-developed routines. For example our implementation of Beier and Mommsen's grouping procedure can be called from a menu list consisting of other novel grouping methods. Selecting this menu item results in the appearance of a dialogue box of the type shown in figure 7. The specific details of this dialogue box are less important than the ease with which it can be created from within S-Plus. However, the three points outlined below are worth making.

1. The dialogue box controls input to the grouping routine by specifying the names of the S-Plus objects containing the data and the measurement errors if known.
2. The dialogue box also controls output from the grouping routine by specifying the names of an S-Plus report file and an S-Plus output object. The latter contains numerical information that results from the statistical analysis and the former provides a user-friendly format for this information by embedding it in a specially designed written report.

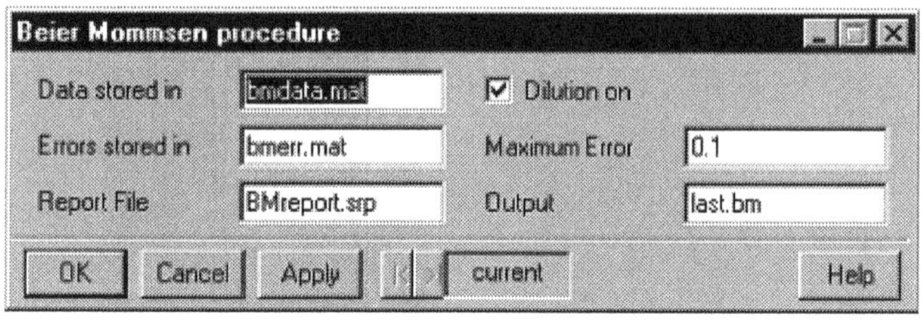

Figure 7: Dialogue box for calling our implementation of Beier and Mommsen's grouping procedure.

3. This dialogue box consists of only three basic user-interface elements: (a) text input via four editable boxes, (b) numeric input via a single editable box and (c), a single "check box" to define whether the dilution correction should be turned on or off. However, there are several other elements that can be used to construct such dialogue boxes. These include sliders, tabbed pages, push buttons, radio buttons, pop-up menus, scrollable lists etc. (MathSoft 1999).

4.2. The S programming language

As a brief example to show the power of the S programming language, we illustrate in figure 8 a simple S code listing for a routine to calculate the Mahalanobis distance between a point, $\underline{x}$, in p-dimensional space, and a group with mean $\underline{y}$ and p by p covariance matrix S. Recall that this is given by equation (1). Reconsidering equation (1), we can write

$$d_{xy}^2 = (\underline{x}-\underline{y})^T \Sigma^{-1} (\underline{x}-\underline{y}) = (\underline{x}-\underline{y})^T \underline{v} \qquad (2)$$

where

$$\underline{v} = \Sigma^{-1}(\underline{x}-\underline{y}).$$

Mathematically, calculating explicitly the inverse of the p by p covariance matrix S is an ill-conditioned process and not to be recommended. However, multiplying both sides of the above equation by S gives

$$\Sigma \underline{v} = \underline{x} - \underline{y}.$$

In other words the vector $\underline{v}$ required in the definition of the Mahalanobis distance (2), is the solution of a linear system of p equations. This system is solved in the third line of the S routine illustrated in figure 8, using the *solve.default(Sigma, x-y)* command. The result is stored in the vector *v*. However note that in this simple example we have made no attempt to trap errors caused by the properties of the matrix stored in *Sigma*. In particular, this matrix will be rank deficient if the group size is less than p, the number of variables. S-Plus has many other built-in commands for carrying out complex mathematical operations on matrices. This makes it an ideal platform for the development and implementation of algorithms in the field of multivariate statistics where the data is naturally represented as an n by p data matrix.

The fourth line of the code calculates d^2_{xy} as defined by equation (2) and stores the result in the variable *d2*. Here the *%*%* operation represents matrix multiplication. The fifth line of the code signifies that it is the value of *d2* that is to be returned as the output from the routine. Given S-Plus variables *x, y* and *Sigma* storing respectively two p by 1 vectors, $\underline{x}$ and $\underline{y}$ and a p by p

```
1. Maldist <- function(x, Sigma, y)
2. {
3. v <- solve.default(Sigma, x-y)
4. d2 <- t(x-y) %*% v
5. return(d2)
6. }
```

Figure 8: S-Plus code listing for a routine to calculate the Mahalanobis distance between a point, $\underline{x}$, in p-dimensional space, and a group with mean $\underline{y}$ and p by p covariance matrix S. Note that line numbers are given only for ease of reference and are not required as part of an S-Plus code.

covariance matrix S, the command *d2 <- maldist(x, Sigma, y)* calculates the Mahalanobis distance between $\underline{x}$ and $\underline{y}$ and stores the result in the variable *d2*.

Finally it is worth noting that this example is chosen simply for illustration, as S-Plus has a built-in command, *mahalanobis*, that is called in a very similar manner to our example. However, as we have previously pointed out in section 2.2, Beier and Mommsen's grouping procedure makes use of modifications to the definition of Mahalanobis distance, given in equation (1), to take into account uncertainty of measurement and dilution effects. These modifications require additional programming effort, especially when we take dilution effects into account.

5. Summary, conclusions and future work

In this paper we have:

1. Shown that otherwise complex methodology can be implemented in S-Plus with relative ease. In particular, we have implemented a powerful but complex approach to the statistical analysis of ceramic compositions developed by Beier and Mommsen.
2. Demonstrated that traditional methods of exploratory multivariate analysis can be used alongside, or in combination with, a technique designed specifically for grouping ceramics by chemical composition.
3. Reported on the development of a user-friendly interface for our S-Plus routines.

Our implementation has been successfully applied to the analysis of chemical compositional data consisting of 27 elemental concentrations for each of 100 ceramic samples from Nichoria in the Peloponnese.

Looking ahead, our involvement in the GEOPRO project has the wider aim of developing an S-plus library of user-friendly routines for grouping ceramics using chemical and/or mineralogical data. In particular, over the duration of the project and beyond, our work will involve the evaluation, development and implementation of many more statistical techniques. A particular difficulty is that other authors have implemented advanced statistical methods for the analysis of mixed-mode data in the programming language FORTRAN. In the interests of widening access to such methodology, some progress has been made in interfacing FORTRAN-based software with S-plus. For similar reasons, we will be giving careful consideration to the provision of support materials (for example documentation and tutorials, both paper and web based). The final collection of routines will be packaged as an S-Plus library and made freely available, via the Internet, to the archaeometric community.

Acknowledgements

We are particularly grateful to Thomas Beier, Hans Mommsen and Anno Hein for fruitful discussions on this work. There is no implication that they necessarily agree with our interpretations and any misunderstandings and errors are entirely our responsibility. Hans Mommsen is thanked, additionally, for providing the data used throughout. This work forms part of the GEOPRO Research Network funded by the DGXII of the European Commission, under the TMR Network Programme (Contract Number ERBFMRX-CT98-0165).

References

BAXTER, M.J., 1994. *Exploratory Multivariate Analysis in Archaeology*. Edinburgh: Edinburgh University Press.

BAXTER, M.J., PAPAGEORGIOU, I., CAU, M.A., DAY, P.M. and JACKSON, C.M., in press. Integrating geochemical and mineralogical data in studies of ceramic provenance: some statistical issues (the GEOPRO project). *Proceedings of CAA99*, Dublin. Oxford, BAR.

BAXTER, M.J. and JACKSON, C.M., in press. Variable selection in artefact compositional studies. *Archaeometry*.

BEARDAH, C.C., PORCINAI, S. and BAXTER, M.J., in press. Clustering with KDEs: art historical and archaeological applications. *Proceedings of CAA99*, Dublin. Oxford, BAR.

BEIER, T. and MOMMSEN, H., 1994. Modified Mahalanobis filters for grouping pottery by chemical composition, *Archaeometry* 36: 287-306.

MATHSOFT 1999. *S-Plus 2000 Programmer's Guide*. Seattle: MathSoft, Inc.

MOMMSEN, H., BEIER, T. and HEIN, A., in press. The Berkeley Neutron Activation Analysis data on Mycenaean pottery - A complete chemical grouping. In Kilikoglou, V. and Maniatis, Y. (eds.), *Proceedings of the 5th European Meeting on Ancient Ceramics*. Athens, 18-20 October 1999. Oxford, BAR.

VENABLES, W.N. and RIPLEY, B.D., 2000. *S Programming*. New York: Springer-Verlag.

Why the Application of a Gaussian Curve and Seriation Programs can be Detrimental

Andrej Pleterski

Institute of Archaeology, Scientific Research Centre of the Slovenian Academy of Sciences and Arts

Gosposka 13, 1000 Ljubljana, Slovenia

e-mail: andrej@zrc-sazu.si

Abstract

Certain attributes are indicative of "chronological groups" while others are indicative of other phenomena. Any attribute may be characteristic of a specific group or not. It cannot, however, be slightly less or more characteristic of a group. Recognition of this situation highlighted problems with the application of the Gaussian curve within seriation programs. The problem of classification is no longer mathematical, but rather hermeneutic. The Bajuwarian cemetery at Altenerding proves that chronologically sensitive attributes occur more or less simultaneously. Here we are faced with a continual process of change that is sometimes faster and at other times slower: sometimes changes are more frequent and at other times sporadic. This is also why algorithms for the seriation of material finds, which anticipate that chronological groups do exist, and, as a rule, follow successively one after another within the context of their respective chronological types, cannot provide adequate results.

Key words: Gaussian curve, seriation program, chronology, archaeology

The practice of resolving chronological questions with the aid of seriation programs, whose algorithms frequently postulate a distribution of material finds through time in the form of a Gaussian curve, is still quite popular. Whilst analysing the Bajuwarian cemetery at Altenerding (Sage 1984), I chanced upon results that impugn the use of the Gaussian curve and seriation programs for such purposes.

The 1530 graves of Altenerding could have been classified chronologically with the aid of those material artefacts that are temporally well defined, such as arched bow fibulas in female graves and belt buckles and belt fittings from male graves. On the other hand, I was tempted to establish, with the help of stratigraphic tables, an entirely independent relative chronology, which would ultimately be furnished with absolute dates. The two principal objects used within the analysis were beads, the most frequent artefact in female graves, and belt buckles, the most frequent artefact in male graves. As the latter are also present in female graves, a correlation is proffered in the composition of beads from female graves. From the database, I extracted the attributes of both types of artefacts and initiated their classification through the KOR seriation program, written specially by Tomaž Zwitter (Pleterski and Zwitter 1993). While resolving questions of the delimitation of seriated groups, we also discovered that the groups in the combination tables and the tables of the seriation program in no way demonstrated simply a seriation of the artefacts through the course of time, but much more.

The problem was reconsidered on this basis. It was felt that the phenomenon of "non-chronological groups" was the consequence of a poor initial selection of attributes. That is, certain attributes are indicative of "chronological groups" while others are indicative of other phenomena. Any attribute may be characteristic of a specific group or not. It cannot, however, be slightly less or more characteristic of a group. Recognition of this highlighted problems in the use of the Gaussian curve upon which seriation programs are based. The problem of classification was felt to be no longer mathematical but hermeneutic (cf. Pleterski 1996). A table of stratigraphic relationships facilitated chronological interpretation. I decided to use 270 graves with absolute dates, which Hans Losert attributed using the chosen analogues, for additional control.

An unexpected result derived from this process. My six groups of belt buckles were indeed bounded quite narrowly chronologically, and they were thus truly "chronological groups" in this respect. However, they were not arranged in chronological succession. The groups chronologically overlapped in part, or were even entirely contemporaneous. This observation was confirmed by the subsequent dating process, after which I was able to determine 154 chronologically sensitive attributes.

It is evident that chronologically sensitive attributes may occur more or less simultaneously. Change and novelties arise continuously over time. Consequently, division into chronological levels is unrealistic and misleading, particularly as this leads to the creation of artificial divisions which simply confirm the preconceived slices of the chronological "sausage".

The graphs presented demonstrate that the majority of novel chronologically sensitive attributes appear between 550 and 600, and that the majority disappear between 600 and 650. The reason for this is that the majority of attributes from this period correspond to multi-part belt fittings. Were the attributes classified differently, the graphs, with respect to the appearance and disappearance of chronologically sensitive attributes, would also be altered. However, the partial contemporaneity of these attributes would remain.

If I were to provisionally define chronologically classifiable artefacts as "chronological types", then it follows that chronological types do not exist within the context of a single "chronological level" simultaneously, or successively (one after another) in further "chronological levels". In reality, we are faced with a continual process of change that is sometimes faster and at other times slower. Sometimes change is more intense and at other times it is sporadic. However, this is all it is. Each "chronological level" is actually an artificial definition. This is why all algorithms for seriation of material finds which anticipate that chronological groups

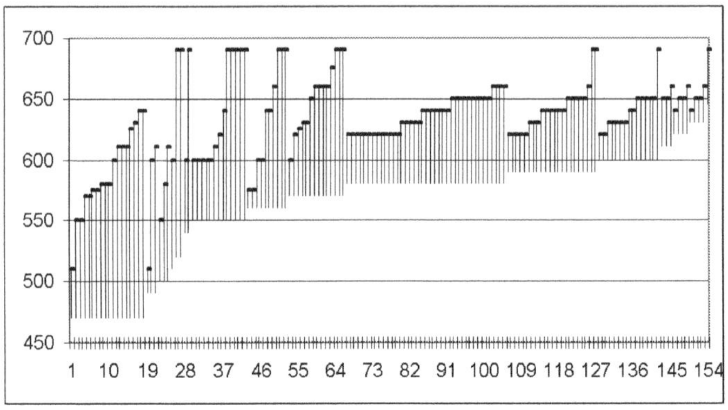

Figure 1: Chronologically sensitive attributes, in order of their succession, showing the temporal range of appearance of attributes within graves.

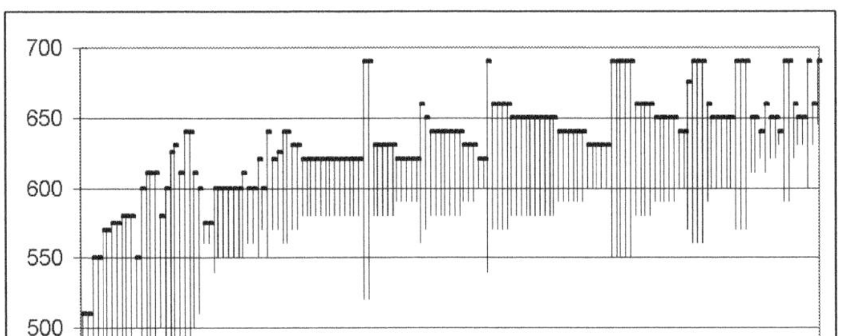

Figure 2: Chronologically sensitive attributes, order by succession and by median temporal range of attributes within graves.

do exist and, as a rule, follow successively one after another within the context of their respective chronological types, cannot provide adequate results. Occasionally analyses may apparently succeed. Despite this, such results could still not be used as a reliable tool. However, they do remain useful tools for searching for congruent structures. I rest assured that I am not alone in my thoughts on the basis of a lecture by François Djindjian; who emphasized

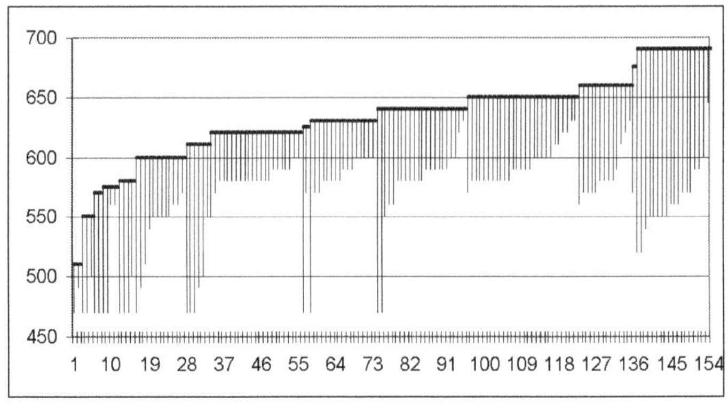

Figure 3: Chronologically sensitive attributes, in order of their succession, showing the temporal range of the disappearance of attributes from graves.

that the understanding of an artefact is a question of proficiency and not a characteristic of computer analysis, that semiotics can be more valuable than taxonomy, that seriations are not necessarily a reflection of chronology and that chronological structures must be separated from other relevant factors (Djindjian 2000).

References

DJINDJIAN, F., 2000. Artefact analysis. – *CAA 2000: Computing Archaeology for Understanding the Past*, April 18. – 21. Ljubljana (lecture).

PLETERSKI, A., 1996. Zur Methode der strukturellen Analyse von Gräberfeldern. In Bialeková, D., Zábojník, J. (eds.), *Ethnische und kulturelle Verhältnisse an der mittleren Donau vom 6. bis zum 11. Jahrhundert*, Symposium Nitra 6. bis 10. November 1994, 15 – 20. Bratislava.

PLETERSKI, A. and ZWITTER, T., 1993. The KOR Seriation Program and its applicability in Archaeological Research. *Arheološki vestnik* 44: 269 – 293.

SAGE, W., 1984. *Das Reihengräberfeld von Altenerding in Oberbayern I.* Germanische Denkmäler der Völkerwanderungszeit, Serie A, Band XIV. Berlin.

Quantities, Possibilities and Probabilities: Some Experiences from the Research of the Roman Age in Slovenia

Iva Mikl Curk

Cimpermanova 5, 1000 Ljubljana, Slovenia

Abstract

The paper considers the state of Roman ceramic studies in Slovenia. It questions the legitimacy of current approaches to ceramics, outlining problems in procedure and placing contemporary practise within its historical context. The range of archaeological uses for ceramics is considered and suggestions are made for the extension of ceramic studies using data from the urban area of Emona and a number of Roman cemeteries in Slovenia.

1. Introduction

Knowledge of how often we are able to detect specific objects or features within an individual archaeological record, within a particular findspot or within a specific (larger or smaller) geographical unit, has always been a very important component of archaeological research. It is often the basis on which conclusions are reached. In the language of theoretical archaeology, it could be said that the statistical method of previous developmental phases in archaeology permitted the development of archaeology. The computer has opened undreamed-of possibilities for contemporary research procedures. It would therefore be advisable to recall experience of previous decades. Such examples could help, to some degree, to establish what types of questions should be framed and what the product of research might be. I am, of course, discussing this matter on the basis of my personal experiences, and the literature cited might appear one-sided at first glance - but I am listing here only those of my reports which substantiate these observations.

These experience relate to the Roman Period in Slovenia. This period is, more or less, identical to the duration of the Western Roman Empire, as it is generally known. It is, to a certain extent, a specific period; a range of written sources provide evidence concerning the circumstances of the Roman Empire. Consideration of the data enables the construction of a material world with a number of common features, allowing the application of analogies between distant places. On the other hand, this huge state retained many local characteristics. Moreover, substantial distinctions existed among the small but diverse regions of our country, located in what was, at the time, an important Northern Adriatic - Eastern Alpine passage to the continent. These distinctions are noticeable from the time of annexation. From the time of the late Republic the littoral zone became a part of Gallia Cisaplina. Most probably the majority of the western part of Illyricum became a passable corridor at the time of Augustus' first expeditions into the Adriatic hinterland. The Alpine part, however, was not formally or legally organised as a province before the time of Claudius. These, along with many other undisputed facts, influence and continue to influence, the problems we wish to research.

2. The pottery

This is the field in which I have most experience. The statistical analysis of ceramics, although previously conducted on the basis of piles of hand written slips of paper subsequently became a component part of the new systematic excavations undertaken after World War II (Horvat M. 1999). The opportunity to publish all finds, along with the entire process of analysis, were far more limited in the past comparison with the situation today. Yet even today the result of much analysis can be presented only in short reports at professional meetings, frequently only to specialists. J. Horvat's (Horvat J. 1999:233) important discussion of analyses based on numeric studies of ceramics, was published in RCRF Acta. Here she comments on the lack of evaluation of criteria and methods used for obtaining comparative data, and this paper serves as an important introduction to this report.

On any occasion when experts from various specific fields meet, it would be in order to dedicate a few words to our established practice of estimating the *number of individual vessels*, from fragments found in individual archaeological contexts or groups. This practice has been introduced, with slight modifications, into our country following its general international use and has, in my view, contributed to the construction of a basis for statistical investigation. I am convinced that this practice is still applicable, all the more so since its completion is accelerated by the use of computers. Over and over again we find ourselves confronted by the task of "translating" the number of fragments into numbers of vessels. Any two results are hardly ever identical and an entirely reliable method is still not available. Even today, none of the detailed research (technical, chemical, etc) of substance and shapes, aimed at achieving this purpose, has been adequately published. Macroscopic analysis, which had been in use from the very beginning of modern archaeological research, still forms the basis for this work - but it demands considerable practice (few hours of practical work in a restoration workshop is the precious source of necessary experience!) and, of course, the time to ponder over it. Almost 90 % of the Roman ceramics found in our country are made on the potter's wheel (in 80 % of cases on a fast wheel) and fired in closed kilns. This manufacturing technique gave the products a unity of surface; but not without exception. Although the product of specialised manufacture, each vessel is to a large extent the result of handicraft and these vessels almost never have the same texture around the entire circumference. They are not equally fired, etc. We must point out that the final, or specially moulded parts of vessels, like the handle or the rim, are almost always slightly different in texture, since intense kneading extracted the larger particles and inclusions from the clay. The question whether several fragments belong to the same or to a different vessels can most easily be resolved in cases of fragments of moulded terra sigillata: this knowledge forms the basis of all the analyses aimed at following the spread of sigillata from southern, middle and eastern Gaul. Work with atypical surface fragments of plain sigillata is only slightly more difficult (a higher percent of errors might occur, although with experienced researchers these errors do not exceed 10 %), if accompanied by a sufficient number of distinctive parts of vessels, stamps, sections of rims and legs. The same holds for all sorts of semi-industrial pottery, manufactured in larger workshops. In comparison , the error margin for handmade local pottery can reach up to 30-40 % of studied groups!

We never place enough emphasis on the importance of possessing sufficient knowledge of other, previously excavated material from the same site or at least from near-by sites. It is easier to define *shapes* of particular vessels, and these are usually the first to be determined. We therefore arrive, relatively quickly, at the conclusion that amphorae, jugs, plates, pots, etc. can be found within a specific group. The decision on what *function* the individual artefact actually served usually demands additional interpretation: there are particular shapes for specific uses of pottery, which more or less serve the same function at various locations, and then there are other shapes which might have been used for a range of functions, although of course, we cannot exclude the possibility that an old and damaged sigillata cup could have occasionally been used for other purposes (as is often the case today!), for instance for mixing paint or glue. My experience tells me that the conclusion about the use of the vessels from an individual archaeological complex, can in most cases only be reached towards the end of the study. Nevertheless, the terra sigillata or other distinct earthenwares, as well as amphorae and mortars, represent exceptions, due to the fact that their function is usually very clearly defined.

The result of any investigation is therefore determined, above all, by the character of the archaeological group and by the stratigraphic record of the ceramics. For this reason I feel it important to emphasise the necessity (even within a large team, where tasks are very clearly divided among experts) for professionals dealing with the investigation of particular materials to have a good overview of the situation in the field. I believe that I was able to form my conclusions about pottery from several excavation sites within the city walls of Emona because I recorded most of the ceramics immediately as they appeared from the ground and was thus able to examine details while the trench was still open. Even when first recorded it was clear from where material originated, e.g., from the stratum later covered by the pavement of the next building phase and from garden humus. These data were later augmented by all previous data and by the overall results of the excavations, corresponding buildings or other components of the settlement structure.

The material records stored during the excavation used to be the quickest procedure for *dating* an individual stratum or archaeological complex. I suppose it is unnecessary to waste words about this procedure for this particular group of readers. We are all well acquainted with the fact that specific types of ceramics have a general and indisputable dating value. In such cases only the youngest object is significant, quantities play a minor role. From this angle, the most informative record for ceramic study, were for me, those from several excavations of a single stratum within the walls of Emona, a stratum which is almost always present and is the deepest archaeological layer, cut by the foundations of the first built structures and covered by the oldest mortar pavement. Fragments of Iron Age pottery can also be found here. However, the majority of the fragments originate from the Augustan period; several individual fragments are reliably made of sigillata dated to the late Augustan period and during the first years of Tiberius's reign (Mikl Curk 1973, 1977). It is thus evident that the formation of this stratum ceased at the moment when house construction inside the walled and planned town area was completed.

However, quantification plays a very important role in dating an individual unit where the ceramics (as well as other materials, of course) have a relative chronological value. At this point it is of key importance how the features of any particular period within a complex obtain their dating value, or to what extent those features are disappearing. As an example, let me present some of my observations regarding coarse Roman pottery, partially hand-moulded or moulded on the slow potter's wheel. This kind of pottery is somehow "softer" to the touch, inclusions are often limestone, grains have rounded edges, so that it appears as if the clay had been carelessly kneaded. But quite a lot of coarse pottery made on the fast wheel has been found as well. This type represents the second group of coarse kitchen pottery in Slovenia. It is usually very "hard" to the touch, so well fired that it gives a distinct sound, inclusions are often finely ground, sharp, small grains of flint. The most frequent coarse pottery form is the oval pot, although there exist a number of other objects made of such a ware. The place of manufacture can only rarely be identified, but the majority of these objects are presumed to be of local origin, although we are aware that this pottery was sometimes sold far away from the place of its manufacture. It is most difficult to ascertain how many vessels we are dealing with in an individual context from the recovered fragments of pottery. The possibility of error is greatest in these cases. Undoubtedly, the huge difference in manufacturing techniques cannot be attributed only to chronological causes. The most primitive and the most accomplished types appear side by side in the same situation and in the same closed context. Each individual technique produced vessels suitable for specific purposes –vessels with walls porous enough for food storage, vessels resistant to open fires, etc. Mould details with a wavy line on the surface appear throughout the entire Roman period and at various sites throughout the country. But particular features are not present at all locations or in equal frequencies or equally represented at all periods. It is therefore important to determine which particular features on the ceramics prevail within an individual archaeological group. Accordingly, I presume we could consider, for instance, unevenly kneaded pottery, decoration with a belt of wavy horizontal lines drawn by a comb, as characteristic of some specific environment and of a specific period (near, but not as exactly defined as in Rodrigez 1977, 1992), or for large pots with a surface that gives an impression of being impregnated by wax, as the characteristic of central Slovenia from the same time or even from the Migration Period – the Dark Ages (Mikl Curk 1975, 1992a). Or furthermore, the oval, roughly shaped pots of the early Middle Ages, often defined by an irregular texture and decorated with lines and prints, all of which testify to the use of a more robust potter's tools. However, I must point out, that one individual fragment of a specific kind of pot, is never enough to play a decisive role in determining the archaeological record. The most important statement in the record is: in what percentage is the described feature represented. Due to the modest numbers of finds in the later strata of Slovenia, a single fragment can represent up to 10 % of the total!

The frequency of individual types of ceramics of known origin, provides a snapshot of the *intensity of trading* and indirectly of all kinds of *other connections*. A number of researchers are engaged in studying this topic both in our country and abroad, so that the working method is well known as is the probability of error. An achievement of the study of ceramics in Slovenia is the study of WHY only particular kinds of terra sigillata are represented in specific archaeological groups, while many of other contemporary products cannot be found. The assumption that the volumes of material transported by sigillata merchants was often quite large (several sets of services) and that therefore the merchants delivered only the sigillata from a single or at least from one smaller group of manufacturers to one place at any one time, has not yet been disputed (Mikl Curk 1987, 1992).

The frequency of finds of vessels, sorted by function, enables us to draw conclusions concerning the activities that took place within specific buildings, *a conclusion about the purpose of a specific building*. Of course these conclusions demand special attention. Stratigraphic data is especially important. All the tools and vessels which had been used in a building do not necessarily remain there after any change, even where a pavement has been left intact. To illustrate: numerous remains of former drinking vessels were left all over the first century AD paved floors of small rooms in the forum of Emona. It is quite probably evidence of the sale and consumption of all kinds of liquids. But at one end the earthen joints used as support for vessels when placed in kilns were also found among the fragments. This testifies that empty vessels were sold as well. The irony is that the early nineteenth century gardens owned by a Ljubljana faïence manufacturer were located at the same site where the Emona forum once stood. For some inexplicable reason the ashes and other remains from the kilns of this faïence factory, which was documented at the other side of the town, ended up there. Ceramic plugs similar to the Roman ones (although not identical in colour and texture) and which were used in kilns as a support for pots were mixed with the top, humus layer, over these remains! On the other hand, practically no finds were discovered in the sandy strata of the forum plateau. Similarly, only a few pot fragments were found in the stratigraphy of the large forum buildings, which L. Plesničar Gec believes to have been built in the third century. This is understandable. The walls and stone paving of these buildings were still partly preserved at the excavation site. In a regulated settlement daily waste, whatever this might have been, was usually removed from site. Moreover, this data in a way supports the presumption that activities in which pottery was used did not take place in the town basilica. Large impregnated pots were found in the stratum covered by these remains (outside the buildings); a discovery which leads me to suggest that this kind of pottery belongs to the period when larger constructions were still standing but were no longer serving their original function. After all, we could hardly imagine that in a regulated town organism the most common food was being prepared and stored or that various common and everyday activities took place close to the external walls of the court basilica or the church portico. The dating of similar ceramics from the late Roman settlement at Vranje in the late fifth century (Knific 1979) confirmed this presumption.

For some other kinds of ceramics dated to the same period and originating from contemporary as well as earlier layers it can sometimes be concluded, that even after changes to the buildings within the constructed area of Poetovio, the ceramics remained close to the location where they originally served their purpose. We thus proposed, on the basis of two distinct locations within the area of Roman Poetovio, an undoubtedly bold and as yet unconfirmed assumption, that the middle Gallic sigillata had been reserved for the upper, mainly officer class of the town, while Rheinzabern pottery was available to the common inhabitant (Mikl Curk 1990).

3. The use of arable land

In our country there are only five, or perhaps six, extensive plains on which were situated all the Roman towns or major settlements. The Vipava valley (Vipavska dolina) which widens in the hinterland of today's Ajdovščina – Roman Castra. The Sava alluvium which stretches beneath the southern foothills of Karavanke Alps from the Sora plain (Sorško polje) to the higher regions of the Sava plain (Savsko polje) north of Ljubljana in the direction of Kamnik. These plains are the wider hinterland of the town of Kranj – the late Roman Carnium, but in the distant past they were mainly the hinterland of Ljubljana – Roman Emona. The Savinja valley (Savinjska dolina) is dominated by the town of Celje – Roman Celeia, the Krško plain (Krško polje) was in Roman times dominated by Drnovo near Krško, now an insignificant settlement, but in those times Neviodunum. The Drava and Ptuj plains (Dravsko and Ptujsko polje) were in Roman times certainly exploited by today's Ptuj – Roman Poetovio. The smaller plain around Slovenj Gradec belonged to the territory of the Roman Colatio, then located at Stari trg near Slovenj Gradec. These plains differ slightly, with respect to the pedological and partly hydrological characteristics, so that reliable conclusions about the extent of cultivation in Roman times cannot be made. Due to the consistent cultivation of these plains, any remains of actual Roman fields are not recognisable even on aerial photos. It is also unknown to what extent the Romans were interested in their agrarian potential. It certainly wasn't their primary interest. Above all, the territory of Slovenia attracted them for its strategic and communication potential, and to a certain degree as a source for material wealth. The existing roads which have been in use for a long time, and the cadastral border-lines, in some places indicate the shape of fields which might have been formed in Roman times (Mikl Curk 1984); yet all these facts are still far from being confirmed. However, we do have a fair knowledge of the traces of settlements and roads constructed during the Roman period on all the above mentioned territories (Mikl Curk 1993a). These traces do not inform us of the cultivated areas, but the plains are measurable and this information can be used for basic calculations. Poetovio was a military camp throughout the whole first century AD. The fact that a legion needed at least 600 tons of corn or equivalent quantity of other crops to pass the winter months, plus a less well known quantity of supplementary foodstuffs. In the eighteenth century in these regions, approximately 150 – 175 hectares of fields would be required to provide such a quantity of corn. For the Ancient World we have to approximate the minimum harvest per hectare. From this we can conclude that a legion could have needed as much as 10 – 15 % of all cultivable land within the reach of the town (Mikl Curk 1993). Of course, we have no knowledge of how many of these areas were already cultivated before the Romans came to the area, nor whether the legion actually supplied itself from the products produced in the surroundings of castra hiberna. Perhaps it might soon be possible to complete and extend this entirely hypothetical study.

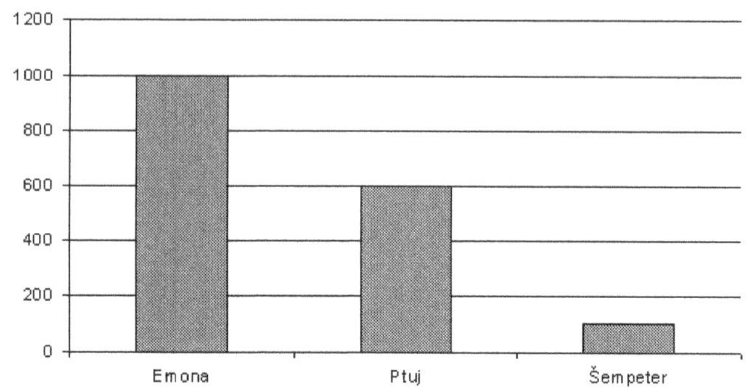

Figure 1: The number of analysed graves.

4. The cemeteries

Funeral gifts and aretefacts placed in tombs are in each civilisation characterised by numerous, specific and complicated symbology. In a prosaic sense, they signify the result of an instantaneous occurrence. In almost every cemetery individual tombs differ, although they have common features. The possibility to document varieties and similarities is obvious, as also is the wish to understand their causes. Regarding research on the Roman period in Slovenia (figure 1), the possibilities for such analysis became a reality after the comprehensive studies of the Emona cemeteries were published and after recent research of a large part of these cemeteries (Petru (1972) published the old data and Plesničar Gec (1972) the recent excavations). The first analyses of these and later cemetery data have certainly condensed the chronological aspect and at the same time revealed the popularity of specific methods of burial, with objects in the tombs characterising the deceased by sex (figure 2). Such synoptic tables have already been included in the publications of Emona material. Chronological and typological questions were treated by Plesničar Gec 1977, and several other authors in Arheološki Vestnik 30, 1979. The criteria for determining the sex from "male" and "female" jewellery, parts of clothes, knifes, mirrors, phials of scents, small chests, needles and similar, are still unreliable; although the results obtained so far are quite good for cremation graves. Experiments with selection of objects and their origin from a specific civilisation (the origin of the type and not the individual product, whether from a Mediterranean or from a local tradition) gave identifiable results concerning the, simplified, ethnical and social affiliation of the deceased (Plesničar Gec 1985, Mikl Curk 1985, 1996). These results are not numerically insignificant. Throughout this work we were aware, and would remain so, that at the starting-point of such studies one meets a range of unexplained facts. It should therefore be taken into consideration that the clear relationships of each human community with the objects it uses in a specific superior or symbolic function (as their funeral use is) are possible only as long as the community can be clearly distinguished from the other communities with which it lives; e.g., for as long as the Italic colonists could be clearly distinguished from the natives, during the first period following Roman occupation. In the next generations these traces would already have lessened.

A fair number of similar problems still wait their turn to be closely studied, and the investigation of those is our future task.

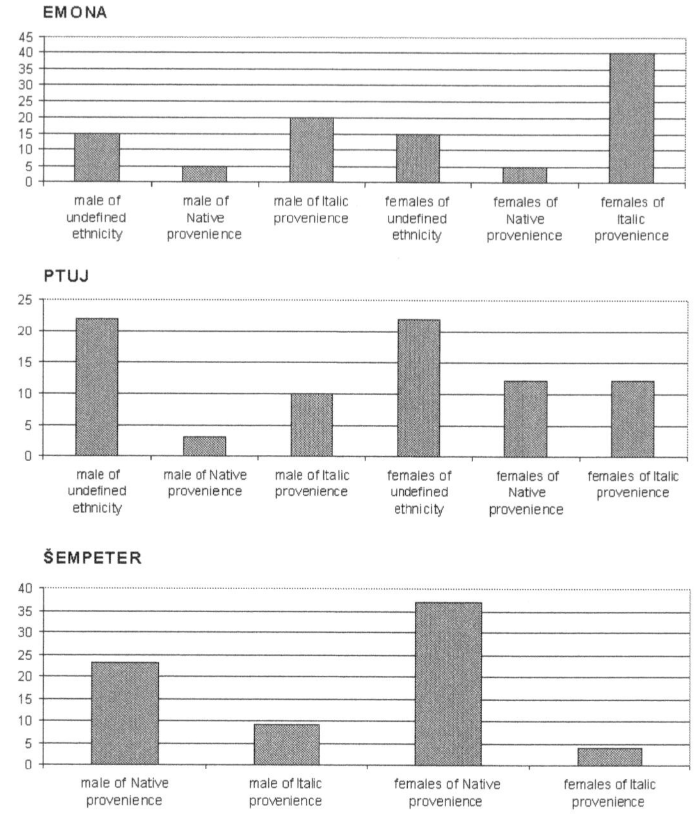

Figure 2: The number of graves, from the same sites, possibly defined by the sex of the deceased but separated upon the groups of men of undefined ethnicity, men of so-called Native and men of Italic provenience, and groups of women of undefined ethnicity, women of the so-called Native and women of Italic provenience.

References

HORVAT, J., 1999. Roman Provincial Archaeology in Slovenia Following 1965, Settlement and Small Finds, *Arheološki vestnik* 50: 215-257.

HORVAT, M., 1999. Keramika - tehnologija keramike, tipologija lončenine, keramični arhiv - eng. abstract, *Razprave FF Ljubljana.*

KNIFIC, T., 1979. Vranje pri Sevnici. Drobne najdbe z Ajdovskega gradca (1974), *Arheološki vestnik* 30, 273-385, pict. 50, 51, 76, 78, 84, 85, 100.

MIKL CURK, I., 1973. Zur Datierung der Keramik der aeltesten Siedlungsschichten in Emona, *Acta RCRF* 14-15: 125-127.

MIKL CURK, I., 1977. Einiges ueber die padanischen Sigilllaten von Emona, *Acta RCRF* 17-18: 81-84

MIKL CURK, I., 1985. Pridatki zah. nekropole v Ptuju kot dokaz socialne strukture - ..ein Zeugniss der sozialen Struktur der Bevoelkerung (sum.), *Materijali* 20, Beograd: 169-175.

MIKL CURK, I., 1984. Sledovi rimskega zemljemerstva na Dravskem polju?, *Arheološki vestnik* 34: 353-360.

MIKL CURK, I., 1987. Buying and selling pottery - An evidence from NW Yugoslavia, *Acta RCRF* 25/26: 403-408.

MIKL CURK, I., 1990. Welche Bevoelkerungsschichten haben vorwiegend Sigillaten aus Rheinzabern gebraucht, *Acta RCRF* 27/29: 29-33.

MIKL CURK, I., 1992. La ceramica italica e i trasporti verso il nord-est, *Acta RCRF* 31-32: 465-474.

MIKL CURK, I., 1992a. Raw ceramics in Slovenia, a brief survey. In Santoro-Bianchi, S., Faila, A, Guermandi, M.P. (eds.), *Archeometria della ceramica*, Bologna: 99-108.

MIKL CURK, I., 1993. Petoviona v sožitju z bližnjimi in daljnimi kraji - Poetovio in coexistence at home and abroad (sum.), *Ptujski arheološki zbornik*, Ptuj: 205-218.

MIKL CURK, I., 1993a. Roman traces in landscape... In Pavuk, J. (ed), *Actes du XIIe congres international des sciences pre- et protohistoriques*, Bratislava 1991. IV: 419-423.

MIKL CURK, I., 1996. Arheološki vir k vlogi ženske v rimskem Ptuju - Zur Rolle der Frau (sum.), *Ptujski zbornik* VI, Ptuj: 158-187.

MIKL CURK, I., 1997. Territoriale Unterschiede der groben Hauskeramik in Slowenien - neue Beobachtungen und neue Fragen, Il contributo delle analisi archeometriche allo studio delle ceramiche grezze e comuni, Bologna - Studi e scavi 4: 110-116.

PETRU, S., 1972. Emonske nekropole, *Katalogi in monografije* 7.

PLESNIČAR GEC, L., 1972. Severno emonsko grobišče, *Katalogi in monografije* 8.

PLESNIČAR GEC, L., 1977. Keramika emonskih nekropol, *Diss. et monogr*. 20, Ljubljana - Beograd.

PLESNIČAR GEC, L., 1985. Emonske nekropole - Necropolises of Emona, ethnic and economic-social aspects, *Materijali* 20, Beograd: 151-168.

RODRIGUEZ, G., 1992. Bemerkungen zur relativchronologischer Gliederung der SO Alpinen spaetroemisch - spaetantiken Gebrauchskeramik, Il territorio tra tardoantico e altomedioevo, Firenze: 159-178.

RODRIGUEZ, G. 1997. Die Zeit vor und nach der Schlacht am Fluvius Frigidus (394) im Spegel der suedostalpinen Gebrauchskeramik, *Arheološki vestnik* 48: 153-177.

Image Quantification as Archaeological Description

Juan A. Barceló, Jordi Pijoan and Oriol Vicente

Universitat Autónoma de Barcelona, Quantitative Archaeology Lab, Dept. of Prehistory, Faculty of Letters

Building B, 08193 Bellaterra, Spain

e-mails: ilphd@blues.uab.es, jordi.pijoan@uab.es, oriolvicente@teleline.es

Abstract

This paper deals with the use of microscopic images as a way to define textures, and with the statistical analysis of quantitatively described textures. Digital image processing is a normal technique in archaeology. Archaeological images range from the microscopic to the macroscopic, and a diverse toolbox of computer techniques is available to process such archaeological data. However, the very nature of images as archaeological data has not been evaluated. Images are not primary data, but a transformation of empirical reality, translated into a language of luminance contrasts. Images are therefore the result of a goal-oriented modification. But how may this modification alter the reliability of the analysis? Very few studies have been published as regards this topic. Our goal in this paper is to integrate different archaeological applications of microscopy (use-wear in lithic tools, and pottery archaeometry) in order to define the observational category we are dealing with: texture. If the texture is the complex set of surface properties in an artefact, how can we describe it? What kind of archaeological, historical information do we obtain from the analysis of texture? A related problem is that of image sampling. Digital image techniques have been applied in disciplines where the assumption of surface homogeneity is valid. However, the modified surface of an archaeological artefact is always discontinuous. Different images can be obtained from the same artefact, and all of these images may be different. Statistical sampling is therefore a basic problem in archaeological image processing, and very few studies have been made. We explore the use of neural networks and related approaches to deal with this problem.

Key words: image processing, use-wear, archeometry, lithic analysis, pottery analysis, microscopy, neural networks

1. The concept of texture

Work modifies matter. As a result of human action, matter is exposed to changes and modifications, the result of which we call *artefacts*. David Clarke defined an *artefact* as anything modified as a result of human action: a tool is an artefact, in the same sense as a house, a pit, a burnt bone or a landscape (Clarke 1978). Social Sciences study how humans modify nature by creating artefacts, and these artefacts should be described in terms of human induced modifications on natural resources, and the sequence of changes across time.

Consider a lithic tool. It is made of stone, consequently, we should explain its *cause* as a human modification on a natural resource (flint), producing as an *effect* a *product*, an artefact whose properties are the result of the modification process or *production*. The same idea is valid for a vase. Here production can be described as: obtaining the resource (clay), obtaining other resources (temper, water, fuel), and processing them. In both cases, artefacts should be considered as nature modified by humans.

The goal of Archaeology (or one of its goals) should be the analysis of these processes, that is to say, the study of how humans modify natural resources in some specific historic circumstance. We study a *cause-effect* relationship, i.e. how social activity *causes* observable modifications in nature. Therefore, we should process a set of observable properties in order to be able to identify material effects of human work. Although the list can be very long, we consider that observable variability can be reduced to: shape, size, composition, texture and location.

Shape and size are the most commonly analysed properties of artefacts (see, for a theoretical introduction Small 1996), as it also holds true for location analysis. There is a lot of research on how to calculate the shape, size and location of lithic tools, pottery artefacts, metallic elements, etc. The shape of bones (human and animal), for instance, has been studied intensively in order to obtain taxonomic information. Composition is also a rather standard domain, especially in recent years: archaeologists are able to decompose any artefact into its compositional elements, both at a formal level (a house is composed of walls, a wall is composed of bricks), or on a physical-chemical level (archaeometry). But not many studies were performed on how human work modifies the surface properties of artefacts.

Artefacts have surface properties because of the way they have been made, or the way they have been used. In this paper, we analyse *archaeological textures*, that is surface properties, in pottery and lithic artefacts. In the first case, texture is a result of manufacture: different ways of producing and using a vase give a thin-section with a characteristic texture. In the second case, use changes the physical characteristics of the flint surface, producing a distinctive texture wear for each use. We explain how texture can be defined using different attributes, such as coarseness, contrast, directionality, line-likeness, regularity and roughness.

2. Observing textures: the creation of images

Observation is a process by which the human brain transforms light intensities into "visual" models. What we usually call *data* are not primary inputs, but a transformation of sensory information into an explanatory model of it. Observation is a 3 stage process: Perception, Recognition, Description (Bunge 1981, Hacking 1983), in which "perception" is only the first. Only once our brain recognises sensory information according to prior experience, we begin "seeing" reality around us. *Data* is the result of the final stage: description – that is, translation of recognised sensory inputs into a specific language.

Observation is a mode of knowledge acquisition, and it is used as a test mechanism for evaluating the reliability of already acquired knowledge. That is, observation is always an intentional act, guided by specific goals. As an intentional act, observation, even the so called "scientific" observation is always affected by available knowledge and can be direct or indirect, precise or wrong, misguided or even fraudulent. This fact leads some authors to reject the scientific method because without "objective" observation, there is no "objective" knowledge. However, given that observation is an "intentional" act, we can build observation in order to be objective. To do that we need is that observation results (data) be made public or collective, and not limited to a single observer.

"Objective" observation is produced by externalising perception, and by formalising recognition and description. We should impose control on:

- the object of observation,
- the observer and his/her/its perceptions,
- the circumstances of observation,
- the means of observation (senses, auxiliary instruments and procedures),
- a knowledge base relating to all above elements.

The means of observation, together with the related knowledge base are *the instruments of observation*. When we "externalise" observation in order to produce objective knowledge of the world, we mechanise the perceptual phase. That is we substitute human senses with a microscope, for instance. Nevertheless, this is not really a substitution. Instruments are necessary for perception, but not sufficient for observation, because nothing can be detected without an observer. The world is not data, but a set of perceptual information waiting for an observer to impose order by recognising an object and by describing it. What we are doing when we use the microscope is trying to avoid perceptual misinformation: a microscope allows two different observers to agree on the perceptual basis of information, but they can disagree on the recognised objects and how to describe the recognised world. Therefore, a fact can be observed if an agent *a* (the observer) is able to record some perceptual information *p* using an instrument *r*, under some circumstances *y*. The instrument is as important as the circumstances of observation, which includes the goals - the knowledge to be tested.

Images are not something to be captured, because they are not a part of reality. They are data, that is formal descriptions of something that exists. Light and colour are properties that really exist in the world, and they can be captured using special devices which transform light into electric or chemical signals, which should be manipulated in order to create a representation (an image).

Data, which is the result of the observational process, is only a model, a representation of some aspect of reality. Given that images are a kind of data, they are not a manipulation of reality, but a guided and intentional explanatory representation of some regularities existing in that real world. They are "real" only in the sense that they are true, that is, they coincide with the real world. Consequently, shape, size, texture, etc. are properties of a perceptual model of reality. Any "visual model" is only a spatial pattern of luminance contrasts that explains how the light is reflected, and it is composed of visual bindings which can be divided into sets of marks (points, lines, areas, volumes) that express position or shape, and retinal properties (colour, shadow, texture) that enhance the marks and may also carry additional information (Foley and Ribarsky 1994, Astheimer et al. 1994). Points and areas connected by the same plane or surface have not the same values. This variation is called *texture*, and it is used to understand those geometric properties that are based on local features. Each surface appearance should depend on the types of light sources illuminating it, its physical properties, and its position and orientation with respect to the light sources, viewer and other surfaces.

A microscope is not a device producing data, but it is used as a perceptual mechanism, whose output is the input of our model. A picture is not primary data, but a visual model of some real world properties, among them also *texture*. Thus (see Marr 1982, Watt 1988, Gershon 1994, Wadnell 1995, Barceló 2000):

- a pattern of changes in light wavelength and surface-reflectance, should be translated into a model of *colour,*
- a pattern of changes in edge orientation (Curvature), where an edge is an abrupt change on luminance values, should be translated into a model of *shape*,
- a pattern of changes in luminance variations in a scene with non uniform reflectance, should be translated into a model of *texture*,
- a pattern of discrimination between edges at different spatial positions, should be translated into a model of *topology,*
- a pattern of discrimination between edges at different spatial-temporal positions should be translated into a model of *motion.*

Although humans readily recognise a wide variety of textures, they often have difficulty describing the exact features that they use in the recognition and description processes. In this paper, our goal is to explain how to create a visual model of texture, using geometry as the formal language for recognition and description of microscopic visual inputs.

3. Describing texture: measuring images

An image is not a surrogate for reality, it is a directed and intentional transformation of reality in order to extract some relevant information. The microscope is not a device for observing some aspect of reality, but for capturing some initial input (luminance perception), which should be translated into observed data by a human agent using a visual model. What we are looking for in that image is the patterning of luminance values across all pixels. This is not the texture of the image, but we should *recognise* texture patterns in it, and build a geometric model of it. This can only be done with the help of prior knowledge as regards the concept to be modelled. The way of obtaining that knowledge is relatively simple: by comparing different images observed in experimental conditions.

Our main assumption is that different artefacts have different textures because they have been altered by different work activities. Consequently, the geometrical model of luminance patterning in each microscopic image should be different, if the activity performed by that artefact was different. We should create a *prototype* model of texture produced by a specific activity, quantifying also different sources of variability within that model, and maximising the variability between models for different activities.

The texture of different images should allow us to discriminate between image groups with some characteristic pattern of luminance variation. Nevertheless, the problem of luminance pattern variation is a complex one. When we see a picture, we *recognise* some differential features (striations, polished areas, scars, particles, undifferentiated background). These features are then a consequence of our prior knowledge, although in some way, they exist in the image. *Recognition* is a subjective procedure if we follow our individual criteria. However, this stage can be formalised, using an algorithmic approach: if we can reduce the amount of irrelevant variation in luminance patterns, the result is a formal representation of relevant features. Of course, what is relevant or irrelevant must be strictly defined. That is, we should distinguish two kinds of texture, one of them is inherent to the artefact surface, and the other one is the result of modifications on the surface generated by work activities. Furthermore, we should also distinguish luminance variations produced during the perceptual stage as a consequence of microscope functioning. Given that generated texture modifies inherent texture, a formal procedure of deleting random variation should allow the extraction of "dominant" or relevant features. We should not look for "meaningful" features, but we should describe formally (quantify) relevant variation measured in a experimentally controlled situation in order to define variation patterns regularly associated with each experiment.

Once relevant features have been extracted ("recognised"), the construction of a geometrical model of their relationships is a fairly straightforward task.

Consequently, analysing archaeological textures is not a single comparison of images, but a comparison of geometric models. Each model is a generalisation of surface properties "observed" through a microscope.

3.1. Quantifying texture

We should take into account that properties of any visual model are expressed as intensity values of colour variation, light and reflectance over surface (Sonka et al. 1994, Ebert et al. 1994). Therefore, a digital image of texture properties is a two dimensional mapping of points (p_i, q_i) with a specific luminance value (r_i). The resulting function is then $p_x q_x r$.

Texture is then described as the relationships of luminance values in one pixel with luminance values in neighbouring pixels. These values can be modelled as forming a set of regions, consisting of many small sub-regions, each with a rather uniform set of luminance values. In our case, these values are defined as grey levels. A group of related pixels can be considered as a texture minimal unit, sometimes called *texel* - texture element - (Sonka et al. 1994). Texture patterning in an image should be described as associations between *texels*.

A two-dimensional measure of texture is based on co-occurrence matrices, which show how often each grey level or luminance value occurs at a pixel located at a fixed geometric position relative to another pixel. For instance, an (3. 17) entry in a co-occurrence matrix means the frequency (or probability) of finding grey level 17 immediately to the right of a pixel with grey level 3. Each entry in a co-occurrence matrix could be used directly as a feature for classifying the texture of the region that produced it. Each different relative position between the two pixels to be compared creates a different co-occurrence matrix (Gose et al. 1996).

The first task in texture description is the *segmentation* of zones with the maximum contrast of luminance (texels). This task can be approached by calculating the *texture gradient* in the image - that is, the direction of maximum rate of change of the perceived size of the texture elements, and a *scalar measurement of this rate* (Sonka et al. 1994). This texture gradient describes the modification of the density and the size of texture elements and so regularity patterns in luminance variation can be determined. A *convolution filter* can be designed so that each pixel in the original image is transformed according to the following function:

$$g(x,y) = \mathrm{G}\left[f(x,y)\right] = \left|\frac{\delta f / \delta x}{\delta f / \delta y}\right|$$

that is to say, each pixel (with *x, y* co-ordinates) is transformed according to the median of the derivative of its pixel neighbours. This is called a *gradient operator.* Its magnitude is defined by the following expression:

$$mag\left[\mathrm{G}\left[f(x,y)\right]\right] = \left[(\delta f / \delta x)^2 + (\delta f / \delta y)^2\right]$$

This operator increases luminance values in areas with sharp luminance and brightness contrasts, and decreases the values in areas with soft luminance and brightness contrasts. As a result, isolated areas are segmented whose shape, size, texture, composition and position may be measured (Pijoan et al. 1999).

3.2. Real image, segmented image and "texel map"

Once texels have been extracted, we should calculate their formal and relational properties, using their variables of shape, size, composition, texture and position. Among others we should measure:

- Area measurements (number of pixels *within* a texel).
- Perimeter measurements (number of pixels around the edge of a texel).
- Perimeter shape. Measured as a pattern of changes in edge orientation.
- Convex Hull: the smallest region which contains the texel, such that any two points of the region can be connected by a straight line.
- Euler-Poincaré characteristic: difference between the number of regions (texels) and the number of holes within them.
- The Frequency and Entropy of Brightness within a texel (histogram of grey levels).
- The Frequency and Entropy of Contrast: local change in brightness (ratio between average brightness within the texel and the background brightness – neighbouring texels).
- Topology of Texture. A pattern of discrimination between the edges at different spatial positions, distance and adjacency relationships between different texels. Among them:

    Degree of Coarseness: edge density is a measure of coarseness. The finer the texture, the higher the number of edges are present in the image,

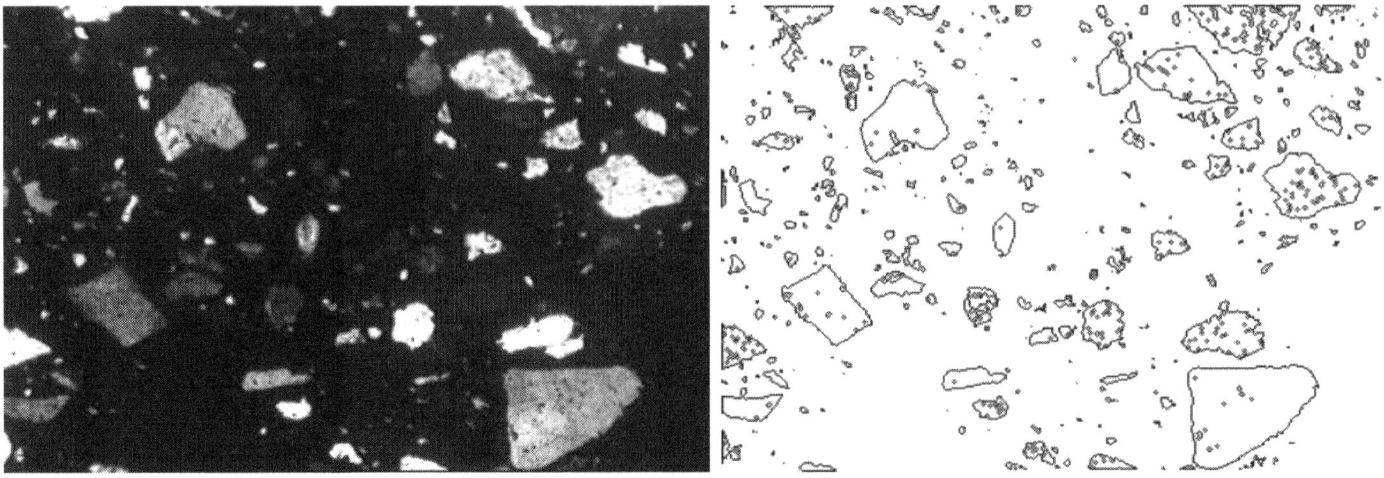

Figure 1: Describing mineral particles in a pottery thin-section by means of texel extraction.

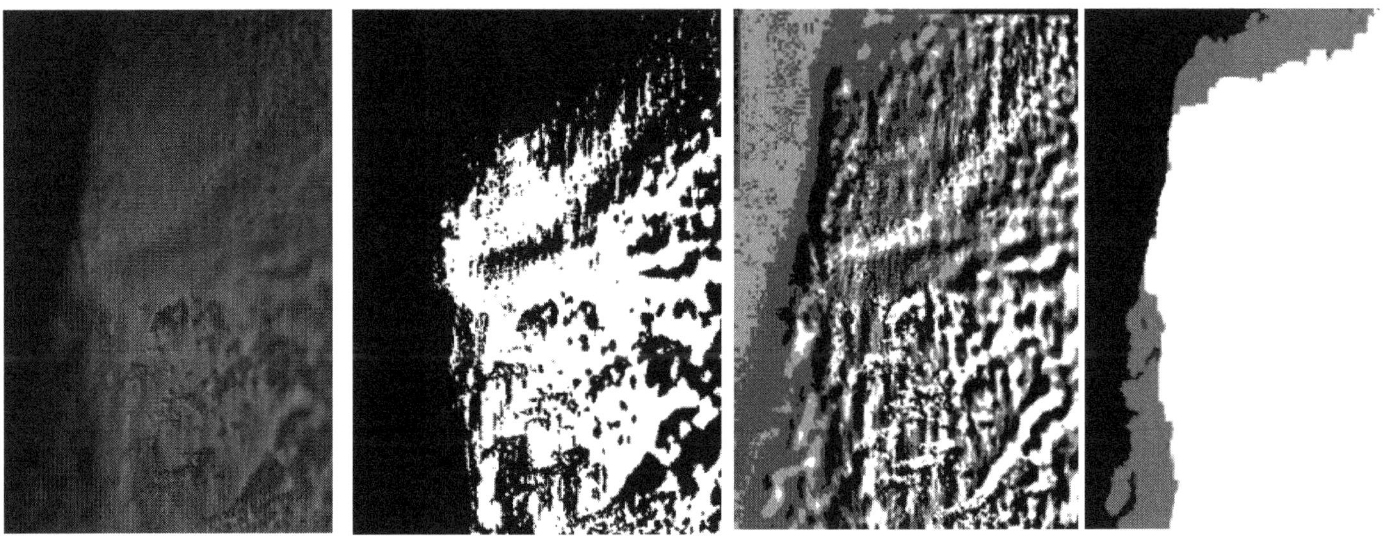

Figure 2: Extracting use-wear areas by means of texel extraction.

Contrast: high contrast textures are characterised by large edge magnitudes,

Randomness: may be measured as entropy of the edge magnitude histogram,

Directivity: entropy of the edge-direction histogram. Directional textures have an even number of significant peaks, direction-less textures have a uniform edge-direction histogram,

Linearity: is indicated by the co-occurrence of edge pairs with the same edge direction at constant distances,

Periodicity: texture periodicity can be measured by co-occurrences of edge pairs of the same direction at constant distances in directions perpendicular to the edge direction,

Size: texture size measures may be based on the co-occurrence of edge pairs with opposite edge-directions at a constant distance in a direction perpendicular to the edge directions.

In the long term, this approach should be directed to the generation or *synthesis* of texture from a program or model, rather than just a digitised or painted image (Musgrave 1994). We are looking for a "procedural" approach where the analysis of properties of observed textures is expressed as a statistical model which should be able to reproduce the textures from statistical data (Ebert et al. 1994).

3.3. Two case studies

The thin-section samples of pottery are utilised in petrography for the description of some petrographic attributes of the vessel fabric mineralogical composition. They are the results of a specific work process, where the productive agents modify the clay status of fabrics through a thermal alteration. These different petrographic attributes are the result of the natural formation of clays and the deliberated human alteration of them. We try to study thin-section samples in the most objective way. In this way we try to assign a series of numeric values to a thin-section microscopic image in order to quantitatively describe the sample. The purpose of this study is the description of mineral particles that compose the fabric of a vase. We represent each particle as texels defined against a general background (clay). In this way, mineral particles can be measured according to luminance intensity, shape, size, etc. The goal is to distinguish different vases (fabrics) from the different characteristics of particles contained in the fabric. These differences could be explained in terms of the manufacturing processes.

In lithic use-wear, we compare different images created as visual models of specific experiments. Each image is not a photograph of an artefact, but a model of the generated texture on a flint surface when an agent makes a longitudinal movement with that tool on a fresh wood material. Our goal is to distinguish between the visual model of that texture and the visual model of the generated texture on a flint surface when an agent makes a transversal movement with the same tool on a leather material. The extracted texels should be recognised as micropolish, microscarring and linear features. Each one should be considered a different kind of texel. We should distinguish these features produced by the movement of a lithic tool done on an specific matter, from the macro and microscopic traces characteristic of the lithic surface alone.

We have reduced the original complexity of microscopic images into grey scale pictures. In this way, it is easier to recognise texels. Once recognised, texels must be described. We use geometry to describe the shape, size, texture, composition and location of micropolished areas, scars and linear features detected on the lithic tool surface, or to describe the particles detected in the microscopic picture of a pottery vessel thin section. Shape is a property that can be used for the differentiation of texels, whereas size can also be used to differentiate between texels generated in different experimental conditions. Density measures give information about the texture (homogeneity within a particular texel) and the position. The following variables are those that the use-wear and thin section analysis techniques have used to describe the differences between cases, and to discriminate those between different groups:

Area measurements: The total number of pixels with the same luminance or range of luminance. The edge is defined by the proximity of a grey level. Normally a simple threshold operation is enough to define the area or areas of a discrete texel. In use-wear analysis we utilise the area measurements to extract the extension of micropolish, the size of microscars_and the striations length. Pottery thin-section analysis is used for measuring the size of each mineral particle in the fabric.

Texels perimeter: We took the information as regards the size of a mineral particle or the length of the striation. This variable is used for calculating different ratios of the variables related with the perimeter shape.

The *Euler-Poincaré* characteristic is used for measuring the ratio in the microtopography and the micropolish spread. This variable is not necessary in the thin-section analysis.

The *frequency* and *entropy of brightness* within a texel is calculated using the histogram of grey levels.

The *frequency and entropy of contrast*: local change in brightness (ratio between average brightness within the texel and the neighbouring texels) is used as an intermediate calculus to describe coarseness.

Perimeter shape and *orientation*: To introduce the category of shape we use the natural geometric shapes as indicators, in order to define the pattern of the geometric model of the sample.

Circularity: the degree of circularity of a texel. I.e. how similar is this texel to a circle. Where 1 is a perfect circle and 0.492 is an isosceles triangle. This shape is expressed by:

$$\frac{4\pi s}{p^2}$$

s: texel area
p: texel perimeter

Quadrature: the degree of quadrature of a texel, where 1 is a square and 0.800 an isosceles triangle. This shape is expressed by:

$$\frac{p}{4\sqrt{s}}$$

Irregularity: measurement of the irregularity of a texel, calculated as the relationship between its perimeter and the perimeter of the surrounding circle. The minimum irregularity is a circle, corresponding to the value 1. A square is the maximum irregularity with a value of 1.402. This shape is expressed by:

$$\frac{p_c}{p}$$

Elongation: the degree of ellipticity of a texel. A circle and a square are the less elliptic shapes. This shape is expressed by:

$$\frac{D}{d}$$

D: maximum diameter within a texel
d: minimum diameter perpendicular at D

All shape measurements are used in use-wear and thin-section for the study of tendencies in the geometric pattern, both for describing the orientation and shapes of the micropolish and the striations in the use-wear analysis, and mineral particles in thin-section pottery analysis.

Orientation: the orientation given by the angle of the detected linear features with the tool's edge is used in use-wear analysis to define the direction of the movement *made* with the tool.

Topology of texture: these measures are measured from relationships and associations between texels (and not at each texel).

Randomness: entropy of the number of texels within a modified surface. It can be used in use-wear for distinguishing the area of the micropolish from the background.

Linearity: linear features can be represented using linear equations: $y = a + bx$, where y and x are co-ordinates, and a and b linear coefficients. We use both coefficients as quantitative variables in our study. We can also include some other numerical attributes such as the quantity of lines, and their longitude. The width of linear features can be measured on a three-dimensional representation, and included in the image quantification.

Directivity: entropy of the edge-direction histogram. Directional textures have an even number of significant peaks, direction-less textures have a uniform edge-direction histogram. This can be used in the description of striation orientation.

Size: number of pixels corresponding to each contour in the image. It allows the study of micropolish topography.

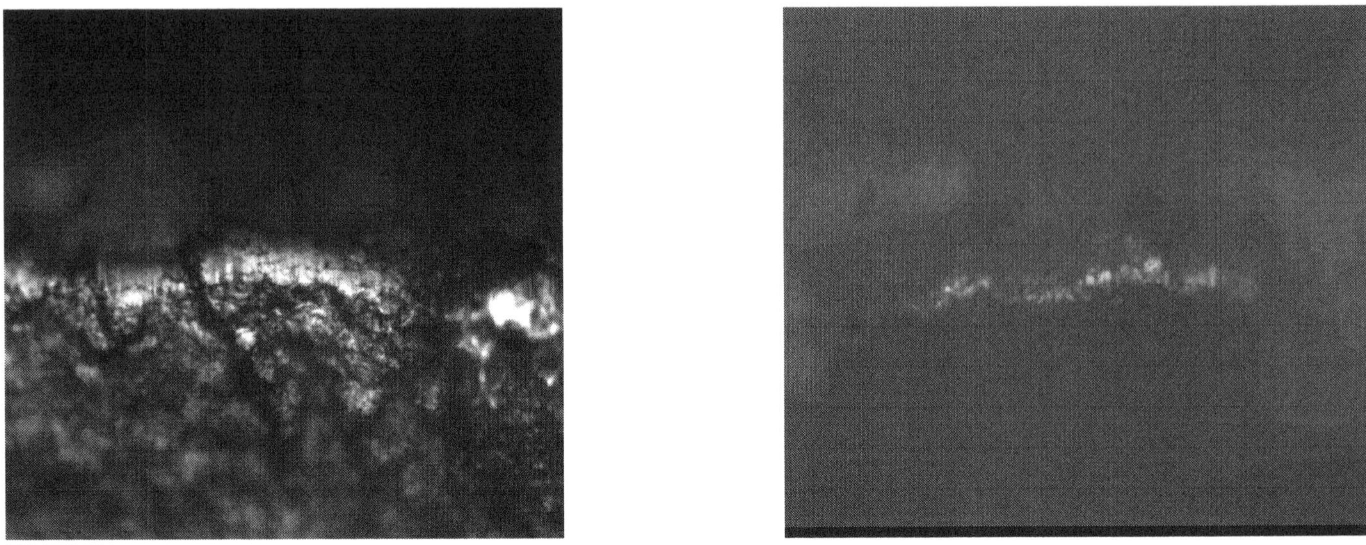

Figure 3: Variations in texture due to focus adjustment. Only the edge of the tool is visible.

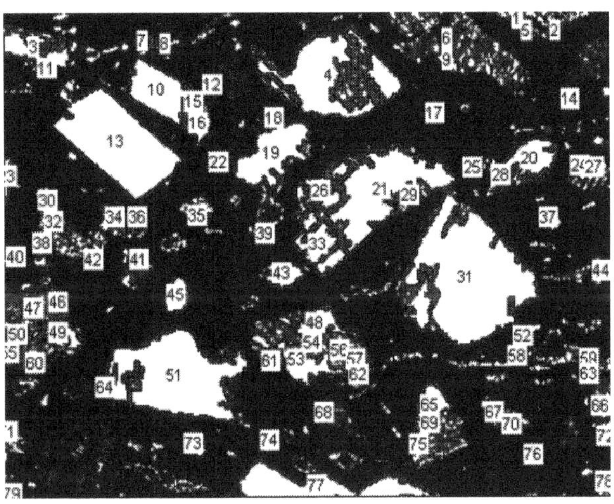

Nº	Area	Circularitat	Quadratura	Elong.	Irregula
1	13	1,05	0,87	1,09	1,07
2	25	0,38	1,44	2,87	1,04
3	36	0,76	1,02	1,82	1,26
4	77	0,28	1,68	2,61	0,85
5	28	0,77	1,01	2,13	1,28
6	13	0,96	0,91	1,79	1,31
7	19	0,75	1,03	1,90	1,19
8	11	1,02	0,88	1,37	1,18
9	46	0,73	1,04	2,17	1,26

Figure 4: Measuring texels.

4. The variability of textures: statistical sampling

A single microscopic image is not a good prototype to determine the artefact's texture properties. This assumption would be correct if all the artefact surfaces were modified in exactly the same way. In the case of use-wear in lithic tools this is clearly not the case, because modified texture related to the work activity appears only in some areas of the surface. Consequently, a geometric model of textures cannot be built as a generalisation of a single image. Sources of variability are too important, and should be taken into account.

Measurement error is the most obvious source of variability. Features we "observe" in a microscope image come from a 3D reality, but the picture is a 2D model. As a result, any image should be considered as a modification of perceptual reality, because it cannot maintain the same focus for the entire field of vision. It is impossible to give the same sharpness to the complete observed surface, because it is not on the same level. As a result, microscopic images are characterised by narrow observation plans, that can be wrongly considered as discrete texels. This is a case of measurement error, and the only way to deal with it is by not using primary images, but modified visual data. That is, texture data should be composite images made of microscope pictures taken with different focus levels. By merging all levels into one, and by posterising them, we obtain a visual model of a sharp-equalised image.

Colour and shadow are also some sources of measurement error. They are the consequence of light reflection across the artefact's surface, and, in a reflected light microscope – used in material sciences – this reflection depends on the angle between the light beam and the observed layer. In these circumstances, we can get a paradoxical situation where a polished texel (more light reflection) seems darker than an unpolished texel, which is less reflective, but light reflects at a perpendicular angle in respect to the source of light. Observing coarse areas (texels with a large number of minor variations of luminance contrasts) is then a matter of light orientation, and not only of surface parameters. To appreciate this, it is only necessary to consider a regularly patterned object viewed in 3D - two effects would be apparent; the angle at which the surface is seen would cause a perspective distortion of the texture element, and the relative size of those elements would vary according to the distance from the observer. The best way of dealing with this source of measurement error is by controlling all observation parameters, and maintaining all of them fixed during the entire procedure. Among these parameters, we can find the following: the distance from the observer slant, the angle at which the surface is sloping away from the viewer (the angle between the surface and the line of sight), and tilt, the direction in which the slant takes place (Sonka et al. 1994). The control of observation parameters is not an easy task, because there is not a single perceptual plan that is useful for all kind of observations. Some features are best seen with perpendicular beams of light, while other can be discovered only using fast horizontal beams.

These are some of the sources of error measurement, and there is a long tradition of dealing with them and reducing their effects. Less known are the sources of variation that prevent the simple generalisation of perceptual images. In our case, the main problem is that the microscope field of vision is too limited for our purposes (from a 4x4 cm. field in the easiest case, to 0.001x0.001 mm or less, if we use electronic microscopes). Without further investigation we cannot accept the assumption, that a reduced frame contains all the elements that characterise the complete surface. We need more than one single image to correctly represent all texture variation present in the surfaces of the artefact.

Consequently, sampling questions are of great importance. In this research we have used a series of images to investigate the variability of texture within an artefact, before using the resulting geometric model to explain differences between artefacts. The problem is to merge different files containing shape, size, composition, texture and position of individual texels identified in all images of different artefacts.

We have used the following approach:

4.1. *Within-artefact* description

We have considered the processing of all observed texels in different images of the same artefact. The number of images depends on the complexity of texture and the position of modified surface patterns. In our research we have selected three or four images for each artefact, in order to look for differences among all texels produced in the same experimental conditions in the same artefact. These texels are described using the variables defined above, and within-artefact variation is then analysed, using standard statistics.

The purpose is not only to describe variation, but also to define prototype values for relevant features. For instance, we have used mean and standard deviation of area and perimeter measures, as well as skewness and kurtosis measures. It is not the absolute value of these prototypes that interests us, but the range of variation each texel may adopt within an artefact.

4.2. *Between*-artefact description

Of course, central-tendency measures are relevant only if within-artefact variation is approximately normal, and this assumption should be tested in each case. However, even when within-artefact variability is not normal, measures of dispersion can be used to compare textures produced by different activities. In some cases, there is not any identifiable texture pattern associated with some specific activity, but a greater or lesser dispersion of values than others. For instance, micropolish in use-wear analysis should be understood not as a discrete texel, but as an area with a low degree of texture variation due to friction, and given as a result a specific luminance value due to light reflection on that homogeneous surface.

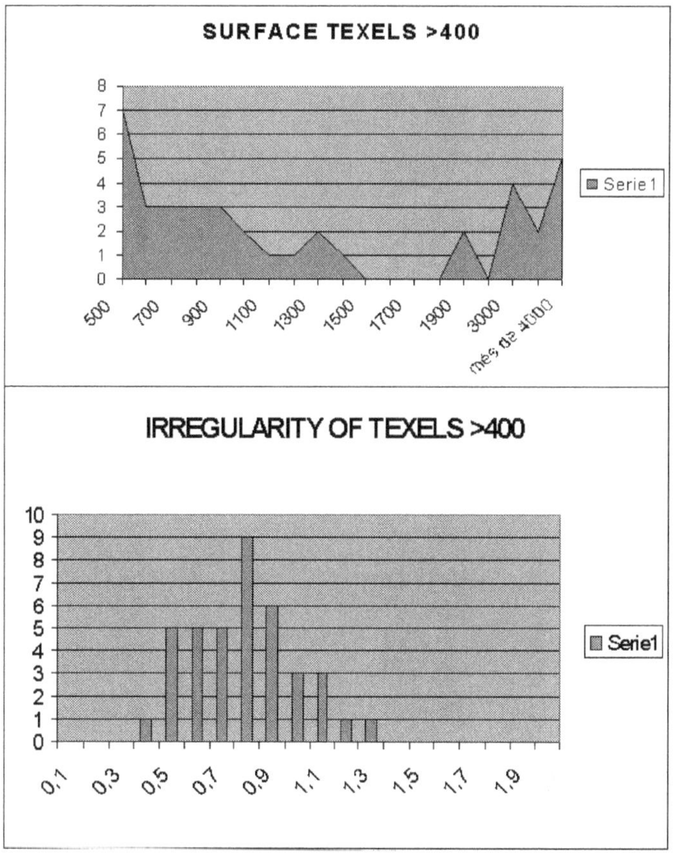

Figure 5: Statistical analysis of some properties (Surface, Irregularity) of the biggest particles (more than 400 pixels surface area) from all four microscopic pictures of the same object.

In such circumstances, specifically when normality is not assumed, between-artefact variability is very difficult to discriminate. In order to perform this task we have used a neural network approach (Barceló 1996, Barceló et al. 2000).

The system we want to build is a diagnosis machine that predicts the probability of any artefact (a lithic tool, a pottery vase) to be it used or produced in any way, given a set of inputs (a quantitative description of macro- and microscopic texels extracted from a number of different images of the same artefact). This prediction does not follow a rigid algorithm in producing an answer based on given inputs, but it is actually learned through training examples.

The network consists of many simple, but individual processing elements ("nodes") arranged in one or more layers and a system of connections. These connections transmit the signals, which the nodes manipulate. A transfer function contained in each node governs this manipulation. The nodes add weight adjusted inputs, and a bias value, and finally they pass the result through an activation function (also called a transfer or squashing function) to be used by other neurones or offered as an output. A learning process is usually performed in the network of connections. Although a network's transfer functions usually do not change, the connection strengths change during the learning process. These changes result from the network making predictions on training examples, which contain known outputs based on real inputs. In our case, training examples are pairs of archaeological experimentation results, that is, the descriptive features observed in those lithic tools that were used for some specific activity in the laboratory.

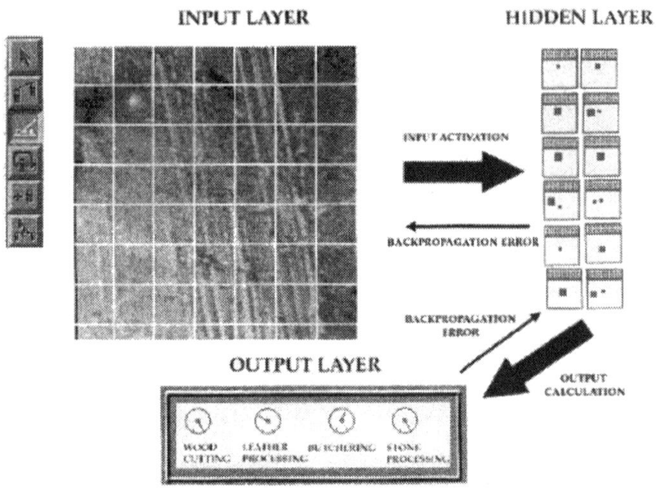

Figure 6: A Neural Network scheme for the analysis of use-wear microscopic pictures.

Observable information (image data) is the input to the first layer which then propagates through the structure of connections and nodes. When the input values finally reaches the output nodes in the final layer, these units produce an answer (a number reflecting the intensity of the function in each unit), which is the network's prediction of the output based on the given data input. The predicted output at every node in the final layer is then compared to the correct (known) output at every node. Errors are generated as the difference between the correct output and the network's predicted output. These errors propagate backwards through the network, modifying the connections' weights based on a mathematical equation that defines what is described as the learning rule. This process continues until the user is satisfied with the accuracy of the network's predictions.

Once the network is satisfactorily "trained", it is put into actual use. The network is fed only input data, preferably data it has never seen before. Feeding the network the exact same data that was used during training only tests the network's ability to "memorise" data. A useful network can accurately predict output to data it has never seen before.

The real challenge in developing a useful neural network system is the training process. We are comparing different supervised learning algorithms where the "training" of the network is an iterative process based on large numbers of data samples representing the traffic flow within a certain region. Using standard connection weights, the network computes a set of outputs, and compares this set of outputs with the input values by calculating a root mean square difference (or global error) and modifies the connection weights to displace the outputs toward the expected values. If the training is successful, the global error is reduced. In over-simplified terms, gradient descent works to optimise a system by minimising a given function. In the case of backpropagation, network error is minimised by optimising the weights values of the connections among nodes. The total network error is minimised by following the gradient (actually followed down towards a *minimum*, hence descent).

Since many indicators appear to be relevant at first glance, we should perform sensitivity analysis with respect to the different inputs. This involves noting the percent change in the output caused by a specific percent change in one of the inputs, keeping all the other inputs the same. But we have also included the possibility of non-linear interactions, that is, changes to two or more inputs in tandem can have a different effect from that of changes to one input alone. Redundancy has not been deleted, because it was one of the goals in our analysis, that is, to evaluate if classificatory results are affected by redundancy. We have carried out only a preliminary sensitivity analysis, in order to drop features that do not produce enough information.

5. Conclusions

The way neural networks process redundancy and irrelevant variation is the reason we have selected this approach. It is important to realise, however, that an erroneous understanding of image processing has confused the fuzzy nature of image descriptions, even at a quantitative level. We think that redundancy, error measurement and within-artefact variability exists at the level of perceptual input, that is, they are inside the images we want to compare. Any experimental approach is nothing more than a "supervised-learning" framework, where it is assumed that between-artefact variation is greater than within-artefact variation, and its patterning can be distinguished. Most image analysis in archaeology and other disciplines neglect the sources of within-artefact variability and error measurement. In this paper we have proposed an approach to deal with this problem.

Acknowledgements

The authors would like to thank Assumpció Vila, Xavier Clop and Ignacio Clemente, for initiating the project and for their continued support during its development. We also express our gratitude to the Center for Computer Vision at the Universitat Autònoma de Barcelona for all analysis carried out using their facilities. This research has been made possible thanks to Project DGICYT PB97-1161 from the Spanish Ministry of Education and Culture. J. Pijoan is a bursary of The Universitat Autònoma de Barcelona, and he thanks this institution for his Grant. J. Pijoan and O.Vicente also received a special bursary from CAA2000. We would also like to thank the Ljubljana organisers for attending the conference. We acknowledge all the aforementioned financial supports.

References

ASTHEIMER, P., DAI, F., GÖBEL, M., KRUSE, R., MÜLLER, S., ZACHMANN, G., 1994. Realism in Virtual Reality. In Thalmann, N.M. and Thalmann, D. (eds.), *Artificial Life and Virtual Reality.* New York: John Wiley Publ.

BARCELÓ, J.A., 1996a. Arqueologia Automatica. Inteligencia Artificial en Arqueologia. Sabadell: Ed. Ausa, *Cuadernos de Arqueologia Mediterranea*, No. 2.

BARCELÓ, J.A., 2000. Visualizing what might be. An Introduction to Virtual Reality in Archaeology. In Barceló, J.A., Forte, M. and Sanders, D. (eds.), *Virtual Reality in Archaeology*. Oxford, ArcheoPress (BAR International Series).

BARCELÓ, J.A., VILA, A., GIBAJA, J., 2000. An application of Neural Networks to Use-Wear Analysis. Some preliminary Results. In Lockyear, K., Sly, T.J.T., Mihailescu-Birliba, V. (eds.), *CAA96. Computer Applications and Quantita-*

tive methods in Archaeology. Oxford, ArcheoPress (BAR International Series).

BUNGE, M., 1981. *La investigación Científica.* Barcelona, Ed. Ariel.

CLARKE, D.L., 1978. *Analytical Archaeology* (2. edition) London, Methuen.

EBERT, D.S., MUSGRAVE, F.K., PEACHEY, D., PERLIN, K. and WORLEY, S., 1994. *Texturing and Modelling. A Procedural approach.* Boston: Academic Press Professional.

FOLEY, J., RIBARSKY, B., 1994. Next-generation data visualisation tools. In Rosenblum, L. et al. (eds.), *Scientific Visualisation. Advances and Challenges.* New York, Academic Press: 103-127.

GERSHON, N., 1994. From perception to visualisation. In Rosenblum, L. et al. (eds.), *Scientific Visualisation. Advances and Challenges.* New York, Academic Press, pp. 129-139.

GOSE, E., JOHNSONBAUGH, R., JOST, S., 1996. *Pattern Recognition and Image Analysis.* Prentice Hall. Upper Sadler River, NJ.

HACKING, I., 1983. *Representing and Intervening.* Cambridge University Press.

MARR, D., 1982. *Vision*, San Francisco: W.H. Freeman and Co.

MUSGRAVE, F.K., 1994b. Procedural Fractal Terrains. In Ebert, D.S., Musgrave, F.K., Peachey, D., Perlin, K. and Worley, S. (eds.), *Texturing and Modelling. A Procedural approach.* Boston: Academic Press Professional.

PIJOAN, J., BARCELÓ, J.A., BRIZ, I. and VILA, A., 1999. Image Quantification in use-wear analysis. *Computer Applications in Archaeology '99.* Dublin (in press).

SMALL, C.G., 1996. *The Statistical Theory of Shape.* New York-Berlin, Springer-Verlag.

SONKA, M., HLAVAC, V., BOYLE, R., 1994. *Image processing, Analysis and Machine Vision.* London: Chapman and Hall.

WANDELL, B.A., 1995. *Foundations of Vision.* Sunderland (MA): Sinauer Ass., Inc.

WATT, R., 1988. *Visual processing: Computational, Psychophysical and Cognitive Research.* Hillsdale (NJ): Lawrence Erlbaum Ass.

The SHAPE Lab: New Technology and Software for Archaeologists

Frederic F. Leymarie
Division of Engineering, Brown University, The SHAPE Lab
Box D, 182-4 Hope St., Providence, RI, 02912 USA
e-mail: leymarie@lems.brown.edu

David B. Cooper
Division of Engineering, Brown University, Box D
e-mail: cooper@lems.brown.edu

Martha Sharp Joukowsky
Center for Old World Archaeology and Art, Dept. of Anthropology, Brown University, Box 1921
e-mail: martha_joukowsky@brown.edu

Benjamin B. Kimia
Division of Engineering, Brown University, Box D
e-mail: kimia@lems.brown.edu

David H. Laidlaw
Dept. of Computer Science, Brown University, Box 1910
e-mail: dhl@cs.brown.edu

David Mumford
Division of Applied Mathematics, Brown University, Box F
e-mail: mumford@dam.brown.edu

Eileen L. Vote
Dept. of the History of Art and Architecture, Brown University
e-mail: vote@lems.brown.edu

Abstract

The SHAPE Lab was recently established (1999), with a grant from the United States National Science Foundation, by Brown University Departments of Engineering, Applied Mathematics, Computer Science and The Centre for Old World Archaeology and Art and Department of Anthropology. It is a significant interdisciplinary effort for scientific research with a direct application to important problems in the analysis of archaeological finds and artefacts. We present the concepts that will underlie a 3D shape language, and an interactive, mixed-initiative system, for the recovery of 3D free-form object and selected scene structure from one or more images and video. This work has impact by providing new practical tools. It also provides an effective testbed for 3D shape reconstruction and recognition, more descriptive local and global models for working with 3D shapes and performing free-form geometric modelling, and for extracting 3D geometry from one or more images and video, as well as associated computational complexity issues. As applied to the field of archaeology, this technology provides, specifically, new ways to analyse and reconstruct pottery, compare objects from different sites and reconstruct sculpture and architecture.

Key words: *Shape language, 3D free-form modelling, ridges and valleys for shape, implicit polynomial models, medial axes, skeletal graphs, 3D object reconstruction and analysis, Great Temple of Petra, archaeological recordings*

1. Introduction

The SHAPE[1] Lab was recently established for the synergistic study of three dimensional (3D) free-form in the disciplines of mathematics, computer graphics, computer vision and archaeology. It answers the needs that arise in the domain of archaeology, but which are also generic across a range of applications. Specifically, we are investigating the following:

1. 3D free-form modelling for surface and volume representation, via the design of a shape language.
2. Geometric information extraction from either: (i) passive optical systems, such as obtained via a single image, many images, or a video stream, (ii) active data-capture systems, such as laser camera systems or structured light systems, and (iii) a combination of data obtained via (i) and (ii) together with auxiliary data when available (e.g. floorplans, survey data).
3. Human/computer interaction (HCI) for facilitating the model building and geometric information extraction, as well as to provide an interactive virtual system for archaeo-

Figure 1: Typical fragments of head sculptures from Petra, to be analysed. These fragments have considerable detail although some parts are missing and the surface is badly eroded.

logical analysis with site features, topography, architecture artefacts and special finds.

4. Decision-directed machine estimation for automatic model choice and geometric information extraction, with an emphasis on the important archaeological problem of stitching together fragments constitutive of an original object, such as obtained from pottery sherds.

These research topics are to be explored through an integrated effort because, on the one hand, the archaeology applications drive the 3D modelling and 3D from images research, and, on the other hand, the research on 3D free-form provides archaeologists with tools to be able to conduct research hitherto impossible or impractical. In addition, in light of the fact that the three disciplines of mathematics, computer graphics and computer vision view 3D free-form in various perspectives, ranging from theoretical to practical, this project provides a unique opportunity to conduct the study in a comprehensive way. Finally, we benefit from the direct involvement of our team of archaeologists at the site of the Great Temple in Petra, Jordan (see an aerial view of the Great Temple in Vote et al. this volume, figure 1) where on-going excavations have been conducted for many years, and from which a large database of artefacts is currently being built (Joukowsky 1998).

1.1. Archaeological research problems

We focus here on the development of a generic technology for the recovery of 3D models which is investigated in the context of archaeology. Archaeologists are typically faced with a series of bottlenecks, including the following ones, which this research aims to alleviate.

1. Excavators want to be able to register the location of artefacts in situ in order to maintain an accurate archaeological excavation record (see section 2). Our proposed technology will allow archaeologists to use relatively inexpensive equipment to expedite excavations and maintain more comprehensive, accurate and accessible records of artefact *geometry* and find locations.
2. Currently artists assist in work on site by documenting the artefacts found and positing reconstructions of broken artefacts, thus leaving archaeologists out of the process with additional delays and much added cost. Our proposed technology allows the archaeologist to use shape models and computer graphics to document and interactively reconstruct artefacts.
3. A significant problem in archaeology is the inability to compare many artefacts stylistically, which requires substantial physical information (e.g., see figure 1). Relating one artefact to another, perhaps found in another site, is an integral part of discovering its role, age, responsible artisan or community, etc. The expression of artefacts in a *shape language* will advance possibilities for interactive or automatic quantitative and qualitative comparison.

We rely mainly on image analysis and photogrammetric methods in order to reconstruct and measure the 3D structure of objects. The use of a passive optical data acquisition technology, in contrast to active scanners, is of interest in order to:

- acquire data on-site at low cost, without imposing hard constraints on the size of objects or the ambient lighting, and without slowing down the excavation campaign;
- use existing image databases from previous excavation campaigns and from other sites.

However, we also make use of active data acquisition techniques, such as laser scanning, structured light and computerized tomography (CT, see figure 2) systems, in order to:

- provide "ground truth" measurements upon which we can gauge passive reconstruction techniques;
- rapidly acquire 3D data in order to conduct our other research objectives;
- maintain an expertise in using both types of systems, and keep track of their evolving differences.

The last point is emphasized by the fact that digital photogrammetry has yet to become automated, while laser camera remain relatively expensive, structured light systems have limited applications in the field (i.e., constrained lighting conditions and limited field of view), and tomography is not a portable technique. See also the work of Pollefeys et al. (2000) who advocate the use of both passive and active systems for different purposes in documenting archaeological sites.

1.2. Shape modelling research problems

Our premise is that the use of more powerful 3D shape representations than the classical points, straight lines, planes, triangles or splines, can lead to practical solutions for the preceding problems, and markedly improve speed, accuracy, and user convenience over most of what is presently possible. The 3D representations we propose to study are hybrid constructions made of *ridges*, *implicit polynomial surfaces* (IPS, i.e., algebraic surfaces) and *skeletal graphs*. They are studied for use individually and in concert, in order to understand their most effective synergy as new hybrid models and algorithms are developed.

Of course, the guiding principle in this work is discovering and understanding the fundamental issues in solving these complex problems in computationally fast, yet user-friendly ways. We seek to identify the most effective ways of handling and processing the huge amounts of data available in the acquired images and video streams, in particular, by identifying tradeoffs between accuracy and complexity. We emphasize that our program for the study of 3D free-form representations for shape:

1. Solves heretofore unsolved problems.
2. Improves speed and user convenience in handling complex problems.
3. Handles huge amounts of data in new and faster ways.

In the remaining of this paper we first describe, in section 2, the present day situation at our main site of excavation, at Petra. Then, in section 3, we give some early results in tackling the previously introduced research objectives. Finally, in section 4, we describe in some detail the basis for our shape language developed to tackle complex archaeological problems.

2. Archaeology at Petra

The Great Temple of Petra, Jordan, is a monolithic structure at the top of a three-levelled precinct measuring 35 meters east-west, and 42.5 meters in length. In unearthing a site such as this, archaeologists want to use the most exact technology to register objects they excavate, and reconstructive technology to help them envision what the building and the objects within looked like (Joukowsky 1980). Our proposed technology aims to help them do both.

The latest archaeological standard for gathering data about finds is to register each object as it is excavated with a costly laser transit station. This requires three people to digitally register the object with the survey equipment; one to shoot the point, another to hold the prism in order to register the location, and a third to label and bag the artefact. Even with this method there is no easy way to correlate the object with the survey. Some archaeologists register an object with one point, indicating the approximate centroid of an object; others take four or five points (or more) per object in an attempt to give additional information about the object's shape and orientation.

In a typical excavation, relevant finds need to be registered with the survey station in different locations at once and excavators wait for the surveyors to register their objects before they can proceed with digging. Furthermore, all of the information regarding the find spot must be retrieved in the field. After the survey station has registered the object, it must go through other phases

Figure 2: Figure made from density measurements of a box of sherds. Densities within the box were imaged volumetrically using a Rhode Island Hospital CT scanner. Iso-density surfaces were then created and rendered using marching cubes (Mortensen and Barrett 1998), a computer graphics technique.

of registration. All artefacts are hand measured, drawn by a site artist, photographed and then put into a database of objects. All these steps must be completed at or near the site because artefacts cannot be taken home. Archaeologists require the ability to digitally register an object's orientation, detailed shape and other physical characteristics quickly and either on site, or a posteriori, when the data acquisition method allows it.

The database for the Great Temple excavation contains already more than 115 000 artefacts, recorded since 1993 (Joukowsky 1999). Unfortunately, the full potential of archaeological databases is rarely realized. Most archaeologists are not able to analyse the geometric characteristics of artefacts and their spatial relationships with other elements of the site (Crescioli and Niccolucci 1998).

Our methodology encapsulates all of the above recording steps in one process. For example, 3D objects can be registered in the field via photogrammetric means (Leymarie et al. 1996). Our proposed technology will also permit archaeologists to reconstruct of broken or eroded fragments. Once 3D information is gained about artefacts and architectural fragments while objects are being initially registered, it will be possible to better exploit reconstruction possibilities. A series of pot fragments (figure 2 and 6) can be interactively, and eventually automatically, reconstructed, eroded sculpture reconditioned to understand the original features and surface, a wall rebuilt without having to lift heavy fragments, and many elephant-head column capital trunks reconsolidated. In many cases, archaeological artefacts go uncited as historically significant because they cannot be interpreted and referenced with other like examples. Our proposed technology allows archaeologists to understand and reference objects within a historic framework and also permits visualization that has, in the past, been unavailable or too costly.[2]

Figure 3: Examples of our perspective-based reconstruction using simple geometric primitives (images from Petra). On the left are shown single snapshots taken with an ordinary camera with the addition of fiducials. On the right are shown the results of drawing CAD-like primitives as overlays on the basis of perspective cues.

3. Early results

3.1. Gestural and verbal user interfaces

User interfaces developed for production environments have not evolved significantly since the introduction of the windows, icons, menus, and point-and-click (WIMP) interface metaphor over two decades ago. Despite the advantages of WIMP interfaces (e.g., ease of use, short learning curve, and wide applicability), they greatly under-utilize the real-world capabilities and skills of users by limiting input and output to a keyboard, mouse, and monoscopic display. While effective for 2D desktop productivity applications, WIMP interfaces are not the ideal solution for intrinsically 3D applications. In this context, we have undertaken the study and evaluation of the next generation of post-WIMP interfaces that leverage application-specific knowledge and human skills to realize a more powerful, natural, and task-efficient user interface. Gestural and voice-driven user interface is being used for interactively modelling 3D objects in a virtual environment. In figure 3, simple geometric models here are overlaid on a photograph from the site at Petra.

Examples of our recent work are 3D widgets (Brook Conner et al. 1992), free-form deformations (Hsu and Hughes 1992) and gestural interfaces (Zeleznik et al. 1996). 3D widgets demonstrate how parameters can effectively be represented by 3D geometry and embedded in a 3D dataspace. Our system *Sketch* (Zeleznik et al. 1996) is a gestural interface for 3D geometric conceptual design which demonstrates that 2D drawn gestures can specify rich, context-sensitive commands to realize a powerful interface without relying on 2D WIMP user interface mechanisms.

Our most recent and on-going effort, the ARCHAVE[3] project, on the development of a multi-platform interactive virtual environment for archaeological analysis within the context of an accurate reconstruction of the site, both in space and time, is presented elsewhere (Vote et al., this volume).

3.2. Three dimensional reconstruction from a single image

We have conducted preliminary work in order to extend our current *Sketch* system (Zeleznik et al. 1996) to interactively generate and edit free-form 3D shape models in a sequence of images. Figure 3 illustrates our first generation system which makes use of single viewed perspective images (Williamson and Brill 1990) together with basic geometric primitives. Our approach in this stream is to maintain a functional system that is fully interactive using, in the early phases, our current knowledge of 3D shape and scene recovery, and incorporates novel shape models and automated shape recovery algorithms, as they become available in the later phases of this project.

Figure 4: Illustration of photogrammetric 3D reconstruction from multiple images for architectural forms; more details in Vote (1999, in press).

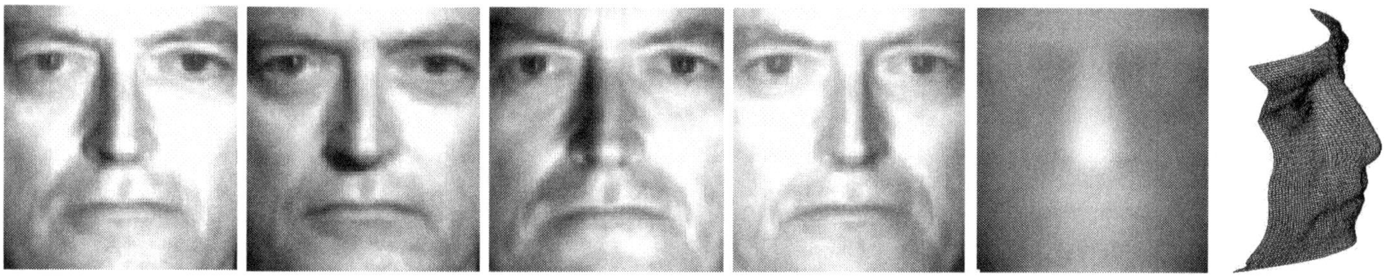

Figure 5: Six images (4 of which are shown) of the face of one of the co-author, D. Mumford, are used to determine its shape. The pose is constant, while the lighting varies. The resulting photometric variation reveals the shape, as shown in the depth map and wire-frame (fifth and sixth pictures).

3.3. Three dimensional reconstruction from multiple images

In order to establish the geometry of a scene and its objects, a number of correspondences (e.g. feature points) need to be recovered between N images of a sequence (Leymarie et al. 1996). In the context of archaeological scenery, corner detectors combined with a model-based approach (for position refinement), prove useful (Blaszka and Deriche 1994). For a video sequence, one can take advantage of the "continuity" of the sequence, by using robust tracking techniques (Leymarie et al. 1996). The correspondence problem is harder to solve for a set of photographic snapshots taken from a-priori unknown positions. This stage is user-driven in classical photogrammetry (Leymarie and Gruber 1997). To automate this task, one can make use of classical (window) correlation-based techniques combined with relaxation methods in an optimisation stage. Given the intrinsic camera parameters, we can then establish the calibration of the sequence (also called "exterior orientation" in photogrammetry). Alternatively, one can recover the full set of parameters, via robust estimation techniques (Faugeras et al. 1995). Once calibration is solved, more feature points can be acquired and matched to generate a cloud of 3D points. Finally, a triangulation can be obtained thanks to methods retrieving the connectivity (topology) of the bounding surface. Such methods permit to obtain realistic renditions of a statue (e.g. at roughly a ±5 mm accuracy in surface deviation). We expect our shape models, to be presented below, to constrain the reconstruction process and greatly simplify the final representation of such free-form objects while maintaining good accuracy.[4]

Similar techniques were applied to photographs taken at Petra and are illustrated in figure 4 where we performed some detailed wall reconstruction under user supervision (Vote, in press). We have also experimented with a "dual" method to photogrammetry, where the camera and object positions are fixed, and, instead, the light source is moved to known positions. This is based on the work of Belhumeur et al. (1996). Note that, this technique by-passes the problem of calibration. However, such a setup provides for excellent accuracy to scan small objects in a constrained environment where lighting conditions can be controlled (see figure 5) and is, thus, comparable to structured light techniques such as used in (Pollefeys et al. 2000).

3.4. Fragment representation and reassembly

The series of detailed head statuary in figure 1 need to be reconstructed by filling-in missing sections, fusing related fragments, reconditioning eroded surfaces and, finally, comparing the shapes of the different heads with others found in the region of Petra (Joukowsky 1998). A similar problem we have solved using IPS models (see section 4.2) by matching 3D fragments of an Egyptian bust (Blane et al. 2000). The use of ridges and skeletal graphs for the same purposes represent on-going work, and more details about these methods are given in section 4.1 and section 4.3 below.

3.5. Site content discovery via 3D geometric history

For analysis, it is essential to maintain the artefacts in their architectural and topographical context. Following what Forte (2000) proposed, we have started exploring how Geographical Information System (GIS) (Kofler et al. 1996) and Virtual Environments (VE) can be useful in helping archaeologists understand their data to develop new conclusions and hypotheses about the history and evolution of the Nabataean culture (Vote et al., this volume).

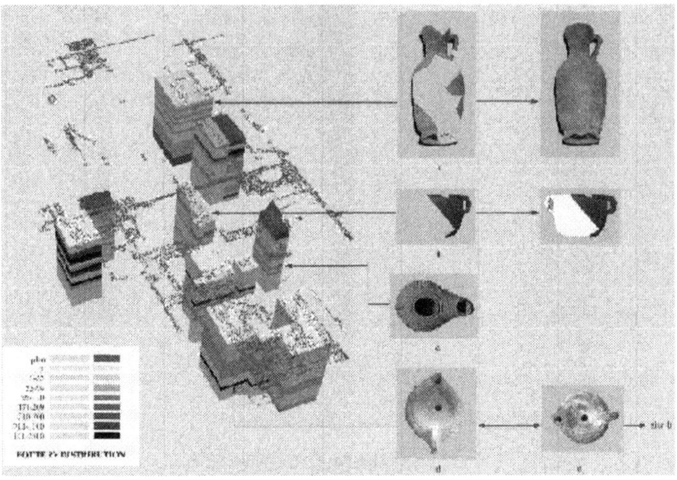

Figure 6: GIS application at Petra's Great Temple. Artefacts referenced and defined in 3D will allow archaeologists to reconstruct objects (a and b), do advanced spatial analysis, link artefacts between sites (For example, tracing lamps between sites in the region will allow us to trace trade routes.) (c, d and e), and maintain more comprehensive, accurate and accessible records of artefact geometry and find locations.

Figure 6 shows a GIS application with a 3D view of trenches. Colour represents the concentration of pottery fragments found in each locus or layer of excavated material. Unfortunately, the "traditional" GIS cannot represent in 3D the in situ or find position of individual artefacts or allow reference to specific finds on site or in other sites around or outside of Petra. With the ability to reference the location and geometry of artefacts, archaeologists will have a more dynamic data set that can be used to reconstruct, link objects for analysis and maintain spatial information for future generations. This is explored in our ARCHAVE project which is described in detail elsewhere (Vote et al., this volume).

4. Three dimensional free-form shape modelling

We have been investigating the use of 3D distinct representations for shape, i.e., ridges, implicit polynomials and skeletal graphs. Our premise is that these representations are intimately connected (e.g. see figure 10b), and we propose a joint, integrated, and comprehensive investigation of these, which shall lay the foundations to establish a complete and formal shape language for general use in archaeology and beyond. Briefly, *ridges* are a representative of curve loci on a surface, a one dimensional construct in space, e.g., a break curve where a sherd was broken from another piece. *Implicit polynomial surfaces* (IPS) are a representative of entire surface loci, a two dimensional construct in space, e.g., the outer and inner surfaces of a pottery sherd. *Skeletal graphs*, also called *Medial Axes* (Blum 1973), are a representative of volumetric features, a three dimensional construct in space, e.g., the main axis of a pot and its symmetric relations with the pot surfaces. These three elements are constitutive of a vocabulary classification for shape. Their relationship via a hyper-graph structure, will define the equivalent of a syntax for 3D shape. In the remaining of this final section we detail each vocabulary class.[5]

4.1. Curvilinear modelling through ridges

What are ridges? It is simplest to define them in 3D by analogy with a 2D case. The boundary of a 2D shape is a simple closed curve which can be divided into convex and concave portions, separated by points of inflection where the curvature of the boundary vanishes. In addition, there are special points on the boundary where the curvature has a local maximum or minimum. The most important of these are the "vertices": local maxima in convex segments, and local minima in concave segments, which are analogs of the vertices of polygons. In particular, each endpoint of the medial axis or skeletal graph (see section 4.3) of the shape is the centre of the osculating circle at a convex vertex (Leyton 1992) (see figure 9a and 9d). The psychologist Attneave proposed that these were the most perceptually salient and informative points on the contour (Attneave 1954).

What happens in 3D? The situation is more complex. Instead of merely convex and concave pieces, the boundary of any 3D shape is divided into three kinds of pieces: (i) the convex parts with both principal curvatures positive, (ii) the parts where both principal curvatures are negative, i.e., the surface is strictly concave, and (iii) the hyperbolic saddle-like parts where one principal curvature is positive, the other negative. Instead of local max and min points for the principal curvatures, one looks for curvilinear collections of points where the larger of the two principal curvatures has a local max on its corresponding *line* of curvature (figure 7a), and points where the smaller curvature has a local minimum. The *ridges* in the *convex* parts of the surface are smooth analogs of the convex edges of a polyhedron and are perceptually salient as the prominent lines where the surface *protrudes*. Likewise, "ridges" in the *concave* parts of the surfaces look like the bottom of *valleys* where the surface is *creased* (Cipolla et al. 1995).

One goal is to use these features to describe 3D shape in an intuitive way. In figure 7b, ridge computations on a sherd surface data obtained via CT scanning (cf. figure 2) are depicted as different shades of grey corresponding to different measures of extremal curvature. We have developed an interactive algorithm to extract ridges and valleys based on this curvature map; this is illustrated in figure 7c. The user clicks a starting point and goal (which may be identical, to close a loop), decides whether a ridge or valley is needed, and then lets the computer rely upon an implementation of a 3D active contour to seek an optimal path (Leymarie and Levine 1993). Such an active contour model tends to minimize a cost function based on an integral of the curvature measures along a path as well as on measures of elastic tension along the contour. Because such features as ridges and valleys correspond well with (human) intuitive curvilinear descriptors for free-form shapes, we believe they will provide a very effective tool for manipulating shape for interactive modelling as well as for indexing and searching databases of shapes, and delimiting break surfaces of sherds.

The next stage in our research program is to explore the use of ridges/valleys on a variety of free-form shapes, bodies and a range of artefacts as well as faces, animals, humans, sculptures of various types, furniture and tools, etc. There has been psychophysics on the human perception of ridges (Phillips 1997) and an additional goal is to characterize how *stable* ridges are for shape modelling.

Figure 7: (a) Concept of a ridge as a line of maximal principal curvature on a surface. (b)Local curvature computations on a sherd from its recovered 3D surface. (c) Result of an interactive ridge and valley computation using a 3D active contour model (see Andrews 2000 for more technical details).

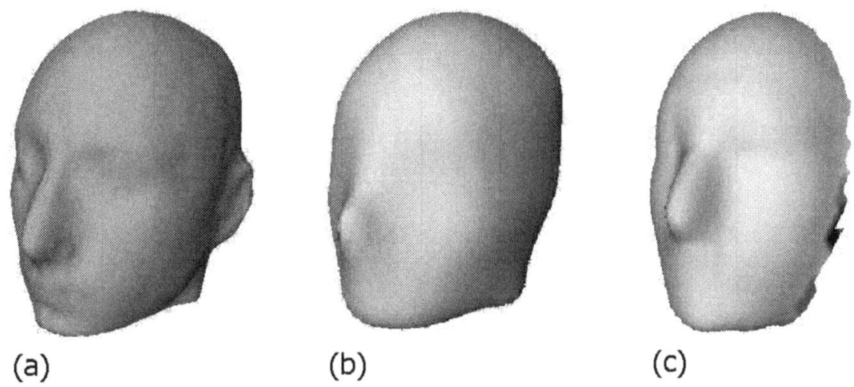

Figure 8: (a) Original triangulated head data set. (b) 10th degree fit (ears discarded). (c)Reconstruction via 12 patches, using 4th degree IPS for each patch.

4.2. Surface modelling through implicit polynomials

Multivariate Implicit Polynomials provide a powerful and rich representation for 2D and 3D curves and surfaces (Bloomenthal et al. 1997). For example, a trivariate d^{th} degree Implicit Polynomial Surface (IPS) is the zero set of a dth degree explicit polynomial, i.e., the set of points (x,y,z) where the explicit polynomial is zero, $f(x,y,z) = \sum_{i+j+k \leq d} c_{ijk} x^i j^j z^k = 0$. These surfaces are generalizations, to more complicated shapes, of the conics, e.g., a hyper-ellipsoid, a cylinder with hyperbolic cross section, etc.

For example, the set of points (x,y,z) for which $(x-x_0)^2+(y-y_0)^2+(z-z_0)^2-R^2 = 0$ is the equation of a sphere of radius R having centre $x = x_0$, $y = y_0$, $z = z_0$. These IPS are useful for representing blobby closed surfaces, open patches as in figures 1, 4, 6 and 8, surfaces attached to prominent ridges, generalized cylinders (Naeve and Eklundh 1994), and other shapes, e.g., free-form shapes with holes. IPS can be used in at least two interesting ways: (i) as a coarse, but smooth, approximation or (ii) as a close fit to the data. As a low resolution approximation to a complex surface, an IPS can be used to extract coarse geometry useful for shape recognition, crude assembly of fragments into reconstructions, etc. On the other hand, a single high degree IPS or a number of patches made of IPS of more modest degree, can be used for a high resolution representation. Some of these uses are illustrated in figure 8.

A goal of our research program is to explore the use of ridges (section 4.1) and skeletal graphs (section 4.3) for the optimal placement of IPS patches so that low order fits, and thus fewer parameters, can be used. Fitting to data is fast, repeatable, and robust, since the fitting is linear least squares (thus resulting in an *explicit* expression for the estimated coefficient vector), it is regularized by our 3L fitting (Blane et al. 2000), and is further regularized by the use of *ridge regression* (Blane et al. 2000). Note that, the principal computational cost is in computing only a scatter matrix of monomials based on the (x,y,z) data points. Once this is done, a refitting to subsets or unions of data point sets, or a modification of surfaces through human interaction, requires orders of magnitude less computation and is possible in real-time.

Our approach to human interaction with shape, when using IPS, is to modify the surface much as a sculptor or a designer might: by specifying a position, or a position and a tangent cut, or a position, a tangent cut and two bendings (e.g. via principal curvatures) that we want the deformed surface to satisfy approximately (soft constraint) or exactly (hard constraint), such that the surface is not modified much away from the position of interest. More generally, we can specify a number of points, or a curve in 3D, or a surface attached to a ridge that we want the deformed surface to approximate.

Our next step will be to investigate a hybrid model by interpolating with an IPS exactly (hard constraint) or approximately (soft constraint) by specifying some surface properties (e.g. curvatures, tangents, etc.) in-between the ridges, where these latter properties could be specified through stochastic processes or through probability distributions. For elongated surfaces like an arm, perhaps an upper torso, an elephant trunk, etc., a *generalized cylinder* (Naeve and Eklundh 1994) can be realized by computing the skeletal axis of skeletal sheets (see below), and then sweeping a cross-sectional planar IP curve along the axis, where the plane is

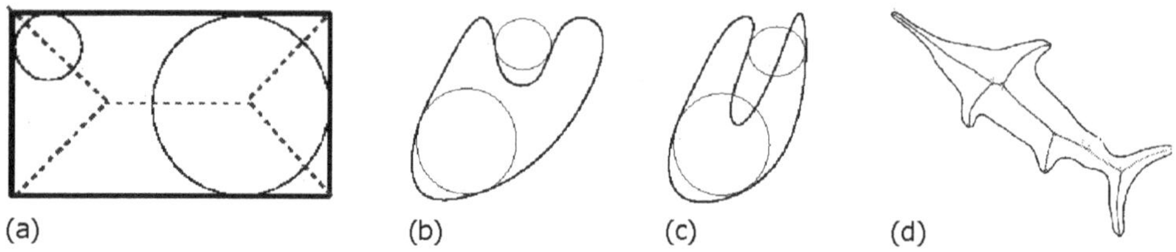

Figure 9: 2D skeletons: (a) Skeleton of a rectangle with two examples of maximally inscribed circles; (b) example of two bitangent circles which are maximal and (c) two which are not, since they cross the boundary; (d) More complex skeleton for a swordfish outline - note that at each curvature extremum of the boundary corresponds the end of a skeleton branch.

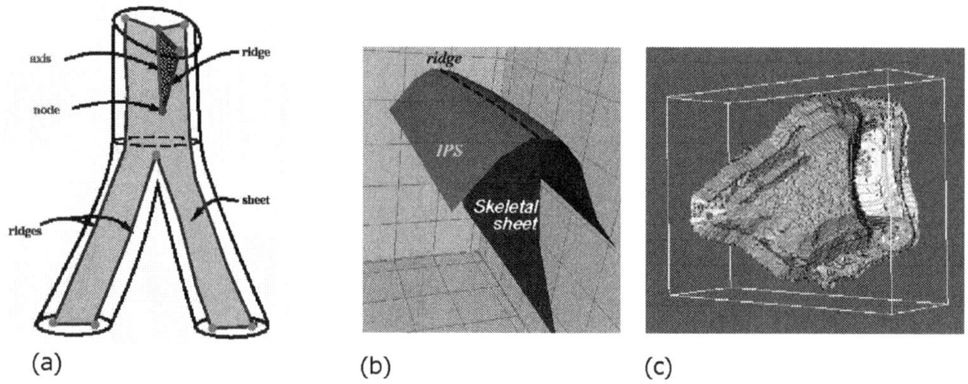

Figure 10: (a) Sketch of the 3D skeleton of a branching shape: the skeleton is made of sheets, axial and ridge curves, and nodes. (b) Concept of skeletal sheet and its relation to IPS and ridges. (c) Computed skeletal sheets of a carpal bone.

orthogonal to the axis and has a local coordinate system determined by the skeletal sheet.

4.3. Volumetric modelling through skeletal graphs

The skeletal graph of a 2D shape, or its medial axis, is the locus of centres of *maximally inscribed circles* (see figure 9). It is an intuitive and efficient representation for the recognition of 2D shapes, since variations in shape often leave the graph structure intact (Blum 1973). This medial axis graph structure, however, maps to a variety of shapes (i.e., by varying its associated radius function) and thus is not sufficiently constrained to reveal qualitative shape. A dynamic view of the skeletal graph as the singularities (shocks) of wavefronts propagated from the initial boundary (Leymarie and Levine 1992) defines a notion of velocity and direction of flow for each medial axis point and thus leads to a finer partitioning of the skeletal branches at points where the flow is reversed. The resulting shock graph, when stripped of radius information, is more discriminating in that it reveals qualitative shape (Kimia et al. 1995).

Shape can be fully reconstructed from the medial axis and the corresponding radii as the envelope of circles centered on the axis. The local nature of this intimate connection between shape and skeletal graphs, however, is not explicit in the envelope reconstruction. In (Giblin and Kimia 1999) the differential geometry of the boundary, i.e., tangent and curvature, is derived as a function of the differential geometry of the medial axis and of the dynamics of shock propagation on the axis, i.e., velocity and acceleration. It is shown that the shock graph, together with curvature and acceleration descriptions for each link, is a complete description of shape.

For 3D shape, the medial axis is the locus of *maximal bitangent spheres*. The wave propagation approach again leads to a dynamic view of shocks propagating on the skeletal locus (Leymarie and Kimia 2000). The points of medial axis (and shock set) have been classified resulting in a hypergraph representation (Giblin and Kimia 1999) consisting of skeletal sheets with associated flow fields, which end either at a boundary corresponding to *ridges* or at curves shared by three medial sheets, much like the central axis of a *generalized cylinder* with a triangular base (Naeve and Eklundh 1994). These curves interact only at special points, namely when they intersect each other at nodes. These are the only generic possibilities (figure 10). The skeletal hypergraph describes the connectivity among symmetries of each portion of the shape (see Leymarie and Kimia (2000) for more details).

The next step in this research is to investigate how the skeletal hypergraph can be matched against other medial axis representations in a pre-stored database of similar objects, in analogy to 2D matches (Sharvit et al. 1998). Also, partial matches of skeletal axial curves should prove useful to solve the difficult problem of automatically stitching together different sherds to recover a full pot (Ucoluk and Toroslu 1999).

5. Conclusions

The SHAPE Lab has been created with the goals of: (i) introducing new geometric modelling and 3D surface and structure recovery from images; (ii) improving human/machine interaction tools for facilitating human input of geometric information to the machine and then visualizing the results in real time; (iii) developing new tools to facilitate reconstructing large geometric structures (e.g., walls of buildings) and smaller objects (e.g., columns and their capitals, and at more detailed levels, with statues and artefacts) from free-form fragments scattered about a site. These objectives require considerable domain-specific knowledge and are central in providing material for analysis in archaeology but also can be used extensively in architecture and architectural history, and ultimately in many other disciplines where the design and manipulation of free-form 3D shapes is required.

In order to fulfil this ambitious program, a key component is the development of a shape language for 3D free-form objects. We have reported in this paper on our early success in putting together a vocabulary based on three classes of elements: *ridges*, to model perceptually significant surface curves, *implicit polynomials*, to model surfaces of various complexity, and *skeletal graphs* to model volumetric features and, furthermore, provide the "glue" to relate together the three classes.

Difficult and interesting challenges still remain ahead of us. There is clearly a *continuum* from the ridges on polyhedra which are most precise as well as most salient and those in near planar or near spherical parts of the surface, and this "scale-space" for ridges needs to be studied (Mumford et al. 1999). A second question is how to *approximate* a 3D shape using ridge and skeletal data. In the plane, an old idea going back to Attneave is to approximate any 2D shape by the polygon joining its vertices. What analogs of this construction can we make in 3D? An essential step in the HCI part of this research (section 3.1), is to be able to estimate an entire shape roughly based on the user marking approximate ridges and local planes of symmetry, and then let the computer position and select implicit polynomial models of the surface patches bounded by such ridges. Another question concerns the location of ridges using reflectance data gathered from one or more images of an object. The basic idea is that since the tangent plane is changing rapidly at ridge points, images of the surface will have rapid changes in intensity along ridges. In addition, specularities "cling" to ridges and with elongated light sources, may even make the whole ridge shine. We want to make these ideas precise and integrate them in the reconstruction of 3D shape from multiple images with varying illumination as a constraint used in the recovery of shape (figure 5).

In addition to developing this approach to shape representation, we want to apply it to object recognition based on shape. It is broadly recognized that one of the most effective techniques for object recognition is the use of Bayesian statistical methods (Cernuschi-Frias et al. 1989). In order to apply this method to free-form shapes, we need priors of the space of such shapes (Mumford 1996). For example, the shapes we find are often built out of parts which may be generalized cylinders or rectangular parallelepipeds; or they may have limbs like a statue, a human or a tree in winter, etc. The approach we want to take is to model stochastically the generic features of shapes, their skeletal graphs and ridges and decomposition into parts. In 2D, Zhu and Yuille (1996) have constructed stochastic models of shapes based on the medial axis. In 3D the development of such priors, involving the explicit representation of ridges and skeletal graphs, is needed.

Acknowledgments

Major funding for this project comes from the USA NSF/KDI grant #BCS-9980091. Frederic Leymarie was partially supported by a Ph.D. fellowship from IBM. Besides the co-authors, a number of students and researchers have been involved in the first year of the SHAPE Lab and have had a major impact on the work reported in this paper. These include graduate students Daniel Acevedo, Stuart Andrews and Jill Baker, Dr Peter Giblin (Liverpool) and Dr Jean-Philippe Tarel (Paris).

References

ANDREWS, S., 2000. *Interactive generation of feature curves on surfaces*. Master's thesis, University of Toronto, Ontario, Canada.

ATTNEAVE, F., 1954. Some informational aspects of visual perception. *Psychological Review*, 61(3):183-193.

AYERS, M.R. and ZELEZNIK, R.C., 1996. The lego interface toolkit. In *Proceedings of the ACM Symposium on User Interface and Software Technology (UIST'96)*:97-98.

BELHUMEUR, P. and KRIEGMAN, D., 1996. What is the set of images of an object under all possible illumination conditions? In *Proceedings of CVPR*: 270-277. San Francisco, CA, IEEE CS Press.

BELHUMEUR, P., KRIEGMAN, D. and YUILLE, A., 1997. The bas-relief ambiguity. In *Proceedings of CVPR*:1060-1066. San Juan, PR, IEEE CS Press.

BLANE, M.M., LEI, Z. and COOPER, D.B., 2000. The 3L algorithm for fitting implicit polynomial curves and surfaces to data. *IEEE Trans. on PAMI*, 22(3):298-313.

BLASZKA, T. and DERICHE, R., 1994. Recovering and charactering image features using an efficient model based approach. Research Report RR-2422, INRIA, Sophia Antipolis, France.

BLOOMENTHAL, J., BAJAJ, C., BLINN, J., CANI-GASCAUEL, M.P., WYVILL, B. and WYVILL, G., 1997. *Introduction to Implicit Surfaces*. Morgan Kaufmann Pub., San Francisco, CA, USA.

BLUM, H., 1973. Biological shape and visual science. *Journal of Theoretical Biology*, 38:205-287.

BROOK CONNER, D., SNIBBE, S.S., HERNDON, K.P., ROBBINS, D.C., ZELEZNIK, R.C. and Van DAM, A., 1992. Three-dimensional widgets. *Computer Graphics (1992 Symposium on Interactive 3D Graphics)*, 25(2):183-188.

BRUCE, J.W., GIBLIN, P.J. and TARI, F., 1996. Ridges, crests and sub-parabolic lines of evolving surfaces. *International Journal of Computer Vision* 19:195-21.

CERNUSCHI-FRIAS, B., COOPER, D.B. and HUNG, Y.P., 1989. Toward a model-based Bayesian theory for estimating and recognizing parametrized 3-D objects using two or more images taken from different positions. *IEEE Trans. on PAMI*, 11(10):1028-1052.

CIPOLLA, R., FLETCHER, G. and GIBLIN, P.J., 1995. Surface geometry from cusps of apparent contours. In *Proc. Int. Conf. on Computer Vision (ICCV'95)*:858-863. IEEE Computer Society Press.

COHEN, J.M., MARKOSIAN, L., ZELEZNIK, R.C., HUGHES, J.H. and BARZEL, R., 1999. An interface for sketching 3D curves. In *Proceedings of 1999 Symposium on Interactive 3D Graphics*, Atlanta, ACM.

CRESCIOLI, M. and NICCOLUCCI, F., 1998. P.E.T.R.A.-Data: An integrated environment for archaeological data processing. In *New Techniques for Old Times: Computer Applications and Quantitative Methods in Archaeology. Proc. of the 26th CAA Conf.*, BAR International Series 757:133-134.

DEBEVEC, P.E., TAYLOR, C.J. and MALIK, J., 1996. Modeling and rendering architecture from photographs. In *Computer Graphics Proceedings, SIGGRAPH'96*. ACM.

FAUGERAS, O. et al. 1995. 3D reconstruction of urban scenes from sequences of images. Research Report RR-2572, INRIA, Sophia Antipolis, France.

FORTE, M., 2000. About virtual archaeology: Disorders, cognitive interactions and virtuality. In Barceló, J.A., Forte, M. and Sanders, D.H. (eds.), *Virtual Reality in Archaeology*. BAR International Series 843:247-259, Oxford, Archaeopress.

GIBLIN, P.J. and KIMIA, B.B., 1999. On the intrinsic reconstruction of shape from its symmetries. In *Proc. of CVPR*:79-84.

GIBLIN, P.J. and KIMIA, B.B., 1999. On the local form and transitions of symmetry sets, and medial axes, and shocks in 2D. In *Proc. of ICCV*:385-391.

GIBLIN, P.J. and KIMIA, B.B., 2000. On the local form of symmetry sets, and medial axes, and shocks in 3D. In *Proc. of CVPR*. IEEE Computer Society 2000.

HAGER, G. and BELHUMEUR, P., 1996. Real time tracking of image regions with changes in geometry and illumination. In *Proceedings of CVPR*. IEEE CS Press.

HSU, W. and HUGHES, J.F., 1992. Direct manipulation of free-form deformations. In *Computer Graphics Proceedings, SIGGRAPH'92*:177-184.

JOUKOWSKY, M.C.S., 1980. *A Complete Manual of Field Archaeology: Tools and Techniques of Field Work for Archaeologists*. Prentice-Hall, Englewood-Cliffs, New Jersey. (Eight printing).

JOUKOWSKY, M.C.S., 1996. *Early Turkey: An Introduction to the Archaeology of Anatolia from Prehistory through the Lydian Period*. Kundall-Hunt.

JOUKOWSKY, M.C.S., 1997. Brown University 1997 excavations at the Petra Great Temple. *Annal of the Department of Antiquities of Jordan*, XLII:293-318.

JOUKOWSKY, M.C.S., 1997. The Petra southern temple: The fourth season 1996. *American Journal of Archaeology*:101:339.

JOUKOWSKY, M.C.S., 1998. The Great Temple at Petra. *American Journal of Archaeology*, 102(3):593-596.

JOUKOWSKY, M.C.S., 1999. *Petra Great Temple - Volume 1: Brown University Excavations 1993-1997*.

KIMIA, B.B., TANNENBAUM, A.R. and ZUCKER, S.W., 1995. Shapes, shocks, and deformations, I: The components of shape and the reaction-diffusion space. *International Journal of Computer Vision*, 15:189-224.

KOFLER, M., REHATSCHEK, H. and GRUBER, M., 1996. A database for a 3D GIS for urban environments supporting photo-realistic visualisation. In *International Archives of Photogrammetry and Remote Sensing (XVIII ISPRS Congress)*, volume XXXI:198-202.

LEI, Z., TASDIZEN, T. and COOPER, D.B., 1998. PIMs and invariant parts for shape recognition. In *Proc. of International Conference on Computer Vision (ICCV'98)*:827-832. IEEE CS Press.

LEYMARIE, F. and GRUBER, M., 1997. Applications of virtual reality: CyberMonument and CyberCity. In *Proc. of the 3rd European Digital Cities Conference*, Berlin, Germany. European Commission.

LEYMARIE, F. and KIMIA, B.B., 2000. Discrete 3D wave propagation for computing morphological operations from surface patches and unorganized points. In Goutsias, J., Vincent, L. and Bloomberg, D. (eds.), *Math. Morphology and its Applications to Image and Signal Processing*, volume 18 of *Computational Imaging and Vision*:351-360.

LEYMARIE, F. and LEVINE, M.D., 1992. Simulating the grassfire transform using an active contour model. *IEEE Trans. on PAMI*, 14(1):56-75.

LEYMARIE, F. and LEVINE, M.D., 1993. Tracking deformable objects in the plane using an active contour model. *IEEE Trans. on PAMI*, 15(6):617-634.

LEYMARIE, F. et al., 1996. REALISE: Reconstruction of reality from image sequences. In Delogne, P. (ed.), *Proc. of IEEE International Conference on Image Processing (ICIP'96)*, volume III:651-654.

LEYMARIE, F., 1997. Exploitation of 3D georeferenced datasets in a GIS. In *Proc. International Workshop on High Performance Computing Networking Exploitation of Multimedia Databases*, number EUR 17349 EN:10-29. European Commission.

LEYMARIE, F., KIMIA, B.B. and GRUBER, M., 1998. *Direct camera pose evaluation from a single view and a map*. Technical Report LEMS-174, Brown University.

LEYTON, M., 1992. *Symmetry, Causality, Mind*. MIT Press, Cambridge.

MARKOSIAN, L., COHEN, J., CRULLI, T. and HUGHES, J., 1999. Skin: A constructive approach to modeling free-form shapes. In *Proceedings of SIGGRAPH'99*.

MORTENSEN, E.N. and BARRETT, W.A., 1998. Interactive segmentation with intelligent scissors. *Graphical Models and Image Processing*, 60(5):349-384.

MUMFORD, D., 1996. *The Statistical Description of Visual Signals*. Akademie Verlag.

MUMFORD, D., GIBLIN, P., GORDON, G., HALLINAN, P. and YUILLE, A., 1999. *Two and Three dimensional Patterns in Faces*. A.K. Peters.

NAEVE, A. and EKLUNDH, J.O., 1994. Generalized cylinders - what are they? In Arcelli, C., Cordella, L.P. and Di Baja, G.S. (eds.), *Aspects of Visual Form Processing*, 2nd International Workshop on Visual Form:384-409. World Scientific.

PHILLIPS, F., 1997. *Geometric structure, frames of reference, and their implication in the localization of features on smoothly curved surfaces*. PhD thesis, Ohio State University.

POLLEFEYS, M., PROESMANS, M., KOCH, R., VERGAUWEN, M. and Van GOOL, L., 2000. Acquisition of detailed models for virtual reality. In Barceló, J.A., Forte, M. and Sanders, D.H. (eds.), *Virtual Reality in Archaeology*. BAR International Series 843:71-77. Oxford, Archaeopress.

SCHMITT, G. et al., 1995. Toward virtual reality in architecture: Concepts and scenarios from the architectural space laboratory. *Presence*, 4(3):267-285.

SHARVIT, D., CHAN, J., TEK, H. and KIMIA, B.B., 1998. Symmetry-based indexing of image databases. *Journal of Visual Communication and Image Representation*, 9(4):366-380.

STECKNER, C., 2000. Form and fabric, the real and the virtual. In Barceló, J.A., Forte, M. and Sanders, D.H. (eds.), *Virtual Reality in Archaeology*. BAR International Series 843:121-128. Oxford, Archaeopress.

STREILHEN, A., 1994. Towards automation in architectural photogrammetry: CAD-based 3D-feature extraction. *ISPR Journal of Photogrammetry and Remote Sensing*, 49(5):4-15.

SUBRAHMONIA, J., COOPER, D.B. and KEREN, D., 1996. Practical reliable Bayesian recognition of 2D and 3D objects using implicit polynomials and algebraic invariants. *IEEE Transactions on Pattern Analysis and Machine Intelligence*, 18(5):505-519.

SUBRAHMONIA, J., KEREN, D. and COOPER, D.B., 1994. An algebraic and Bayesian technology for object recognition. In Mardia, K.V. (ed.), *Advances in Applied Statistics*:165-205. Carfax Publishing Co., Abingdon, Oxfordshire.

TAREL, J.P., CIVI, H. and COOPER, D.B., 1998. Pose estimation of free-form 3D objects without point matching using algebraic surface models. In *Workshop on 3D Modeling, ICCV98*:13-21.

TAREL, J.P., GUICHARD, F. and AUBERT, D., 1999. Tracking occluded lane-markings for lateral vehicle guidance. In *Proc. of IEEE International Conference on Circuits, Systems, Communications and Computers (CSCC'99)*.

UCOLUK, G. and TOROSLU, I.H., 1999. Automatic reconstruction of broken 3-D surface. *Computers and Graphics*, 23(4):573-582.

VOTE, E., in press. Using desktop photogrammetry to document archaeological remains: The Great Temple at Petra, Jordan. In Cooney, G. and Masterson, B. (eds.), *Proc. Int. Conf on Computer Applications in Archaeology (CAA'99)*.

VOTE, E., FELIZ, D.A., LAIDLAW, D. and JOUKOWSKY, M.S., 2000. ARCHAVE: A virtual environment for archaeological research. In *Proc. Int. Conf on Computer Applications in Archaeology (CAA2000)*. This volume.

WILLIAMSON, J.R. and BRILL, M.H., 1990. *Dimensional Analysis through Perspective - A Reference Manual*. Kendal/Hunt Publishing co.

ZELEZNIK, R.C. and FORSBERG, A.S. and STRAUSS, P.S., 1997. Two pointer input for 3D interaction. In *Proceedings of 1997 Symposium on Interactive 3D Graphics*:27-30.

ZELEZNIK, R.C. and FORSBERG, A.S., 1999. Unicam - 2D gestural camera controls for 3D environments. In *Proceedings of 1999 Symposium on Interactive 3D Graphics*.

ZELEZNIK, R.C., HERNDON, K. and HUGHES, J.F., 1996. SKETCH: An interface for sketching 3D scenes. In *Computer Graphics Proceedings, SIGGRAPH'96*, Annual Conference Series:163-170.

ZHU, S.C. and YUILLE, A.L., 1996. FORMS: A flexible object recognition and modeling system. *International Journal of Computer Vision* 20(3):187-212.

Notes

1 SHAPE: SHape, Archaeology, Photogrammetry, Entropy; a multi-disciplinary project established in the Fall of 1999; visit our website at: www.lems.brown.edu/vision/extra/SHAPE/.

2 N.B., we also plan to relate and compare objects and aspects of our site with other sites within Petra and other Nabataean sites like Medain Saleh.

3 ARCHAVE: ARCHAeology with Virtual Environment systems; see (Vote et al., this volume).

4 Such a model-based constraint paradigm is similar to Debevec et al.'s, earlier DARPA community's and others' approach to 3D reconstruction from images. However, in these other approaches, much simpler set of models is used. These simpler models are typically made of regular primitives to represent simple architectural shapes (Streilhen 1994).

5 The syntaxic properties of our shape language will be reported elsewhere; see (Giblin and Kimia 1999) for early theoretical investigations.

An Experimental Method for the Analysis of Attributes of Flint Artefacts Using Image Processing

Sorin Hermon

Archaeological Division, Department of Bible and Ancient Near Eastern Studies, Ben-Gurion University of the Negev

Beer-Sheva, Israel

e-mail: sorinh@bgumail.bgu.ac.il

Marco Petrone and Luigi Calori

CINECA, via Magnanelli, 6/3, 40033 – Casalecchio di Reno (Bo), Italy

e-mails: m.petrone@cineca.it, calori@cineca.it

Abstract

The article describes an experiment that explored the applicability of image processing to lithic attribute analysis. Following this experiment, we developed an interface for the application of a computerised method of attribute analysis of lithic artefacts using image processing. The main purpose of the experiment was to examine the potential of image processing methods which would be used for the automation of attribute analysis of flint artefacts, using relatively non-expensive devices, increasing the accuracy of the process and reducing time consumption typical for the manual process. The developed interface is able to recognise the contours of an object, extract its sizes (length and width) as well as the shape of its edges, and store all information in a data-base (measurements and image of the object). Further developments will include classification of edge-shapes (according to a priori defined criteria) and number of arises on the dorsal side of the objects.

Key words: attribute analysis, flint artefacts, image processing, automatisation, VTK, attributes extraction, data-base

1. Introduction

One of the tasks archaeologists are often involved in is the reconstruction of social and economical organisation of past cultures. These reconstructions rely heavily on the material culture retrieved through excavations. For prehistorians, such reconstructions sometimes depend on a scrupulous examination of the material culture, mostly flint artefacts. This data-base is manipulated by minute classification into various types, mostly using morphological and metrical characteristics of chosen items. This process of examination, known as "attribute analysis" (e.g. Clay 1976, Patterson 1983), involves the minute examination of lithic artefacts, including measurements of various sizes and ratios between them, as well as morphological observations of attributes, such as scar patterns, types of butts, modification patterns, etc. (see Inizan, et al. 1992). Thus, through these observations, the prehistorian hopes to reconstruct the mode of flint tool production, which will further be used for his/her reconstruction of models from the past societies:

> "…Even the simplest techniques of any…society take the character of a system that can be analysed, in terms of a more general system. The techniques can be seen as a group of significant choices that each society has been forced to make, whether they are compatible, or incompatible with other choices…" (Lévi-Strauss 1976:11).

Techniques involve the manipulation of materials, sequences of action, tools and a particular knowledge (know-how, manual skills, procedures, etc). Different techniques form a system - each technique is a locus of multiple interactions and constant adjustment among its elements: without the action animating it and without the knowledge of its effects, the tool is nothing. Action itself is constantly adapted to transformations in material, to the characteristics of the tool, to the evolution of know-how; technical knowledge takes into account the available tool, its effective action, the material worked upon, and so forth (Lemmonier 1986).

The basic notion is that technology embraces all aspects of the action performed upon the material as being "a social production" (Lemonnier 1990). Furthermore, technique is "a socialised action on matter…explained by three orders of facts: suites of gestures and operations (technical process), objects (means of action) and specific knowledge (*connaissance*)" (Schlanger 1994:145). Therefore, the archaeologist's task is to describe in which way a given technology is a social production. The knapper is also limited by the knowledge (*savoir faire*) of knapping methods. The possibility that these constraints emerged not only from the nature of the environment, functional needs, and knowledge of the knapper, but also from the social system within which the technique was practised, cannot be excluded. This approach is inspired by a "conception of techniques as social products: the "objectification" of what are socially elaborate thoughts" (Lemonnier 1990).

The current approach to the analysis of lithic assemblages takes place by reconstructing "production systems" (Karlin and Julien 1994). The process of knapping involves an intellectual operation: abstraction, anticipation, working out a solution to the encountered problems, construction of models. Pelegrin (1991) distinguishes a scale of "conceptual knowledge" (*connaisance*) acquired through the memorisation of concepts ("the mental representations of forms that are considered ideal and also the materials involved") and memorisation of modes of operation and procedural knowledge (*savoir faire*), that can either be ideational, arising from intelligence and memory, or physical movement, presupposing bodily skills. A distinction can be made between conceptual, abstract knowledge (*connaissances*), and practical or pro-

Figure 1: Image of sickle-blades used for the experiment, as acquired by the camera, before processing.

cedural know-how (*savoir-faire*). Both of these are present in an "elaborate flint knapping activity" (Pelegrin 1990).

Thus, a detailed analysis of material culture remains, and in this case, flint artefacts can be used for reconstructing various aspects of past cultures and their socio-economic organisation. In recent years, several techniques were developed for accurate investigation methods of attribute analysis of artefacts. These methods also include a broader use of computers and their accessories (e.g. Barceló et al. 1999).

The main goal of the experiment described below was to explore the potential of image processing in the use of identifying various attributes of flint artefacts and their classification, in order to automatise, speed up the process of attribute analysis and improve its accuracy through the elimination of human subjectivity (as for measurements or classification of morphological attributes). Therefore, an interface was developed for these purposes. The main component of this interface is a camera attached to a computer. The camera captures the image (which is the artefact we want to analyse) and the interface reads the image and analyses it. The results are then stored into a data-base.

The case study chosen for the experiment was a collection of sickle-blades from Israel, belonging to the Chalcolithic period (approx. 4500-3500 BC). These items are well defined by morphological characteristics (see Gilead 1995). Their polythetic definition is:

1. Narrow, but relatively thick, elongated blade obtained from a prismatic or pyramidal blade core, mostly of banded flint (see below).
2. One steeply backed side edge. The cross-section is rectangular or triangular, with a perpendicular side formed by backing.
3. Mostly straight, but sometimes finely denticulated working edge.
4. Truncations on the proximal, on the distal or on both ends are common (figure 1).

Sickle-blades are among the tool-types recognised to respond (to a higher degree than other tool-types) to chronological/cultural variations in proto-historic and historic assemblages of southern Levant. Moreover, the recognition of some specialisation is reflected by the discoveries of workshops for producing sickle-blades at various sites (Gilead and Hermon in press). Therefore, the importance of a detailed attribute analysis of sickle-blades can reveal important socio-economic aspects of Chalcolithic cultures in the study area. Thus, the interface developed was designed to identify the sizes of sickle-blades (length and width, sensitive to cultural/chronological and regional changes), the pattern and number of scars on their dorsal face as well as the edge shapes.

2. Description of hardware and software

While developing the interface, it was necessary to maintain the following features: all the processes were to be automatised to the highest possible level, thus reducing human interference to the minimum (such as artefact positioning), the analysis process was to be substantially faster than the human analysis and the required equipment (both hardware and software) as simple and cheap as possible.

The equipment used was:

a. Cameras

Hyper Had Standard PAL black and white Sony camera (spt – m102ce)

Computer TV lens 1:1.4 , 3.5 – 8 1/3 inch CS

Auto luminosity electronic (ccd iris)

b. Acquisition board

Matrox Meteor II professional frame grabber (the acquisition board), with a resolution of up to 768x576 pixels with PAL video signals. This board is used to digitise the signal from the camera in order to obtain an image, i.e. a matrix of points each one with its own grey value. The use of the professional board allows maximum use of the PAL video signal resolution. A lower cost solution could use a commercial acquisition board (e.g. the one integrated in the Matrox Marvel); this would probably require reducing the resolution of the acquired images to reduce the jitter noise (e.g. using 384x288 images).

c. P.C. computer (basic requirements)

Intel PII 400 Mhz, with 128M RAM, and Microsoft NT OS.

d. Table with arms for lights

Standard stand for reproduction (Mini Repro/2 1192), with column height 70 cm and base board 40 x 50 cm. Micrometric elevation with screw locking. Universal attachment stud, ¼ inch. Complete with two light arms each equipped with one movable lamp holder.

The program used for developing the interface was VTK – Visualisation Tool Kit. One of its characteristics which should not be neglected is that it is a freeware program, available for any user, running both on Unix or PC systems. VTK is a true visual system; it supports a broad range of algorithms (scalars, vectors and tensors), texts as well as volumetric methods. The program supports several techniques for building 3D models (including mesh smoothing, polygon reduction, implicit modelling, contouring, cutting). The language is Tcl (Tool command language), in this case used for the experiment phase and the development of the graphic user interface, and C++ (to expand VTK with new objects) (for more information, see Schroeder et al. 1998).

3. Description of the experiment

3.1. Extraction of the information from the object

One of the tasks explored was how to determine the best mode of extraction of the contour of the analysed objects. It was noticed that an important influential factor is the position (location) of the light source, which can be used to good effect in order to enhance the morphology of the objects. If the light source comes from the top, it does not allow to distinguish the number of surfaces (scars of previous removals on the dorsal face of the objects). It was decided that the image was to be taken with the light source from the top, in order to enable us to find the contour of a single object, due to the contrast between the background and the object itself. In this way, we also reduced the problem of shade and we could focus on the area of interest (the object itself) - the histogram of the object pixels is not corrupted by other information arising from the background. A simple sheet of glass supported by a black box was used as a table. The black box was placed underneath the glass sheet in order to avoid shading problems and was thus used to obtain a dark background and simplify the thresholding process. In order to avoid reflection from the camera itself the camera was covered by a dark shield.

In order to determine the surfaces of the object, at least two images had to be taken, each lighted from an opposite angle, so all information regarding the morphology of the object could be obtained. The light angle i.e. the angle between the camera axis and the direction of the source of light (sharp angle) should range between 45-90° - according to the best image obtained on the screen (the best image being the one where the surfaces are most visible, one side is almost completely dark, the other with emphasised surfaces). The luminosity should be constant as should the distance of the light source to the object (20 cm). While taking a sequence of images (up to 40 sec at two frames per second), we moved the light from one side to the other, in an arch above the object, in order to find the best angle for the camera.

3.2. Ambient luminosity

The problem of the ambient source of light (surrounding light) is that it must be diffused, in order not to create an unwanted shadow, particularly when we try to extract the contour of the object. In addition, if the ambient luminosity is too strong, it is of no use to change the angles of the light sources, since there will be no strong changes in the enlightenment of the object's surfaces. Therefore, we had to either create a constant shadow, or work in a dark environment, with a low ambient luminosity. Thus, the chosen light source was a normal 60 W light-bulb with a diffused light.

3.3. Process of image acquisition

In order to fit the object into the frame the distance from the ccd to the board was 16 cm, that is with the above specified lens. The positioning of the object was with the longer side parallel to the angle of the picture, but precise definition of the object location was not required. Zooming apparently increases the accuracy of the image (higher precision is translated in terms of higher resolution – one pixel corresponds to a smaller surface on the piece; maximum zooming depends on the distortion of the edges, due to the camera lens). A possible solution to this problem would be developing an algorithm of sub-pixel precision (measures), which would virtually increase the information per pixel.

In order to obtain a sufficient number of images of a single object, a sequence of 40 sec (2 frames/sec) were taken from various angles. This was performed in order to identify the morphology of the object – to observe how the angle of the light source affects the acquirement of the image (the lumination of surfaces). Finally, it was decided to use three frames: one with the light source from the top and two frames with the light source from opposite angles (the latter served for surface identification purposes).

It should be noted that several objects could be placed on the main board at any one time, and they could be chosen sequentially. The number of items on the board should be between 3 to 6 in each sequence.

3.4. Saving the images

The images were saved in a pgm format (format for grey images, which can be read by VTK, Paint Shop Pro – read/write). This format saves the image in a raw format (400kb), which is an appropriate format, for it does not lose any information about the image during the process of compression. Pgm format is also useful because it can be read by a number of different graphic programs. The information regarding morphological characteristics of the object is stored in a database (accessed by Office – Access program, or compatible).

3.5. Algorithm description

The main stages of the program are described in figure 2. Following is a description of the main stages of the program development:

Thresholding is a labelling operation on a grey scale image. It distinguishes pixels that have higher grey values from pixels with lower grey values. This method is used for automatic determination of the threshold value. Since the threshold separates the dark background from the bright object (or vice-versa), the separation could be done if the distribution of dark and bright pixels was known. The threshold value could then be determined as that separation value for which the fraction of dark pixels (binary labelled with 1) equals the fraction of bright pixels (binary labelled with 0). The difficulty is that the independent distribution of dark and bright pixels are often not known, since what is known is the image histogram, which tells how many pixels are associated with any grey value in the mixture distribution. In the event that the distribution of dark pixels and bright pixels are widely separated, the image histogram will be bimodal (corresponding to dark and bright pixels, respectively).

At this stage of our research, the method implied for the development of the interface is manual thresholding: a window with a cursor scale, showing the grey scale is set; the operator changes the grey scale until the required image (the contours) is acquired, and the object is separated from the background (figure 3). However, the thresholding required is minimal, since the contour is found at a low level of thresholding in any case. A future development of the interface will include the implementation of automatic thresholding, using either the Otsu method (Otsu 1987), or the Kittler and Illingworth method (Kittler and Illingworth 1985, 1986).

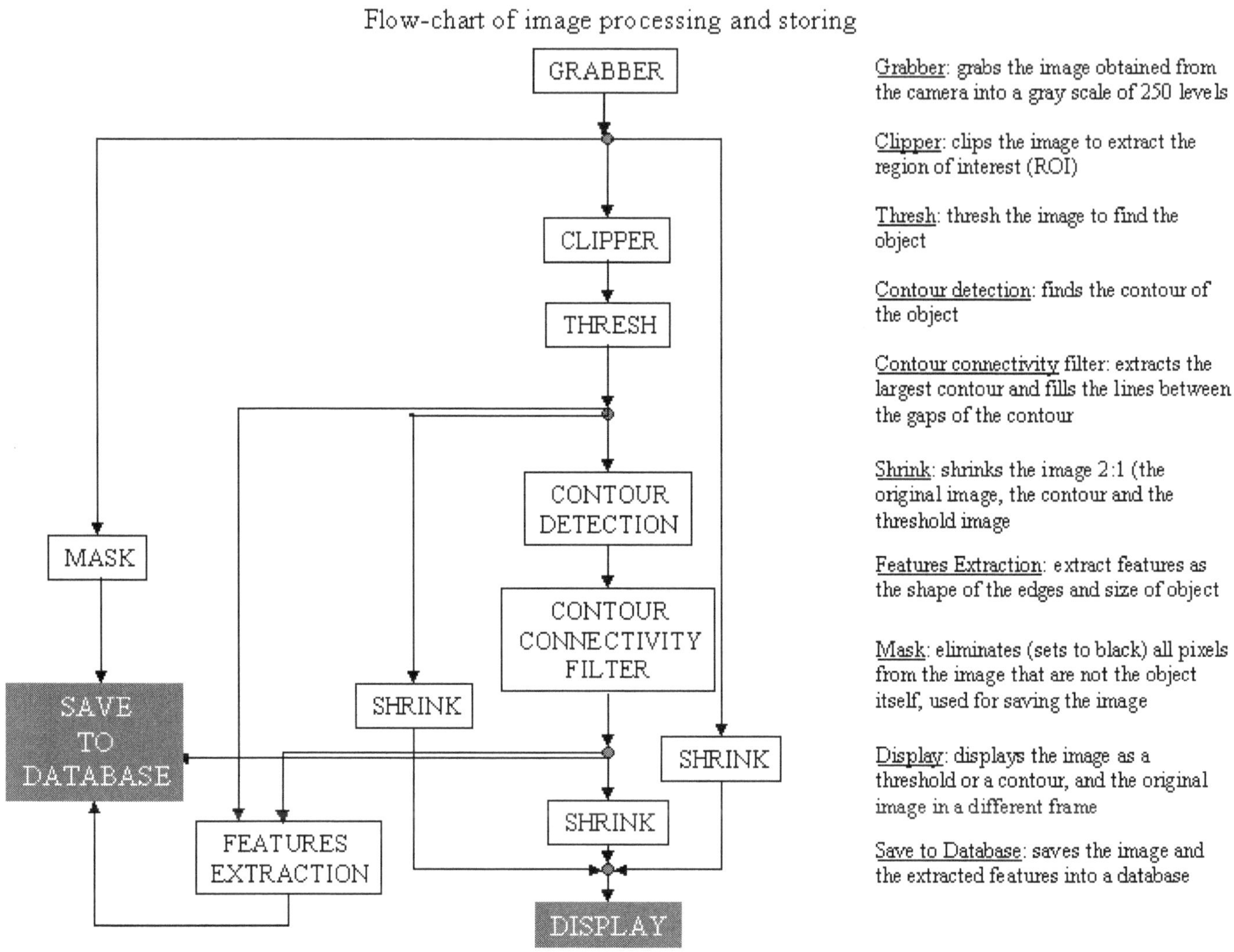

Figure 2: Flow Chart describing the main stages of the developed algorithm.

Contouring is a method of constructing boundaries between regions of similarly coloured areas. Contouring begins with selecting a scalar value (or contour value), which corresponds to the generated contours values. To generate the contours, some form of interpolation must be used, usually a linear interpolation along the edges. Once the points on the cell edges are generated they can be connected into contours. The method implied for contouring was the implementation of a 2D algorithm of *marching squares* (Schroeder et al. 1998:159). The basic assumption applied by this method is that a contour can only pass through a cell in a finite number of ways.

The method of extracting the sizes of the object (length and width), are based on the following formulae, which define the second-order spatial moments of the region (the shape of the object):

Second-order row moment:

$$M_{rr} = \frac{1}{A} \sum_{(r,c)\in R} (r-\bar{r})^2$$

Second-order mixed moment:

$$M_{rc} = \frac{1}{A} \sum_{(r,c)\in R} (r-\bar{r})(c-\bar{c})$$

Second-order column moment:

$$M_{cc} = \frac{1}{A} \sum_{(r,c)\in R} (c-\bar{c})^2$$

where r is the row; c is the column, R is the contour and A is the number of points in the contour.

In order to obtain the lengths of the major and minor axis and their orientation from the second-order moments, four cases are considered:

1. $$\begin{cases} M_{rc} = 0 \\ M_{rr} > M_{cc} \end{cases}$$

The major axis is oriented at an angle of - 90° counter-clockwise from the column axis and has a length of $4M_{rr}^{1/2}$. The minor axis is oriented at an angle of 0° counter-clockwise from the column axis and has a length of $4M_{cc}^{1/2}$.

2. $$\begin{cases} M_{rc} = 0 \\ M_{rr} < M_{cc} \end{cases}$$

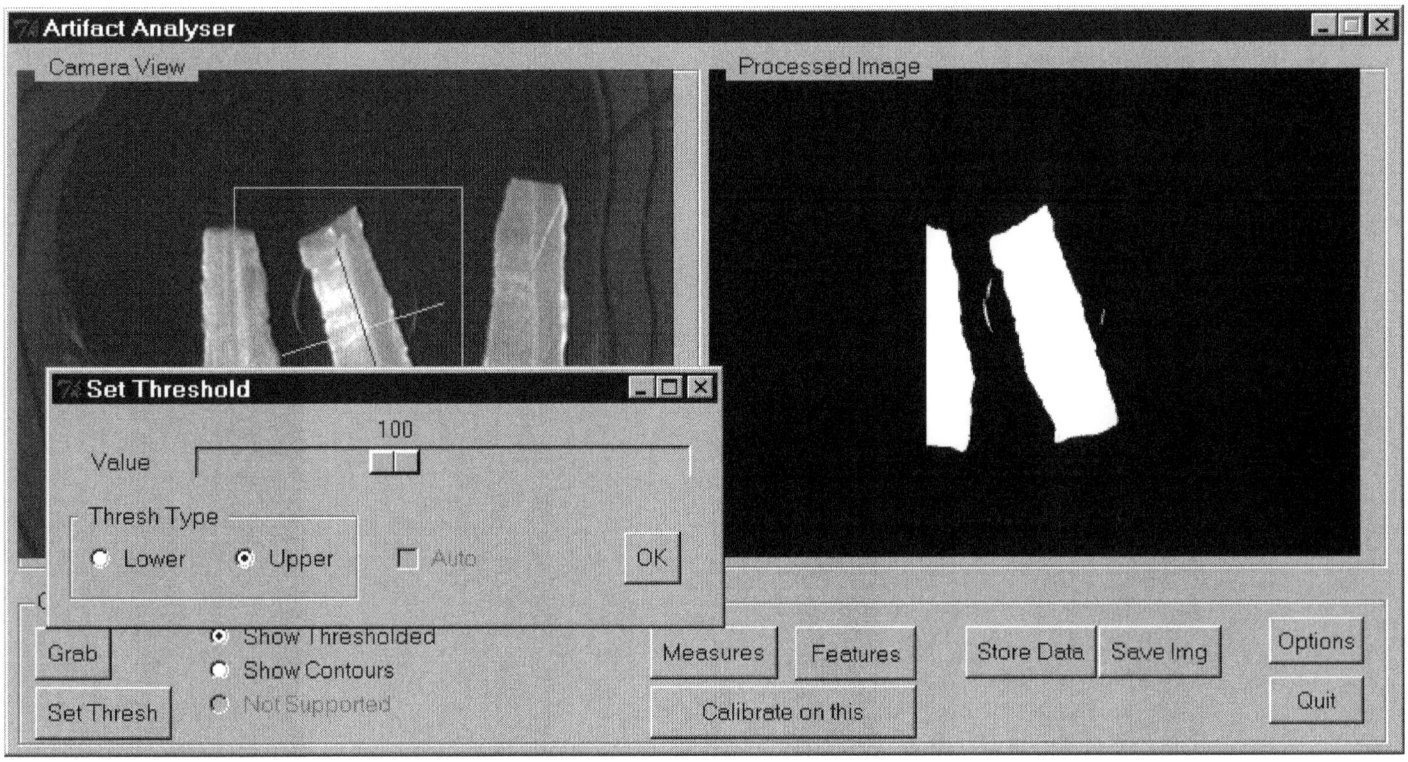

Figure 3: Image of the interface, showing the processing of the thresholding function.

The major axis is oriented at an angle of 0° counter-clockwise from the column axis and has a length of $4M_{cc}^{1/2}$. The minor axis is oriented at an angle of - 90° counter-clockwise from the column axis and has a length of $4M_{rr}^{1/2}$.

3. $\begin{cases} M_{rc} \neq 0 \\ M_{rr} \leq M_{cc} \end{cases}$

The major axis is oriented at an angle (between the orientation of the object related to the camera) of:

$$\alpha = \tan^{-1}\left[-2M_{rc} / \left(M_{rr} - M_{cc} + \sqrt{(M_{rr} - M_{cc})^2 + 4M_{rc}^2}\right)\right]$$

counter-clockwise with respect to the column axis and has a length of

$$\sqrt{8\left(M_{rr} + M_{cc} + \sqrt{(M_{rr} - M_{cc})^2 + 4M_{rc}^2}\right)}$$

The minor axis is oriented at an angle of 90° counter-clockwise from the major axis and has a length of:

$$\sqrt{8\left(M_{rr} + M_{cc} - \sqrt{(M_{rr} - M_{cc})^2 + 4M_{rc}^2}\right)}$$

4. $\begin{cases} M_{rc} \neq 0 \\ M_{rr} > M_{cc} \end{cases}$

The major axis is oriented at an angle of:

$$\alpha = \tan^{-1}\left[\left(\sqrt{M_{cc} + M_{rr} + \sqrt{(M_{cc} - M_{rr})^2 + 4M_{rc}^2}}\right) / 2M_{rc}\right]$$

counter-clockwise with respect to the column axis and has a length of

$$\sqrt{8\left(M_{rr} + M_{cc} + \sqrt{(M_{rr} - M_{cc})^2 + 4M_{rc}^2}\right)}$$

The minor axis is oriented at an angle of 90° counter-clockwise from the major axis and has a length of

$$\sqrt{8\left(M_{rr} + M_{cc} - \sqrt{(M_{rr} - M_{cc})^2 + 4M_{rc}^2}\right)}$$

4. Results

The results of the experiment are shown in the development of an interface (figure 4) that enables us to extract features such as contour and size of the object, as well as store data in a database, together with the image of the object itself. The entire process is almost fully automatised, in the sense that the person's task is only to place the object on the table, threshold the image and press the required buttons on the interface. It should be stressed that there is no special requirement for a specific position of the items to be analysed (the only requirement is that they are in the camera frame). The entire process does not take more than a few seconds, thus drastically increasing the acquisition process of attribute analysis. Moreover, the data is acquired by means of measurements taken by the developed program, and thus has an increased accuracy, and at the same time it is stored in a database available for general use. Thus, the operator's tasks are to place the objects to be analysed on the board, choose the object by framing it in the rectangle (in the camera view window), set the thresholding on the processed image window to the required size (until the contour is clear) and click on the appropriate buttons in order to obtain the required information. The measurements given on the processed image are translated into the metric system (millimetres)

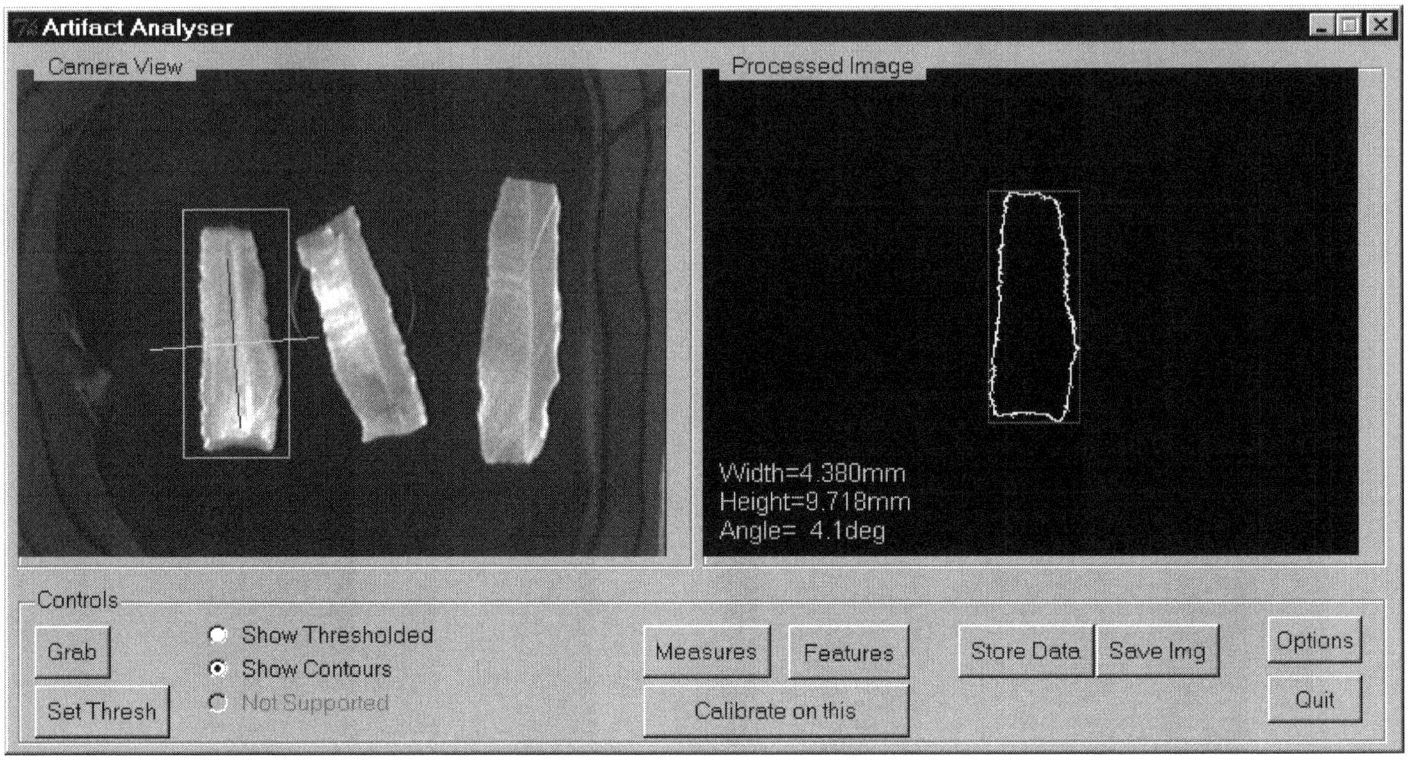

Figure 4: Image of the developed interface.

and stored in a database, which can be reached through any standard database program.

Further developments of the program will be total automatisation of thresholding, recognition of edge shapes and a more vigorous relation with the database. Even though it seems that the results obtained so far are modest from an archaeological point of view, the importance of this experiment lies in its nature itself, i.e. in the exploration of the application of image analysis to the technological interpretation of flint artefacts, or, in a wider view, to any object. The algorithm developed can be used for any object, no matter its nature, and can be adjusted for the needs of any final user. Thus, following this experiment, the potential and limitations of using image processing and analysis in the field of archaeological artefacts is better understood. Further developments will hopefully improve and expand the performance of the interface, making it more applicable from an archaeological scientific point of view.

References

BARCELÓ, J.A., BRIZ, I. and VILA, A., 1999. C*omputer Applications and Quantitative Methods in Archaeology*, BAR International Series 757, Oxford.

CLAY, B.R, 1976. Typological classification, attribute analysis and lithic variability. *Journal of Field Archaeology* 3: 303-311.

GERO, J., MAZZULLO, J., 1984. Analysis of artefact shape using Fourier series in closed form. *Journal of Field Archaeology* 11: 315-322.

GILEAD, I., 1995. *Grar A Chalcolithic Site in the Northern Negev.* Ben-Gurion University Press, Beer-Sheva.

GILEAD, I. and HERMON, S., in print. *The Flint assemblages of Abu Matar and Safadi.* CRFJ, Jerusalem.

INIZAN, M.L., ROCHE, H. and TIXIER, J., 1992. *Technology of Knapped Stone.* C.R.E.P., Meudon.

KARLIN, C., JULIEN, M., 1994. Prehistoric technology: a cognitive science? In Renfrew, C. and Zubrow, E.B.W. (eds.), *The Ancient Mind*, University Press, Cambridge: 152-165.

KITTLER, J., ILLINGWORTH, J., 1985. On threshold selection using clustering criteria. *IEEE Transactions on Systems, Man, and Cybernetics.* Vol. SMC-15: 652-655.

KITTLER, J., ILLINGWORTH, J., 1986. Minimum error thresholding. *Pattern Recognition vol.* 19: 41-47.

LEMONNIER, P., 1986. Anthropology as technical systems. *Journal of Anthropological Archaeology* 5: 147-187.

LEMONNIER, P., 1990. Topsy turvy techniques. Remarks on the social representation of techniques. *Archaeological Review of Cambridge* 9: 27-37.

LEVI-STRAUSS, C., 1976. *Structural anthropology*. Basic Books, New York.

PATTERSON, L.W., 1983. Criteria for determining the attributes of man-made lithics. *Journal of Field Archaeology* 10: 297-307.

PELEGRIN, J., 1991. Les savoir - faire: une tres longue histoire. *Terrain* 16: 106-113.

SCHLANGER, N., 1994. Mindful technology: unleashing the chaine operatoire for an archaeology of mind. In Renfrew, C. and Zubrow, E.B.W. (eds.), *The ancient mind*, University Press, Cambridge: 143-152.

SCHROEDER, W., KUNA, M., LORENSEN, B., 1998. *The Visualization Toolkit.* Prentice Hall, New Jersey.

Documents available on the World Wide Web:

www.bgu.ac.il
www.cineca.it
http://minos.cineca.it

3. National and Regional SMR

Transforming Diversity into Uniformity – Experiments with Meta-structures for Database Recording

Torsten Madsen

University of Aarhus, Dept. of Archaeology

Moesgaard 8270 Hojbjerg, Denmark

e-mail: farktm@moes.hum.aau.dk

1. Introduction

In the sixties a high-spirited optimism spread throughout many parts of archaeology. The advent of the digital computer and the promise of storage and handling of information on a grand scale had an impact on archaeologists. As recalled by Scholz and Chenall the prospect was to create "A framework or model for recording any site, site feature, artefact, or archaeological situation, in the form of a highly structured, though expandable, code for the descriptive attributes of function, form, material, technique of manufacture, surface treatment, and design" (Scholz and Chenhall 1976:92).

However, the optimism did not last for very long. In the mid seventies Scholz and Chenall noted "that a generalised data bank is not very useful for research purposes". Moreover it seems evident that a preliminary description of data is necessary to formulate ideas concerning the choice of variables, which need to be observed and scaled for any specific purpose. The time to record data on the computer would appear to be only after this preliminary descriptive step - i.e., when actual procedures and variables have been defined for testing specific hypotheses. Data categories for observation, and conventions for recording data cannot be chosen independently of problem orientation" (Scholtz and Chenhall 1976:92).

The lesson learned by then was that there is no path leading to a unified description of archaeological materials. Descriptions are bound to particular research problems, and as these are ever changing, so will descriptions. Over the last 25 years the interest for databases in archaeological research has been marginal. Many researchers have created specialised databases for recording particular materials. As soon as the results of the research have been published all interest in the databases and their content is lost.

On the other hand, administrative bodies associated with archaeology have systematically tried to develop the use of databases over the last 30 years. Their answer to the Tower of Babel problem of database recordings has been - standardisation. If only we can agree to the same description system, there will be no problem at all, they argue. The administrators have increasingly been able to agree, and increasingly lost contact with the realities of research.

On previous occasions I have argued that standardisation of content is a non-issue in archaeological research (Madsen 1998, 1999a). In the years to come we may find an increase in large-scale standardised and centralised administrative databases, but at the same time we will find an explosion in small local databases with their own unique structure serving the research efforts of individuals, projects or small organisations. Wherever research is carried out there will be a need for individually designed recording systems that will contain the base material for this research. The content of these databases, however, are in grave danger of being lost. As soon as the tasks for which they are created have been accomplished, they merge into oblivion.

One possible remedy for this inherent anarchy in data recording and consequent loss of data is to design a database structure that will literally encompass all other structures. A database designed in such a way that it does not represent any particular part of reality, but rather represents the way we model reality for the purpose of database recording using, for example entity-relationship modelling. What we are looking for is a meta-structure for database recording. One benefit of such a database is that accessing data can always happen through the same application interface no matter what particular data structure is actually at hand, and it will thus be infinitely easier for users to access data. Furthermore, time could be invested in creating efficient and powerful ways of searching and presenting data, because the investment would not apply to just one database instance, but to all instances.

Over the last three years I have experimented with a system called GUARD (the name possibly an acronym for General Utility Archaeological Recording Database). Its background lies in the IDEA project from the mid-nineties (Andresen and Madsen 1992, 1996a, 1996b). This project, carried out together with Jens Andresen, was aimed at creating a flexible database solution to excavation recording. By the end of the project, which was partly a success, we used the experiences gained to outline the principles for a meta-structure for database recording. Although the design has been modified considerably from what was originally conceived, it is these principles that I have used now to implement a working system.

2. The meta-structure design

The meta-structure I have chosen is not very complicated, but as the level of abstraction is high it may not be easy to grasp how it works at the first glimpse. However, if you keep in mind that what is implemented is more or less the entity-relationship model, it is not all that difficult. I have used six basic building blocks named *Entity types*, *Classes*, *Entities*, *Attributes*, *Entity Attribute Values* and *Relationship Attribute Values*. These are the entity types of the meta-structure (figure 1).

The reason for including *Classes* as a separate entity type may not be self-evident and has in fact caused much debate between Jens Andresen and I. Why separate *Classes* from *Entity Types?* For instance, if you take artefacts, these will surely constitute a basic Entity Type in any specialist's database aimed at artefact recording. However, the artefacts will also be qualified according to elaborate class-hierarchies, and different classes are likely to have

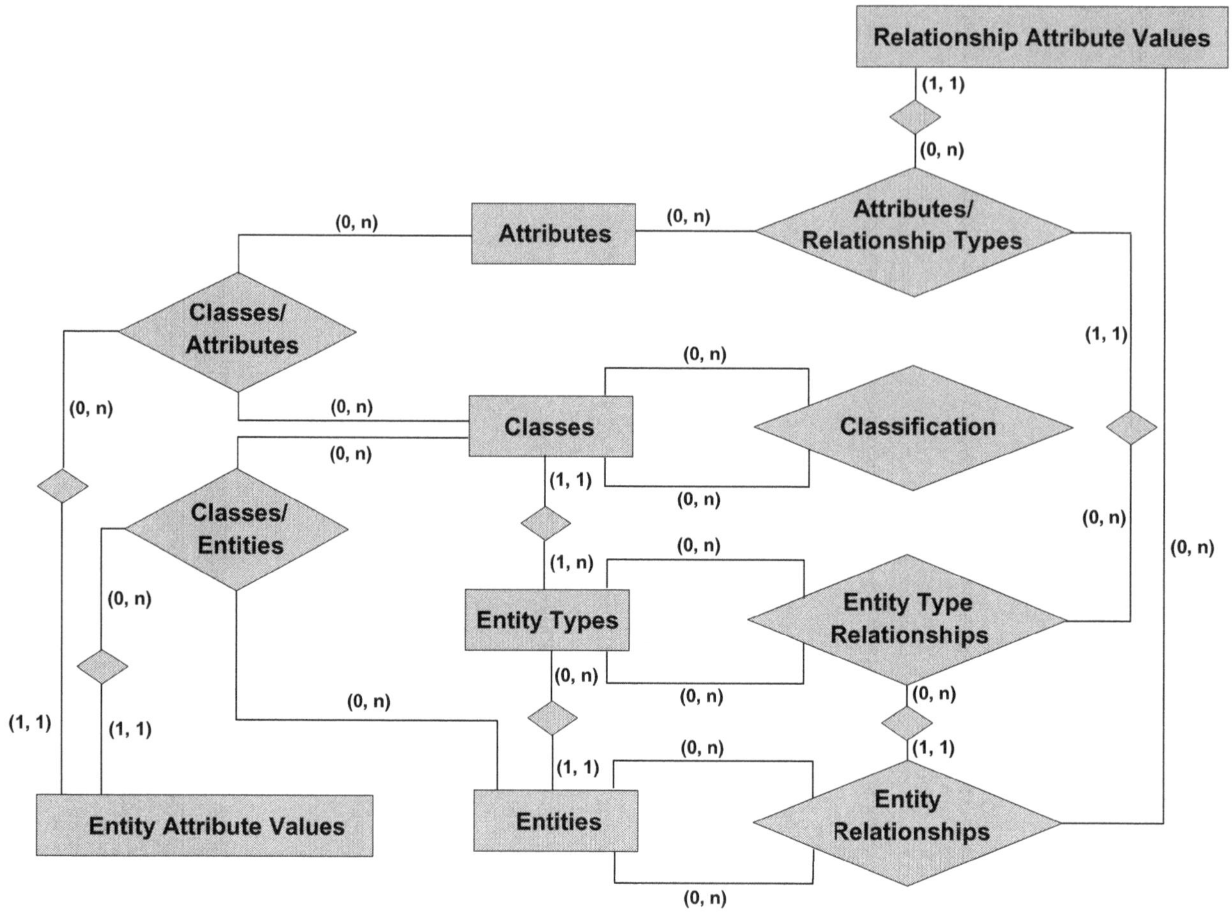

Figure 1: Entity-relationship model for GUARD.

different descriptive attributes. According to the standard database theory entities with different attributes should also be of different entity types. Ultimately, this would mean that each class could end up as its own entity type with a separate unique identification number, which of course would not agree with the way we normally use classifications.

To counter this problem *Classes* has been introduced as a kind of entity type that does not posses identification numbers – well it does, but the user is kept unaware. Otherwise it has all the qualities of entity types. For practical reasons it has become *Classes* and not *Entity Types* that have attributes associated with them. Thus any instance of *Entity Types* has at least one class called the root. If attributes are to be associated directly with an instance of *Entity Types* this will happen through its root class. In addition, any number of classes may be associated with an instance of *Entity Types* and each class may have its own set of attributes. Obviously, a relationship exists between *Entity Types* and *Classes* so that an instance of *Classes* can only exist if linked to an instance of *Entity Types*, and an instance of *Entity Types* must have at least one instance of *Classes* associated with it. Further, a relationship exists between *Entity Types* and *Entities* so that an instance of *Entities* can only exist if linked to an instance of *Entity types*.

Six named relationship types tie the six basic entity types of GUARD together. Three of these - *Entity Type Relationships*, *Classification* and *Entity Relationships* - are used to create an internal structure for *Entity Types*, *Classes* and *Entities* respectively, with a many to many cardinality between their entities.

The other three relationship types named *Classes/Entities*, *Classes/Attributes* and *Attributes/Relationship Types* are used to tie different entity types together with a many to many cardinality. Thus *Classes/Entities* makes it possible for an instance of *Entities* to be associated with many instances of *Classes*, and an instance of *Classes* to be associated with many instances of *Entities*, while *Classes/Attributes* makes it possible for an instance of *Attributes* to be associated with many instances of *Classes*, and an instance of *Classes* to be associated with many instances of *Attributes*. Together these two are then linked to the *Entity Attribute Values* entity type making it possible to assign a value to a unique combination of *Entities*, *Entity Types*, *Classes* and *Attributes*. The relationship type *Attributes/Relationship Types* makes it possible for an instance of *Attributes* to be associated with many instances of *Entity Type Relationships* and an instance of *Entity Type Relationships* to be associated with many instances of *Attributes*. *Attributes/Relationship Types* together with *Entity Relationships* are linked to the *Relationship Attribute Values* entity type making it possible to assign a value to a unique combination of *Attributes*, *Entity Type Relationships* and *Entity Relationships*.

The Entity-Relationship diagram of GUARD is thus fairly straightforward, and the same is true if we look at the table structure (figure 2). The only slight complication concerns the *Entity Attribute Values* and the *Relationship Attribute Values*. In both cases it is not just one table that is needed but as many tables as there are different data types involved. Each record in a table stores one value and that value is of one type only for that specific table. Thus, there are separate tables for text, double, integer, memo, etc.

Many questions probably spring to mind when confronted with a structure like this. Will it work at all? And if it works will it be

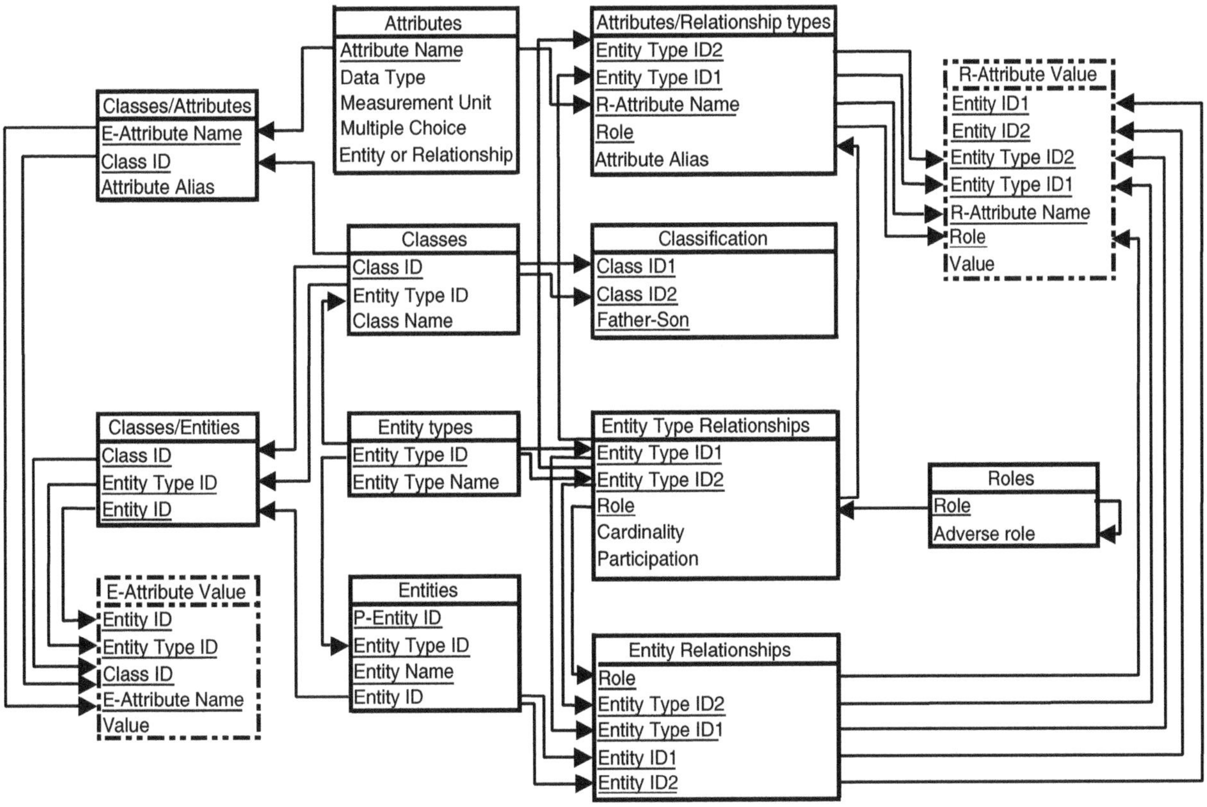

Figure 2: The central table structure of GUARD. Fields in primary keys are underscored. Tables drawn with dotted lines indicate a number of identical tables each holding values of specific data types.

painstakingly slow? Is it possible to create an intelligible user interface to this spaghetti of cross-referenced ID-numbers? Can you efficiently access and operate data, once it has been entered?

First of all it works. Surprisingly well, actually. It would be foolhardy to claim that I can depict any database design in GUARD, but I have tried it out on many different designs, and so far I have had no problems. In fact, database designs in archaeology are fairly simple, and I believe that the ones that will not fit will be few and far between.

Secondly, is it slow? In the beginning I feared this very much, but it does not seem to be the case. The potential problem stems from the way that every bit of information is atomised across many tables, and at the same time being overloaded with identification numbers. The latter surely has its price, as the size of any given database will be much larger in GUARD than in its original form. We are talking of a factor 3 or 4 in size. However, the numerous identity numbers, which are all indexed, help to maintain speed in searches. In fact the majority of searches associated with the user interface are done on individual tables and all searches are on indexed fields. The result is that the response time does not seem to grow significantly with the growing amounts of data. I have had the complete Danish SMR loaded on to GUARD taking up some 600 MB of space. In the *Entities* table there were more than one million entries, yet the result was merely a slight decrease in response time. Slowness, where it appears, is foremost related to the user interface itself and not to the table structure.

3. Design of the user interface

The user interface is the most complex part of GUARD. It is self evident that the abstraction level of the table structure does not allow forms to be drawn up in the usual straightforward way. It is necessary to write fairly extensive pieces of code to make it work. This is where the time investment in GUARD lies, but then of course the interface will work with any database design implemented.

Being a meta-structure GUARD has no inherent recording structure. A new database cannot record data before it has been structured. The structuring elements are themselves data, and they have to be entered before "real data" can be recorded. Hence the user interface has a number of forms through which it is possible to define entity types, relationship types, class structures, variables, lookup lists as well as setting links between variables and classes. To define a recording database in this way is very challenging and instructive. You have to be explicit in all your choices.

The main entry form presents you with the basic elements of GUARD in a straightforward way (figure 3). In the upper left-hand control the entity types are found. You only see those entity type relationships that have an existence dependency between them - shown as hierarchies in the control. The example in figure 3 is the Danish SMR with a county ("Amt"), district (Herred), parish ("Sogn") and site number ("SB Nummer") breakdown, and various entity types related to the site number. At present you cannot see if a relationship type exists between say "event" and "objects", but if it has been defined you can open a form through which you can link instances of "event" with instances of "objects" using the relationship type.

In the lower left-hand control the entities are found. When we have a dependency structure as the one just described, we will also have dependency among the entities, where a number of entities will be the offspring of one entity of the independent entity type. Selecting an entity type thus sets a selection path only. To

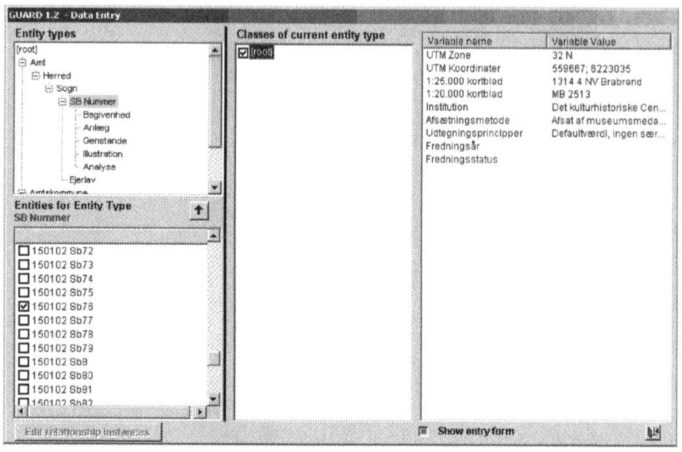

Figure 3: The main form of GUARD to access the data structure for data entry and editing.

reach a particular entity of the selected entity type we also have to traverse the hierarchical structure of the entities. Ideally this can be done in a tree view control as the one used for the entity types, but since the number of entities typically will run into thousands, the time it will take to build up the content of this control would be devastating. Instead a list view control has been chosen, where you start at the bottom of the entity type selection path, and from where you can select your way through the entity hierarchy by double clicking an entity, which will give you access to the next level, etc.

In the centre of the form the class structure of the current entity type is shown. In the example only the root class is present, but there might have been an extended hierarchy class present. In the rightmost control the attributes (here called variables) of the current class is shown, and if values have been recorded for the attributes these are shown as well.

You can enter and edit data in a form based on the attribute settings for a particular class (figure 4). The input form is highly standardised, and may appear rather primitive in its appearance. The problem is that there is no way of knowing in advance, what variables, and thus what data entry fields are needed, not only from database application to database application, but indeed from entity type to entity type, and from class to class. The solution is to use a form containing a large number of data entry fields related to the different data types available, organised into pages by way of a tab control. Whenever a class is selected the proper number and types of entry fields are activated, and their record sources are set to the proper slots in the database, while all unused fields and unused pages of the control are hidden. This all-enclosing data entry form works quite well despite its appearance.

From the main form you may call up a form in which to set entity relationships between the current entities and other entities (figure 5). In the actual example we are about to set this is represented by the institution responsible for recording the position of the current site. We have chosen the entity type "Institution", we have chosen the action or role "has been marked by" and we can now choose the institution and link it. If there are attributes defined for the entity type these will appear in the lower right-hand control of the form, where they can be edited. The entity relationships that may be created between entity types can be of one to one, one to many or many to many cardinality. Naturally, relationship types within individual entity types may also be established.

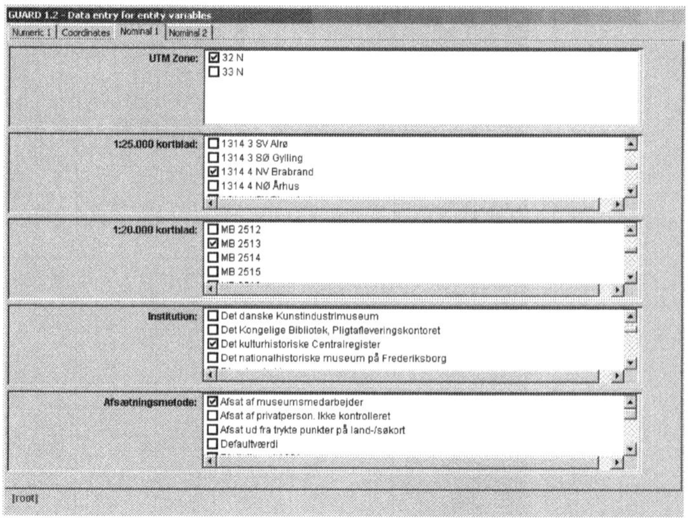

Figure 4: Pop-up form for data entry and editing values for entity variables.

Thus an intelligible and usable interface can be created. The same applies to search procedures whether for tabular output or for reports. Various experiments have been carried out and a standardised user interface for data search is currently being developed. It is too early, however, to say exactly how this interface will be organised.

4. Prospects of GUARD

GUARD may be used directly as a "production" database, or it may be used as a repository for already existing databases. My experience so far shows that it is fairly easy to migrate data from an existing database into GUARD. To do so you must be familiar, however, with the table structure in GUARD, and most of the migration process has to be run from VBA code modules. It will be possible to write some standard procedures and functions, but in each specific case it will be necessary to stitch them together according to the specific structure of the database we wish to import. Most standard research databases can be migrated with less than a day's work, which is not very much if you consider the benefits gained from the common table structure. In the future it will be easier for the curators of these databases, and it will also be easier for future users to access the data, when all data-sets are accessed in exactly the same manner irrespective of their structure.

GUARD may thus be well suited for archival purposes, but the reasoning behind its development never was archival purposes. Basically, it was to counter the tendencies to dictate data standards in archaeology. I have previously argued that we may seek standards of form, which is exactly what GUARD does, but never standards of content (Madsen 1998, 1999a). We need our recording systems to vary according to the issues and problems. Research carried out according to predetermined common standards of what should be recorded and how it should be recorded will soon cease to be research. It will be mere reproduction.

Thus GUARD is aimed at making archaeological recordings flexible and versatile. A system that will allow users to create powerful database solutions with little effort and allow them to redesign at least part of the recording structure on the fly, as recordings progress, has great research potentials. At the same time, as the

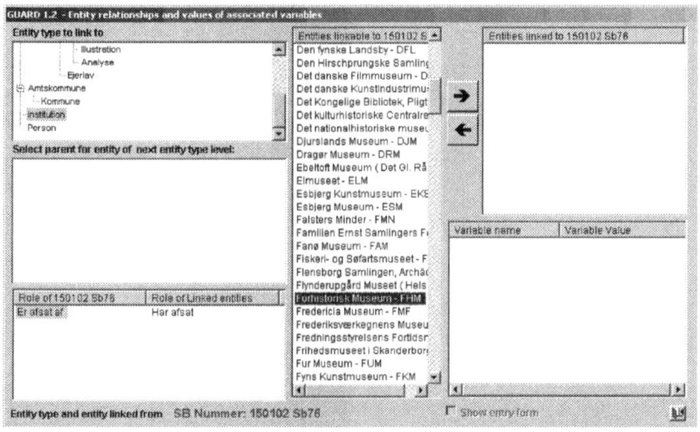

Figure 5: Pop-up form for setting entity relationship links, and to enter and edit associated relationship type variables.

structure of the database remains the same regardless of the structure and content of the recordings, we will hopefully be able to shut the mouths of those who claim that the only way to achieve compatibility is through conformity.

In developing GUARD for research purposes, special attention was paid to the potentials of using proper classification systems in connection with entity types. In most current applications, classification is a matter of choosing one class from a list of alternatives. Thus an entity can have a class assigned to it from a set of classes that are all at the same level. This is obviously not satisfactory. In GUARD a classification can consist of a hierarchical tree-structure with one root and as many branches and levels as needed. Classifications are thus of a classic monothetic divisive nature. At the moment I stick to this because, whereas the table structure allows polythetic classifications, I have so far found no satisfactory way to implement a polythetic structure in the user interface.

Within any classification tree you may choose to assign only one class to an entity, or as many classes as you wish. Further, you may operate with as many parallel classification trees as you wish. When, in the design phase, you assign an attribute to a class, all classes in the branches of the tree structure above it will automatically inherit this attribute. In this way it is ensured that all more specialised classes will hold the basic attributes of their parent class apart from what ever more specialised attributes they may posses themselves. If you do not want a class to share attributes with its parent class you may of course remove the attributes from it.

Classifications are operational. That is, when searching for data you may use the position in the classification tree as a parameter in you searches. You can specify that you want all of this class only, or all of this class and all classes lying in branches above it in the classification tree.

Another area that I have given some consideration to is the possibility of using the *Entity Type Relationship* and *Entity Relationships* to depict complex relationships. It is mostly relationships between entities of the same entity type that are of interest. Archaeology is full of such relationships. One example could be contexts with contexts in an archaeological excavation leading to statigraphy and the use of Harris Matrices to solve the complexity of this. Another example could be the fitting together of artefacts like pottery and flints and the use of graphs to elucidate the refitting patterns. A third example could be the association of decorative elements with each other leading to composition patterns. Recently I have used GUARD in my work with the latter problem, and I found a clear potential for systematic studies within an area that traditionally has been considered very difficult to handle (Madsen 1999b).

GUARD is a general utility system. Its primary aim is to enhance the quality of descriptive practice in archaeology. It offers a tool that gives the researcher a better chance of creating a description and recording system in accordance with the complexity of the problems at hand. The database solution attained, however, is one of a far greater scope than just providing flexible recording. The use of a meta-structure design points towards a standardisation of form in recording systems, detached from content, and hence it points towards the area of archival standards. I do not intend to suggest that GUARD is *the* solution, but I think it is an important step in the direction we need to take if we do not wish for everything to end up in either rigid conformity dictated by administrators on the one hand, or chaotic *ad hoc* use of databases on a primitive level and with a severe loss of information on the other.

References

ANDRESEN, J. and MADSEN, T., 1992. Data Structures for Excavation Recording. A Case of complex Information Management. In Larsen, C.U. (ed.), *Sites and Monuments. National Archaeological Records*. The National Museum of Denmark: 49-67

ANDRESEN, J. and MADSEN, T., 1996a. IDEA – the Integrated Database for Excavation Analysis. In Kamermans, H. and Fennema, K. (eds.), *Interfacing the Past. Computer Applications and Quantitative Methods in Archaeology CAA95*. Analecta Praehistorica Leidensia 28, Leiden: 3-14.

ANDRESEN, J. and MADSEN, T., 1996b. Dynamic classification and description in the IDEA. III International Symposium on Computing and Archaeology. *Archeologia e Calcolatori* 7: 591-602.

MADSEN, T., 1998. Digitaliseret registrering af udgravninger – er der brug for datastandarder og fælles strategier? Hansen, H.J. and Ødegaard, V. (ed.), *De Nordiske Museer og Informationsteknologien – rapport fra en konference 1.-3- december 1996*. TemaNord 1998: 513, Nordisk Ministerråd, København: 167-177.

MADSEN, T., 1999a. Digital recording of excavations: Do we need data standards and common strategies? Hansen, H.J. and Quine, G. (eds.), *Our Fragile Heritage. Documenting the Past for the Future.* København: 131-138.

MADSEN, T., 1999b. Coping with Complexity. Towards a formalised methodology of contextual archaeology. *Archeologia e Calcolatori* 10, 1999: 125-144

SCHOLTZ, S., CHENHALL, R.G., 1976. Archaeological Data Banks in Theory and Practice. *American Antiquity*, Vol 41, no 1: 89-96.

Archaeological Applications of Fuzzy Databases

Franco Niccolucci

University of Florence

Via Pisana 56, 50143 Firenze, Italy

e-mail: niccolucci@unifi.it

Andrea D'Andrea

CISA – Istituto Universitario Orientale

V.tto I, S. Maria ad Agnone 8, 80139 Napoli, Italy

e-mail: dandrea@iuo.it

Marco Crescioli

Unirel SRL

Via Volturno 12, 50019 Sesto Fiorentino (FI), Italy

e-mail: marco@unirel.it

Abstract

This paper deals with problems concerning statistical data (e.g. deriving from archaeometry) in an archaeological database, when some unsuspecting application may lead to erroneous conclusions. A new model is proposed for these cases, using fuzzy logic to assign a reliability coefficient to imprecise attributes. Considering a case study, we generalize the assignment of age, gender and chronology to burials. The procedures are general and can be fruitfully used also in other investigations. To manage these fuzzy attributes we personalized a free Relational Database Management Systems (RDBMS) and created a WWW interface to ease data consultation and allow remote access.

Key words: fuzzy set, fuzzy database, excavation database, necropolis, Etruscan

1. Quantitative applications and archaeological theory

In a recent paper (Barceló 2000), J. Barceló wisely pointed out that computer applications in archaeology have reached an elevated level of complexity, often characterized by sophisticated and expensive techniques; and yet such resources are still not fully exploited for their investigative potential, notwithstanding the goals achieved especially in spatial technologies and virtual reality applications. For the Spanish scholar, the sparse use of these advanced computer technologies in archaeological research derives from the fact that we are not able to ask questions complex enough for such complex instruments. Consequently, archaeological results are still lacking.

Pursuing the application of the most recent hardware solutions in archaeology, as well as the most promising software developments, produced by research or by the market, technical systems, which are efficient and reliable but are not accompanied by an adequate level of theoretical and methodological reflection, are often generated.

A preoccupying trivialisation is often hidden behind a shiny technological apparatus; it is caused by the absence of reflection on the impact of the use of advanced technology on the process of historical knowledge. However, critical elements pervade the archaeological use of virtual reality and emerge also towards intersite GIS systems, oriented only to environmental variables and therefore deterministically biased. Stating the importance of the connection between the improvement of computer applications and archaeological research, Harris and Lock pointed out that a GIS system is not impartial or neutral: it

> "represents the social reproduction of knowledge and, as such, the development of a GIS methodology cannot be divorced from the development of the theory needed to sustain it" (Harris and Lock 1995:355).

Very recently, similar attention to a correct definition of the correlation between techniques and interpretative processes seems to characterize also mathematical and statistical applications. For a long time these have represented the main quantitative application in archaeology; and now, after a decline resulting from the crisis of the processual approach, which had represented their theoretical and methodological basis (Moscati 1996), they seem to be undergoing a new growth.

The recent contribution of quantitative methods (Buck et al. 1996, Delicado 1999), such as non-parametric statistics, Bayesian statistics and fuzzy theory, certainly helped to invert the negative trend that had characterized quantitative archaeology around the middle of the 1980s under the influence of post-processual criticism (see, for instance, Hodder 1982). Perhaps the negative reaction to the use of statistics to support interpretation may have generated a new relationship between archaeology and mathematics.

A new quantitative approach is based on evaluating the impact of statistical-mathematical models while carrying out archaeological research (Moscati 1996, Voorrips 1996, Wilcock 1999), not only as far as data analysis and classification are concerned, but also in formalizing procedures and in the use of statistical sampling techniques. Thus, the post-processualist image of the computer as a neutral instrument sides with the New Archaeology vision of it as an objective meter of historical and human facts and behaviours: both these approaches, only apparently opposing, in fact converge to the same negation of the importance of computers in archaeological theory and method. Regarding the deep connection between computational methods and their impact on archaeological theory leads inevitably to a cul-de-sac: the blind pursuit of the "discovery" of "innovation" and novelty without understanding the function of the proposed solution in the process of historical and archaeological investigation, and the consequent inability to bypass the proposal of toys, which so often are as expensive as they are useless.

Hopefully, a different approach from the one we suggest for fuzzy theory may represent a useful step for foreseeing a new and more promising relationship between archaeological theory and practice, as well as the use of models deriving from other disciplines (Crescioli et al. 2000). In our opinion, it is not correct to choose a quantitative technique only because it seems to correspond better with the current investigation; this attitude inevitably produces a mechanic and unreflective application of quantitative techniques that may lead to erroneous conclusions. By choosing a technique, we must bear in mind that we are thus making a cognitive choice, which will reflect on data and results. Fuzzy theory continuously reminds us that during an investigation we make choices that are determinant to formalize data, but they leave no sign in the interpretative process, so that raw data and hypothetical or reconstructed information become inseparable: the more the formalism used for data analysis is hidden, as in computer applications and, in particular, in database applications, the bigger the risk of overwhelming the original information content of data with the subjective meaning of interpretation.

2. Databases and archaeological theory

The huge amount of data that characterize any archaeological investigation and the pervasive presence of computers in every aspect of present life have ultimately led to a generalized use of DBMS's (Data Base Management Systems) for managing excavation data, as well as any other kind of archaeological records. Nowadays, it appears quite natural to store and search for archaeological data in the memory of a computer, due to the highly structured nature of forms used to record them, a condition that perhaps precedes the advent of computers but certainly is enforced by their use. These tools undoubtedly serve a great purpose in easing archaeological data management and the synthesis process, so that nowadays even the most conservative educational institutions can no longer exclude database training from archaeologists' curricula. Using DBMS has thus become a part of the current archaeological practice and little attention is therefore paid to its implications on the correctness of data. On occasion, this is due to excessive confidence in automatic processing, while sometimes it is the ignorance of simple statistical laws concerning error propagation that may induce false conclusions; moreover, these very conclusions have the aspect of indisputable truth, since a machine, which by assumption makes no mistake in computations, produces them. After they have been recorded into a database, archaeological records lose any element of uncertainty and subjectivity and become as trustworthy as the computer itself.

This consideration should not imply a luddite rejection of computers, which by the way are not guilty of such erroneous results, but simply the awareness that computations based on uncertain data follow rules that differ from ordinary ones, with or without a computer. Even the simple act of counting is no longer the same. In other words, since archaeological data have an intrinsic uncertainty, any conclusion drawn on their foundation cannot ignore elementary statistical rules, including the paradox that 1 + 1 does not always make 2.

In fact, every time you recognize something there is some uncertainty in the attribution. And in repeating this process several times, as occurs for instance when classifying archaeological finds, errors associated to each item add up, producing a total error that may be unacceptable in certain cases.

In most cases one can ignore this feature, because the error is so small that deterministic rules and statistical rules in practice do not differ. However, this should not be taken for granted in every case.

This reasoning has particular implications when using a DBMS to record data. Usually, checking boxes or filling fields according to standardized dictionaries accomplishes this, and no space remains for uncertainty or doubts. One has to decide to cross the box labelled "black" or the one labelled "white", with no possibility of grey. Then *alea iacta est*, the die is cast, and that choice will forever obliterate the real archaeological record and will be processed with many similar ones, possibly thousands of them, as it happens when managing finds from an excavation. The computer, in its cold assurance, will keep no track of the archaeologist's human hesitation.

Thus the subjective attribution is unconsciously objectified and disparate levels of reliability are equalized to absolute certainty by the magic of computers. Should we not introduce a warning that some of the data are "more subjective" than others, including even the archaeologist who originally interpreted them and trusted them at different degrees? Probably yes. And it is a common practice to mark less reliable attributions with question marks. But interrogation marks are difficult to process, and in no way are they supported by DBMS. So our proposal aims at introducing some attributes that make the reliability of data evident, as well as a few simple and transparent rules to process them.

Ignoring the problem of data reliability is still worse when they are derived from statistical processing. This happens when archaeology uses the results of other scientific techniques, as in archaeometry; our case study will illustrate one such example.

In conclusion, databases are very useful for recording archaeological data, and using them in everyday archaeological practice is an achievement that need be not discussed. But a naive usage may lead, in certain circumstances, to incorrect conclusions; this can be prevented with some simple technical improvements. Our contribution hopefully moves towards this perspective by simply quantifying (in an absolute subjective way) how much the compiler of the database trusted the data, and consequently, by giving some reasonable rules to process this reliability coefficient through all the computations for which the database is used. It must be pointed out that the numeric nature of this reliability coefficient should in no way be interpreted as an "objective" measure of the uncertainty, rather only as an expression of the compiler's reliability regarding subjective evaluation. Therefore, the meaning of different numeric values should be clearly stated in the accompanying documentation, as well as how they are computed when the coefficient derives from computation, as it will happen in our case study.

This approach, together with other practices, such as the generalized disclosure of archaeological databases to the public, will further contribute towards guaranteeing the correctness of application of the scientific method, which necessitates the possibility of backtracking, at least in theory, and the inference of results from data beyond the "black box" of the database.

3. The case study

The present paper considers the data resulting from a sample of burials discovered in the cemetery of Pontecagnano, an important Etruscan-Campagnian settlement situated about 70 km south of Naples. The funerary area, extending below the modern town centre, produced in over forty years' investigations more than 100 burial nucleuses and more than 8000 tombs dating between the First Iron Age (9th century BC) and the Hellenistic period (beginning of the 3rd century BC). To manage the enormous quantity of finds, the archaeological team is carrying on a GIS project lasting already a few years (D'Andrea 1999). The project consists of a cartographic database, implemented with Mapinfo, whose main function is to exactly position the ancient remains on modern cartography and to store topologic, spatial and alphanumeric data regarding each tomb and burial area.

The burials examined in the present paper pertain to the most recent phases of use of the Etruscan-Campagnian cemetery. Serritella (Serritella 1995) edited these burials in a volume, which includes a philological study of the grave goods, an analysis of the most significant pottery production and, above all, a reconstruction of the ancient community of Pontecagnano during the 4th and 3rd centuries BC, starting with an analysis of funerary customs. The tombs studied by Serritella are distinct from the remainder of the cemetery and are situated in free areas that were not occupied in previous periods, thus constituting a privileged observatory for studying the society of the Hellenistic period. Of all the tombs, 65 % revealed grave goods from within, while the remaining 35 % did not, including about 7 % which had certainly been violated already in ancient times.

In order to examine funerary behaviours, the author uses the analysis of pottery and burial typology, as well as the results obtained - concerning the gender and age of the deceased - by classical anthropometric methods (Scarsini and Bigazzi 1995, Petrone 1995). These are based on statistical values that may be obtained using different procedures, resulting with a variety of numeric coefficients. In particular, the gender coefficients vary within a range of +2 and –2: positive values refer to male gender, the negative ones to female. Unfortunately, most of the values do not reach these extremes: only 11 are outside (–1, +1), that is about 20 % of the cases, for which an osteological coefficient can be evaluated and 13 % of all cases. So in most cases, the level of uncertainty is rather high.

Notwithstanding the uncertainty of the palaeo-anthropological results, obtained with a statistical computation applied to the dimensions of each skeleton, the tables that compare grave goods, gender and age, so as to reconstruct the horizontal stratigraphy (age classes) and the vertical stratigraphy (social status) of burial areas, do not show the variability of anthropological determinations. So the statistical information turns into certain data.

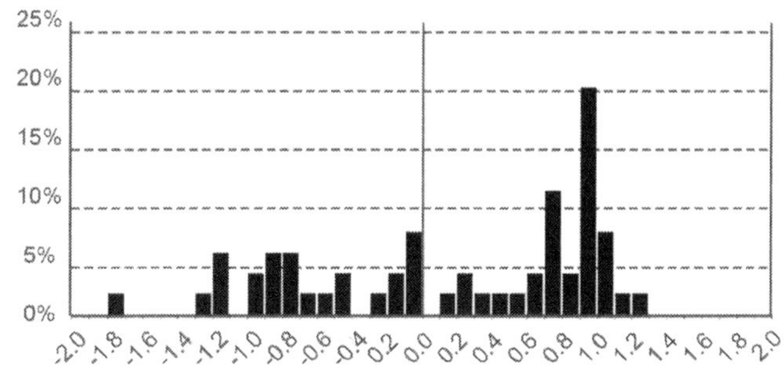

Figure 1: Frequencies of gender coefficients.

Correctly, the publication provides all the details of the anthropological analysis so that the reader may check the scientific results; but all this is irreparably lost when data are stored in a database.

To circumvent this drawback, our proposal suggests the use of statistical information already available, by creating special attributes and showing how to process them, so as to keep the coefficient variability within the data structure.

This being our goal, we have analysed the frequency distribution of the osteological coefficients obtained in the case study, dividing [-2, 2] into intervals of length 0.1. A bi-modal distribution, with peaks corresponding to the two most frequent values denoting males and females, is anticipated. The histogram of this frequency distribution is shown in figure 1.

As can be easily verified in figure 1, the distribution of gender coefficients is only roughly bimodal: the male modal value is +1, while female coefficients have no mode. Moreover, the total does not reach 30 %, even when adding the frequencies of modal values and the nearest neighbours.

Presumably, such characteristics may be influenced by the choice of the interval width: using 0.2 instead of 0.1 gives a better double-bell-shaped curve (with still low frequencies of the modal values). However, it confirms that discriminating the gender by means of osteological coefficients is not a straightforward task.

4. Fuzzy set concepts

We shall not go into detail here regarding fuzzy theory; further details are attainable in Crescioli et al. (2000) and the bibliography included. Let it suffice that, given a set A, a *fuzzy entity* is the couple formed by a variable X having values x in A and a function f_X from A to [0, 1]. Hence, to any instance x of X, a number $f_X(x)$ in [0, 1] is associated, which can be interpreted as the degree of reliability of x, and will henceforth be named the *(fuzzy) reliability coefficient* attached to x, while f will be named the *(fuzzy) reliability function.* So, a fuzzy entity extends the concept of any variable by adding these reliability coefficients.

In particular, a *fuzzy label* is such a couple, the first one assuming nominal values (the labels). For instance, fuzzy gender is a fuzzy label with the nominal values "male" and "female"; each one has a number attached, which represents the fuzzy reliability coefficient of the assignment.

A *fuzzy value* is another kind of fuzzy entity, in which the first element of the couple, the variable, has a numeric range. Fuzzy age is such, being formed by a possible range of ages, each one having a corresponding fuzzy reliability coefficient.

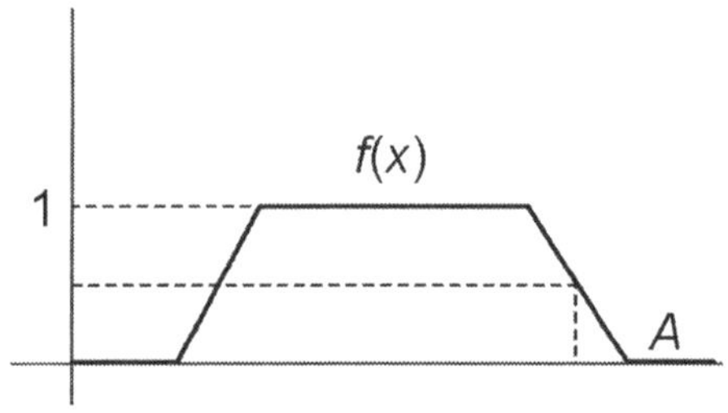

Figure 2: Graph of a trapezoidal fuzzy reliability function f.

Fuzzy labels can be fully described as arrays, where the labels are set in the first column and the corresponding reliability coefficients in the second.

Fuzzy values can be represented in the same way if the range of possible values is finite; otherwise, a function from A to [0, 1] is needed. A typical form of this function is trapezoidal, as shown in figure 2.

The concept of equality also needs an extension to operate with fuzzy entities.

We first introduce the *similarity* $s(x)$ between the fuzzy entities X and Y, respectively with fuzzy reliability functions f, g, defined over the same domain A, which is the (numerical) function

$$s : A \to [0,1] : s(x) = \min(f(x), g(x)), x \in A.$$

Shown in figure 3 is a graphical representation of $s(x)$, in which it is assumed that X and Y are fuzzy values so that A is a numerical set, and both f and g have a very simple, trapezoidal form.

To globally compare the two fuzzy entities X and Y, the maximum of s over A is taken: thus we define a *fuzzy operator*, that is, a function associating a number in [0, 1] to each couple of fuzzy entities. We shall use the symbol ~ to denote this operator, which is called the *fuzzy equal*. In the previous picture, the value of $X \sim Y$ is given by the ordinate of the marked point in figure 3.

The rationale of this definition depends on the interpretation of the fuzzy reliability function: taking the minimum of the two functions means that for each possible value in A we consider the worst condition for each fuzzy entity. However, speaking globally, the most likely situation corresponds to the greatest of these values. The equality of the two items may derive from their being both "male", or both "female". So the reliability of the equality, regardless of what case determines it, is larger than the reliability of each single case, which adds credibility to the overall reliability. A prudent approach would, for the reliability of each case, establish the minimum of the two reliability coefficients, and the overall reliability would equal the greater of the two, with no additional contribution from the other. For another example, consider two disjoint age intervals X and Y. Strictly speaking, there is no equality between them, since the parts in which they have a reliability of 1 are disjoint; but both have overlapping tails in which they are less likely, however not impossible. The common value where they have the highest likelihood is the point marked in figure 3.

The definition of fuzzy equality is an example of generalization to fuzzy entities that are of familiar concepts, such as equality, counting, adding, averaging, and so on. We are not going to deal with these concepts any more: this paper shall only approach the counting of fuzzy quantities. To count occurrences, that is, to compute frequencies, we need to generalize the familiar operation of

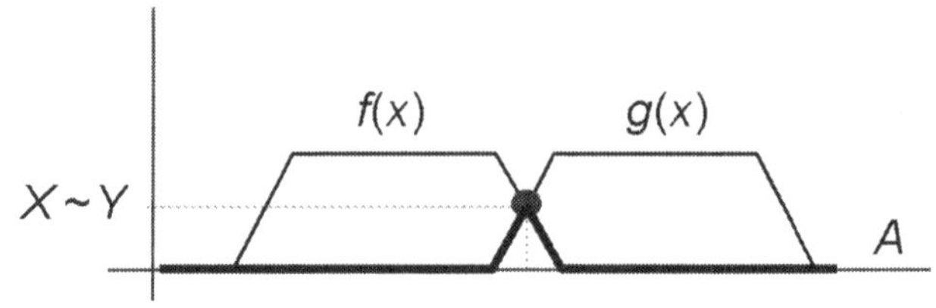

Figure 3. Graph of similarity function s *(heavy line) and the value of* X ~ Y *(marked point).*

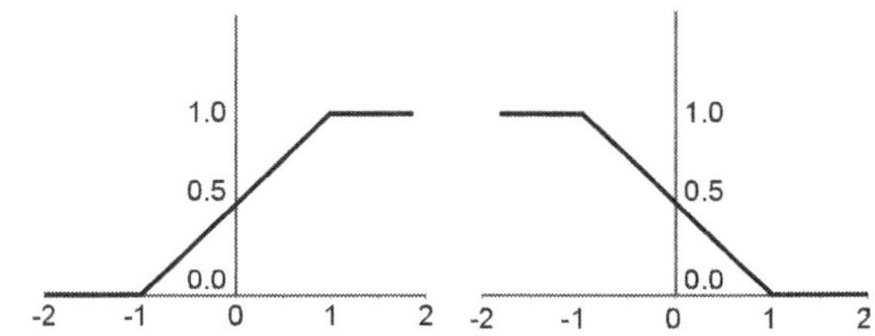

Figure 4: Derivation of m *(left, ordinates) and* f *(right, ordinates) from* k *(abscissa).*

counting: adding one when the desired result comes out (for instance, "female" when counting gender occurrences) and oppositely, adding zero when it does not (that is, the result is "male"). In our generalized model, we shall total the fuzzy coefficients for each case, so that the count of each possible outcome will be the sum of the fuzzy coefficients. This is in accord with common sense weighing in average evaluation and furthermore, also represents a particular case of the more general "Extension principle" (see Yager and Filev 1994:16-18).

5. Fuzzy entities in the case study

Three attributes have been recognized as fuzzy entities in the case study: gender and age of the deceased, burial in a tomb, and the chronology of the burial.

Gender may be considered a fuzzy label, as stated before, while age and chronology are fuzzy values. For each one of these fuzzy entities we shall briefly explain how to evaluate the second member of the couple, the reliability coefficient. For fuzzy gender, this will imply the evaluation of two numbers: one for each gender, based on the osteological coefficient. The other two attributes, however, require the definition of a function, as shown below.

There are several (in fact, infinitely many) possible ways to assign numerical values to the gender coefficients. The ones we chose are based on the following considerations:

- In this case study, few osteological coefficients (less than 20 %) go beyond +1 or –1; this can be considered the best possible result in these conditions.
- When the male coefficient gets the highest value, the female one should get the lowest, and vice versa.
- When the osteological coefficient varies between –1 and +1, the corresponding fuzzy gender coefficient increases (or decreases) uniformly.

So, denoting the osteological coefficient by k and the male and female corresponding gender coefficients by m and f respectively, to derive m and f from k we can build the function shown in figure 4.

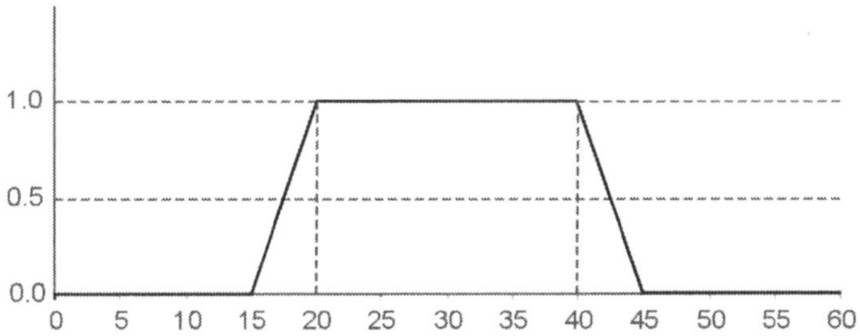

Figure 5: Fuzzy age coefficient (example for the age range 20 – 40).

$$f(k)=\begin{cases}0 & \text{per } k \le \mu-1.5\sigma \\ 1-\dfrac{(\mu-\sigma)-k}{\sigma/2} & \text{per } \mu-1.5\sigma < k \le \mu-\sigma \\ 1 & \text{per } \mu\text{-}\sigma < k \le \mu+\sigma \\ 1-\dfrac{k-(\mu+\sigma)}{\sigma/2} & \text{per } \mu+\sigma < k \le \mu+1.5\sigma \\ 0 & \text{per } \mu+1.5\sigma < k \end{cases}$$

Figure 6: Fuzzy coefficient f *as a function of the mean* μ *and the standard deviation* σ *of the estimated osteological age.*

Thus we are able to obtain the fuzzy gender coefficients for each item from the value of its osteological coefficient. The resulting fuzzy gender attribute will be an array, as already noted. For instance: {(male, 0.8), (female, 0.3)}.

Notice that, even if there is, in general, no mutual dependence between *m* and *f*, the definition we chose implies that $m + f = 1$.

There are some cases in which the value of *k* remains undefined, since there were not enough elements to apply the osteological method. In these cases, our choice is to assign a value of 0.5 to each of the gender coefficients, that is $m = f = 0.5$. This assignment is based on the fact that the gender of the deceased is undecided regarding the known elements; hence, both gender are equally likely (or unlikely). The disadvantage lies in that the difference, if any, between the case of $k = 0$ and *k* not computable is lost; so one might prefer a different assignment as, for instance, $m = f = 0$. We believe that there is no relevant loss of information and furthermore, the method applies with both choices; so the one we adopted will have no consequence on the model's validity.

The osteological determination of age ranges is based on two different methods, in several cases producing conflicting results; for instance, in the case of tomb 4046 (Scarsini and Bigazzi 1995:139, pl.1a), for which the two estimates are 50 ± 2 years and 20 – 25 years respectively.

While the results of the second method are given in the form of a range with no other information included, the authors cited for the first one (Scarsini and Bigazzi 1995, 139-140), the central value *m* and the standard deviation *s*. So it is reasonable to assume that estimated ages have a Gaussian distribution, with mean *m* and standard deviation *s* as presented in the paper. Since the area below the Gaussian curve between $m - s$ and $m + s$ equals 68% of the total area (see any text on statistics, for instance Mood, Graybill and Boes 1979), we may conclude that the estimated age values between $m - s$ and $m + s$ are the most probable and hence the most reliable, with a tail on both sides, having a lower probability. In terms of reliability of the result, we may therefore assume that this is the highest for the values within $[m - s, m + s]$, decreasing to zero outside; to keep things simple, the usual trapezoidal shape may be used, so that the reliability coefficient will be 1 within $[m - s, m + s]$, going to 0 at $m - s - q$ and at $m + s + q$, *q* being a positive number. In order to estimate *q*, the table of the normal distribution demonstrates that when $q = 0.5s$, 7% of the area is left out on each side; for $q = 0.64s$, the remainder is 5%; for $q = 1.33s$ it is 1% and so on, the last two values being those normally used for confidence intervals in hypothesis testing. The choice among these possibilities is subjective, and we cautiously chose $q = 0.5s$, considering those cases that have a probability less than 0.07 as unreliable. Since the age range has a width of $2s$, this means that in our trapezoidal approximation, we allow for a slack of 25% of the length of the estimated age interval, on each side of it, assigning a fuzzy function with value 1 on the interval determined by osteology, which descends to 0 on both sides with constant slope. The same rule will be applied to the second osteological method.

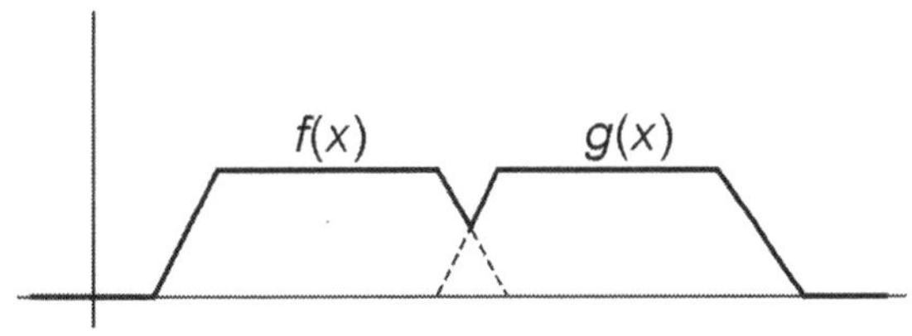

Figure 7: Graph of fuzzy OR (heavy line).

For instance, an osteological estimate of the age range as 20 – 40 will correspond to the fuzzy age represented by the function shown in figure 5.

The general formula for *f* in terms of *k*, *m* and *s* can easily be computed from the above rule and results as shown in figure 6:

When the osteological investigation renders two distinct, not overlapping ranges for the age, they will correspond to a fuzzy reliability function that is built taking into consideration the two separate parts as distinct reliability functions $f(x)$ and $g(x)$, and defining their fuzzy OR as follows:

$$f \text{ OR } g(x) = \max(f(x), g(x))$$

with the graph shown in figure 7.

This set of rules implies that a greater uncertainty corresponds to wider age intervals and consequently, a larger slack, and has other consequences that are worth considering.

Let us consider two cases. The first case incorporates an osteological age range of 22 – 30 years and the second a range of 23 – 39. Which one should be considered "younger"? Intuition says the first one, but this is not true. Since the ranges have a statistical nature, there are tails for both: our cautious assumption determines 2 years for the first one and 4 years for the second one (traditional statistical assumptions would have fixed them at 2.56 years and respectively 5.12 years, at a confidence level of 5 %, and even larger for a confidence level of 1 %). This implies that the complete age range (with tails) is 20 – 32 for the first case and 19 – 43 for the second. Now who is "younger"?

This simple example shows that our common sense reasoning may be fallacious and new categories need to be introduced, even for

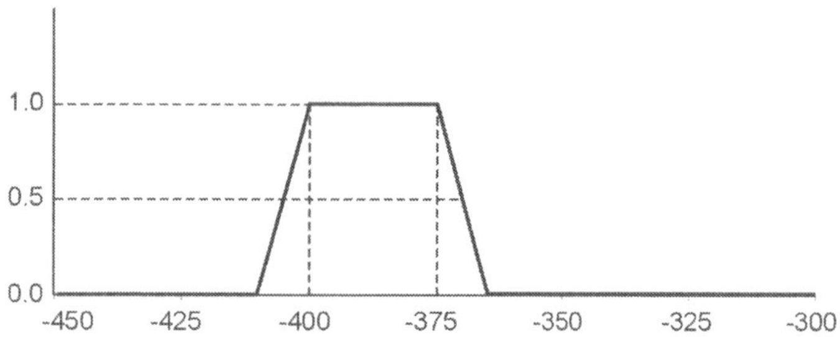

Figure 8: Fuzzy chronology with a slack of 15 years (example).

simple comparisons, which lose any significance when applied to statistical data.

As far as chronology is concerned, nominal constants were used in this context, as usual; for instance, the "first quarter of the 4th century BC" meaning the time interval [–400, –375].

Each of these has therefore been converted into a fuzzy value, with a fuzzy coefficient given in the usual trapezoidal for each time range; that is, equal to 1 for the corresponding time interval, and a tail *d* on each side, on which the fuzzy reliability coefficient varies uniformly from 0 to 1, or vice versa.

Again, several choices are possible for *d* and the candidates are 6.25 years (25 % of the range as for age), 8 years (32 % of the range, corresponding to a statistical confidence level of 0.1) and 16.625 years (66.5 % of the range, corresponding to a statistical confidence level of 0.01). We chose 15, the rounded value closest to the latter, to express the high degree of indeterminacy, in numerical terms, of the chronological traditional assignment; this leaves more space for tail values outside of the interval. It would be equivalent to assume a slack of 60% of the time range length. Figure 8 demonstrates an example.

It might be argued that wider intervals lower the confidence level in each single value of a fuzzy value. For instance, an age interval such as 20 – 40 would make any single age, 35 for instance, less credible. If this is the case, and confirmed by the osteological method used to determine the age interval, the top value of the reliability function should be lowered accordingly. However, provided with no other information, then all values should have equal and top reliability factors, as we suggest.

This argument possibly derives from a misunderstanding between the probability of a single value, which is lower if more cases are possible, and the fuzzy reliability of each single case, which is not influenced by their number since there is no constraint when adding up to 1. This substantiates our use of this approach as opposed to a probabilistic one: from a probabilistic point of view, the probability of an exact value is zero, as it is well known, even if an exact value, perhaps unknown to us, should have existed. So in this context archaeology does not need to deal with low probability figures, rather with "exact" figures having a different reliability, so as to establish archaeological inference knowing how reliable they are.

Concluding this paragraph on the evaluation of fuzzy reliability coefficients, we want to emphasize that the subjective character of their choice has ultimately emerged. In our opinion, however, this is their strength and not their weakness. It has been proved beyond doubt that any attempt to construct a completely deterministic model, intended to provide archaeological results only by means of computations, would be fallacious. Subjective, in this meaning, differs from arbitrary, and is strictly related to the concept as used in De Finetti (1970) or in Savage (1972), who prefers the term "personalistic" to denote this approach.

6. Operating in the database with fuzzy entities

To store data in a database, it is necessary to create special data types corresponding to fuzzy entities, which are fuzzy labels and fuzzy values, and to set the rules to process them. Since these have already been defined in Crescioli et al. (2000), we refer to that paper for a more detailed description.

As noted before, a fuzzy label is an array of couples formed by a label, that is, a nominal element and a number in [0, 1]. The nominal elements are chosen from the common domain, so different instances of the fuzzy label consist of the same nominal elements, possibly with different coefficients. If we agree to always put the nominal elements in the same (not relevant) order in the array, then the fuzzy label is characterized by the domain: that is, the common set of *n* possible labels and an *n* multiple of numbers, the values of which are different for each instance of the label. In our case, the fuzzy gender is characterized by the domain as follows: the two labels "male" and "female", and a couple of numbers as (0.3, 0.7), having conventionally agreed that the former refers to "male" and the latter to "female".

The definition of each member of the data type FUZZY_LABEL, and likewise also FUZZY_GENDER, thus requires that it store the domain somewhere, being {"male", "female"} in this case. It then consists of a one-dimensional array of real numbers.

Regarding fuzzy values, the required function is approximated by a piecewise linear function, so that only the corner points need to be stored. Previous models used only "trapezoidal" functions as the ones shown in the above figures, but this limits the field of application, as shown in Crescioli et al. (2000). Our models set no such restriction.

So, the FUZZY_VALUE data type consists of a two-dimensional array of real numbers. For example, the age interval 20 – 40 is represented as the array {(15, 0), (20, 1), (40, 1), (45, 0)} according to the assignment of the fuzzy age function stated in paragraph 5; these values are, in fact, the coordinates of the corner points of the graph portraying the reliability function.

It may be useful to define constants for any type of fuzzy data, which are stored in a separate table. Due to SQL naming rules, and in order to make the use of constants in queries easier, they are denoted as functions with no parameters, for instance YOUNG().

To determine the values of these fuzzy constants, we refer to the commonly used ranges by anthropologists; so that, for instance, YOUNG() means an age included within the range of 15 and 20 years, with a slack of 2 years before and after. Naturally, constants may be modified or be defined with other values, so long as their value is clearly stated.

Finally, we need to define the operator fuzzy equal, denoted with ~. This follows the definition given in paragraph 4, and the result is a number in [0, 1]. Therefore, the comparison between two homogeneous fuzzy entities or a fuzzy entity and a constant will produce a list of numbers corresponding to the values of fuzzy equality for different instances. An example: the comparison be-

Age interval (osteological)	Possible age range (fuzzy)	Young constant	Result of FUZZY_AGE ~ YOUNG()
10 – 18	8 – 20	15 – 20	1
22 – 30	20 – 32		0.5
40 – 48	38 – 50		0
22 – 38	18 – 42		0.666
22 – 22	22		0
...			...

Table 1: Results of FUZZY_AGE ~ YOUNG() (example).

tween the FUZZY_AGE attribute and the fuzzy constant YOUNG() will give a list of numbers, each representing the similarity of the fuzzy age of each record to the constant (fuzzy) value chosen for YOUNG(). A dummy result for the query FUZZY_AGE ~ YOUNG() is represented in table 1, where the descriptions are presented instead of the values of fuzzy entities in order to ease readability.

The apparently counter-intuitive result that 22 – 30 is "less similar" to YOUNG than 22 – 38 is a consequence of the statistical nature of age ranges and is perfectly coherent with the fuzzy treatment of data, as already noticed in paragraph 4.

Fuzzy counting, as defined in paragraph 4, does not require any special function; it simply uses SUM.

7. Implementing the fuzzy database

Implementing the fuzzy database requires an extensible DBMS. We chose PostgreSQL for this, a RDBMS available under a Linux operating system. Furthermore, being fully relational, it is free software and is customisable in the sense that new data types, functions and operators can be added to the standard ones.

PostgreSQL can be queried within a terminal window using psql, an SQL environment command line with standard features. In our case, we used psql to create the new data types, to define the database structure and to load the data, which were available and had previously been typed, verified and converted to text format. Any software can be used for this, and we did not develop a graphical interface because we did not need it to for data input, as direct conversion was quicker. The data were then manipulated to give expressions such as the following:

INSERT INTO TOMB VALUES(85, 4012, 'Maisto', 'Cappuccina', 'FALSE', 'TILES', 40, 216, 70, 'SE-NW', 'FALSE', 'M','', '{0.9, 0.1}', 'A', '46-52; 40-45', '{{38.75, 0}, {40, 1}, {45, 1}, {45.45, 0.64}, {46, 1}, {52, 1}, {53.5, 0}}', '1st quarter 3rd cent. BC', '{{-285, 0}, {-300, 1}, {-275, 1}, {-260, 0}}', 'TRUE', 'SUPINE', 'INHUMATION');

The above SQL expression assigns values to all the fields of the record, most of which are not fuzzy and have not been dealt with in the present paper. TOMB is the name given to the table, values in bold refer to fuzzy attributes and italics to the corresponding "original" expression, which are kept in the database for comparison. In this case, the (osteological) gender was M for "male"; the (osteological) age, according to the two different methods, consisted of the two distinct and not overlapping intervals of 46 – 52 and 40 – 45 (a contradiction, if manipulated with traditional methods, and also impossible to manage with previous fuzzy database models), and the chronology was rendered as explained above.

Creating the table TOMB necessitated the previous definition of the fuzzy data types, which was accomplished thanks to the psql command CREATE TYPE that allows the definition of personalized data types.

The GRAVEGOODS table was created in a similar manner with data concerning grave goods.

The constants were inserted into the CONSTANTS table and then they used to create functions and operators, such as fuzzy equality ~ . This operator is based on the function f_equal(x,y), the only piece of software written in C to make computation quicker, and it is computed according to the definition given in paragraph 4. It was introduced in order to allow an expression in the form f_age ~ ADULT(), which is much easier to understand than the equivalent (and cumbersome) expression -

f_equal(f_age,'{{16,0},{21,1},{40,1},{45,0}').

9. Archaeological application to the case study

At this point, it might be reasonable to ask what is the impact of the machinery set-up presented in the previous paragraphs on archaeological research? Even if the present paper aims at contributing at a methodological level, we propose to present in this last paragraph, some evidence of the results that can be obtained when using fuzzy models.

Among the information obtained from the "archaeology of death", data concerning demography are traditionally considered as "natural" or biological. However, as d'Agostino (1985:52) noted, anthropological information must also be interpreted within an archaeological framework, since the definition of age classes and male or female roles, only apparently objective, must always be referred to within the social context of their origins. Thus, demographic data cannot be taken into account mechanically, basing solely on sociological assumptions. On the contrary, they should always be compared with information derived from the analysis of funerary customs, in order to avoid the imposition of categories deduced from the "community of the living" upon the "world of the dead".

Within the framework of demographic applications to cemetery interpretation, two criteria are particularly significant for the verification of how representative the funerary sample actually is (d'Agostino 1990).

The first one is based on the ratio between the number of adults and the number of children, and on the ratio between the number of males and the number of females: each of these ratios, for pre-industrial societies, should be approximately equal to 1 (Weiss 1973). When one of the sampled values is substantially different from this model, it may be concluded that the funerary sample is not representative of all the components existing in the community: in this case, the sample of tombs may reflect the adoption of discriminating burial practices.

Another important criterion for cemetery analysis is based on funerary variability (O'Shea 1984). It is based on the principle that if social statuses of the deceased are present uniformly, then the tomb sample represents only one social class of the community. Interpreting funerary variability presents more complex problems than the above ratio criterion: indeed, hierarchy may be emphasized more or less according to the economic and social struc-

Land property	Maisto-Boccia	Rossomando	Tascone-Di Dato	Cemetery
Age ratio (Serritella's data)				
Infant	22.9%	33.3%	16.0%	22.7%
Children – Young	0.0%	6.7%	0.0%	1.3%
Adult – Elderly	54.3%	53.3%	76.0%	61.3%
Indeterminate	22.8%	6.7%	8.0%	14.7%
Total	100.0%	100.0%	100.0%	100.0%
Ratio A/(I+C)	2.37	1.33	2.75	2.56
Age ratio (Fuzzy model)				
Children	43.9%	45.3%	28.5%	41.0%
Adult	56.1%	54.7%	71.5%	59.0%
Ratio A/C	1.28	1.21	2.50	1.44
Gender ratio (Serritella's data)				
Male	47.4%	75.0%	63.1%	58.7%
Female	42.1%	25.0%	31.6%	34.8%
Indeterminate	10.5%	0.0%	5.3%	6.5%
Ratio M/F	1.13	3.00	2.00	1.69
Gender ratio (Fuzzy model)				
Male	53.1%	70.4%	57.6%	57.2%
Female	46.9%	29.6%	42.4%	42.8%
Ratio M/F	1.13	2.38	1.36	1.34

Table 2: Age and gender ratio corresponding to land property (percentages). Derived from Serritella (1995: 116, 121, 123) and computed from the fuzzy model.

ture of the community, becoming inadequate when egalitarian ideologies prevail.

In the past, applications of statistical methods to the study of funerary custom have been based on purely quantitative logic. Mathematical models were used to measure - "objectively" - the richness of a tomb, to determine the funerary variability or to identify social hierarchy on the basis of "energy expenditure" (see Cuozzo 1994:268, Cuozzo 1996). Contrarily, we suggest the use of fuzzy logic (using palaeo-anthropological demographic data) to estimate how much a funerary sample is representative of the community. A few examples are given below, and we consider this perspective a potentially substantial contribution for the execution of an analysis that aims to determine horizontal and vertical stratigraphy. Since the age and gender coefficient result from statistical computations on osteological parameters, as shown above, an "improper" use of them, or the mechanical assumption that they represent an "objective" truth, may lead to an unconscious variation of the true ratios of children to adults or males to females, thus turning awry any deduction derived from the sample.

Consider the first block of table 2, derived from those published in Serritella (1995: 116, 121, 123), counting the occurrences of age categories and shown as percentages of the total (blocks are identified by heavy lines). The tombs are grouped by modern landowners, which gives a rough indication of their space position in the cemetery. In other words, land property, represented in the database with the name of the modern owner, is a simple and approximate, but effective clustering of the tombs. Serritella demonstrated that this grouping reflects significant differences in chronology or rite.

Indeterminate gender assignments add up to 15 % of the total and consequently, they may significantly alter the confidence level of the sample. For the second group, assigning all the indeterminate cases to infants or children gives a ratio of 1:19, turning this sample into a highly representative one.

Using fuzzy coefficients, the fuzzy age "Infant or Child" class, ranging from 0 to 20 years, needs to be introduced, as well as the "Adult or Elderly" class, ranging from 21 upwards. The data then needs to be compared with the two new constants. We then compare Serritella's results with those shown in the second block of table 2, which are easily obtainable by means of an SQL query on the database. The "Rossomando" and "Maisto-Boccia" groups fit the model, while the "Tascone-Di Dato" does not. The latter shows a prevalence of adult burials. Considering the male to female ratio, the values shown in the third block of table 2 are from Serritella's work and the ones shown in the last block of the same table are from our database. From this it follows that the "Maisto-Boccia" group portrays a representative sample, the "Tascone-Di Dato" is less so and the "Rossomando" group is not.

After combining the two tables, only "Maisto-Boccia" remains as a representative sample, while the others present some discrimination: "Tascone-Di Dato" for age, "Rossomando" for gender. An explanation for this discrimination, in terms of social status and age and gender roles, can derive only from an investigation of the grave goods; that is, to understand the underlying cultural model of social representation. However this reaches beyond the aim of the present paper. Comparing our results with Serritella's, what she suggested is confirmed by our own research; however, with a much higher level of reliability, due to the absence of indeterminate cases, which in her tables should suspend every conclusion.

10. Conclusions

If the proposed model is accepted, at least in a negative sense, if not positive, then greater caution should be instigated when using statistical data in archaeological investigations. It will no longer be possible always to take the reliability of archaeometric data for granted; at least, not when these data are critical for the validity of an interpretation model. Nevertheless, we hope that our model, together with the computer tools we made available (possibly improved by future work) will help research. The user-friendly aspect of computers and the extensive availability of software tools together have the positive effect of spreading the advantages of their applications, while at the same time their usage occurs unawares. Perhaps this contribution will indeed augment the awareness of archaeologists that even when using databases, the quickest solution is rarely the cleanest one.

References

BARCELÓ, J.A., 2000. Visualizing what might be: an introduction to virtual reality techniques in archaeology. In Barceló, J.A., Forte, M., Sanders, D.H. (eds.), *Virtual Reality in Archaeology*: 9-36. Oxford: Archaeopress (BAR International Series 843).

BUCK, C.E., CAVANAGH, W.G. and LITTON, C.D., 1996. *Bayesian Approach to Interpreting Archaeological Data.* New York: Wiley (Statistics in Practice).

CUOZZO, M., 1994. Patterns of organisation and funerary customs in the cemetery of Pontecagnano (Salerno) during the orientalising period. *Journal of European Archaeology* 2, 2:263-298.

CUOZZO, M., 1996. Prospettive teoriche e metodologiche nell'interpretazione delle necropoli: la post-processual archaeology. *AION ArchStAnt n.s.,* 3:1-39.

CRESCIOLI, M., D'ANDREA, A. and NICCOLUCCI, F., 2000. A GIS-based analysis of the etruscan cemetery of Pontecagnano using fuzzy logic. In Lock, G.R. (ed.), *Beyond the Map: Archaeology and Spatial Technologies*, European University Centre for Cultural Heritage, Ravello, Italy, October 1-2 1999. Amsterdam: IOS Press.

D'AGOSTINO, B., 1985. Società dei vivi, comunità dei morti: un rapporto difficile, *DialArch* 1.3, III s.: 47-58.

D'AGOSTINO, B., 1990. Problemi di interpretazione delle necropolis. In Francovich, R. and Manacorda, D. (eds.), *Lo scavo archeologico dalla diagnosi all'edizione, III ciclo di lezioni sulla Ricerca applicata in Archeologia* Certosa di Pontignano (Siena), 6-18 novembre 1989: 401-420. Firenze: All'insegna del giglio.

D'ANDREA, A., 1999. Il GIS nella produzione delle carte dell'impatto archeologico: l'esempio di Pontecagnano. *Archeologia e Calcolatori* 10:227-237.

DELICADO, P., 1999. Statistics in Archaeology: New Directions. In Barcelò, J.A., Briz I. and Vila, A. (eds.), *New Techniques for Old Time, Proceedings of the CAA98 Conference*: 29-37. Oxford: Archaeopress (BAR International Series 757).

DE FINETTI, B., 1970. Teoria delle Probabilità. Sintesi introduttiva con appendice critica, Torino: Einaudi. English translation: Probability theory: A Critical Introductory Treatment, New York: Wiley, 1974.

HARRIS, T.M. and LOCK , G.R., 1995. Toward an evaluation of GIS in European archaeology. The past, present and future of theory and applications. In Lock, G. and Stančič, Z. (eds.), *Archaeological and Geographical Information Systems: a European Perspective*: 349-365. London: Taylor & Francis.

HODDER, J., 1982. *Symbols in action.* Cambridge: Cambridge University Press.

MOOD, A.M., GRAYBILL, F.A. and BOES, D.C., 1979. *Introduction to the Theory of Statistics*, New York: McGraw-Hill International Edition.

MOSCATI, P., 1996. Archeologia Quantitativa: Nascita, sviluppo e "crisi". *Archeologia e Calcolatori* 7: 579-590.

O'SHEA, J., 1984. *Mortuary variability.* New York: Academic Press.

PETRONE, P.P., 1995. *Analisi paleodemografica e paleopatologica delle tombe in proprietà Rossomando.* In Serritella 1995, Appendix I: 129-134.

SAVAGE, L.J., 1972. *The Foundation of Statistics.* New York: Dover (second edition).

SCARSINI, C. and BIGAZZI, R., 1995. *Studio antropologico dei resti umani.* In Serritella 1995, Appendix II: 135-148.

SERRITELLA, A., 1995. *Pontecagnano. II.3. Le nuove aree di necropoli del IV e III sec. a. C.*, Annali del Dipartimento di studi del Mondo Classico e del Mediterraneo antico dell'Istituto Universitario Orientale, Quaderno n. 9, Napoli.

VOORRIPS, A., 1996. Information Science in Archaeology: a Short History and Some Recent Trends. *Archeologia e Calcolatori* 7: 303-312.

YAGER, R.R. and FILEV, D.P., 1994. *Essentials of Fuzzy Modeling and Control*, J. Wiley & Sons, New York.

WEISS, K.M., 1973. *Demographic Models for Anthropology*, Washington: Memoirs of the Society for American Archaeology 27.

WILCOCK, J.D., 1999. Getting the Best Fit? 25 Years of Statistical Techniques in Archaeology. In Dingwall, L., Exon, S., Gaffney, V., Laflin, S., and Van Leusen, M. (eds.), *Computer Applications and Quantitative Methods in Archaeology 1997*: 19-27. Oxford: Archaeopress (BAR International Series 750).

A Metastructure for Thesauri in Archaeology

Martin Doerr

Institute of Computer Science, Foundation for Research and Technology – Hellas

Heraklion, Crete, Greece

e-mail: martin@ics.forth.gr

Demetrios Kalomoirakis

Greek Ministry of Culture, Athens, Greece

e-mail: demetrios.kalomoirakis@damd.culture.gr

Abstract

Controlled terminology is becoming increasingly important for archaeological documentation and information retrieval in wide area networks. Even though there is plenty of literature on thesaurus creation, the logical organisation, the "metastructure" of a thesaurus is still rather unreflected, in particular in the more abstract layers. So-called facet indicators are distributed intuitively in order to group a set of sibling terms by an implicit property. They can be regarded as potential elements of a metastructure. In co-operation with ICS-FORTH the Greek Ministry of Culture is developing a general-purpose thesaurus for archaeology. An extensible, systematic metastructure for terminology in archaeology is proposed, based on contextual and phenomenal notions. This seems to be more comprehensive than traditional approaches. This paper discusses the benefits of such an approach. In particular, the proposed structure seems to provide a better understanding of the implicit criteria intrinsic to object classification terms and the relation between compound expressions and traditional terms.

Key words: archaeological terminology, facetted classification, thesauri, qualia, subsumption hierarchies, concept formation

1. Introduction

Controlled terminology is becoming increasingly important for electronic documentation of archaeological object description and its consistent retrieval in wide area networks. In particular thesauri, which organises terms or concepts in semantic networks, is becoming a mandatory tool for users to find their way through the rapidly growing electronic information flood and for unambiguous communication through automated intermediaries. Even though there is plenty of literature and an established practice on creating thesauri (see e.g. publications of the Getty Foundation, the British Arts and Humanities Data Service AHDS, and numerous national standards), their logical organisation, the metastructure of a thesaurus is still rather unreflected, in particular in the more abstract layers. In this paper, we will try to clarify some concepts and present the logical organisation of a general purpose thesaurus for archaeology developed at the Greek Ministry of Culture in co-operation with ICS-FORTH. Finally we will point out some research topics, that came up with this work.

2. What is a Thesaurus

The notion of a linguistic thesaurus has been introduced over one hundred years ago by Peter Mark Roget. In 1852, he published his famous "Thesaurus of English Words and Phrases", intended to help authors find synonyms or more precise terms. Such thesauri are ever since quite popular, and Roget's one is still being edited [Encarta].

With the raise of electronic communication, and in particular with the search for information in database records, the need to refer to some concepts with standardised expressions also arose, otherwise entries made by curators and requests entered by users would never match. The currently most successful approach to standardise expressions are the so-called thesauri in the sense of computer science.

At this moment two directions can be distinguished: a linguistically oriented and a conceptually oriented. The "linguistic" one regards the "term" as the key element of the thesaurus, coming from a dictionary world. Terms are represented as the expert would use it in speech: potsherd, vessel, column, pottery etc. Besides textual descriptions of the term meaning, linguistic information is also added, such as part-of-speech, sort keys etc.

Respective data standards have been promoted by the Text Encoding Initiative [TEI], the latest developments being ISO 12620, MAchine Readable Terminology Interchange Format (MARTIF), and the Virtual HyperGlossary (VHG) (Murray 1998:34). Terms are typically organised in mono hierarchies (i.e. one term belongs exactly to one broader term) as a kind of associative decision trees leading from higher to lower abstraction levels of conceptually close terms. These hierarchies serve mainly as searching aids for the user to find a term rather than expressing abstraction levels. Often information is added to create a nice lay-out of indented lists in printed reports.

These thesauri might be helpful for individuals looking up and understanding terminology as an encyclopaedia, for linguistic work, translations and natural language search. However, there are severe drawbacks for indexing and classification in databases. The naïve assumption that a term uttered by an expert is a sufficient surrogate for the item we are describing or looking for does not hold true from the following reasons: terms are context dependent, be it true homonimity ("contrastive ambiguity" (Pustejovsky 1995:27)), shift of meaning ("complementary polysemy"), or because the term expresses not all aspects of an item relevant for the one or other future user. A "vessel" can be a number of different things, "pink" is not always a colour etc. A "temple" from the Greco-Roman period may have different associations and abstractions from a Protestant one. "Neo-classical building" and "school house" can be equally relevant characteristics of the same object. Further, users often query at a higher level of abstraction than the most detailed classification of an object, e.g. "black vases from Attica of the third and fourth century BC". Hierarchies that are not simply associative, but that are built on true abstraction, can help the software to resolve such queries automatically through the thesaurus.

Such considerations lead the middle of the twentieth century library scientists to the development of classification languages based on concepts rather than terms. As C. Welty (Welty and Jenkins 1999:155) pointed out, only since shelving books on unique places is no more the major organisation principle, assignment of multiple subjects/terms were considered. So on one hand, the problem was addressed by the rules to combine "elemental" concepts either already in the terminological system (pre co-ordination) or at time of classification (post co-ordination). For this purpose the classification vocabulary was grouped into so-called "facets", i.e. sets of elemental concepts, going back to Ranganathan's Colon Classification System (Ranganathan 1965). On the other hand, these "facets" were internally organised by conceptual relations. Such systems form *thesauri* in the sense we will use from here on. As Pollitt (Pollitt 1997a) stated:

"Problems in the Verbal Plane brought about the introduction of regularisation and structure into the language used in post co-ordinated systems via thesauri." Vickery (Vickery 1960:181) reports how he first encountered the word thesaurus:

> "Speaking at the Dorking conference on classification, Helen Brownson said 'The problem (of information retrieval), as some investigators see it at least, is to transform concepts and their relationships, as expressed in the language of documents, into a somewhat more regularised language, with synonyms controlled and syntactic structures simplified. Now it is reasonable to think that the further we can go in routinising and mechanising the techniques of translating ordinary language into a regularised language and of coding for machine manipulation, the more we will be likely to achieve economically feasible machine searching on a large scale...[Some investigators] have come up with the thought that the best answer ... may be the application of a mechanised thesaurus based on networks of related meanings.' In the same paper Vickery (1960:185) cites Bernier and Heumann (1957) as proposing the introduction of an organised vocabulary in the form of a thesaurus."

This "conceptual direction" of thesauri, typically elaborating on the semantic links as defined in ISO2788, ISO2709 and ISO5964, regards the "concept" as a key element of the thesaurus (see e.g. Svenonius 1989:82). These concepts can be alluded to by a series of linguistic expressions, but the concept itself has no name. In order to improve the orientation of the user, one term may be selected arbitrarily as an *identifier* for each concept (often called "*descriptor*"), eventually giving the impression as if the thesaurus were still about the term. This idea is also well explained in the explanations to the Art and Architecture Thesaurus AAT (Getty AHIP 1994), whereas ISO2788 still has no clear notion of separating between terms and concepts. WordNet (Miller 1993) identifies the concept (here called "meaning") by the combination of *all* appropriate terms. Miller et al. make clear the excessive relations between words or terms and concepts.

Finally it should be mentioned, that there is an automatic generation of the so-called "thesauri" (e.g. Hazewinkel 1996, Chen 1993:25). By statistical means, measures for a conceptual distance between terms or expressions are derived based on the co-occurrence *in corpora* and by grammatical criteria. These methods do not give much insight into the semantic nature of the derived relations, and should be regarded more as retrieval aids than thesauri. Nevertheless, in the future such methods may be quite useful in the creation of a true thesauri.

3. About concepts

Cognitive psychology scientists have proposed several definitions for the notion of "concept" (see also Doerr 1998). According to the first point of view, a concept is a person's conceptualisation of the notion of categories (the class or set of entities which are grouped together on the basis of some criterion or rule (Michalski 1993). According to this point of view, there would have to be as many concepts for the same category as there are different cognitive representations for it. For thesauri we do not adopt this approach since the thesauri incorporated are collections of terms with generally agreed on semantics and not individual definitions or representations of a concept.

According to the second point of view, a concept is perceived as a set of entities, called "concept instances" by a common agreement rather than formal reasoning on the properties which may make an entity the instance of the concept. In other words, we regard a concept as defined within a group of people, as long as the people agree which items are characterised by the concept. This does not exclude the use of rules, however it is not based on it. It differs to the first point of view in the fact, that we do not argue about what people really conceptualise, but what they can agree to on the outside. This is actually the basis for a semantic thesaurus construction. It is further a principle, which can be easily applied to the classification of material objects. Eventually we are forced to communicate such concepts to each other by a verbose account of examples and contextual information, called (in the context of thesauri) "scope notes".

According to the third point of view, a concept is identified by a classification rule. This means, that the concept is further analysed into a logical expression as regards its properties. However, quite often conscious rules cannot easily be formulated even for obvious concepts, e.g. the optical characteristics of an aquarelle, and the rules themselves cannot be further analysed. Nevertheless, this principle is the basis for the so-called "Terminological Logic" or "Description Logic"(Borgida 1995:671), [DL], which can be regarded as the most advanced extension of conceptual thesauri. It has been proven quite successfully in medical applications (Rector 1997) as well as other domains. In our opinion it is more appropriate for cases where distinct overt features dominate, like in scientific objects and technical artefacts. As I will show below, it is quite useful to complement an agreement-based body of concepts by the more dynamic rule-based expressions.

Advantages of the conceptual approach are, that the conflicts of homonymity, as well as cultural and contextual differences can be overcome or made explicit. What counts is the definition of the concept rather than the term itself. Consequently, the AAT team has created expressions like "pink (colour)", which are not natural, but resolve ambiguity. Further, we can identify clear abstraction hierarchies based on semantic inclusion (generalisation, super-class, subsumption, hyponomy and whatever synonym term may be in use), that allow to expand more abstract database queries into sets of specific terms. E.g. "dime" is a "coin", "baby doll" is a "recreational artefact", "kachina" is a "figural work" etc.

3.1. About objects and subjects

As we can see from the historical considerations above, library science, and in particular the problem of subject classification, has been the driving force behind these developments. Museums and archaeology, dealing to a very large extent with material objects, have relatively unreflectedly adopted the same method to classify their objects instead of the literature about them. At this point let us make a few propositions as regards the differences:

- Physical objects are not "abstract". An "Introduction to Biology" and "The Behaviour of Geese" can be the distinct subjects of two different books. In case of objects, there is no single "living being" e.g., without being something concrete, like "goose".
- Consequently, a physical object is truly classified by any higher abstraction of its type, but only the lowest has instances. My dog is a dog, a carnivore, an animal, a living being, a material object. Any existing animal however must be of the most specific type, whereas a book about animals may be as generic as the concept "animal".
- The lowest abstraction level is relative, a function of our attention to more or less characteristics. It depends on the size and constitution of the collection at our disposal. The relatively well-defined level of a "species" in biology has only weak analogies in archaeological objects.
- An object may be classified by an open number of relevant views, according to the investigators objectives. These views may be completely uncorrelated as e.g. "exchange item", "caouri shell", and "property of an Oba", "bead", "element of a preserved string", "slightly worn". The "subject" on the other side may be seen as *one* such view, valid for literature objects, but mostly absent in material objects.
- Even though any object can give raise to a subject in the sense of "talking about..." (Welty and Jenkins 1999:155), the "parallel hierarchy" (Soergel 1995:369) generated by objects viewed as subjects may differ. E.g. talking about "bridge construction" is a kind of talking about "bridges". "Bridge construction" itself is not a kind of "bridges", but a kind of construction, either a design or an activity or both.

Of course, any piece of literature as a whole has its own unique individuality similar to a museum object (Welty and Ide 1999). We however agree here with (Welty and Jenkins 1999:155), who prefers to see the subject of a certain book not as a parallel individual associated with each book, but as a collective concept applicable to more than one book. He further points out that in libraries fiction itself is mostly not classified with subjects. The focus of libraries and museums is obviously quite different, and give raise to endless misunderstandings between both communities.

To summarise, we are under the impression, that conceptual hierarchies for material objects can be and need to be stricter defined with respect to the notion of abstraction (hyponomy, broader/narrower terms) than usual in library subject catalogues, and that they need more structure reflecting different aspects of investigation. E.g. we cannot declare, "Top: Arts: Classical Studies: Journals" and "Top: Arts: Classical Studies: Academic Departments" as narrower terms of "Top: Arts: Classical Studies" (see the DMOZ project [DMOZ]), or regard bridge construction as specialisation of bridges (LCSH), but must separate entities of the real world by their deeply distinct nature.

3.2. Descriptive and characteristic concepts

The examples above suggest, that we may be able to separate concept hierarchies by the aspects they express, about people for instance by sex, race, intellectual properties, etc. This can be easily performed with concepts, which represent more or less verifiable properties of the one or the other kind, which *describe* basically an object by observational criteria imposed by the researcher. See e.g. the definition of "knives (weapons)" in the AAT: "Weapons designed primarily for cutting, consisting basically of a single-edged, pointed blade and a grip mounted asymmetrically in relation to the axis of the blade, closer to the back edge". The concept of "weapon" itself is purely functional, and bare of any assumption about its physical form. A dynamic (post co-ordinated) term like "railway museum" expresses no more than what it says (see the AAT).

However, we also have to simultaneously operate with historical concepts of things *characteristic* for a kind of "design model" of a culture, like an "ushabti" or "kachina doll". These concepts comprise of a series of aspects and implications. "Persian rugs", e.g. exhibit the subtlety that not all rugs from Persia are Persian Rugs, nor are all Persian Rugs from Persia, even though at least a "genuine" one should be. Obviously the provenance "Persian" does not imply that a rug necessarily has the characteristic design observed in Persia, and the characteristic design may be copied or truly followed outside of Persia. The latter has a history, evolution and social context.

In this sequence let us talk about *descriptive* object concepts, as those, that analyse our observations about the objects, that have come upon us from the past, be it on features that can be at present verified on the item itself, or features that originate in historical knowledge.

In contrast, let us talk about *characteristic* object concepts, as those, that designate a "design model" of a culture, which has specific functions, designs and contexts of use altogether, has characteristic traditional names (e.g. "alcazar"), and varies over regions and evolves over times. The "ushabty" e.g. was initially a surrogate for the mummy, and later became a servant for the afterlife. It has a characteristic span of forms and sizes. "Ushabty" (ushebti, ushabti, shawabty etc.) are transcriptions from the original Egyptian terms for that concept. In other cases, like with Neolithic objects, those are lost, and researchers "coin" modern terms as surrogates.

As we see, our object concepts come from different perspectives of cognition, with more complex relations between each other and the objects they classify, than simple hierarchies can appropriately capture. The results of pressing those into a simple hierarchy are a disorientation of the user trying to find concepts and poor or even questionable conclusions from the given semantic relations.

Ranganathan (Ranganathan 1965:38), characteristically states:

> "...that the work gets resolved into three different planes: the Idea, the Notional, and the Verbal Planes. Looking back from this position into the work done before these three planes of work were clearly seen and separated, it is found that much of the difficulty arose out of frequent, listless, and unconscious change from one plane to another. Another cause of difficulty was the inhibition of work in the Idea Plane by the limitations of the Notational plane, and by the inherent defects of the Verbal Plane." (see also Pollitt 1997a).

In this paper, we want to investigate some properties of the Idea Plane, the world of concepts and meanings, and their relation to objects of our investigation, free of limitations of the Notational plane, and in sequence propose some improvement in the Notational plane.

4. About facets

We have already mentioned the so-called "faceted classification". The term "facet" alludes to the fragmented surface of a diamond, suggesting that the same thing can be seen from different sides. Faceted classification and further developments in the form of Description Logic are becoming increasingly popular in the dealing with the complexity of precise capturing of information contents (e.g. Prieto-Diaz 1987:6, Constantopoulos 1995, Welty and Ide 1999 and Pollitt 1997a). Actually, one may distinguish three slightly different notions of facets with at least two different applications.

Following (Prieto-Diaz 1987:6): "The faceted method... relies... on... building up or synthesising from the subject statements of particular documents. By this method, subject statements are analysed into their component elemental classes, and these classes are listed in the schedule. Their generic relationships are the only relationships displayed."

This is the first notion of facets: *Elements for synthesising* complex (post co-ordinated) terms or statements by enumeration. In our opinion, this kind of analysis results more in grammatical elements, such as subject, predicate, object, etc. (see e.g. Constantopoulos 1995), than in "elemental classes". They become elements of an indexing language used to create an open number of potential "compound terms". Typical library facets are: Topics, periods, places, genre.

Whereas statements often summarise a scientific paper quite well, material objects are better characterised by noun phrases. Soergel (Soergel 1995:369) discusses in much detail the use and limitations of an indexing language (the AAT), which foresees only one generic relation between each facet. An example for such a compound is "nineteen century Massachusetts wood chairs". Please note, that this denotes no more than a chair. The AAT distinguishes the facets: "Object, Agents, Activities, Styles and Periods, Materials, Physical Attributes, Associated Concepts". Description Logic is the natural extension of such indexing languages, as it enriches the set of relationships allowable between the concepts and puts them in a well-defined logical framework (see e.g Welty and Jenkins 1999 and Bechhofer 1999 as regards the utility of DL).

The separation of our concepts into *disjoint cognitive categories* gives raise to a slightly different interpretation of "facet". Under this aspect, we regard as facets notions of objects, actors, events, measures, time, space, etc. "Facets" are seen as the building elements of our conceptualisations of a domain, without an immediate syntactic purpose. Such elemental classes are often referred to as "*major facets*". Ranganathan talks about "the five fundamental categories PMEST (Personality, Matter, Energy, Space, Time)", and the AAT facets seem to be of that kind. The difference as regards the above interpretation becomes obvious, when someone regards grammatical subjects and objects, which are identical in nature, but different in their role within a statement. E.g. a physical object can be the subject and the object in a grammatically correct phrase.

For an archaeological thesaurus, we regard the principle of "major facets" as appropriate to bring the first order into all potential concepts of the domain. The ontology of the CIDOC Conceptual Reference Model (Doerr 1999) is an attempt to provide a formal standard model used to describe objects and their history, mainly to be able to mediate and transform between different data and metadata formats. It distinguishes: *Temporal Entities* (including periods, events and activities), *Actors, Physical Objects, Conceptual Objects, Place* and *Time*. Recently the IFLA and the Dublin Core community came up with almost identical categories (see: the Indecs project, http://www.indecs.org/results/model.htm). The obvious similarity encourages standardisation. As Steven Pinker (Pinker 1994) suggests innate categories of human perception are most probably hidden behind this. We would however not dare say, to which degree nature itself dictates those. It is easier to derive facets for synthesising complex terms for different applications from these categories, than from purely syntactically motivated categories.

There is however yet another interpretation of "facet", which is, in our opinion completely different. Prieto-Diaz (Prieto-Diaz 1987:6) continues in the same paper: "Facets are sometimes considered as perspectives, viewpoints, or dimensions of a particular domain", and continues with an example about animals classified by habitat versus genealogy. Actually this is the notion that Ranganathan originally introduced, which indeed reminds us of facets of a jewel. The MDA Archaeological thesaurus [MDA97] e.g. introduces a term "armour by construction", and below "scale armour", etc, "armour by form", and below "cuirass", etc, "armour by function" and below "parade armour". Such terms, that announce the criterion by which the subsequent analysis is carried out, are called node-labels (ISO2788), *minor facets* or guide-terms (AAT). Frequently, the values these criteria can take on are taken from another "major facet". As e.g. in the term "cutting sword", the value "cutting" appears in the major facet "Activities" of the AAT.

4.1. The minor facets in the AAT

Ranganathan observed, that an open number of concept arrays can be used to provide the criteria for a facet on another array. Similarly, the AAT has systematically introduced guide terms (facet indicators) in a great number, to be precise 2840 in the 1998 edition. From a scientific and philosophical point of view, this situation is not very satisfactory. In the first statistical investigation, we have identified that there are a number of predominant criteria, that can be used to create a metastructure.

For that purpose, we have separated the guide terms of the object facet. We separated approximately 1640 of them (some concepts are not really objects, such as "illustrations", "museums"). From these, 615 actually refer to the criterion with the characteristic connector "by" like "swords by function". These could easily be separated, and became the object of our analysis. A more thorough analysis could well be done on the total, but at this stage we are only interested in a qualitative result:

318 guide terms out of the 615 fall in three clearly distinct dominant groups:

...by form:	129
...by function:	121
...by location or context:	68.

The remaining have frequencies below 15. The notion "location or context" is not easy to interpret. It is often a geometric relation of a part to a kind of whole, but also of independent smaller elements to a larger structure such as landscape to sea, lamps to parts of a room, etc. The other categories are clearly defined. In practice, form and function can overlap in the form of "functional forms". There are 15 guide terms "...by form or function", like "furniture by form or function" and others.

This analysis is only linguistic. Actually a closer look on the actual narrower terms appearing under the less frequent kinds of guide terms reveals more of the major categories, such as "...by shape", "...by merchandising practice", "...by position". The AAT has never tried to standardise these terms. Further, a large multitude of differently named categories can be characterised as construction criteria, be it the internal structure or the process of creation of the internal operation principles of machinery. We have not tried to go into higher detail, nor have we tried to distinguish event-like functions like "cutting" from activity-like functions like "dwelling". Construction and form sometimes overlap, in the case of technically necessary forms of machinery we have preferred to classify the criterion as constructive.

We have arbitrarily added the group "form or function" to "function". There is a small group of criteria about the social context of creation or use. Surprising enough, only 5 times the "subject type" appears in the proper sense, e.g. for sculpture, as a term-generating criterion! For example we have classified the "subject" of a museum under "function". Here are the statistics of our interpretation of the AAT object guide terms with the "by" connector:

Criteria of form:	34 %
Criteria of function:	29 %
Criteria of relative location:	15 %
Criteria of construction:	15 %
Criteria of social context:	5 %
Criteria of subject type:	1 %
Criteria of naming (like coins):	1 %.

From the impression these statistics give we derive the hypothesis, that there are a few predominant aspects to analyse objects, and not an arbitrary, unpredictable mass. These aspects are reflected in the terms themselves, but may be as well explicit criteria (distinct data base fields) of classification (function, form, material etc). Some British museums have made the functional aspect explicit: they classify objects with SHIC, (Social, Historical and Industrial Classification) [SHIC93] only by their relevance to human activities - a clear indication of the relevance and utility of the "function" and "social context" aspects. In this practice, the actual "function" is always explicit, and not hidden in some traditional term.

The above analysis gets an unexpected confirmation from modern linguistics. Pustejovsky (Pustejovsky 1995) has developed a theory to cater for the dynamic change of word meaning in new contexts. He shows, that nominals in general, and in all languages imply multiple meanings, which are "activated" in a sentence by the appropriate context, e.g. a verb that refers to a certain aspect. An example he gives is:

> "He walks through the door" (function, opening), and "He paints the door" (object).

He talks about the "Qualia Structure" of nominals, which he analyses in the following categories (referring also to Aristotle's notion of modes of explanation):

- Constitutive: The relation between an object and its constituents. (Material, weight, parts and component elements)
- Formal: That which distinguishes the object within a larger domain. (Orientation, magnitude, shape, dimensionality, colour, position)
- Telic: Purpose and Function of the object. (The purpose the agent has in performing an act. Built-in function or aim which specifies certain activities)
- Agentive: Factors involved in the origin or "bringing about" of an object. (Creator, artefact, natural kind, causal chain)

The similarity is striking. If eventually, common sense in understanding language and classification in archaeology will lead to the same or at least similar categories, will be shown in future research. By sure, both disciplines can learn from each other.

5. A metamodel for object thesauri

The fact that traditional terms silently imply one or the other aspect can bring confusion into the classification. Some data about an object may be classified under the above aspects with descriptive concepts, especially if a characteristic term for it is missing. Another object may be classified with traditional characteristic terms that imply a descriptive concept shared with another object. We shall not be able to retrieve both under the same terms.

For example, let's look at the AAT Terms "armor", "parade armor" and "garnitures". The scope note of "garnitures" says: "Sets of armor, ...for use in warfare or tournaments. They were invariably made for important persons according to a precise design for specific types of events." Neither the aspect of protection in combat nor the aspect of social representation are explicit from the wording or from the broader terms given. On the other hand, a "pure descriptive" classification with the above criteria is neither efficient nor always appropriate. Archaeologists at the Greek National Archive of Monuments and others [PrivCom] have repeatedly referred to us with the problem, that the characteristic terms imply various properties not accessible in a database search. On the other hand, explicit reference at each instance is not economic, and implicit properties may change with the period even for the same term.

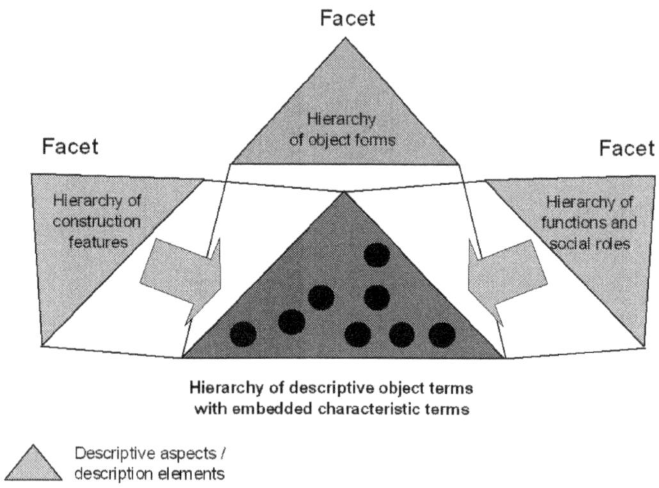

Figure 1: Conceptual Scheme of a multi faceted Hierarchy.

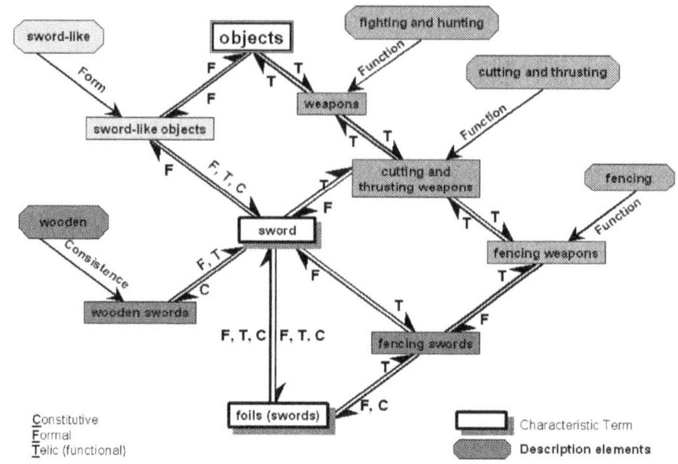

Figure 2: Sample hierarchy with typed broader/narrower relations.

Under this consideration we propose to systematically classify characteristic terms by descriptive terms. With this practice, direct classification of objects with descriptive terms and classification with the correct characteristic term can become consistent and compatible. As we have seen above, a thesaurus can serve as a language indexing device and as an information source as regards the conceptualisation of a domain. For the latter, systematic classification of descriptive terms by all aspects (qualia) is a remarkable finding aid for users, as detailed in section 6.

An object clearly identified as a model of its culture, like an ushebti, is of course best characterised by its vernacular, historical term. Only that will render all relevant aspects. An object, which cannot clearly be identified as one or the other, should better be classified by all possible aspects, in order to provide future researchers with all potential candidates for their requests (ensure recall). Similarly, a term with multiple potential meanings should be generalised (i.e. assigned broader term relations) by all possible meanings. Equally prominent in archaeology are cases, were we try to identify unknown models of a culture, in particular of prehistoric ones. In such an event, we may use combinations of descriptive terms and eventually invent a new characteristic term, a "coined term", when our knowledge about the kind of items seems to stabilise.

This is not in contradiction to defining its proper scientifically established meaning by respective scope notes. This double practice not only helps to explain vernacular terms or their equivalents to the interested scholar, but also provides the correct generalisations for query expansion in the use as an indexing language. Even more, shifts in meaning could be expressed by specialising a term into its local and temporal variants, and generalising them differently. For example the most ancient Egyptian items that we have encountered are burial items, but their later equivalents are not [PrivCom], (contemporary secular equivalents are typically not preserved).

5.1. Examples

In order to illustrate this theory, let us in the following assume, that the relevant qualia are: function, form, construction. To create the proposed metastructure, we may provide an initial vocabulary of expressions of function (like SHIC), of forms, and of constructive features relevant to the set of objects under consideration. From these, we create the set of classes of objects, which are characterised by precisely one feature of the initial vocabulary, e.g. "objects for warfare", "objects for eating and drinking", "rectangular objects", "built objects". If the initial vocabularies are organised by hyponymy (broader/narrower terms), this procedure creates a "parallel hierarchy" (Soergel 1995:369) for each aspect (quale). These would be the primitive descriptive concepts. Each hierarchy could be headed by a node label like "objects by function" etc., even though the practice is to enumerate only a flat list under each node label. Any characteristic object concept should be describable as a narrower term of some combination of these classes. In this theoretic picture, characteristic concepts appear dispersed in the "pool" of all possible combinations of the primitive descriptive concepts, as figure 1 suggests.

In practice, two factors cause severe distortions to this picture: Properties of the formal aspect and the relations between the characteristic concepts. The formal aspect or quale seems to play a dominant role in the discrimination of characteristic concepts. Pustejovsky refers to it as "basic", without further justification. We observe a basic difficulty in the defining of expressive sets of shapes without the reference to objects types, however we easily derive shapes from object types: we may talk about a cross-shaped ground plan, a hammer-like fish ("hammerhead shark"), etc. This does not hold true for the other qualia. This is further confirmed by the dominance of form criteria in the AAT. As described in chapter 6, the Greek National Archive of Monuments has introduced the formal aspect as «ειδος», literally translated as "kind, model or article".

It is beyond the scope of this work to analyse the deeper reasons behind that. It seems economical to take characteristic concepts themselves per default as sufficient to represent their formal aspect. Consequently, an independent vocabulary of form is only created for those forms not captured by the characteristic terms themselves, and the total of characteristic terms is seen as a source to generate respective form expressions, e.g. with a "like" operator.

Beyond that we observe, that characteristic terms of some higher level of abstraction may be specialised themselves by one or the other aspect, or all aspects together. In the AAT, the guide terms appear characteristically at the top and then at any level without a recognisable system. Our interpretation is that actually any characteristic term is potentially specialised by all aspects at any level, however, these potential places in the polyhierarchical grid may not always be occupied with actual established terms. The current practice to introduce a node label in order to announce the aspect of making narrower terms (NT) has two limitations: firstly, it cannot be continued on the next level. Another node label is needed. Secondly, it does not say, which aspect of the inversion, the broader term relation (BT), entails. This is actually not of a trivial nature. There are combinations of all aspects between the two directions of a BT/NT relation. The broader term relation is actually the one that classifies the characteristic concept, whereas the narrower term relation is the one we need to systematically create the thesaurus top to down and to expand higher concepts into individual classification concepts.

Instead of node labels, it would be better for us to introduce different types of BT/NT relations, a real metastructure feature, which goes beyond ISO2788. The idea is illustrated in figure 2, around the characteristic concepts "swords" and "foils (swords)". A type determinator in brackets accompanies the latter term, in order to disambiguate the concept from homonyms. The right branch (weapons...) shows part of an object hierarchy induced by activities "fighting and hunting", "cutting and thrusting". These pure functional concepts become specialised by the characteristic term "sword", which adds a morphological and constitutive aspect. The latter is again specialised by the functional aspect "fencing", and in turn by the characteristic concept "foils (swords)".

On the left side, we see the morphological aspect, starting from "sword-like", and penetrating through the concepts "swords" and "foils (swords)". "Fencing swords" takes a hybrid position. Further, we show a constitutive aspect, the "wooden swords". In the hierarchy here, wooden swords appear as weapons, and to our knowledge at least in medieval Japan, wooden swords were in use as respectable weapons. Alternatively, one could put wooden swords as a formal/constitutive hybrid under "sword-like objects", and distinguish between "wooden swords (toys)" and "wooden swords (weapons)". From this picture, we can easily recognise how the generalisation or specialisation has a direction towards the formal, the constitutive, and the functional aspect. A similar graph can be seen in Pustejovsky (1995:145).

"Weapons" is an established, "historical" term and concept, which is purely functional. Obviously, we cannot assign any prototypical shape to "weapon", as we do to "sword". As such, we regard it as descriptive and not characteristic. However, in general, non-formal terms are more frequently compounds than the others. In particular concepts which complement gaps in the lattice of traditional terms in order to cover relevant feature combinations compounds with terms from the participating aspects can be noticed, as expected from the faceted classification schemes presented above. The difference of our position here is that we cannot replace a part of the hierarchy by purely descriptive compounds as propagated by current indexing systems, because characteristic terms and traditional terms penetrate such hierarchies. A solution to embed these in a comprehensive way must be found, in a scheme of dynamic feature combination, as indicated in this paper.

6. Polydeykis, a practical experience

POLYDEUKIS is a general purpose, ontological Thesaurus aimed at consistently classifying cultural and archaeological terms. It is being elaborated by the Directorate of the Archive of Monuments of the Greek Ministry of Culture in co-operation with ICS-FORTH. It has reached its first state of completion, where the basic principles can be verified and further refinement can be determined. Its appellation was coined to honour Polydeukis, a second-century BC Alexandrian who authored a Lexicon of terms of the Attic dialect. The specific activities of the Directorate of Monuments have instigated a thematologically multi-faceted logical organisation of cultural and Archaeological terms contained in Polydeukis. Indeed, the Directorate handles an archive of extensive and multi-faceted data concerning cultural monuments. These are the material traces of coherent human activity spanning over several millenniums from a geographical area considerably larger than the Modern Greek State. The influences exerted by the Hellenic culture over other cultural environments with which it held converse and still does, are very broad and remarkably fertile.

Consequently, all objects under investigation are characterised by an exceptional variegation of terminological designation and distinction into characteristic concepts. The analytical overview of concepts in Polydeukis follows an anthropocentric logical structure, that is to say, it observes closely, albeit up to a certain degree, the relationship which man (as a living being) has with his fellow humans, and with the creations created by himself or nature. Every human being, for instance, is a biological entity designated by his parents (creators), the time and place of birth, name, physiognomy and bodily form, personality, biological structure, as well as his activities, and the ambient wherein he lives and works.

In analogy, the logical organisation of Polydeukis establishes a conceptual anthropocentric structure for every type of object/ monument, to which all characteristic historic terms designating and specifying it are linked through a system of guide terms. The analysis of characteristic concepts conducted so far yielded the formulation of a relatively repetitive conceptual structure for their every set.

Let us illustrate the above by the example of the concept "temple" (in Greek "Ναος"). This concept has two designated uses according to Polydeukis: a descriptive and a characteristic historical one. The descriptive concept "temple" designates every single building of religious function, regardless of the religion or the religious tradition it belongs to. As such, it is a specialisation of a morphological descriptive term "building", and the functional term "object from religious life". "Temple" is further specialised into characteristic historic terms by the particular religious tradition it belongs to, i.e., "sanctuary, temenos, temple, dominicium, mosque" (temple appears again in its narrower sense). Associated with it are characteristic concepts of temple "elements", functional parts like: "altar, nave, opisthodomos, diakonikon, prothesis, choir, narthex, royal doors, mihrab, minbar" and structural parts like: "capital, architrave, chancel barrier, frieze, pediment, mukarnas, etc.". In a similar way, each characteristic concept is embedded in a conceptual structure capturing its social, functional, morphological and stylistic significance with the corresponding terminology.

The conceptual structure underlying all terms contained in Polydeukis also follows the above-mentioned anthropocentric model. The distinctions of concepts in Polydeukis are purely methodological: they are meant to convey not the levels of the very nature of things, but the levels of man's cognition according to the specific standpoint by which he searches and understands the

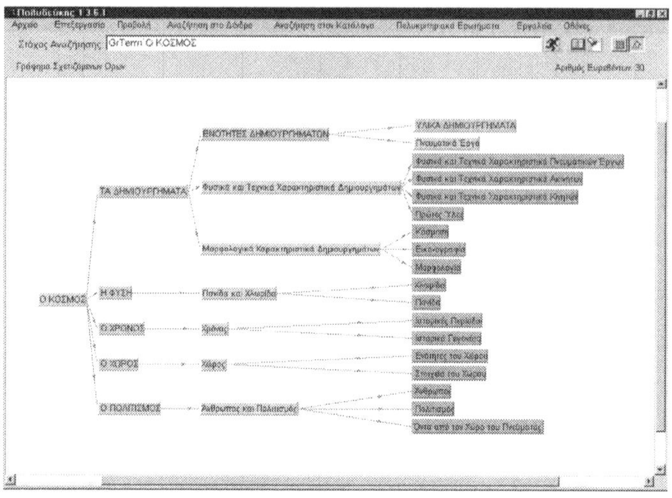

Figure 3: Screen dump from Polydeykis, "The Kosmos".

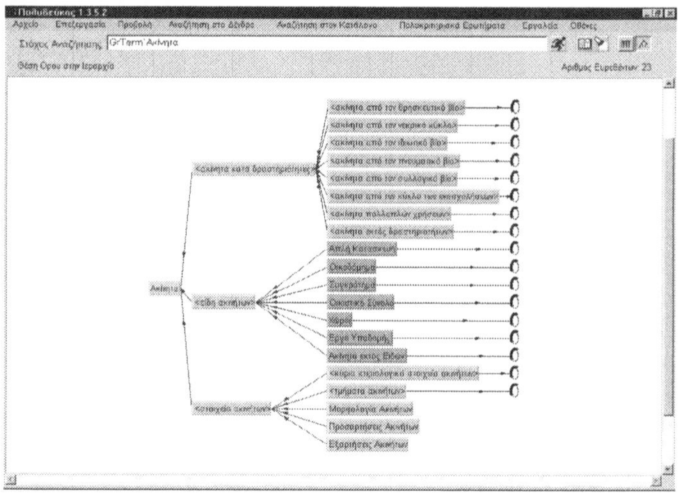

Figure 4: Screen dump from Polydeykis, immobile objects hierarchy.

nature of things. For that purpose, Polydeukis uses the following major facets (see figure 3):

- Kosmos, the world as subject
- Living Nature, as historical subject
- Culture and Civilisation
- Space
- Time
- Creations, the man-made world:

  Material creations
  Conceptual works
  Associated concepts: Stylistic, physical and technical characteristics.

Hereby "Kosmos" includes other facets as subjects of representations. The "Living Nature" is organised by the relation to man, habitat and feeding habits, rather than genealogy. The man-made objects (as a formal aspect "ειδος"), their associated concepts (constitutive aspect) and relevant social/functional terms from the Culture and Civilisation facet are combined systematically between each other in two levels of hierarchy of descriptive terms into a kind of regular grid. The combinations are announced group wise by guide terms like "functional parts of immobile objects". Under these, the characteristic terms are introduced, and the hierarchy continues on demand with the multiple criteria. E.g. "immobile objects" are split into "simple constructions", "buildings", "built complexes", "housing areas", "open spaces", "other immobile objects", whereas activities are split into "religious life", "burial rites and functions", "private life", etc. (see figure 4). The combination of "buildings" and "religious life" leads to, for example, "temple", as seen above.

This kind of hierarchical layout allows for the formation of a common structural environment for the descriptive and the historical/characteristic level. According to such an organising principle of the Polydeukis material, the higher we are in the hierarchy of terms, the more we function on a level of descriptive categorisation. On the other hand, the more we descend from the terms at the top, the more we are apt to encounter characteristic historic terms. The main characteristic of this superintending view of facets is that the underlying synthetic structuring of terms follows a flexible policy of conceptual integration which is capable of registering the whole spectrum of variegations of concepts whether descriptive or historic/characteristic. Furthermore, it allows for the formation of broader or differentiated descriptive categorisations for the same characteristic concepts.

The first experience with users is quite encouraging. The systematic structure of the upper levels and the explicit relation to descriptive concepts provide a remarkable orientation for the user, and we got enough positive responses. Characteristic terms seem to "find a natural place" in the hierarchy. On the other hand, the creation of combinations leads to a kind of "combinatorial explosion", and is not easy to be maintained. Also, the introduction of guide terms at all levels makes the presentation "heavy". Speaking with Ranganathan, we have successfully ignored the limitations on the notational plane, overcome the verbal plane by introducing compound terms, and satisfied the idea plane. Now, the next step is to introduce a suitable notation, and to improve the compound term creation.

7. Conclusions

The experience with Polydeykis, the analysis of AAT guide terms and the linguistic theory of qualia structures makes us confident, that the analysis of a thesaurus by placing object terms into a limited number of aspects and the respective logical and notational organisation of thesauri is feasible and can considerably improve user orientation, precision of classification and reasoning during retrieval. We have introduced the notion of the "descriptive" versus "characteristic" concept, which seems to be a viable distinction able to account for certain phenomena in the classification. An initial distinction into "modern" and "historical" was soon abandoned, as we could find all kinds of transitions between historical and modern terms with respect to their significance. We would be interested to learn if similar notions have been developed in philosophy or cognitive science.

We have encountered limitations of the current notational practice. We believe, that these can be overcome by mechanisms of virtual pre co-ordination based on Description Logic that we have conducted our first experiments with (Ntoas 1999). Eventually also new modes of graphical presentations need to be developed. However, several open questions about thesaurus structuring remain, for example:

1. Should there be a relation between certain qualia and the level of abstraction, or between descriptive and characteristic terms in a practical thesaurus? E.g. do functional terms tend to be more abstract than characteristic or formal ones (as psychological studies suggest)?
2. Are the qualia of common sense language the same to those of a specialist vocabulary? Are the latter similar, a part of the latter or can they be completely alien, e.g. in some technical fields?
3. A formalisation of multi-typed hyponymy for thesauri, i.e. broader/narrower term relations specific to qualia, should be developed.
4. To what detail should sense distinction in thesauri be made in order to account for historic concept shifts? Where is the optimal precision/complexity?

Finally, this approach may be useful to merge or make the transition easier between multiple thesauri in networked environments, as it can give a better account for reasons and aspects of broader/narrower term distinctions.

References

BECHHOFER, S., GOBLE, C.A, 1999. *Classification Based Navigation and Retrieval for Picture Archives*. IFIP WG2.6 Conference on Data Semantics, DS8 , Rotorua, New Zealand.

BORGIDA, A., 1995. Description Logics in Data Management. *IEEE Transactions on Knowledge and Data Engineering*, 7(5): 671-682.

CONSTANTOPOULOS, P. and DOERR, M., 1995. Component Classification in the Software Information Base. In Nierstrasz, O. and Tsichritzis, D. (eds.), *Object-Oriented Software Composition*. Prentice-Hall.

CHEN, H., LYNCH, K., BASU, K. and NG, D., 1993. *Generating, Integrating, and Activating Thesauri for Concept-based Document retrieval*. IEEE Expert, 8(2): 25-34.

[DMOZ] DMOZ Open Directory Project http://dmoz.org/

[DL] More information can be found for instance on http://www.ida.liu.se/labs/iislab/people/patla/DL/index.html

DOERR, M. and FUNDULAKI, I., 1998. *A proposal on extended interthesaurus links semantics*. Heraklion - Crete, Greece: FORTH, Institute of Computer Science - Technical Report FORTH-ICS/TR-215.

DOERR, M. and CROFTS, N., 1999. *Electronic Esperanto: The Role of the Object Oriented CIDOC Reference Model*. Proc. of the ICHIM'99, Washington, DC.

[Encarta] Microsoft Corporation: "*Mircrosoft Encarta Reference Suite 2000*", CD SET X04-99037, 1993-1999.

GETTY ART HISTORY INFORMATION PROGRAM AHIP, 1994. *Introduction to the Art & Architecture Thesaurus*. Published on behalf of The Getty Art History Information Program, Oxford University Press, New York.

HAZEWINKEL, M., 1996. *Enriched Thesauri and their Uses in Information Retrieval and Storage*. First Delos Workshop, An Overview on Projects and Research Activities in Digital Library Related Fields. INRIA - Sophia Antipolis. *ERCIM Workshop Proceedings* - No. 97-W001.

[MDA97] Museum Documentation Association, English Heritage & Royal Commission on the Historical Monuments of England, "*mda Archaeological Objects Thesaurus*", 1997, http://www.open.gov.uk/mdocassn/archobj/archcon.htm

MICHALSKI, R.S., 1993. *Beyond Prototypes and Frames: The Two-tiered Concept Representation*. In Van Mechelen, I., Hampton, J., Michalski, R. and Theuns, P. (eds.), *Categories and Concepts, Theoretical Views and Inductive Data Analysis*, Academic Press, San Diego.

MILLER, A., GEORGE, BECKWITH, RICHARD, FELLBAUM, CHRISTIANE, GROSS, DEREK and MILLER, K., 1993. *Introduction to WordNet: An On-Line Lexical Database*.

MURRAY-RUST, P. and WEST, L., 1998. Knowledge, Language and Semantics: XML and VHG (TM). *ASLIB Managing Information* vol.5, no.4: 34-36.

NTOAS, D., 1999. *Economy and consistency in Thesauri*. Heraklion - Crete, Greece: FORTH, Institute of Computer Science -Technical Report FORTH-ICS-TR-262.

PINKER, S., 1994. *The Language Instinct*. New York: W. Morrow and Co.

POLLITT, S., 1997a. *Interactive Information Retrieval based on Faceted Classification using Views-Knowledge Organisation for Information Retrieval*. Proceedings of the 6th International Study Conference on Classification, University College, London 16-19 June 1997. http://www.hud.ac.uk/schools/cedar/dorking.htm

POLLITT, S., 1997b. *The key role of classification and indexing in view-based searching*. IFLA '97 Copenhagen Aug 31 - Sept 3 1997. 63rd IFLA General Conference Booklet 4. Section on Classification and Indexing Session 95 Paper 009-CLASS-1-E.

PRIETO-DIAZ, R. and FREEMAN, P., 1987. *Classifying Software for Reusability*, IEEE Software: 6-16.

[PrivCom] Private Communication with Tanya Szraiber, British Museum.

PUSTEJOVSKY, J., 1995. *The Generative Lexicon*. MIT Press. ISBN 0-262-16158-3.

RANGANATHAN, S.R., 1965. *A descriptive account of Colon Classification*. Bangalore: Sarada Ranganathan Endowment for Library Science.

RECTOR, A., BECHHOFER S., GOBLE, C., HORROCKS I., NOWLAN, W. and SOLOMON, W., 1997. *The GRAIL Concept Modelling Language for Medical Terminology*. THE GALEN Project, Medical Informatics Group, Department of Computer Science, University of Manchester, England.

[SHIC93] The SHIC Working Party, "Social, Historical and Industrial Classification", The Museum Documentation Association, Cambridge 1993, ISBN 0 905963 91 1.

SVENONIUS, E., 1989. *Design of Controlled Vocabularies*. Encyclopaedia of Library and Information Science, New York, Marcel Dekker: 82-109.

[TEI] Text Encoding Initiative (http://www.uic.edu/orgs/tei), Chapter 13, "Terminological Databases" http://www.uic.edu/orgs/tei/p3/doc/p3te.txt.

VICKERY, B.C., 1960. Thesaurus - A new word in documentation. *Journal of Documentation*, 16:4: 181-89.

WELTY, C., and JENKINS, J., 1999. Welty, Formal Ontology for Subject. *Knowledge and Data Engineering*. 31(2): 155-182. Elsevier.

WELTY, C., and IDE, N., 1999. Using the Right Tools: Enhancing Retrieval from Marked-up Documents. *Journal of Computers and the Humanities*, 33(1-2). Kluwer.

SOERGEL, D., 1995. The art and architecture thesaurus (AAT): A critical appraisal. *Visual Resources*, X: 369-400.

SMR in New Clothes: The Danish National Record of Sites and Monuments on the Verge of a New Era

Lars Bagge Nielsen and Henrik Jarl Hansen

The National Archaeological Record, The National Museum of Denmark

Ny Vestergade 11, DK-1471 Copenhagen K, Denmark

e-mails: bagge@natmus.dk, jarl@natmus.dk

Claus Dam

The National Archaeological Record, c/o Dept. of Prehistoric Archaeology

Moesgaard, DK-8270 Hojbjerg, Denmark

e-mail: farkcd@moes.hum.au.dk

Abstract

The Danish National Archaeological Record of Sites and Monuments (DKC) is at the moment in a transitional phase - in two respects:

One is that DKC, as an institution, is changing from its traditional SMR function of registering information to providing a wider service for use of cultural historical information technology.

The other is that the DKC-database is progressing from a traditional SMR-base to become a Cultural-Historical Atlas of Denmark – a portal to digital knowledge about the cultural history from past to present.

This paper provides a brief presentation of these two developments.

Key words: SMR, database, cultural-historical atlas, GIS, internet mapping, digital excavation plans, multidisciplinary projects, multimedia

1. DKC in a transitional phase

The Danish National Record of Cultural History (in short: DKC) is a national database of archaeological sites and monuments. The register, which is housed at the National Museum, was founded in 1984.

The original basis for the Central Register is the so-called "Parish Record" which comprises comprehensive text and map data concerning archaeological monuments and finds.

During the years from 1873 to 1937 archaeologists from the National Museum's staff visited the whole of Denmark. The result was a systematic nationwide record of Danish ancient monuments and locations for many of the objects in the present National Museum's collections. In 1982 DKC started the digitisation of the "Parish Record", a task which has not yet been completed. The database can be accessed from the Internet with full access only for professional users.

During the last decade a number of new developments have occurred at DKC. One of the reasons for this is, of course, the rapid development in computer technologies, a development that is enabling the implementation of many ideas that before were not truly viable.

DKC is at the moment in a transitional phase - in two respects:

- DKC, as an institution, is changing from its traditional SMR function of registering information to providing a wider service for use of cultural historical information technology.
- The other is that the DKC-database is progressing from a traditional SMR-base to become a Cultural-Historical Atlas of Denmark – a portal to digital knowledge about the cultural history from past to present.

This paper will provide a brief presentation of these two developments.

2. Status: The DKC today

The main task for DKC is maintaining the national database of archaeological sites. As of now the database contains more than 140.000 sites at three different levels of registration (figure 1). It is clear from this illustration that care is necessary when making distribution maps at a national scale.

3150 new site reports were received from the Danish museums during the year 1999 – a fairly typical annual average.

The current procedure for a new site report is shown in figure 2: Information from a new archaeological location is brought to the museum where a new site report form is sent through the mail system to DKC. The information is then typed into the database.

2.1. DKC online

DKC online is a presentation of the data in the DKC database on the net – it is the DKC front to the public (http://www.dkc.natmus.dk). There are several entry points to the online database.

First the user is shown a map of Denmark with a geographical search procedure. By clicking on an area, a new menu appears with a choice of various maps at different scales. The public can

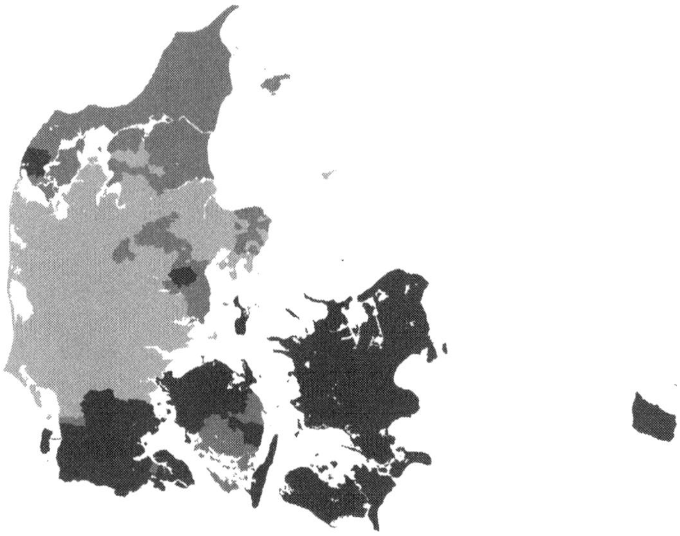

Figure 1: Registration status by parish. Dark parishes are fully registered, meaning updated with all information from the original "Parish Record", supplied with information from other National Museums archives. A check of the geographical position has also been made. The medium gray marks are partly registered parishes, which contain information from the original Parish Record and a site classification (dating and category) based on "Parish Record" information only. Light gray denotes parishes, where the location of the sites is registered, but only a provisional classification has been made.

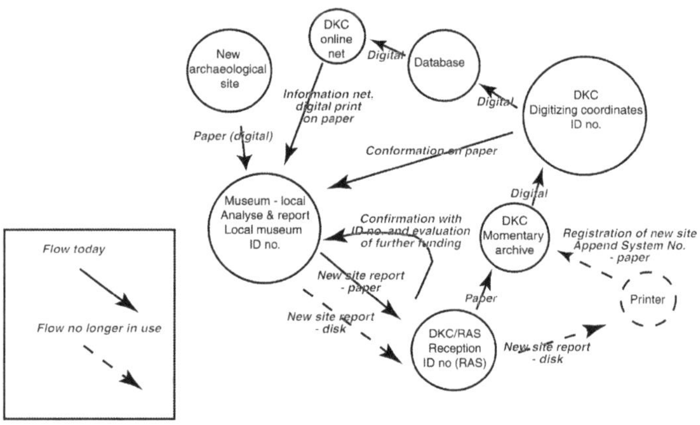

Figure 2: The current procedure for the processing of new site reports.

view the distribution on maps in 1:200,000, but to access 1:100,000, 1:50,000 or 1:25,000 scales (figure 3) a password is needed, thereby limiting the possibility for the public finding the exact location of a find. The user can click on a dot representing a find and further restricted information can be obtained.

The other approach is topographic (figure 4). Here the user is supposed to have more information about the database. Information on a find in the form of an illustration can be presented, for instance the burial shown in figure 5.

3. Non-registration tasks

As mentioned in the introduction, the main role of DKC is the electronic registration of various cultural historical data. This, of course, is still the DKC's function, but there are increasing expectations in the museum community that DKC will function as a

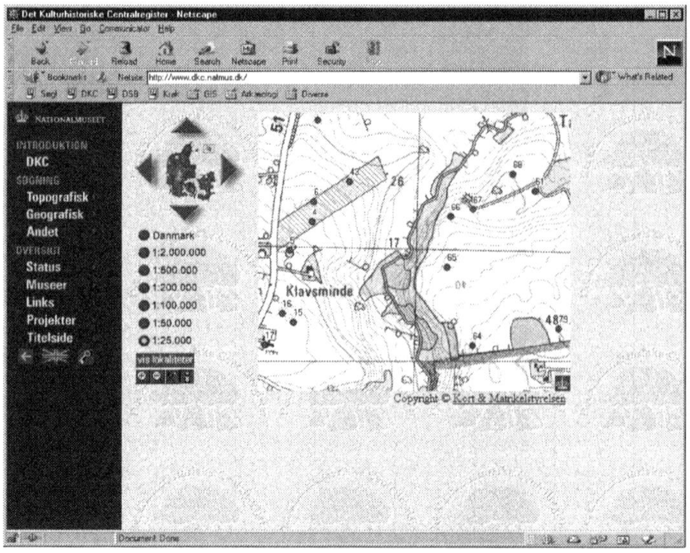

Figure 3: DKC online: Geographic search.

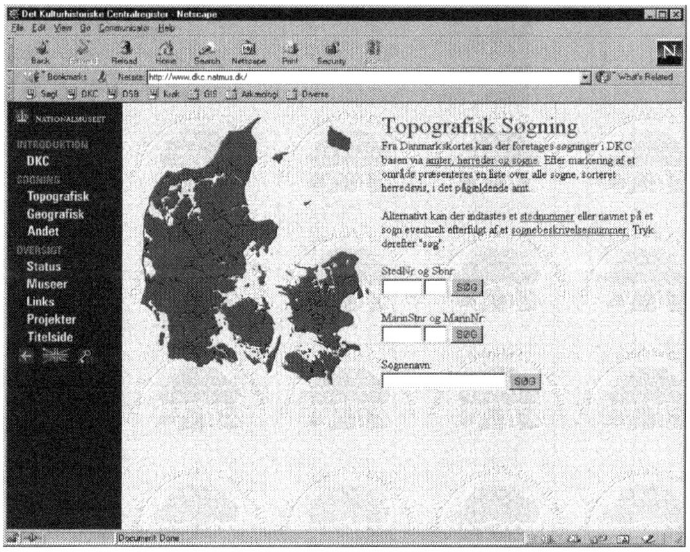

Figure 4: DKC online: Topographic search.

centre for cultural historical IT knowledge capable of any task involving IT and cultural historical data.

Examples of these tasks are the production of distribution maps and other types of illustration for publication, digitising map data, GIS analyses, programming tasks etc.

We can do all of this, of course, but with limited staff and no signs of extra funds in the near future, it is a difficult situation. So there is no alternative other than to charge for these extra services, but even in this situation it can be difficult sometimes to find someone with time to do a job.

4. Introducing GIS

One of the recent initiatives by DKC has been to arrange introductory courses in MapInfo GIS for representatives from the museums. A number of these sessions have already taken place, and more are still needed. In fact we were somewhat surprised to see the introductory courses become such a success. It is inspiring to see the willingness among the museums to accept new technologies – and to realise the potential in these technologies. So far

Figure 5: "A barrow from ancient times at Raklev on Refsnaes" by Thomas Lundbye (1839). Thorvaldsens Museum.

about 75 percent of the archaeological museums in Denmark have purchased a MapInfo license.

A substantial number of man-hours have been invested in the running of the introductory courses. But as a long-term investment there is no doubt it pays off for the archaeological community as a major leap forward in persuading the museums to "go digital".

Courses for experienced users will take place at a later date.

Why MapInfo? In our opinion it is the best GIS in terms of being easy to learn, easy to use –the fact that we struck a very fair deal with the Danish MapInfo distributor KampSax also helped.

This program of courses in MapInfo GIS is the direct result of a wish to expand the subjects recorded within DKC to include digital excavation plans. By using GIS the documentation and data will be more easily accessible in the future. In an online perspective, it offers interesting possibilities.

Imagine clicking your way through map levels, starting at the national level with your site represented as a dot, zooming through regional levels where the site perimeter is represented as a polygon and ending up at a local level with the plan of your Iron Age village shown against the background of a contemporary landscape as shown in a high resolution digital orthophotograph.

5. A National Record of Social History

So far DKC has mainly recorded archaeological information, but we are now facing the challenge of dealing with modern ethnological information. The background for this is the fact that we have more than 100 local, but professionally managed, museums in Denmark. The majority of these museums are working in the field of ethnology and social history, and they are concerned with the period after 1660. Most of them have their own electronic records of things, archives, photos etc. in their possession. By and large they have conceded to a common structure of recording this information, more than half of the museums using the same software program specially developed by an institution under the Danish Ministry of Cultural Affairs.

In the wake of general IT development the idea is to try to gather all these local databases, merge them into one central, national record of social history and present this comprehensive record on the Internet. The idea has materialized in so far as we are now running a pilot project in small scale. We hope to be able to present a preliminary result of a Cultural-Historical Information System on the Internet by the end of this year.

The illustration in figure 5 represents the various approaches one can adopt when approaching cultural history. The archaeologist will define the monument as a burial with a dating etc., while the ethnologist sees a national romantic representation. Both have to be incorporated in the structure of DKC and perhaps within a common database. The aim for the DKC at present is to deal with these widely variable perceptions.

6. DKC as a provider of digital maps

Another of the more recent functions of DKC is to provide digital map material to the Danish museums. Through some very favourable deals with The National Survey and Cadastre and Kampsax, the two largest geo-data agencies in Denmark, it is now possible to offer the Danish museums a wide range of historical and modern digital map material as well as digital orthophotos at a reasonable price.

These data are also available to the museums in paper-form as map sheets with archaeological locations plotted in.

7. Project involvement

At the moment DKC is involved in two large multidisciplinary projects: "Changing Landscapes" and "Agrar 2000". The two projects have a number of similarities, focusing as they do on the cultural landscape and with a strong emphasis on strategic aspects, one of the aims being to develop methods and tools of use in the management of the landscape. Both use MapInfo GIS.

7.1. Changing Landscapes

The DKC subproject in Changing Landscapes (http://www.ou.dk/Hum/ForandLand/English/Index.htm) deals with the development of a Cultural Landscape information System - or CLIS in short.

The aims of the CLIS-project are:

- To develop an information system containing digital information about the cultural landscape in the past as well as in the present - information being presented in the form of geo-data and metadata about the geo-data.
- To generate data for the information system through digitisation of a series of analogue maps from the oldest topologically accurate maps of 200 years ago through to the maps of today.
- To give access to existing DKC-data in a way that is easier to interpret, i.e. where data points have been transformed into areas.
- To produce thematic layers with a predictive content.
- To make information available to landscape managers and researchers on the internet and as thematic layers for use within a GIS context.

There is a real need for an Information System containing factual as well as hypothetical information about the location of sites in the landscape. It is not only a question of which archaeological sites exist in a given area, but also their presumed extent and what type of site is likely to be found in that area.

DKC is frequently consulted at the planning stage of major public projects, including the construction of new railway lines or highways, with questions about the number and location of sites within or near the project area.

One of the more important issues is to develop methods to transform the point data to surfaces via analysis of landscape characteristics, thus providing a more adequate representation of a prehistoric reality.

The use of historical as well as modern map material makes it possible to study the changes in land use over time and compare this information with the archaeological record in order to assess the significance of archaeological sites.

7.2. Agrar2000

The other multidisciplinary project involving DKC is Agrar2000 (www.natmus.dk/Agrar2000). The title of the DKC-subproject is "*The agrarian Landscape: Land-use and regional variation during the Roman Iron Age, Early and High Middle Ages*". The aim of the project is to develop methods to refine the information in the DKC database in order to enable a qualitative improvement of these data and increase data homogeneity, so that the database can more confidently be used in analyses aimed at elucidating the development of the agrarian landscape and prehistoric settlement patterns.

The research is taking place at 22 different locations spread around the country. The distribution of lakes, which provide the potential for pollen analysis, was the principal determining factor within this project. Collection and critical evaluation of the archaeological evidence was limited to within the pollen catchments around chosen lake sites.

The end result of the sub-project is expected to be a set of tools which can provide plausible definitions, and clarifications, of archaeological data in the DKC-database, and which can be used in mapping the interaction between human populations and agrarian landscape.

8. Multimedia project involvements

In 1997 the Danish National Museum made its first electronic publication *Gods and Graves* (http://www.natmus.dk/kulturnet) available on the Internet as part of the pilot project *Culture Net Denmark*. This publication, aimed at the general public, provides information on some of the more important sites from the Early Bronze Age in Denmark, and is unique in as far as it is the first time that the two large databases the DKC-base and the National Museum's Object Register of the Department of Prehistory and Early History were combined and presented on the internet. Since the initial publication, *Gods and graves* has been updated to include the Late Bronze Age.

9. Danish Museums online

Another *Culture Net Denmark* project involving DKC is *DMOL* - Danish Museums OnLine (http://www.dmol.dk/). *DMOL* will provide a common gateway to information on individual museums – cultural-historical museums as well as art museums. *DMOL* itself

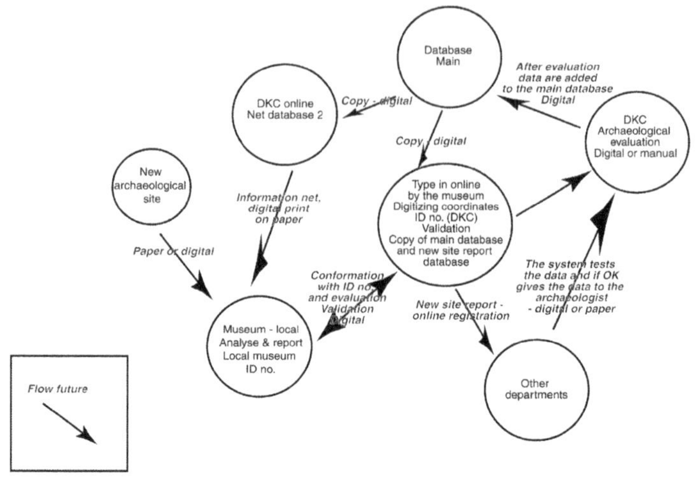

Figure 6: The future procedure for the processing of new site reports.

can be considered a virtual museum, containing key information on other museums and presenting some of the major artefacts or works of art from each institution. Essentially these will represent the highlights of each museum.

Altogether 150 museums will be represented in *DMOL*, with links to their own home pages, enabling the public to get more detailed information.

Apart from these web-projects, DKC has been involved in CD-ROM production in co-operation with other departments at the National Museum and the National Survey and Cadastre.

10. Online improvements

The DKC-developed GIS used with DKC online is about to be replaced with MapInfo Corporation's MapXtreme mapping server. This will represent a quantum leap forward in that it will give the professional users of DKC online a whole new suite of possibilities. At present, as was mentioned above, there are several ways of searching the database, but with MapXtreme it will be possible not only to search but also to spatially analyse DKC data on interactive maps. Increased flexibility, thematic mapping and full layer control are some of the possibilities. But one of the main advantages is the possibility to show not only point data, but also line and polygon objects.

MapXtreme is also a sensible choice in respect of new site reports. The complicated flow of information connected with these was shown in figure 2, but with this new mapping server software it will be possible to have online registration of new sites with entry fields for information and a map window for point and click marking of the location - as a point, a line or a polygon (figure 6).

11. Where are we going from here?

DKCs involvement in the Changing Landscapes and Agrar2000 projects is nearing its conclusion, but steps are presently being taken to start new initiatives and to develop these multidisciplinary projects still further.

We have discussed a number of new initiatives in this paper:

- The creation of a database and structure within the framework of DKC to register and analyse ethnological data through to modern times.
- The GIS courses which are helping museums to enter the digital era.
- The registration of digital excavation plans.
- The introduction of new map server software enabling the implementation of a whole new degree of flexibility and possibilities for digital archaeology on the web.
- DKC as a provider of digital map material.
- Involvement in multimedia projects.
- Participation in multidisciplinary projects.
- The transformation of point data to surfaces.

With this whole suite of initiatives, DKC has taken several steps towards its planned transition in function and content.

To the Danish museum community, we are firmly established as a cultural historical IT knowledge centre, although the organizational aspects still remain to be sorted out. With the transformation of point data to areas within the DKC-database and the introduction of ethnological data, we are also rapidly moving away from being just a traditional SMR and towards the goal of being more of a Cultural-Historical Atlas of Denmark and a Cultural-Historical Information System. In this way a firm foundation for DKC's new role as a portal to digital knowledge about cultural history from past to present has been laid down.

National Registries of Sites and Monuments in Norway – Developing GIS-based Databases

Evy Berg

Directorate of Cultural Heritage

Box 8196 Dep., N-0034 Oslo, Norway

e-mail: evb@ra.no

Abstract

The Directorate for Cultural Heritage of Norway is the owner of several national registries of sites, monuments and buildings. This is a presentation of some of the work in progress aimed at establishing these databases on a GIS-platform. Among these databases is the Registry of Archaeological remains; the Foundation for Nature Research and Cultural Heritage Research is responsible for updating and developing it into a GIS-based database in collaboration with the Directorate. Another important new database is the National Registry of Buildings. This paper presents the status for the work done so far, and includes a short description of plans for the future.

Key words: national databases, registries, GIS, archaeology, buildings

1. Introduction and historical background

For more than a century the sites, monuments and buildings of cultural-historical value have been recorded in various registries and by different institutions. The five Archaeological museums located in the cities of Oslo, Stavanger, Bergen, Trondheim and Tromsø (figure 1) concentrated their efforts towards recording prehistoric sites and monuments while the Medieval towns, churches and buildings were covered by Riksantikvaren, the forerunner of the Directorate of Cultural Heritage. The institution became a Directorate ten years ago, and took on a broader set of cultural heritage management responsibilities, including the management of prehistoric sites and monuments as well. It is now one of five environment directorates; the others are the Norwegian Pollution Control Authority, the Directorate for Nature Management, the Norwegian Mapping Authority and the Norwegian Polar Institute. The directorates reside under the Ministry of Environment. In addition, a new regional body of cultural heritage management on the county level was set up. Currently there are 19 regions, each with at least one archaeologist and one buildings specialist, who do the bulk of new registrations and perform the planning and management duties. The Archaeological museums now concentrate their efforts mainly towards excavation of prehistoric sites, while the Foundation for Nature Research and Cultural Heritage Research (NIKU) do the various kinds of registrations and conduct excavations in the Medieval towns.

2. Legal points of importance

The Norwegian Cultural Heritage Act states that all Cultural remains from before the Reformation, which means from before 1537 AD, are automatically protected by law. The same rule applies to Sami remains older than 100 years. The law thus distinguishes between two different age divisions concerning automatic protection, one static and one dynamic. The legal protection for both kinds automatically includes a safety zone stretching 5 meters out from the border of the site. This zone is easy to define around clearly delimited site types, but difficult to define in other cases. Figure 2 shows examples of this. The data used comes from two different registrations. The runic R on the map marks objects recorded as part of the Economic Mapping registrations in 1968, while the dwelling-sites and find-spots were recorded during a survey of plans for new railway tracks in 1995 (Berg 1995). The area between the two collections of grave mounds is cultivated land so the registration method was field walking. Dwelling sites are identified as collections of artefacts found within a restricted area, while find-spots are single artefacts with no immediately identifiable context. Represented among the find-spots on the map are an amber pearl and a broken whetstone. The security zone around a grave mound or a collection of grave mounds is relatively easy to define. Invisible sites like hunter/gatherer dwelling sites from the Stone Age or farming settlements from the Iron Age are far more difficult to define spatially, since their extensions are not known so well. These sites are defined on the basis of artefact finds in test-pits or in modern-day fields, and most often simply marked by a point on the map. Find-spots where single artefacts are found are not automatically protected, but are mainly used for prediction purposes at evaluating an area.

Changes in the Cultural Heritage Act, which are due to take effect in 2001, provide that registration of an automatically protected object or area shall be regarded as such until proven otherwise. The Directorate is the sole authority with the right to search for and record cultural Heritage remains according to this section of the Act, but may delegate this authority to the regional level, NIKU or the Archaeological museums, authorising them to carry out the registrations. Furthermore, the field registration must be followed by an entry into an *official, national registry*. A formal procedure for entry and deleting the records in the Registry must be worked out, because of the legal implications of entering an object into such a registry. This is a reversal of current legal procedures. In the past, when an object was partially or wholly destroyed, the cultural heritage authorities had to prove that it was automatically protected. Now the Directorate will be given the authority to decide whether an object is automatically protected or not.

The changes in the Act mean that we have to work out better procedures for informing owners of land with automatically protected sites about the location and type of object. Laymen are not expected to recognise sites like Stone Age dwelling sites with no visible markers above ground, or to understand that a pit in the

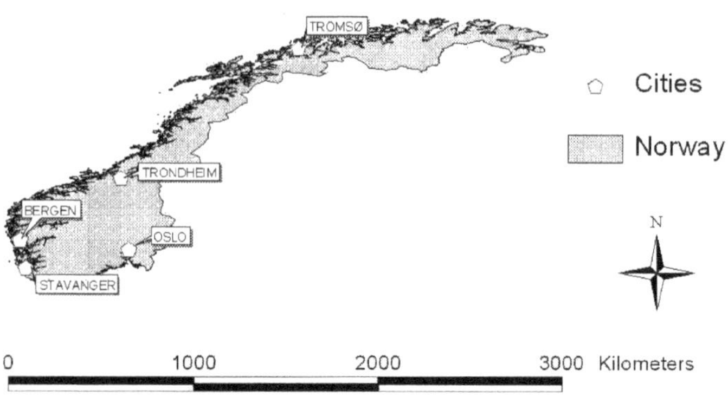

Figure 1: Map showing the location of the five cities with Archaeological museums responsible for recording Archaeological sites and monuments.

ground really is a Viking Age - Medieval coal-pit used in iron extraction.

More precise placement of the location of objects and areas by coordinates. The most widely used method of placement of registrations is still by marking the location on a large-scale map, and digitising it later on. With the increasing use of GPS to determine precise locations, this procedure will be reduced to one stage instead of two. Another important document is the Norwegian Mapping authority's new standard for property information (1999). This document clearly states that coordinates and a representation point must delimit the location of the areas protected by the Cultural Heritage Act. The decision on how the location of single objects is to be represented is left to the Cultural Heritage authorities. Here we will have to define the margin between object and area. Since we no longer map at a specific map scale, single objects, such as big grave-mounds, may cover an area large enough to be represented as a coordinate - enclosed area, not as a single point in the case of large map scales 1:1000, 1:500 or bigger.

Improving the quality of location data for existing sites and monuments already in the registry. There are several examples that this data is not always precise enough, and sometimes completely wrong. This work must be done mainly in connection with the planning of the work, new surveys or registrations in an area, by checking the location against the map and by new GPS measurements.

3. Existing databases

The table 1 shows the databases kept by the Directorate of Cultural Heritage. Of these only the National Registry of Buildings, the Archaeological Remains Registry and the Rock-Art database are available on-line for management authorities on different levels. The rest are strictly internal registries, and the Directorate must be contacted if anyone wants to get information about objects in them. This situation cannot continue as such, it will be essential to convert these registries into a more coherent and usable form.

3.1. National Registry of Buildings

This registry is a novelty that came into being February 2000. The contents are a merger of two building registries. One is the Buildings part of GAB (short for property, addresses and buildings in

Elvestad farm, Råde municipality, Østfold county

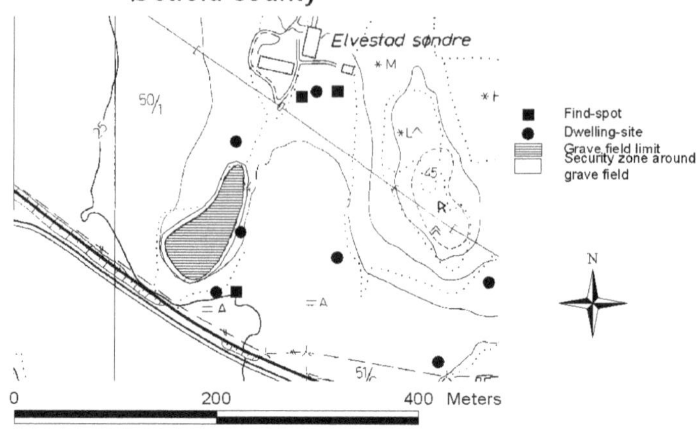

Figure 2: Elvestad farm, examples of various site types and security zone.

Existing databases	
Buildings and Installations	Archaeological remains
National Registry of Buildings	Archaeological remains Registry
Registry of legally protected Buildings and Installations	Rock-Art database
Church Registry	Medieval ruins Registry
Project databases:	Medieval cemetries Registry
Light-house Registry	
Industrial sites Registry	Post-Medieval sites not
Historical Gardens Registry	automatically protected by law
20th century Architecture	

Table 1: List of databases kept by the Directorate of Cultural Heritage

the Norwegian language) maintained by the Norwegian Mapping Authority. This registry contains technical information used by local administrations on the municipality level, mainly for younger buildings. Local authorities update this registry on a daily basis. However, for older buildings, the information is inconsistent. The second registry was the result of a 25-year programme called SEFRAK, to record all buildings from before 1900 AD in southern Norway. In the northernmost part of the country the limit was 1940, as large parts of northern Norway were burnt at the end of the Second World War. The Directorate of Cultural Heritage and the regional authorities on the county level maintained this registry. It was difficult to properly update it, especially the information on ownership changes. By merging the two essentially identical registries, in the sense that both were concerned with standing buildings, resulted in a single registry, which is far easier to update and maintain. All the cultural and historical information collected during the SEFRAK-buildings registration is found in the new National Buildings Registry. It is still operated by the Norwegian Mapping Authority. The regional cultural heritage management authorities will update cultural and historical information, while technical and ownership information relating to buildings will continue to be updated by municipal authorities. This registry is primarily a building registry, not a cultural and historical database. Of the 500,000 buildings older than 1900/1940 probably less than 50 % are of a cultural or historical value. Those of such value will need a place where broader information

Archaeological Remains Registry		
Data sources	Site-types recorded	Coverage
Economic Mapping registrations from 1963	Automatically protected sites,mainly visible types like grave-mounds	Populated areas and agricultural landin all counties
Topographic registrations predating Economic Mapping	Almost exclusively visible, monumental types	Populated areas and agricultural landin some counties
Project registrations	All kinds of automatically protected sites, visible and invisible	Interior and coastal areas affected by large-scale construction projects
Small-scale registrations andsurveys connected with zoning	All kinds of sites both automatically protected and post-Medieval sites visible and invisible	Nationwide, mostly populated andagricultural areas
Heritage management projects national and regional level	All kinds of sites both automatically protected and post-Medieval sites visible and invisible	Areas not well represented otherwise: mostly large interior woodland areas

Table 2: The Archaeological Remains Registry most important data sources.

than what the National Buildings Registry can provide will be stored.

All the data on the buildings in the registry is in digital form. During the year 2000 a small number of legally protected buildings registered in the internal database at the Directorate will be transferred to the National Buildings Registry and thus be digitised. These buildings were not a part of the SEFRAK programme registrations, and therefore could not automatically be included in the merging of the two registries.

3.2. The Archaeological Remains Registry

Table 2 shows the most important data sources that make up the Archaeological Remains Registry. The majority of registration work has been done in populated areas though some of it is associated with hydro-electrical projects, concentrated along river valleys in the mountainous parts of the country. Large parts of Norway's interior are therefore not covered by any kind of survey or registration.

Representation of different types of sites and monuments varies greatly. Visible monuments like grave mounds have the best geographical coverage, while invisible types like dwelling sites from the Early Mesolithic are to present day highly underrepresented. Site types common in woodland and outlying areas are also seriously underrepresented. They are mostly known from big projects where the registration methods covered all kinds of sites, not primarily the visible ones. This is mainly due to economic reasons: It is obviously more time consuming to look for invisible sites then to look for visible ones. At accepting or rejecting a construction plan the regional authorities, tend to use modern zoning registrations. This is an attempt to cover all kinds of sites, both automatically protected features, and the Sami remains. Large-scale registration projects connected with construction plans are often undertaken by NIKU, and also aim at including all kinds of cultural remains, regardless of visibility or age.

4. The purpose of the Database

The most important function of the database is Cultural Heritage management. The site type, age and location are prioritised. The information about the physical state of a site and whether or not it is included in an upkeep programme is recorded. Any legal dispensations must also be recorded. When a dispensation is granted,

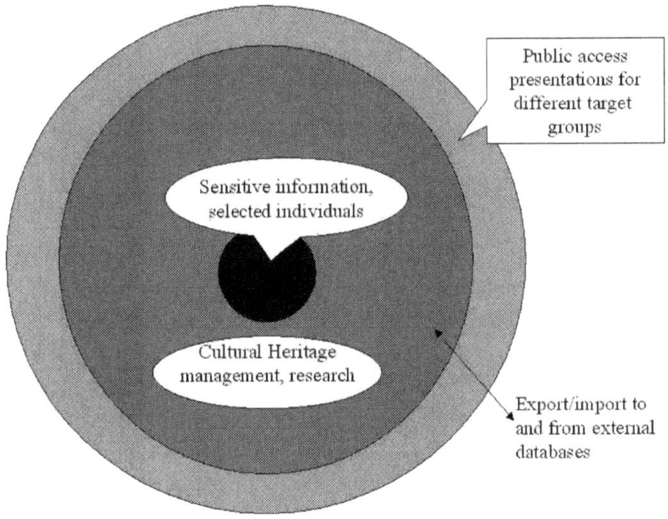

Figure 3: Principle of access to the Archaeological Remains Registry.

it is usually under the condition that excavations are conducted at the site and documented before construction works may commence. The Act contains several ways by which dispensations can be obtained. The most common way is in connection with zoning, but it also provides for finds occurring during construction work and for cases where there is no zoning plan in place. Another paragraph deals with research - excavations. The database must contain information about which paragraph and which section of that paragraph grants the dispensation. The status "automatically protected" is removed from the site, but the site itself is not deleted from the database. All information is kept for historical and research-purposes, but it no longer figures on the maps as a protected object or area.

5. Who will have access to the Database

Figure 3 shows the principle of various degrees of access to the database. The outer ring represents information directed towards the public in the form of presentations aimed at specific target groups. The second level contains the bulk of the information in the database, except for extremely sensitive information. Access is provided by password. The primary users will be the Cultural Heritage authorities, Archaeological museums and university teaching establishments, and the Foundation for Nature Research

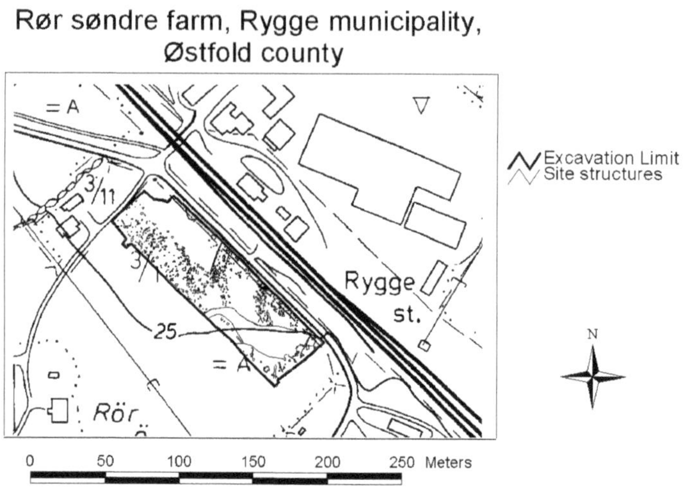

Figure 4: Rør søndre farm with excavated area and structures.

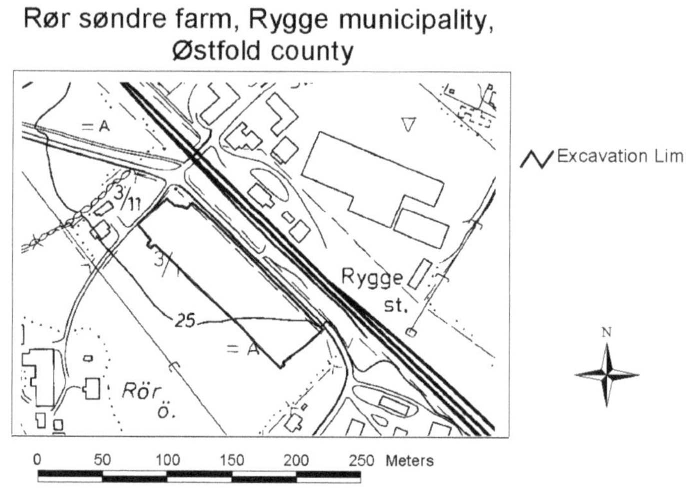

Figure 5: Rør søndre farm with excavation boundary shown.

and Cultural Heritage Research (NIKU). The innermost ring represents information restricted to select individuals in cultural heritage institutions. What will be placed on this strict level must be determined in co-operation with several parties, especially the Sami cultural heritage authorities. Some of the Sami site types are sensitive in terms of the spreading of information.

The database will be linked to GAB to which it will transfer key information about automatically protected sites, including their coordinates. In return it will import updated information about ownership and about changes of the property boundaries. The last factor is important since the rural landscape is currently undergoing massive changes and farms are being merged into bigger units for economic purposes. GAB will also provide a good opportunity for solving the problem of notifying owners of the legal status of their property. Since all the updated information about owners, addresses and property is kept here, it is easy to mail information to the right people. GAB is the most important tool for municipalities regarding owners, addresses and property - and the information is regularly updated.

The general public will not gain access to the database in the form it is now. Presentations directed at different user-groups, for example schools, are planned but have not yet been initiated. This is due mainly to the fact that the work to establish the GAB connection is a priority task. After this link is well established, the public education orientated tasks will be undertaken.

6. What information the Database contains, and what it does not

Automatically protected sites, mainly prehistoric and medieval, outside of the Middle-Age towns, are the most numerous site types. The cultural layers in the medieval towns form a very complex situation when it comes to placing them into a database together with objects and areas far simpler in structure. Most of the information about medieval urban cultural layers will therefore be found elsewhere. A fair number of Sami cultural remains can be found here, but big databases are also operated by various Sami institutions. We must therefore solve the problem of what to incorporate in the database and what should be located elsewhere in co-operation with the Sami cultural management authorities.

There are major differences with respect to the site types covered by various kinds of registration and surveys. These differences are reflected in the database. Representation is biased, focusing mostly on visible and monumental sites, while dwelling sites and other invisible site types are seriously neglected. This situation can only be improved through time, with new registrations being added to the database. In addition, there are several projects trying to tackle the problem of prediction, to compensate for the fact that it is impossible to register all invisible sites, and all the sites in large woodland areas, visible or invisible.

The most severely underrepresented category are the post-medieval sites that are not buildings or installations. No national programme for surveying and recording such sites has ever been initiated. The SEFRAK project did attempt to include such sites, but had to give it up due to time and financial limits. About 10 000 objects were recorded by SEFRAK, with large variations between the counties. These records are currently difficult to access, since they could not be included with the SEFRAK-buildings in the merger with GAB, but they will be incorporated into the Archaeological Registry. Some post-medieval objects were recorded during the Economic Mapping registrations, and a steady stream of new sites is coming into the registry from zoning registrations on the county level.

As of June 2000 roughly 73,000 sites have been recorded in the Archaeological Remains Registry. Of these, nearly 46,000 have been prepared for digitising and 29,000 of these are completed. Over the next couple of years the number of digitised objects will rise due to more effective procedures of digitising on raster-maps (NIKU 1. Half year report 2000). The most important work for the rest of 2,000 is to add these coordinates into the database, making it fully functional for GIS use.

7. The interface between the archaeological museums' topographic databases and the registry

After a rescue excavation conducted by one of the Archaeological museums or NIKU is completed, the quantities of documentation relating to finds and structures are amassed. This information belongs to the excavating unit, but it is also public in that it is the result of an official decision. The report is sometimes published, most often not since most excavations are too small to be published alone. The report is a public document, and some of the information in it must be sent to the registry. The example used here in figures 4 and 5 is an excavation related to the building of a new railway line and new station in Rygge municipality, Østfold county, conducted by the author in 1996 (Berg 1997,1998).

7.1. Proposal of a division between management and museum's databases

This site contained the remains of a Roman Iron Age farm in the form of five house foundations, but also a series of traces of older activities. An early Bronze Age cultural layer/refuse area with remains of cereal threshing, a votive find of flint flakes from the Late Neolithic - Early Bronze Age and post-holes dated to the Late Neolithic, cooking-pits from Pre - Roman Iron Age, indicating that this site had been in use for a very long time, and that the discovered farm-houses represent the location of a mobile farm at a particular point in time. The excavation boundary is the extent of the zoning plan for the new railway station and line. There is every reason to believe that the site stretches further into surroundings on all directions. Construction and buildings have not destroyed the land only in the south-west direction towards the modern farm buildings. The Registry therefore needs the information about the exact location of the excavation boundary, the site type and the date. Coordinates should represent the boundary, the rest must be extracted from the documentation kept in the museum's electronic and manual archives. So, from this situation, the registry will need the *boundary coordinates*, but not the coordinates of each structure inside the area. These are stored in the museum's database with the rest of the documentation. Also the museum's artefact number will be stored in the Registry, to provide a link between databases.

7.2. Another such division is between finds and find-spots

If the find-spot of an artefact can be located well enough for it to be represented by a point coordinate, the information about the place and a short description of the artefact and dating information will be found in the Archaeological Remains registry, together with a reference in the form of the museum's artefact number. The traditional description of the find will then be available in the museum's database. Concerning finds which cannot be located precisely enough to give them a point coordinate, the information about artefact can only be obtained in the museum's database.

8. The future

Table 1 shows the numerous databases in existance. Now that we have finally managed to gather together the Archaeological databases and merged them into one, and that we have only one Buildings registry, the next goal for The Directorate of Cultural Heritage of Norway is to expand the new Archaeological database to a comprehensive cultural-historical database about all kinds of monuments, buildings and sites. A plan and a rough data model for this merger has been approved (Steinnes et al. 2000), and will be forwarded to the regional level and the research institutions for comments and suggestions during the second half of 2000. The expanded database is expected to be in operation by the end of 2002. This new database is to be based on GIS and every object in the database will be digitised. By collecting all the various sites together in a GIS-structure the new possibilities for analysis and presentation will expand dramatically.

References

BERG, E., 1995. *Høyhastighetsprosjektet Oslo - Kornsjø. Konsekvensutredning fase II - Automatisk fredete kulturminner*. Parsell Haug, Råde - Seut, Fredrikstad. IAKN, Registration Report. Universitetets Oldsaksamling. Oslo 1995.

BERG, E., 1997. *Gård og grav på Rør i Rygge, Østfold. Dobbeltsporprosjektet. Arkeologiske undersøkelser 1994-1996.* VARIA 47. Universitetets Oldsaksamling. Oslo 1997. Excavation Report. Abstract in English.

BERG, E., 1998. Nye funn i Østfold - over og under markoverflaten. In Østmo, E. (ed.), *Fra Østfolds oldtid. Foredrag ved 25-års jubileet for Universitetets arkeologiske stasjon Isegran.* Universitetets Oldsaksamlings Skrifter nr. 21: 35-54. Ny rekke. Oslo 1998. Abstract in English.

The Cultural Heritage Act. Act No. 50 of 9 June 1978 Concerning the Cultural Heritage with later Amendments. The last amendment is dated 03.03.2000. The Ministry of Environment.

Foundation for Nature Research and Cultural Heritage Research: Half-Year Report on the work with the Archaeological Remains registry, dated 01.06.2000.

Norwegian Mapping Authority: Proposal for Standard for registration of Property Information. Hønefoss 1999.

STEINNES, S., BERG, E., LONG, C., HJØNNEVÅG H., 2000. Proposed data model for the new, aggregated Cultural Heritage database. Planning document. Directorate of Cultural Heritage. Oslo.

A GIS Driven Regional Database of Archaeological Resources for Research and CRM in Casco Bay, Maine

Matthew Bampton and Rosemary Mosher

Geography-Anthropology, University of Southern Maine

Gorham, ME, USA

e-mail: bampton@usm.maine.edu

Abstract

The needs of research and cultural resource management (CRM) are frequently at odds with one another. In the first case maximum accuracy of all records and open access to complete data sets is essential. In the second case differential access to data for the general public and resource managers is a regrettable necessity for protecting archaeological resources. The potential offered to researchers and resource managers by the availability of cheap, powerful, and portable computers and GPS units is offset by the potential for abuses of these same tools to exploit archaeological sites for personal gain. The legal and moral issues raised for a free society by this dilemma are not amenable to a technical fix. However, careful design of a GIS database offers an interim solution to the problem by simultaneously providing accurate information to the research and management community, while rendering sensitive information in a protected form for the broader audience.

Key words: GPS, GIS, database, accuracy, security

1. Introduction

For the past six years the GIS Laboratory at the University of Southern Maine has been exploring the potential of GIS as a tool for archaeological research and cultural resource management (CRM) by developing a case-study data base of the Casco Bay area on the coast of Maine, USA. In previous work some of the parameters within which this work has been done, and some of the difficulties faced, have been outlined (Bampton and Hamilton 1996, 1997). Rapid developments in the technology, an exponential increase in the amount of data available, and a significant increase in the number of actual and potential users have all contributed to a significant development in the approach used in the Casco Bay GIS.

Further developments have been prompted by the growing number of models for multi-purpose GIS data sets presented by workers in other areas of the world. Of particular note is the model of integrated archaeological and geographical data linking field and archival materials currently used in Sweden by the State Archaeology Group (Riksantikvarieämbetet Avd för Arkeologiska Undersökningar). In the USA discussion of the problems and potentials of archaeological GIS is now becoming more common (e.g. Allen et al. 1990, Harris and Lock 1995, Esser and Shaw 1997). In this paper we discuss some the evolving needs of the constituencies using the Casco Bay GIS and some of ways in which we have tried to address these needs.

2. Problem statement

To be useful an archaeological data set must meet two necessarily contradictory conditions: It must precisely record the location, extent, and contents of sites for a professional audience, and; simultaneously preserve the integrity of sites by concealing their location, extent, and contents from a broader public audience. Astride this contradiction sit cultural resource managers, charged with serving both the scientific community and the public (for examples see Lynott and Wylie 1995, Speiss 1978, 1994).

The increasing use of GIS, GPS and total stations by archaeologists in the past ten years has brought this problem to the fore. What was previously a pipe dream is now becoming a reality: Sub-meter accuracy GIS data-bases incorporating all available digital environmental data, and all archaeological field data within a single data structure. Websites such as that at the University of Rhode Island now offer many of these data as downloadable GIS coverages — though with some restrictions placed on archaeological site location data (URI 2000). As these technologies become more widely applied there is growing discussion of the role of GIS in both research and CRM (e.g. Hurlbert 2000).

From the perspective of scientific analysis this means that scholars can now combine the results of field work with a broad range of environmental data sets generated by workers in other fields. Regional scale analyses of patterns in space and time can be developed, as can speculative models of past habitats, combining such factors as sea level change, soils, vegetation coverage, aerial photography, remotely sensed imagery, and cartographic records.

From the perspective of cultural resource management (CRM) an accurate accounting of archaeological resources can be combined with a cartographic representation of threats to the record in the form of environmental changes such as erosion, and human social changes, such as development. When combined, the scientific and the CRM GIS data sets can be used strategically, to identify resources that are of particular value or are unusually vulnerable, and can provide supporting information to decision-makers, who must decide what and when to preserve, excavate, or even sacrifice.

There are two groups of challenges confronting anyone attempting to produce a GIS data base that serves both research and CRM functions. The first group of challenges are technical, centring on the unique data demands placed on a GIS by both research and CRM users. The second group of challenges are operational and centre on the precision and accessibility of the GIS.

3. Technical challenges

Archaeologists - and most other field scientists - have long recognized the importance of gathering spatial data sets, and have proven themselves capable surveyors using standard tools such as optical instruments and topographic maps. And like most field scientists they allow a fairly generous error factor in their surveys, historically tending to ignore such arcane issues as projections, reference surfaces, datums and changing magnetic declinations.

In respect to this tendency GIS has proven to be rather unforgiving. If the GIS is simply used as a cartographic device slight failures of registration, location, orientation, or projection can be "fudged" by juggling colours, line widths, symbols, and as a last resort, by using correction fluid or a pair of scissors and a bottle of glue. However, when the data are intended for use in any analytical application minor survey issues can have disastrous consequences. Incorrectly specified projections and datums do not simply create skewed images, in some of the most widely employed GIS packages they can quite literally collapse the data structure.

Imported database files used to generate point coverages can also create problems. A convention used by some researchers to enter latitude and longitude values of 0°0'0" by 0°0'0" for site locations they have not yet calculated resulted in the sudden appearance of several dozen prehistoric Native American middens 425 km due south of Accra, Ghana. Accidental transposition of values for Casco Bay — actually about 43°40' by 70°10' - placed a cluster of sites in central Greenland, 500 km northwest of Godhaven (we have assumed that these database entries were errors, though it is possible that some particularly diligent field crews have discovered hitherto undreamed of cultural links).

While the spectacular anomalies are fairly easy to spot, the more subtle ones present greater problems. In Maine all nautical charts are referenced to North American Datum 83 (NAD83), all topographic sheets are registered to NAD 27, most hand-held GPS units default to World GPS Datum 1984. The three datums produce results that have, on average, a difference of between 200 and 2000 m in our study area.

Serious analytical problems can result from these types of survey or mapping errors. A simple procedure such as generating a point in polygon overlay - a response to a query such as "how many archaeological sites are located on sandy soils"- can be rendered meaningless by any of the issues outlined above, or by the locational inaccuracies resulting from the size of the dots drawn with a .05 mm mechanical pencil on a 1:25,000 map (the standard 7.5" USGS topographic sheet).

When a diversity of other data sets are combined with the archaeological field data the problem of deficient survey work is exacerbated by errors and inconsistencies in these source materials, whether digital or in some other form.

Data cleaning, and the resolution of issues of projection, registration, and scale have all proven to be as important as data gathering in constructing the GIS. Once the archaeological data are cleaned the question of registration with other data sets must be addressed. Much of the time sacrifices of precision and accuracy are made in order to preserve compatibility with the growing number of publically available environmental data sets. Finally, even when data are cleaned and corrected they can frequently prove problematic, as both graphical and database conventions that appear correct to field workers not familiar with GIS can be unsuitable for application in many standard GIS packages. Spreadsheet fields with long narrative entries are essentially unsearchable in many applications.

4. Operational challenges

The Casco Bay GIS is designed to serve two constituencies. On one level it functions as a secure repository of technical information, and on another level it serves as an open-access public resource. Presenting the professional community with accurate site maps, and precise site locations has become fairly straightforward if survey-quality GPS and total station equipment is used. Once the data are acquired, it is simply a question of presenting them in a PhD-proof software package, such as ArcView. Security can be ensured by combining good data management practices with fairly exclusive system password privileges, the internal data hiding capabilities of the software, and the storage of data on a removable medium, such as a CDROM.

The public access component of the system presents greater problems. Although at some future date some information concerning selected sites may be made available to the general public as an educational service, at present both legal and practical concerns prevent this application from being explored in any detail (Arthur Speiss, pers. comm. 1992-2000).

At present the primary public access application is site preservation, and is designed to serve the planning community. To preserve sites from vandalism and looting, and in some cases to preserve human remains, precise locations must be kept secret — this is becoming increasingly important as the advent of cheap hand-held GPS units has made it very easy for the curious or malicious person to get to within 30 m of any point for which latitude and longitude values are available. However to preserve sites from destruction during construction activities it is necessary to make property owners, town planners, and building inspectors aware of site locations. This becomes particularly important in such as Casco Bay where sites can, at times, occur every 300 m along a shoreline.

It is fairly easy to use a graphical device such as a disproportionately large symbol to conceal site locations. However all point data are still stored in the accompanying data base by their exact latitude and longitude location. Rudimentary system skills will enable an unauthorized user to find the location of any point. It is possible to automatically generate buffer zones around known sites and so conceal precise locations within a large polygon, and then remove the point data from the public access component of the data set. However the system centres buffers on point locations, so if you find the middle of the circle, you can find the point location of the site.

To present planners, and the general public with a data set that will alert them to the proximity of a site if, and only if, they are working in its immediate neighbourhood, it is necessary to map all known sites but to present them nested inside randomly shaped buffer zones. These zones must be small enough to prevent extended areas from being unnecessarily blanketed with historical preservation orders, and yet large enough to encapsulate sensitive places. They must be discernable only when the user requests information on a single location. And they must be physically separated from the original point data from which they were generated.

The site locations were randomised by imposing a regular grid over the entire region. Cell size was half of the desired protection radius. That is, a 50 m cell size was used if a 100 m protection radius was required. The grid was then reclassified as a binary coverage site/no site. All cells not containing sites were discarded, and all of the original point locations were discarded. The cells were then buffered to the desired protection radius, and the cells discarded. Finally all parts of the buffer falling on ocean were discarded. Thus the final polygons were irregularly- shaped, and surrounded a point falling somewhere unspecified no less than the desired protection radius from the boundary but no more than 1″ times the desired protection radius from the boundary.

In Casco Bay it was decided to identify two classes of sites, "small" and "large" — more complex archaeological classification schemas proved to be beyond the powers of the GIS lab team. Small sites were protected with a 200 m buffer, large sites were protected with a 500m buffer. Sites having particular archaeological value, as defined by the Maine Historical Preservation Office (MHPC), were identified as "Protected", sites with either unknown value, or limited value were defined as "Maybe Protected", all other areas were defined as "Unprotected".

The whole coverage was coloured in a single shade, with no visible differentiation between these three classes of protection. A coverage depicting roads was added to allow users to orient themselves. The whole project was presented in a heavily customized ArcView desktop. Users were not granted access to database files, printing capabilities, customizing tools, and zoom capabilities were severely limited.

A user activates the cursor, and defines an area of interest - perhaps drawing in the four corners of a lot that he or she wishes to build on. An information screen appears stating that the area falls into one of the three classes defined: "Protected", "Maybe Protected" or "Unprotected". In either of the first two cases the user is alerted to the fact that they must contact an MHPC officer before proceeding with any development activities.

5. Conclusions

Creating secure data structures for archaeological data sets is a necessary evil. The increasing sophistication and the decreasing cost of powerful computing and navigational tools mean that the general public are more likely than ever to be able to find, record, and return to sensitive archaeological sites. As the pressures of development remain as strong as ever, the net result is an increasing pressure on a diminishing resource. We believe that appropriate use of GIS technology can, in some measure, mitigate the negative effects of these trends by providing a powerful tool for planners that may help to preserve archaeological resources by aiding planners and preservation officers to co-ordinate their efforts to protect archaeological sites. And while the diminishing costs and increasing power of computers and GPS units may pose a threat, they also present an opportunity. Well-designed databases can serve to provide researchers with a new set of analytical tools. The research potential of such complex multi-dimensional data is only now starting to be explored.

Acknowledgements

Our thanks are due to Dr Nathan Hamilton of the University of Southern Maine Department of Geography-Anthropology and Dr Arthur Spiess of the Maine Historic Preservation Commission: Both generously shared unpublished data with us and provided valuable insights and commentary during the preparation of this paper.

References

ALLEN, T.M.S., GREEN, S.W. and GREEN, E.B.W., 1995. *Interpreting Space: GIS and Archaeology*. Taylor and Francis: London.

BAMPTON, M., 1997. Archaeology and GIS: the view from outside. *Archeologia e Calcalatori* 7:19:23-45.

BAMPTON, M. and HAMILTON, N.D., 1996. GIS, Education and Public Stewardship in Casco Bay Estuary, Maine. *International Union of Prehistoric and Protohistoric Sciences XIII Congress Pre-Conference Publication.*

BAMPTON, M. and HAMILTON, N.D., 1997. Archaeology, CRM, and the Casco Bay Database. In McLaren N. and Jonson I. (eds.), *Archaeological Applications of GIS*. Sydney University Archeological Methods Series, No.5.

ESSER, K and SHAW, S., 1997. *A public viewing tool for archeological data.* ESRI Conference Proceedings 1997. http://www.esri.com/library/userconf/proc97/proc97/to250/pap213/p213.htm

HARRIS, T.M. and LOCK, G.R., 1995. Towards an evaluation of GIS in European Archaeology: the past, present, and future of theory and applications. In Lock, G.R. and Stančič, Z. (eds.), *Archaeology and Geographical Information Systems*. Taylor and Francis, London.

HURLBERT, D., 2000. GIS as a Cultural Preservation Tool in Shenandoah National Park, http://www.nps.gov/gis/newsletter3/shen.htm

KELLOGG, D.C., 1982. Environmental Factors in Archeological Site Location for the Boothbay, Maine Region, with an Assessment of the Impact of Coastal Erosion on the Archaeological Record. MA Thesis Quaternary Studies, University of Maine at Orono.

LYNOTT, M.J. and WYLIE, A., 1995. Stewardship: the central principle of archaeological ethics. In Lynott, M.J. and Wylie, A. (eds.), *Ethics in American Archaeology: Challenges for the 1990s*. Society for American Archaeology Special Report: Washington DC.

SPIESS, A., 1978. *Conservation Archaeology in the Northeast: Towards a Research Orientation.* Peabody Museum Bulletin No. 3. Harvard University. Cambridge: MA.

SPIESS, A., 1994. CRM Archaeology and Hydroelectric Relicensing in Maine. In Kerber, J. (ed), *Cultural Resource Management: Archaeological Research Preservation, Planning, and Public Education in the Northeastern United States*. Bergin and Garvey Westport CT.

URI 2000. Archaeological sites listed by the RI. Historic Preservation Commission masked to 20 acre grid cells. http://www.edc.uri.edu/rigis-spf/statewide/state.html#cultural

Using a Relational Database Management System for the Recording of Ancient Settlements and Sites in the Vrachneika Territory in Western Greece

Vangelis G. Tsakirakis

Landscape Archaeology Group

Amfitrionos 33, 118 54, Kato Petralona, Athens, Greece

e-mail: bm-ecoikk@otenet.gr

Abstract

During the study of the local history and archaeology of western Achaia, Greece, by the workgroup of Landscape Archaeology Group (LAG), an Access 97 database application was used for the recording of archaeological sites throughout the territory of the modern villages of Vrachneika and Monodendri. The application was characterized by its advanced capabilities regarding the types of data entry, such as topographical maps of the sites, pictures of the finds, indexing of the related bibliography, etc. Other components include the capacity to implement a select data research, for example of sites with the same chronology, or sites with similar finds. Some quantitative remarks concerning the numerical data were also attainable with the use of statistical graphs. Finally, all the data entries of the sites may be printed in reports or accessed through other software applications (Word, Excel etc.) for advanced processing.

Key words: Access, relational database, database application, landscape history, historical sites, landscape archaeology group, cultural data management and retrieval

1. Preface

Since 1995, a non-profitable scientific society in Greece, called the Landscape Archaeology Group (L.A.G.), has been investigating the history of the western part of the Achaia province in the northwest Peloponnese.[1] The Landscape Archaeology Group was founded in an attempt to accomplish the following four goals:

1. To contribute to the preservation of the historical landscape and the historical memory.
2. To awake the historical memory and consciousness of the native population.
3. To encourage interdisciplinary co-operation, not only among Group members but also in the wider research field, for the advancement of archaeological research.
4. The exploration, promotion and support of modern techniques and, in general, of a new strategy that will correspond to the research notion of our times.

The means used by the Group to achieve its goals are field surveys, implementation of historical, archaeological, anthropological, environmental and social studies and the publication of a scientific Newsletter, as well as individual studies, such as monographs and articles in scientific journals.

2. LAG's background and present studies

Our work in Achaia began as early as 1991, with the participation of some Group members in an interdisciplinary study of the historical topography of Achaia, a project supervised by the National Hellenic Foundation of Research and the local branches of the Archaeological Service (Papagiannopoulos and Zahos 1994, Zahos et al. 1995, Tsakirakis 1996, Andrinopoulos 1998, Simoni and Papagiannopoulos 1998, Andrinopoulos and Simoni in press, Zahos and Papagiannopoulos in press). After this initial step and the foundation of the society, the Landscape Archaeology Group started to develop an interest in the recording of the local history of western Achaia. This interest was partly in response to the concern of local mayors and community council members regarding the establishment of historical documentation of their towns and villages. Since 1996, our society has given lectures and organised exhibitions relating to the cultural identity and history of the local communities. From the beginning we have focused on the diachronic unity of a geographic area subjected to human intervention. This is our Group's main methodological approach, as we believe that the landscape consists of a combination of natural and man-made characteristics. During successive historical periods, settlements and human exploitation leave marks of cultural intervention on the natural landscape. When looking into the human past of a specific area, attention must be paid not only to historical testimonies and archaeological finds, but also to the study of the natural landscape in all its aspects. This can be attained with the contribution of the natural and environmental sciences, like geology, chemistry, forestry and geography. When such a combination of disciplines studies all landscape components, local history can be apprehended in detail.

At the present time our workgroup for the study of the cultural landscape of western Achaia has completed its work regarding two communities, Agiovlassitika and Kareika. The result of these studies was the writing of two books of history (Argyropoulos et al. 1998, Papagiannopoulos and Chronopoulos 1998). Since 1998, the city council of Vrachneika, a newly founded municipality in western Achaia, has assigned to our society's workgroup an ambitious project for the documentation of the diachronic history of the area within the modern town's administrative boundaries. An essential part of the project was the study of human settlement and intervention in this area, a task partly undertaken by the author.

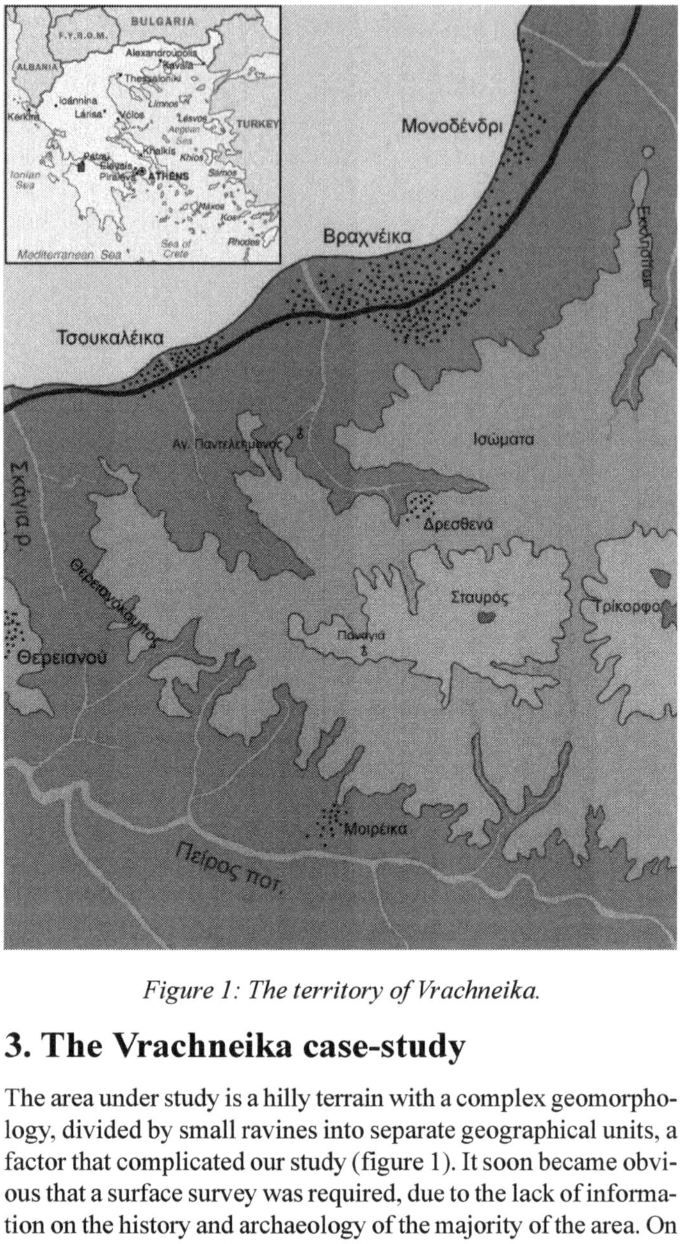

Figure 1: The territory of Vrachneika.

3. The Vrachneika case-study

The area under study is a hilly terrain with a complex geomorphology, divided by small ravines into separate geographical units, a factor that complicated our study (figure 1). It soon became obvious that a surface survey was required, due to the lack of information on the history and archaeology of the majority of the area. On the other hand, smaller areas near the coastline and in a ravine had already been surveyed by archaeologists of the local branches of the Archaeological Service and members of the project for the historical topography of western Achaia (Petropoulos and Rizakis 1994). These areas are well documented in the archaeological and historical bibliography as places evidencing important settlement activity during the Late Bronze Age (Papadopoulos 1979:26, Petropoulos and Rizakis 1994:190, table 1, site no. 21) and the Roman period (Petropoulos and Rizakis 1994:190, table 1, site no. 22). Consequently, there were two stages of research: selecting all published evidence of human habitation during prehistoric, historic, medieval and post medieval times in the area, then surveying the terrain by fieldwalking in order to evaluate the published information, in the light of new evidence and finds. In the first stage, we used the testimonies of ancient and modern travellers and some published excavation reports from the last 50 years. In the second stage of research, a wider range of information was collected, including topographical maps, place-names, environmental and geomorphologic evidence, survey drafts, survey fieldnotes and photographs. As a result, a large amount of information was gathered and we had to utilize it in the best way in order to formulate the final interpretations and conclusions of our work. This was achieved by creating a relational database application, using Access 97 software (Tsakirakis 1997-98).

4. Database application

4.1. General description

The database of the Vrachneika project, soon to reach its final form, is the first computer application in the Achaia province for the documentation of the local cultural heritage. It consists of 20 tables with a total of 87 fields of data and more than 50 select and crosstab queries in Structured Query Language - SQL. Most of the queries are embedded in forms and reports. Data can be recorded, viewed and edited via the input and output forms and subforms, which also contribute to a better user interface. Macros and Visual Basic for Applications – VBA routines and procedures manage automation of the application. Tables are bound to each other with "one-to-one", "one-to-many" and "many-to-many" relationships. The overall size of the application was about 5 MB with no data and 60 MB when all information of the project was stored in the database. Eighty-seven historical sites were recorded.

4.2. Table structure and methodology

The overall structure of the Vrachneika database was determined by the research methodology of our workgroup. Initially, it was necessary to decide on the categories of information that should be recorded and processed. Our methodology necessitated a primary data table, where the basic characteristics of each site would be recorded. To create the primary table of sites we took into account a definition of the term "site" as an area of concentrated human intervention in the past, in relation to the historical evidence and finds of its proximate surrounding geographical space. This helped us define the "site" as the basic record unit of our application. Modern topographical parameters and a description of the natural landscape of each site were important, as well as information about landowners and the place-names. All these fields of data, along with a field for the general description of the site, were included during the layout process in the primary table (table 1).

The table of sites, associated with the other tables, plays a pivotal role in the application. For each recorded historical site of the primary table, a number of fields provide additional information in supplementary tables about topographical site maps, site finds, surface survey ceramics and bibliographical quotations. These tables are bound to the primary site table with "one-to-many" relationships. Some of their fields support Object Linking and Embedding - OLE, in order to accept picture entries: this is the case in the fields of site topographical maps, survey ceramics and finds tables.

As mentioned before, archaeologists of the Archaeological Service had previously investigated some sites. This provided us with a bibliography of published finds, such as pots and necklace beads from two Late Bronze Age cemeteries of rock-cut chamber tombs (Papadopoulos 1979:66 and 141, Papapostolou 1977:485-490). According to the archival classification system of the local branches of the Archaeological Service, these finds are indexed as "movable" finds, in contrast to the category of architectural finds, like ancient tombs or building walls. This classification, which serves an administrative purpose, was at first adopted for

FIELD NAME	DATA TYPE	FIELD SIZE	VALIDATION RULE	DISPLAY CONTROL	ROW SOURCE (SQL)	KEY	INDEX
SiteID	AutoNumber	Long Integer	-	-	-	Primary	Yes (no dublicates)
SiteName	Text	255 bites string	No zero length	Text Box	-	-	Yes (no dublicates)
Village	Text	100 bites	Is Not Null	Text Box	-	-	No
Settlement	Text	100 bites	No zero length string	Text Box	-	-	No
Toponym	Text	255 bites	No zero length string	Text Box	-	-	No
Description_ of_Location	Memo	-	No zero length string	Text Box	-	-	No
Category_of_Size	Number	Integer	-	Combo Box (2 Columns, 1st Bound, 2nd Visible)	SELECT SizeCategories_ of_Sites.* FROM SizeCategories_ of_Sites ORDER BY SizeCategories_ of_Sites.Size;	-	Yes (dublicates OK)
Natural_Landscape	Memo	-	No zero length string	Text Box	-	-	No
Landowner	Memo	-	No zero length string	Text Box	-	-	No
General_Site_Description	Memo	-	No zero length string	Text Box	-	-	No

Table 1: Sites Table Layout.

the database classification of site finds. During the first layout procedure of the database, all finds were indexed as "architectural" or "movable" in two different tables, in order to make our application compatible with possible future databases of the local branches of the Archaeological Service. But later on, when it was realized that such a prospect was not immediately forthcoming, we decided to merge all the finds in one table, to avoid duplicate table fields and refine our database.

Except for the aforementioned primary and supplementary tables of the application, other tables with a few data fields, conventionally named "list tables", were created for feeding in the selection lists and combo boxes of the forms with information. This is the case with the tables for researcher names, site categories of human use, categories of size, historical periods of human use, titles of the relevant bibliography, typologies of finds and surface survey ceramics. The researcher names were required for the documentation of site surveys. The site category of human use is a term for specifying what function a site had during its historical lifetime, for example, if it was a settlement, a farmstead or a cemetery. We had to take into account the fact that many sites had different uses in different historical periods, especially in the case of the multi-period sites. Size category is a subjective criterion for assorting sites according to their size on the surface and was measured in square meters. We assigned three basic size categories. The purpose of this classification was to compare the site categories of human use with the average size of sites and conduct explanation models for contemporary sites that could not be ascribed a clear use, due to lack of sufficient archaeological evidence. As for the historical periods of human use, each site could be represented by more than one record. This proved to be very convenient in the case of multi-period sites, where human activity occurred at the same place during different periods of time.

4.3. User interface

Having arranged all issues concerning table structure and the overall layout of the application, we proceeded to design the user interface. A switchboard form with navigation buttons was created to provide access to the recorded information in the database (Tsakirakis 1997-98: figure 1). From here the user can load the sites primary form (figure 2).

This form retrieves all data related to each site from the primary sites table (table 1). Through the sites form, the user can view or edit each site's maps (figure 3), finds, fieldnotes and bibliographical citations, stored in the supplementary tables. These categories of information are accessible via graphic buttons that open other forms, by using VBA routines and macros. Access to the same information is also available through the switchboard form of the application. All forms have combo boxes, which retrieve their data from the "list tables" of categories, typologies, historical periods and bibliographical titles. The records of the combo boxes can be selected by scrolling lists. At any time and by request of the user, these lists are automatically updated with more records, as VBA procedures add new entries in their source tables. These procedures are activated after specific events, for example the user double clicks on a combo box field to enter a new record on its list.

From the main switchboard form, the user can proceed to search for information about a site by using the overview form (Tsakirakis 1997-98: figure 7). In this case, all recorded data of a site is retrieved from the related tables and presented synoptically. These data can also be viewed and printed in report format.

Another function of the application is to search the sites categories of information for common records, for example, by filtering all site entries via output forms, to locate the records of a specific find or ceramic category and select all sites dated in a specific historical period. The output data of these forms can be viewed and printed in report format, or exported to another application.

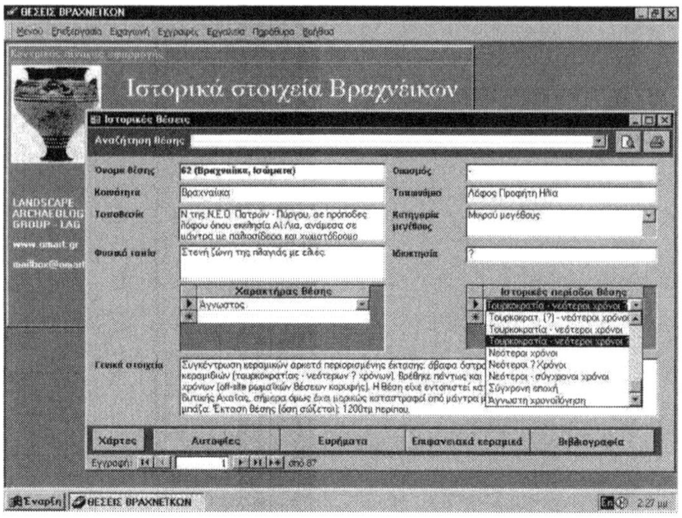

Figure 2: The sites primary form.

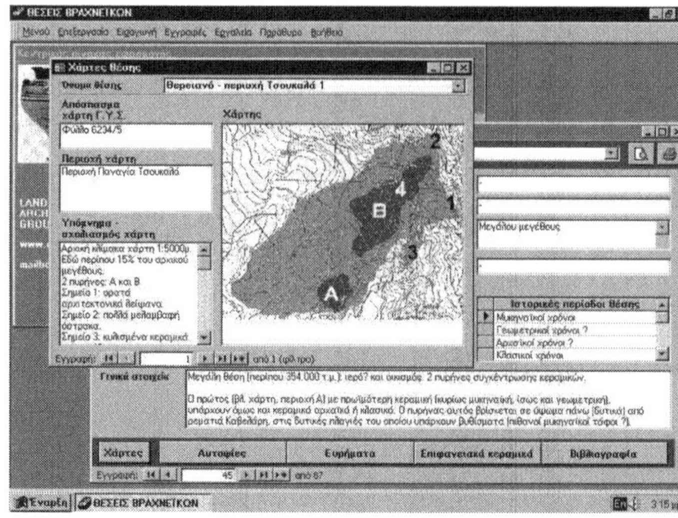

Figure 3: The sites map form.

Besides the retrieval of data, some graphs of quantitative ratios of the site data categories are available via the switchboard form of the application. The user can view graphs displaying the ratio of the total number of sites during historical periods (figure 4), the frequency of the site categories of size and the categories of human use. These graphs, though of no special statistical interest, helped the workgroup to arrive at useful conclusions regarding the history of settlement and the kinds of human intervention during historical periods. Graphs can also be viewed and printed in report format.

Finally, an important feature of the application is the automatic generation of reports of the recorded information. By using macros and VBA procedures, a report of a selected site can be viewed, printed or sent to another application for advanced processing (Tsakirakis 1997-98: figure 6).

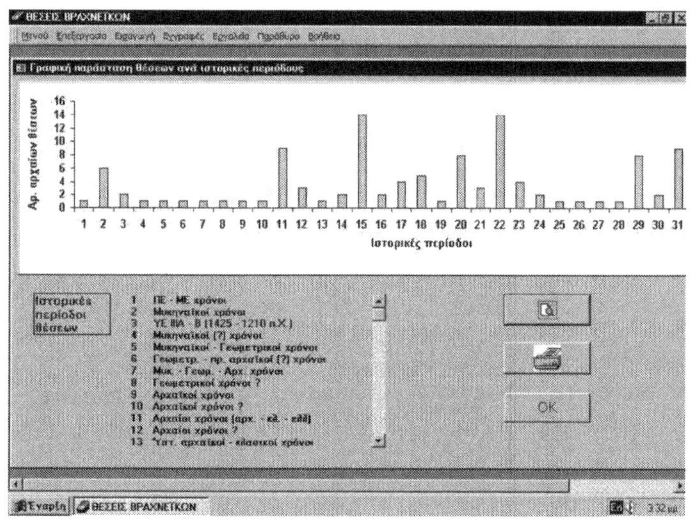

Figure 4: Graph form displaying the ratio of the number of sites during historical periods.

5. Conclusions

I would like to mention again that our application is the first non-administrative database in Achaia for the recording of historical sites throughout the landscape. Our attempt is basically experimental and by no means a finished endeavour. Presumably some things may need to be changed and improved in order to satisfy the demands of our research. In the future our work in western Achaia will continue and we intend to create a similar database application for every community we study. In the long-term, we plan to link all these databases within a central relational database management system. Thus, all the historical information will be accessible to any individual, from any of the local communities, interested in studying the cultural heritage of this area.

Acknowledgement

I would like to thank Aristotelis Koskinas, Helen Simoni, Jean Mertzanakis and Betina Spirou for their valuable advice.

References

ANDRINOPOULOS, A., 1998. Ancient and Contemporary Land Use in West Achaia, Peloponnesus, Greece: A GIS Application. In Peterson, J. (ed.), *Cost Action G2; Paysages antiques et structures rurales; The use of Geographic Information Systems in the study of ancient landscapes and features related to ancient land use*: 37–42. Norwich, UK: School of Information Systems, University of East Anglia.

ANDRINOPOULOS, A. and SIMONI, H., in press. G.I.S. application on the archaeological survey of western Achaia. In Rizakis, A. (ed.), *Proceedings of the A' International Conference on Ancient Dyme, Dymaia – Vouprassia*, Kato Achaia, 6-8 October 1995.

ARGYROPOULOS, B., PAPAGIANNOPOULOS, K. and CHRONOPOULOS, B., 1998. *Kareika of Achaia: community and identity*. Patras: study submitted to the Community Council of Kareika.

PAPADOPOULOS, A., 1979. *Mycenaean Achaea*. Studies in Mediterranean Archaeology LV:1. Göteborg: P. Åströms Förlag.

PAPAGIANNOPOULOS, K. and CHRONOPOULOS V., 1998. *Agiovlassitika; steps through time*. Patras: Community of Agiovlassitika.

PAPAGIANNOPOULOS, K. and ZAHOS, G., 1994. Intensive Survey. In Petropoulos, M. and Rizakis, A., Settlement Patterns and Landscape in the Coastal Area of Patras, Preliminary Report. *Journal of Roman Archaeology* 7:187 – 189.

PAPAPOSTOLOU, I., 1977. *Collection of antiquities from Achaia*, Praktika tis en Athinaes Archaeologikis Etaireias 1977 B:485 – 490.

PETROPOULOS, M. and RIZAKIS A., 1994. Settlement Patterns and Landscape in the Coastal Area of Patras, Preliminary Report. *Journal of Roman Archaeology* 7:183 – 207.

SIMONI, H. and PAPAGIANNOPOULOS, K., 1998. Project for the Topography of Ancient Achaia, Greece: Quantitative Analysis and Visualisation of the Results of the Intensive Surface Survey at Kamenitsa. In Peterson J. (ed.), *Cost Action G2; Paysages antiques et structures rurales; The use of Geographic Information Systems in the study of ancient landscapes and features related to ancient land use*: 43–56. Norwich, UK: School of Information Systems, University of East Anglia.

TSAKIRAKIS, V., 1996. Statistical notes on archaic, classical and early hellenistic pottery from the intensive survey at Western Achaia. *Pyxida* 2:37 – 43. Patras: Landscape Archaeology Group.

TSAKIRAKIS, V., 1997-98. Database application for historical–archaeological research: recording of historical sites. Preliminary report. *Pyxida* 3: Web edition at www.omart.gr/pyx3gr.html#Pyxida3_MeletiBT. Landscape Archaeology Group.

ZAHOS, G., PAPAGIANNOPOULOS, K., SIMONI, H. and THANASSOURAS, N., 1995. Experimental application of intensive survey of Therianou area, Achaia. *Pyxida* 1: 24 – 34. Athens: Landscape Archaeology Group.

ZAHOS, G. and PAPAGIANNOPOULOS, K., in press. *The intensive surface survey in Achaia; a different approach.* In Rizakis A. (ed.), Proceedings of the A' International Conference on Ancient Dyme, Dymaia – Vouprassia, Kato Achaia, 6-8 October 1995.

Notes

[1] Cf. at the LAG's website, http://www.omart.gr/.

Vienna Archaeological GIS (VAGIS): A Short Outline of a New System for the Stadtarchäologie Wien

Wolfgang Börner

Magistrat der Stadt Wien, Geschäftsgruppe Kultur – Stadtarchäologie

Friedrich-Schmidt-Platz 5/1, 1080 Wien, Austria

e-mail: bor@gku.magwien.gv.at

Abstract

The first archaeological systematical research in the city of Vienna was conducted by Josef Nowalski de Lilia and Friedrich von Kenner. Their work on mapping the archaeological sites in Vienna set the fundations for the archaeological map of Vienna. Their data was upgraded after a century of archaeological research in Vienna and presents the basis for the present sites and artefacts database called VAGIS (Vienna Archaeological GIS). In this paper we present how artefacts locations and plans of sites registered on historical maps from the year 1819 were transferred to the modern database.

Keywords: Vienna archaeological GIS, Franziszeischer Kataster, Nowalski de Lilia, Stadtarchäologie Wien, Vienna

1. Introduction

Cities are never "finished". Even if they came into being according to a development plan, they are constantly in "statu nascendi" – they are constantly adapted to the needs of their inhabitants (Artner/Krenn 1995:45).

The growth of the 2 million population of the City of Vienna has increased many times over the past decades. An increasing number of green areas have been designated as development areas. Over 1000 ha have been designated for new residential, infrastructure and roads to be constructed during the next decade. (Jakob 2000:76)

One of the tasks of the Stadtarchäologie Wien is to develop strategies, which help to prevent uncontrolled destruction of our cultural heritage.

1.1. Morphological and historical background

For those unfamiliar with the characteristics of the Viennese landscape and its archaeological record, we will give a short introduction here.

A political slogan coined during the 1990s stated that: "Vienna is different." This statement is also applicable to the city's geography and to its history.

Geographically, Vienna can roughly be divided into 3 parts: the area north of the Danube, the terraced landscape south of the Danube and the wooded hills of the Wienerwald to the west.

The Vienna Basin provides an excellent environment for settlements and has therefore attracted settlers since prehistoric times.

The Danube served as an east-west-highway for water transport, fords by which the branched-out Danube could easily be crossed provided an ideal north-south-communication.

The area south of the Danube with its terraces formed during the ice ages and by rivers, in particular, offered the greatest incentive to settle. Settlements dating back to the Neolithic can be found at the heights and slopes of these hills.

After the Vienna area was attached to the Roman Empire, the Romans found a relatively flat platform, protected by brooks and rivers in what is today the first district. The location was ideal for building a fort (Mosser 1998:74). The civilian town (Zivilstadt) was situated in the third district (Mader 1998:89). Roman Villas and smaller towns could be found near other watercourses (Wölfl 1998:113).

The northern part of Vienna, districts 20 to 22, is less favourable to settlements because of its morphological features. The area is not terraced and the height-differences are negligible. Until its regulation the constantly changing bed of the Danube and the formation of new branches have largely destroyed all evidence of settlements. Most designated development areas are situated in those districts, which will therefore increasingly draw the attention of our archaeological research.

2. Fundamentals

Archaeological research has a long tradition in Vienna. The first Roman finds were mentioned in the 15th century. Until the 19th century collection was the top priority. Serious research of archaeological remnants did not begin until late in the 19th century. The fall of the bastions, the fortifications that hemmed in Vienna's growth for centuries, triggered an impressive building boom beginning in 1870. At this time Josef Hilarius Nowalski de Lilia, a Polish emigrant (Wielowiejski 1989:89), arrived in Vienna and started systematically recording archaeological evidence and collecting finds from the various building sites, particularly from the city centre. His notes and sketches have been preserved to a large extent and form the basis of our site database (Stipanits 1998:67).

2.1. Database

Before we were given the idea for a Geographical Information System, our first step was to sift through the existing notes on finds and known sites. Eight years ago the first database was constructed, first with the program Superbase, then with Access, as it saves a lot of work.

Each site is given a number, the site code, consisting of the year the excavation took place and a consecutive number. In the meantime about 2000 numbers are in use. All other numberings, e.g.

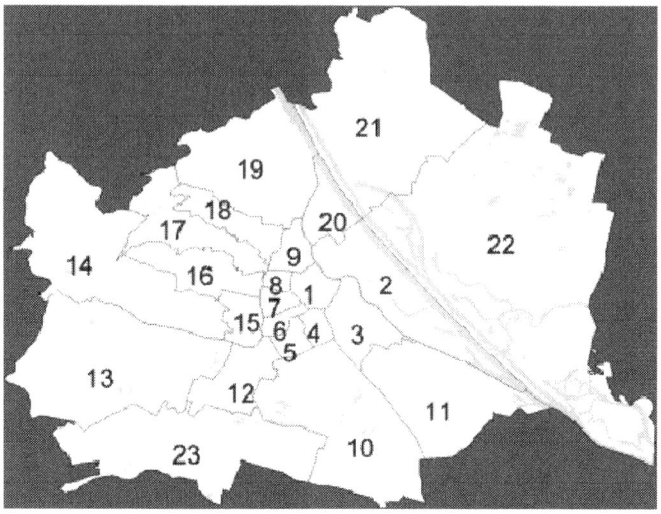

Figure 1: The districts of Vienna.

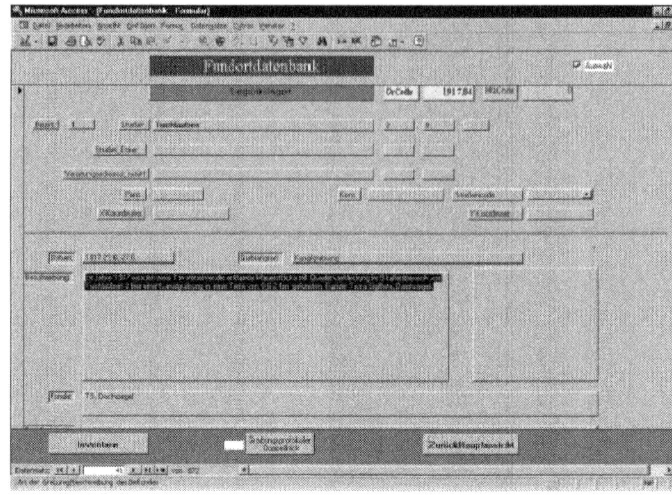

Figure 2: Site-Database of the Stadtarchäologie Wien.

for artefacts, photos or context sheets, are based on this code (Börner 1998:52, Turecek 1999:55).

The first database created was the site database of Stadtarchäologie Wien listing all sites in the urban area. For each site there is a short characterisation and description. This main database is linked by the site code with all other databases, like inventory, photo, excavation and so on. Additionally to the site database, an excavation database has been set up and is at disposal to our field teams.

Because we wanted to include as many earlier excavations as possible, different methods of excavation had to be taken into account at developing this database. The integration of the Harris-Matrix in the database management system is planned for the near future.

3. CAD

Shortly after we began developing this set of databases, it was decided to transfer all existing plans of the contexts and sketches into a CAD System. This proved to be quite easy for excavations that were already linked to a system of coordinates.

Mr. Nowalski's sketches of sites were more problematical. In the past years our experiences have shown that the sites depicted in these sketches have been surveyed very accurately. Yet, it was difficult to establish a link to today's city map. Luckily Nowalski also drew property borders or the walls of houses. All we had to do was to find a map of Vienna from before the building boom. The "Franziszeische Kataster" turned out to be especially useful. (Börner and Mosser 1997:9)

3.1. The "Franziszeische Kataster"

Emperor Franz I. of Austria succeeded in abolishing legal and technical obstacles and in establishing a complete change. With his land tax patent of the 23 December 1817 he founded the Austrian Land Register. On the whole, the guideline of this patent is still valid today.

The land register is also called "Franziszeischer Kataster" after his creator, or stable land register. "Stable" because the rates of net proceeds used for calculating taxes were meant to be invariable, that is without consideration for possible higher profits as a result of exceptional diligence – with the exception of the total or partial loss of the soil's fertility due to natural disasters (Fischer 1995:39f.). The land survey was based on scientific principles.

Before this land register was digitised and transformed by today's city map its accuracy was reviewed.

3.1.1. Review of the land register's accuracy

The terms of reference were an analysis of the accuracy and the transformation of the "Franziszeischen Kataster" for the third district. For the first district a transformation of the "Franziszeischen Kataster" was carried out in summer 1996. On account of the high number of identical objects, a transformation „house after house" could be made. Non-identical areas were transformed together with identical objects. In this case identical means objects existing in nature today, which were already in existence at the time the "Franziszeische Kataster" was made and were shown on the plans.

This method didn't work with all land register plan in question, because the number of identical objects were lower than in the plan of the first district, furthermore identical objects were not situated as close to each other.

Since completely different methods of survey were in use at the time the maps were made, one cannot say the land register plan is inexact. Rather it is to be admired how precise the depiction of Vienna in maps of those days really is. Therefore we refrained from a bit-by-bit transformation to preserve the geometry of these old maps. Determination of the scale was done according to the Mekenkamp method and the transformation of identical points was done after Helmert.

3.1.2. Determination of points

First of all each sheet of the land register was divided at random into zones and columns, to guarantee an indisputable identification of points.

3.1.3. Mekenkamp method

By this method identical coordinates are taken from both the digitised "Franziszeischen Kataster" and today's digital map of Vienna.

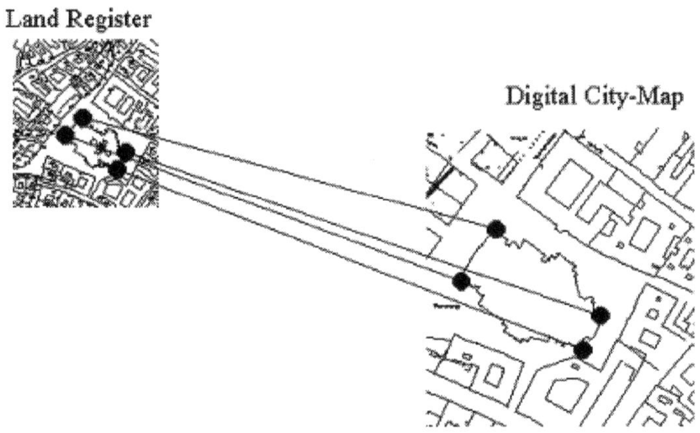

Figure 3: Scrutiny of the land register's accuracy (Identical Points - Transformation house after house).

After that all possible distances between all points are calculated for both bodies of maps separately. Matrices are drawn up and an average scale is calculated. Furthermore a standard deviation expressed in percent can be calculated for each point. With this method exceedingly deviating identical points were found.

3.1.4. The Helmer-transformation

The Helmertransformation is a transformation from one system of coordinates into another with several identical points (coordinates). When solving the problem a shift, a distortion and a mean change of scale has to be calculated between the two systems of coordinates. The Helmertransformation works according to the principle of the balance account.

3.1.5. Results

	Scale	Deviation	Deviation for 100 m
Initially intended	2880	0	100.00 m
Sheet XVII	2924	44	101.53 m
Sheet XXIII	2904	24	100.84 m

A paper warped by 1 mm equals an object shift of 2,9 m!

Results of the Helmerttransformation:

	Scale	Rotation [gon]	Linear translation parameter - Gk	
Sheet XVII	2924	0.332199	3023.40	338079.01
Sheet XXIII	2904	0.317054	3033.73	338081.17

(Twaroch 1998)

With the help of these fundamental maps roughly 95% of all sketches could be fitted in. This land register, however, does not serve only one purpose, for in the past years it turned out to be even more valuable.

After the basics had been created, it was time to think about linking all data.

The task of Stadtarchäologie is not only to react to reports of finds, but also to set preventive measures.

That was the hour of birth of Vienna Archaeological GIS.

The first theoretical considerations dealt with the question of which thematic layers should be included.

It was decided to create 3 groups of thematic layers:

1. The Archaeological Layer (plans of contexts, maps of the distribution of find a.s.o.)
2. Historical Layer (Franziszeischer Kataster and earlier city maps, Map of the Danube with contour lines from the year 1849 - that is from before the regulation of the Danube)
3. Auxiliary Layer (modern digital city map, geo-referenced aerial photos, geological maps, mains system register a.s.o.)

3.1.6. Archaeological layer

Every archaeological site shown on the map is marked with the site code. Each entry is cross-referenced to a record in the site database. This reference is multilayered. The topmost layer contains information about the excavation and a description of the site. All other layers contain further information about the archaeological contexts of the site.

3.1.7. Historical layer

At the moment the main focus of our work is on digitising and analysing historical maps.

As mentioned before, the "Franziszeische Kataster" is not only used to localise earlier sites by tying them to specific coordinates, but also to recognise the courses of small rivers, which have either completely disappeared or have been regulated and put underground.

Since the land register was drawn up as an aid for taxation and written tax assessments also exist, it is possible to deduce the quality of the soil.

Of course, historical settlement areas can be made out on this body of maps, which comprises 180 maps for Vienna (Banik-Schweitzer 1998:19).

By combining the "Franziszeische Kataster" with other maps, e.g. the "Donauplan" of 1849 with aerial photos, and with the digital terrain model of today's citymap, and of course, all known archaeological sites, areas favourable for settlements can be determined.

Particular attention will be focused on the districts north of the Danube, because several large building projects are planned for this area for the near future.

3.1.7. Auxiliary layer

The digital map of Vienna is of outstanding importance for positioning our sites.

The digital Topographical Database of the City of Vienna is better known under the misleading name "Mehrzweckstadtkarte" (MZK) (Wilmersdorf 1995:46). Since all available information of the topographical database is never shown on one and the same series of maps, the name is not quite suitable.

The Topographical Database is saved centrally and contains detailed information about all objects on the surface like buildings, fences, property borders, landscape features, vegetation, road network and other details.

The data acquisition is made tachymetrically within the road areas and photogrammetrically within the building complexes.

The accuracy of the tachymetrical measurements is around ±0,01 m, those of the photogrammetrical measurements around ±0,25 m. Therefore the absolute accuracy of 3D-models of the city is limited to ±0,25 m. For the generation of the terrain model (DTM resp. Digital Terrain Model) only tachymetrically calculated heights should be used in any case (Halmer 1999:43).

The Topographical Database contains records for the entire urban area with the exception of the wooded areas at the outskirts of the city. It is updated every three years.

The GIS-Software Arc-Info is used for the administration of the data described in the lecture, that is the thematic layers and the archaeological finds and context data, as well as for the photographic archive.

As Stadtarchäologie Wien does not have this GIS-software at the moment, the preliminary work is first carried out in an AutoCAD Map and later transferred into ArcInfo with the help of the division of graphic data processing.

A prototype of the Vienna Archaeological GIS will be presented to a wide public on the occasion of an event organised by the city of Vienna and called "Global Village", whose theme will be the remote-controlled city, in June 2000.

4. Outlook

Based on these fundamentals, we would like to develop a strategy for the preservation and research of our historical heritage in co-operation with the city-planning bodies, as well as with private developers (Kobermaier et al. 1998:37).

Considering that it is impossible to be present at each and every building site, it is necessary to construct and maintain a body of maps that will not only show all the known archaeological sites but will also help in identifying the broader areas in which archaeological implications are most likely to occur in the course of their development.

Besides processing the material from earlier excavations this will be our task in the forthcoming years.

We cannot prevent building sites, but we can raise peoples' awareness of our past through public relations and the publication of appropriate maps. This process should begin in school.

For this reason Stadtarchäologie Wien runs two projects besides the research prescribed by law: the Junior and the Senior Archäologie. Both initiatives actively involve Viennese citizens in the archaeological exploration of the city (Strohschneider 1998:38, 44).

References

ARTNER, G., KRENN, M., 1995. Stadtarchäologie im Raum Krems-Stein. In Krenn-Leeb, A. (ed.), *Perspektiven, Zum Werdegang von Krems und Stein. (Festgabe zum 10000-Jahr-Jubiläum der Stadt Krems)*, 40-56. Wien: Archäologie Österreichs (Sonderausgabe).

BANIK-SCHWEITZER, R., 1998. *Rekonstruktion von Erschließungs und Bautypologien mit Hilfe von Baualterplänen.* In Forschungsgesellschaft Wiener Stadtarchäologie (ed.), Workshop 2 – Archäologie und Computer 1997:19-36.

BÖRNER, W., 1998. Archäologie und Computer. *FWien* 1:52-56.

BÖRNER, W., MOSSER, M., 1997. *Von der 100-jährigen Karteikarte zum Plan des Legionslagers.* In Forschungsgesellschaft Wiener Stadtarchäologie (ed.), Workshop 1 – Archäologie und Computer 1996:9-20.

FISCHER, K., 1995. *Streiflichter auf die Wien-Kartographie der letzten beiden Jahrhunderte.* In Fischer, K. (ed.), *Das ist die stat Wien*: 38-45, Wien: Wiener Geschichtsblätter (Beiheft 4).

HALMER, A., 1999. *Teilautomatisierte Generierung digitaler 3D-Modelle des Wiener Stadtgebietes.* In Forschungsgesellschaft Wiener Stadtarchäologie (ed.), Workshop 3 – Archäologie und Computer 1998: 43-53.

JAKOB, A., 2000. Digitale Siedlungsflächenbilanz Wien. In M. Schrenk, M. (ed), *Informationstechnologie in der Raumplanung*: 75-80. Wien: Computergestützte Raumplanung (CORP 5).

KOBERMAIER, F., KREPPENHOFER, A. and MATOSEK, J., 1998. *Das neue Schutzzonenmodell für Wien.* In Forschungsgesellschaft Wiener Stadtarchäologie (ed.), Workshop 2 – Archäologie und Computer 1997:37-44.

MADER, I., 1998. Die EDV-gestützte Erfassung der archäologischen Fundstellen im 3. Wiener Gemeindebezirk. *FWien* 1:89-97.

MOSSER, M., 1998. Das Legionslager Vindobona - EDV - gestützte Erfassung alter und neuer Grabungen. *FWien* 1:74-88.

STIPANITS, U., 1998. Über 100 Jahre handschriftliche Fundmeldungen und ihre EDV-gestützte Erfassung. *FWien* 1:67-73.

STROHSCHNEIDER-LAUE, S., 1998. Über die Unbezahlbarkeit freiwilliger Helfer. *FWien* 1:38-43.

STROHSCHNEIDER-LAUE, S., 1998. Initiative Juniorarchäologie oder: Wie viel Archäologie braucht die Schule. *FWien* 1:44-47.

TURECEK, C., 1999. *Archäologische Datenverwaltung in 3 Dimensionen. Aufbau eines umfassenden Geoinformations-systems für die Stadtarchäologie Wien.* In Forschungsgesellschaft Wiener Stadtarchäologie (ed.), Workshop 3 – Archäologie und Computer 1997:55-62.

TWAROCH, F., 1998. *Genauigkeitsanalyse und Transformation des Franziszeischen Katasterplanes.* Technischer Bericht für die Forschungsgesellschaft Wiener Stadtarchäologie (not published).

WIELOWIEJSKI, J., 1989. Uzdial Józefa Nowalskiego w Ratowaniu Rzymskich Monet na Terenie Wiednia (Josef Nowalski's contribution to preservation of Roman coins in Vienna). *Wiadomo'sci Numizmtyczne* XXXI. 1987:88-97.

WÖLFL, K., 1998, Ausgrabung Unterlaa – Ein Siedlungskomplex im Hinterland des Limes. *FWien* 1:113-116.

NARS - Nabunken Aerial Photograph Retrieval System – A Way to the GIS

Susumu Morimoto

Nara National Cultural Properties Research Institute

2-9-1, Nijo-cho, Nara, 630-8577, Japan

e-mail: susumu@nabunken.go.jp

Abstract

The Nara National Cultural Properties Research Institute (Japanese abbreviation: Nabunken) houses an important historic collection of aerial photographs. The collection originates after World War II when private companies began to take aerial photographs for mapping or exploratory purposes. Unfortunately, private companies do not have the capacity to preserve this data, so Nabunken archives these records for the whole of Japan. This collection does not include aerial photos taken by the government for national mapping. The total number of these photos is about 1,800,000 and Nabunken maintains a database of the entire archive. The original film is 10 inch format and difficult to treat. Microfilm is made from the original. About 900 thousand pictures have been processed to date. Whilst the data entry system provides coordinate data for the map image, the retrieval system was separated from the entry system. However, a new system is now being created incorporating the digital map image, the digitised image of each aerial photo and data entry and data retrieval. The system also acts as an intranet and as data entry and digitisation proceeds the data will become available for use within true GIS analyses.

Key words: Nabunken, aerial photo, microfilm, database, retrieval system, digital map, high resolution

1. The origin of the collection

Nabunken (The Nara National Cultural Properties Research Institute) is a national organization for the study of archaeology in Japan. It was founded in 1952 at Nara, the ancient Japanese capital city of the 8th century. Nabunken operates as part of the Agency for Cultural Affairs under the Ministry of Education.

Nabunken possesses an important collection of aerial photographs. After World War II private companies began to take aerial photographs for cartographic purposes. However, these private companies did not have a enough space to preserve this data and it was intended that the photographic archive should be disposed of in c. 1970.

Nabunken believed that that these aerial photos were cultural property and that their loss should be prevented. The photographs have considerable historic value. They are material evidence for past land use and provide evidence for a variety of sites and monuments. In Japan there are very few ancient structures made of stone or brick, and evidence for past settlement is rare. Despite this, through the aerial photographic record we are able to distinguish, for example, ancient tombs from 3rd to 7th century AD and the reorganisation of agricultural land from the 7th century to the medieval periods. This is possible because these aerial photos were taken before the development and the demolition of the historical landscape.

Nabunken therefore decided to preserve these primary records for the whole of Japan. The total number of air photos is about 1,800,000 and comprises c. 25,000 rolls of film. Nabunken provides an air-conditioned storage facility specifically for these films. It should be emphasised that this collection does not include aerial photos taken by the government office for national mapping (the Geographical Survey Institute).

2. The history of the database

NARS Generation 0 (before 1981)

Prior to his retirement in 1998, Mr. Taisaku ITO developed a computer aided system - NARS (Nara national cultural properties research institute Aerial photo Retrieval System). Within this system, he arranged the original documentation and, in-between 1973 to 1977, placed a location mark for each picture on a paper map. This was not a convenient system. Sometimes there were too many pictures on the same map. Consequently, around 1979, Mr. Taisaku ITO tried to introduce a database system. Unfortunately, there were so many errors in the original documentation that the photos themselves had to be re-checked (figure 1).

NARS Generation 1 (figure 2)

As the original photographic film is 23 cm square negative roll film format it is difficult to use. To provide imagery that is accessible this is transferred to positive 16 mm microfilm from which it is possible to check individual aerial photographs. One roll of microfilm includes approximately 4,500 pictures. About 900 thousands pictures have been processed to date. Some original films had attached data about the picture, but about a half of them had no original documentation, and the image itself is the only data source. We check each microfilm to determine the place where the photograph was taken. Detailed features, such as the direction of shadow, lines of railways or roads, vegetation etc. assist in this task. This process is difficult and requires expert involvement.

The text database of the aerial photo has the following fields for each picture:

- orientation map ID,
- title,
- photograph date,
- photograph scale,

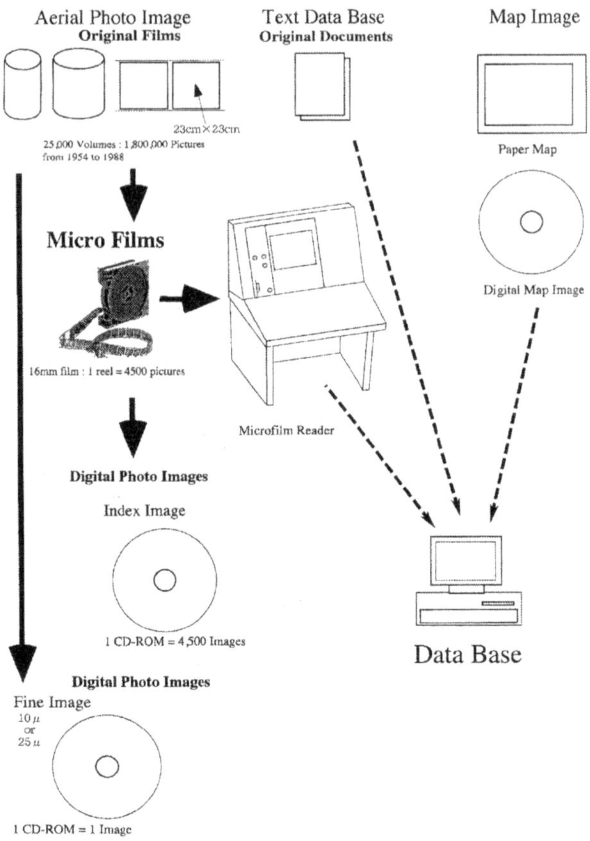

Figure 1: Data entry of the database.

- photographic company,
- type of photosensitive materials,
- microfilm magazine number,
- original roll number,
- number of points in the photographic course,
- photograph point list,
- photograph point number,
- longitude of the point,
- latitude of the point.

The data entry system includes coordinate data from the map. For this a paper map was digitised and the coordinate taken from a PC. The data retrieval system was originally separated from the data entry system. The host computer, an IBM mainframe, was situated at the National Museum of Ethnology at Osaka and was connected to Nabunken by an exclusive line and modems. Aerial photographic images were checked via a microfilm reader.

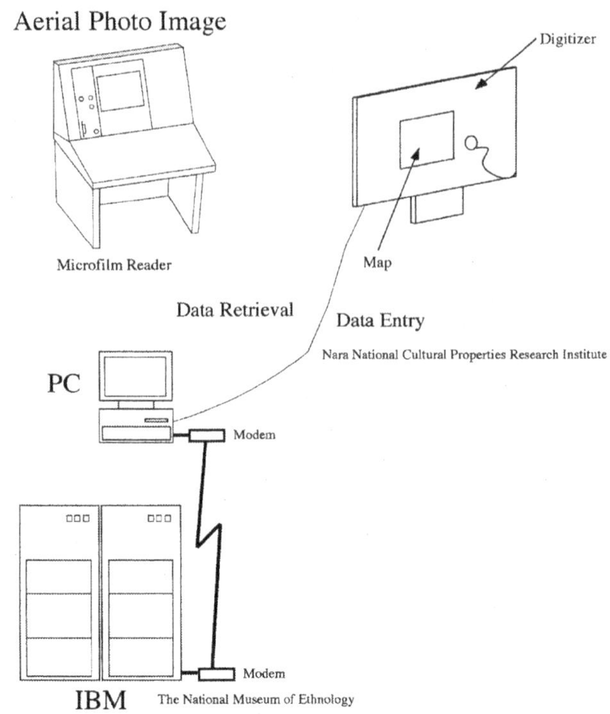

Figure 2: NARS Generation 1.

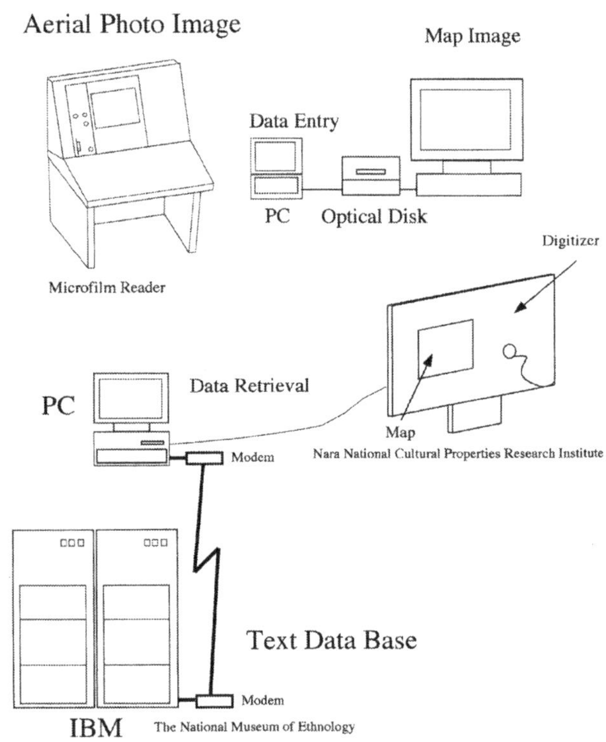

Figure 3: NARS Generation 2.

NARS Generation 2 (figure 3)

Each aerial photographic image was checked to determine location. The data entry system was on a stand alone PC and the map image was stored on optical disks. 1291 tiles of the national map drawn at a scale of 1 to 50,000 for the whole of Japan were digitised as black and white images and stored on 3 optical disks. Map images were connected to each other and a seamless map image for 1/3 of Japan was provided on a single optical disk. Positional data from each map was selected via a mouse on the display: the software then calculated the longitude and the latitude of the point and stored them in the PC which then sent the data to the host computer.

For data retrieval, we used a digitiser to select coordinate data object point in the same manner as for NARS Generation 1. The text database was stored on the host computer was situated at the National Museum of Ethnology.

NARS Generation 3 (figure 4)

The microfilm reader was still used as the primary tool to view aerial photographic data. Data entry was achieved through a stand

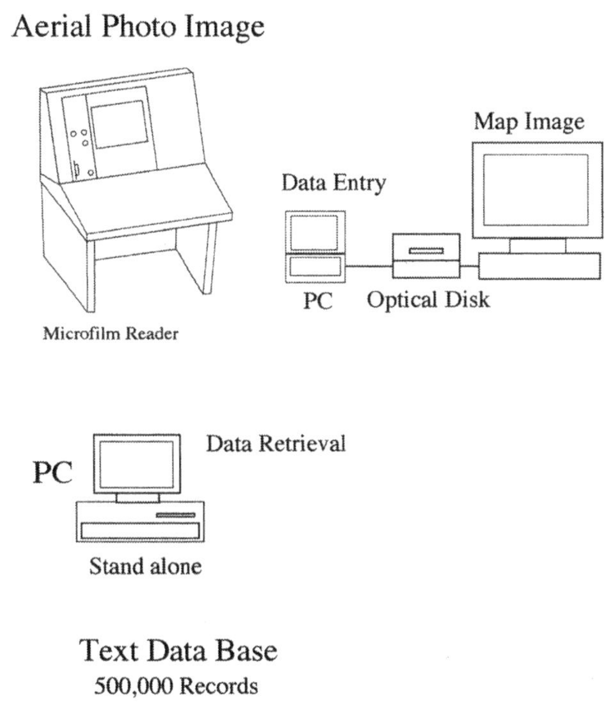

Figure 4: NARS Generation 3.

alone PC with map images on the display. There was no real change here from NARS Generation 2. Data retrieval system was at this time achieved via another stand-alone PC with a text database of about 500,000 records stored on a local hard disk. The mainframe was abandoned, as it was not quick enough. After NARS Generation 3 all the components of NARS were stored and maintained within Nabunken itself. Unfortunately, there were problems with this new system. There was still no on-line connection between the data entry system and the data retrieval system. The data retrieval software was also designed for more general use and was not truly convenient for our purposes.

NARS Generation 4 (figure 5) in 2000

Today we use NARS Generation 4. To check and confirm aerial photo images, we always use the microfilm reader. This part of the system still uses analogue data.

We have a database server within the Institute which provides the topographical map image, based on a national map scale: at 1:25,000 scale the system uses 4356 maps, and also stores the text data of each aerial photograph. This system is also intranet compatible. The user accesses the system via a PC for data entry and also for data retrieval. The server computer is a PC running Windows NT. The data retrieval software is Cach (Inter Systems), and the application software has been specially written for NARS. Using this system we can at last locate photographs quickly and easily from any PC in the Institute and at any time. NARS Generation 4 allows us to make efficient use of our aerial photographic collection for the first time.

NARS Generation 5 (figure 6)

Recently we have begun to digitise the aerial photo collection for use as index images within the database. Images from microfilm are stored as index images requiring c. 200 kBytes in JPEG format. To date some 120,000 photos have been digitised in this manner. The NARS system will be totally reconstructed using these images. Ultimately, three types of data; the aerial photograph, the map and the text data will be incorporated within the computer system within the next couple of years. Storage in microfilm is expensive so an all-in-one PC system is a necessity.

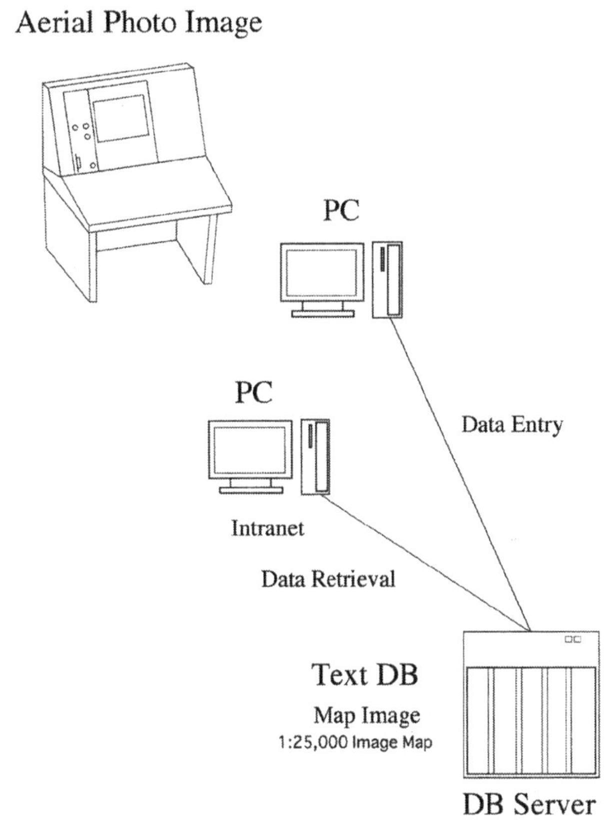

Figure 5: NARS Generation 4.

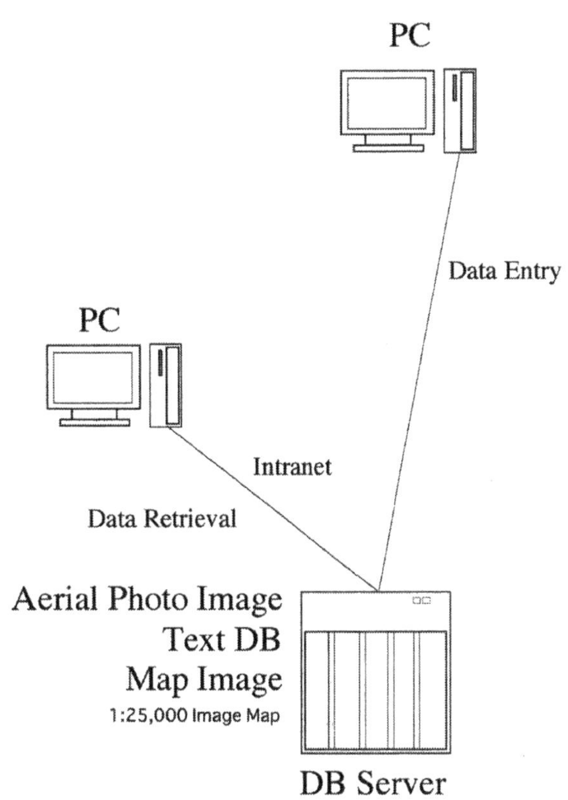

Figure 6: NARS Generation 5.

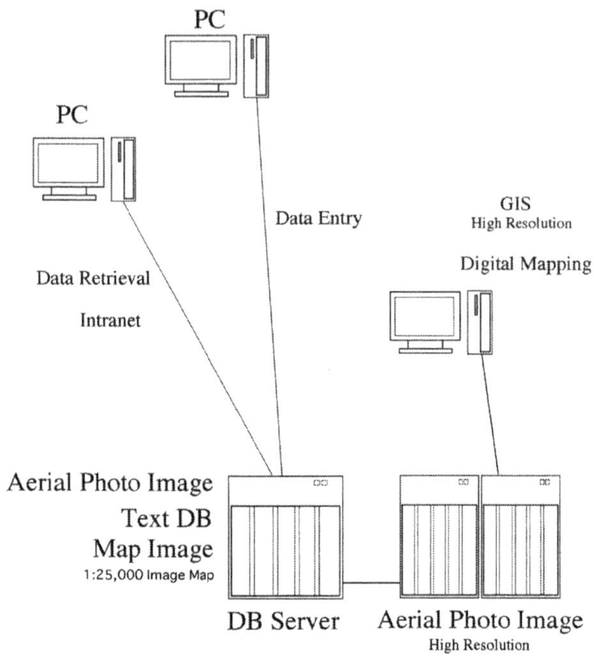

Figure 7: NARS Generation X.

NARS Generation X (figure 7)

Within our collection are photographs which cover the greater past of the Nara basin covering 13 km from east to west and 30 km from north to south. Within this area is the ancient capital as well as a series of palace sites, many imperial and powerful clans' tombs and also prehistoric sites. This series of photographs, numbering c. 2,000 in total, was taken in 1962 just before the urbanization of the area. Recently we have begun to digitise and make an orthophoto of the series. To date we have digitised the original film. Scanning resolution is about 25 microns. The scale of the original photographs is 1 to 6,000 and the resolution is 15 cm. Soon we will be able to use the whole of this photographic series to study this important historical region. When stored digitally, the entire landscape will be available for study within a true GIS-based analysis.

4. Intra Site Spatial Analyses

Formalizing Fact and Fiction in Four Dimensions: A Relational Description of Temporal Structures in Settlements

Mads Kähler Holst

Department of Prehistoric Archaeology, University of Aarhus

Moesgaard 8270 Hojbjerg, Denmark

e-mail: farkmh@moes.hum.au.dk

Abstract

Long lasting settlements with physically overlapping structures often contain a vast amount of observations, which can be taken to indicate specific relative chronological relations between different structures. Our concept of the temporal significance of the observations can be translated into a formal logical expression. This allows the construction of a detailed graph of the temporal structures of the settlement. The procedure will be demonstrated on a section of a Danish Iron Age settlement. Compared to the traditional handling of temporal structures the suggested formalized approach gives a more detailed picture of the development of the settlement with the potential of conducting formal analysis on the relational structures. Furthermore, it presupposes the explicit formulation of the preconditions on which the temporal sorting is based, revealing a problematic mix of fact and fiction in the traditional analysis of Danish Iron Age and early Medieval settlements.

Key words: relational description, networks, Harris matrix, relative chronology, settlement analysis

1. Introduction

When settlements occupy the same area for several centuries traces of buildings and activities may accumulate to such an extent that original distinct patterns and structures become totally obscured.

Northern European Iron Age and Early Medieval villages may be taken as an example from this kind of settlement. The buildings were constructed of posts dug into the subsoil, leaving durable and archaeologically easily recognizable traces. The buildings themselves, on the other hand, had a relatively short life span of one or two generations leading to frequent changes in the layout of the settlement over several centuries. Consequently the excavation plans are often characterized by an overwhelming confusion of postholes and other structures (Hvass 1979, 1980).

Normally, it is possible to identify the individual buildings, first and foremost the longhouses but also the smaller out-houses as well as the fences surrounding the buildings. Furthermore, different types of physical connections between the buildings point to specific relative chronological relations. Buildings may, for instance, cut each other indicating that the cutting structure was constructed after the demolition of the structure being cut. Another example could be that one fence had been added to another. Here the added fence must have been erected after the construction but before the demolition of the other fence. A final example is structures which overlap although without a clear stratigraphy we only know that the involved buildings did not exist at the same time.

When the connections between the different structures form a closely interwoven net, it should, in principle, be possible to generate a detailed relative chronological model of the development of the settlement. Temporal sorting of a number of Iron Age and Medieval settlements have also resulted in seemingly clear and unambiguous phases of contemporary structures, but the analyses are informal and often seem somewhat intuitive. The combination of complex data and the lack of an explicit, formalized approach to the sorting makes it very difficult to test the results and evaluate what preconditions they are based upon.

2. Relational description of the temporal structures

A formalized method for the interpretation of temporal structures of settlements must take its starting point in the character of the data. We can consider settlements to be a relational system consisting of a number of entities, i.e. the buildings, which are related to each other by physical connections with temporal implications (figure 1) (Dallas 1992, Holst 1999).

The connections refer to the construction and demolition of the structures and to the life span of the entities, which can be understood as the time between construction and demolition. If we define demolition as the time where physically the building no longer influences subsequent structures, we may consider both construction and demolition as points in time, notwithstanding that abandonment of the building might have been a gradual process. In a relational description each structural entity in this way has two nodes: A starting point and an end point. If, based on this, we explicitly formulate the life span as the open interval between the two extreme points, we can furthermore distinguish between continuity, where the end point of one structure is contemporary with the start point of another structure, and discontinuity where the end point of one structure is earlier than start point of another structure. The temporal sorting is purely a relative chronology, and as we are dealing with points in time, in principle only three types of relations are possible: Contemporary with, earlier than and later than.

In this way the relational network is rather similar to the one found at stratified excavations, where a formal, graph based representation of the relations has long been used in the form of the Harris Matrix (Harris 1975).

However, it should be stressed that the analysis of temporal structures in settlements differs from the handling of stratigraphy for a number of important reasons. First and foremost, stratigraphical observations are relatively objective, whereas it is quite evident that attempts to uncover the temporal structures of the settlements deal with interpretative constructs and are based upon our subjective evaluation of the significance of connections between different constructs.

Another difference is that in settlement analysis the structural entities consists of two nodes, and the connection between two entities might consequently be a set of relations rather than just one relation as in stratigraphical analysis, making it possible to describe the temporal consequences of the connection very specifically. This level of detail is necessary, because settlements contain a wide range of different types of connections with different temporal implications, whereas stratigraphical analysis principally only operate with two connections i.e. above and below.

A final characteristic that distinguishes settlement analysis from stratigraphical analysis are the ambiguities and uncertainties of the temporal connections. As the data are a fragmented and partial reflection of the past, we are often left with several possible relations. An example is the overlapping structures without cuttings. Here we know that either the start of structure X is later than or contemporary with the end of structure Y, or the end of structure X is earlier than or contemporary with the start of structure Y. To describe this kind of ambiguity it is necessary to introduce an either-or-connective.

We may now describe the temporal relationship between structures in a settlement by relating the start and end points of the individual structures with the relations: earlier than, later than and contemporary with and by combining these relations with the logical operators *and* and *either-or* (Holst 1999). The relations can be represented formally using the symbols:

=	for the relation	*contemporary with*
<	for the relation	*earlier than*
>	for the relation	*later than*
≤	for the relation	*either earlier than or contemporary with*
≥	for the relation	*either later than or contemporary with*
AND	for the connective	*and*
XOR	for the connective	*either or*

The example of the overlapping structures without cuttings would formally be expressed as:

X(start) ≥ Y(end) XOR X(end) ≤ Y(start)

3. The procedure in practice

Having established the general principles of description, attention can now be directed towards the practical use of the formalized handling of the temporal structures.

The first step is to identify the different types of connections between structures of a settlement. Secondly, our interpretation of the temporal significance of the connection types must be formulated, and thirdly, this interpretation has then to be expressed formally according to the principles outlined above.

The types of connections can be classified into 13 classes each with their specific temporal implications as seen in figure 2.

Having defined the connections and expressed their temporal implications formally the next step is the actual recording of the temporal structures in the settlement. This process can be demonstrated on a small part of the 3rd to 7th century AD Nørre Snede settlement situated in Mid Jutland (figure 3) (Hansen 1988). The

	Time	Entities	Links	Interpretation
Data level	Span of time	Structures	Connections	Temporal implication
Formal representation	Points in time	Nodes	Relations	Formal expression

Figure 1: The basic concepts in the analysis of temporal structures related to a distinction between a data level and a formal representation.

Partial asynchronism

a) $X(start) < Y(start)$ XOR $X(start) > Y(start)$

Full asynchronism

b) $X(start) \geq Y(end)$ XOR $X(end) \leq Y(start)$

General synchronism

c) $X(start) < Y(end)$ AND $X(end) > Y(start)$

Specific synchronism

d) $X(start) = Y(start)$

e) $X(end) = Y(end)$

Full synchronism

f) $X(start) = Y(start)$ AND $X(end) = Y(end)$

Asymmetrical synchronism

g) $X(start) \leq Y(start)$ AND $X(end) \geq Y(end)$

General diachronism

h) $X(end) < Y(end)$

i) $X(start) < Y(start)$

k) $X(start) < Y(end)$

Full diachronism

l) $X(end) \leq Y(start)$

General continuity

m) $X(end) = Y(start)$ XOR $X(start) = Y(end)$

Directional continuity

n) $X(start) = Y(end)$

General discontinuity

o) $X(end) < Y(start)$ XOR $X(start) > Y(end)$

Directional discontinuity

p) $X(start) > Y(end)$

Combined expressions

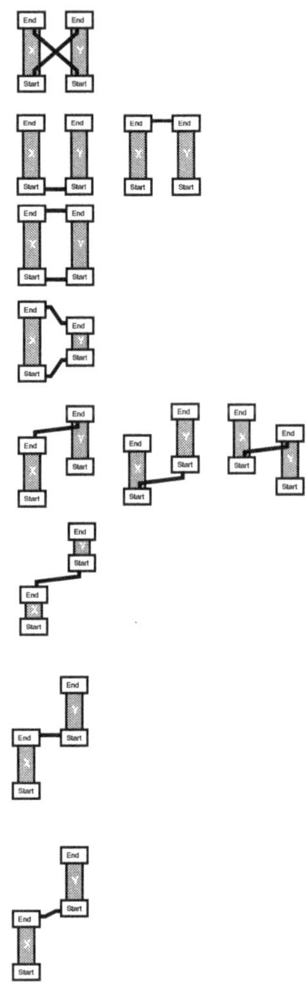

Figure 2: Classification of the temporal implications with graphical representation, where the implication is unambiguous.

area shown consists of two farms, which can be divided into a number of minor structural entities, i.e. fences, longhouses, minor houses and barns. The defined structures are listed in an m x m matrix, figure 4, where the temporally significant connections between the structures are shown. If the individual connections are substituted with their formal relations between start and end-nodes of the structural entities, a relational network representing the temporal structure of the settlement is achieved.

Figure 3: The analysed area of the Nørre Snede excavation showing the identified structures in the settlement (scale 1:500).

4. The ambiguities of the relational network

The relational network of the Nørre Snede settlement can be represented graphically as shown in figure 5. It offers a very detailed picture of the temporal structure, where we can distinguish between contemporary replacements of buildings and fences in one of the farmsteads and gradual changes in the other. The level of detail makes it possible to give a very precise description of the character of changes in the settlement.

However, the network is also a highly flexible structure and very much open to interpretation. The partial character of the archaeological record means that only a few elements are related to each other and hence there are a limited number of constraints on the relative positions of the elements. They are seldom locked, but can be moved relative to each other, making it difficult to point out exactly which elements existed simultaneously. Furthermore, due to the either-or relations several different ways of sorting the elements will often be logically consistent and thus equally probable. Consequently, it is not possible to identify unambiguously which structures existed simultaneously.

What then is the basis for the often very detailed segregation of phases of contemporary structures seen in the publications of the North European Iron Age and Medieval settlements? Apparently the chronological sorting is based on the same types of connections as in the network analysis presented here, but there seems to be important differences in the concept of their temporal implications. In the construction of the network at least 13 classes of different temporal implications were used, whereas the more traditional analyses only seems to distinguish between full synchronism, full diachronism and full asynchronism, even though it is rarely expressed explicitly.

This is a generalization, which to a wide extent ignores the ambiguities of the archaeological data, and makes the temporal sorting appear more certain than it actually is. Furthermore, it may have far reaching consequences for the interpretation of the organisation and social structure of the settlement, as it can impose false structures on our picture of the development of the settlement. For instance, the exclusive use of the three mentioned types of temporal implication, will make changes in the settlement appear contemporary even if they occurred gradually, and these apparently contemporary changes have been used to argue that the settlements were strongly regulated. Again this influences our concept of power structures and the character of the power in the Iron Age and Medieval society.

Another reason why traditional chronological sorting results in a relatively inflexible picture of the temporal structure of the settlement can be found in the introduction of a number of implicit

	Longhouse I	Longhouse II	Longhouse II	Longhouse IV	Longhouse V	Longhouse VIa	Longhouse VIb	Longhouse VII	Fence 1	Fence 2	Fence 3	Fence 4	Fence 5	Fence 6	Fence 7	Fence 8	Fence 9	Fence 10	Fence 11	Fence 12	Fence 13	Fence 14	Fence 15	Fence 16	Fence 17	Other 1	Other 2	Other 3	Other 4	Mindre 1	Farm 1	Farm 2
Farm 2	.	.	.	g	g	g	g	g	.	.	.	.	.	c	g	g	g	g	g	g	g	g	g	g	g	.	.	.	g	g	.	..
Farm 1	g	g	g	.	.	.	.	.	g	g	g	g	g	c	.	.	.	.	.	.	.	.	.	.	.	g	g	g	.	.	..	
Mindre 1	.	.	.	.	.	.	.	.	.	.	.	.	.	.	.	.	.	.	.	.	.	.	.	.	.	.	.	.	.	..		
Other 4	.	.	.	.	.	.	.	.	.	.	.	.	.	.	.	.	.	.	.	.	.	.	.	.	.	.	.	.	..			
Other 3	.	.	.	.	.	.	.	.	.	.	.	g	.	.	.	.	.	.	.	.	.	.	.	.	.	.	.	..				
Other 2	.	.	.	.	.	.	.	.	.	.	g	.	.	.	.	.	.	.	.	.	.	.	.	.	.	.	..					
Other 1	.	.	.	.	.	.	.	.	.	g	.	.	.	.	.	.	.	.	.	.	.	.	.	.	.	..						
Fence 17	.	.	.	.	.	.	.	.	.	.	.	.	.	b	bb	.	c?	.	.	.	.	.	.	.	..							
Fence 16	.	.	.	.	.	.	.	.	.	.	.	.	.	.	.	.	.	.	.	c	.	.	i	..								
Fence 15	.	.	.	.	.	.	.	d	.	.	.	.	.	.	.	.	.	b	b	b	b	b	..									
Fence 14	.	.	.	.	.	.	.	.	.	.	.	.	.	.	.	.	.	b	b	b	bn	..										
Fence 13	.	.	.	.	.	.	.	.	.	.	.	.	.	.	.	.	.	b	b	bn	..											
Fence 12	.	.	.	.	.	.	.	.	.	.	.	.	.	.	.	.	.	b	bn	..												
Fence 11	.	.	.	.	.	.	.	.	.	.	.	.	.	.	.	.	.	b?n	..													
Fence 10	.	.	.	.	.	c?	.	c?	.	.	.	.	.	c?	.	.	.	..														
Fence 9	.	.	.	.	.	.	.	.	.	.	.	.	.	.	b	biln	..															
Fence 8	.	.	.	c	.	.	.	.	.	.	.	.	.	.	bim	..																
Fence 7	.	.	.	.	.	.	.	.	.	.	.	.	.	bln	..																	
Fence 6	.	.	ci	.	e	.	.	.	.	.	.	c	c	..																		
Fence 5	.	.	.	.	.	.	.	.	.	.	.	c	..																			
Fence 4	.	.	g	.	.	.	.	.	.	m	.	..																				
Fence 3	.	.	g	.	.	.	.	.	.	.	..																					
Fence 2	l	f	.	.	.	.	.	.	m	..																						
Fence 1	cf	.	.	.	.	.	.	.	..																							
Longhouse VII	.	.	.	bin	bi	bi	bi	..																								
Longhouse VIb	.	.	.	b	bl	bln	..																									
Longhouse VIa	.	.	.	b	bln	..																										
Longhouse V	.	.	.	bin	..																											
Longhouse IV	.	.	.	..																												
Longhouse III	b	bln	..																													
Longhouse II	bbn	..																														
Longhouse I	..																															

Figure 4: Matrix representing the connections between the structures of the analysed area of the Nørre Snede settlement. The letters in the matrix refer to the codes used in figure 2.

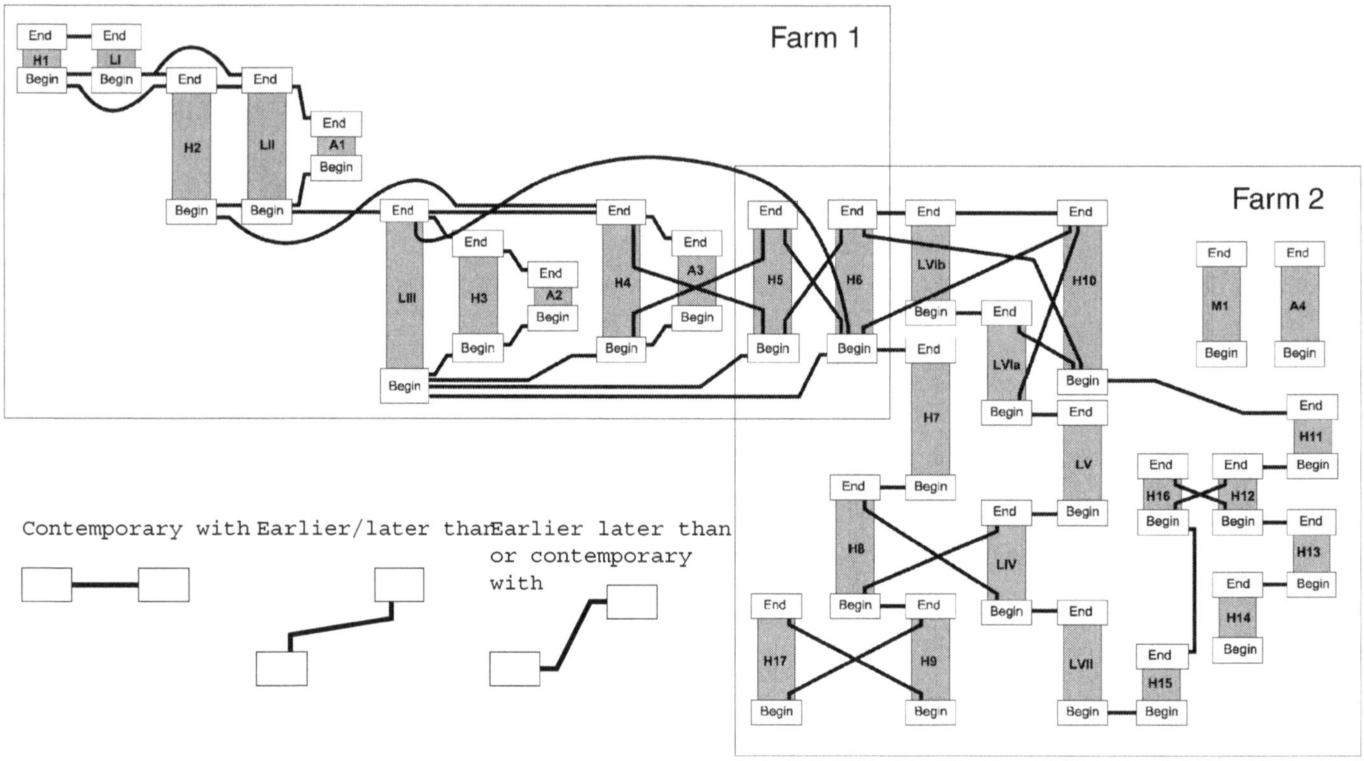

Figure 5: Graphical representation of the temporal structure within the analysed area of the Nørre Snede settlement.

preconditions. For instance, it is often more or less consciously assumed that all the structures of the settlement have approximately the same life span. Furthermore, continuity is often supposed, where it is not directly opposed by dating finds. These preconditions may be useful and often also reasonable, but they are not directly supported by the data, and should be treated carefully and explicitly as they may introduce false structures, as do the generalizations of the temporal implications.

Both the generalizations and the implicit preconditions are obviously a product of a non-formalized, non-computer-aided approach. The vast amount of data necessitates a reduction of the complexity and the ambiguities, if they are to be handled purely intellectually, but the by-product is an unfortunate mixture of fact and fiction, where it is often the fiction which creates those structures in the data upon which we build our interpretation of Iron Age and Medieval society.

5. The use of the relational network

Even though the relational network may not result in an immediate, unambiguous, relative chronological sorting of the structures of the settlement, it must be stressed that it does produce a very detailed picture of the temporal structure of the settlement. Consequently it ought to be possible to obtain valuable information through studies and analysis of the network. The question is simply how.

One way may be to introduce the implicit preconditions of the traditional sorting of the settlements as explicit parameters in the formal analysis of the temporal structures. In this way we reduce the number of possible solutions by introducing structuring principles, which might seem reasonable, but which are not directly supported by archaeological observations. For example, the principle that the structures of the settlement have had approximately the same life span. By making the preconditions explicit and only allowing formal operations, it should be possible to estimate the consequences preventing circular arguments and allow for appropriate modifications of the conclusions.

Another approach is to accept that due to the partial nature of the data, there will not be one definite answer, but several equally probable solutions. We can then formulate queries examining the possibilities of specific temporal structures existing within the temporal network. For instance we can examine whether the introduction of relations representing simultaneous changes in the settlement cause logical inconsistencies, and so get closer to answering the question if the movements of the villages were the result of organized restructuring of the whole settlement or a gradual translocation of individual farms. Other examples could be a query of the possible maximum and minimum number of contemporary farms within a given area, or a study of the possibility of strongly structured layouts of settlement.

6. Conclusions

The work with the relative chronological structure of Iron Age and Medieval settlements is very much a work of interpretation. It is based upon ideas of the temporal consequences of a wide range of connections, and it presupposes a specific syntax for the layout of the farms and a specific syntax for their changes. This relative chronological sorting involves both fact and fiction.

However, the prominent subjective element of the work does not prevent a formal approach. On the contrary the need almost becomes even more pressing. It is necessary to clearly state the premises of the temporal sorting, so that they can be openly discussed and their effect can be studied. Thereby we can evaluate the conclusions on their premises, and we can choose to discard those which are based upon premises we cannot accept, and adopt those conclusions which are based upon what we conceive as reasonable premises. Furthermore as the interpretation often involves very complex structures, a formalized approach will enable us to achieve a much higher level of detail in our analysis.

If the potential of the detailed description of the temporal structures is to be satisfactorily exploited and a better understanding of Iron Age and Medieval settlements is to be achieved, it requires further development of computer applications. In this work, two areas are of special importance.

Firstly, a database system, where the complex data structures can be recorded and retrieved, has to be developed. This is a precondition for analysis of the temporal structures and some of the suggested questions concerning structures in settlements could in principle be easily answered through querying in a database. The GARD-system seems to be an ideal solution (Madsen 1999, this volume).

Secondly, tools for the sorting and analysis of the network structures have to be developed. The ability to handle the complicated logical structures with the possibility of describing ambiguities is essential. A graph theoretical approach seems the obvious solution to this problem, and this area will consequently take a central role in the future development of the analytical methods.

References

DALLAS, C., 1992. Relational description, similarity and classification of complex archaeological entities. In Lock, G. and Moffett, J. (eds.), *Computer Application and Quantitative Methods in Archaeology 1991*: 167-178. British Archaeological Publications (International series S577).

HANSEN, T., 1988. Die Eisenzeitliche Siedlung bei Nørre Snede, Mitteljütland. *Acta Archaeologica* 58: 171-200.

HARRIS, E., 1975. *The Stratigaphic sequence: a question of time. World Archaeology,* vol. 7 no. 1: 109-121.

HVASS, S., 1979. Die völkerwanderingszeitliche Siedlung Vorbasse, *Mitteljütland Acta Archaeologica* 49: 61-111.

HVASS, S., 1980. Vorbasse. The Viking-age Settlement at Vorbasse, Central Jutland. *Acta Archaeologica* 50: 137-172.

HOLST, M., 1999. The dynamic of the Iron-age village. A technique for the relative-chronological analysis of area-excavated Iron-age settlements. *Journal of Danish Archaeology* 13: 95-119.

MADSEN, T., 1999. Coping with complexity. Towards a formalised methodology of contextual archaeology. *Archeologia e calcolatori* 10: 125-144.

Introspective Sitescaping with GIS

Dora Constantinidis

Athens University

Thyatiron 28, Nea Smirni 17121, Greece

e-mail: dorac@postmaster.co.uk

Abstract

The application of GIS to archaeological research has primarily facilitated landscape analyses up until now. In the process of applying a GIS to facilitate an intrasite analysis it was observed and subsequently proposed that GIS processing concepts applied to landscape analyses can also be mapped *to "sitescape" analyses. Emphasis is placed in either case on the importance of defining questions about the available data, and a set of spatial GIS analyses to answer the questions. A small sub-set of questions and possible analyses are presented for the site of Akrotiri on Thera. The relatively well preserved state of this site offers a number of research options for the spatial organisation of the site. This paper presents preliminary results of a metrical analysis of the site's plans, utilising MapInfo in order to determine what, if any, units of measurement were used for constructing the buildings at the site.*

Key words: sitescape analysis, mapping processes, metrology, Minoan foot

1. Introduction

Archaeological data across a landscape or within a site have a spatial element associated with them. Geographic Information Systems (GIS) facilitate the management and analysis of spatial data. Therefore it was only a matter of time before GIS were applied to the multi-faceted management and analytical requirements of archaeological research. Over the last ten years, GIS have facilitated many projects and applications (Allen et al. 1990, Lock and Stančič 1995). Patterns of GIS use became apparent with an overview of the various applications of GIS to archaeological research, and from this new research directions have been sought.

GIS can be utilised at any level of the archaeological record. The two fundamental levels are those of landscape and intra-site, or as proposed here, "sitescape" archaeology. This paper presents the idea of mapping GIS processing concepts from one level of archaeological research to the other. Some of the underlying concepts of GIS and how these can be mapped to either level of archaeological research will be presented. For instance the buildings can be thought of in terms of the physical surroundings of the artefacts, where the artefacts are "sites" within a "landscape" of buildings.

1.1. The importance of defining which data are to be analysed

For either level of analysis the main question that needs to be asked is, what specific data should be utilised? For landscape analyses the data mainly consist of site locations in relation to the topography and geology of an area. In the case of sitescape analyses, it is usually artefact locations within a site. In most cases this choice of data leads to a distributional analysis, whether of sites across a landscape, or artefacts within various buildings.

The basic issue behind all research is to choose an area to focus on. Projects can range from analysing settlement patterns to the distribution of objects within a room. What issues should an archaeological investigation focus on? A set of questions has to be derived from the available data before an analysis can proceed. Since sites vary by region and period, this results in different questions being asked of the available data. The introduction of GIS to archaeological research makes it even more necessary to constantly re-assess the questions posed on the available remains across a landscape or at a site. For instance how does the distribution of artefacts provide an indication of a culture's behaviour? Colin Renfrew makes a pertinent point that "... very often the form of the question determines the form of the answer." (Renfrew 1984:9). So it is necessary to ensure that GIS are not being used to mould the data to provide for predetermined answers.

Before an analysis can take place, therefore, careful attention should be placed on what questions are asked of the data. Bertrand Russell's remark on the importance of defining questions is very applicable to archaeology. By substituting the word philosophy with archaeology it reads: "In *archaeology* what is important is not so much the answers that are given, but rather the questions that are asked." (Russell 1959:17). This set of questions should guide an investigation towards the provision of information about a culture and eventually lead investigators to understand an aspect of the culture under consideration. Many factors influence the process of working towards a set of questions from the available data. Ultimately a researcher's interests or field of expertise often decides upon what aspect of an archaeological site or sites are to be investigated.

1.2. Case Study: The site of Akrotiri on Thera

In order to better appreciate some of these issues a practical application of a sitescape analysis was made for the site of Akrotiri on the island of Thera. The site of Akrotiri provides a good case study since it offers many avenues for GIS data management and analysis.

Akrotiri can be sitescaped for a variety of analyses such as the density and distribution of artefacts and samples within a building or across the site. Furthermore since many of the buildings are preserved up to the second storey, 3D visualisation is another possibility. There are also a number of potential analyses for the buildings' architectural and dimensional aspects which have not been exhausted.

This paper presents a small sub-set of questions that can be applied to some of the available data. The underlying question is whether a culture can be identified and understood from the spatial organisation of its buildings, down to their sub-division into rooms. The GIS functions of MapInfo were used to digitise and store the site-plans, from which a metrical analysis of the walls of the buildings was performed to determine if standard units of measurement were used at the site.

2. Mapping GIS processes

Since GIS is a tool which facilitates the quantification of the spatial relationships of data, and to some degree through time (Langran 1992), the object of most archaeological analyses is to measure the dimensions, distances, density, and distribution, of sites, buildings, and artefact types. Comparative studies may be made across different regions or periods to determine, for example, how settlement patterns differ from one region to another or from one period to the next. Another example may be analyses between site-types and the various buildings and artefacts found at each site.

It appears that the types of analyses that are commonly undertaken for landscape archaeology can also be applied to sitescape analyses. These remain essentially the same since spatial elements (for example distances) are quantified or measured in each case. So one type of analysis at the landscape level can *map* to a corresponding sitescape analysis. Therefore I propose to define the term "mapping processes" since many of the concepts behind spatial analyses can be *mapped* to either level of landscape or sitescape research. The ability to map processes from one level of archaeological research to another is what allowed GIS to be applied to archaeology in the first place.

2.1. Landscapes versus sitescapes

Both levels of landscape and sitescape archaeological research are basically comprised of two "data units" each. For landscapes, the units are *sites* and *land* where sites are related to the land. For sitescape archaeology, *objects* such as artefacts are related to *site-areas*. These *four* units can be analysed either amongst themselves or in relation to each other. An example is illustrated in figure 1. where artefacts are related to either (P) public or (D) domestic building types, and (S) small (M) medium and (L) large sites are related to an environment. For either level, the spatial relationship between the two units is what links the two levels together. This link allows the underlying GIS management and analytical processing concepts to be interchangeably *mapped* or transferred from one level to the other.

Landscape analyses have primarily focussed on examining the spatial organisation of sites across a geographic area (Gaffney and Stančič 1996). Often geo-morphological elements of the environment are related to the sites and their locations within a landscape. An environmentally deterministic approach has characterised most research projects given the slant of the investigation and the available data (Gaffney et al. 1995). In some cases the less evident and tangible factors of cultural, social and religious elements are considered, although it is difficult to avoid an environmentally deterministic approach given the fact that GIS developed in order to manage and analyse land-use. This deterministic approach occurs since most GIS focus on descriptive data con-

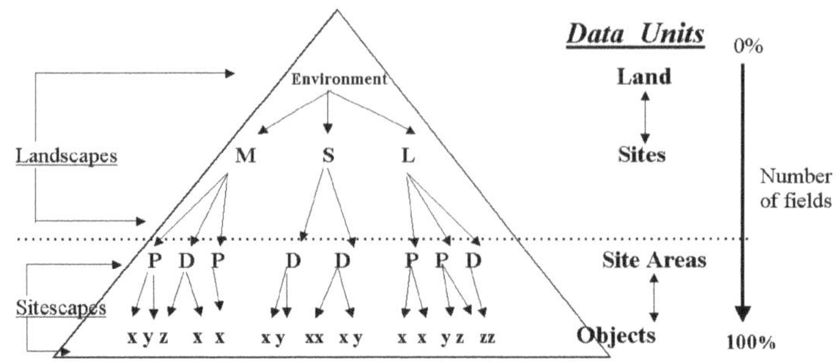

Figure 1: Schematic diagram of archaeological data units in relation to landscapes and sitescapes.

cerning the environment that the sites are located in. Similar issues and concerns also need to be dealt with and addressed when a GIS is used for sitescape analyses. It is possible that building types may be used to influence and subsequently determine ideas about certain types of artefacts that are associated with them, and vice-versa.

In landscape archaeology, for example, certain site types may correspond to particular environments, where a certain environment-type has a higher percentage of one type of site. On the other hand for sitescape archaeology certain site-areas may have a higher proportion of certain objects. In this case (figure 1) artefacts are related to either public or domestic building types, and small, medium and large sites are related to an environment. The resulting analyses may be influenced by the way the buildings of a site and their related objects are categorised. Correspondingly the way an environment and site-types are categorised can also influence the interpretation of results. As shown in figure 1, proportionally more records are contained as the lower levels of analyses are reached. That is the number of data fields related to each of the units (*land* to *sites* and *site-areas* to *objects*) increases as we move down to sitescape analyses. Hence one of the differences between the two levels of archaeological research is the amount of data fields which are available for analysis and management.

All the capabilities of GIS, especially for sitescapes, have not yet been fully exploited. In the case of sitescapes, up until now most attention has focussed on using a GIS for the analysis of excavation data to investigate artefact distribution across a site or between the various phases and stratigraphy of a site (e.g. Biskowski 1994:115-134). Limited research has been done to investigate all the possible permutations and combinations of the numerous types of analyses that can be made within and between the two levels of archaeological research, and amongst the data units of which they are comprised. For now only a broad overview of some of the capabilities of GIS and how these have been applied to archaeological research so far, follows.

2.2. Mapping GIS processes from landscapes to sitescapes

First of all GIS applications in archaeology can be subdivided into the two broad categories of *managerial* and *analytical* applications. Overall for both levels of landscape and sitescape archaeology there are *four* main types of GIS applications which *map* to either level, see figure 2.

1. *Data management*, where a monuments or sites register *maps* onto an artefact or building register. The storage of site names and locations across a landscape provides a Sites

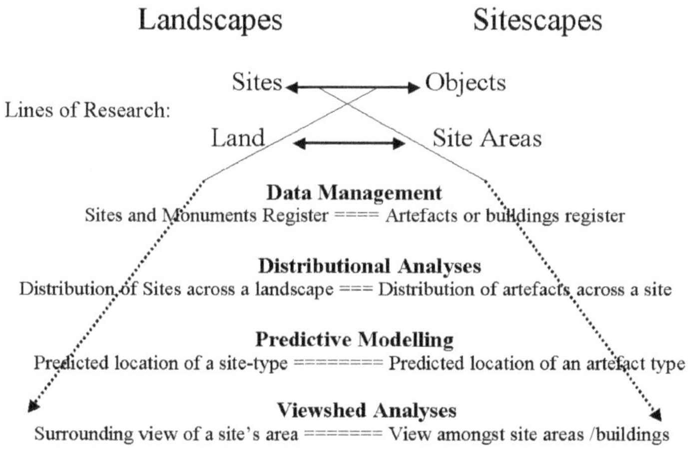

Figure 2: Chart of GIS applications which "map" from landscape to sitescape archaeology.

and Monuments Register. This managerial capability of a GIS can also apply or be *mapped* to an excavation where a GIS can be utilised as a tool which assists in centralising excavated data into one archive. This digital archive can be easily updated and accessed for information about various data that have been excavated at a site. As each monument or site has a specific location and set of attributes within a landscape, so too each artefact or building has a specific location and set of attributes within a site.

2. For a *distributional analysis* the spread of sites across a landscape would *map* to, for example, the distribution of an artefact-type within a sitescape.
3. For *predictive modelling* if the location of a site-type can be predicted in a landscape then these processing concepts can be *mapped* to predict the location of an artefact-type within a site.
4. For a *viewshed analysis* around various sites the process or concepts used would *map* to creating views around a building within one site.

These lines of research correlate to either level, and *mapping* is possible because the underlying data elements are spatial. For instance a set of points can be spatially related to each other and to external features. So in the case of landscapes these points may be considered to be sites relating to each other and to the environment. In a sitescape the location of objects can be related to each other as well as to the buildings or areas of a site. Taking this even further, artefacts scattered within a room can also be managed or analysed along similar lines for "roomscapes".

3. Sitescaping Akrotiri

More archaeological research utilises GIS for landscape than sitescape analyses, to be expected considering that GIS developed in order to manage and analyse land-use over large territories, and subsequently the initial application of GIS to archaeological research was influenced by this. Most of the earliest archaeological studies which used GIS were those investigating sites in relation to their environment, and to other sites (Allen et al. 1990). However the effective management of daily data collected at a site can also be facilitated by the application of a GIS. When it comes to analysing various aspects within one site limited methodologies have developed so far, so, re-thinking what and how a GIS can be applied to a smaller area is important.

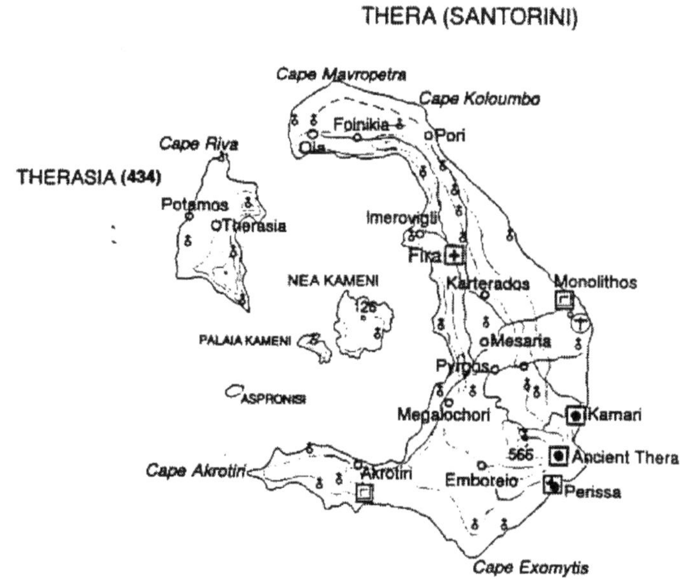

Figure 3: Island of Thera in 1:250,000 scale (Atlas 1999:84).

Therefore, the process of applying a GIS to sitescape analyses was undertaken for the site of Akrotiri to investigate some possibilities, keeping in mind the ability to *map* some of the already developed methodologies or processing concepts from landscape analyses.

3.1. The site of Akrotiri (Thera)

Approximately 3500 years ago the cataclysmic eruption of the volcano on the island of Thera buried the site of Akrotiri under a layer of volcanic ash 30 to 60 m in depth. The volcanic eruption created a geological phenomenon - the largest caldera in the world with cliffs that rise steeply above the level of the sea. On the other hand this destructive event buried the remains of a culture. Akrotiri has been referred to as the Pompeii of the Aegean, because just as at Pompeii the buildings and artefacts have been relatively well preserved (Doumas 1984). Systematic excavations commenced in 1967, by Professor Marinatos, and after his death in 1974 excavation and conservation at the site has continued under Professor Doumas.

Akrotiri, situated on the south coast of the island of Thera, faces Crete which on a clear day is visible on the horizon, see figure 3. Not surprisingly evidence of many elements of Minoan culture are present at the Late Cycladic (LC) site of Akrotiri. Various artefacts, architectural features, wall painting motifs, and local pottery display the close links Akrotiri had with Minoan Crete during the Aegean Bronze Age (Castleden 1990).

It seems that people had enough time to load awaiting boats in the near-by harbour with their most precious possessions since relatively few luxury items and metal objects have been uncovered so far. However what was left behind provides a wealth of information to archaeologists with many of the buildings relatively well preserved, up to two storeys in some cases. The most spectacular discovery was the wall paintings: many of the rooms were decorated with wall paintings depicting various scenes including animals, plants, landscapes, as well as people engaged in various activities, from sea-faring to gathering saffron (figure 4 - from building Xeste3, room 3a). The variety of themes and subjects which have been depicted offer a unique glimpse into Aegean Bronze Age life and culture at Akrotiri.

Akrotiri had been inhabited since the late Neolithic. The Early Bronze Age or the Early Cycladic period lasted during the fourth to the third millennium BC. A major earthquake ruined many of the Middle Cycladic (MC) or early 17th century BC buildings (Marthari 1984). The LC city of Akrotiri was built on top of the MC phase, and many of the buildings are modifications of the MC buildings. Evidence of rebuilding is apparent throughout the site. In many areas ruined walls were demolished and new walls were built on top of the MC rubble, raising the overall LC level of the site.

The earthquake gave the inhabitants the opportunity to rebuild and refurbish their homes according to their functional requirements. New styles from Minoan Crete influenced the reconstruction. Popular Minoan architectural features such as ashlar masonry, timber reinforcements, stone bases for columns and orthostats were introduced. Palyvou (1984) notes that more ground floor rooms existed prior to their filling-up with debris from the earthquake. Furthermore some of the ground floor rooms became basements since the level of the streets increased when the rubble was spread outside. It also seems that some of the house entrances were also altered to accommodate the changes the debris layer created.

Up until now approximately 10,000 m^2 of the site have been excavated, revealing various building complexes. Only a small fraction (5%) of the estimated 200,000 m^2 area of the site has been revealed to the present moment without there being any indication as to the actual extent of the settlement (Doumas 1984:45). It will take many more years, perhaps generations of archaeologists to uncover all the available data at this site. So far the amount of available data offers numerous possibilities for GIS analyses at the sitescape level alone. So what are some of the possible GIS analyses for the data which have been found to date?

3.2. Some possible sitescape analyses at Akrotiri

GIS functions can facilitate a variety of spatial analyses. There are a number of possible analyses for the wall paintings alone. For instance the location of wall painting themes across the site can be related to the size of a room. Do certain recurring themes appear in larger or smaller rooms? This would mean that the wall paintings would have to be categorised according to the subject matter and then these categories would then be related to the size (in m^2) of the room they are located in.

It would also be of interest to see if there is a pattern regarding which walls, in terms of orientation, the paintings are located on in each room. Were the paintings located on certain walls, and if so were they on those walls to avoid harsh sunlight from damaging them (Doumas, pers. comm.)? Another major project is to investigate the distribution of artefacts and various samples within the buildings around the site. Can the distribution of artefacts be used to determine the functional purpose of each building or room?

A GIS can also allow for the analysis of the buildings at the site. Some questions that can be asked of the buildings focus on aspects of (1) density and distribution of the buildings within the area of the site which has been excavated to date, (2) design of the buildings, and (3) wall orientation. Some possible investigations relating to the buildings are listed below.

3.2.1. Building layout and design

The study of the relatively well preserved buildings offers many challenges. The density, distribution, general orientation and layout of the buildings throughout the site can be analysed. This may result in patterns appearing where certain building-types are grouped in the vicinity of particular areas at the site. Patterns of distribution of certain building-types across the site may reveal functional or social hierarchies within a site. The dimensional aspects of each individual building may reveal how it was designed based on the spatial organisation of the rooms. Are larger rooms to be expected for larger houses? Can the functional purposes of the rooms be determined by the design?

These types of investigations can be easily catered for by digital plans. Digitised site plans facilitate the easy measurement of the rooms, walls, doors, windows and stairways. Each room has been digitised as one entity allowing for the area of the room to be made available. All these dimensions can now be utilised to provide various statistics, such as averages and ranges of the number and size of rooms, doors, windows, and stairways per building.

These digitised dimensions can be sorted and grouped to enable observations to be made for the way buildings were planned and organised into areas and rooms. Certain patterns may be revealed by considering:

- the dimensions and number of doors and windows per room in relation to the size of each room,
- the total number of doors, windows and stairways per building in relation to the size of the whole building complex,
- the number and size of the rooms for each floor of the buildings,
- the number and average size of rooms in a building in relation to the total area of a housing complex,
- the internal area of the buildings as opposed to the external areas.

The list goes on … The quantification of the dimensions of the buildings by using a GIS allows for a more structured approach to studying the buildings in greater detail than was possible by using only analogue site plans.

3.2.2. Wall orientation - environmental considerations?

The application of a GIS to site plans can also facilitate the investigation of the orientation of the walls. For instance how are the longest walls orientated? Was the orientation of the walls based on cultural or practical requirements? In the case of practical requirements major environmental considerations would be the sunlight, shadows and wind factors on the building throughout the day and the seasons. Are the density and general orientation of the buildings the way they are to provide for optimal shade in summer and to reduce wind speed in winter? The position and orientation of the doors and windows on the buildings can also be examined in light of these climatic factors which can be incorporated into a GIS analysis.

Climatic factors can be easily examined by visualizing the seasonal prevailing winds or sun's rays on a digitised map in relation

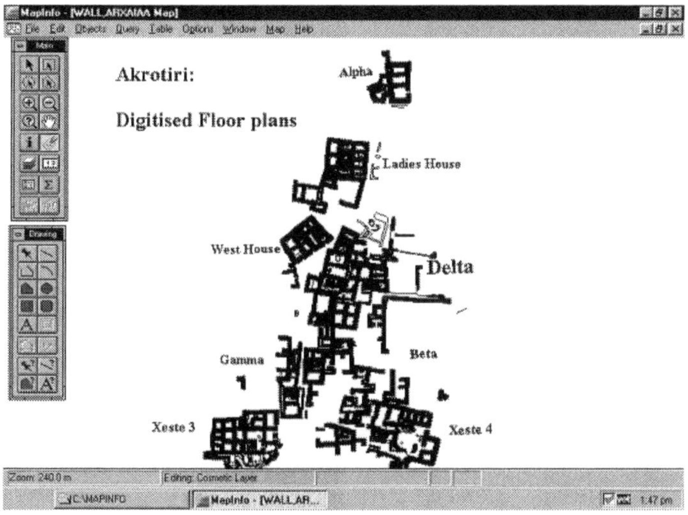

Figure 4: Akrotiri: Digitised ground and first floor plans.

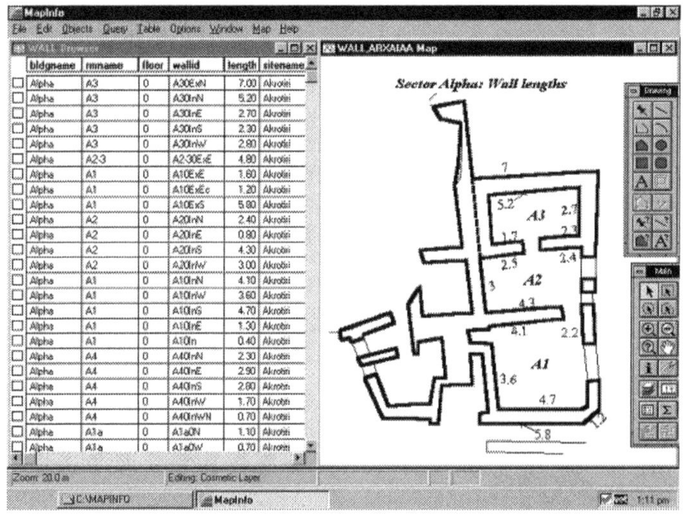

Figure 5: Building Sector Alpha: Digitised plan and "wall" table browser.

to the position of the highlighted doors and windows located on the walls throughout the site. Were the doors and windows orientated in such a way in order to keep the building warm in winter and cool in summer? Do the doors and windows face or avoid the direction of the sun and wind? The orientation of doors and windows can also be considered to determine whether there was some other pattern. Do most doors or windows face the direction of the harbour and coast, or some other feature in the landscape? Or do the doors simply face the streets in between the houses?

This is only a small sample of ideas for sitescape analysis at Akrotiri. For this paper it was decided to undertake a metrical investigation of the buildings at Akrotiri.

4. A metrical analysis using a GIS: Methodology

The aim behind the present analysis is to determine if whole units of measurement were applied to the construction of the buildings at the site. Jones (2000:73) suggests that identifying units of measurement which were utilised at the time may lead to a better understanding of the design of ancient buildings.

The assumption behind this investigation was that people constructed the buildings by using a set of standard units of measurement. On first appearance the layout of the buildings does not suggest that any measurements were used, however, the investigation had not been carried out before. The sheer number of walls had meant that this would have been a very time-consuming task with the possibility of no results, other than to note that no units of measurement were used. Now that the wall lengths are available in a digital format, numerous investigations can be facilitated more conveniently, one of which is a metrical investigation of wall dimensions.

4.1. GIS and Akrotiri

The EPET[1] "Archaeotool" research project has applied the GIS functions of MapInfo in the development of an excavation tool which would essentially archive the daily data collected at Akrotiri.

The present author took part in this project during 1996 and 1997, and subsequently MapInfo was adopted as the tool for this present metrical investigation since the functions it provides are adequate for the current research requirements. Even though MapInfo is a desktop mapping facility, it has many of the GIS functions found in more expensive and less available GIS programs (Johnson 1996:2). MapInfo can be applied to a variety of archaeological research projects which range from the analysis of archaeological sites in relation to the geography of a region, to the analysis of excavation data. These projects can provide further insight into the spatial distribution patterns of archaeological sites across the landscape with possible reasons for their locations, or for patterns of artefact distribution within the various phases and stratigraphy of a site, respectively (Snow 1994:143-148). In the case of Akrotiri the numerous buildings and their associated wall dimensions can now be more readily studied for metrological investigations.

4.2. Data input

The original plans of the site of Akrotiri are in analogue form. The site's architects are currently using AutoCAD to facilitate the updating of plans as new data becomes available. The latest (1997) architectural site plans, supplied by Dr T. Salli, were imported from AutoCAD format and provide a base map from which the walls, rooms, doors, windows, and wall painting locations were digitised by using the on-line digitising facility provided by MapInfo (figure 4).

Separate layers were created for various features such as walls, rooms, doors, and windows. All the dimensions of the walls of the buildings (over 2,300) are stored in the associated MapInfo table named "wall". Because some of the buildings have been preserved up to the second storey, a floor level (0, 1 or 2) field was entered into the wall table in order to analyse each of the walls of the floors separately. Taking a closer look at the building Sector Alpha (figure 5) provides an example of some of the fields contained in the wall table, and depicts details of the wall lengths for some of the walls.

In this case seeking out units of measurement entails metrical studies which can be easily facilitated by a GIS. The process of investigating the digitised plans was to analyse the wall dimensions in the "wall" table for possible units of measurement.

Unit length [cm]	33	27	30	20
Building		% of whole units		
Alpha	0	4	22	38
Beta	1	3	46	50
Gamma	2	5	34	42
Delta	3	3	34	52
Ladies House	3	5	40	44
West House	1	3	36	54
Xeste 3	1	7	40	45

Table 1: Results of the initial metrical analysis: Buildings category, Walls over 90 cm in length.

4.3. Data analysis

Digitised site plans can be used to display and, most importantly, analyse the lengths of walls or other architectural features. Evidence of the use of standard units of measurement is available from some early societies such as Egypt. The process of determining what the length of the measuring unit is when it is not known from written sources, is by dividing the length of the walls by a number of possible units. It is easier to determine units that may have been used to plan buildings at a site if some benchmark from other sites is available. In order to determine a measuring unit used by a society that is being studied some notion of an already known unit of measurement must be sought from related societies.

Since Minoan influence is evident at Akrotiri, it can be assumed that the Minoan foot was used to lay out the walls of the buildings. From previous research, particularly that of Graham (1960:335) who studied wall dimensions at sites on Crete, the Minoan foot ranges from 27.8 to 33.4 cm and averages 30.36 cm in length (Graham 1987:225). However, other units of measurement may also have been used at Akrotiri. So apart from investigating whether the Minoan foot was used, other unit lengths were also used to determine if building standards were utilised.

Graham (1987) assumes that the more regular and uniform a building is, the more likely that a measuring unit was used to construct it. That is, if buildings have accurate right angled corners, it is more likely that units of measurements were used to lay them out. Even if the unit of measurement is not known at a particular site, it is assumed that it will be easier to determine what it was for walls which were built in a regular fashion with neat walls with straight corners.

Unfortunately most of the internal walls at Akrotiri do not have accurate 90° angles, in fact, the corners of most of the internal rooms of the buildings are mainly curved and not squared. Therefore many of the dimensions of the walls are difficult to determine precisely given the fact that many of the internal walls were not constructed with clear-cut corners and did not follow straight lines. To make matters worse some of the early archaeological restoration work at Akrotiri was done before the original state of the walls or features such as doors and windows were accurately recorded. So some of the measurements in the wall table are a reflection of these reconstructed architectural features.

This leads to a certain degree of approximation of the length of the walls on the plans. The schematised architectural building plans were used where the curved corners of many of the rooms have been angled. Furthermore in many cases the walls were distorted by the impact of volcanic debris. Therefore, for the wall table it was decided to initially use a rounded-off figure of the wall lengths for analysis. The length field in the "wall" table was updated by dividing it by a number of possible units and the results were then sorted in search of the most frequently occurring whole unit of measurement that may have been used for constructing the buildings.

Initially all of the wall dimensions of the buildings at Akrotiri were divided by the average Minoan foot of 30.36 cm to see what proportion of the walls may have used this as a whole unit. The ranges of the Minoan foot length were also applied to the wall table. As expected the results were not convincing; taking into consideration the approximation of many of the wall lengths it was decided to base the investigation on units of 30 and 40 cm as well as using the rounded-off units of the extreme ranges of the Minoan foot, that is 27 and 33 cm. These last two units resulted in very few whole units being found. It was noticed that many of the lengths contained half 40 cm units, so it was then decided to group all of the whole and half 40 cm units into a 20 cm unit group.

5. Results and observations

This present paper presents preliminary results of a metrical investigation of the dimensions of the walls of the buildings at Akrotiri. The digitised site plans and associated database provide a central archive of the dimensions of all the buildings excavated at the site to date. The results of this present analysis are limited to some degree by the fact that the total data from the complete area of the site still remains to be uncovered, with only approximately 1/20th of the site revealed so far. Hence the following results, though promising, may be irrelevant when the site is completely excavated!

5.1. Preliminary site results

Tables 1 and 2 present a summary of the results of the metrical investigation for the buildings at Akrotiri. Of all the buildings only the West House can be studied as a completely excavated unit. This is then followed by the Ladies House, Sector Beta, Gamma, Delta and Xeste 3. Only the ground floor of Sector Alpha has been preserved so any observations made relate to that. This is in fact reflected by the results in table 1 below, where a proportionally lower percentage of whole units of measurement are present for Sector Alpha. Table 1 is categorised by building, where the percentage of each of the proposed units of measurement, which is divisible as a whole unit into the length of the wall, is out of the total number of walls over 90 cm in length for each of the selected buildings.

Table 2 is categorised by wall length and the percentage of each of the proposed units of measurement, which is divisible as a whole

Unit length [cm]	33	27	30	20	
Wall length			% of whole units		#of walls
over 20 m	0	0	50	25	4
10 - 20 m	0	0	33	41	17
8 - 10 m	0	12	23	30	26
6 - 8 m	7	0	34	55	44
4 - 6 m	0	4	30	52	186
3 - 4 m	10	0	43	46	231
2 - 3 m	0	10	31	47	359
1 - 2 m	0	0	24	58	400
e.g. The longest wall is 23.4m = 78 x 30 cm units.					

Table 2: Results of the initial metrical analysis: Walls category, Walls over 1m in length.

unit, is out of the total number of walls over 1 m in length for each of the wall length categories.

Therefore, for the over 20 m wall category two of the walls are divisible as whole units of 30 cm and one of the walls is divisible by 20 cm.

This summary of the results in both tables 1 and 2 suggests that 20 and 30 cm units (in approximation) may have been used concurrently when laying out the foundations of the buildings at Akrotiri. The concurrent use of two units of measurement at a site is not unheard of, and occurs elsewhere. Graham (1987) noticed that the use of two or more units of measurement is evident at many ancient sites.

5.2. Conclusions

Since all archaeological data are spatially referenced, GIS can facilitate a multitude of analyses. As new GIS applications in archaeology develop at one level it will be possible to map them to the other level. The spatial aspect of the data allows for the processing concepts behind the two levels of landscape and sitescape archaeological research to be mapped to either level. The ability to map GIS processes from one level to another enables methodologies to be formulated for new research projects. An example of this has been refined for the sitescape analysis of the dimensional aspects of the buildings at Akrotiri. Unfortunately the underlying assumptions and objectives which guide all research means that only a small subset of the numerous possible analyses has to be selected for a project to test the hypothesis at hand. In this case it was data for a metrical analysis. The process of determining whether any standard units of measurement were used in constructing the buildings at one site may *map* to investigations of standard units of distance between sites of one period in a particular area.

Regardless of what data are used, the analytical process requires that the data are entered into a GIS program which will then facilitate the quantification and categorisation of spatial elements. The range of data that can be investigated come from either beyond a site's boundaries, or introspectively from within one site. GIS assists the spatial analysis of the chosen data as well as providing a way of visualising data from a database directly on digital plans. This inspires new analyses to be formulated and allows the process to start all over again. Ultimately new theories are generated which may open up new doors for understanding our past.

Acknowledgements

For the opportunity of taking part in the EPET research project, I would like to thank Professors C. Doumas and T. Sellis and Dr A. Agrafioti.

References

ALLEN, K.M.S., GREEN, S.W. and ZUBROW, E.B.W. (eds.), 1990. *Interpreting Space: GIS and Archaeology*, Taylor and Francis, London.

ATLAS 1999. *Archaeological Atlas of the Aegean: From Prehistoric times to Late Antiquity*, Ministry of the Aegean, University of Athens, 84, Map 49.

BISKOWSKI, M., 1993. GIS and the study of grinding tools in the Teotihuacan valley, Mexico. In Johnson, I. (ed.), *Methods in the Mountains: Proceedings of UISPP Comm. IV Meeting*, Mt. Victoria, Australia, Sydney University Archaeological Methods Series 2 1994: 115-134.

CASTLEDEN, R., 1990. *Minoans: Life in Bronze Age Crete.* Routledge, London.

DOUMAS, C.G., 1984. *Thera, Pompeii of the ancient Aegean.* Thames and Hudson, London.

DOUMAS, C.G., 1992. *The Wall-Paintings of Thera.* Thera Foundation, Nomikos, Athens, 152: 116.

GRAHAM, J.W., 1960. The Minoan Unit of Length. In *American Journal of Archaeology* 64: 335-341.

GRAHAM, J.W., 1987. *The Palaces of Crete.* Princeton Uni Press, NJ.

GAFFNEY, V., STANČIČ, Z. and WATSON, H., 1995. The impact of GIS on archaeology: A personal perspective. In Lock, G. and Stančič, Z. (eds.), *Archaeology and Geographical Information Systems: A European Perspective.* Taylor and Francis, London: 211-230.

GAFFNEY, V., STANČIČ, Z., 1996. *GIS Approaches to Regional Analysis: A case study of the Island of Hvar,* Znanstveni Institut Filozofske Fakultete, Ljubljana.

JOHNSON, I., 1996. *Understanding MapInfo - A Structured Guide.* Archaeological Computing Laboratory, University of Sydney.

JONES, M.W., 2000. Doric Measure and Architectural Design 1: The Evidence of the Relief from Salamis. In *American Journal of Archaeology* 104: 73-93.

LANGRAN, G., 1992. *Time in Geographic Information Systems.* Taylor and Francis, London.

LOCK, G. and STANČIČ, Z. (eds.), 1995. *Archaeology and Geographical Information Systems: A European Perspective.* Taylor and Francis, London.

MARTHARI, M., 1984. The destruction of the town at Akrotiri, Thera, at the beginning of LCI: Definition and Chronology. In MacGillivray, J.A. and Barber, R.L.N. (eds.), *The Prehistoric Cyclades: Contributions to a workshop on Cycladic Chronology*. Department of Classical Archaeology, Edinburgh: 119-133.

PALYVOU, C., 1984. The destruction of the town at Akrotiri, Thera, at the beginning of LCI: Rebuilding Activities. In MacGillivray, J.A. and Barber, R.L.N. (eds.), *The Prehistoric Cyclades: Contributions to a workshop on Cycladic Chronology*. Department of Classical Archaeology, Edinburgh: 134-147.

RENFREW, C., 1984. *Approaches to Social Archaeology*. Edinburgh University Press, Edinburgh, 9.

RUSSELLL, B., 1959. *Wisdom of the West*. MacDonald, 17.

SNOW, D.R., 1994. Using MapInfo to map archaeological data. In Johnson, I. (ed.), *Methods in the Mountains: Proc. UISPP Comm. IV Mtg.*, Mt. Victoria, Australia, Sydney University Archaeological Methods Series 2: 143-148.

Note

[1] EPET was funded by the general secretariat of Research and Technology, Greece.

A GIS Solution for Excavations: Experience of the Siena University LIAAM

Vittorio Fronza, Alessandra Nardini, Federico Salzotti and Marco Valenti

Dipartimento di Archeologia e Storia delle Arti – Università di Siena

Via Roma, 56 – 53100 Siena, Italy

e-mails: scarpazi@mac.com, alessandranardini@virgilio.it, freeco@virgilio.it, valenmar@unisi.it

LIAAM website: http://www.archeo.unisi.it/archeologia_medievale/welcome.html

Abstract

It is our intention to present the experience accumulated in the last decade by LIAAM at the Department of Archaeology and Art History - University of Siena, Italy. During these years we specialised in developing solutions managing every kind of information produced by an archaeological project. We work on different scales (from regional surveys down to detailed records of all the finds). All the data is administered within a system made up of three components: different GIS platforms, an alphanumerical database and a media database; these are linked by a system level application called OpenArcheo, *directly engineered and developed at our Laboratory. The basic concepts of our system are the multidirectional links between information types (which allow the user to query and retrieve all the information related to a feature starting from any of the components mentioned above), modular organisation of architecture in order to implement the ever changing variables and detail levels of archaeological research that suit the specific needs of every single project, and of course, user-friendliness making the management of complex data possible to anyone who has a basic knowledge in the use of computers.*

In short, what we try to do is optimise the management of information produced by archaeological projects and allow the archaeologist to have all the different kinds of data at hand in real time. Such a system fits perfectly with the application of landscape and intrasite spatial and statistical analysis tools, allowing us to build predictive models of landscapes and settlements, orientating every campaign of our projects and providing simulations of the parts we cannot investigate.

We also take particular care in editing our web pages (our web-site is made up of more than 1000 pages constantly updated, and it also provides a portal with more than 7000 links; http://www.archeo.unisi.it/archeologia_medievale/welcome.html), 3D modelling and hypermedia production, in order to spread the information we produce to the widest range of people.

Key words: excavation GIS, excavation DBMS, intrasite analysis, Medieval archaeology, OpenArcheo, global archaeological data management

1. Introduction

The Siena University LIAAM (Laboratorio di Informatica Applicata all'Archeologia Medievale – Laboratory of Information Technology Applied to Medieval Archaeology) made its first steps in the field of computing archaeology at the end of the 1980s (regarding the change in our working philosophy, the evolution of the last years and the new frontiers towards which we are heading, see: Francovich 1990, Valenti 1998a, Francovich 1999, the Italian state of art is described in Francovich and Valenti 2000). The target of our experiments, always aiming towards the global running of archaeological data, remains the production of working tools which allows integrated consultation of all gathered data, free of any preliminary interpretations: not only a supporting instrument but a really useful tool for the *in fieri* research. As already stated our working philosophy consists of six main points:

- technology as an instrument of research in the field;
- technology as a recording instrument for lab files;
- technology used to query and investigate the data we have in order to produce knowledge;
- great attention is paid towards any technological innovation on the market;
- the whole process is managed by archaeologists who, without reaching the "know how" typical of computer scientists, acquire a high end user level;
- ability to use the computer as an advanced household appliance, to be ordered (through programming) to perform what we want it to.

Throughout the years we have acquired skills in processing aerial photographs, GIS manipulation of excavations and landscape projects, advanced scanning and editing, 3D modelling, photorealistic rendering, animation and morphing procedures, electronic video documentation, digital photography, CAD techniques, programming languages, multi-media productions, QuickTime Virtual Reality applications, the production of web pages. OpenArcheo, the system of global data manipulation, is conceived through hierarchical recording levels, each of them characterised by different levels of detailed information; at present they can be divided into: macro level (Tuscany), semi-micro level (areas under survey) micro level (excavations).

Our global data management system, called OpenArcheo, is conceived on hierarchical levels, implying different detail classes; we can actually distinguish three different scales: macro or regional (entire Tuscany), semi-micro or local (areas under survey), micro (excavations). OpenArcheo represents an integrated and open solution for the manipulation of archaeological evidence;

thanks to it all kind of data (cartographic, planimetric, alphanumerical, graphical, multi-medial, etc.) can be queried and visualised in a multidirectional way. The system revolves mainly around two fundamental issues: documentation (what kind of documents do we want to obtain?) and the relation keyword (what is the research key through which we want to gather the requested documents?). Let's assume we are working on the GIS of an excavation and have a need to study the ceramics related to the period of the selected evidence: documentation will concern pottery finds, the relation keyword will be the period and the link will originate from the GIS to the DBMS of the excavation.

The usefulness of such a manipulation is clearly perceivable; this is especially true if we take into consideration the multidirectional possibility of the links (any single type of data interacts with all the others), the wide range of information gathered from the macro to the micro scale (that means from a regional view down to the record of a single pottery piece) and the promptness and easiness of the connections. It is a sort of application that allows the best kind of monitoring of the archaeological evidence: that has been our main purpose during the last few years.

Within the creation process of our system we have concentrated especially on experimenting with GIS platforms for archaeological excavations. We have already written about the way we conceive the close connection between information sciences and archaeology, the standards we have been working on, and some aspects of stratigraphical data management (Valenti 1998a, Valenti 1998b). At this point we would like to deal with the problems connected with the building of a data model for excavations (as to the more recent LIAAM contribution on the subject see Francovich and Valenti 2000, Valenti in press, Fronza in press, Nardini in press); we will also try to demonstrate how useful a computer can be in the production of information and in the verification of it afterwards (computer as a relevant instrument for research and GIS as a medium to interpret and understand what we come across). The lab experiments have mainly focused on the excavations at Poggio Imperiale in Poggibonsi (SI) where we have been investigating a long-term medieval settlement (see Valenti 1996, Valenti 1998c, Valenti 1999a, Valenti 1999b). Since the very beginning of the project, in 1991, we wanted to match new investigating strategies and new technological media. One of the main points supported by the Poggio Imperiale project is that the whole excavation is digitally recorded and managed. All the evidence is included within a GIS platform related to an articulated system of files and records; they contain all the information concerning the excavation and allow the development of new interpretation methods and the planning of the enlargement of the excavation and transformation into an archaeological model.

2. Characteristics of the excavation GIS

The GIS management of the archaeological excavations is intended as an instrument of research and not as an illustration of something already completed. Our proposal can be defined as a GIS solution meaning a series of platforms and files, which interact among them and allow the consulting of all the documents through a wide range of interrelations. It is an instrument that allows immediate manipulation of all the data, the building of various information levels in real time, the creation of new questions according to the gathered answers, as well as the formation of predictive and interpretative hypothesis. In order to achieve all that we had to create 3 applications and place them in a dynamic relation: the excavation platform, the alphanumeric DBMS (the stratigraphical unit file system, finds, etc.) and the picture and multimedia DBMS (the system of photographic, graphic and movie documentation). This and the OpenArcheo system have made complex questions easier and have also made elaboration times shorter, thus accelerating the construction of interpretative models.

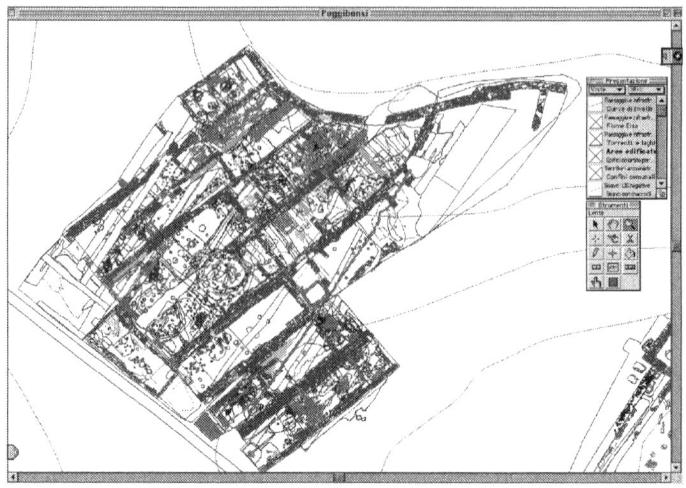

Figure 1: Poggio Imperiale GIS platform. Area 2: composite plan (updated to July 2000).

The GIS platform – At the time we started using GIS software (autumn 1995) we had already graphically planned the Poggibonsi excavation within CAD software, and realised that the tool did not suit our needs (Nardini in press). The necessity of being able to work with an open and functional instrument meant, in our intentions, to reproduce graphically the exact situation of an excavation: a vast composite plan, continually updated, formed by non-interpreted graphical objects. This is the real innovation when compared to other experiences; in fact this is beyond a certain logic (well supported by intrinsic characteristics of CAD applications) which considers the digital manipulation of excavation plans as complementary and subordinate to the one of alphanumerical files (figure 1).

The excavation GIS is conceived as a platform containing the recording of all operations and research campaigns conducted on the site. The Poggio Imperiale hill, for example, has been vectorised completely and placed in its landscape and settlement context (figure 2). The main parts of our data are the detailed geological map, the work of the geoarchaeologist Antonia Arnoldus (geological sections of the hill, archaeological probability map, a theory concerning a possible water gathering system), the preliminary investigations over the territory (field walking projects in 1991 and 1992) and interpretation of different scales of digitalised aerial photographs (the cartographic regional aerial photographs, historical sites photographs, photographs from a tourist plane and from a balloon). Eventually the entire excavation and dumping areas in their progressive movement (they are an integral part of the history of the hill) and the stratigraphical data (from humus to the virgin soil), are complete; the reality of the various archaeological levels is fully represented (the GIS base, counting only the 12 hectare extension of the hill, conducts to a census of 60,112 vectorialised elements at the end of 1999).

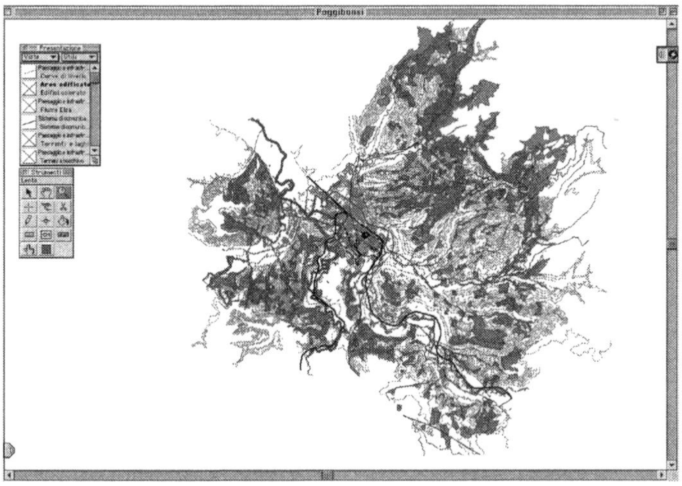

Figure 2: Poggio Imperiale GIS platform. Landscape of the Poggibonsi commune with land cover; in the centre lies the excavated area.

The construction of the "data model" - The choice of software is very important and should not tend to products supported by a rigid structure (like some of the standards for landscape projects; for example ArcView). That is why we have chosen MacMap (made in France), a software programme which allows an *ex novo* construction of data through the creation of a model, organised in types and subtypes, defined both geometrically (surfaces, lines, and points) and graphically. This peculiarity allows to build up a single data base (and not many different thematic layers) where all vectorialised graphs are gathered and classified at the moment of input, according to prefixed logic schemes (which can always be changed).

The data model has been organised in order to gather (in a stratified way) the whole information set of an excavation: the morphology of the site before the beginning of the archaeological work, the non destructive investigation (survey and crop-marks analysis), the excavation, particular research topics (paleopedological survey, geomagnetism and resistance test, anthropological examination of skeleton finds, etc.), future musealisation (project of an archaeological park) and predictive elaboration on the basis of already acquired information. All of this data has been classified into 13 types (for a detailed list see Nardini in press) based on the typology coherence and geometric identity, and avoiding redundancies. The basic concept is grounded on the intrinsic characteristics of the evidence and not on the basis of subjective interpretations. For example, having to vectorialise the road system, it would be incoherent to insert systems already in use and those under planning in different types of the model: they are typologically similar from a cartographic point of view and therefore belong to the same type.

The DBMS (Data Base Management System) - The system of graphical and multimedia files sees the use of databases created for the manipulation of images, films and sounds (in our case Canto Cumulus Desktop Plus); documents are represented in a gallery of miniatures (which can be seen full size with a simple double click) and consist of a descriptive space and a series of keywords and categories (corresponding to precise identifications in our system) which allows for the consultation of different media and subjects. The keywords associated to each record correspond to the number of represented stratigraphical features, the excavation areas, the stratigraphical and the interpreted US definition, the

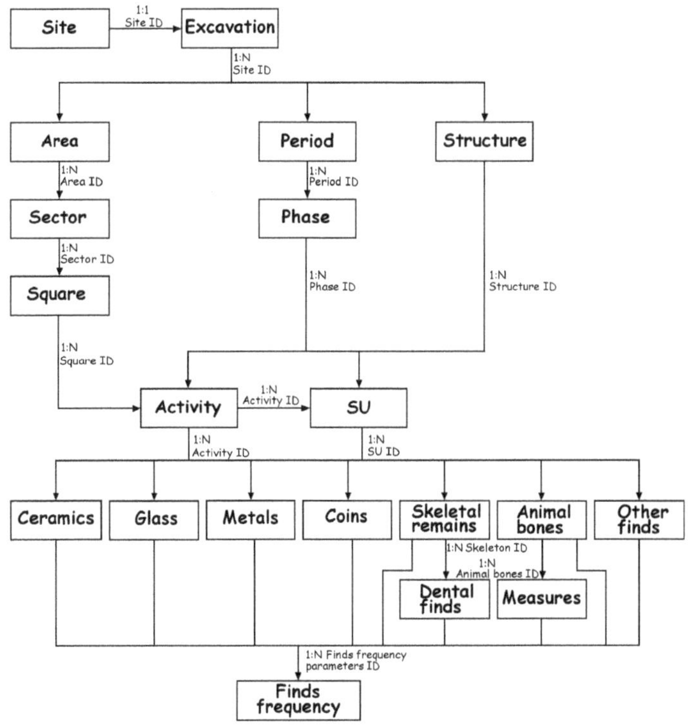

Figure 3: Archaeological excavation DBMS. Diagram of relational architecture.

interpreted structure, the periodisation, etc. So far 5700 documents regarding the Poggio Imperiale project have been recorded (including pictures, movies 3D renderings, QTVR documents, animations).

The alphanumerical database represents a substantial point in developing an efficient solution of comprehensive excavation data management. Quality, usability and availability of information depend heavily on this projectual moment. Our file system was conceived as a hierarchical relational database including detailed stratigraphical and finds records (figure 3). If we consider only the Poggio Imperiale project, 19,425 records (comprehending all different tables) have been entered so far.

Regarding all kind of DBMS applied to archaeology we consider it essential to primarily refer to all of the specific requirements linked to archaeological research, and subsequently elaborate a data model which can join the strict (and often abstract) logical principles of computer science with the possibilities of the hardware/software platforms available (in our case a LAN with a Macintosh server, and commercial DBMS applications largely and easily available). We have to point out that it is sometimes hard for cognitive procedures aiming at the production of historiographical models to match the methods of information science; the problems usually solved by computer analysts concern definitive and irrefutable data, which scarcely need updating. The same cannot be said of archaeological research (Fronza in press). During the planning stage of a database (and, more generally, of a global computer solution), functional for research, we have to be very careful as regards two points: the creation of an open structure which can be integrated at any time with new information types and the precise definition of a detail level in recording data. Not taking into consideration these two aspects might end up with the creation of inefficient or partial solutions. An open structure is, in fact, strictly connected with the very essence of archaeological research. This is often dynamic and evolves according to the

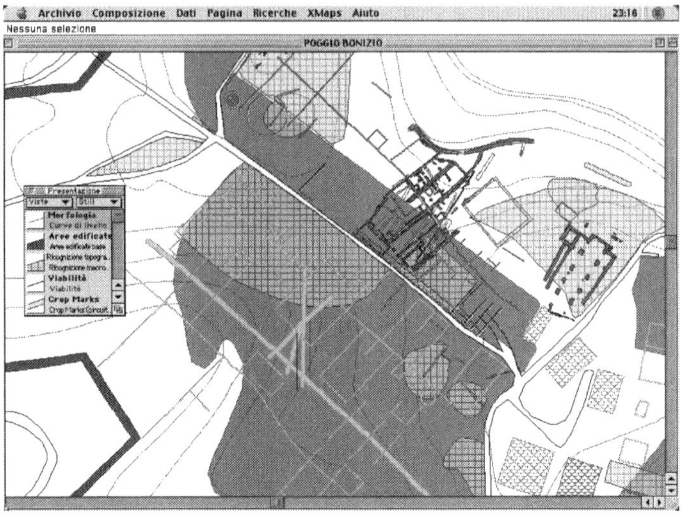

Figure 4: Poggio Imperiale GIS platform. Crossed information levels in order to produce an urbanistic hypothesis of the mid 12th century settlement.

targets of the project; investigations sometimes require a deeper analysis as the excavation goes on. The consistency of the information detail is directly connected with the efficiency of the database; the ideal solution should match the two different needs: detailed study of particular aspects of the project and profitable availability of data.

3. Visualisation of data

The recorded data has been organised into different "views" worked out (elaborated) with the help of visualisation parameters, which can integrate and combine any element according to queries. As we have said before the system is open and temporary views are continuously added to the main views during the use of our platform.

A preliminary view locates and frames the hill of Poggio Imperiale into the surrounding territory and describes its present natural, morphological and monumental features: it shows land use, land cover, contour lines, monumental emergencies, the road system, the new building areas, etc. A further study is recorded and it (through various different colours) shows the different geological phases according to the geological map of Italy. Another layout shows the changes that occurred after the opening of the excavations; the various layers of dumped soil, the location of the huts, and other structures connected with the security of the yard and the consolidation of the found evidence and monuments. Preliminary investigations (surface survey and crop-marks analysis), characterised with different colours and patterns, are organised either in single views or overlapping the wall features excavated during the intensive research in order to verify the correspondence between preliminary data and excavation reality (figure 4). The stratigraphical units (corresponding to three different types: positive stratigraphical units (SU) like surface objects, negative SU like linear objects and characterisations like linear objects) show different colours according to their subtype: a sort of introduction to the vast composite map, from which it is possible to obtain results easily and quickly. Three more views show all the SU according to their main chronological periods: late antique, early and late Middle Ages. Within them some macro views show the single phases of the settlement. Other views show the evolution of the excavation project (the digging areas according to the various campaigns), the perspectives concerning the transformation of the area into a museum (vectorialized, reproducing the master plan of the archaeological park) the paleopedological research (aiming at the creation of a map of archaeological probabilities) and the contextualization in the surrounding landscape (the communal territory of Poggibonsi). The GIS platform, besides allowing a general and detailed survey and analysis of the whole area, the various and different calculations (averages, extensions, distances etc.) the printing of subject maps (periodisation, distribution of finds etc.) allows simulations concerning the structures that have not yet been excavated. This new type of investigation, mainly based on processing excavation data combined with the results of preliminary investigations, gives detailed and interesting suggestions in order to interpret the settlement and orientates the opening of new areas. Other analytical studies concern the early medieval huts (Fronza and Valenti 2000), the early medieval necropolis (Francovich et al. 2000), the distribution of the finds of animal bones (Nardini and Salvadori 2000) and the reading of preserved wall features (Bianchi and Nardini 2000). At present we have just started working on spatial and statistical analysis in order to build predictive models of intrasite settlement patterns, especially dealing with the planning of the small medieval town (which extends on all of the 12 hectares of the hill of Poggio Imperiale; the first results are presented in Nardini in press). This kind of analysis can be considered as the main point of applying GIS technology to an excavation.

References

BIANCHI, G., NARDINI, A., 2000. Archeologia dell'architettura di un centro storico. Proposta per un'elaborazione informatica dei dati su piattaforma GIS bidimensionale. In Brogiolo, G.P. (ed.), *II Congresso Nazionale di Archeologia Medievale*: 381-388. Firenze.

FRANCOVICH, R., 1990. Dalla teoria alla ricerca sul campo: il contributo dell'informatica all'archeologia medievale. *Archeologia e Calcolatori* 1: 15-27.

FRANCOVICH, R., 1999. Archeologia medievale e informatica: dieci anni dopo. *Archeologia e Calcolatori* 10: 45-63.

FRANCOVICH, R., NARDINI, A., VALENTI, M., 2000. La piattaforma GIS dello scavo nella gestione di un'area cimiteriale. In Brogiolo, G.P. (ed.), *II Congresso Nazionale di Archeologia Medievale*: 28-36.

FRANCOVICH, R., VALENTI, M., 2000. La piattaforma GIS dello scavo ed il suo utilizzo: l'esperienza di Poggibonsi. In Brogiolo, G.P. (ed.), *II Congresso Nazionale di Archeologia Medievale*: 14-20.

FRONZA, V., in press. Il sistema di gestione degli archivi nello scavo di Poggio Imperiale a Poggibonsi (Insegnamento di Archeologia Medievale dell'Università di Siena). Una soluzione all'interno della "soluzione GIS". *Archeologia e Calcolatori* 11.

FRONZA, V., VALENTI, M., 2000. L'utilizzo delle griglie di riferimento per lo scavo di contesti stratigrafici altomedievali: elaborazione di una soluzione informatica. In

Brogiolo, G.P. (ed.), *II Congresso Nazionale di Archeologia Medievale*: 21-27.

NARDINI, A., in press. La piattaforma GIS dello scavo di Poggio Imperiale a Poggibonsi (Insegnamento di Archeologia Medievale dell'Università di Siena). Una soluzione all'interno della "soluzione GIS". *Archeologia e Calcolatori* 11.

NARDINI, A., SALVADORI, F., 2000. La piattaforma GIS dello scavo e i modelli distributivi di manufatti e reperti osteologici animali. In Brogiolo, G.P. (ed.), *II Congresso Nazionale di Archeologia Medievale*: 38-45.

VALENTI, M. (ed.), 1996. *Poggio Imperiale a Poggibonsi (Siena)*. Dal villaggio di capanne al castello di pietra. I. Diagnostica archeologica e campagne di scavo 1991-1994. Firenze: Biblioteca del Dipartimento di Archeologia e Storia delle Arti. Sezione Archeologica. Università di Siena, 1.

VALENTI, M., 1998a. La gestione informatica del dato; percorsi ed evoluzioni nell'attività della cattedra di Archeologia Medievale del Dipartimento di Archeologia e Storia delle Arti-Sezione archeologica dell'Università di Siena. *Archeologia e Calcolatori* 9: 305-329.

VALENTI, M., 1998b. Computer Science and the management of an archaeological excavation: the Poggio Imperiale Project. *Archaeological Computing Newsletter* 50 (Spring): 13-20.

VALENTI, M., 1998c. La collina di Poggio Imperiale a Poggibonsi. Uno spaccato di storia insediativa toscana tra tarda antichità e basso medioevo: ipotesi e modelli diacronici (aggiornamento 1997). In *Atti della giornata di studio "I castelli della Valdelsa. Storia e archeologia"*. Miscellanea Storica della Valdelsa 104: 9-39.

VALENTI, M. (ed.), 1999a. *Carta Archeologica della Provincia di Siena*. Vol. III, La Val d'Elsa (Colle di Val d'Elsa e Poggibonsi). Siena.

VALENTI, M. (ed.), 1999b, *Il progetto Poggio (Poggibonsi - SI)*. Guida allo scavo archeologico (aggiornamento 1999). Poggibonsi.

VALENTI, M., in press. La piattaforma GIS dello scavo nella sperimentazione dell'Insegnamento di Archeologia Medievale dell'Università di Siena. Filosofia di lavoro e provocazioni, modello dei dati e "soluzione GIS". *Archeologia e Calcolatori* 11.

Data Integration and Intra Site Spatial Analysis of the Castellaro del Vhò

Manio Pessina

Archeosys snc., Via Valmaira 15, 20162, Milano, Italia

e-mail: manlau@iol.it

Abstract

The main objective of this paper is to present an intra site GIS application of the Bronze Age site Castellaro del Vhò in the municipality of Piadena, province of Cremona, Italy. The site is known since 1890; in 1995 the Archaeological Museum of Milan started the excavations and field survey. The specific goals of the analysis presented in this paper are the integration of archaeological field survey data with geomorphological prospecting, aerial photography and excavation data in order to evaluate the impact of agricultural activities on this site which extends on an area covering over 7 hectares.

Key words: Terramare, agricultural works, aerial photos interpretation, DEM, survey, Middle and Late Bronze Age, core borings, 1890-1910/12 trenches

1. Introduction

The site Castellaro del Vhò is located in the municipality of Piadena, province of Cremona, Italy. The site dated to the Bronze Age was discovered in 1890 (Castelfranco 1911, 1916). Since 1995 intensive archaeological excavations were carried out by the Archaeological Museum of Milan. The most distinctive features of this site are the embankment, the ditch and several houses built on aerial scaffoldings. The site can be considered as a typical representative of Middle to Late Bronze Age settlements in the Po plain best known as the "Terramare" culture (Frontini 1997, Degasperi and Frontini 1997a, Degasperi and Frontini 1997b, Degasperi and Frontini 1998).

The main objective of the project was to integrate all the data from previous excavation campaigns, intensive field surveys and geomorphological prospecting in order to make it easier for us to understand the site formation. The specific goals of the analysis were to evaluate the impact of agricultural activities on the site as regards the extensive soil removal in 1992, analyse the state of preservation of the site and understand the internal structures of the site without any additional archaeological excavations. The data used in the analyses was mostly based on geomorphological prospecting, archaeological field survey data, aerial photography and excavations from the end of the 19th century/beginning the 20th century.

The Castellaro del Vhò site covers around 72,800 square metres. The site is limited on its south and east side by the modern road Molino Vecchio which leads to the River Oglio; on the north and west the site is terminated by two small intersecting ditches.

2. Comparison of data from aerial photographs

The need for this analysis became especially urgent after extensive agricultural works took place on this site. Despite the fact that the land on this site was ploughed and used for crop production over numerous centuries, the 1992 levelling work was particularly destructive. At this time the entire surface of the site was altered by heavy machinery. These alterations were most problematic in sectors F-P 5-12, where the surface was lowered by approx. 19 cm. The soil from this area was deposited in sector G-Y 15 –16 (figure 1). However, digital elevation models of the site prior to this destruction could have been produced on the basis of properly recorded elevations of the site.

The aerial photos of this site were taken on two occasions, prior to and after the 1992 levelling work. On the 1988 aerial photograph (figure 2) one can clearly see the ditch (A) as a dark, 20-30 m wide soil mark. The embankment (B) is visible only in the eastern part of the site, where some traces of the ditch-embankment-ditch sequence can be seen. Lighter areas marked as C and D are interpreted as riverbeds. Other features on this aerial photograph might also have archaeological importance: Feature marked as E could represent the traces of a second ditch, while F could be another embankment. The feature marked as G is a light area, not interpretable, with sub-circular shape. The feature marked as H is a sub-elliptic brown zone not interpretable, save as depressed zone. The field survey showed that this is a zone with dense anthropic elements.

On another aerial photograph taken after the agricultural works (figure 3), one can clearly see the ditch (a) and the embankments (b1, b2 and b3). C1 and c2 are possibly traces of other ditches, while d1 could represent an entrance to the enclosure. D2 was interpreted as possible remains of the embankment. The reason for all the other soil marks is as yet not known.

On the basis of the comparison of the two aerial photographs one can fairly accurate determine the shape, location and width of the ditch as well as the later disturbance of the ditch by a watercourse (feature C in the figure 2) on the eastern side of the enclosure.

The embankment can be seen much better on the photograph from 1992. If we compare the location of the riverbeds we could be lead to the conclusion, that these are a result of later changes of the river location.

All remaining soil marks are hard to interpret and demand additional testing in the field. However, it is clear, that the soil removal which took place in 1992 destroyed traces of riverbed soil marks C and some yet unidentifiable features, especially E, F; feature H on figure 2 still remain (it is feature marked "e" in figure 3). All the other soil marks are visible on the air photograph from 1992.

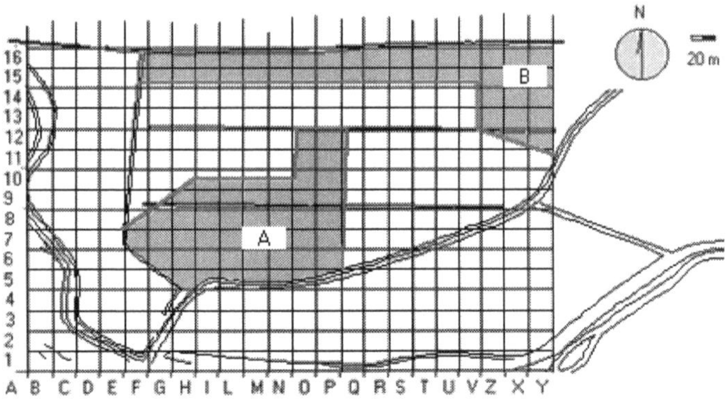

Figure 1: 1991 deeper areas of agricultural works.

3. Intensive archaeological field survey and interpretation

The main objective of the 1998 field survey was to determine the major archaeological activity areas on the site. More precisely, we wanted to determine the number and the density distribution of the prehistoric pottery appearing in the field as well as record the structures which were visible on aerial photographs. We took attention at the datable material too.

After the initial processing of the field survey data, it was realised that the site distribution has to be analysed in two parts. The first one represents the area which has been to a large extent destroyed by agricultural activities, while the second part is represented by the area which was only slightly damaged. The agricultural activities have influenced the surface artefact distribution in such an extent, that if they were analysed as a single area, no relevant information would be gained.

The processing was fairly simple and straightforward. The arithmetic mean of the number of shards per quadrant was calculated for the northern and the southern part of the site. In figure 4 we have displayed the density of shards in all quadrants which have a higher density than the mean density in the analysed zone. One can clearly see that the high density areas coincide with the structures on aerial photographs. In the north part, these are the built-up area, the ditch on the western side of the site and an area in the eastern part of the site that correspond to the feature "e" in figure 3 and "H" in figure 2. But in the western side the agricultural works were intensive, not so intensive in the eastern side. In the south part the shards density correspond to the structures, but we haven't high density areas, because here there are many negative structures. The archaeological stratigraphy that grows in a negative structure normally stoops down, and for this reason the agricultural works less easily damage them.

The DEM analysis and the 1995 core borings show that the built-up area was built on a hillock with a slope in S direction more marked than today, so we can suppose that the south zone was more preserved from agricultural works (at least untill 1992). We can conclude that, in spite of all the activities of ploughing, sowing, and levelling works imposed on the Castellaro del Vhò during the last 50 years, the surface data can still offer useful information for the understanding of the buried stratification.

During the field survey 45 datable artefacts were collected. They were grouped in the following periods:

- Middle Bronze 1 initial and Middle Bronze 2 initial,

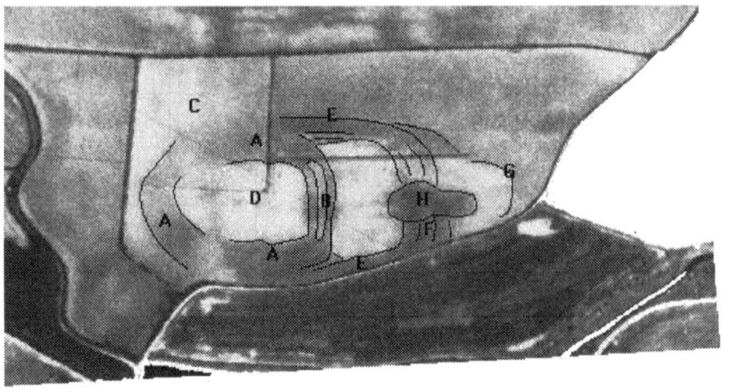

A :ditch
B:embankme.
E:ditch ?
F: emb.? G: elev.area
H:area antr.
C: channel
D: palaeoalveo
N
20 m

Figure 2: Photo '88: The Castellaro field prior to the agricultural works: interpretation of tracks (photograph by: Dr J.Tirabassi, of Museo Archeologico di Reggio Emilia, filtered with histogram equalisation by M. Pessina, interpreted by J.Tirabassi and M. Pessina).

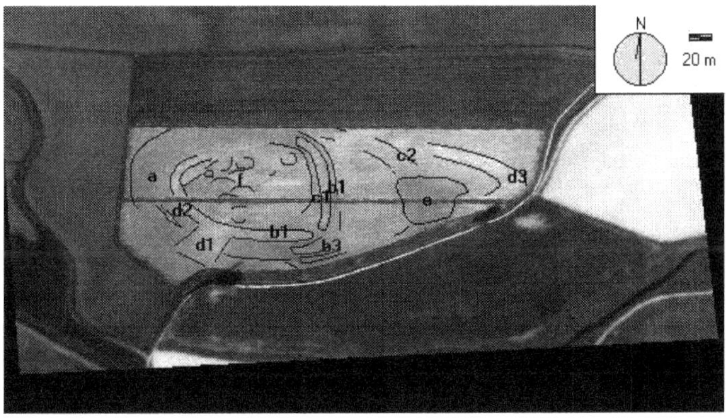

a: ditch
b1,b2, b3: emb.
c1,c2, c3: dit.?
d1, d2, d3: elev. areas
e: area antr.
f :tracks inside the site

Figure 3: Photo '91: The Castellaro field after the agricultural works: interpretation of tracks (photograph by: Dr J.Tirabassi, of Museo Archeologico di Reggio Emilia, interpreted by J.Tirabassi and M. Pessina).

- Middle Bronze 2 and Middle Bronze 2 middle,
- Middle Bronze 2 middle-advanced and Middle Bronze 2 advanced,
- Middle Bronze 2 late and Late Bronze.

The positioning of each quadrant containing datable artefacts was correlated to the features identified on the aerial photographs. The Middle Bronze 1 initial and Middle Bronze 2 initial quadrants are concentrated in the central part of the site and areas where the agricultural levelling works were highly destructive (figure 5). Only 3 evidences dated Middle Bronze 2 and Middle Bronze 2 middle are in the south field zone, where the levelling works were more incisive.

The others are in the north zone, 3 in the built up area, 5 eastwards.

Middle Bronze 2 middle-advanced and Middle Bronze 2 advanced artefacts were most prominent in the northern and north-eastern part of the site. Finally all the Middle Bronze 2 late and Late

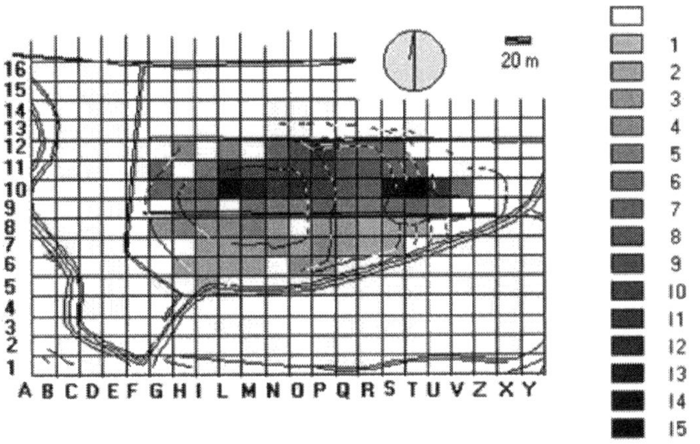

Figure 4: Medium and maximum values of pottery distribution in the southern and northern area and tracks on aerial photographs: the evidences have been sorted on a scale of 15 levels (each level corresponds to 100 shards).

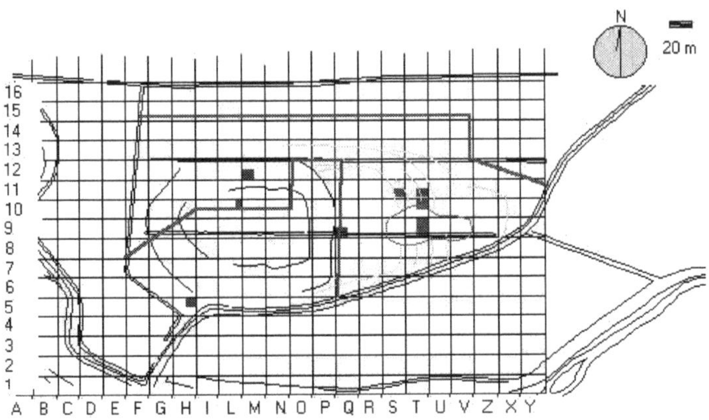

Figure 5: Distribution of datable Middle Bronze 2 initial and Late Bronze artefacts with the boundaries of deeper areas of agricultural works and the tracks on aerial photographs.

Bronze artefacts were located in the eastern part of the site, in the area marked "e" in figure 3 and H in figure 2.

This analysis marks that the evidences of the first two groups are concentrated in the zone where the agricultural works were more incisive and where the ancient hillock was more exposed: here the archaeological stratification is almost completely destroyed.

The third group in the north zone is out of the zone where the agricultural works were more incisive, in the south instead is in this zone maybe because here the presence of more negative structures allowed a better the stratification preservation.

The most interesting group is the fourth because the recent evidences are grouped in an area where:

- the agricultural works haven't been incisive
- there weren't archaeological excavations
- there has been found an archaeological stratification *in situ* during sections cleaning of a modern small ditch
- there is the highest distribution density of prehistoric pottery.

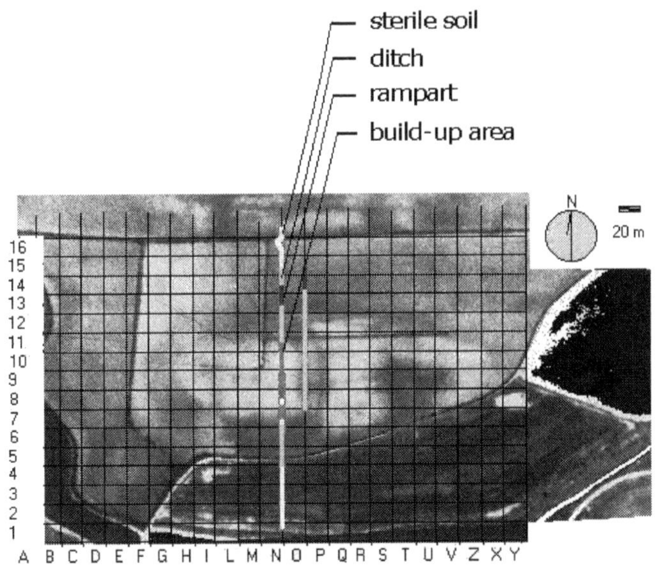

Figure 6: 1995 core borings; line N-O, -260 cm elevation; line O-P –260 cm elevation.

4. Geo-archaeological core borings

Since we wanted to gain some insight into the sub-surface geology and the relation between archaeological features and geology, we decided to make a series of core borings on the site. So, in 1995, a set of core borings were made, crossing the Castellaro del Vhò in the S-N direction.

The first analysis (Bassetti 1997), made without aerial photographs, identified the palaeo river bed presence, flowing in the E-W direction, which later on moved from the southern zone to the north. This change is marked by the meander cut causing marshy conditions which were later on dried out by the modern canal. Core borings proved the presence of a hillock (marked by the sands layer) with a slope in the N-S direction. Over this hillock the prehistoric built-up area was set. We also managed to determine the presence of negative evidences in the stratification sections with core borings.

Core borings enabled detail insight into the interpretation of the stratigraphy of the site, especially when the results are compared to the aerial photographs.

The highest elevation of both cross sections measuring 210 cm and 220 cm below the arbitrary reference point represents the plough soil. Deeper, at –260 cm traces of the ditch are already visible in the core borings (figure 6). Here we can see a sequence of the ditch and rampart on both cross sections with the built up area in its centre. The longer section is concluded by sterile soils.

If we look deeper (figure 7), at -300 cm elevation the most significant change is the substitution of the built up area in the centre by sterile soils. We can still follow the ramparts and the ditches in both cross sections. The clay soils in the southern part of the section are probably not the ditch remains but most probably belong to the palaeo riverbed.

At lower elevations of -380 cm and –350 cm sterile soils start dominating both cross sections. Only remains of the palaeo river beds can be traced. This is repeated for all deeper elevations up to -440 cm where we reach sterile soils along both cross sections.

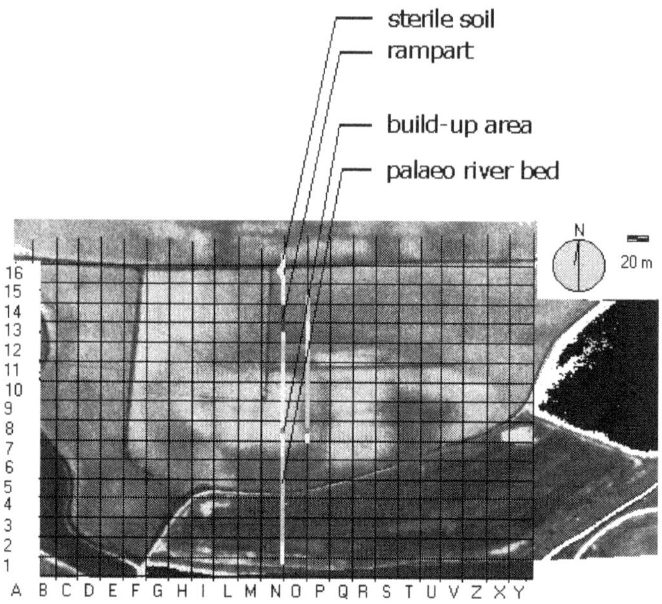

Figure 7: 1995 core borings; line N-O, -300 cm elevation; line O-P –300 cm elevation.

The core boring analysis helped in the interpretation of the aerial photographs, especially of the ditch and the built-up area inside the enclosure and it also aided the explanation of the negative tracks seen in the aerial photographs. However, for a higher detail analysis, archaeological excavations will be needed.

5. Data from archaeological excavations in 1890-91 and 1910-12

Since we wanted to obtain a more detailed insight into the interpretation of the site we decided to incorporate data also from the previous excavations. Unfortunately, integrating this data into our work was not as easy as first thought. The original location of the excavations in 1890-91 is not known (Castelfranco 1911, 1916). The reports mention the presence of built structures as well as traces of a ditch and an embankment. The grey sterile sands were 150 cm below the topsoil. The later reports also state that these excavations were located on the opposite side of the 1910 excavations: which can lead us to the conclusion that the 1890-91 excavation trenches were located in sectors O-R 8.

With the data from the excavations conducted in 1910-12 we had much less of a problem and this data was incorporated in our interpretations without any major difficulties. When we positioned the 1910-12 trenches onto the aerial photographs, we obtained a better understanding of some conclusions drawn by the archaeologists and we were able to add new data to the site analysis. Despite the fact that seven trenches measuring several hundred meters were made, and the centre of the built up area was identified, the ditch-embankment sequence was never detected. The reason for this was that all trenches made for detecting ditches were located outside the enclosure or in the ditch, where the sequence isn't visible; the only one was located between the built-up-area and the ditch was in a area very confuse by the Oglio river floods. The presumable absence of the ditch lead the archaeologists to not interpret the Castellaro del Vhò as a terramara.

6. Conclusions

Integration of various data, ranging from digital elevation data through aerial photography, archaeological field survey and excavation data to geo-archaeological core borings enabled us to have a detailed interpretation of the Castellaro del Vhò site. By using digital elevation models we could analyse the relation between paleo geomorphology and the original terrain prior to the major disturbance of the site in 1992. On the basis of the comparison of the archaeological surface survey data, aerial photographs and the digital elevation model the extent of the 1992 disturbance can be defined. This can also help us to establish the chronological period of the layers removed by this activity.

The interpretation of the tracks which can be seen in the aerial photographs has proved the existence of a large site surrounded by a ditch and an embankment. The presence of additional structures can also be seen on aerial photographs. From the structures on the eastern part of the site (outside the settlement surrounded by the ditch-embankment-ditch sequence and a recent fluvial canal crossing) special attention should be paid to the centre of the site. The existence of settlement traces outside the central part of the site was additionally proved by positioning the location of trenches excavated in 1910-12 onto aerial photographs. Last but not least, the analysis of the layers sequence in the core borings compared with the aerial photographs has evidenced the tracks of a structure inside the central part of the site as well as tracks of the palaeo river beds. We also hope that this case study did not only lead to a better interpretation of the Castellaro del Vhò site, but has also demonstrated how geographical information systems can be used for integrating various data types in the intra site analyses.

Acknowledgements

I want to thank everybody who enabled and contributed to this work: Zoran Stančič, Patrizia Frontini of Civiche Raccolte Archeologiche sezione preistorica di Milano, who directed the excavations and all studies at Castellaro del Vhò, J.Tirabassi of Museo Archeologico di Reggio Emilia, who took and interpreted the aerial photos, as well as Nicola Degasperi, Michele Bassetti, Mauro and Stefano Rossi, Enrico Faccio, Adele Buccarelli, Andrea Sangiorgi, Manuela Montagnari, Ginevra Scotti and Roberta Gianadda.

References

AA.VV.,1998. Methodological trends and future perspectives in the application of GIS in archaeology In *Archeologia e Calcolatori* 9, 1998

ALLEN, K.M.S., GREEN, S.W. and ZUBROW, E.B.W. (eds.), 1990. *Interpreting space: GIS and archaeology*, Taylor and Francis, London.

ANDRESEN, J., MADSEN, T. and SCOLLAR, I. (eds), 1993. Computing the past, *CAA 92 Computer applications and quantitative methods in archaeology*, Aarhus University Press, Aarhus.

BASSETTI, M., 1997. Il contesto geomorfologico. In Frontini, P. (ed.), *Castellaro del Vhò. Campagna di scavo 1995*, Milano: 27-38.

BERNABO, B.M., CARDARELLI, A., CREMASCHI, M., 1997. *Le Terramare - La più antica civiltà padana, catalogo della mostra tenutasi a Modena*, Electa 1997.

CASTELFRANCO, P. and PATRONI, G., 1916. La stazione palustre di Campo Castellaro presso il Vhò di Piadena. In *Monumenti Antichi, Regia Accademia dei Lincei*, XXIV,1: 309-347.

CASTELFRANCO, P. PATRONI, G., 1911. Vhò di Piadena. In *Notizie degli Scavi*: 14-17.

DEGASPERI, N. and FRONTINI, P., in press. Castellaro del Vhò, Piadena, Cremona. Abitato dell'età del Bronzo. In *Notiziario della Soprintendenza Archeologica della Lombardia 1998.*

DEGASPERI, N. and FRONTINI, P., 1997a. Castellaro del Vhò, Piadena, Cremona. Abitato dell'età del Bronzo. In *Notiziario della Soprintendenza Archeologica della Lombardia 1995-1997*: 24-25.

DEGASPERI, N. and FRONTINI, P., 1997b. Castellaro del Vhò, Piadena, Cremona:l'abitato; Castellaro del Vhò, Piadena, Cremona: i materiali. In Bernabo, B.M., Cardanelli, A., Cremachi, M., *Le Terramare- La più antica civiltà padana, catalogo della mostra tenutasi a Modena*, Electa 1997: 228-331.

FRONTINI, P. (ed.), 1997. Castellaro del Vhò. Campagna di scavo 1995, Milano.

GAFFNEY, V. and STANČIČ, Z., 1996. *GIS approaches to regional analysis: a case of study of the island of Hvar*, Znanstveni institut filozofske fakultete, Ljubljana.

HODDER, I. and ORTON, C., 1976. *Spatial analysis in archaeology*, Cambridge University Press, Cambridge.

LAFLIN, S. (ed.), 1986. *Computer applications in archaeology 1986*, University of Birmingham.

LOCK, G. and MOFFETT, J. (eds.), 1992. *Computer applications and quantitative methods in archaeology 1991*, Tempus Reparatum, BAR international series S577.

LOCK, G. and STANČIČ, Z. (eds.), 1995. *GIS and archaeology: a European perspective*, Taylor and Francis, London.

PETRIE, L., JOHNSON, J., CULLEN, B. and KVAMME, K., 1995. *GIS in archaeology.* An annotated bibliography, Sidney University Archaeological Methods Series 1, Sidney.

RAHTZ, S.P.Q (ed.), 1988. *Computer and quantitative methods in archaeology*, British Archaeological Reports International Series 446, vol.2: 435-451.

STANČIČ, Z., GAFFNEY, V., OŠTIR, K. and PODOBNIKAR, T., 1996. GIS Analysis of Land-use, Settlement Patterns and Territories on the Island of Brač. In *Archaeological Applications of GIS: Proceedings of Colloquium II, UISPP XIIth Congress*, Forlì, Italy.

STANČIČ, Z. and KVAMME, K., 1999. Settlement pattern modelling through Boolean overlays of social and environmental variables. In Barcelo, J.A. (ed.), *New techniques for old times, CAA98*, Barcelona, BAR International Series 757.

5. Archaeological Regional Spatial Analyses and Predictive Modelling

Ancient Roads and Fields in Northwestern Gaul – A GIS-Based Analysis

Frank Vermeulen

Ghent University, Department of Archaeology and Ancient History of Europe

Blandijnberg 2, 9000 Gent, Belgium

e-mail: frank.vermeulen@rug.ac.be

Marc Antrop

Ghent University, Department of Geography

Krijgslaan 281- S8, 9000 Gent, Belgium

e-mail: marc.antrop@rug.ac.be

Beatrijs Hageman

Ghent University, Department of Archaeology and Ancient History of Europe

Blandijnberg 2, 9000 Gent, Belgium

Torsten Wiedemann

Ghent University, Department of Geography

Krijgslaan 281- S8, 9000 Gent, Belgium

Abstract

This paper is a summary of the methodology and the main results of a recent (1997-1999) geo-archaeological project concerning the ancient landscapes of parts of northwestern France and Belgium. The primary goal of this project was to reconstruct and interpret, the Roman road system and land organisation in this area, as well as other ancient (pre-medieval) linear features. To understand the structural organisation in this poorly preserved ancient landscape a full battery of techniques has been deployed: e.g. aerial photography, field survey, selective excavation, pedologic and palynological analysis, regressive examination of cartographic material, etc. All data gathered was digitised and integrated in a GIS, which allowed the spatial analysis of all cartographic and archaeological information and was very helpful in locating and explaining protohistoric and Roman roads and field patterns. Important methodological problems that needed to be solved concerned the use of oblique aerial photographs and cadastral map data when studying ancient patterns in a GIS-environment.

Key words: GIS, landscape analysis, ancient roads, field systems, aerial photography

1. The basic methodology

The methods and research results presented in this paper are part of a geo-archaeological project undertaken between 1997 and 1999 at Ghent University, with the support of the Fund of Scientific Research (Flanders). In general this study was concerned with the reconstruction of the landscapes in Western Belgium during the Roman occupation, essentially the first 4 centuries of our era. But as the Roman landscape cannot be isolated from earlier evolutions, it is the pre-medieval landscape as a whole that was looked at.

The primary goal of this project was to reconstruct and interpret the Roman road system and land organisation in this area, as well as other ancient (pre-medieval) linear features using a combined set of geographical and archaeological methods. This work should ultimately lead to some understanding of the human behaviour behind the landscape structure, in particular concerning the processes of the acculturation during the Roman period. Local answers to political, economic, technological and ideological stimuli should then be evaluated in relation to the Roman influence on autochthonous land organisation.

Another objective of the study was to explore the potential of new methods of palaeo-landscape analysis in a geographical area of Northern Gaul where, until recently, very little was known about the protohistoric and Roman landscape.

The study area is the *Civitas Menaporium* (figure 1). This former administrative unit of the Roman Empire covers the coastal plain, the flat sandy area and the loamy, hilly lowlands of Interior Flanders up to the river Scheldt. It stretches from north-western France to the south-eastern Netherlands. Even more interesting than its variety of natural landscape types is an unusually clear division during Roman times into a more romanized villa landscape with a classical urban centre Cassel in the south, and a more indigenous, less exploited rural area in the north (Vermeulen 1992).

At first sight, the chosen region appeared not to have known any pre-medieval landscape structures. The area is situated just south of the regions where large protohistoric field systems of the Celtic field type have been discovered. Also, the study area is a little too far to the north to expect any profound Roman influence on the land organisation. The well-structured and essentially military Roman road network only stretched as far as the southern fringes

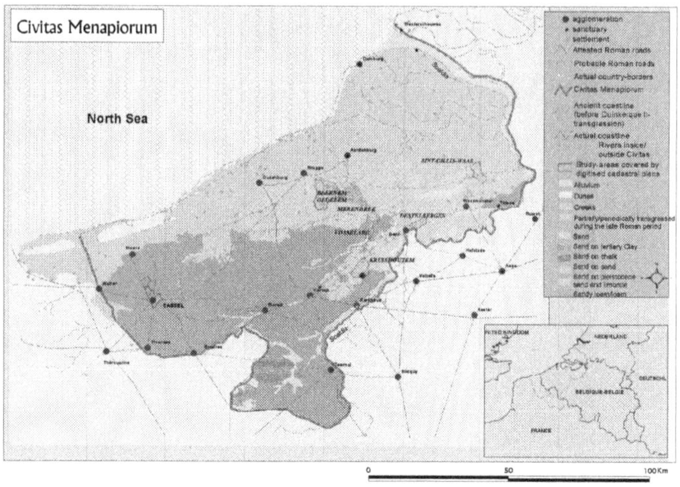

Figure 1: Localisation of the case studies and the main natural landscape types in the Civitas Menapiorum during Roman times.

of this region and only vague traces indicate a possible existence of an orthogonal Roman land division (centuriation).

Except in the surroundings of Roman capital Cassel (*Castellum Menapiorum*), most research was concentrated in the northern part of the *Civitas Menapiorum*, which is now called the Sandy Flanders landscape region. It is a sandy and lightly sandy-loamy, generally flat area situated between the river Scheldt and the coastal plain. The main reason for choosing this region as study area is the high level of objectivity and quality of the available archaeological information obtained through systematic aerial photography and numerous field observations carried out in the past 15 years.

Although the rural landscape in the study area is extremely varied and intensively used, the results of the systematic oblique photographical survey are impressive (Bourgeois et al. 1995). Currently, more than 50,000 aerial photographs, representing several thousands of settlement sites and burial places, have extended the archaeological database of the region in an almost "alarming" way for a detailed archaeological analysis (see figure 2a). Moreover, an intensification of flights since the late eighties, combined with a growing geographical approach, stimulated the discovery of many off-site features. Ancient roads, field-systems, wells, pools, cult-places and other isolated traces in the landscape, most obviously of pre-medieval date, are some of the features that are now registered on a regular basis too. The discovery and interpretation of such phenomena offer an important additional resource to systematic archaeological research concerning the human landscape organisation of the periods concerned here.

Initially, the archaeological study consisted mainly in building a computer inventory of the "fossil" roads and fields based on the oblique aerial photographs of the Department of Archaeology of Ghent University (Vermeulen et al. in press). After charting all the relevant data, the central issue became to relate the detected linear patterns with archaeologically datable elements (see figure 2b). These fossil structures, which consisted of more than 2000 probable ancient roads and field boundaries, had to be differentiated from more recent and modern elements in the landscape and had to be interpreted and dated correctly. This meant historic-geographical and essential archaeological research at a detailed scale was necessary. The historic-geographical research involved classic regressive examination of cartographic and vertical photographic material, while the archaeological research demanded fieldwork (Vermeulen and Antrop in press).

This fieldwork included first a survey by field walking and auguring and second the excavation of selected sites. During the field campaign, several hundred selected ancient road traces and field systems were surveyed. Drillings with a simple Edelman-auger enabled us to make a detailed study of a selection of road-like features and a few supposed field ditches. Some 30 double line traces that were clearly visible on aerial photographs were verified systematically. This technique allowed us to draw some general conclusions regarding the nature of the traces, and more than once they confirmed our assumptions that these were indeed fossil roads. In general, it was fairly easy to retrace these ditches, to determine their location and to specify their dimensions (width, depth) and backfill. Sometimes a thin hardened stratum of soil constituting the original earthen road surface between the drainage ditches, was found. However, these drillings could provide only sporadic dating evidence.

Only by excavating the traces it is possible to determine their precise nature and date. Within the framework of this research, two methods were used. First a systematic and focussed study of some carefully selected traces by means of narrow sample trenches, and

(a) (b)

Figure 2: (a) Ancient land divisions and Bronze Age burial monuments at the archaeological site of Aartrijke - Gemene Veld (prov. of West-Flanders). (b) Ancient field system at Koekelare "Bergbeek" (both aerial photographys: J. Semey, Ghent University).

Figure 3: General plan of the Roman traces (black) of roads, field ditches and farms found during the 1996-97 excavation campaign at Sint-Gillis-Waas.

second a large-scale excavation of entire systems in an open area, often as part of current rescue archaeology going on in the region.

The first method involved the digging of a series of limited sample trenches cutting across a number of fossil roads and field boundaries selected from the aerial photographs.

This type of fieldwork confirmed the existence of pre-medieval, mostly Roman, tracks and of some fields that were more or less simultaneously parcelled out. They are reassuring evidence of the fact that a considerable but as yet indeterminable part of the archaeological data consists of road and land division structures that were used in the Roman period and sometimes even before.

The second method, the open area excavation, was only applied within the scope of a number of large-scale rescue projects within the study area. Four excavations were the direct result of indications on aerial photographs, over these areas which were threatened with destruction. These *case* studies made a very important contribution to our research. They made it possible to excavate several hundred metres of Roman roads and in one particular case (Sint-Gillis-Waas) research could be done on the excavated parcel boundaries (figure 3).

Integrated with this fieldwork, colleagues from Ghent University were able to do quite a number of palaeobiological and pedological analyses. These pollen analyses, the study of macro-plant remains and detailed pedological observations, allowed us to reveal the environmental context of the line patterns detected with the help of the aerial photography, as well as the function and evolution of the various field structures.

2. GIS and spatial analysis

To be able to develop efficient methods of spatial analysis all collected data need to be integrated in a geographic information system or GIS, using a common cartographic reference base (Vermeulen 2000, Johnson et al. 1997, Jones 1997). In this study Arcview GIS 3.1 from ESRI formed the core for the integration of data from different sources and the spatial analysis was mainly performed with Arcview's extensions Spatial Analyst 1.1 and 3D Analyst 1.0. For extraction and exploding geometrical data, procedures from ArcInfo 3.5.1. and AutoCad LT were used as well and also Microsoft Excel 97 offered interesting possibilities. Data management was achieved with Microsoft Access 97 and statistical analysis was performed with Microsoft Excel 97 and Statistica.

First, as many as possible relevant primary sources, such as maps and vertical aerial photographs were stored in this GIS, in vector or scanned versions. For the region considered this means that geographical information had to be derived from very different data sources provided by three countries: France, Belgium and the Netherlands. The integration of these data was not a simple task, as we had to deal with a multitude of sources with different data quality, scales and especially different projections. The following map-layers were integrated in the GIS: scanned topographic maps, digital orthophotographs, waterways, municipality-borders, geology, soils, a scanned version of an 18th century military map of the Low Countries and various vectored map-figures from archaeological literature. For some case study-areas we also digitised and georeferenced some 19th century cadastral plans. As the three-dimensional landform information is fundamental in all spatial analysis, we also extracted and transformed contour and height information from the scanned topographical map to build a digital elevation model (or DEM) of the same study micro-areas.

Secondly, the GIS comprised a database of all relevant archaeological information on the protohistoric and Roman periods in the *Civitas Menapiorum*. This not only includes the Roman settlement sites and other ancient findings known from earlier publications or recent fieldwork, but also all vectored excavation plans of relevant archaeological fieldwork in the area. Furthermore, the archaeological database includes all processed oblique aerial photographs, of which a majority is now scanned and stored on CD ROM. The main purpose is to store this archaeological information in a suitable way to examine vertical and horizontal relations between different archaeological sites and other elements of the landscape, like soil type, land form, geographical situation, terrain, rivers, etc. in time and space.

The oblique aerial photographs are an important data source. Several procedures had to be developed to optimise their manipulation and interpretation.

Each photograph is described in a relational database. The description contains, besides characteristics of the features, the metric co-ordinates of one point, which is clearly visible on the image and can be located easily on the topographical base-map. These reference points were entered in Arcview as a point theme that allows the creation of hot links between the GIS-maps and the photographic database. Thus, the oblique aerial photographs can be visualised easily by clicking on its map position. A first interpretation can be made by comparing the image directly with other thematic themes, such as known Roman settlements, roads and excavation-plans. Also, photographs can be selected from the database using map queries.

The photographic quality of the most interesting images was improved using imaging-software such as *Corel PhotoPaint*. Techniques such as contrast enhances, edge detection and filtering were used to improve the recognition of the marks.

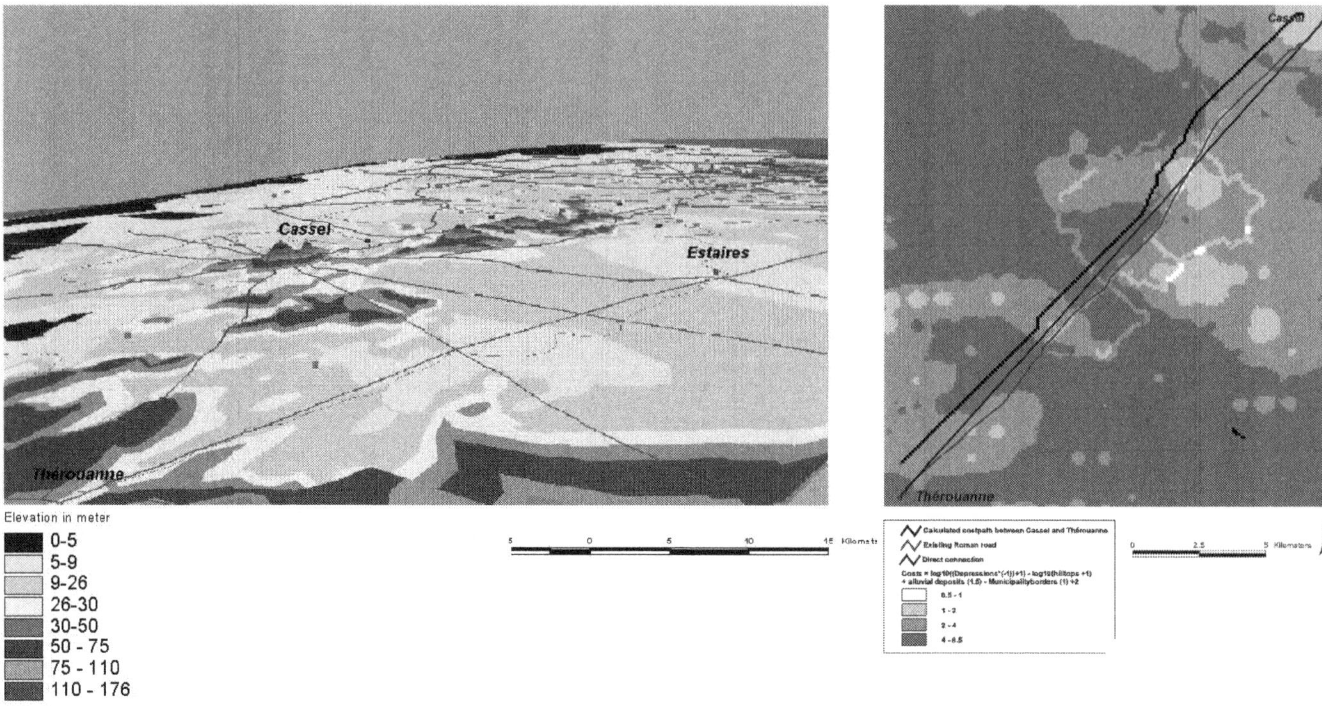

(a) (b)

Figure 4: (a) 3D-visualisation looking from the SW to the NE, with known and supposed Roman roads and settlements overlaid of the area around Mount Cassel. (b) Cost-path analysis of the Roman road between Thérouanne and Cassel based upon the digital elevation model, the geological map and the viewshed analysis.

For the selected linear traces observed on oblique aerial photographs a method had to be developed for restitution and georeferencing these upon a common topographical base. This was achieved using the warping procedure in Arcview's Spatial Analyst. It applies a polynomial transformation of grid themes using control points (tic or tie points) that are common on the distorted image and the rectified topographical base, such as orthophotomaps. Useful tic points are the corners of the field that contains the marks.

So much for the basics of methodology. We now have a closer look at some of the analyses and results concerning the ancient roads and land divisions, which were studied. As space is limited here we will simply present a short overview of some of the most important approaches. A full publication of the results of this project is in preparation (Vermeulen and Antrop in press).

3. Ancient roads

An attempt has been made to examine the common characteristics of the limited number of well-attested and archaeologically known major Roman roads within the Civitas Menapiorum. The purpose is to derive a model to reconstruct possible connections between large Roman settlements where no exact road is known so far. The characteristics of archaeologically known Roman roads can be evaluated by map–overlaying, spatial analysis and statistical testing.

Cost surface analysis is one of the methods of spatial analysis available in GIS that was applied to study the Roman roads. It allows the modelling of route directions considering different features of the landscape and parameters of movement through the geographical environment. The connection between two places, a source and target place, is directed by the effort or "cost" to move through the landscape. The CostPath-algorithm in Arcview's Spatial Analyst searches the route with the lowest accumulated "cost" between the source and target place. As many Roman roads have characteristic straight line connections, this method allows to evaluate the deviations between the theoretical and shortest straight line connection and the actual route. Deviations are compared to the local terrain conditions in order to explain them and understand the reasoning of the road builders. The results of this type of analysis can stimulate and help further research in areas where the major Roman roads are still hard to trace in the field.

This *cost path analysis* was applied on the Roman capital site of Cassel and its connection with the neighbouring major town of Thérouanne (ancient *Tarvenna*), located south of it. Roman Cassel was situated on a hilltop and became the core of a whole network of well-identified Roman roads. The connection between both cities follows the interfluve between the alluvial lowlands of the river Aa basin in the north-west and the river Lys in the south-east (figure 4a).

The present road reflects in its major part an almost straight line connection between these two *Civitas* capitals and there is a lot of evidence that it corresponds partially to the original Roman road trace.

If we compare the Roman road with the straight line connection, several areas can be detected where the difference is more than 500 m. One of them is situated nearby Bavinchove where the reason for the choice for this particular trajectory might have been the existence of a little isolated hill, which allows better vision of the landscape. Visibility analysis using the same data in the GIS, confirmed that the visibility is much better from here than from the straight line connection (figure 4b).

The resulting cost-path, calculated from Cassel as source to Thérouanne as target, shows that the existing road fits quite well to known remnants of the Roman road. The model seems to indi-

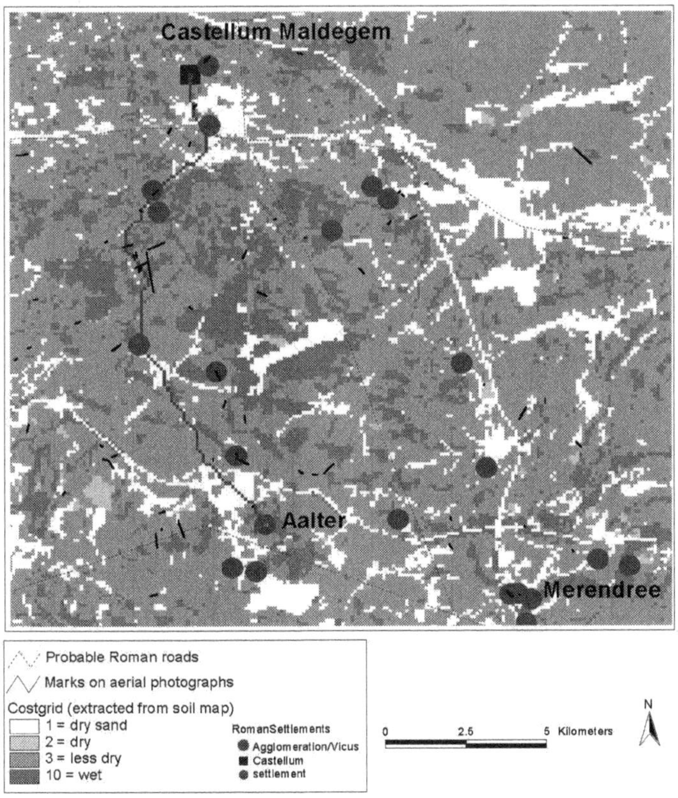

Figure 5: Cost path analysis between the Roman camp at Maldegem and the stone quarry area of Aalter in the north of Sandy Flanders.

cate that especially topographic depressions and the later developed municipality borders, often remnants of ancient roads in this region, are the main factors that explain the differences between straight line connection and the existing route.

A comparable model, based on similar parameters, was applied in the northern part of the study area where archaeological proof still lacks for the possible reconstruction of major Roman roads between the central settlements. This analysis met with some success. In some instances there seems to be a real connection between the hypothetical road trajectory and the location of recently discovered Roman sites and of fossil roads detected on aerial photographs. This kind of spatial analysis and GIS-modelling can be extremely helpful to detect corridors of high probability for major Roman road trajectories (see figure 5). It can become a valuable tool to define micro-regions where possible Roman roads have to be searched for in areas where archaeological and historical proof still lacks.

The application of these models needs detailed data with a high degree of certainty. This is difficult to obtain for the localisation of the large number of secondary and local roads. The irregularity of these roads, interconnecting minor settlements or leading from Roman or protohistoric farms towards their fields, is notorious. For the time being only a combination of systematic prospection with aerial photography and selected fieldwork can help to locate these.

Our excavation work in Sandy Flanders has shown that, although before the 18th century only earth roads and tracks seem to exist here, these minor connections can be investigated quite well. The fieldwork campaigns indicate so far that a large part of the double-linear features visible on aerial photographs can be dated in

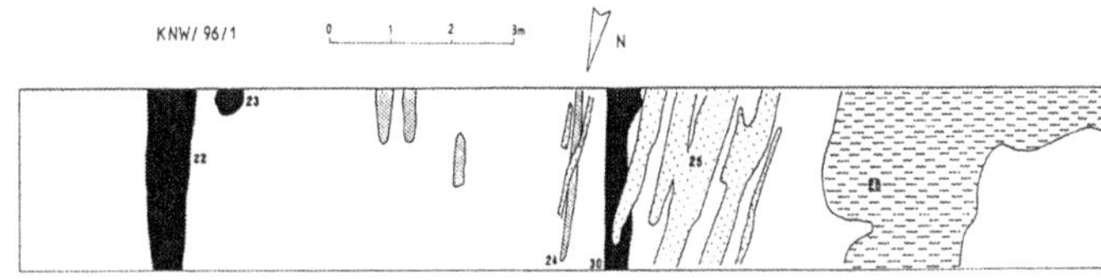

Figure 6: Knesselare: plan of the excavated pre-medieval road with ditches (22-30) and cart tracks (24) and of the cart tracks of a medieval road (25).

the Roman period, although precise dating is not easy, as the ditch fillings seldom contain numerous artefacts (figure 6). The excavated traces of these Roman roads may vary:

- Sometimes only the drainage ditches are preserved and the road surface has been completely eroded.
- When this road surface survives somehow, then a bundle or a couple of wagon tracks can be distinguished. Pedological processes such as iron precipitation often stress their presence.
- In other cases, the Roman road is succeeded by a medieval one, and a slight shift in location or orientation can be recognised.
- When the Roman road is excavated in an open area then its connection with adjoining and contemporary field-ditches could be noticed.

It is clear, however, that our selection of traces checked in the field by way of excavations is somewhat biased. Most of them already displayed surface indications, such as scatters of pottery, or the nearness of a known Roman settlement, pointing towards a possible Roman date. Therefore, it is also much too soon to draw statistically meaningful conclusions from these field checks.

There can be no doubt that many of the unexcavated road traces have a medieval or even a post-medieval date. A part of these can be isolated using techniques which analyse the relation of the old road traces seen from the air with cultural elements in the landscape. A buffering of municipality borders, which often coincide with medieval roads, is one of the possibilities. In figure 7a we see that the abandoned roads at the northern rim of the municipality of Aalter are very much related to the municipality border between Aalter and Ursel. A buffering zone of 200 m around this border allows a selection of those linear traces that are situated very close by. Several fragments of single and double lines with a mainly E-W orientation appear over a distance of some 2 km. The ditches are probably the remains of an early medieval (?) connection between Aalter and Ghent and are situated mainly on slightly elevated parts in the landscape.

However, not all traces of abandoned roads seen from the air are Roman or medieval. The use of different spatial analyses upon the GIS-database seems to indicate that some of the ancient roads and tracks might have a prehistoric date. A simple example from the northern edge of Sandy Flanders demonstrates this. A clearly visible lining of some 30 Bronze Age barrows has been discovered near the municipality border of Maldegem and Oedelem (Bourgeois et al. 1995). The barrows are spread over an area of more than 25 ha and they all are situated on dry to moderate wet sand. In the immediate vicinity of the barrows (figure 7b), a large number of field borders and a few fragments of ancient roads appear. The field borders do not show one main orientation, which may indi-

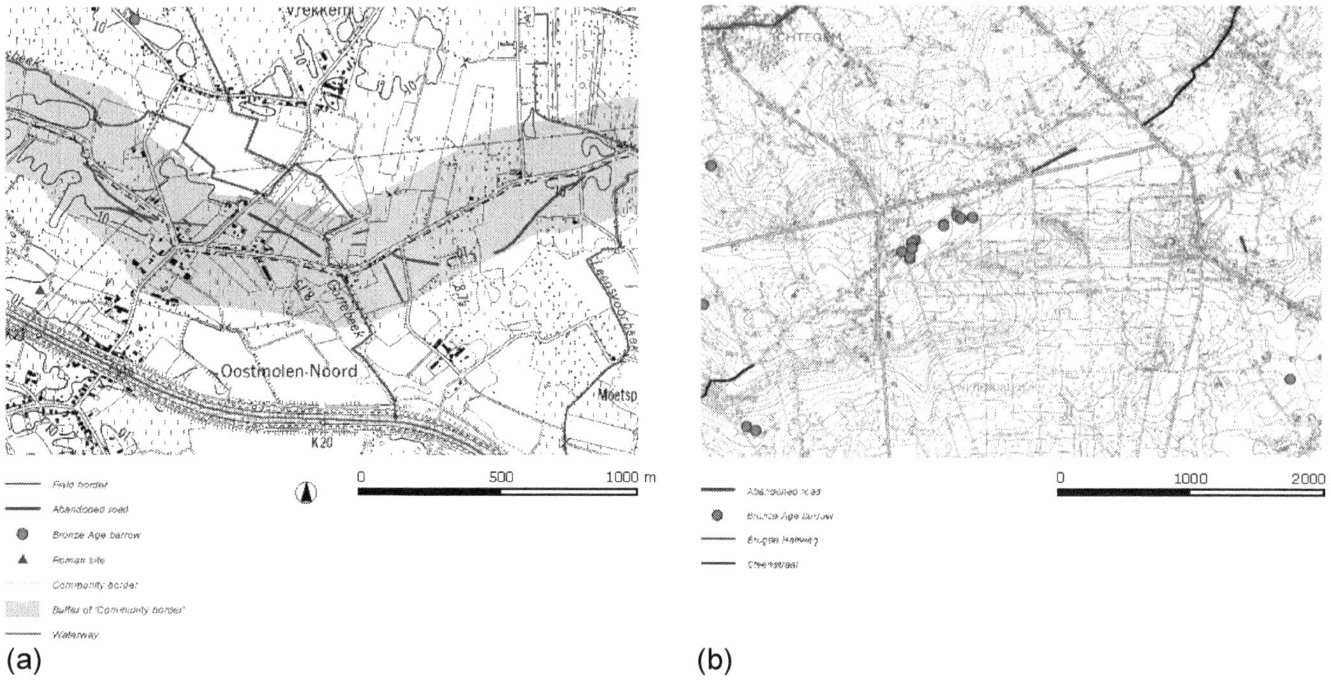

Figure 7: (a) The municipality borders between Aalter and Ursel and associated linear traces visible on aerial photographs. (b) Fragment of an ancient road connecting two parts of the "Roman" Steenstraat and other ancient features visible on aerial photographs.

cate their different chronology, but the orientation of the longest ancient road segment might suggest contemporaneity with the Bronze Age cemetery.

To conclude this section on the ancient roads, it must be clear now that the implementation of the results of aerial photography in the GIS environment, opens up a whole set of possibilities for studying so-called "fossil" roads, routes and tracks. The overlay of the archaeological traces with different thematic maps and geographical information and the use of spatial analysis in relation to physical and cultural aspects of the landscape offer great opportunities. Although confirmation in the field remains essential, firm hypotheses about the date and interrelation of a wide range of road traces can be constructed.

4. Field systems

Most of the detailed research concerning field systems was carried out in smaller micro-regions in which we had recent and reliable information. Six of these test areas lie within the Sandy Region, one in the south around the Roman urban centre of Cassel.

Different types of analyses were carried out, which will be summarised briefly here.

In the surroundings of the town of Cassel, spatial analysis in the GIS-environment was used to check the possible existence of a Roman normative land division (Antrop et al. in press). Several authors suggested recently that the rural landscape in a wide area to the south-southwest of Cassel and possibly even in other directions shows clear marks of Roman centuriation (Jacques 1987, Roumegoux and Termote 1993, Malvache and Pouchain 1994). In particular the hypothesis put forward by François Jacques, concerning "classic" 710 by 710 m module land divisions with orientations ranging from 31° NW (north of Cassel) to 35° NW (south of Cassel), has many followers. This work, mainly based on the visual analysis of 19th and 20th century topographic maps, has never been tested in the field.

We checked the centuriation idea, not yet supported by field research, using digital overlays of different maps and by directional filtering the major landscape structures in the area. The analysis shows the classification of the orientation of field boundaries overlaid upon the presumed centuriation grid for the area around Cassel. The most significant concentrations of line segments are situated here in the intervals -25° to -35° and 55° to 65°, confirming a dominant north-west orientation of 31° (+/- 5°) as proposed by Jacques. Interesting to note is that some orientations agree rather well with the direction of the Roman roads, such as the ones towards the salt producing sectors on the edge of the former coastal area. However we must recognise that some orientations are also parallel to the drainage flow of the nearby Peene river, which could well have influenced the direction of field systems in Roman times.

In certain areas a more detailed observation of the cadastral field boundaries selected for directions between 26° and 36° shows much more "centuriation-oriented" field edges than published by F. Jacques in 1987. This illustrates well the potential of GIS based spatial querying on digitised ancient cadastral documents. It offers not only more detailed and flexible approaches for testing of centuriation hypotheses, but it also delivers basic maps for precise testing in the field.

After an inventory had been made of all ancient field systems on oblique aerial photographs in the region of Sandy Flanders, and the relevant traces were charted, the problem of interpreting and dating the structures became crucial. The surveyed micro-region is almost completely covered with sandy soils that facilitate agriculture, requiring less manual labour input. As these soils provide a good basis for agriculture, there is a high probability they have been used for longer periods during history. Light soils also reveal many features in aerial photography. However, these are seldom registered complete in one observation. Their visibility strongly depends upon meteorological conditions, crop type and phenology of the vegetation. Hence, many observations in the most diverse conditions are necessary to obtain some coherent view. Also, all features, dating from different periods, are seen superposed, which makes their interpretation difficult.

The perceivable ditches might most probably also be a combination of multi-periodical land use (figure 8). Thus, it is impossible to assign a chronology to these features without excavating them.

It is even very difficult to establish a typology of the traces (Boissinot and Brochier 1997). In this study a purely morphological typology of the linear structures was attempted, for which it is still impossible to attach any chronological meaning.

Besides this purely morphological approach of linear ditches and field borders, the relationship of these features with surrounding elements is more significant in many cases (Chouquer 1997). These context elements can consist of environmental features, such as a river or a difference in elevation, or cultural features caused by human intervention, such as a road or other ancient structures.

In some cases both can offer good possibilities to collect some evidence about the chronology of the field structures. Two examples illustrate this.

- In the close vicinity of Ghent, several areas between 15 and 70 ha could be selected where the occurrence of traces could be associated clearly with environmental conditions. The traces indicated Bronze Age burial mounds and field ditches in close relation to slightly higher and drier sandy soils and could indicate of the nearness of the supposed settlement sites and surrounding fields. The zone shown here (figure 9a) is divided by the community border and consists of land sections which form a well preserved half-circle truncated by a brook. The total zone covers 70 ha, providing more than enough space for living and burying the death.
- The relation between the attested field borders and parallel traces or roads has also been examined. This example is situated in the municipality of Zingem where a regular arrangement of a field system was detected. The field borders, spreading over an area of about 13 ha are situated in the immediate vicinity of a recently discovered Roman settlement, cemetery and coin hoard. The ensemble consists of a framework of regular lines with a parallel or perpendicular orientation towards the presumed protohistoric and Roman road between the rivers Lys and Scheldt (figure 9b). The intermediate distance between the ditches is rather small, the estimated plot size varies between 0.6 and 1 ha. The field system is clearly the result of an organized construction, but the supposed Roman character still has to be proven by means of excavation.

Field excavation has been used in the more romanised landscape of the south of Sandy Flanders in a number of cases, such as with the field system discovered in Kruishoutem. Between two regional centres several field borders and a possible local road were found perpendicular to the Roman road. Like the major road they were dated in the 1st and 2nd centuries AD. Pedological and palaeo-botanical evidence suggest that the fields adjacent to the road system were mainly in use as meadows.

The discovery and precise dating of these systems gave us the opportunity to do some further GIS-analysis in the surroundings

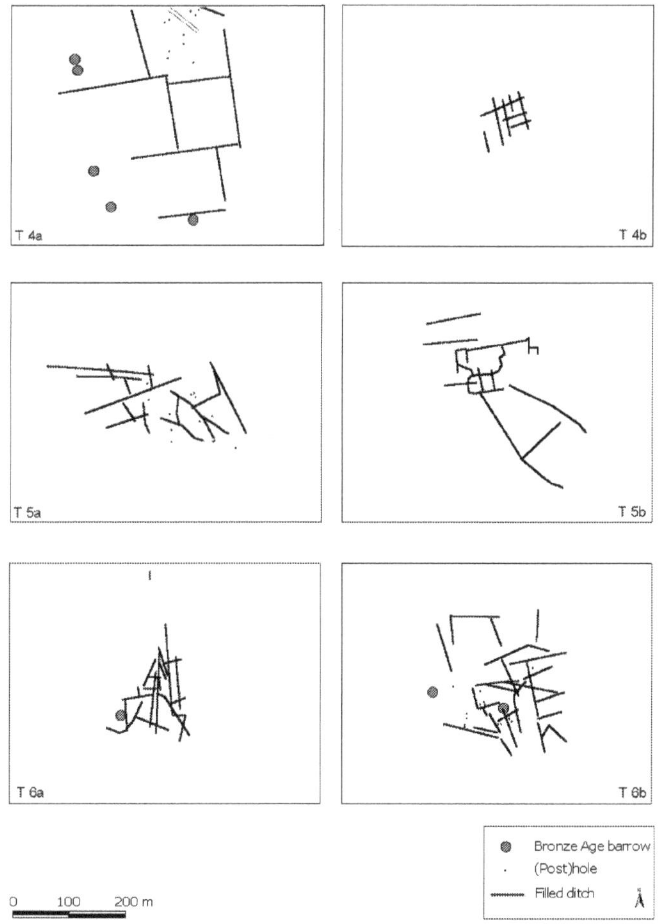

Figure 8: A selection of types of ancient field systems on aerial photographs in Sandy Flanders.

of these ancient features. The digitized cadastral plans of the beginning of the 19th century, allowed us to compare the orientation of the fields dated as Roman with the former land division of the whole municipality.

On the map representing the field border orientation four different concentrations of one dominant orientation can be recognised (figure 10a). These correspond to four different geomorphological land facets having different slope aspect, drainage conditions and soil types:

- The centre is situated on the most elevated zone of this study-area, with a flat terrain and where sandy soils dominate. The main orientation is north-east and the fields are quite large. The Ferraris map of 1770 shows that the landscape was characterised then by open arable land.
- In the Northwest on the valley slope of the river Lys, the same dominant northeastern orientation is found, which corresponds to the orientation of the waterways and the slopes, which can be up 5%. In the surroundings of the Gaverbeek where the terrain is flat, the field system is oriented perpendicular to this river, showing a main north-north-eastern orientation. This orientation is also found in small clusters in the immediate surroundings of the modern centre of the settlement of Kruishoutem, as well as close to the west of the Roman *vicus*.
- In the East sandy loamy soils dominate and the slope aspect is Northeast. The hydrography is oriented east-north-

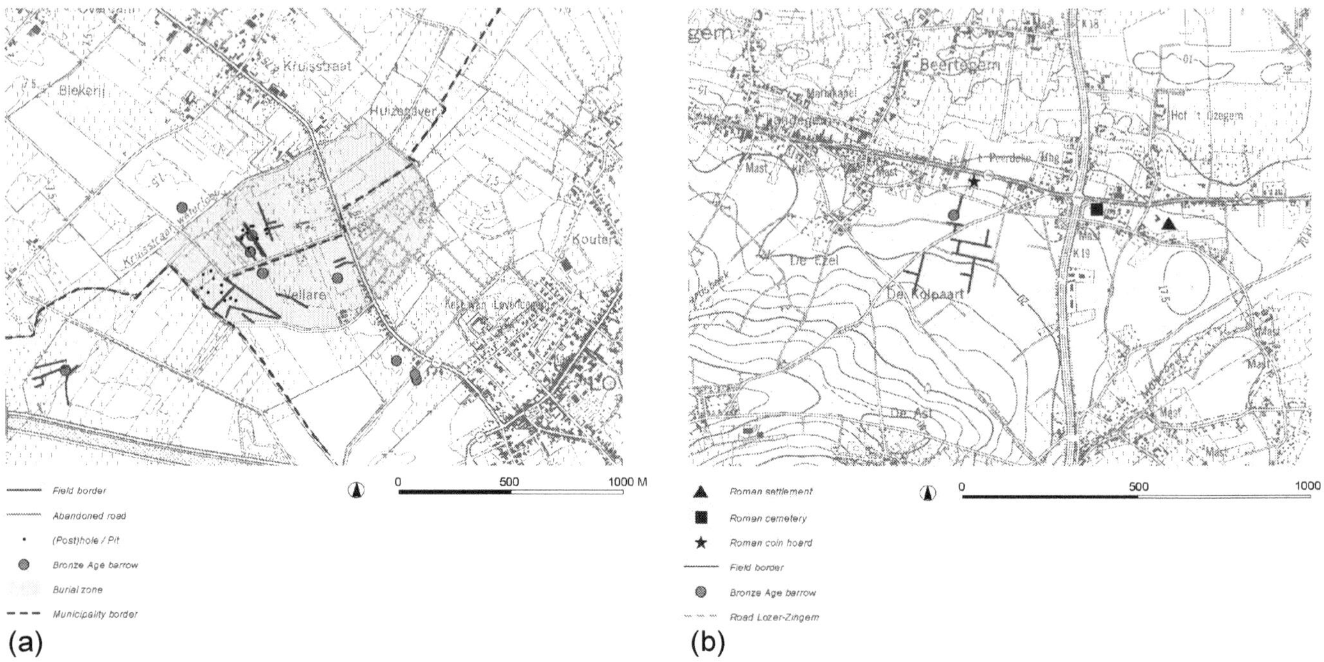

Figure 9: (a) The possible burial zone on the communities of Zomergem and Lovendegem. (b) A possible framework of Roman field boundaries near Kruishoutem.

east and the major field borders have an orientation perpendicular to the waterways.

- In the South the field orientation is less concentrated than in the other parts of the study area. The dominant orientation is east-north-east, but disturbed by the second dominant north-eastern orientation.

Further analysis shows that the marks of the Roman cluster are correlated with a cluster of north-north-eastern and west-north-western oriented field systems of the 19th century cadaster map. Further research must confirm whether these might also be relics of the ancient field systems.

A comparable case study was achieved in the municipality of Sint-Gillis-Waas. Here the study area is situated far from known important Roman roads and settlements at the northern border of the *Civitas Menapiorum* where the romanisation of the settlements is rather low. No Roman roads or settlements were known here until very recently when two Roman field systems, some Roman houses and local roads were discovered by excavations. As the Roman road is not very straight and situated far from the main network of Roman roads in the Civitas, only the orientation of the ancient field systems has been examined here.

As the vectored cadastral map shows (figure 10b), the majority of fields are oriented between -30° (ENE) and 10°(N). This can be explained by the orientation of the drainage which is west-east oriented and most of the fields are perpendicular to the hydrography. In the south of the study area the orientation of the field system seems to be perpendicular to the main road (Reepstraat), which turns to the Southwest in the western part of the area and to the Southeast in the eastern part. The corresponding orientations of the field system range from -40° (NE) in the west to 10° (N) in the east.

In the centre and in the north, where sandy soils dominate, the main field orientation is northeast. Here, where the study area was covered by woods in 1770 as shown on the Ferraris map, the fields are much larger than in the rest of the study area.

The most complex cluster of marks is concordant with the 19th century cadastral divisions and absolutely parallel to the excavation plan of Roman fields near the Reepstraat. Because the marks complement the Roman fields very well we may suppose that the excavated Roman field system could once have been be part of a much larger system of land division. However, no causal proof for this observed matching can be determined as many factors and chance might be implied.

A part of the study area is situated in the polders of the river Scheldt where a concentration of east-north-eastern oriented fields can be recognised.

Generally, the field systems are oriented perpendicular to the actual natural drainage system, which most likely had the same general direction during the Roman period.

5. Some conclusions

The greatest challenge in landscape analysis is the investigation of the *cultural landscape* (Vermeulen and De Dapper 2000). Certain aspects of the cultural landscape can be interpreted very easily, such as the location of cities, cemeteries and rural settlements. More complex archaeological and historical landscape features, such as segments of roads, networks of tracks and whole field systems, demand more effort and care to get integrated in a GIS and their spatial analysis and interpretation asks for sophisticated geographical skills. Typical problems concern the understanding of "black box" procedures available in a GIS, dealing with uncertainty and error propagation and the critical assessment of the historical meaning of ancient and actual landscape structures as revealed by aerial photographs and maps.

However, the project on the *Civitas Menapiorum* showed some convincing results of the application of GIS and spatial analysis for the efficient processing and integration of the most diverse data sources that are significant for study of the ancient land or-

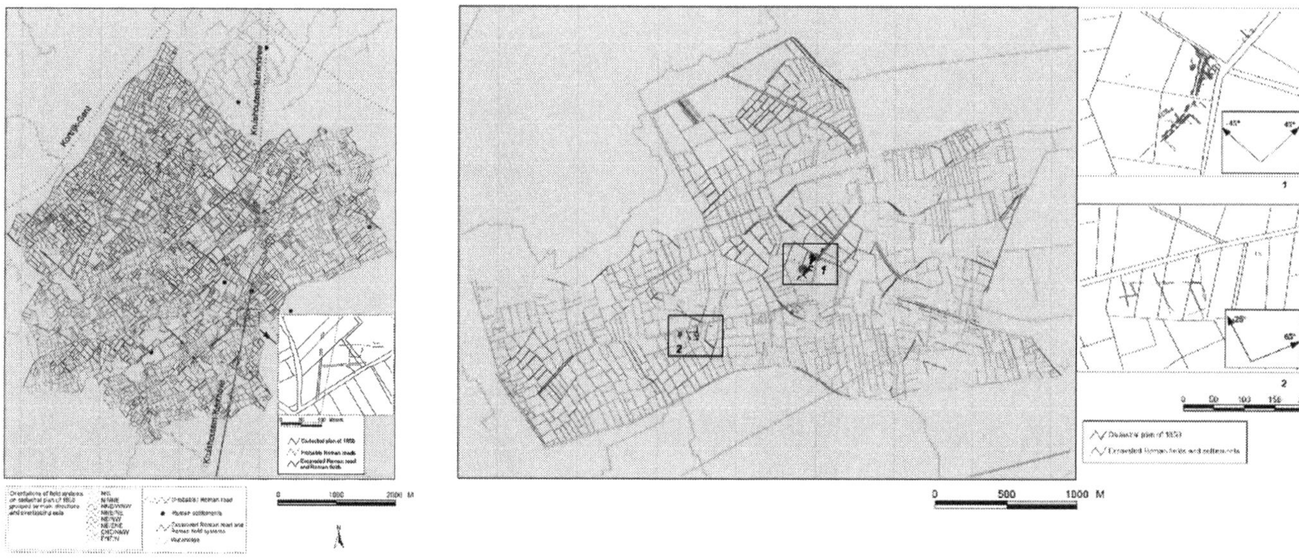

(a) (b)

Figure 10: (a) Kruishoutem: cadastral plan of 1850 and excavation plan of a Roman road and possible Roman fields. GIS-filtering of the field boundaries according to the major orientations. (b) Sint-Gillis-Waas: GIS-filtered cadastral plan of 1850 and excavation plans of Roman field systems.

ganisation. The advantages are multiple and only some will be summarised here:

- data about the physical environment (topography, landform, soils) can be well integrated in a GIS-environment and spatial analysis can be shown to be a valuable tool for explaining Roman and other ancient roads and field patterns;
- a more complete use of oblique aerial photographs when studying such patterns is now possible;
- the cadastral map data, which are so important because of their historical dimension, can be used more efficiently for the spatial analysis of ancient field systems.

The examples shown here demonstrate, however, that detailed archaeological surveying and in particular stratigraphic excavations will remain essential to support even the best hypotheses. This is especially true when confronted with a type of archaeological features, *in fact just some lines in the landscape*, which present a whole range of problems concerning their historical interpretation.

References

ANTROP, M., VERMEULEN, F. and WIEDEMANN, T., in press. Une organisation cadastrale dans la partie méridionale de la Civitas Menapiorum (Gallia Belgica). In Clavel-Lévêque, M. (ed.), *Atlas Historique des Cadastres d'Europe II.*

BOISSINOT, PH. and BROCHIER, J.E., 1997. Pour une archéologie du champ. In Chouquer, G. (ed.), *Les formes des paysages. 3. L'analyse des systèmes spatiaux.* Paris: 35-65.

BOURGEOIS, J., CROMBÉ, PH. and SEMEY, J., 1995. L'archéologie aérienne en Flandre sablonneuse. Initiatives pluridisciplinaires de l'Université de Gand. In *Actes du Colloque international d'Archéologie Aérienne.* Amiens 15-18 octobre 1992.

CHOUQUER, G. (ed.), 1997. *Les formes des paysages.* 3. L'analyse des systèmes spatiaux. Paris.

JACQUES, F., 1987. Témoins de cadastres romains dans la région de Cassel. *Revue du Nord* 69 (nr. 272), 101-108.

JOHNSON, I. and MACLAREN, N. (eds.), 1997. *Archaeological Applications of GIS. Proceedings of Colloquium II, UISPP XIIIth Congress,* Forli, Italy September 1996. Sydney.

JONES, C.B, 1997. *Geographical Information Systems and Computer Cartography.* London.

MALVACHE, E. and POUCHAIN, G., 1994. La recherche des cadastres antiques dans la région Nord-Pas-de-Calais (état provisoire de la recherche). *Revue du Nord – Archéologie* 76 (no. 308), 83-98.

ROUMEGOUX, Y. and TERMOTE, J., 1993. Op de rand van een imperium de Romeinen in de Westhoek. *Westvlaamse Archaeologica, jg.* 9 (2), 61-80.

VERMEULEN, F., 1992. Moderate Acculturation in the Fringe Area of the Roman Empire: Some Archaeological Indications from the Civitas Menapiorum. *Bulletin de l'Institut Historique Belge à Rome* LXII, 5-41.

VERMEULEN, F., 2000. The potential of GIS in landscape archaeology. In Slapšak, B. and Stančič, Z. (eds.) *On the good use of GIS in ancient landscape studies.* Proceedings of the International Worshop held in Ljubljana (December 1998). Ljubljana.

VERMEULEN, F. and ANTROP, M. (eds.), in press. *Ancient Lines in the Landscape.* Gent.

VERMEULEN, F. and De DAPPER, M. (eds.), 2000. Geoarchaeology of the landscapes of Classical Antiquity. Proceedings of the congress in Gent (october 1998). *Babesch Supplementa* 5. Leuven.

VERMEULEN F., HAGEMAN B. and WIEDEMANN T., in press. Photo-interprétation et cartographie des systèmes spatiaux anciennes: l'archéologie des routes et des parcellaires en Belgique. In Clavel-Lévêque, M. and Vignot, A. (eds.), *Atlas Historique des Cadastres d'Europe* II. Besançon.

VERMEULEN, F., HAGEMAN, B. and WIEDEMANN, T., in press. Organisation du paysage et traitement d'images et de photo-aériennes dans le cadre d'un SIG: le cas du Civitas Menapiorum (Gaule Belgique). *Paysages Agraires et Structures Anciennes*. Rapport intermédiaire de l'action COST G2.

An "Integrated Space" Approach for the Interpretation of a Medieval Stronghold in Middle Pomerania, Poland

Rafał Zapłata

Institute of Prehistory, Adam Mickiewicz University in Poznań

Św. Marcin 78, PL – 61-709 Poznań, Poland

e-mail: rafalz@amu.edu.pl

André P. Tschan

Institute of Archaeology, University of Oxford

36 Beaumont Street, GB – OXFORD OX1 2PG, United Kingdom

e-mail: andre.tschan@archaeology.ox.ac.uk

Abstract

Polish archaeology, at present, adopts traditional approaches with the result that, by and large, there are only utilitarian explanations for the subject of medieval strongholds. In an attempt to open new analytical avenues, this work combines post-processual concepts applied in landscape archaeology together with the power of geographic information systems (GIS). The goal is to engage in holistic investigations of archaeological remains beyond the current emphasis on artefacts and features as isolated phenomena that are, seemingly, devoid of any large-scale contextual setting. The site of Wrześnica is therefore not just a self-contained example of an early medieval stronghold but, rather, represents an integral utilitarian and *symbolic part of the middle Pomeranian cultural landscape and its overall spatial organization. The specific research applies a viewshed analysis based on the strong suggestion that visibility and intervisibility played a highly important role with regard to the area's spatial symbolism.*

Key words: geographic information systems, landscape archaeology, medieval strongholds, Pomerania, post-processual archaeology, space, symbolism, viewshed analysis

1. Introduction

A major focus of post-processual archaeology is the concern with the meaning of space and landscape. For the following study, a fundamental notion is the belief that both individuals and communities "create" space as a social construct. Embracing this idea also defines space as an all-inclusive manifestation that is full of commonly recognized and shared symbolic content. Consequently, there is an added interpretation potential with regard to past cultures by investigating the symbolism of given spaces and/or the objects contained within. Ideally such research can supplement some of the otherwise limited archaeological assessments that are preoccupied with the identification of mainly physical and functional characteristics.

In particular, the traditional interpretation for early medieval strongholds of "the tribal period" (sixth to tenth century AD) in Poland, including the site of Wrześnica in middle Pomerania (figure 1), concentrates mainly on militaristic functions. Other potential purposes such as seats of power or cult places are, overwhelmingly, treated in a secondary fashion. Recent developments rooted in post-processual archaeology, on the other hand, provide the crucial means for a different approach with regard to the role of these strongholds. The essential impetus comes through the incorporation of space and landscape as integral parts of the archaeological record; thereby allowing strongholds to be investigated in an inclusive and more holistic manner. Further inspiration for this work is the result of anthropological research and humanistic geography (Hall 1959, Hall 1966, Tuan 1977, Cosgrove 1998) that stresses individual and collective human experiences with regard to the role of space. In other words, the topic of people and their environments is examined through the relations of meaning, value, will and emotion, and not only as rational or intellectual relationships (Grzybowski 1981:623).

The specific research tool of choice for this paper is a viewshed analysis based on the strong suggestion that visibility/intervisibility played an important role at Wrześnica micro-region. An "integrated space" approach using GIS visibility maps, in combination with a post-processual interpretation of the wider archaeological context, forms an enhanced and intriguing dataset that potentially achieves a greater understanding of the past. There is therefore a good likelihood that this stronghold, by virtue of its clear visual prominence, also exhibited a significant symbolic relationship with much of the surrounding cultural landscape features including settlements and burial mounds.

2. Landscape archaeology

In traditional settlement archaeology, the relationship between humans and their environment is understood in physical terms by the way space is exploited and utilized (Tilley 1994:7). For post-processual landscape archaeology, on the other hand, the fundamental issue is conceptualisation of space by virtue of perception, through its content in the form of environmental and cultural phenomena. However, at present, there is some considerable debate regarding what landscape archaeology should really represent (Fisher and Thurston 1999). Tilley (1994), in particular, deals with landscapes as anonymous sculptures that are perpetually created by human activity and, thus, it is never a complete process because it continually develops. He also suggests that the relation-

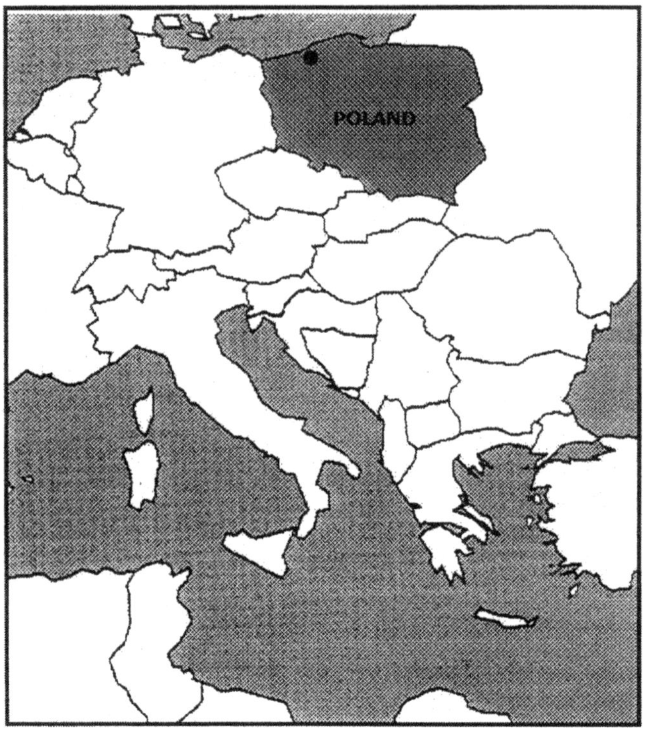

Figure 1: Location map of the Wrześnica stronghold, Poland.

ship between humans and the landscape is a dialectic and constant practice of structuralisation. The landscape is therefore a medium or intermediary for activity as well as an activity itself and is an experiential setting for the people living their everyday lives. With regard to this paper the term landscape archaeology is used to describe:

> "...a broad, inclusive, holistic concept created intentionally to include humans, their anthropogenic ecosystem and the manner in which these landscapes are conceptualised, experienced and symbolized." (Fisher and Thurston 1999:630).

3. The creation of space

It is critical to understand that based on different perceptions of the world, individual communities and groups possess and create social spaces that are unique to them in the form of an unrepeatable arrangement of spatial meanings and values. Every society develops its own space and has ways of acting and partaking within it through practices that finally lead to the creation of human spaces:

> "Man feels and understands all space from the centre of his or her own space, which is always that space created to a significant degree by man himself. Anthropomorphisation of space on the part of man cannot be completely avoided." (Cackowski 1998:41)

Znaniecki (1938), feeling that any researcher of culture should treat space with a humanistic coefficient (i.e., considering the way it is experienced by those human subjects whose culture is investigated), has specifically examined the issue of social space. In his opinion, human subjects never come across some universal, objective, unqualified, boundless and infinitely divisible space. On the contrary, the participants of a culture experience countless "spaces" that are qualitatively varied, limited, indivisible, changeable and either positively or negatively valued. Space is therefore never a uniform or indifferent setting for all the varied forms of economic, social and cultural life in which people engage (Tilley 1994:14-117). Thus, space plays an active, often-conflicting role by perpetually creating or modifying phenomena and processes.

3.1. Space in early medieval culture

Contemporary categories of space will have nothing or little in common with other periods in the past and, consequently, we need to first address those factors that affected the creation of social spaces. As part of a whole plethora of cultural categories this would include the natural environment, technological developments, prevailing power structures and a common value system, to name a few, that in combination form the respective worldview for a given society.

According to Guriewicz (1976:168), during the early middle ages spaces would have been established and recognized through the variety of human activities (e.g., economic, religious, etc.). In this way, the experiential environment in which life took place hardly left any blank spots on a unique landscape map of events and geography. This was clearly not a static process inasmuch as "...matters blending into the landscape would blur after some time to be replaced by new meanings; they would, however, sometimes last for ages" (Banaszkiewicz 1979:988). The actual process is one where people gave names to the specific phenomena occurring within a given environment and thereby "spiritually" appropriated the natural landscapes in which they moved and resided. This allowed humans to understand the world by superimposing meaning onto space and to simultaneously create a unique cultural record for a community.

3.2. Space and value

One of the aspects emphasized as part of this research, is symbolism or, more specifically, symbolic space. Halbwachs (1970:132) argues that:

> "...when a particular group is connected to a certain section of space... then each aspect, each detail of this place has its own meaning and can be interpreted only by the members of that group."

In other words, humans create their own space and shape its function and meaning. These spaces, in turn, create the conditions for the behaviour of the people themselves. Clearly, space has not only a physical and functional character but it is also full of individual/communal symbolic references and content.

For the purpose of this work, one needs to embrace this notion that human efforts perpetually transform natural spaces into social spaces. In other words, through engagement with our surroundings we create space and bestow value-based cultural significance upon it. The mechanism for doing so depends on the specific worldviews of a given culture and the applied system of valorisation (models that change in the course of time). Niczyporuk (1998) identifies three types of valorisation pertaining to space (table 1).

Magical valorisation requires appropriate "safety measures" such as magic in order to establish one's place while, at the same time, no single place may exclusively belong to anyone. The modern type, on the other hand, is not necessarily relevant for this study

Magical (tribal societies)	Space is considered qualitative, non-homogenous and concrete (i.e., of changeable quantity, unstable, and uncertain).
Modern	Treats the human world and nature separately using categories of spheres for cultural spaces (e.g., economic, religious etc.).
Religious (magical-religious)	Space is divided into two distinct categories pertaining to the *sacrum* and the *profanum*.

Table 1: Valorisation Types for Space.

because of the functional separation into many distinct and often task-specific spheres. Religious valorisation, however, is critical for the subsequent interpretation of the Wrześnica site inasmuch as the analysis focuses specifically on the relationships between places of potentially high ritual and symbolic significance and secular areas that are not designed or used for these purposes.

3.3. Religious and secular space

Based on religious valorisation, the *sacrum* is a holy space that is separated or removed from its surroundings to form a centre of religious functions. This also means that any such site will exhibit a transcendental or a "not of this world" quality (often the location for the appearance of divinity and the centre of the world itself) while acting as a bridge between the human world and transcendence. The *sacrum* is understood as a sacred and symbolic area where only those behaviours are permitted that communicate content by "explaining" or "interpreting" the world in line with the prevailing religious and ritual beliefs. Thus, in universalistic religions, the *sacrum* is not differentiated inasmuch as the purpose of each location is identical. And, that function is to organize space and giving a comprehendible order to a culture (Niczyporuk 1998:48).

The *profanum*, on the other hand, belongs to the people and comprises the places of work and shelter – it is, essentially, the secular space where day-to-day activities are undertaken. These areas are clearly qualitatively differentiated since they vary from one to another by virtue of unique purposes and characteristics. The *profanum* also involves spaces that are generally safe while keeping in mind that this may not always be a constant state due to new and alien elements that may introduce themselves and thereby encroach upon these "mortal" areas (Ibid.: 48).

Ethnologists have for a while supported the importance of beliefs and magical practices in preliterate communities, particularly in matters requiring direct contact with nature. Bystroń (1939) generally sought to determine and conceptualise the magical factors present with regard to settlement processes:

1. Magical-religious division of space,
2. Positive/negative evaluation of taking a given direction,
3. Determining/evaluating bounds,
4. The value of a given place,
5. Identifying locations for burial sites,
6. Evaluation of the landscape and its features,
7. Presence/absence of animal and plant species.

With regard to this research, the overall degree to which the religious element in the form of superstition or magic (e.g., omens) was a decisive factor in accepting/rejecting a given site location is difficult to establish even when employing the most modern archaeological methods and techniques. Nonetheless, any magical practices would have been intended to separate the inhabited area from the natural environment in an attempt to create a safe family abode, protected by supernatural forces. Based on works where it has been shown that religion permeates all aspects of (Bystroń 1939:32), religious valorisation of space would seem to be of great importance in the case of the early medieval period in Pomerania including Wrześnica.

Most importantly, though, is the fact that religious valorisation clearly demarcates the *sacrum* and *profanum* in spatial terms (Niczyporuk 1998). This work therefore looks at the specific arrangements that relate to areas identifiable by their potential for "inner mysteries" and ritual symbolism in contrast to the "uninitiated" spaces representative of everyday life. However, the emphasis is on avoiding a singular preoccupation with each site based on the separation into religious and secular locales since the goal is to holistically investigate and interpret the specific interrelationship of the *sacrum* and *profanum* for the entire cultural landscape of the Wrześnica micro-region.

4. The Wrześnica case study

Hall (1966), when researching the individual and communal spaces for various cultures, pointed to the existence of many different forms of space based on the diverse means of sensory perception (i.e., sight, sound, smell, taste and touch). With regard to the early medieval site at Wrześnica, perceptual recognition through sight provides a new set of information pertaining to the overall role for this stronghold. Based on the intriguing nature of immediate as well as distant visual spaces this important archaeological site was deemed ideal for a detailed GIS viewshed investigation. The specific approach incorporates arguments that, beyond current interpretations (Rączkowski 1999), suggest Wrześnica was not only a centre of social organization but also fulfilled an additional and highly symbolic purpose due to its prominence as a spatial focal point within the surrounding cultural landscape.

To achieve this goal the above notions with regard to the role and function of space have to form the basis for the Wrześnica case study while the main concern is with visibility or visual spaces. Clearly, humans not only identify the physical nature or structural components for a specific setting, but also establish and perceive the symbolism of a place as part of overall cultural developments. However, the perception of space is not just the consequence of what can be seen but also relates to things hidden from us either physically or conceptually because the cultural significance of a particular place may not be recognized (Hall 1966:28). Within this context, the question relates to what medieval peoples could or could not have "seen" and the effect that this might have had on the symbolic definition or appropriation of the landscape and its contents.

4.1. The setting

The general area under investigation is situated on the Sławno uplands, adjacent to the Słupsk uplands of middle Pomerania

(Pomorze), and comprises the broader geographic area of the southern Baltic seaboard of northern Poland. The Wrześnica micro-region is approximately 35 km^2 where many sites have been recorded including, in addition to the aforementioned stronghold, several settlements and burial mound cemeteries. For this study, all the available archaeological research data was consulted beginning with excavation and field survey results from 1982 onwards and, to-date, the corpus of the collected information is based on the culture histories and environmental records for the early middle ages (Rączkowski 1998).

According to the literature, the culture that occupied the site during the early middle ages would have belonged to the group of Pomeranian tribes comprised of the north-western Slavs (Leciejewicz 1989:58). Current research suggests that the Wrześnica micro-region was used by just one of the tribal units (Sułowski 1960) that settled in the basins of the central and lower runs of the Wieprza, Łupawa and Łeba rivers. These peoples are commonly identified as the Słowińcy based on the nomenclature of Wulfstan (Wenden) (Łosiński 1982).

Important archaeological remains obviously pertain to the site of Wrześnica located within the floodplain of the Wieprza river valley and they have been dated to the ninth to tenth century AD. The maximal height of its embankments or ramparts is approximately three meters from the interior surface and four meters above the current water level of the river. It is a "valley ring" type stronghold (Olczak and Siuchniński 1976) and kidney-shaped with a concave open space that encompasses a total area of approximately 0.6 hectares. It therefore puts the Wrześnica site into the category of average-sized strongholds ranging from 0.5 to 1.0 hectares.

4.2. Viewsheds

A GIS viewshed analysis was selected in an attempt to study and evaluate visual spaces and their potential symbolism with regard to the wider cultural landscape of Wrześnica. The overall goal was the interpretation of cultural phenomena by virtue of a spatial visibility/intervisibility context beyond the information available through limited culture and period-specific distribution maps. In other words, the application of this GIS tool should allow an examination of the archaeological data from a different perspective and through this process provide potentially new insights regarding life during the early middle age. As a standard viewshed application, there are limitations inherent with this tool particularly when used for archaeological research (Gillings and Wheatley 2000). It is therefore our belief that any such shortcomings should be explicitly mentioned as part of a study.

For one, the visibility maps that were created use a centrally situated viewpoint for each of the sites selected. Clearly, this approach disregards actual areal extents and thereby limits the capacity to generate viewshed results representative of an entire location. However, it has to be stressed that the issue of what constitutes a site is generally based on interpretation or specific archaeological practices and as a result concrete boundaries may vary considerably. With regard to the Wrześnica region the settlements and cemeteries in particular have been subject to different conceptual spatial definitions and it was deemed safer to select single viewpoints as a respective centre for each site.

In addition, the overt emphasis on the topographical layout of the Wrześnica micro-region obviously ignores other important environmental data such as vegetation. The problem of viewshed analyses applying barren landscape approaches that omit tree coverage has already been identified specifically (Tschan et al. in press). However, some understanding of early medieval culture together with present-day ecological circumstance does allow for further relevant information to be incorporated within the final interpretation. This study identifies an agrarian society that predominantly cultivated the western banks of the Wieprz river valley. The fact that the eastern sides of the valley are composed of old growth forest may therefore remedy to some degree the absence of other data for the generated results.

There is furthermore a general difficulty in reconstructing accurately the characteristics of significant archaeological structures (i.e., building dimensions, heights of barrows, etc.). And, it should be mentioned that the viewsheds produced depend on surface data. In other words, no raised elevations based on man-made structures were involved aside from the most probable maximal vertical extents of the Wrześnica stronghold ramparts set at a total of five meters. The next step involved the issue of a viewer's height and it was decided to add a realistic, albeit assumed, 1.75 meters above ground for each site. The visibility maps also use a value of 8.6 kilometers to define the field of view and this figure was derived from the maximum distance diagonally across the total study area (5x7 km).

In all of this, the inherent assumption is that the physical shape of the landscape around Wrześnica has not changed to a point where it is now solely indicative of current circumstances, while bearing no relationship to past characteristics. From preliminary geomorphological investigations, it is clear that the overall profile for this area has not altered to the extent that modifications to the dataset would be required.

4.3. Analysis

The actual computer tools included AutoCad to digitise the data available from 1:10,000 scale maps and the Idrisi GIS package to process the captured digital information. The basic spatial data consists of archaeological sites and contour lines with intervals of 1.25 meters. The final scale in Idrisi employs an individual raster cell size that represents a ten square-meter section of the real world. From the topographical information a digital elevation model (DEM) was created and, based on the specific site data, individual viewsheds were generated using the dedicated routine available in Idrisi.

The results concerning the archaeological content were then visually and statistically analysed in order to establish the number of sites that can be seen from each selected location respectively. At the same time, this work does not address any climatic conditions (e.g., precipitation, fog, variable lighting, etc.) that actually would have a dramatic impact on visibility especially with regard to distant background areas. We therefore accept that for our purposes the capacity for human sight concerning early medieval peoples at Wrześnica has been modelled in a relatively idealized fashion.

In order to examine the visual spaces and in particular the detailed intervisibility relationships, a total of 19 viewshed maps were generated. The archaeological data included five burial mound cemeteries, fourteen settlement sites (figure 2) as well as the stronghold of Wrześnica itself and the results for each were subsequently compiled (table 2). From this tabular arrangement

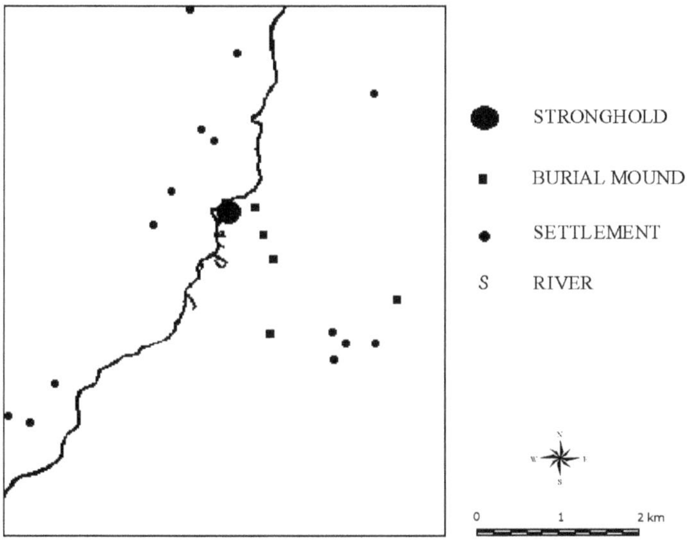

Figure 2: Sites in the Wrześnica region.

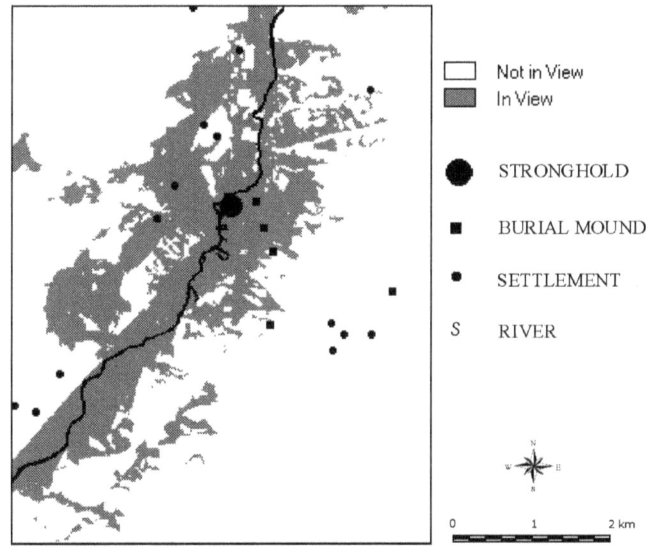

Figure 3: "Viewshed map" of the Wrześnica stronghold.

		Burial Mounds					Settlements													
Sites	S	49	50	57	58	59	4	5	7	10	12	15	25	30	38	46	52	54	55	60
S		X	X	X				X	X	X	X	X		X						
49	X							X		X	X	X	X	X	X					
50	X				X			X	X	X	X	X	X	X	X	X	X	X		X
57	X	X				X		X		X	X		X	X						
58																X	X	X		
59	X			X				X	X	X	X		X	X						X
4			X						X								X			
5	X	X	X	X					X		X									
7	X		X	X			X	X		X	X	X	X				X			
10	X	X	X	X		X			X		X	X		X						
12		X	X	X				X	X	X		X		X						
15		X	X	X					X	X	X									
25		X	X	X		X			X						X			X		X
30	X	X	X	X					X	X	X	X								
38			X							X			X	X				X		X
46					X	X											X	X	X	
52			X		X				X				X			X		X	X	
54			X		X											X	X			
55																X	X	X		
60		X	X			X							X		X			X		

Table 2: Site Visibility. S = Stronghold, X = visible, □ = not visible.

and the actual maps, it rapidly becomes clear that the stronghold incorporates the largest visible areas including the regions mainly to the east and west in addition to a substantial portion to the south of Wrześnica along the Wieprz river valley (figure 3). The cemeteries seem to have a similar visual preference as the stronghold inasmuch as the western parts and also much of the southern river valley are in clear view. In contrast, settlements seem to have no explicit visible association with specific areas and/or other sites. Thus, based on this preliminary assessment, the conclusions drawn would support the idea that the stronghold of Wrześnica from which the greatest number of settlements and burial mounds can be seen plays an important symbolic and unifying role as part of the wider cultural landscape.

5. An interpretation of Wrześnica

The preliminary interpretation of the results for this ongoing study is based on the preferred concepts of space outlined above. Clearly, the stronghold appears to be the singularly most unique feature of this cultural landscape because it allows a superior view throughout the region. This also means that it is possible to examine its role not just as a dominant place in terms of central physical and visual prominence, but also with regard to the meaning that the people of Wrześnica would have bestowed upon it.

The fact that Wrześnica represents such a significant vantage point within the landscape means that people coming along the river valley clearly saw the site from afar. Similarly, anyone residing within the stronghold would have been able to recognize human movement throughout the surrounding landscape even at great distances. This extensive capacity for visibility may have been a necessary means of security since trouble could have been recognized at an early stage and it would have formed part of the physical preparations and refuge mindset required in defending the stronghold. In principle it therefore fulfilled a practical and symbolic separation from the world "outside". As an advantage specific to the stronghold's inhabitants the same would not have been available for everybody else living in the settlements.

Another main idea is that the stronghold seems to connect places such as settlements and cemeteries as part of a wider social and symbolic structure. This is particularly evident when considering that the symbolic burial grounds lie outside of the visual range for the inhabitants living in the settlements on the western side of the river valley. Thus, people would not have had the opportunity to directly observe the ritual burial mounds on the eastern side of the river valley but would have been able to do so from the stronghold. The stronghold might therefore have acted as a specific intermediary to commune between these two areas through its role as a visual go-between.

In other words, the symbolism relates to the bridging of worlds on either side of the river, involving the world of the living and the one of the dead. This link might also be extended inasmuch as the stronghold could have acted as a gateway from one world to the next based on a religious definition of the human life cycle and the specific ritual practices associated with it. This connection allowed the symbolic crossing of boundaries while in reality these would not have been bridged until death and internment in one of the cemeteries had actually occurred.

Based on the location of the stronghold the suggestion is that physical access to the afterlife was restricted while at the same time it allowed for a visual participation. Thus, it provides the means to be present in both worlds simultaneously. The question of whether or not this was the sole privilege of the inhabitants of the stronghold may, therefore, also reflect potential economic role divisions as well as socio-political structures with regard to the different cultural spaces for Wrześnica and its surroundings. Clearly, continued research is required that specifically investigates these and other space-related issues in order to extend upon these early results obtained from this viewshed analysis.

6. Conclusions

The initial assessment of this work provided the opportunity to combine post-processual landscape ideas with a GIS-based viewshed analysis. In effect, it has created the essential preconditions for what will ultimately be a holistic interpretation of the entire Wrześnica region, where different spaces are not just identified by their physical and functional properties.

"Seeing" specific places continually stirs the memories of people as part of the recognition of space and these spaces in turn form the permanent memory of its inhabitants. The available results of this study are therefore very encouraging, considering that the stronghold's role within the wider cultural landscape clearly involves visual spaces that were full of a meaning known to and transformed by the people that lived at Wrześnica during the early medieval period. We therefore suggest that Wrześnica was not only a place of habitation or shelter but, moreover, fulfilled an important role with regard to socio-cultural and magical-religious creation and use of space.

Acknowledgements

The authors would like to thank Dr Wlodek Rączkowski for help in the creation of this paper and for his invaluable input.

References

BANASZKIEWICZ, J., 1979. Fabularyzacja przestrzeni. Średniowieczny przykład granic. *Kwartalnik Historyczny* 86/4:987-999.

BYSTROŃ, J.S., 1939. Czynniki magiczno-religijne w osadnictwie. *Przegląd Socjologiczny* 7/1:25-46.

CACKOWSKI, Z., 1998. *Osobliwość przestrzeni ludzkiego świata*. In Symotiuk, S. and Nowak, G. (eds.), *Przestrzeń w nauce współczesnej*: 31-44. Lublin: UMCS.

COSGROVE, D.E., 1998. *Social formation and symbolic landscape*. Madison: The University of Wisconsin Press.

FISHER, C., THURSTON, T., 1999. Dynamic landscapes and socio-political process: the topography of anthropogenic environments in global perspective. *Antiquity* 73/281: 630-631.

GAFFNEY, V. and STANČIČ, Z., 1991. *GIS approaches to regional analysis: a case study of the island of Hvar*. Ljubljana: Znanstveni institut Filozofske Fakultete.

GRZYBOWSKI, J., 1981. Z filozoficznych zagadnień geografii. *Przegląd Geograficzny* 53/3:620-625.

GURIEWICZ, A., 1976. *Kategorie kultury średniowiecznej*. Warszawa: Państwowy Instytut Wydawniczy.

HALBWACHS, M., 1970. *La morphologie sociale*. Paris: A.Colin.

HALL, E.T., 1959. *The Silent Language*. New York: Doubleday.

HALL, E.T., 1966. *The Hidden Dimension*. New York: Doubleday.

LECIEJEWICZ, L., 1989. *Słowianie zachodni. Z dziejów tworzenia się średniowiecznej Europy*. Wrocław: Ossolineum.

LLOBERA, M., 1996. Exploring the topography of mind. GIS, social space and archaeology. *Antiquity* 70:612-622.

ŁOSIŃSKI, W., 1982. *Osadnictwo plemienne Pomorza (VI-X wiek)*. Wrocław: Ossolineum.

MASCHNER, H.D.G. (ed.), 1996. *New Methods, Old Problems. Geographic Information Systems in Modern Archaeological Research*. Carbondale: CAI.

NICZYPORUK, D., 1998. Przestrzeń w kulturowym obrazie świata. In Symotiuk, S. and Nowak, G. (eds.), *Przestrzeń w nauce współczesnej*: 45-60. Lublin: UMCS.

OLCZAK, J. and SIUCHNIŃSKI, K., 1976. Typologia wczesnośredniowiecznych grodzisk Pomorza środkowego. *Slavia Antiqua* 23:111-152.

RACZKOWSKI, W., 1998. Między programem a przypadkiem: badania osadnictwa w dorzeczu środkowej Wieprzy. In Dworaczyk, M., Krajewski, P. and Wilgocki, E. (eds.), *Acta Archaeologica Pomoranica I: XII Konferencja Pomorzoznawcza*: 145-156. Szczecin: SNAP.

RACZKOWSKI, W., 1999. Osadnictwo i proces urbanizacji wczesnośredniowiecznych społecznosci Pomorza Środkowego. In von Zitzewitz, L. (ed.), *Zeszyty Kulickie, 1: Początki miasta Słupska. Nowe wynik badawcze z Niemiec i z Polski*. 63-88. Kulice: Europäische Akademie Külz-Kulice.

SUŁOWSKI, Z., 1960. Słowiańskie organizacje polityczne nad Bałtykiem. *Roczniki Historyczne* 26:55-88.

TILLEY, C., 1994. *A Phenomenology of Landscape*. Oxford: Berg.

TSCHAN, A.P., RACZKOWSKI, W., LATAŁOWA, M., 2000. *Perception and Viewsheds: Are they mutually inclusive?* In Lock, G. (ed.), *Beyond the Map: Archaeology and Spatial Technologies*. NATO Science Series, IOS Press Amsterdam: 28-49.

TUAN, Y.F., 1977. *Space and Place: the perspective of experience*. London: Arnold.

WHEATLEY, D., 1995. Cumulative Viewshed Analysis: A GIS-based method for investigating intervisibility, and its archaeological application. In Lock, G. and Stančič, Z. (eds,), *Archaeology and GIS: A European Perspective*. London: Routlege: 171-187.

WHEATLEY, D., GILLINGS, M., 2000. Vision, perception and GIS: developing enriched approaches to the study of archaeological visibility. In Lock, G. (ed.), *Beyond the Map: Archaeology and Spatial Technologies*. NATO Science Series, IOS Press Amsterdam: 1-28.

ZNANIECKI, F., 1938. Socjologiczne podstawy ekologii ludzkiej. *Ruch Prawniczy, Ekonomiczny i Socjologiczny* 18/1:87-113.

Interpreting Field Survey Results in the Light of Historic Relief Change: The Fogliano Beach Ridges (South Lazio, Italy)

Hendrik Feiken and Martijn van Leusen

RPC Project, Groningen Institute of Archaeology / N.W.O.

Poststraat 6 9712 ER Groningen, The Nederlands

e-mails: h.feiken@let.rug.nl, p.m.van.leusen@let.rug.nl

www.let.rug.nl/RPC/

Abstract

*This paper details the authors' feasibility study into the use of historic elevation data for the mapping and correction of biases in the archaeological record. Detailed elevation maps produced by the Italian Military Topographic Institute (IGM) in 1927/8, in preparation of large-scale land reclamation works (*Bonificá*) in the Agro Pontino, were digitised and compared with more recent commercially available DEMs produced by the IGM. After the removal of a number of extraneous causes of differences between the two DEMs, the remaining differences could be correlated both to the historic activity that took place during the Bonificà, and to the effects of intensive agricultural land use in the subsequent 70 years. A brief discussion of the results of field surveys conducted in the area in 1998/9 demonstrates the need to include historic relief change as a significant biasing factor in archaeological land use models.*

1. Background

The Pontine region, a low-lying and partly marshy area along the coast of south Latium, was taken up into the Roman power sphere in a slow process completed only midway through the 4th century BC, which is why the preceding period 500 - 350 BC is named "post-Archaic" rather than "early Republican" (Attema 1993). Much of it appears to have been marginal to the major political and economic developments of the time, and this translates itself into the relatively low density of surface ceramics reported in field surveys (Attema et al. in press, Van Leusen 1998). Low finds densities present us with particular interpretation problems because of the relatively large influence of biases in the type and amount of research conducted, the visibility of the surface, statistical effects, and geopedological circumstances (Van Leusen in press). It is the effect of the latter factor that we have attempted to study in more detail, using the landscape of ancient beach ridges around the Fogliano lagoon as our study area (figure 1). However, a second area, below Sezze on the margins of the Lepine mountains, was also studied for the same purpose, but was later dropped from the analysis because of problems with the quality of the available data.

This feasibility study into the use of historic elevation data for the mapping and correction of biases in the archaeological record was conducted as part of the Regional Pathways to Complexity (RPC) project[1] in the Pontine region (south Lazio, Italy). The general aims of the RPC project are, first, to understand indigenous versus externally-induced growth in complexity (especially urbanisation), and second, to conduct detailed surveys in marginal areas to understand the scope and nature of dynamics that are mostly known from urbanised sites. Specifically, the case study presented here is part of the project's methodological focus on GIS approaches to the detection of spatial patterning in the archaeological record. One reason for concentrating on the Fogliano area is that it has been subject to two major surveys in the last two decades (Voorrips et al. 1991, Attema et al. in press), and the surface record is therefore relatively well known; the other reason is that the Pontine plain as a whole was subject to major restructuring in the late 1920s and 1930s by the Italian government in a so-called *Bonificá*. These works have changed many parts of the landscape, and must therefore be taken into account if we are to interpret and understand our survey data. A detailed elevation survey of the Pontine plain was made in 1927 by the Italian Military Topographic Institute (IGM), allowing us to compare the post-World War II relief to the relief that was present before the Bonificá.

Before we proceed to describe the feasibility study itself, a brief overview of the settlement and land use history of the area is in order. The landscape of ancient beach ridges around the Lago di Fogliano must have attracted humans from earliest prehistory, and flint artefacts dating from the Middle Paleolithic onwards were found during the surveys. This material seems to concentrate mostly along the banks of the larger water courses and the lake itself, and one can imagine the water rich environment being very well suited for fishing and fowling. However, throughout this period and into the late 2nd millennium BC human presence would appear to have been quite rare and impermanent, the earliest indications for agricultural activity and the use of ceramics dating to the Bronze Age.[2] Figure 2 shows the probable settlement locations for the protohistoric period (running from the Bronze Age up to the beginnings of Roman influence in the Pontine region, around 500 BC) as derived from the finds densities of the RPC survey, on the background of the 1928 DEM. The absolute number of finds from this period tends to be very low for most of the area surveyed - on the order of 1 to 5 finds per hectare. Settlement in this period seems to be concentrated on relatively well-drained capes and banks along the larger streams, where access to natural resources would have been easiest and preconditions for paleotechnic agriculture were positive (Kamermans 1993: 100-4; see also Attema et al. in press). By the end of this period (6th century BC), the number of settlements begins to grow.

Figure 3 shows the post-Archaic, Roman Republican, and early Imperial sites identified by the RPC survey, again covering a period of approximately one thousand years (500 BC to AD 500). The number of settlements has greatly increased, but this occurs mostly between 200 BC and AD 200 - the late Republican and

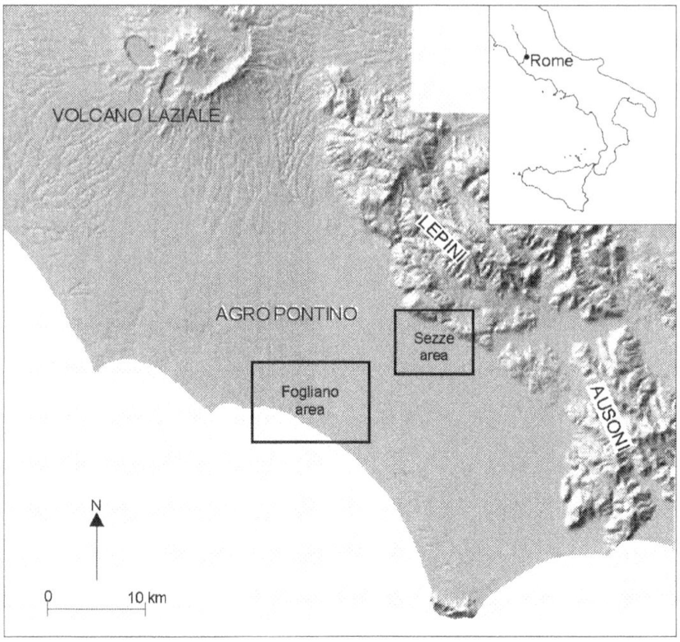

Figure 1: Location of the Fogliano study area within the Pontine region (Lazio, Italy).

Fogliano proto

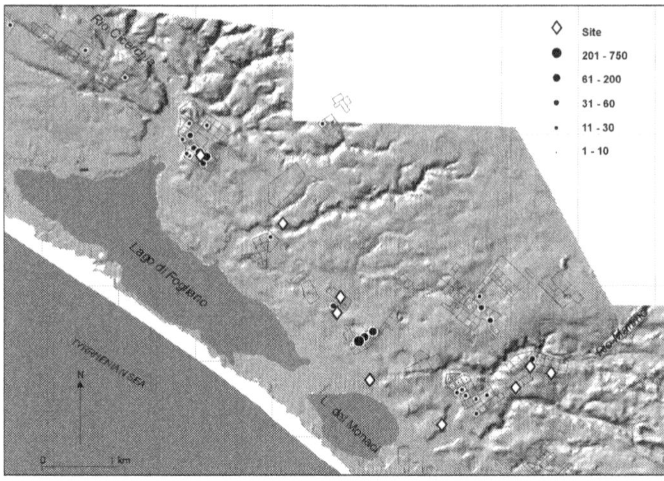

Figure 2: Results of the 1998/9 RPC survey in the Fogliano area, protohistoric period. Find densities per hectare corrected for surface visibility. Background: 1927 shaded relief map and km grid.

early Imperial period. Settlement is concentrated on the relatively level and agriculturally usable area to the east of the Lago di Fogliano, where a late Republican village seems to have sprung up. We can tentatively relate this development to the inclusion of the area in the wider economy of Roman society. Roman agricultural technology (drainage of lower-lying areas, heavy ploughs) enables more land to be farmed; Roman hydrocultural technology (regulation of lake levels) enables commercial fisheries to be established in the lagoons; Roman roads and markets enable the commercial exploitation of the clay beds along the nearby Astura river; from the 1st century BC rich Romans even built their summer palaces along the banks of the coastal lagoons of south Lazio. All this activity makes it likely that the growing number of small farms became dependent on a few large specialised rural villae. However, this system collapses from the 2nd century AD onwards, as a wetter climate[3] led to the return of marshy conditions and the expanding Roman empire found its supplies elsewhere.

2. Tracking Historic Relief Change

For the purposes of studying historic relief change in the Fogliano area, we have used historic maps of land use and land form. We know from historical research that the landscape was little changed since the Middle Ages, and historic maps dating to the 17th century show us an approximation of the landscape as it was in the late Roman period. Such maps may even be a better guide to interpreting the proto- and early historic archaeological record than the modern ones, which were made after major restructuring of the region during the Bonificá.

When we digitised and compared a very detailed digital elevation model (DEM) from the 1:5,000 scale maps prepared during the Bonificá (figure 4; IGM 1927) with a commercially available DEM of 1 arc second (~25 m) resolution (figure 5),[4] some major differences became immediately apparent. Whilst obtaining a map of the differences between the two mappings was easy – all we needed to do was to subtract the two DEMs from each other – most of our study was concerned with the identification and removal of various mapping errors obscuring the "real" relief changes that might have taken place during and following the Bonificá. Since it is usually one of the most important layers in a regional archaeological GIS, deficiencies in the DEM can cripple much analysis. These sources of error, and our attempts at removing them, are discussed in section 3; our interpretation of the archaeological evidence in the light of the "cleaned" elevation data follows in section 4.

3. Extraneous sources of differences between the two DEMs

The first point we need to make is that any DEM is a *model*, that is, a simplified version of reality. The type and amount of simplification that can be supported in any analysis depends on the questions asked, and any analysis relying on DEMs should be explicit about its limitations. Secondly, the *comparison* of two DEMs forces us to be even more precise in our description of the data. Differences between two DEMs may be due to real changes in the morphology of the terrain they represent, or to errors committed in the process of producing the digital elevation data, or to the precision with which the data were recorded, or to the use of different projection parameters and coordinate systems. The following list summarises seven distinct sources of differences we discovered between the two DEMs, which are NOT due to an actual (real) change in the land form.

Cause of difference/discussion

Scale:

The scales of the topographic map sheets, from which our two DEMs are derived, differ. The 1927 DEM derives from a 1:5,000 scale map; the more recent DEM derives from a 1:25,000 scale map. The reduced mapping scale implies that features are simplified and smaller features may even be lost. When comparing DEMs of different scales, such features will stand out as differences.

Resolution:

Both the horizontal (X and Y) and the vertical (Z) resolution of our two DEMs differ. (1) The horizontal resolution of the 1927

Fogliano roman

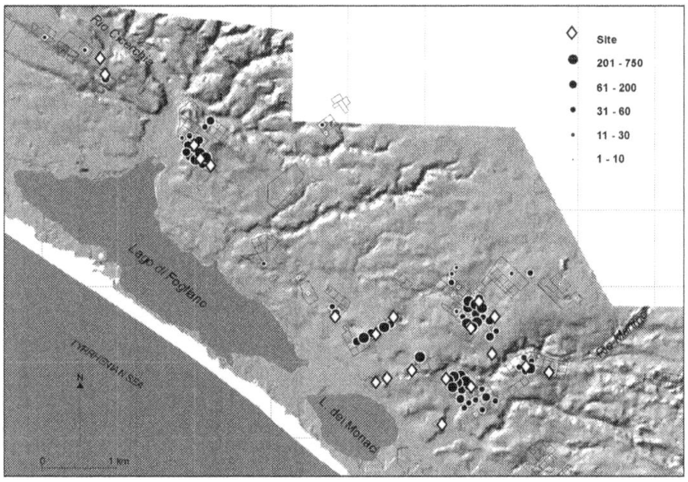

Figure 3: Results of the 1998/9 RPC survey in the Fogliano area, Roman period. Find densities per hectare corrected for surface visibility. Background: 1927 shaded relief map and km grid.

DEM is approximately 5 by 5 m; that of the more recent DEM is 1 arc second (approximately 25 by 31 m at the latitude of Italy). This means that an area represented in the latter by a single elevation value, is represented by approximately 30 (5 by 6) values in the former. Unless that area happens to be level, most of those values will be different from the single elevation value provided by the low resolution DEM – they will lower downslope, and higher upslope (see figure 6a), giving rise to the "banded" appearance in some areas of the raw differences map (figure 7a). The amplitude of the differences is related to the terrain slope and the difference in resolution of the DEMs: at any given percent of slope, the error is directly proportional to that difference. Because the error is systematic, we can devise a formula for deducing it: $E = \delta R * S / 2$ (where E = Error; δR = resolution difference; S = percent slope). (2) The vertical resolution of the 1927 DEM is 0.1 m; that of the more recent DEM is 1 m. Since the latter must round any values to the nearest whole meter this may lead to differences of up to 0.5 meters compared to the former.

Mapping errors:

Mapping errors may (and will) occur both during the original recording of elevation measurements, and during the subsequent cartographic process. Proper control procedures are needed to minimise the occurrence and effects of such errors. While we could not check the quality of the primary cartographic data, we were able to compare both DEMs with the elevation data in the original map sheets, and found some major mapping errors. One of these can be seen in figures 4 and 5 (area marked 'A') where a small valley that existed in the 1928 DEM had mysteriously changed into a hill by the 1950s. Whilst such obvious errors can be found and corrected fairly easily, there certainly remain many less obvious mapping errors in our DEMs - a very worrying situation...

Interpolation:

Our two DEMs were created by interpolation from digitised contour lines and elevation points, and this has led to the introduction of three different kinds of errors in the data. Since many properties of the resulting DEM are determined by the interpolation method (e.g., inverse distance weighting or thin plate splining; see Hageman and Bennett 2000), it is important to study its effects: (1) While the interpolation method used in creating the 1927 DEM is known (a "flood fill" algorithm provided by GRASS GIS), no such information was available for the more recent DEM. However, it appears from the data that some sort of inverse distance weighting using a low number of data points was used, and this has led to a large number of visible artefacts in the latter. Since the two DEMs were created using different interpolation algorithms, the interpolated values will also differ; (2) Some softwares are unable to handle '0' (zero) as a real elevation value, in which case such values are ignored during interpolation. The method used to create the more recent DEM apparently suffers from this error which, for example, has caused the present dunes to "smear out" across the lagoons because the water's edge was digitised as a zero contour (see figures 4 and 5, area marked 'B'). The elevations here are clearly in error, and the area cannot be used in any further analysis; (3) Many interpolation algorithms that start from an input of digitised contour lines (including the ones used to create both our DEMs) do not resample the input values. Since contour lines always represent cardinal elevations, these end up being overrepresented in the resulting DEMs - the map histogram has a "saw tooth" appearance, with the distance between "teeth" depending on the vertical resolution of the original map. While

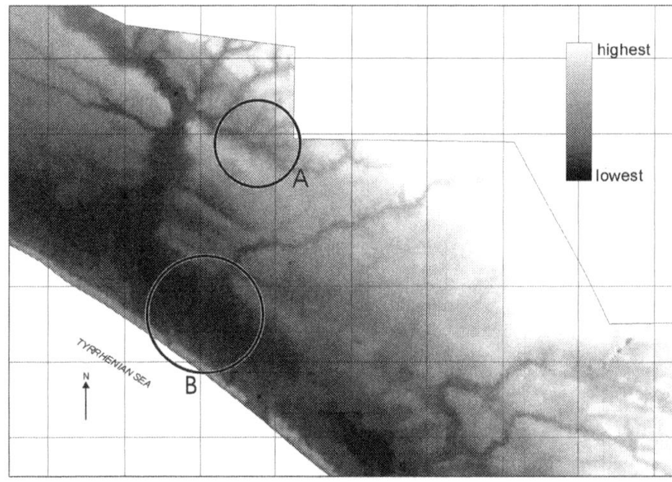

Figure 4: Digital Elevation Models of the Fogliano area, derived from the 1927 1:5,000 IGM maps.

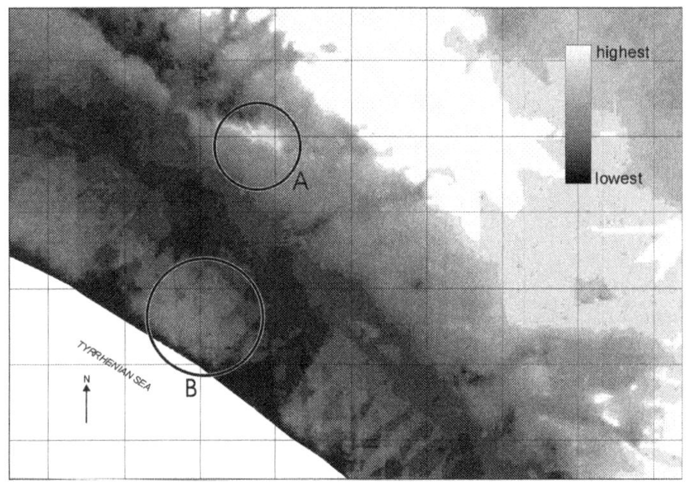

Figure 5: Digital Elevation Models of the Fogliano area, from 1" numeric cartography derived from the 1:25,000 IGM topographic map series (permission IGM 26/05/98, no. 4805). A: digitising error; B: interpolation error.

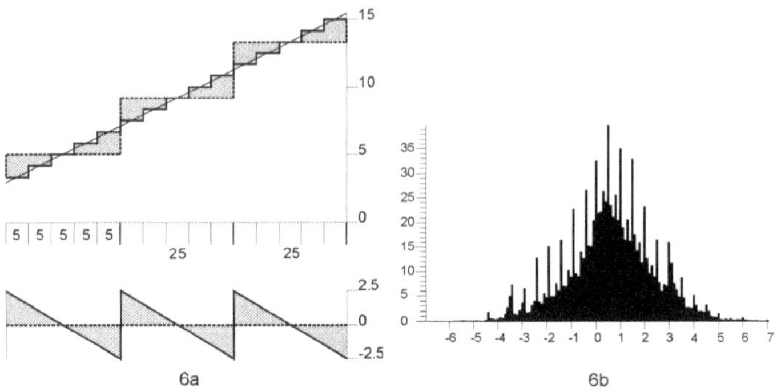

Figure 6: (a) On a slope represented by the thin continuous line, the differences between two DEMs of differing horizontal resolution (here, 5 and 25 m) alternate between positive and negative (shaded areas). The maximum difference depends on the terrain slope and the difference between the resolutions of the two DEMs. All units in m. (b) The histogram of a typical DEM shows the systematic "saw tooth" shape caused by overrepresentation of cardinal values. Horizontal axis: elevation in m; vertical axis: number of cells x 1000.

the 1927 DEM used contour lines every 0.5 m, the more recent DEM used contour lines every 5 m. In terms of accuracy, this means that values are only accurate to half the distance between contours, i.e. to 2.5 m in the case of the more recent DEM (see figure 6b).

Datum shift:

One of the parameters of the coordinate system used by both DEMs is its datum (origin). The horizontal datum of Italian topographic maps was moved by 2.53" (about 70 m) on one occasion; we could not obtain any information regarding changes in the vertical datum used in either DEM. We attempted to compensate for this by comparing the recorded elevations of relatively stable landscape features (e.g., buildings) in both DEMs. No vertical datum shift could be deduced from these. Left uncorrected, comparison of the two DEMs using different datums would have led to measurement errors especially in areas of steep slope; in this case, we were able to correct the horizontal datum shift.

The aim of identifying all these errors is, of course, to be able to correct them. Mapping errors (at least those that were discovered by us) were corrected by re-digitising and interpolating from contour lines on the original map sheets, and replacing the incorrect sections with these new data. Other areas of faulty data could only be "corrected" by removing them entirely from the analysis (using "masking" functions in the GIS). But the most interesting type of error was the one that could be estimated or calculated, because these errors could be corrected by first mapping them and then using them as "noise filters" when interpreting the differences between the two DEMs. Thus, the error formula referred to under "Resolution" in the list given above was used to calculate a map layer of the estimated amplitude and sign of the error, given the known slope and resolution of the more recent DEM. We then applied the correction by subtracting this calculated error from the observed difference between the two DEMs.

Figure 7 shows the differences between the 1928 and 1950s DEMs, both (a) before and (b) after the corrections were applied. The general effect is one of producing a much less extravagant map, whose values do not exceed 5 meters. We were now ready to evaluate these differences in terms of natural and human processes that occurred in the area, rather than in terms of data error.

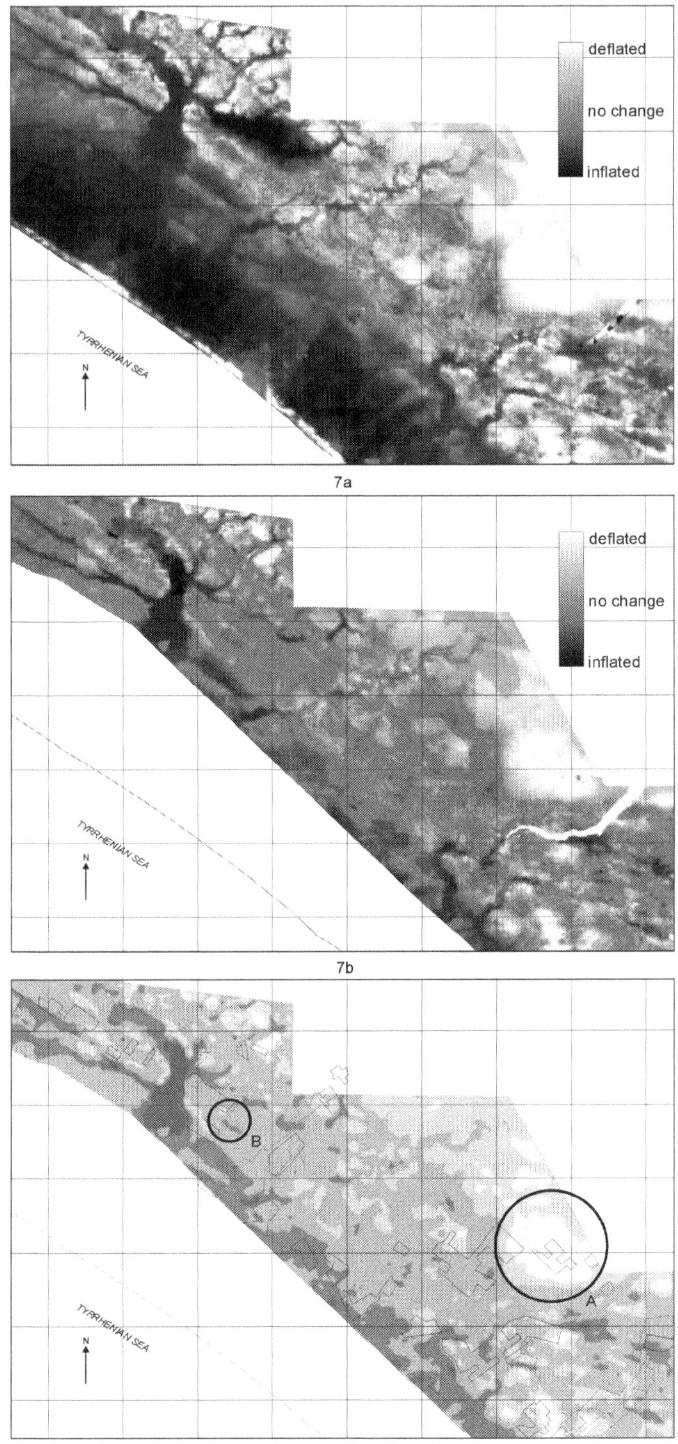

Figure 7: Interpreting the differences between the two DEMs. (a) uncorrected; (b) corrected; (c) simplified corrected differences between the two DEMs, with the RPC (continuous lines) and APP fieldwork areas (dashed lines) overlaid. A: area where large-scale sand removal may have taken place; B: area where farmer has levelled his field.

4. Interpreting the evidence

Any real differences between the two DEMs might be the result of either natural or human causes, or to a combination of both. In the simplified corrected map of differences between the two DEMs (figure 7c), light zones have become up to 7 meters lower, dark zones up to 6 meters higher since 1927, and these zones may therefore betray works carried out during the Bonificá. However, we must also consider other processes such as plough-induced erosion and settling of zones with a clayey subsoil.

With the help of a soil scientist familiar with the region we began by evaluating the potential size of the effects of natural causes. We could see that the lower-lying areas (especially the small valleys) had all gotten considerably higher since the late 1920s, whereas some areas to the north-east had been reduced by 3 meters or more. The former is probably due both to intentional dumping of material into marshy places and to the effects of ploughing the slopes of the adjacent ancient beach ridges. The latter may well be partly caused by soil settling after the construction of the canal system during the Bonificá,[5] but it is unlikely that a difference of more than one meter could be explained that way. We have some historic evidence that a top layer of aeolian sands may have been removed from some terrains in order to provide material for the founding of the new capital of the area, at Latina, and sand and gravel pits were still in use in the northwestern part of the study area in the early 1980s and recently (Sevink et al. 1984:28); archive research may reveal further evidence as to which terrains were historically involved in such activities.

Certainly many parts of the study area were levelled in order to facilitate access and workability for modern farming equipment, but such works were not often recorded (with the exception of the local, often marshy, hollows which were mapped on the 1920s land cover maps), so we have to rely on the recollections of the local farmers and on the occasional evidence from soil cores. Raised terrains also occur along the banks of the coastal lagoons, and we know from historical sources that these were deepened and the material used to raise the surrounding land enough to prevent the formation of marshes. In fact, this was one of the major aims of the Bonificá – to reclaim the Pontine marshes for agriculture by destroying the habitat of the malaria mosquito.

Figure 7c shows that some of the fields surveyed in 1998/1999 by the RPC project (and in the 1980s by the Agro Pontino Project) lie in zones affected by serious deflation or inflation. For the fields surveyed by the RPC project, the relation between the average amount of change in the elevation and the average density of ceramics found (see figure 8) gives rise to the suspicion that limited deflation (of ca. 0.5 m) is helpful in bringing archaeological material to the surface, while strong deflation (more than 0.8 m) destroys the record altogether. Conversely, even weak inflation tends to hide any archaeological material that might be present on the surface. A closer look at the observations made in the field confirms the general picture, and suggests ways in which our archaeological intrepretations may be improved:

- Areas with strong deflation (over 1.5 m) were observed during the survey to have no soil profile at all and to be archaeologically sterile (area marked 'A'); Sevink et al., in their account of the soils of the area (1984:30), suggest that subsoil was brought from elsewhere, but this is flatly contradicted by the evidence for deflation from our study. It is therefore quite possible that the top layer of soil, including any archaeological remains, was removed from this area and used elsewhere in construction work.[6] The late Republican village mentioned in the introduction may therefore have extended further to the northwest than the results of our survey suggested.

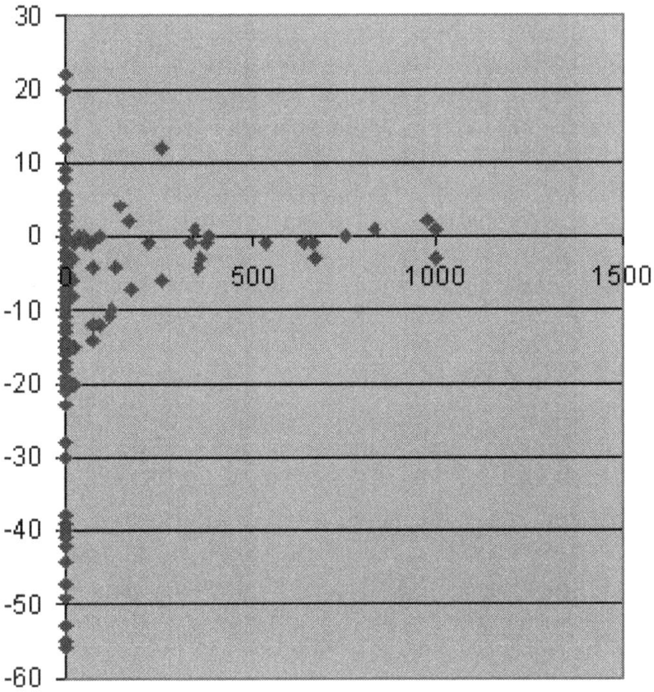

Figure 8: Scatterplot of average deflation/inflation (in dm) (vertical axis) versus average find density per surveyed unit (horizontal axis) of the Fogliano survey (Attema et al. in press).

- The reduction of local relief through ploughing or levelling can be traced in many fields:

1. Areas that were observed during the survey to have an unusual soil colour or material often correlate with small hollows and valleys mapped in 1927/8 and afterwards levelled;
2. The raised banks of canals dug during the Bonificá on occasion also contain archaeological material, which must presumably have come from the immediately adjoining stretch of the canal;
3. We have several examples where our study confirms that sites located on hillocks were, along with the soil, "smeared out" across the surrounding fields by tillage;
4. One hollow, which the owner informed us he had filled in using soil (containing Roman ceramics and building materials) from elsewhere in the same field, coincides exactly with a patch of inflated soil (area marked 'B').

These observations confirm the importance of landscape history as a factor biasing the results of field surveys, and the need for a structured "source criticism" so that we can trace and correct such biases.

5. Conclusions

This case study has demonstrated the feasibility of employing GIS to extract and interpret land form changes from historic elevation and land cover data. Although it is generally recognised that the interpretation of survey results requires knowledge of local geopedology and landscape history, workers have not yet gone very far with this approach. The present case study shows that, provided certain requirements regarding data quality are met, historic elevation data can be used to track one of the most important factors biasing the results of field surveys today – changes in land form caused by human agricultural and construction activity. By comparing historic to recent elevation data, using GIS, maps can be made of the location and approximate amount of deflation/ inflation influencing the presence and visibility of the archaeological record. Such maps can be used both to target future surveys to areas that are likely to have survived undisturbed, and to re-interpret the results of older surveys.

References

ATTEMA, P.A.J., 1993. *An Archaeological survey in the Pontine Region, A contribution to the early settlement history of South Lazio (900-100 BC)*, Groningen (PhD thesis).

ATTEMA, P.A.J., BURGERS, G.J., KLEIBRINK, M. and YNTEMA, D.G., 1998. Case Studies in Indigenous Developments in Early Italian Centralisation and Urbanisation: a Dutch Perspective. In *European Journal of Archaeology*, Vol. I:326-381.

ATTEMA, P.A.J., van JOOLEN, E. and van LEUSEN, P.M., in press. A Marginal Landscape: Field work on the beach ridge complex near Fogliano (South Lazio), *Palaeohistoria* 41/ 42.

HAGEMAN, J.B. and BENNETT, D.A., 2000. Construction of Digital Elevation Models for Archaeological Applications. In Westcott, K.L. and Brandon, R.J. (eds.), *Practical Applications of GIS for Archaeologists*: 114-127. London: Taylor & Francis.

IGM 1927 (Year 5). *Bonificazione Pontina*. Topographic map in scale 1:5,000, published by the Istituto Geografico Militare on behalf of the Consorzio della Bonificazione Pontina. Firenze: IGM.

IGM 1996. Catalogo 1996. Produzione e Prestazioni dell'IGM. Firenze: IGM.

KAMERMANS, H., 1993. *Archeologie en landevaluatie in de Agro Pontino (Lazio, Italië)*. Unpublished PhD thesis, University of Amsterdam.

SEVNIK, J., REMMELZWAAL, A. and SPAARGAREN, O., 1984. *The soils of southern Lazio and adjacent Campania* (Reports of the Fysisch Geografisch en Bodemkundig Laboratorium 38). Amsterdam: University of Amsterdam.

Van LEUSEN, P.M., 1998. Archaic Settlement and Early Roman Colonisation of the Lepine Foothills. In *Assemblage* 4 (1998), http://www.shef.ac.uk/~assem/4/.

Van LEUSEN, P.M., in press. Marginal Landscapes: Survey and interpretation biases in low finds density regions in Italy. In Gojda, M. and Darvill, T. (eds.), *Landscape Archaeology at EAA 1999* (BAR).

VEENMAN, F.A., 1996. *Landgebruik in de Pontijnse regio (1900-1100 BC). Een paleobotanisch en archeologisch onderzoek.* Unpublished MA thesis, University of Groningen.

VOORRIPS, A. LOVING, S.H. and KAMERMANS, H., 1991. The Agro Pontino Survey Project. *Studies in Prae- and Protohistorie* 6. Amsterdam: University of Amsterdam.

Notes

1 The RPC project was conducted jointly from 1997 to 2001 by the Groningen Institute of Archaeology and the Archaeological Institute of the Free University of Amsterdam. It studies protohistoric landscape and settlement dynamics in three Italian regions - the Pontine region, the Salento Isthmus, and the Sibaritide. Processes of centralisation, early urbanisation and colonisation are its main themes (Attema et al. 1998).

2 The evidence for this comes from an unpublished pollen core from Fogliano (pers. comm. E. van Joolen), and from the unpublished finds database of the Agro Pontino Project (Voorrips et al. 1991).

3 As deduced from pollen cores analysed by Haagsma - cited by Attema (1993:253) and Veenman (1996:59).

4 A matrix of numeric data, containing one value for each 1" by 1" square, and based on the digitised contour lines and elevation points of the 25 and 25V map series produced in the 1940s and 1950s at a scale of 1:25,000 (IGM 1996:13 and Table 42).

5 Evidence of such soil settling is plentiful in the Pontine region. The approaches to many of the small bridges built in the late 1920s have had to be lengthened, and the concrete bottom of the channel below them broken through, because the surrounding land had sunk 70 cm or more.

6 It is known from historical records that large amounts of soil were brought in to found the new capital of the Agro Pontino at Latina.

Understanding the Neolithic Landscape of the Carnac Region: A GIS Approach

Corinne Roughley

Department of Archaeology, University of Cambridge

Downing Site, Cambridge, CB2 3DZ, United Kingdom

e-mail: cfa1001@cam.ac.uk

Abstract

The Neolithic landscape of southern Morbihan is archaeologically very rich, but topographically subtle. This combination provides an interesting and important challenge for GIS approaches. This paper discusses the applicability of a range of techniques for exploring and analysing the relationships between monuments and their locales. Various of approaches to the description of the shape of the terrain are investigated. The views of and from monuments are affected by landform over a range of scales. This forms an important part of the experience of approaching, being in, and leaving a locale.

Key words: Neolithic, monument, Carnac, GIS, topography, viewshed

1. Introduction

The Neolithic monuments of the Carnac region of Brittany are the focus of an internationally important archaeological landscape. For a study of the monument locations and their inter-relationship, the region poses considerable challenges for traditional Geographic Information Systems (GIS) approaches due to the gently undulating nature of the topography and the large number of monuments. An appreciation of landform and context is crucial to understanding this landscape and the place of the monuments within it.

Simple GIS topographic measures (for example, the traditional elevation, slope and aspect) are inadequate for describing a monument location. In this paper, methods for characterizing the shape of the terrain are presented. When the positions of the Neolithic monuments are compared with other non-site locations, they are found to occupy specific and definable types of location.

Landform shape has an important effect on the perception and experiencing of a location. The approach to a monument by a person needs to be considered. Views to and from each monument are investigated at a range of scales, and the influence of visibility on site location examined. The relationship between the orientation and size of monuments and their local landscape context has also been investigated. In this approach, the inter-correlation of visibility measures with other topographic characteristics needs to be understood. However, the effectiveness of any study such as this rests on the resolution and accuracy of the digital terrain model in its representation of the Neolithic landscape.

1.1. Archaeology of Morbihan

The Neolithic archaeology (approximately 4500 to 2500 BC) of southern Morbihan has been studied extensively over the last 150 years. Around 150 monuments have been excavated in the coastal area alone (although few using modern excavation methods) and the famous Carnac Alignments have been much debated. There are also a large number of isolated standing stones, smaller alignments and stone circles. There is little settlement evidence known in the area.

The precise chronological relationships between the monuments are still problematic (Boujot and Cassen 1993, Le Roux 1999), but it is likely that the earthen long mounds (fr. tertres tumulaires) and decorated menhirs are amongst the earliest monuments in the region. The passage graves (fr. dolmens à couloirs) were probably built over a long period of time, with some of the more complex forms and the gallery graves (fr. allées couvertes) belonging to the later Neolithic period. The dating of the alignments and Carnac Mounds remains controversial (Boujot and Cassen 1993, Le Roux 1999), but the arguments need not be elaborated further here.

The Neolithic monuments of Morbihan have, in the past, been studied in isolation as individual sites or as part of a typological classification, rather than as part of a complex landscape. However, the importance of the locales chosen by the Neolithic builders has long been recognized, including mention in pilot manuals (Lecornec 1994) and folk history. The relationships between monuments are complex, with locales being re-interpreted by successive generations. Recent work by Boujot and Cassen has begun to incorporate an appreciation of landscape into studies of the Neolithic of this region (e.g. Boujot et al. 1998). This paper is part of a wider-ranging attempt to understand the complex landscape which was created and inhabited by Neolithic people in southern Morbihan, and to assess the ability of GIS to contribute to that search.

The combination of extensive excavation and recent survey work (Desdoigts 1978, Boujot and Cassen 1995, Gouezin 1995) has enabled a substantial database of site information to be constructed containing over 1000 records, of which about half of the sites have coordinates. The expansion of tourism in the region has led to the destruction of some sites; others are inaccessible in woodland or are in private ownership.

For the purposes of this analysis, sites have been categorized into earthen long mounds, passage graves, later passage grave forms (which for the purposes of this study includes lateral entrance graves angled gallery graves (fr. allées coudées), and gallery graves, which are not technically passage graves), menhirs, and alignments, cromlechs and other groupings of multiple standing stones. There is one other class, unidentified tumuli, which includes mounds which are probably of Neolithic date, that have either not been excavated or adequately recorded and are now damaged or destroyed.

1.2. The study area

This paper considers the coastal area of Southern Morbihan which includes the Carnac area and the Golfe du Morbihan. This is part of a larger study area covering some 2000 square kilometres, including a dissected plateau which poses different analytical and interpretative challenges. The topography of the coastal area is generally very subtle, with gently rolling small hills and river valleys. In the north-east corner of the coastal study area the land rises up to 120 m. The Golfe has been flooded by rising sea level (approximately 5 m) in the intervening four thousand years; in the Neolithic it was probably a river valley and estuary. The Neolithic monuments appear to be divided between the two areas, with fewer sites in between. However, the lack of detailed survey work directed at this area may be contributing to this impression and biasing the data.

2. Approaches to describing landscapes

Landscapes have been the focus of considerable archaeological debate in recent years. Post-processual approaches have become widespread and challenge the way in which we consider and analyse landscapes. There has been an increasing emphasis on subjective experience (with quotations from phenomenological philosophers), with narrative replacing more traditional analysis techniques (e.g. Edmonds 1999).

The relationship between GIS approaches and landscape theory has been questioned (Gaffney and van Leusen 1995, Llobera 1996). GIS has been used by many archaeologists in a simplistic way, which has led to an impression among many landscape archaeologists that GIS does not have a part to play in a wider, theoretically-aware research agenda. Information about the relationship between site locations and topographic characteristics has formed a large part of the use of GIS in inter-site studies. The use of topographic measures and spatial distributions to describe some of the characteristics of landscapes must be undertaken with care if environmentally and economically deterministic interpretations are to be avoided, and the complexity of the landscape preserved in the final interpretations. A more data-oriented perspective, in which statements about the characteristics of locales are questioned and evaluated, does have an important role in landscape archaeology. The ability to assess trends (and those sites which do not fit the trends) is an important step towards identifying factors which may have influenced the choices that Neolithic people made.

2.1. Creation of the terrain model

The Digital Elevation Model (DEM) used for this research was created from digitised 5 m contours (with 2.5 m contours in the flatter areas), from the 1:25,000 maps. The bathymetric data are taken from nineteenth century British 1:10,000 Admiralty Charts and the 1:25,000 maps. A horizontal resolution of 25 m was used throughout to ensure comparability of results. The prehistoric sea level is estimated to be about 5 meters below modern sea level, based on sea level modelling (Lambeck 1997).

2.1.2. Simple topographic characteristics

One of the most frequent uses of GIS in inter-site studies has been for comparing elevation, aspect and slope values for sites and for the total study area. This has sometimes been interpreted as providing a reliable assessment of the relationship between sites of different periods and topography. As part of the assessment of the potential contribution of GIS to the study of the Neolithic landscape of the Morbihan, these methods were applied to the study area and the results critiqued. The ArcInfo SLOPE and ASPECT commands were used, rather than the output from CURVATURE, as these are probably more accurate (Wise 1998).

Graphs of elevation, aspect and slope showed little distinction between site types or between site and non-site locations. Although there may be a slight trend towards avoidance of the steepest slopes, a preference for flat ground and the absence of sites from the areas of greatest elevation, there is little of interpretative value here. Two-sampled *t*-tests fail to find any significant differences between site types and between sites and the total area.

The lack of strong relationships between sites and the topographic characteristics analysed above may be due to one (or more) of three possibilities. The first is that the DEM used is too coarse, or insufficiently accurate, to adequately represent such subtle terrain. Work is continuing on the development of a high-resolution DEM from stereo aerial photographs using photogrammetry. Secondly, there may indeed be no relationship in this landscape between site location and topography. The topography is slight, and there are only a few areas in which sites could not be built (e.g. due to too steep a slope or marshy ground). However, accepting this conclusion would be to limit the role of topography in landscape interpretation to an environmentally deterministic one. Topography has an important contribution to the experience and interpretation of landscape. Natural features acquire names and significance. Terrain affects what is seen, it is walked over, changed by monument construction, and reinterpreted. The third hypothesis is that the measures used were not adequate to characterize the subtle topography of the area.

3. Alternative approaches

The unsurprising failure of the simple topographic measures to provide any significant data about site locations led to the search for other, more appropriate, approaches to the characterization of topography. The local topographic features need to be discerned, rather than disappearing into the broader-scale trends.

As the first step towards discerning what a more appropriate characterization might be, many of the sites were visited with photographs taken and verbal descriptions made of the locations of the sites. An impression of the kind of locations in which monuments were located was slowly built up in a qualitative fashion. It included, for passage graves, location on small hills or protruding areas of land. Monuments appeared to be near, but often not at the top of these features. Passage graves were often on the edge of a slightly flatter area, at the top of a steeper slope. Other site types were thought to be associated with different landforms, for example small alignments ran at right angles to the major slope direction.

These descriptive impressions are highly subjective. It is easier to include a location in a category once a trend has been identified. A more objective and quantifiable approach is needed to assess

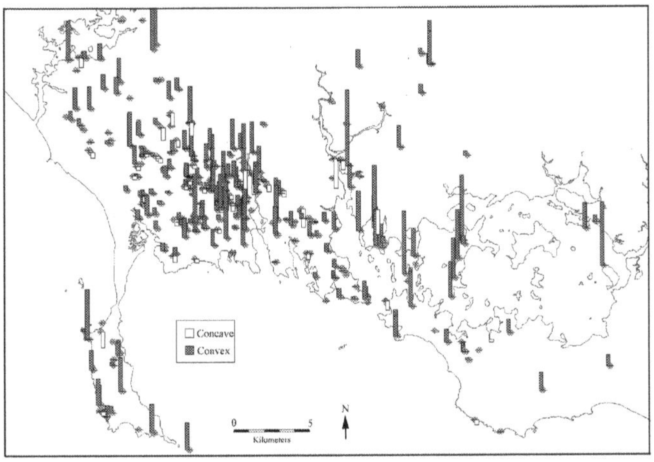

Figure 1: Profile curvature.

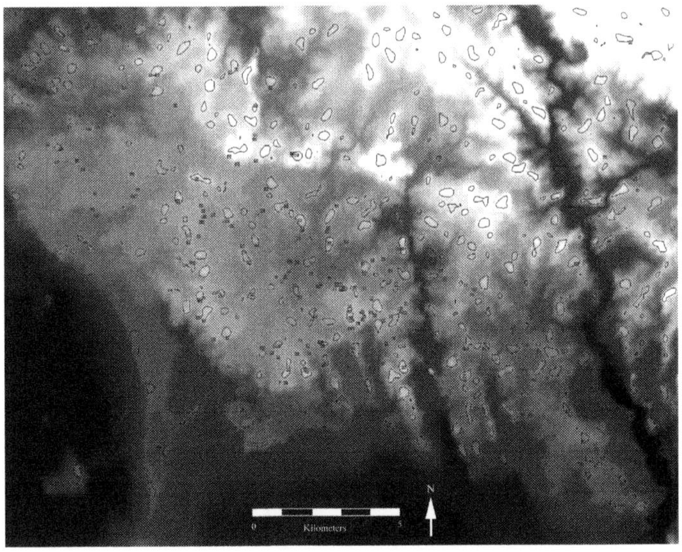

Figure 2: Location of hillocks (outlined in black) and sites.

the accuracy of impressions gathered in the field, which will reliably identify which sites fit a hypothesis, and the probability that such a relationship between site location and topographic characteristics could not have occurred by chance.

3.1. Profiles

To explore whether the DEM showed the characteristics observed in the field, ArcInfo's SURFACEPROFILE function was used. This function draws specified profiles through the DEM at a range of scales. When profiles 1 km in length were drawn through the locations of passage graves, both in the orientation of the site and at right angles to it, many of the sites were seen to be on or near local maxima in the profiles.

Although this method confirmed that features observed in the field are present in the DEM, and do appear to be associated with the monuments, it is difficult to analyse these profiles quantitatively. The profiles are also limited to the selected directions. At an exploratory stage, the interactive facility makes this an easily accessible tool. Creating the standard input lines necessary for analysis is difficult to automate in ArcInfo (a C-program was written for this purpose). Another problem with the technique is that it only describes changes in slope. There is no counterpart for changes in aspect, which form an important part of the shape of a location.

3.2. Curvature

The use of curvature as a characteristic of landform is well established within geomorphology (Wise 1998). Curvature is normally measured in two ways: profile curvature, the change in slope, and plan curvature, the change in aspect (see Wood 1996 for other measures). Profile curvature is normally measured in the down-slope direction, and this convention is followed here, rather than ESRI's description of the up-slope measure.

When the curvature of the DEM was calculated and the values for the locations of each site extracted, it was found that almost all site locations had convex profile (see figure 1) and plan curvature. Many of the sites have a very strongly convex location for both profile and plan curvature when compared with the range of values for the DEM as a whole. The high proportion of convex locations of the monuments can be contrasted with a 50 : 50 convex : concave ratio for the landscape as a whole. This suggests that there may be some quantitative basis to the observations in the field.

Statistical analysis of the results obtained using the curvature algorithm is problematic. The curvature algorithm is highly resolution-dependent. Small changes in pixel size resulted in large changes in output. The choice of interpolator also affects the results. The TOPOGRID interpolator was used here, as it has been shown to provide more accurate results than linear or quintic interpolation (Wise 1998). Comparison with a quintic lattice confirmed the importance of interpolator choice. Errors within the DEM are enhanced using this measurement, as curvature is the second derivative of the surface. Using a low-pass filter on the surface prior to calculating the curvature improves the results (Wise 1998), reducing the impact of erroneous data points. Perhaps most importantly, the curvature values calculated here is only a localized measure of landform (see Wood 1996 for an alternative quadratic scaling method). The algorithm does not capture the overall shape of a landscape feature, being simply a measurement over a 3 x 3 kernel.

Landform characterization in geomorphology has been dominated by algorithms aimed at hydrological modelling. There are few methods available which are suitable for the subtle, rounded landscape of southern Morbihan. The use of a simple kernel-based extraction of peaks and local maxima (e.g. ArcInfo's VIP routine) fails to identify most of the locations of interest.

3.3. Hills

The automated identification of the small hills noticed in the field was considered first. The method used is contour-based. It extracts areas which are less than a specified area (here 100,000 m^2) which are at least 0.5 m higher than the surrounding land on all sides. This differs from buffering a kernel-based extraction of peaks to a particular distance in that the shape of the area defined by the hill-finding procedure reflects the actual land surface rather than the distance to the highest point.

The areas which have been identified by the algorithm are shown in figure 2. Although many sites are located in the identified areas (31 %), there are a significant number which are not. When the results were plotted by site-type, it was observed that over half

Figure 3: Location of promontories (outlined in black), hillocks (outlined in white) and sites.

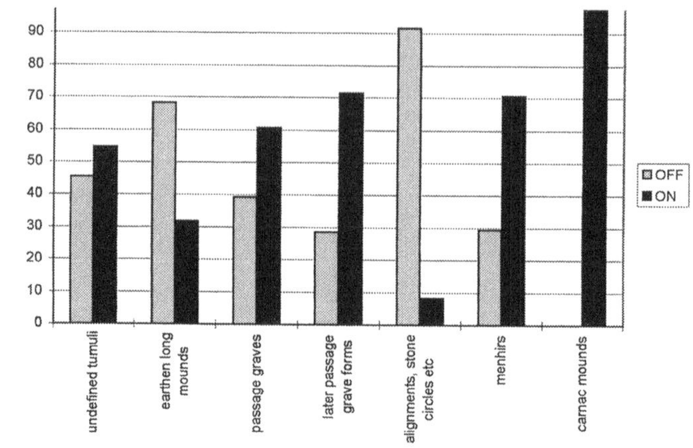

Figure 4: Percentage of sites by type on hillock or promontory.

the passage graves have locations on small hills, which is significantly more than would be expected from the proportion of the total area which has been designated as hilltop (7 %). However, other site-types, particularly the alignments, do not appear to differ from the expected proportion for a randomly distributed set of points.

3.4. Promontories

It was noticed that many sites were located on areas of prominent ground which were not being identified by the hill-finding algorithm. These areas were not higher than their surroundings on all sides, and formed small bulges or promontories (see figure 3). Promontories can be considered as hills which have their shallowest downward slope parallel to the x - y plane. They thus fail to be higher than their surroundings on all sides. By tilting the surface, the slope of the shallowest side is increased, and the area becomes higher than all surrounding locations. In order to extract all promontories, the surface must be tilted in the orientation of each promontory. In this example, 8 tilt directions were used.

For a gently rolling landscape such as this, a tilt angle of one or two degrees appears to be sufficient. Further work is needed to explore fully the relationship between tilt angle and the landscape forms extracted. Software is being written to rotate and tilt the surface, rather than exporting the DEM to other software, rotating and tilting it, and then re-importing to ArcInfo, which is time consuming and not possible to automate.

Preliminary results on a small area of the DEM are encouraging (see figure 4). The proportion of sites which are included in the defined area increased from 39 % to 50 %. 60 % of the passage graves, 71 % of the later passage graves and 70 % of the menhirs have locations on either a hill or a promontory. However, the alignments, cromlechs and other stone settings have a lower than expected proportion on hills or promontories (8 %, compared with 18 % of the total area). The earthen long mounds remain less distinct, having a higher proportion than the background, but with only a third of the locations occurring on a hill or promontory.

The results suggest that these measures may be enabling the locations of some site-types to be distinguished from both the total landscape and the locations of other site-types. However, the sample has been much reduced, and the method needs further refinement. The promontory finding procedure extracts only 6 % of the area remaining after hills have been identified, and includes 19 % of the remaining sites. Although this is only three-quarters of the efficiency of the hill-finding procedure, it nevertheless represents a distinct improvement in the identification of site locations. Optimising the extraction of promontories is required to define as accurately as possible the locations of the monuments. The application of more sophisticated kernel-based methods for extracting ridges (defined as a point on a local convexity that is orthogonal to a line with no convexity/concavity (Wood 1996)) is also being considered.

4. Topography, views and monuments

Identifying landform measures and categorizing the locations of monuments does not in itself explain the landscape context of a site. It is necessary to consider how landform has been used by the builders and participants in the creation and experience of a monument. We will never know for certain what aspects of a particular type of landform were considered important to the choice of monument location, or the reasoning, conscious or subconscious, that led to a decision being made. The cultural context of a location is crucial to understanding it, and the presence of other monuments nearby, or the past history of the locale is an important part of its meaning. However, by examining the characteristics of monument locations, and comparing them with non-site locations, some of the factors which influenced location choice (either directly or indirectly) may be suggested.

4.1. Local scale

Landform shape affects the way a site is encountered at the local scale. As a person approaches a monument, it may appear, disappear, and reappear again. The view from the monument is different from the view towards it. If a monument is unenclosed by a covering mound, with the interior accessible, like a passage grave, a specific view is encountered upon leaving the monument (Tilley 1994).

To explore these ideas, the viewsheds from and to a smaller sample of 43, well-documented, passage graves (including some potentially later sites) and the 43 earthen long mounds within the same area were generated. These were compared with the viewsheds for 500 random points within the same area. To examine the local scale, the viewsheds were buffered to include only the area within 500 m of the monument. Many of the monuments have had most of their mounds removed by erosion, excavation or vandalism, any reconstruction of mound height would in most cases be reduced to speculation.

The limited pollen and soil micromorphology evidence suggests that the area immediately around the monuments was cleared, often considerably before the construction of the monument (Marguerie 1992, Gebhardt and Marguerie 1993). It has therefore been assumed that at the local scale there was no substantial hindrance of view by trees. As local trees have a greater affect than distant ones on visibility, this also provides an important consideration when assessing the models for visibility at larger scales. The presence of modern trees and buildings seriously impedes the assessment of the local viewsheds in the field.

The local viewsheds created give an indication of the local prominence of the locations of the monuments, and the extent to which a view of the immediate surroundings is possible from the site. The local viewsheds from the passage graves are smaller than those from the random sample, suggesting that maximising local visibility was not of primary importance. However, the tertres have greater local visibility than either the passage graves or the random points.

Hilltops have smaller viewsheds than average over short distances and larger ones over greater distances, whist valley locations have the opposite characteristics (van Leusen 1999). The larger number of passage graves than *tertres* or random points on hills (see section 3.5 above) is therefore causing the smaller values observed for the local viewsheds of passage graves.

4.2. Increasing the distance

The issue of scale in the interpretation of viewshed data has been noted by several authors (van Leusen 1999, Woodman 2000). Van Leusen reminded archaeologists that the topography of a location affects the relationship between visible area and increasing viewshed radius. However, this observation can be taken further. The area of a location's viewshed as distance increases can be used itself to infer something about the topography of that location, and describes an important characteristic of the locale.

Viewsheds for the sample of passage graves used above, the random sample, and the earthen long mounds in the same area were buffered at 5 distances. The viewshed areas of the sites were compared to the non-site locations, and are expressed as a percentage of that value (as opposed to the total possible area, e.g. Woodman 2000). This provides a measure of how the site locations differ from the random points within the landscape. The results are shown in figure 5.

The passage grave locations appear to show different characteristics to the earthen long mounds. Having started with a smaller than expected viewshed over the shorter distances (section 4.1), they have a much larger viewshed over longer distances. The earthen long mounds have a larger than expected viewshed at the local scale, but over longer distances they have only slightly larger

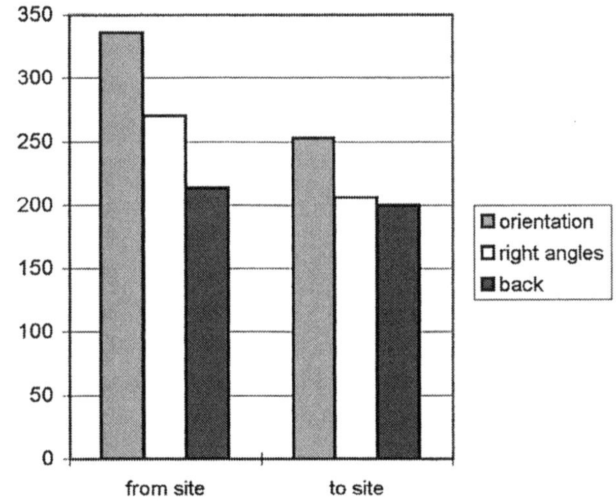

Figure 5: Visibility from and to passage graves by orientation.

viewsheds, much smaller than the corresponding viewsheds for passage graves. This accuracy of these results needs to be assessed once the higher resolution DEM is completed.

The viewshed calculations confirm the observations in sections 3.5 and 3.6 above that the passage graves and earthen long mounds tend to be in slightly different types of topographic locations, with a higher proportion of passage graves occurring on hills or promontories. The passage graves are visible from a wide area, over large distances, but tend to disappear as you approach them, and reappear only when you are closer to them. The earthen long mounds appear to have views of the local area (see Roughley in press), while the passage graves occupy locations with views into the distance.

4.3. Wider perspectives

Passage graves appear from the results above to occur in places which maximize the total viewshed. To investigate further the viewsheds of passage graves, a total viewshed grid (with viewshed values for each cell of the DEM) was constructed for a larger part of the study area (but at a coarser resolution, of 100 m, to reduce processing time).

Interpreting long range views is known to be difficult because of the correlation between viewsheds and elevation (Wheatley 1995). The viewsheds for all sites were correlated with elevation (r-squared 0.30 for views to and 0.25 for views from sites, $p = 0$ for both cases). When the residuals were plotted a clear spatial trend resulted (spatial autocorrelation was confirmed by Moran's I: correlation = 0.33, $p = 0$). Locations nearer the sea have larger viewsheds than locations inland.

The correlation between the Euclidean distance to the Neolithic coast and the residuals of the correlation between visibility and elevation is significant (r-squared 0.18, $p = 0$ for views from sites, r-squared = 0.11, $p = 0$ for view to sites). However, the residuals remain spatially autocorrelated. At this larger scale, the general topography is having a considerable effect on the viewshed statistics, and any interpretation must be made with care. Using an estimate of the prehistoric rather than the modern coastline has a considerable effect on the results, emphasising the importance of modelling the prehistoric landscape rather than utilising the modern one.

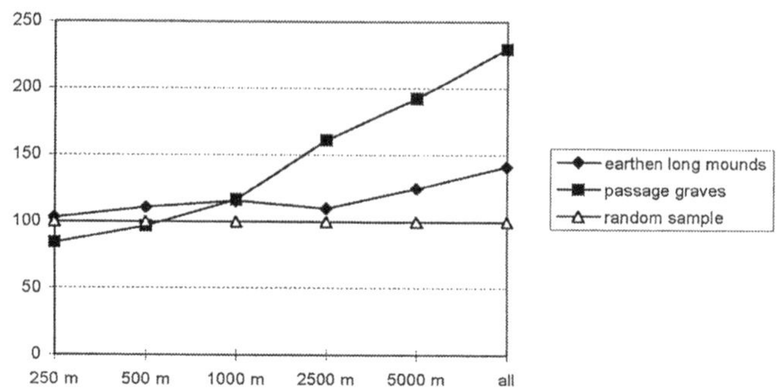

Figure 6: Viewshed from passage graves and earthen long mounds by distance, expressed as a percentage of the viewshed from the random sample.

Surface profiles across the viewshed surface show that many sites are on or near local maxima. It is possible that, within the area in which a passage grave was to be located, the precise choice of location was influenced by a desire to maximize the overall viewshed. Work aimed at comparing views of (and from) land and sea is continuing. The utility of Neolithic sites for navigation markers in more recent times has already been noted, emphasizing the strong impression these monuments make to people on the sea. The Neolithic people used boats extensively, and it is possible that trade was conducted at least partially by sea (for example, for jadeite, variscite and Grand-Pressigny flint).

4.4. Orientation of passage graves

Passage graves, unlike many of the other monuments, have a clear orientation through the axis of the passage. Although the orientation of the passage graves (as defined by the entrance to the passage) is predominantly southeast, there is considerable variation, with directions ranging from northeast to West. This suggested that there might be some local factor which was influencing the choice of passage entrance. Although there has been considerable interest in the alignment of sites in the area on astronomical features (and, perhaps more importantly, each other), the first possibility to investigate was that there was a link between topography and orientation. Sites were not oriented with the passage entrance facing up-slope. However, the steepest down-slope was not always in the direction of the passage entrance. It was surmised in the field that there might be a relationship between orientation and visibility.

The views to and from the monuments (up to a distance of 500 m) were compared in the direction of the entrance (described here as forwards) with views behind the monument and at right angles to it. The views from the site show a strong tendency towards having the greatest unbroken distance in the orientation of the entrance. This means that on leaving the chamber, the nearest horizon is further away than it would be if the monument had been oriented in any of the other directions (see figure 6).

There appears to be less difference in the distances from which a site can be seen uninterruptedly (see figure 6). This is perhaps not surprising, as although orientation may have some connection with the way a site is approached, the person is not as physically constrained in his or her choice of route to the site as the person is upon leaving the monument's passage.

There are some monuments which do not fit the trend observed above. A more detailed examination shows that they tend to be sites which are part of a more complex arrangement of monuments. For example, at Mane Kerioned, there are three passage graves. Two are parallel, and face south, while the third, which lies between the other two, faces east. Although there is insufficient evidence to provide a chronology for this site, a plausible hypothesis is that the two south-facing sites were built first, with the east-facing one then being oriented with respect to the existing sites rather than the direction of furthest first horizon.

The orientation angles have a strong tendency to fall between east and south-south west. As the topography has a general trend downwards towards the coast, it was necessary to investigate whether this had any affect on first horizon distances. The viewsheds for the 500 random points were considered, with the distance to the nearest horizon measured for each of 16 directions. The variation which exists among the different orientations is smaller than that observed between forwards and backwards views for the site locations. It may be inferred from this that the difference in distance to first horizon in the forward and backward direction is not caused by the co-existence of a southerly sloping landscape and the trend towards southerly orientation of passage graves. At larger scales, this relationship breaks down. When the maximum distance considered is increased (rather than buffering to 500 m), the general topographic trends appear to be more important, and there is spatial autocorrelation present in the difference between forward and backward views

4.5. Interpreting viewshed results

Any interpretations must remain speculative at this stage. It is possible that the passage grave locations were chosen for their broader perspective, whilst the earthen long mounds were more important in their immediate surroundings, perhaps a focus for a smaller area, and a smaller number of people? The Irish portal tombs and passage graves may also have been built with different foci in mind: the portal tombs appear to occupy lower ground in river valleys, while the passage graves are on hill tops in highly visible locations (Cooney 2000:147).

The orientation of the passage graves does appear to be important at the local scale, with the local view obtained on leaving the passage perhaps being maximised. It is at this local scale that many of the sites reappear as you walk towards them, having been hidden from view by a local horizon. The results obtained can be associated with two different scales of encounter with the passage grave. They form as a highly visible presence in the distance, both from land and from the sea. The passage graves also use the landscape to structure experience at a local scale, a more intimate encounter with the sacred, ritual or the ancestors.

5. The limits of landform discrimination?

Examining the locations of monuments help us to understand the sites more fully. Techniques have been developed which have enabled new insights to be made into the passage graves and earthen long mounds, revealing differences between these potentially contemporaneous monuments. Application to other monumental landscapes may reveal further uses for these approaches.

The techniques discussed above do not adequately characterize the locations of all site-types. It was observed above that the alignments did not significantly differ from the background population. Fieldwork did give the impression that for the smaller alignments there was a careful use of topography. It has long been observed that in the major alignments the largest stones are on the highest ground. However, for the smaller alignments, the largest stones, on the highest ground, are also the end of the alignment. They are located on ridges running perpendicular to the alignment. As you walk along the alignment, you cannot see beyond the end of the monument until you reach this terminal locale. At this point, the sea becomes visible.

At this point in time, it is difficult to envisage a GIS approach which might adequately assess these impressions gained during fieldwork. The sample size is small, the sites vary in size and are imperfectly preserved, and many locations in the landscape might fit the description given. To investigate the locales of alignments and other larger, complex structures, the sites need to be represented as polygons rather than point or line features. Further work is proposed to consider how best to analyse more the complex relationships of these polygons.

By applying appropriate methodology, it is possible to use GIS to contribute substantially to the study of past landscapes. Techniques need to be sensitive to both the questions asked and the data available. Interpretations need to be grounded within a realistic appreciation of the reliability and limits of the methodology and data and to have a sound theoretical basis. This paper has shown that subtle landscape features are not outside the scope of quantitative approaches. More importantly, it has demonstrated that GIS-based analysis does not have to be limited to the generation of environmentally and economically deterministic interpretations. Although conclusive statements as to the motivation behind the siting of megalithic monuments are not possible, some of the probable influences have been discovered, helping us to appreciate the complexity and subtlety of this important archaeological landscape.

Acknowledgements

I wish to thank my supervisor Dr Colin Shell and my advisor Dr Chris Scarre for their assistance with my research. This research is part of a PhD funded by the NERC. I am grateful to Dr Serge Cassen (CNRS, Nantes) for information on site locations. Hans-Martin Gutmann has helped with writing C-programs, and Dr Rosemary Anderson assisted with the fieldwork.

References

BOUJOT, C. and CASSEN, S., 1993. A pattern of evolution for the Neolithic funerary structures of the west of France. *Antiquity* 67: 477-491.

BOUJOT, C. and CASSEN, S., 1995. *Prospection thématique. Les tertres funeraires néolithiques de la region carancoise.* Service Regional de L'archeologie, Rennes: UMR 153 du CNRS, Laboratoire de Prehistoire Armoricaine.

BOUJOT, C. and CASSEN, S., et al., 1998. Some Abstraction for a Practical Subject: the Néolithization of Western France Seen Through Funerary Architecture. *Cambridge Archaeological Journal* 8(2): 193-206.

COONEY, G., 2000. *Landscapes of Neolithic Ireland.* London:Routledge.

DESDOIGTS, J.-Y., 1978. *Inventaires des monuments préhistorique du pays d'auray.* Rennes: Travaux de Laboratoire Anthropologie Préhistoire Protohistoire Quaternaire Armoricains.

EDMONDS, M., 1999. *Ancestral Geographies of the Neolithic: Landscapes, monuments and memory.* London: Routledge.

GAFFNEY, V. and van LEUSEN, M., 1995. Postscript-GIS, environmental determinism and archaeology: a parallel text. In Lock, G. and Stančič, Z. (eds.), *Archaeology and Geographic Information Systems*: 367-382, London: Taylor and Francis.

GEBHARDT, A. and MARGUERIE, D., 1993. *La transformation du paysage armoricain sous l'influence de l'Homme.* XVIeme colloque interrégional sur le Néolithique, Documents Archéologiques de la France.

GOUEZIN, P., 1995. *Les Mégalithes du Morbihan Interieur Des Landes de Lanvaux au nord du Département.* Rennes, Institut Culturel de Bretagne - Laboratoire d'Anthropologie - Prehistoire (UPR CNRS 403), Universite de Rennes I.

LAMBECK, J., 1997. Sea-level change along the French Atlantic and Channel coast since the time of the last Galcial Maximum. *Palaeogeogrpahy, Palaeoclimatology, Palaeoecology* 129: 1-22

LE ROUX, C.T., 1999. Réflexions autour d'une chrono-typologie du mégalithisme armoricain des Ve et VIe millenaires avant J.-C. In Guilaine, J. (ed.), *Mégalithsmes de l'Atlantique a l'Ethiopie*: 41-56. Paris: Editions Errance.

LECORNEC, J., 1994. *Le Petit Mont Arzon - Morbihan*, Rennes: Association pour la diffusion des recherches archéologiques dans l'Ouest de la France. (Documents Archéologiques de l'Ouest).

LLOBERA, M., 1996. Exploring the topography of the mind: GIS, social space and archaeology. *Antiquity* 70: 612-622.

MARGUERIE, D., 1992. Evolution de la végétation sous l'impact humain en Armorique du Néolithique aux périodes historiques. Rennes: Université de Rennes I.

TILLEY, C., 1994. *A Phenomenology of Landscape.* Oxford: Berg Publishers.

WHEATLEY, D., 1995. Cumulative Viewshed analysis: a GIS based method for investigating intervisibility, and its archaeological application. In Lock, G. and Stančič, Z. (eds.), *Archaeology and Geographic Information Systems*: 171-186, London: Taylor and Francis.

WHEATLEY, D., 1996. The Use of GIS to Understand Regional Variation in earlier Neolithic Wessex. In Maschner, H. (ed.), *New Methods, Old Problems: Geographi Information Systems in Modern Archaeological Research*: 75-103. Carbondale: Center for Archaeological Investigations, Occasional Paper No. 23.

WISE, S.M., 1998. The Effect of GIS Interpolation Errors on the Use of Digital Elevation Models in Geomorphology. In Lane, S., Richards, K. and Chandler, J. (eds.), *Landform Monitoring, Modeling and Analysis*. 139-164. Chichester: John Wiley and Sons.

WOOD, J., 1996. *The Geomorphological Characterisation of Digital Elevation Models*. PhD Thesis, Department of Geography, University of Leicester.

WOODMAN, P.E., 2000. Beyond significant patterning, towards past intentions: the location of Orcadian chambered tombs. In Buck, C., Cummings, V., Henley, C., Mills, S. and Trick, S. (eds.), *U.K. Chapter of the Computer Applications and Quantitative Methods in Archaeology, Proceedings of the Fourth Meeting*: 91-105. Oxford: British Archaeological Reports (International Series 844).

The Hidden Reserve: Predictive Modelling of Buried Archaeological Sites in the Tricastin-Valdaine Region (Middle Rhone Valley, France)

Philip Verhagen

Archeologisch Adviesbureau RAAP

Zeeburgerdijk 54, 1094 AE Amsterdam, The Netherlands

e-mail: ph.verhagen@raap.nl

Jean-François Berger

Université de Nice, UMR Centre d'études Préhistoire, Antiquité, Moyen Age

250, rue Albert Einstein, Sophia-Antipolis, 06560 Valbonne, France

e-mail: jean-francois.berger@cra.cnrs.fr

Abstract

Over the past ten years, a number of archaeological surveys have been carried out in the Tricastin-Valdaine region (Middle Rhône Valley, France). The results of these surveys, which included digging trenches on many locations, indicated that many more sites are hidden below the surface than can be inferred from traditional field-walking data alone. In order to illustrate this effect, a predictive modelling study was carried out, using both the "old" and the "new" data from the area. The paper demonstrates that the use of data on buried sites strongly influences the outcome of the predictive model, as well as the estimation of the actual site quantities involved. This also implies a reassessment of existing theories of site location in the area. Ultimately, the results of the study make a strong case for the use of sub-surface surveying methods in any sedimentary area.

Key words: predictive modelling, χ^2, site-to-area proportions, K_j-method, survey methods, buried sites, Holocene geomorphological evolution, Rhône valley

1. Introduction

The middle Rhône Valley is located at the boundary of the Mediterranean, central European and alpine climate zones. Within the middle Rhône Valley, the Tricastin-Valdaine region is an area of 1086 km^2 located on the eastside of the river Rhône. The area consists of two distinct zones: to the north, the Valdaine Basin comprises the valleys of the Roubion and Jabron rivers, surrounded by Secondary pre-alpine hills. The most important town in this area is Montélimar. To the south, the Tricastin forms a broad transitional zone between the riverbed of the Rhône, the lateral Holocene alluvial fans and the Tertiary pre-alpine hills. The most important towns in this area are Pierrelatte, St-Paul-Trois-Châteaux and Bollène.

Due to its location at a climatic and geological boundary zone, both vegetation and geomorphological processes in the middle Rhône Valley are highly sensitive to climate change and anthropic impact. The area is known to have a complex history of erosion and sedimentation since the beginning of the Holocene (Brochier et al. 1991, Berger 1996, Berger et al. 1997, Berger et al. 2000, Berger in press (1), Berger in press (2), Berger and Brochier in press). Because of this landscape dynamic many archaeological remains are known to be buried below the current surface, especially in the alluvial plain of the Rhône. The purpose of this study is to show that the use of the results of traditional field-walking survey (which will not detect buried sites) for analysing the relationship between site location and landscape characteristics can lead to both a wrong representation of the distribution of the sites with regard to landscape units, and of actual site quantities. This has been achieved by creating a qualitative predictive map of the area, and by performing a quantitative extrapolation of site densities for the sedimentary areas where most buried sites are found.

2. The predictive model

2.1. Introduction

For most predictive modelling studies, the relation of site location to one or more landscape characteristics is inferred by applying an overlay of the known site locations on the cartographic background available. This overlay is then subjected to a quantitative analysis of the observed distribution pattern, an approach also known as inductive modelling (Dalla Bona 1994). In most cases this analysis is done assuming that the known site sample is representative for the total population. However, this is not necessarily true.

First of all, the method of survey determines which sites will be discovered. It is clear that buried sites will not be detected by means of field walking. However, augering and digging trenches are relatively expensive forms of survey, which are not usually available to amateurs, and even professional archaeologists will not use these forms of survey unless there is a clear necessity. In practice, this means that in most archaeological site databases the number of buried sites will be underestimated.

Furthermore, when the size of the area actually surveyed is not known, there is no information available on the absence of sites, which is equally important for the statistical analysis of site location preference.

Thirdly, the area surveyed (and therefore the site sample) is not usually representative of the total study area. This may be a consequence of difficult access of the terrain, for example because of steep slopes, or because of a research bias for certain areas.

Fortunately, the situation for the Tricastin-Valdaine region is different, as we have both detailed records of buried sites, as well as a mapping of the surveyed zones. However, the current study cannot account for a fourth distorting factor, the differential visibility of archaeological surface remains under different types of land use.

2.2. The Taphonomical map: An interpretation of the landscape in terms of sedimentation and erosion

In order to get a grip on the history of sedimentation and erosion of the area, the landscape has to be interpreted in terms of its geomorphological and pedogenetic history. In order to arrive at a map that could be used as a taphonomical base layer, various geological and soil maps have been digitized, and combined in a GIS. They then have been interpreted into in taphonomical terms.

The only base maps available for the whole region are the geological maps 1:50,000 of France. The following sheets have been digitised:

Sheet:	Name:	Publisher:	Year:
842	Crest	BRGM	1976
866	Montélimar	BRGM	1979
XXX-39	Valréas	BRGM	1964
914	Orange	BRGM	1971

These maps have been edited where necessary. The geological map units have then been assigned to one of the 18 taphonomical categories distinguished.

The basic geological information has then been updated with other available information on the geological and pedological conditions in the area. This information comes from three sources:

1. a classified remotely sensed image of the Tricastin area (Tounsi et al. 1997);
2. a delimitation of the main pedological and sedimentary units obtained during fieldwork in the Valdaine (Berger 1996) and Tricastin (Berger et al. 1997); and
3. existing 1:25,000 pedological maps of the area (Bornand 1967, 1971).

Essentially, the remotely sensed image gives detailed information on the location of old riverbeds, alluvial fans, terraces and *cuvettes* in the Tricastin. The fieldwork data provides additional information on the location of colluvial and alluvial deposits (colluvium, alluvial fans, alluvium and *cuvettes*). The pedological maps have been used to find the actual extent of alluvial and colluvial deposits in the Roubion and Jabron valleys, to find the location of stable Pleistocene terraces and alluvial fans in the Valdaine, and the distribution of colluvial deposits and *cuvettes* in the Tricastin. The additional information has been used to update the reclassified geological maps. The final taphonomical map is therefore a patch of several maps of varying scale and precision.

2.3. The archaeological dataset

The archaeological site sample consists of data coming from various sources. Data collected by means of field walking was taken from Beeching et al. (1995) and Berger (1996). Data collected by means of digging trenches was taken from Berger (1996), Berger et al. (1997, 2000), and from the archaeological surveys and excavations for the TGV Méditerranée (the high-speed railway connection between Lyon and Marseille), which were carried out between 1995 and 1998. Of course the latter method of fieldwork will result in the detection of buried sites, and this data therefore forms the most important part of the archaeological database. Data on other sites has been collected from literature, and from the archaeological map of the regional archaeological service of the Rhône-Alpes region.

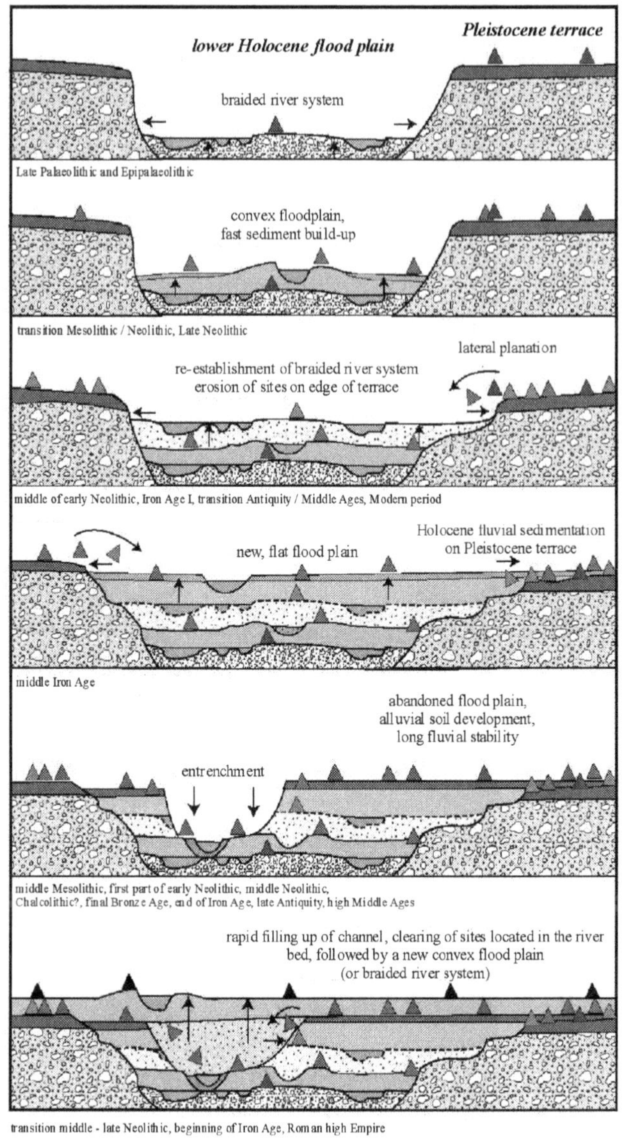

Figure 1: The influence of river dynamics on site location and preservation in the lower alluvial plains of the middle Rhône Valley (adapter after Butzer, 1982). River changes play a predominant role during the Holocene in the occupation and exploitation of the fertile alluvial soils of the lower flood plain. A pattern emerges of phases of fluvial stability coupled to dense occupation, alternating with phases of instability coupled to depopulation.

At the regional scale, the relationship between site location dynamics and geomorphological evolution is currently understood in terms of fluvial systems (Berger 1996, in press (1), in press (2)) and can be schematized for the Holocene period according to figure 1. In the Rhône hydrological system, the fluvial regime of the rivers changes depending on variations in the ratio of liquid and

Period	visible	not visible	total
Epipalaeolithic and Mesolithic	57	8 (12.3 %)	65
Early Neolithic	21	17 (44.7%)	38
Middle Neolithic	57	28 (32.9%)	85
Late Neolithic and Chalcolithic	73	43 (37.1%)	116
Bronze Age	24	87 (78.4%)	111
Iron Age	42	53 (55.8 %)	95
Total	274	236 (46.3%)	510

Table 1: Distribution of archaeological sites over 6 chronological periods, and the proportions of visible and non-visible sites.

Window	E/M	EN	MN	LN	BA	IA	Total
FULL(VIS)	57	21	57	73	24	42	274
SURV(VIS)	46	12	35	41	9	14	157
FULL(ALL)	65	38	85	116	103	95	502
SURV(ALL)	51	23	53	60	45	43	275

Table 2: Distribution of the archaeological sites over the analysis windows. Figures in grey indicate situations where the results of the analysis will be unreliable (n < 40). Windows: FULL(VIS) – whole study region, visible sites; SURV(VIS) – surveyed zones, visible sites; FULL(ALL) – whole study region, all sites; SURV(ALL) – surveyed and trenched zones, all sites. Periods: E/M – Epipalaeolithic/Mesolithic; EN – early Neolithic; MN – middle Neolithic; LN – late Neolithic; BA – Bronze Age; IA – Iron Age.

solid flows. Anastomosed or braided channel systems are unattractive for settlement, except for seasonal activity (Epipalaeolithic, the middle of the early Neolithic, first Iron Age). Flat or convex floodplains, associated with meandering river courses, are more favourable for settlement, but frequent flooding can be a restraint to occupation (late Neolithic, middle Iron Age, Roman high Empire). The most favourable situation is a meandering river system together with a deep entrenchment of the riverbed, which stabilizes the floodplain for a long period of time (middle to late Mesolithic, middle Neolithic, final Bronze Age, second Iron Age). Evidence is found for a cyclic alternation between stability of watercourses and expansion of occupation in the flood plains on the one hand, and instability of watercourses and depopulation of these same areas on the other hand.

From the sources mentioned, a comprehensive archaeological database was constructed, containing the location of 510 pre- and proto-historic sites and probable sites, discovered in different taphonomic contexts (buried in situ, buried in secondary position, washed down slopes; see Berger in press and Berger et al. 2000 for more details). For each site it was documented whether the site was visible at the surface, or if it had been covered by sediment. The term "site" is not synonymous with find spot in this case. A single find spot may have produced evidence for occupation in more than one period; each of these occupation phases has been stored as a separate site in the database. The occupation phases distinguished have then been regrouped into six different chronological periods (table 1). This was done in order to obtain a sufficiently large number of sites per period. The number of sites needed for a reliable analysis of site location in relation to the taphonomical map is approximately 40. For the early Neolithic, this requirement is not met, but the other periods do have sufficient sites to carry out a site location analysis.

2.4. The site sample: Dealing with the problem of representativity

In order to see if non-random sampling influences the site location analysis of the study area, the available site sample was divided into a visible sample and a full sample that also included the non-visible sites. These samples for each period have then been analysed using two geographical analysis windows: the full study region and the area surveyed. In the case of the visible sites, this only encompassed the field walked zones. In the case of the full sample, both the trenched and field walked zones were included (table 2). Because of the small number of sites involved, it was not possible to carry out a separate analysis for the trenched zones alone. However, the chances of finding a non-visible site in the trenched zone are much larger than in zones that have only been field walked. As the trenched zone only constitutes a relatively small portion of the total surveyed zone, the actual importance of non-visible sites may be larger than is suggested by the predictive modelling.

3. The predictive model: Methods applied

In order to analyse the relationships between archaeological site location and the taphonomical map, three separate analyses were undertaken:

3.1. χ^2 test

A χ^2 test is often used as a first step to see if any statistically significant patterns between site location and map units can be observed. The method has first been suggested by Hodder and Orton (1976), and has been applied on a number of occasions in the Netherlands for predictive modelling purposes (Verhagen 1995). However, χ^2 in itself does not say anything about the relative importance of map units for site location, and its application as the only statistical tool for predictive modelling has therefore been criticised on a number of occasions (Wansleeben and Verhart 1992, van Leusen 1996, Kamermans and Rensink 1999).

In order to better comply with the limitations of the χ^2 test (the demand of having at least 5 expected sites per map category, which is in turn dependent on the size of the site sample; see e.g. Thomas 1976) the taphonomical map was reclassified into 9 categories. Even so, in some cases the statistical requirements could not be met. In these cases, Yates' correction has been applied to calculate χ^2. It should however be pointed out that in the case of less than 40 observations, the application of χ^2, even with Yates' correction, should be regarded with suspicion.

3.2. Ratio of site to area proportions

The ratio of site (p_s) to area (p_a) proportions is a simple and straightforward way to look at the importance of certain map categories for site location. This ratio has for example been used in the Netherlands to create the Indicative Map of Archaeological Values (Deeben et al. 1997). However, it does not provide a relative weighting of the categories according to size. This problem is best illustrated by taking the zero site case: a large unit without sites will be less important for site location than a small unit without

1	colluvial deposits (1,14,18)
2	stable Pleistocene alluvial fans and terraces (2,9)
3	unstable Pleistocene alluvial fans and terraces (3,10)
4	recent alluvial fans, terraces and riverbeds (4,5,6,7,8)
5	*cuvettes* (11,12)
6	loess formations (13)
7	resistant rocks (15)
8	intermediate rocks (16)
9	soft rocks (17)

Table 3: Reclassification of the taphonomical map into 9 categories.

sites (in order words, it is statistically more significant). Calculated p_s/p_a values however, will give a value of 0 for both units, thereby attributing them equal importance.

In order to account for this effect, Atwell and Fletcher (1985, 1987) suggested calculating a statistic that is described as a relative weight factor for each map unit. In the case of three map-units a, b and g, the following weights are calculated:

$$A = \frac{a'bc}{a'bc + ab'c + abc'}$$

$$B = \frac{ab'c}{a'bc + ab'c + abc'}$$

$$C = \frac{abc'}{a'bc + ab'c + abc'}$$

where

a, b, c = area proportion of map units α, β and γ,
a', b', c' = site proportion of map units α, β and γ.

This is arithmetically equivalent to dividing each p_s/p_a value found by the sum of all p_s/p_a values, from which it follows that the relative weights calculated with the Atwell-Fletcher method are only normalised p_s/p_a calculations. They will therefore not fully solve the problem of relative weights.

Apart from that, the p_s/p_a calculations do not say anything about the statistical significance of the observed pattern. Atwell and Fletcher (1985, 1987) suggest to test the significance of the pattern by means of comparing the weights to those obtained by simulated site location patterns, a method applied by Wansleeben and Verhart (1992) and Kamermans and Rensink (1999). This analysis depends on the creation of random site distribution maps against which to test the actual pattern. Unfortunately, a random point generating routine is not supplied with ARC/INFO (which was used for the Tricastin-Valdaine model), and time did not permit us to write a separate routine, so the simulation was not performed for this study.

3.3. K_j method

A more complex method of assessing the importance of map categories for site location is the use of the K_j parameter. This parameter was developed by Wansleeben and Verhart (1992), and is defined as follows:

$$K_j = \sqrt{(p_s(p_s - p_a))}\ .$$

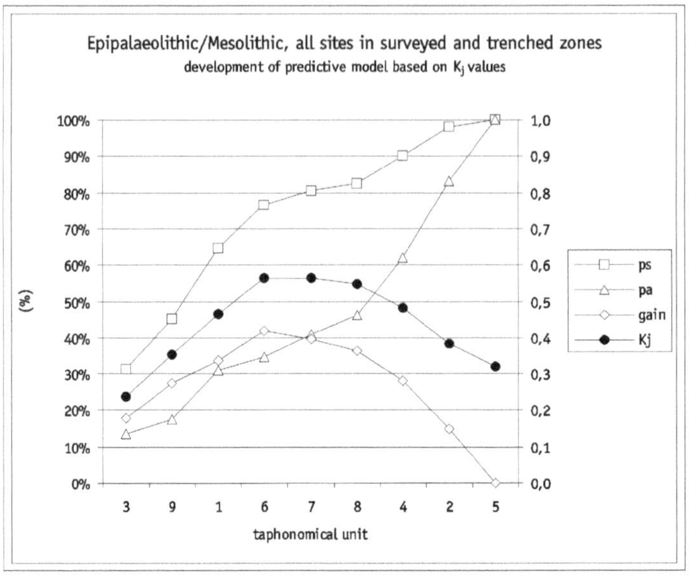

Figure 2: Example of the development of the K_j model for the Epipalaeolithic/Mesolithic period, all sites in surveyed and trenched zones. Units 3, 9, 1 and 6 have a strong positive predictive power, units 7 and 8 are more or less neutral, and units 4, 2 and 5 have a strong negative predictive power.

In the original equation, $p_s - p_a$ is divided by p_w (the proportion of the area without sites); however, this modification is only useful when (hypothetical) site surfaces are used, which is not the case for this model. K_j is calculated for each map category. The category that yields the highest K_j value is considered most successful. In an iterative procedure K_j is calculated again, including the next most successful category in the model until all categories have been included. The utility of K_j is that it takes into account the relative importance of the observed site densities: a small unit with high site densities will not necessarily be considered the most successful. Each time K_j is calculated, the gain ($p_s - p_a$) can be calculated to assess the performance of the model. A model with high predictive power will have high gain values (Kvamme 1989).

Wansleeben and Verhart (1992) state that the actual performance of the model increases as long the value of K_j increases on each consecutive run. However, we find in a number of instances that the gain is dropping while K_j is still increasing. This is because the equation attributes a higher weight to categories that contain a large number of sites. A gain of 40% can be achieved by a model that contains 50% of the sites on 10% of the surface, but also by a model that contains 80% of the sites on 40% of the surface. The K_j method decides that in the latter case the model performs better, although the gains obtained are equal. However, for the purposes of archaeological resources management, it seems that a model based on gain values is more useful, as the total surface to be considered is smaller.

In the cases that were analysed for this study, we often see categories that contribute strongly to the increase in gain and K_j. Other units will only have a limited effect on the total gain, and there are also units that will strongly decrease the gain of the model. In other terms, these groups may be said to have a positive, neutral and negative predictive power (or high, intermediate and low to use a more traditional terminology). These groups may easily be identified by plotting the development of the model in a graph, and one example of these is given in figure 2.

No.	Taphonomical unit	km²	p_a	n	p_s	$K_{j(MAX)}$	rank	$p_{s(CUM)}$	$p_{a(CUM)}$	gain
1	colluvial deposits	6.6436	13.5%	10	0.1961	0.4663	3	64.7%	31.1%	33.6%
2	stable Pleistocene alluvial fans and terraces	10.4040	21.1%	5	0.0980	0.3828	8	98.0%	83.1%	14.9%
3	unstable Pleistocene alluvial fans and terraces	6.6368	13.4%	16	0.3137	0.2372	1	31.4%	13.4%	17.9%
4	recent alluvial fans, terraces and riverbeds	7.8804	16.0%	3	0.0588	0.4809	7	90.2%	62.0%	28.2%
5	cuvettes	8.3500	16.9%	1	0.0196	0.3187	9	100.0%	100.0%	0.0%
6	loess formations	1.7188	3.5%	6	0.1176	0.5660	4	76.5%	34.6%	41.9%
7	resistant rocks	3.1048	6.3%	2	0.0392	0.5637	5	80.4%	40.9%	39.5%
8	intermediate rocks	2.5740	5.2%	1	0.0196	0.5466	6	82.4%	46.1%	36.3%
9	soft rocks	2.0788	4.2%	7	0.1373	0.3519	2	45.1%	17.6%	27.5%
	TOTAL	49.3912	100.0%	51	1.00					

Table 4: Example of the calculation of K_j for the Epipalaeolithic and Mesolithic, using the surveyed and trenched zones with the full site sample. n = number of observed sites.

4. The predictive model: Results of site location analysis

4.1. Epipalaeolithic and Mesolithic (ca. 12,000-6800 BP; Azilian, Sauveterrian and Castelnovian cultures)

The calculated χ^2 values (with Yates' correction) for the Epipalaeolithic and Mesolithic sites indicate that the taphonomical units are significant for site location at the 99.9% probability level in all analysis windows.

When looking at the ratio of site to area proportions, it is clear that the loess formations show very high p_s/p_a values for all analysis windows. For the full study region, the unstable Pleistocene alluvial fans and terraces also have very high p_s/p_a values. However, in the surveyed zones the soft rocks seem more important. Low p_s/p_a values are observed for the recent alluvial fans, terraces and riverbeds, the intermediate rocks and the *cuvettes*. The recent alluvial fans, terraces and riverbeds are clearly less important for the visible sample than for the full sample. Even though only six buried sites have been observed for this period, they do seem to a have a (limited) effect on the analysis results.

The K_j model that was developed for all analysis windows is rather strong. Maximum K_j and gain values decrease when the surveyed zones are used instead of the full region, and the overall performance is weaker when the full sample is included. It is absolutely clear that map units 3, 1, 9 and 4 are the most important ones for site location. The models are less clear about the units with negative predictive power, so most shifts in ranking are observed for these categories. The pattern of site location observed largely conforms to the pattern obtained with p_s/p_a values, with one notable exception: the position of the loess formations is much less dominant than could be expected from the p_s/p_a calculations. This is because the actual gain obtained by including the small unit of loess formations first in the model is less than the gain that can be achieved by including the large unit of unstable Pleistocene alluvial fans and terraces. This clearly demonstrates that the K_j model is able to perform a relative weighting of map categories.

It can be concluded that the known site sample is representative for the area. Neither the restricted analysis, nor inclusion of the non-visible sites leads to drastic changes in observed site location preference. The most important units for site location are the unstable Pleistocene alluvial fans and terraces, the colluvial deposits, the soft rocks and the loess formations. The observed pattern is distinct, as is demonstrated by the high maximum K_j and gain values observed.

It seems that Epipalaeolithic/Mesolithic settlement is strongly concentrated on the intermediate elevations (with the exception of the stable Pleistocene alluvial fans and terraces), avoiding both the humid zones and the hills. This under-representation of settlements in landscape units that are marked by numerous geomorphological events since the end of the Late Glacial is probably the consequence of taphonomical bias. The observed absence of buried sites can be attributed to strong erosion of the recent alluvial fans, riverbeds and the lower reaches of the *cuvettes* between 6400-6200 BP, associated with the first evidence of agropastoral activity in the south of France and an abrupt hydroclimatic event (Berger 1996, Berger and Brochier in press). Furthermore, the Epipalaeolithic/Mesolithic sites found are usually small in size and are characterised by a dispersed lithic scatter, and as such are difficult to detect, even by means of trenching. A geographical bias of Epipalaeolithic/Mesolithic sites is observed for the Valdaine basin. This bias may be the result of selective surveying.

4.2. Early Neolithic (ca. 6500 – 5800 BP; Cardial and Epicardial cultures, and the transition of Cardial to Chassean)

The total number of early Neolithic sites is only 38, which means that calculated χ^2 values are not reliable.

When looking at the ratio of site to area proportions, it is clear that the loess formations show very high p_s/p_a values for the full study region. The unstable Pleistocene alluvial fans and terraces and colluvial deposits also have very high p_s/p_a values. Low p_s/p_a values are observed for the resistant and intermediate rocks. However, in the surveyed zones the position of the loess formations is less dominant. When including the non-visible sites in the sample, the recent alluvial fans, terraces and river beds become much more important, largely at the expense of the loess formations and colluvial deposits. This again illustrates the importance of including these sites in the analysis, even when working with small samples.

The models developed with the K_j method are unstable. Both maximum K_j and gain values are variable. Units 1, 3 and 6 seems to be most important for site location. It is also obvious that by including the non-visible sample, the importance of recent alluvial fans, terraces and riverbeds becomes much larger.

It is difficult to draw conclusions about site location preference for the early Neolithic because of the low number of sites. This low density is in part the consequence of the major erosion phase occurring between 6400 and 6200 BP (Berger and Brochier in press). The K_j model for the full sample indicates one important change compared to the Epipalaeolithic and Mesolithic: the important position of the recent alluvial fans, terraces and riverbeds. Since only the earliest horizon of early Neolithic occupation is destroyed or reworked, this implies that the more recent horizons have been preserved under younger alluvial deposits.

4.3. Middle Neolithic (ca. 5800 – 5000 BP; Chassean culture)

The calculated χ^2 values (with Yates' correction where applicable) for the middle Neolithic sites indicate that the taphonomical units are significant for site location at the 99.9% probability level. For the visible sample however, the number of sites drops below 40 for the surveyed zones, which makes the χ^2 calculation unreliable.

For the full study region, the unstable Pleistocene alluvial fans and terraces and the loess formations have the highest site densities. Very low site densities are found on the resistant and intermediate rocks. For the surveyed zones, the most important units are the soft rocks and the unstable Pleistocene alluvial fans and terraces, and low densities are observed for the *cuvettes,* resistant rocks and intermediate rocks. When comparing the visible sample to the full sample, there is marked increase in p_s/p_a for the recent alluvial fans and riverbeds, again pointing to the importance of the non-visible sites for the analysis.

The calculation of K_j for the visible sites results in strong models with high maximum K_j and gain values. However, when the non-visible sites are included in the models, they are considerably weaker. In spite of this, the importance of units 1, 3 and 9 is very clear for both the visible and full sample. Although the recent alluvial fans, terraces and riverbeds become more important when looking at the full site sample, the effect is less marked than for the early Neolithic.

It can be concluded that the known visible site sample is representative of the area. However, the inclusion of the non-visible sites shows that there is a strong effect of underestimation of the importance of the recent alluvial fans, terraces and riverbeds. A preference can be observed for the unstable Pleistocene alluvial fans and terraces, the colluvial deposits and soft rocks. Compared to the Epipalaeolithic and Mesolithic however, the recent alluvial fans, terraces and riverbeds are more important, at the expense of the loess formations. As the K_j models developed for the full sample for the middle Neolithic are not very strong, this implies that settlement is more dispersed than during the Epipalaeolithic and Mesolithic.

4.4. Late Neolithic (ca. 5000 – 3700 BP; including Chalcolithic)

The calculated χ^2 values (with Yates' correction where applicable) for the late Neolithic sites indicate that the taphonomical units are significant for site location at the 99.9% probability level, with the exception of the visible sample for the surveyed zones.

The p_s/p_a ratios obtained for the full study region show that the loess formations and the unstable Pleistocene alluvial fans and terraces have the highest site densities. No very low site densities are found. In the restricted zones, the most important units are the soft rocks, the loess formations and the unstable Pleistocene alluvial fans and terraces. When comparing the visible sample to the full sample, there is a marked increase in p_s/p_a for the recent alluvial fans and riverbeds for the surveyed and trenched zone, again pointing to the importance of the non-visible sample for the analysis.

The calculation of K_j for the full study region produces a weaker model than for the surveyed zones. It is clear that unit 3 is the most important unit for site location; however, the models differ considerably in attributing a ranking to most other units. For the full sample, the recent alluvial fans, terraces and riverbeds are clearly more important than for the visible sample. In all cases, the *cuvettes* and stable Pleistocene alluvial fans and terraces have a negative predictive power.

It can be concluded that the known site sample is not representative of the area, as the performance of the K_j models differs considerably when looking at the restricted zones. From the available data it can be deduced that the higher elevations (soft rocks, intermediate rocks and resistant rocks) may have been neglected in previous surveys. Furthermore, the inclusion of the non-visible sites shows that there is a very strong effect of underestimation of the importance of the recent alluvial fans, terraces and riverbeds. A preference can be observed for the recent alluvial fans, terraces and riverbeds, unstable Pleistocene alluvial fans and terraces, and soft rocks. As the K_j models developed for the full sample for the late Neolithic are not very strong, this implies that settlement is rather dispersed. In fact, during this period a slight shift in occupation towards the higher elevations is observed.

4.5. Bronze age (ca. 3700 – 2700 BP)

The calculated χ^2 values for the Bronze Age sites are not reliable for the visible site sample, as it only includes 24 sites. For the full sample the calculated value (with Yates' correction when applicable) is significant for site location at the 99.9% probability level for the whole region, but for the surveyed zones it is not.

When looking at the p_s/p_a ratios obtained for the visible sample, the soft rocks clearly exhibit the highest values when looking at the whole region. However, when looking at the surveyed zones, the resistant rocks are most important, followed by the soft rocks and unstable Pleistocene alluvial fans and terraces. Low values are found for the loess formations, intermediate rocks and stable Pleistocene alluvial fans and terraces. When including the non-visible sample, the resistant rocks seem most important when looking at the whole region. Within the surveyed zones, the recent alluvial fans, terraces and riverbeds are most important. The observed patterns seem highly irregular; however, when looking at the two reliable samples, it is obvious that a change in importance can be observed from resistant rocks to recent alluvial fans, terraces and riverbeds.

When looking at the models developed with the K_j method, it is clear that the performance of the models is better when only looking at the visible sample. The inclusion of the non-visible sample results in a more important position for the recent alluvial fans, terraces and riverbeds. When looking at the surveyed zones for the full sample, the importance of the resistant rocks for site location is clearly diminished.

Obviously, the visible site sample is wrongly representing both the quantities of Bronze Age sites, as well as their distribution in the landscape. The visible sample contains relatively more sites on resistant rocks. This can be related to a small amount of cave settlements on this unit near Donzère, which are not included in the surveyed zones. This is due to the history of regional archaeological research, which privileged karstic areas (secondary calcareous formations) until the last decade (Berger et al. 2000). Apart from that, the role of the recent alluvial fans, terraces and riverbeds is clearly underestimated when only looking at the visible sample. When looking at the total sample, a preference is found for site location on recent alluvial fans, terraces and riverbeds, unstable Pleistocene alluvial fans and terraces, and soft rocks. The preference for the river valleys, together with an increasing importance of the *cuvettes* indicates that humid zones become more important for settlement. However, the K_j models developed for the full sample are not very strong. Together with the low χ^2 values found this implies that the settlement pattern is highly dispersed and might even be randomly distributed. This is certainly due to the strong increase in occupation during the final Bronze Age (3200-2700 BP). The original settlement pattern may however have been more strongly concentrated in the alluvial plains. Many Bronze Age sites have been found in secondary position down terrace slopes or in river channels as a consequence of a major erosion phase between 2700 and 2300 BP (known as the first Iron Age hydroclimatic crisis), following a long phase of fluvial stability during the Bronze Age period (Berger et al. 2000). This might imply that many more have been totally destroyed in this period.

4.6. Iron age (2700 – 2200 BP; Hallstatt and La Tène cultures)

The calculated χ^2 values for the Iron Age sites are not reliable for the visible sample in the surveyed zones, as only 25 sites are found there. In the other cases, the calculated values (with Yates' correction when applicable) indicate that the taphonomical units are not significant for site location at the 99.9% probability level for the whole region, with the exception of the visible site sample for the whole study region.

The p_s/p_a ratios obtained for the visible site sample for the full study region indicate three important units: the resistant rocks, the soft rocks and the *cuvettes*. For the surveyed zones, the resistant rocks are clearly the most important. When looking at the full sample for the whole region, very little difference in site density is found. However, in the restricted zones the resistant rocks show the highest p_s/p_a ratios. It is interesting to observe that the importance of the resistant rocks increases when looking at the surveyed zones: this seems to indicate that more sites (fortified *oppida*) may be found on this unit, but may have been overlooked outside the surveyed areas.

The calculation of K_j results in rather weak models, with the exception of the visible sample for the restricted zones. This better performance for these zones is associated with the clear preference for resistant rocks in these windows. A dramatic shift in importance is observed for the recent alluvial fans, terraces and riverbeds when looking at the full sample for the full region and the surveyed and trenched zones.

The known site sample is not representative for the whole region, as the performance of the K_j models is stronger for the surveyed zones. The large differences in K_j and gain values between the visible and full sample indicate that the visible sample is not representative for the actual settlement distribution. From the full sample it can be concluded that the recent alluvial fans, terraces and riverbeds, resistant rocks and unstable Pleistocene alluvial fans and terraces are most important for site location. The important position of resistant rocks is related to the existence of hill forts. Most of the sites found by trenching under alluvium can be identified as small farm sites dated to the second Iron Age.

However, the weak performance of the K_j models for the full sample together with the low c^2 values can be taken as an indication that settlement distribution in the Iron Age is close to random. This is totally contradictory to the existing theories on Iron Age site location before the trenching campaigns started (cf. Odiot 1985).

4.7. Conclusions

The results of the site location analysis for the area point to large differences in reliability of the site samples. Especially for those periods where large numbers of sites have been discovered by trenching (notably the early Neolithic, late Neolithic, Bronze Age and Iron Age) it is clear that the sedimentary areas are much more important for site location than can be deduced from the visible site sample alone. Furthermore, the visible site sample is not always representative for the area, as becomes clear for both the late Neolithic and the Iron Age. Any predictive map to be made for the area will therefore have to include both the extent of the prospected zones as well as the information on buried site locations.

It also seems clear that site location characteristics become less pronounced in the later occupation phases. The strong preference for the intermediate elevations in the Epipalaeolithic and Mesolithic is gradually replaced by a rather dispersed settlement pattern in the Bronze Age and Iron Age. This can largely be attributed to the effect of differential conservation for the various periods. From the analysis results it follows that the definition of zones of Epipalaeolithic and Mesolithic settlements will be much easier than for the later periods. It may be possible that for these later periods, site location preferences are dependent on landscape elements that are not included in the taphonomical map, either because they are too small in size, or because the soil units that were aggregated to create the map are not particularly relevant. A more detailed reconstruction of the (palaeo-)landscape is therefore needed to arrive at an alternative explanation of site location preferences for these periods.

From the point of view of reliability, the best model will be based on the results of the full sample for the surveyed and trenched zones (analysis window SURV(ALL)). The results of the χ^2 test indicate that with the exception of the early Neolithic, sufficient

No.	Taphonomical unit	E/M	EN	MN	LN	BA	IA
1	colluvial deposits	6,2%	8,3%	7,3%	-0,1%	-2,3%	0,5%
2	stable Pleistocene alluvial fans and terraces	-13,2%	-16,7%	-9,7%	-14,4%	-21,1%	-14,1%
3	unstable Pleistocene alluvial fans and terraces	17,9%	12,6%	20,5%	16,6%	8,8%	5,2%
4	recent alluvial fans and riverbeds	-8,1%	18,8%	1,0%	2,4%	19,6%	12,0%
5	cuvettes	-14,9%	-12,6%	-13,1%	-11,9%	-1,4%	-7,6%
6	loess formations	8,3%	0,9%	0,3%	3,2%	-1,3%	1,2%
7	resistant rocks	-2,4%	-6,3%	-6,3%	0,4%	-1,8%	7,7%
8	intermediate rocks	-3,3%	-5,2%	-5,2%	-1,9%	-3,0%	-2,9%
9	soft rocks	9,5%	0,1%	5,2%	5,8%	2,5%	-1,9%

Table 5: Gain development of K_j models for each chronological period (full site sample, surveyed zones).

sites are available in order to construct a model for this window based on taphonomical categories. However, for the Bronze Age and Iron Age significance requirements are not met, so a predictive map based on the χ^2 test for these periods will not be very useful, as the settlement pattern can also be explained by a random distribution of sites. This leaves us with three periods where a useful predictive model based on the χ^2 test can be constructed with a high degree of confidence, the Epipalaeolithic/Mesolithic, middle Neolithic and late Neolithic. These models can then be weighted by means of p_s/p_a ratio, or by using the Atwell-Fletcher method.

It is interesting to observe that valid K_j models may still be developed in cases where the χ^2 test does not meet significance requirements. This is explained by the emphasis placed by the K_j method on the combination of large number of sites and large area units, whereas χ^2 can better be regarded as a measure of concentration of sites. The value of χ^2 depends on the difference between observed and expected sites. Within the total sample, this difference is potentially largest for small area units. In the case where 10 sites are found where only 1 was expected (a difference of 9), the resulting value of χ^2 will be 9.00. If 59 sites are found where 50 are expected however, χ^2 will only be 1.62. In the K_j model such a unit will nevertheless be considered more important than the smaller unit with only 10 sites (as is clearly observed for Epipalaeolithic and Mesolithic, where the strong concentration of sites on loess formations is not reflected in the K_j model).

It remains an open question if K_j models may be used when the number of available sites is very low. The results obtained for the early Neolithic seem to indicate that with low numbers of sites the models become unstable.

A simple method to arrive at a predictive map based on K_j is by plotting the development of the model, either by gain or by the value of K_j itself. Table 5 shows the gain per map category for each chronological period. These individual gain values can be used as a weight factor for each map category. If a qualitative mapping is desired, the mapping becomes a question of deciding on the limits between positive, neutral and negative predictive power (figure 4).

This method of weighting is preferable over the Atwell-Fletcher method. The actual ranking of the units obtained with the K_j method is different, reflecting their relative importance and the "zero site" categories are given different weights, depending on their size. However, the method applied does not say anything about absolute site densities. In order to compare the weights per period, a correction should be applied for the total amount of sites. When

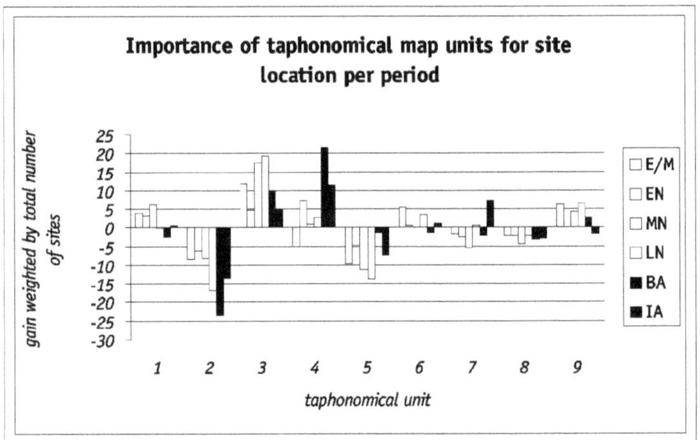

Figure 3: Development of the importance of the taphonomical map units for site location through time, by weighting the gain values from the K_j model by the total number of sites involved.

these weighed values are plotted in a histogram (figure 3) it is immediately clear which units are when important for site presence or absence.

5. Extrapolating site densities

Because of the relatively small number of sites per period in the trenched zones (which only occupy 0.4% of the area), and the fact that the area trenched is not fully representative of the total area, no model has been developed for the trenched zones alone. On the other hand, the trenched zones should give the most reliable estimate of site densities possible, because in theory no sites will escape discovery, whereas field walking will only yield those sites showing significant amounts of archaeological remains at the surface. Within the trenched zones we therefore have the most reliable site sample that can be obtained by means of archaeological survey.

Given this reliable sample, it is theoretically possible to perform an extrapolation of the actual amount of sites per map category. The units where most trenches have been dug are the zones of potential sediment accumulation during the Holocene, i.e. the recent alluvial fans, terraces and riverbeds, the colluvial deposits, the unstable Pleistocene alluvial fans and terraces, and the *cuvettes*. In these units, 0.71% of the total surface has been trenched. For these areas, a cautious prediction can be made of the total number of sites to be found. The total number of sites to be expected is

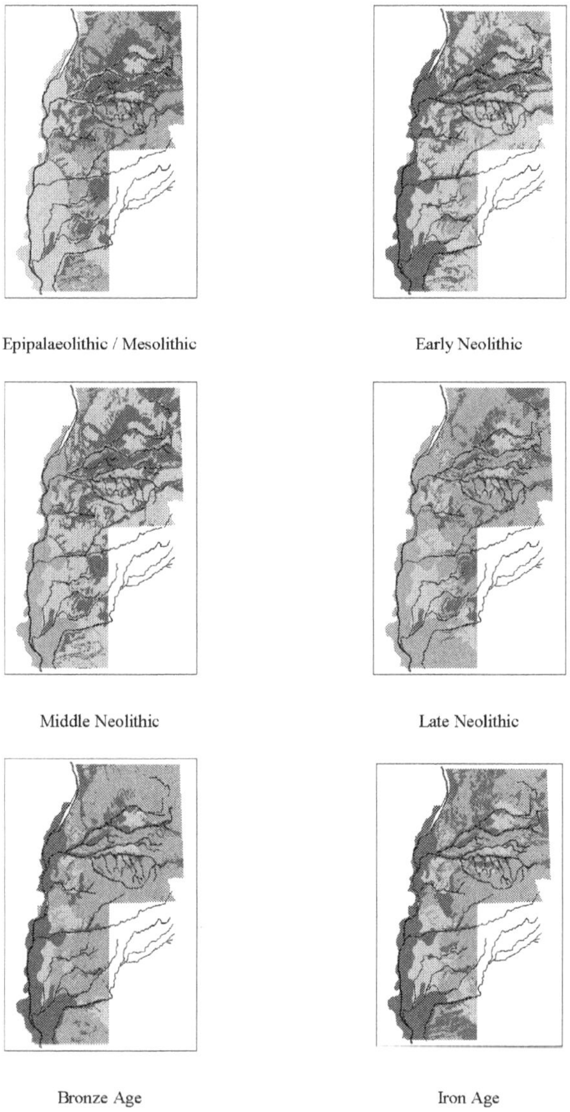

Figure 4: Predictive maps for each of the analyzed periods. Dark grey = positive; medium grey = neutral; light grey = negative.

simply a multiplication of the number of sites found per area unit and the total area:

$$X = N\frac{x}{n}$$

where

X = total number of sites
N = number of area units
x = number of sites found
n = number of area units analysed.

However, this extrapolation is not very useful when the error margin is not known. The standard error of the estimate (Shennan 1988:310) is given by:

$$s_x = N\left(\frac{s}{\sqrt{n}}\right)\cdot\left(1-\frac{n}{N}\right)$$

where

s = standard deviation of the sample.

All probabilistic sampling studies in archaeology depart from the assumption that basic sampling units like survey quadrats (Nance 1990), or even parcels of land (Kvamme 1990) can be defined. Casley and Lury (1982:75) state:

> If the total population of the area is very large, compared to the sample to be selected, the variance of, and hence the precision of estimates calculated from the sample data is a function of the absolute number of sample units, not the sampling fraction.

The basic problem to be solved in order to obtain reliable standard error estimates is therefore defining the size of the sampling units. The smaller these units are, the larger will be the standard errors. In the case of the trenched zones however, the term sampling unit is virtually without meaning, as we can assume that the area trenched has been sampled completely, and therefore no counts per area unit can be performed: the area trenched is equal to one sampling unit. However, with a sample size of 1, the standard errors can not be calculated. The only practical solution – although not an elegant one - is to use the mean surface of the sites as the basic sampling unit. In this case, each individual observation is either a site or a non-site (in statistical terms: we are dealing with a population of ones and zeros), and then the standard deviation of the sample can be calculated with (Shennan 1988:311):

$$s = \sqrt{\frac{p(1-p)}{n}}$$

where:

p = the proportion of interest.

Stratified sampling theory then allows us to narrow down the standard errors for the total sample somewhat by applying the following equation:

$$s_{st} = \sqrt{\sum\left(\frac{N_k}{N}\right)^2 \cdot s^2}$$

where:

s_{st} = standard error for the complete sample
N_k = size of stratum k.

The site surfaces involved can of course not be measured with extreme accuracy. Although surface estimates have been made for most of the sites involved, these are given as ranges from 0.0-0.1 ha, 0.1-0.2 ha, 0.2-0.5 ha, 0.5-1.0 ha and so forth. The surfaces given in table 6 can therefore only be regarded as rather crude approximations (Epipalaeolithic/Mesolithic not included because of lack of buried sites).

When using the site surface estimates from table 6, the standard errors obtained in table 7 are relatively large. Obviously, the low number of "ones" compared to the "zeros" leads to this large standard error. Because of this, decreasing the number of zeros (large site surfaces; middle Neolithic) is more efficient in reducing the standard error than increasing the number of ones (more sites; Bronze Age).

The site surface estimations allow us to perform the same extrapolation for the field walked areas. In general, the estimated surfaces for the visible sites are much larger than for the non-visible sites, because of the spread of archaeological artefacts over the

Period	non-visible sites m²	n	visible sites m²	n
Epipalaeo-Mesolithic	-	-	78438.6	57
Early Neolithic	3029,4	17	73700.0	20
Middle Neolithic	6071,4	28	127590.0	50
Late Neolithic/Chalcolithic	4187,5	40	76890.6	64
Bronze Age	1323,5	85	38948.7	39
Iron Age	3533,3	45	16954.5	11
All periods	3653,5	215	78634.9	241

Table 6: Mean site surfaces for all periods. n = number of sites for which a surface estimate was available.

No.	Taphonomical unit		E/M	EN	MN	LN/C	BA	IA	ALL
1	Colluvial deposits, non-visible sites	X	-	198.7	993.4	0.0	794.7	0.0	1986.7
		s_x	-	197.2	217.9	0.0	393.7	0.0	606.8
	Colluvial deposits, visible sites	X	201.8	80.7	161.4	161.4	20.2	0.0	625.6
		s_x	56.7	37.4	49.6	51.5	19.1	0.0	83.3
3	Unstable Pleistocene alluvial fans and terraces, non-visible sites	X	-	154.6	154.6	154.6	77.3	77.3	618.2
		s_x	-	107.6	53.7	77.8	76.2	76.1	212.7
	Unstable Pleistocene alluvial fans and terraces, visible sites	X	193.5	41.5	193.5	152.0	41.5	13.8	635.6
		s_x	43.2	21.8	39.9	39.3	22.0	12.8	53.5
4	recent alluvial fans and riverbeds, non-visibles sites	X	-	1273.0	763.8	1273.0	1527.6	763.8	5601.2
		s_x	-	562.8	218.2	407.1	618.7	436.9	1141.0
	recent alluvial fans and riverbeds, visibles sites	X	36.3	36.3	72.6	36.3	36.3	0.0	217.7
		s_x	35.1	35.1	49.0	35.1	35.2	0.0	83.5
5	Cuvettes, non-visible sites	X	-	0.0	0.0	59.9	299.2	59.9	418.9
		s_x	-	0.0	0.0	42.5	131.2	58.7	153.8
	Cuvettes, visible sites	X	8.5	8.5	0.0	16.9	8.5	8.5	50.8
		s_x	7.4	7.4	0.0	10.5	7.5	7.5	17.7
	Total, non-visible sites	X	-	1626.2	1911.7	1487.4	2698.8	900.9	8625.0
		s_{st}	-	607.3	313.7	417.6	750.7	448.4	1321.7
		%	-	37.3	16.4	28.1	27.8	49.8	15.3
	Total, visible sites	X	440.0	166.9	427.5	366.7	106.4	22.3	1529.8
		s_{st}	82.0	57.6	82.3	76.5	47.3	15.5	133.8
		%	18.6	34.5	19.3	20.9	44.5	69.5	8.7

Table 7: Estimated number of sites (X) and standard error (s), all periods except Palaeo-Mesolithic.

surface by erosion and agricultural practices (table 6). This obviously means that the number of "zeros" will be substantially reduced, regardless of the number of sites involved.

The extrapolation of the amount of sites for the field walked zones yields the figures in table 7. The total number of sites calculated is much lower than the number obtained for the trenched zones. At the same time, standard errors are much smaller as well. An extrapolation based on the visible sites in the field walked zones is clearly strongly underestimating the number of sites to be found in the sedimentary zones. Furthermore, an extrapolation based on the larger site surfaces provides "false security" when it comes to the accuracy of the estimates.

Theoretically, it is possible to calculate the size of the area that is needed to bring back the standard errors to a more reasonable limit. This is done applying the following equation (Shennan 1988:310):

$$n = \left(\frac{ZsN}{d} \right)^2$$

where:

Z = confidence limit of the estimate in standard deviation units

d = desired tolerance of the estimate.

Z can be used to define a confidence limit for the results of the equation; a Z value of 2.0 equates to a confidence interval of 95.46%. A finite population correction (n / (1 + n/N)) should be applied afterwards obtain the correct values for the required sample size. The third column of table 8 shows the figures obtained when d is set to +/-10% for each single map category. These figures show that it will be necessary to trench about 26 times the area that has currently been trenched in order to obtain a 95.46% reliable estimate within 10% of the total number of sites to be found in the area covered by the four map categories. In general it

No.	Taphonomical unit	d (abs)	d (%)	n (%)	d (abs)	d (%)	n (%)
1	colluvial deposits	198.67	10.00%	15.94%	205.55	10.35%	15.05%
3	unstable Pleistocene alluvial fans and terraces	61.82	10.00%	38.61%	131.72	21.31%	12.17%
4	recent alluvial fans and riverbeds	560.12	10.00%	6.17%	419.79	7.49%	10.47%
5	*cuvettes*	41.89	10.00%	48.25%	105.44	25.17%	12.83%
	total area	862.50	10.00%	18.60%	862.50	10.00%	12.11%

Table 8: Calculated values of n for an accepted tolerance d of 10%, at the 95.46% confidence limit. Columns 1-3 show the results when d=10% for all map units, columns 4-6 show the results when d is weighed according to map unit size.

can be stated that the smaller the number of observed sites in a map unit, the more area needs to be trenched. This means that for statistically reliable estimations of site numbers per map category, the areas with low probabilities of site occurrence should be surveyed more intensively than the areas with high probabilities. In order to reduce the amount of area to be surveyed, while still achieving a tolerance of +/-10% for the whole area, the accepted tolerances for the individual map units may be weighed according to the units' size, as is shown in the last three columns of table 8. Of course trenching of such large areas is not feasible – it will therefore be more practical to combine field walking and augering for such an exercise.

6. Conclusions

We set out to investigate the effect of non-random sampling on the interpretation of site quantities and site location distribution in the landscape of the Tricastin-Valdaine region. The results of the predictive modelling have shown that this effect may be very strong indeed. Especially for the later periods, two effects are observed: firstly, the visible sample is not always representative of the sites found in the field walked zones, which means that certain types of sites are easily overlooked during a field walking campaign.

Secondly, the amount of buried sites is very large for the later periods. These sites are found in landscape units that will not yield a comparable amount of visible sites (this is especially true for the recent alluvial plains and riverbeds). An interpretation of site distribution based on the visible sample alone will therefore strongly underestimate the importance of the sedimentary zones for site location.

The actual quantities of sites extrapolated for the sedimentary zones are very large. Basically, it means that the total number of buried sites to be expected in the sedimentary areas is approximately four to seven times as large as the number of visible sites. The actual reliability of the estimate is difficult to judge, given the crude approximation of mean site surfaces that was used to obtain the sampling unit size. However, these mean site surfaces used are not unreasonable estimates, and obtaining more accurate size data will therefore not drastically change the outcome of the extrapolation because of the effect of the small site surfaces when compared to the total area.

The results of both the predictive modelling and the extrapolation strongly emphasize the need for sub-surface surveying methods in sedimentary areas. This need is long recognised in the Netherlands, where augering has become in integral part of archaeological survey in sedimentary areas, and has led to the discovery of many hitherto unknown buried prehistoric sites, sometimes at considerable depth (e.g. Haarhuis 1995, 1996). The results of the current study indicate that there is no reason to suspect that the situation in France will be very different.

As a last remark, it can be stated that the currently presented predictive model is not very specific for the later periods, as is demonstrated by the lower gain and K_j-values obtained. This means that the model is not very well suited as a tool to guide future surveys, or as an instrument to judge the effect of infrastructural and building activities on the archaeological record. Obviously, the model was not primarily constructed as a tool for archaeological resources management. It served to demonstrate that taphonomy is far more important for site location than was previously thought, and showed the need for a reassessment of both existing site location theory as well as research strategies. A useful model for archaeological resource management is better served by combining elements of the inductive and deductive lines of reasoning, which will probably result in models that make the most of our current archaeological knowledge (Verhagen et al. in press). This might include further research into the palaeogeography of the area, and the analysis of other site location parameters - which may be different for different archaeological periods or parts of the landscape.

Acknowledgements

The authors would like to acknowledge that this work was done in the framework of the ARCHAEOMEDES Project (II), funded by the Directorate General XII of the Commission of the European Union under contract ENV 4-CT95-0159. The authors would like to thank Sander van der Leeuw (Université de Paris I), coordinator of the ARCHAEOMEDES Project, for his support. They would also like to thank the following colleagues and institutions involved in the archaeological surveys in the Tricastin and Valdaine regions:

- A. Beeching, J.L. Brochier and J. Vital (Centre d'Archéologie Préhistorique Valence and CNRS) for the assembly and the interpretation of a substantial part of the archaeological database of the Valdaine basin;
- the TGV Méditerranée archaeological project, financed by the French national railway company SNCF (directed by T. Odiot, excavations carried out by AFAN), which contributed more than 60 excavated settlements;
- and the authors of the archaeological map of the region Rhône-Alpes, J.-P. Daugas and C. Laroche (Direction Régionale des Affaires Culturelles de la région Rhône-Alpes).

Finally, the authors would also like to thank Milco Wansleeben (Universiteit van Leiden) for his valuable comments on the predictive modelling methodology.

References

ATWELL, M.R. and FLETCHER, M., 1985. A new technique for investigating spatial relationships: significance testing. In Voorrips, A. and Loving, S.H. (eds.), *To pattern the past. Proceedings of the Symposium on Mathematical Methods in Archaeology*, Amsterdam 1984 (PACT II): 181-190. Strasbourg: Council of Europe.

ATWELL, M.R. and FLETCHER, M., 1987. An analytical technique for investigating spatial relationships. *Journal of Archaeological Science* 14:1-11.

BEECHING, A., BERGER, J.F., BROCHIER, J.L. and VITAL, J., 1994. De l'espace au territoire en Valdaine, de 9000 à 900 av. J.C. In Beeching, A. and Brochier, J.L. (eds.), *Archéologie spatiale en Vallée du Rhône, espaces parcourus/territoires exploités. Le groupe néolithique et son territoire*: 38-44. Valence: Centre d'Archéologie Préhistorique (Rapport d'ATP). Unpublished report.

BERGER, J.F., 1996. *Le cadre paléogéographique des occupations du bassin valdainais (Drôme) à l'Holocène*. Paris: Université de Paris I. PhD-thesis.

BERGER, J.F., BROCHIER, J.L., JUNG, C. and ODIOT, T., 1997. Intégration des données archéologiques et des données naturelles dans le cadre du TGV Méditerranée. In Bravard, J.P., Chouquer, G. and Burnouf, J. (eds.), *La dynamique des paysages protohistoriques, antiques, médiévaux et modernes. XVIIe Rencontres Internationales d'Histoire et d'Archéologie d'Antibes*: 155-184. Sophia-Antipolis: Éditions APDCA.

BERGER, J.F., MAGNIN, F., THIEBAULT, S. and VITAL, J., 2000. Emprise et déprise culturelle à l'Age du Bronze: l'exemple du bassin valdainais et de la moyenne vallée du Rhône. *Bulletin de la Société Préhistorique Française* 97:95-119.

BERGER, J.F., in press (1). Cycles anthropiques et environnementaux à l'Holocène dans des bassins-versants rhodaniens de rang inférieur (Valdaine et Tricastin, Drôme). In Barrué-Pastor, M. and Bertrand, G. (eds.), *Les Temps de l'Environnement, Toulouse: Geode/CNRS, Actes des Journées du Programme Environnement, Vie et Sociétés PIREVS*, Toulouse, Octobre 1997. Toulouse: Presse universitaire du Mirail.

BERGER, J.F., in press (2). La "dégradation des sols" en moyenne vallée du Rhône et dans le domaine méditerranéen français à l'Holocène (contextes morphoclimatique, paléobotanique et culturelle), chapitre 2: Méthodes, concepts et facteurs de l'érosion, Chapitre 3: Les étapes de la morphogenèse holocène dans le sud de la France. In van der Leeuw, S.E. (ed.), *The ARCHAEOMEDES Project - Understanding the natural and anthropogenic causes of land degradation and desertification in the Mediterranean basin*. Sophia-Antipolis: CRA-CNRS.

BERGER, J.F. and BROCHIER, J.L., in press. Evolution des paysages et des climats dans la moyenne vallée du Rhône et sa bordure préalpine de 13000 à 5000 B.P. In *Actes du colloque international "Epipaléolithique et Mésolithique; les derniers chasseurs-cueilleurs d'Europe occidentale"*, Besançon, octobre 1998. Besançon: Les Belles Lettres.

BORNAND, M., 1967. *Etude pédologique du Bassin Valdainais, 1:25,000*. Montpellier: Institut National de Recherche Agronomique, Service d'Etude des Sols, Centre de Recherche Agronomique du Midi.

BORNAND, M., 1971. *Carte des sols de la région de Pierrelatte – Bourg St. Andéol*, 1:25,000. Montpellier: Institut National de Recherche Agronomique, Service d'Etude des Sols, Centre de Recherche Agronomique du Midi.

BROCHIER, J.L., 1991. Environnement et culture: état de la question dans le Sud-est de la France et principes d'étude autour du Chasséen de la Moyenne Vallée du Rhône. In *Colloque international de Nemours, Identité du Chasséen*, 1989: 315-326. Paris (Mémoire Musée de Préhistoire d'Ile de France 4).

BUTZER, K.W., 1982. *Archaeology as Human Ecology*. New York: Cambridge University Press.

CASLEY, D.J. and LURY, D.A., 1982. *Monitoring and Evaluation of Agriculture and Rural Development Projects*. Baltimore: The Johns Hopkins University Press.

DALLA BONA, L., 1994. *Archaeological Predictive Modelling Project, Ontario Ministry of Natural Resources*. Thunder Bay: Center for Archaeological Resource Prediction, Lakehead University.

DEEBEN, J., HALLEWAS, D.P., KOLEN, J. and WIEMER, R., 1997. Beyond the crystal ball: predictive modelling as a tool in archaeological heritage management and occupation history. In Willems, W., Kars, H. and Hallewas, D.P. (eds.), *Archaeological heritage management in the Netherlands*. Fifty years State Service for Archaeological Investigations: 76-118. Amersfoort: Rijksdienst voor het Oudheidkundig Bodemonderzoek.

HAARHUIS, H.F.A., 1995. *De Waalsprong, gemeente Nijmegen. Archeologisch onderzoek fase A1*. Amsterdam: Stichting RAAP (RAAP-rapport 122).

HAARHUIS, H.F.A., 1996. Gemeente Nijmegen, De Waalsprong. Archeologisch onderzoek fase A/B deel 2. Amsterdam: Stichting RAAP (RAAP-rapport 175).

HODDER, I. and ORTON, C., 1976. *Spatial Analysis in Archaeology*. Cambridge: Cambridge University Press.

KAMERMANS, H. and RENSINK, E., 1999. GIS in Palaeolithic Archaeology. A case study from the southern Netherlands. In Dingwall, L., Exon, S., Gaffney, V., Laflin, S. and van Leusen, M. (eds.), *Archaeology in the Age of the Internet – CAA97*. Computer Applications and Quantitative Methods in Archaeology 25th Anniversary Conference, University of Birmingham. Oxford: Archaeopress (British Archaeological Reports, International Series 750), CD-ROM.

KVAMME, K.L., 1989. Geographical Information Systems in regional archaeological research and data management. In Schiffer, M.B. (ed.), *Advances in archaeological method and theory I*: 139-203. Tucson: University of Arizona Press.

KVAMME, K.L. 1990. The fundamental principles and practice of predictive archaeological modelling. In Voorrips. A. (ed.), *Mathematics and Information Science in Archaeology: A Flexible Framework*: 257-294. Bonn: Holos Verlag (Studies in Modern Archaeology Vol. 3).

LEUSEN, P.M. Van, 1996. GIS and locational modeling in Dutch archaeology: a review of current approaches. In Maschner, H.D.G. (ed.), *New Methods, Old Problems: Geographic Information Systems in modern archaeological research.* Center for Archaeological Investigations, Southern Illinois University (Occasional Paper No. 23).

NANCE, J.D., 1990. Statistical sampling in archaeology. In Voorrips, A. (ed.), *Mathematics and Information Science in Archaeology: A Flexible Framework*: 135-163. Bonn: Holos Verlag (Studies in Modern Archaeology Vol. 3).

ODIOT, T., 1985. Occupations fortifiées de hauteur en Tricastin. In Duval, A. (ed.), *Les Alpes à l'Age du Fer. Actes du colloque "Les Alpes à l'Age du Fer", Chambéry*: 57-72. Paris: CNRS (Revue Archéologique de Narbonnaise, supplément 22).

SHENNAN, S., 1988. *Quantifying archaeology.* Edinburgh: Edinburgh University Press.

THOMAS, D.H., 1976. Figuring anthropology. First principles of probability and statistics. New York: Holt, Rhinehart and Winston.

TOUNSI, I., JUNG, C., BERGER, J.F. and CHOUQUER, G., 1997. Etude de la paléohydrographie et du réseau routier ancien en pays tricastin (France, Drôme-Vaucluse) à partir d'images "Thematic Mapper". In *Journées de Télédétection en Sciences Humaines*, juin 1996: 113-126. Editions Eska (Photo-interprétation 1-2).

VERHAGEN, P., 1995. La carte du potentiel archéologique en Hollande. Une méthode de prédiction fondée sur les données de l'archéologie et du paysage. *Les Nouvelles de l'Archéologie* 61:34-39.

VERHAGEN, P., WANSLEEBEN, M. and Van LEUSEN, M., in press. Predictive Modelling in the Netherlands. The prediction of archaeological values in Cultural Resource Management and academic research. *Archaeological Dialogues.*

WANSLEEBEN, M. and VERHART, L.B.M., 1992. The Meuse Valley Project: GIS and site location statistics. *Analecta Praehistorica Leidensia* 25:99-108.

Archaeological Predictive Modelling for Highway Construction Planning

Zoran Stančič, Tatjana Veljanovski, Krištof Oštir and Tomaž Podobnikar

Spatial Information Centre, Scientific Research Centre at the Slovene Academy of Sciences and Arts

Gosposka 13, 1000 Ljubljana, Slovenia

e-mails: zoran@zrc-sazu.si, tatjanav@zrc-sazu.si, kristof@zrc-sazu.si, tomaz@zrc-sazu.si

Abstract

The main objective of this paper is to present some results of the application and methodology of archaeological predictive modelling in Slovenian highway network planning. Recent analysis of the Slovenian highway construction budget has shown a considerable and constant increase in archaeological fieldwork expenses. According to state concern with cost diminution, we were asked to design a simple and effective methodology for archaeological site prediction. The results of a predictive model were used for consideration of the possible changes of highway location and optimal planning of archaeological fieldwork methodologies.

Key words: archaeological predictive modelling, highway planning, remote sensing

1. Introduction

Predictive modelling is fairly well established in North American archaeology. Interest in predictive modelling in Europe lags behind American trends by about one decade. Generally, predictive modelling in Europe tends to come within the purview of basic pragmatism. While the theory and methodology underwriting predictive modelling is conventional, practical applications of archaeological predictive modelling today can largely be considered within the context of cultural resource management. A great deal of predictive modelling work has already been carried out in the United States (Kvamme 1988, 1992a, Warren 1990, Kohler 1988) and Canada (Dalla Bona 1994), whilst significant applications have occurred in the Netherlands as well (Kamermans and Vansleeben 1999, Van Leusen 1996).

The strength of predictive modelling is that the methodology itself enables exploration and evaluation of locational factors. Therefore the knowledge of possible site location determinants can be tested and improved. What we propose here is that archaeological predictive models can also support highway planning. Integration of predictive models in highway planning is opportune in the early stages of planning (whilst selecting a highway corridor), here the results of predictive models provide immediate information for damage estimates to archaeological monuments. Secondly, predictive models can also be used during the construction stage, their main objective being integrate archaeological fieldwork methodologies with construction. Thirdly, results from predictive models can be used as a consolidated planning tool to define required fieldwork methodologies. Each of these propositions can play facilitate the efficient use of time, and promote cost efficient practice.

The Slovenian national plan seeks to complete and link highways in a network across the country. Prolongation of the building programme affects the public purse detrimentally. A recent examination of the Slovenian highway construction budget has shown a constant and considerable increase in overall costs, but it also disclosed that the principal increase in costs was related to archaeological fieldwork. The reasons for this increase are costs associated with unforeseen archaeological work which cause further delays in construction.

Concern over costs led led to an invitation to design a methodology for archaeological site prediction for a test section of a motorway corridor in the Pomurje region (along the Mura River), in northeastern Slovenia. The corridor measured approximately 11.2 km in length and archaeological prospection had already been carried out in the area including; extensive archaeological fieldwork, geophysical survey and air photo interpretation). The results from archaeological fieldwork could therefore also be used for evaluation of the predictive methodology.

2. Background

2.1. The Pomurje landscape

The Pomurje region lies in northeastern Slovenia, on the western edge of the Pannonian Plain. Here, the surface is gently dissected, with extensive lowlands surrounding the Mura River. The landscape opens to the east; and is interspersed with gentle terraces to the north and small hills towards the southwest. The area could be described as a monotonous landscape and this is reflected in its vegetation. Black alder woods flourish on, mainly, brown alluvial soils, while occasional hornbeam trees are found away from the flood plain. It is noteworthy that a very fertile plain, which is today almost entirely cultivated, lies between the two larger rivers – the Mura and Ledava. Distinguished from the other parts of Slovenia, this region is characterized by its stable geological structure. Non-agglutinate alluvial sediments belonging to the Quaternary and Pliocene compose the drift geology. Consequently, the area is dominated by gravels, sand and clay. Toward the hilly landscapes to the north clay based Holocene sediments emerge, and some light soils can be found in specific areas.

The Pomurje environment most probably always consisted of broad-leaved trees and widely dispersed damp meadows. These native environmental characteristics should be considered when attempting to understand the dynamics of past settlement in a region. Archaeological surveys have demonstrated that both the lowlands and uplands were settled. The distribution of settlements appears to represent a complex and extensive utilisation of the area.

2.2. The archaeology of the Pomurje region

The archaeological heritage of the region was not well known till the mid 19th century. Following this period, chance finds and sites were discovered and recorded in the local archives. Unfortunately, documentation of sites in the area was sporadic, and often lacks geographic coordinate attributes. Archaeological topographical efforts were initiated just after WW II, although a systematic approach to fieldwork was not then implemented. In 1975, the Institute of Archaeology at the Slovene Academy of Sciences and Arts published the first national Register of archaeological sites in Slovenia. Further archaeological research in the 1970s and 1980s confirmed that the Prekmurje region was continuously settled, although settlement intensity was highly variable (Šavel 1991). The national database of archaeological sites ARKAS (Tecco Hvala 1992, Modrian 1994) is currently being updated with recent records. Unfortunately, the level of preservation of the sites is not good. For instance, many prehistoric barrows in the region have been looted, and further damage has also been caused in recent decades by the use of machinery for cultivation.

Some 90 archaeological sites were included within this study. The chronological attribution of the sites ranges from the Bronze Age to the Early Medieval period. The types of sites also vary from isolated finds to settlements. In the Pomurje study, site locations were broken down into five distinct types to allow locational characterisation, and to allow us to recognize diversity in their distribution throughout the landscape.

2.3. The site's predicting methodology

A great deal of information on predictive modelling methodology, concepts and application have already been published (Judge and Sebastian 1988, Dalla Bona 1994, Westcott and Brandon 2000). It is important to emphasize that within the context of the Pomurje study, our goal could not be oriented towards the development of a model that would aid our understanding of the known distribution of the sites. Our aim was to develop a consistent model, which would assist in the process of highway planning.

Instead of presenting a theoretical background on predictive modelling methodology, we prefer to explain briefly what guided us in our choice of the method. Considerable problems often occur when the sample of sites analysed is small, as is the case in our study and are particularly obvious when multivariate statistics are used (Stančič and Veljanovski 2000). Some statistical evidence may be hidden in this process and the significance of specific variables also lost. Another issue issue is that most of the variables used are, naturally, based on the topographic characteristics of the area. An archaeological predictive model is most useful and reliable when it is applied in a region with a diverse natural environment. Surface topography plays a special role in predictive models as a consequence of its association to past land use. As the Prekmurje case study area is almost flat, it would be difficult, and misleading, to extract any rules, or predict potential, solely on the basis of the relief-based distribution of the sites. This situation applies not only to sites that vary in terms of their chronologies, but also between sites of different types. Despite the fact that a powerful predictive model can be made with a limited number of strong location factors,

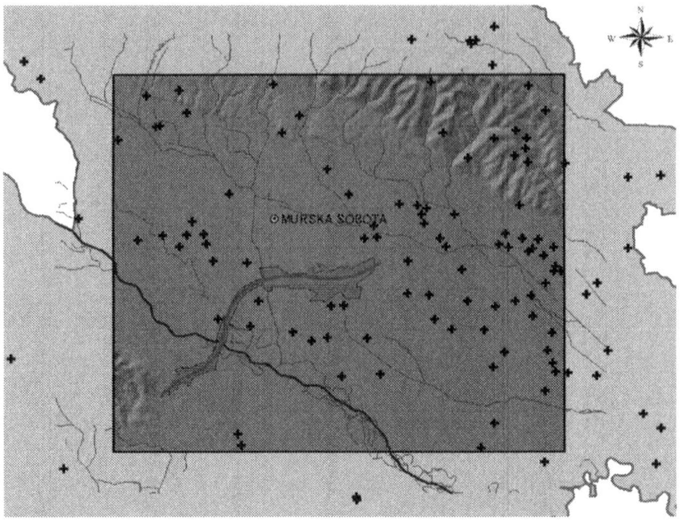

Figure 1: The Pomurje region and working environment for predictive model development together with the distribution of known archaeological sites (from National Database ARKAS) and anticipated location of highway section.

much effort was expended in the search for a range of other, potentially, relevant variables.

With regard to the situation described above and the objective of the case study, complex methods based on multivariate statistics failed to suit our data. Therefore it was finally decided to apply a simple traditional method of Boolean overlay to create a predictive model of site location in the region. As insufficient numbers of variables was a potential pitfall of the analysis, we additionally investigated the potential of remotely sensed data to the case study.

3. Data and modelling

The case study focused on a very flat part of the Pomurje region with little geomorphological change. Elevation across the study area as a whole ranges from c. 165 m in the lowlands to c. 315 m on the upland terraces. In the 11.2 km long highway test section, however, variation in elevation is only 16 meters. The lowlands sweep gently along the Mura River, and the terraces are crossed by a series of gently sloping valleys; some of these are also important sources of water. The study area comprises part of the northern upland terrace and extends about 21 km in an east-west direction and 18 km north-south, all together incorporating about 384 km^2 (figure 1). The basic units of analysis for the development of the model were regular square cells measuring 30 m.

We incorporated elevation, slope and aspect as variables representing the morphology of the landscape. As the natural environment is rather homogeneous, we decided that an accurate digital elevation model (DEM) would prove important, and we created a high-resolution DEM. A high resolution and accurate DEM are required for modelling settlement distribution in flat areas (Stančič and Oštir in press) and a DEM with a cell size of 25 by 25 metres, with a height accuracy of 3 m and a positional accuracy of 10 m was generated using radar images and radar interferometry (Oštir and Stančič in press).

Geological structure and data relating to the quality of soils can also serve as relevant variables representing the natural characteristics of the region. While geological factors are considered trustworthy, soils should be handled with more caution, as the

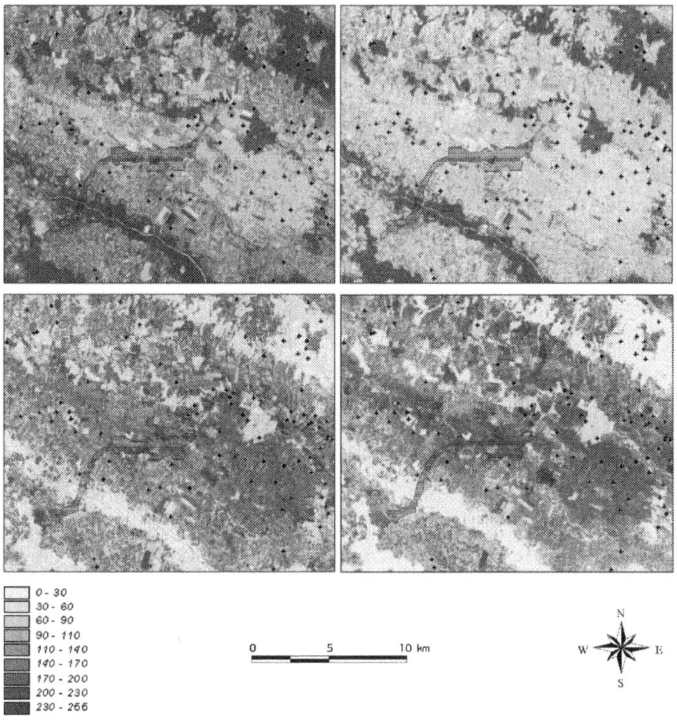

Figure 2: Vegetation index (upper left), Clay index (upper right), Ferrous Minerals index (lower left) and Ferric Oxide index (lower right) with distribution of archaeological sites and anticipated highway section.

data available usually expresses the most contemporary situation in the region. In addition to these data, a remotely sensed land use layer was created. Hydrology is also relevant. Given the case study goals, artificial ditches were not excluded from the hydrology vector layer, and these were considered "streams". Following this the distance to the nearest stream and distance to the main river within the region were calculated. Using these two layers, our aim was to enable an evaluation of the importance of the Mura River as a possible communication route.

The above variables are familiar and have been used in a number of locational studies (Kvamme 1992b). Given that our study area is a large plain, differences observed in variables such as elevation, slope and aspect could never play significant roles. Hence we were primarily concerned the variables that could provide a description of the lowland itself and this we felt was likely to be obtained from remotely sensed data.

Besides land use and a vegetation index, original information relating to surface soil type was attained through the generation of mineral delineation layers. These indices are extensively used in mineral exploitation and vegetation analyses, essentially because they can indicate minute differences between various rock types and vegetation classes (ERDAS Field Guide 1997). In many cases, judiciously chosen indices can highlight and enhance differences which cannot be observed in the original colour bands displays from satellite imagery. Therefore we produced three layers suitable for GIS analyses representing Clay Minerals, Ferrous Minerals and Ferric Minerals (Iron Oxide). Actually, these indices represent the presence of minerals on the surface (figure 2). In calculating these indices we hoped to obtain evidence for significant variation in the lowlands which was not be anticipated from the geomorphic variables.

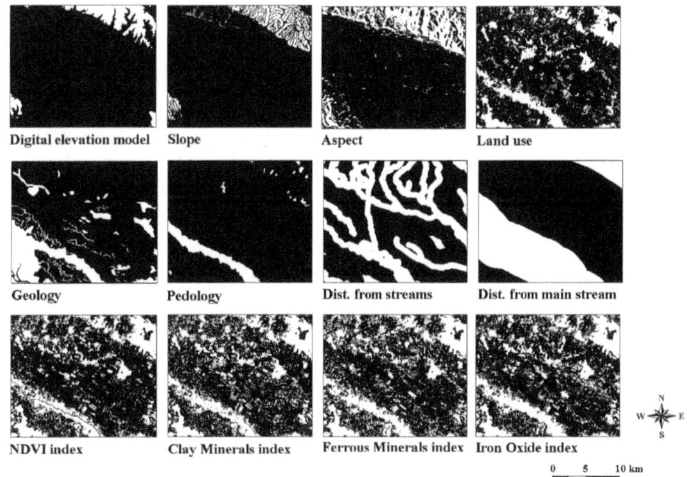

Figure 3: Binary layers representing prehistoric settlements proneness to twelve variables.

All map images were converted to a grid so as to create a series of primary layers portraying elevation, water streams, geology, soils, land use and archaeological site distribution. Some of the primary layers were later manipulated for derivative layers including aspect, slope, distance to the stream and distance to the river, as we hoped that these would help demonstrate how settlement behaviour related to the natural environment.

The majority of these variables used were in ratio scale. Some were later transformed into classes for further correlation testing. Several variables, such as geology, land use and quality of soils, were represented as a nominal scale.

Exploration and statistical analyses were carried out for five archaeological groups: prehistoric settlements (9), Roman and Early Medieval settlements (11), barrows (29), undated barrows (18) and isolated finds (mainly prehistoric axes, 15). A few of archaeological records were omitted from further processing, as it proved impossible to include them amongst these specified groups.

Univariate statistical tests (χ^2 and the Student *t*-test) were used to test the correlation between site locations and each of the variables. The results of univariate statistical tests indicate that different site types in the study area tend to occur in different environmental settings; however, from a statistical point of view, the correlation was quite tentative. Nevertheless, all the variables that were initially examined for significance were later included in the model, if they fulfilled "threshold" conditions. As two scale types were 5nv63ved, two criteria were introduced: a) for continuous variables, the interval between minimum and maximum values for sites was defined, and the circumscribed area considered as having potential, b) for nominal variables the area covered by classes associated with sites was treated as having potential. Twelve binary layers were prepared for each archaeological sample.

The initial model was designed as a simple sum of binary layers. Cells that fulfilled the entry condition for a particular variable were assigned a value of 1 and the rest a value of 0 (see figure 3). As a result, we attained five models for different site types, with a potential ranging from 0 to 12. Evidently, the score was a combined measure of the potential of the cell. For the purposes of overall compliance, the predictive model for the region was then obtained by merging the five model outputs.

On the basis of the preliminary analysis, we decided not to apply linear or logistic multivariate regression to define the relative weights of the variables. The weights are supposed to represent the relative relationship of locational factors within the past settlement system. As opposed to weights usually obtained through complex statistical examination, we believed that the weights most appropriate for a model of this kind should be based simply on the "sum of potential". Criteria were based on the assumption that a good predictor isolates highly specific and, therefore, a close or discrete signal in comparison to the values of the background. If a variable interpreted as a potential indicant also covered a large part of the study area then a smaller weight was assigned to that variable. Consequently, weights were applied in proportion to the observed power of the variable. As three mineral indices and one vegetation index seemed to be highly correlated we decided to give these additional weights. Each of their powers was diminished by a quarter and ultimately they represented – in their combined form – a single variable.

The final model (figure 4) was based on archaeological data attained from the national database and environmental data obtained from maps and satellite images. The social environment was poorly represented, as data , for example with respect to communications, were not reliable enough to allow prediction.

4. Results and testing

The validity of the resulting weighted model was tested by measuring the accuracy of its predicted potentials using two different samples,. The first was a sample of all archaeological sites that were used for the development of the model. This type of testing is known as internal testing and was carried out as a control of the process. Its results indicate that the model's prediction of potential was correct by about 81 %, if "high potential" incorporates areas with potentials ranging from 70 to 100 (figure 5). The remaining 19 % of sites fell within the limits of "medium potential" (40 to 70); no sites fell into the "low potential" range.

The data used for secondary testing were obtained through archaeological survey in the area of the proposed highway corridor. As most of this area has undergone cultivation for some time, survey was restricted to cultivated fields with good ground-surface visibility. The sites discovered were on the basis of artefact scatters located on the surface. Where a scatter was deemed promising the ground was intensively searched for artefacts and other traces of occupation sought through excavation of several 1 x 1m test trenches. Geophysical survey and aerial analysis of the area were also carried out prior to detailed work.

The second subset of archaeological data was therefore an actual independent testing sample as the sample was gained from recent archaeological survey. As these data were not at our disposal until the model was developed and tested with our own data, we can consider this second sample as valid for use for external testing of the model. As no excavations were carried out in the area during the period of development the extent of potential sites was not defined, and as a consequence we could only extract a centroid from the various areas that surface artefact scatters were located. These centroids are presumed to be representative of the these locations. As a scatter's area is sometimes quite extensive, two test samples were made: one with 13 principal centroids from 13 "sites" and the second one with 25 centroids derived from specified artefact densities within 13 "sites". It is important to note that

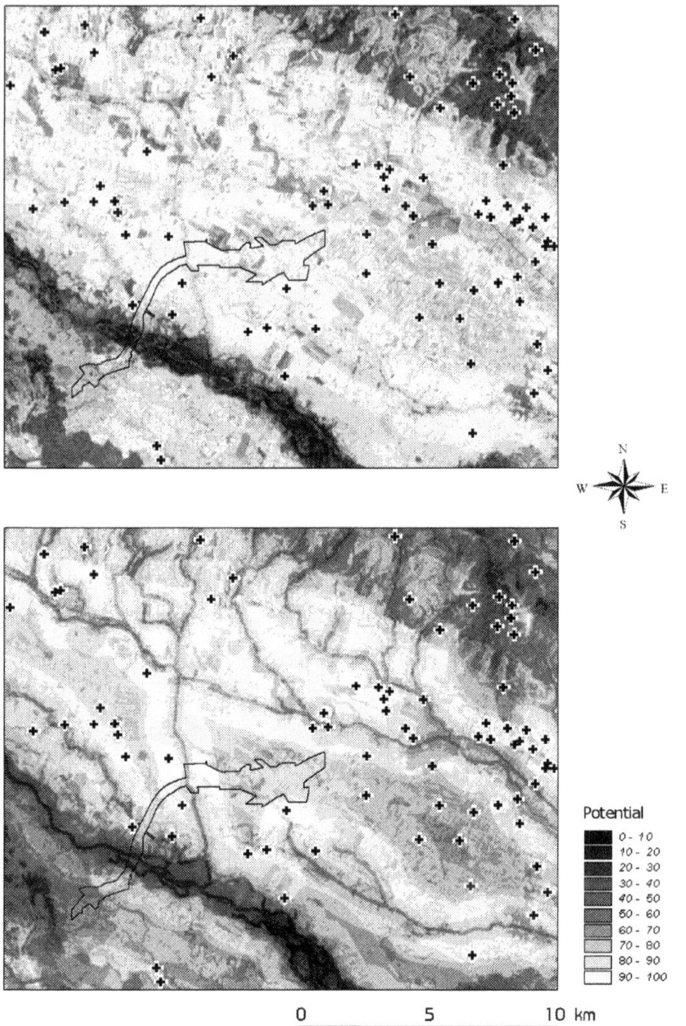

Figure 4: Archaeological predictive model for the Pomurje region: a) unweighted model (upper) and b) weighted model (lower) with the distribution of the sites upon which the model was developed.

the predictive model of the region was "windowed" to extract an area coincident with the independent archaeological survey, which was constrained to the highway corridor itself (figure 5). The results from both tests were similar. In the first case, six out of 13 centroids fell within the zone of highest potential (90 – 100) and the remaining seven just in the lower zone. The second test located 14 centroids within the highest potential zone, 10 centroids within the second highest range, and a single centroid fell within a lower range. A comparison of the statistical parameters within these tests could also be used as a technique to confirm the validity of the process. There is a substantial shift of both "sites" average potential within these tests. The predicted potential of the cells of sample centroids rises to 88 against a background average of 70.

Both tests indicate that the model's predictions are accurate. Although it is difficult to ascertain from the final model why particular sites within the region are situated within a specific area or why the potential of another site is lower, the model still distinguishes between locations where sites are most likely to be present and locations where they are not. The model does not, however, allow us to isolate the most important environmental or social determinants of site locations. In the end what we are providing to

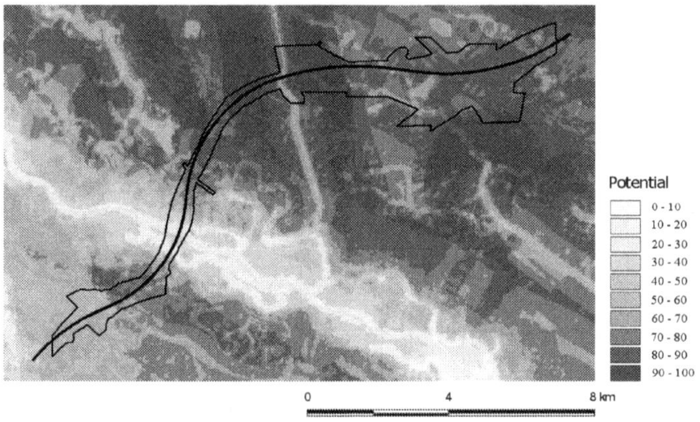

Figure 5: Predictive model of the Pomurje region (weighted model) showing area restricted to the highway location.

highway network planners is simply the expected distribution of archaeological resources.

Figure 5 illustrates the range of potential for site location. In studying this image it is important to note that in this "windowed" image, 63 % of the total land area can be coded as high potential, 26 % as medium potential and 11 % as low potential. This assumes that a range of 70 – 100 is considered as high potential, 40 – 70 is medium, and 0 – 40 has a low potential. Although there are clearly problems in this process the predictive power of such a model should still be useful for highway network planners. Ultimately, the relatively poor performance of the model may simply result from the geographical conditions that are far from suitable for predictive modelling. The manner of determining threshold conditions and the need to generate five smaller predictive models for specific site types may also mitigate against the model's efficient performance.

5. Conclusions

Although the results obtained from the model are good, several issues must be kept in mind when considering the results. The first issue, which is almost impossible to overcome, is the lack of comprehensive and reliable archaeological data. This is particularly problematic when creating an inductive model. The second point is that any generated prediction may be misleading because of the underlying dataset. The predictive model of the Pomurje region is ultimately based on the available, and essentially environmental, data. It does not take into account any of socio-political factors that must have played an important role in past settlement and land use decisions within the area.

Despite this the model does seem to provide highway planners with accurate information relating to areal archaeological potential. On this basis we believe that the model's predictive capabilities could be used in several ways:

- During the early stages of a highway planning (corridor study). Prior avoidance of a high-potential area could result in lower costs both for archaeological fieldwork and in terms of delay in highway construction.
- During the ongoing stages of highway construction, the model could be helpful in determining priority areas for intensive archaeological fieldwork. Although this would have to be in combination with results obtained through previous archaeological survey. The model could also identify low-potential areas where the construction process could be initiated with minimal archaeological supervision.

Acknowledgements

The Slovene National Agency for Highway Construction (DARS d.d.) funded the project backing this paper. The authors would like to extend their gratitude for all data provided concerning the highway section location, and especially for the results from recent archaeological fieldwork analyses, which were carried out by the Ministry of Culture, the National Institute for the Preservation of Cultural Heritage. A significant opportunity for independent testing of the predictive model was enabled by means of these recent archaeological results. Special thanks go also to the Institute of Archaeology, Scientific Research Centre at the Slovene Academy of Arts and Sciences, for accessing the national database of archaeological sites ARKAS.

References

DALLA BONNA, L., 1994. *Cultural Heritage Resource Predictive modelling Project*. Volume 3, Methodological Considerations. Centre for Archaeological Resource Prediction, Lakehead University, Thunder Bay.

ERDAS Inc., 1997. *ERDAS Field Guide TM*. Fourth edition. Atlanta, Georgia.

JUDGE, W., SEBASTIAN, L. (eds.), 1988. *Quantifying the Present and Predicting the Past: Theory, Method and Application of Archaeological Predictive Modelling*. U.S. Government Printing Office, Washington, D.C.

KAMERMANS, H. and VANSLEEBEN, M., 1999. Predictive modelling in Dutch Archaeology, Joining forces. In Barceló, J.A., Briz, I. and Vila, A. (eds.), *New techniques for old times, CAA98*, Barcelona, BAR International Series 757.

KOHLER, T.A., 1988. Predictive Locational Modelling: History and Current Practice. In Judge, W. and Sebastian, L. (eds.), *Quantifying the Present and Predicting the Past*: 19-61. U.S. Government Printing Office, Washington, D.C.

KVAMME, K.L., 1988. Development and Testing of Quantitative Models. In Judge, W. and Sebastian, L. (eds.), *Quantifying the Present and Predicting the Past*, 325-429. U.S. Government Printing Office, Washington, D.C.

KVAMME, K.L., 1992a. A Predictive Site Location Model on High Plains: An Example with an Independent Test: 19-40. *Plains Anthropologist* 37 (138).

KVAMME, K.L., 1992b. Terrain Form Analyses of Archaeological Location Through Geographic Information Systems. In Lock, G. and Moffet, J. (eds.), *Computer Applications and Quantitative Methods in Archaeology*: 127-136. BAR International Series 577.

LEUSEN, van P.M., 1996. GIS and Locational Modelling in Dutch Archaeology: A Review of Current Approaches. In Maschner, H.D.G. (ed.), *New Methods, Old Problems: GIS in Modern Archaeological Research*: 177-197. Center for Archaeological Investigations, Occasional Paper No. 23, Southern Illinois University.

MODRIAN, Z., 1994. Kataster arheoloških najdišč Slovenije (ARKAS) (II. del). *Arheo* 16, Ljubljana.

OŠTIR, K. and STANČIČ, Z., in press. Interferometric generation of DEM for mobile telephone network planning. In *Proceedings of the 'Fringe 99' Workshop on ERS SAR Interferometry*, Noordwijk, The Netherlands.

STANČIČ, Z., OŠTIR, K., in press. Producing Digital Elevation Models with Radar Interferometry. In Computer Applications and Quantitative Methods in Archaeology. *Proceedings of the CAA conference, Dublin 1999.*

STANČIČ, Z., VELJANOVSKI, T., 2000. Understanding Roman Settlement Patterns Through Multivariate Statistics and Predictive Modelling. In Lock, G. (ed.), *Beyond the Map: Archaeology and Spatial Technologies*. NATO Science Series, IOS Press Amsterdam: 147-157.

ŠAVEL, I., 1991. *Arheološka topografija Slovenije, Topografsko področje XX (Prekmurje).* SAZU, Ljubljana.

TECCO HVALA, S., 1992. Kataster arheoloških najdišč Slovenije ali zgodba o nastanku neke računalniške baze podatkov (I. del). *Arheo* 15, Ljubljana.

WARREN, R., 1990. Predictive Modelling of Archaeological Site Location: A Case Study in the Midwest. In Allen, K.M., Green, S.W. and Zubrow, E.B.W. (eds.), *Interpreting Space: GIS and Archaeology*: 201-215. Taylor & Francis, London.

WESTCOTT, K.L., BRANDON, R.J. (eds.), 2000. Practical Applications of GIS for Archaeologists – A Predictive Modelling Kit. Taylor & Francis, London.

6. Future Trends in Spatial Analyses

The Aksum Project (Ethiopia): GIS, Remote Sensing Applications and Virtual Reality

Maurizio Forte

CNR-ITABC, Institute of Technologies Applied to Cultural Heritage

Via Salaria km.29, 300 C.P.10, 00016 Monterotondo St., Roma, Italy

e-mail: maurizio.forte@mlib.cnr.it

Kathryn A. Bard

Boston University

Boston University 675 Commonwealth Ave. Boston, MA 02215, USA

e-mail: kbard@bu.edu

Rodolfo Fattovich, Monica Foccillo, Andrea Manzo and Cinzia Perlingeri

Istituto Universitario Orientale, Dipartimento di Studi e Ricerche su Africa e Paesi Arabi

Piazza S. Domenico Maggiore 12, 80134, Napoli, Italy

e-mail: ybeyene@iuo.it

Abstract

Between May and June 1993 the Istituto Universitario Orientale, Naples (Italy), and Boston University, Boston (USA) archaeological expedition to Aksum (Ethiopia) resumed investigations at Bieta Giyorgis, a hill to the NW of Aksum. Six field seasons have been directed by Rodolfo Fattovich (IUO) and Kathryn A. Bard (BU) (in 1993, 1994, 1995, 1996, 1997 and 1998). The general goal of the IUO and BU Expedition is to study the development of complex societies in Tigrai from late prehistoric (3rd-2nd millennia BC) to medieval times (14th century AD). Earlier investigations have focused on the rise of complex societies in the western Sudanese lowlands (Kassala, Sudan) in the 3rd-2nd millennia BC, and the Pre-Aksumite state on the Tigrean plateau (Yeha, Tigrai) in the 1st millennium BC. The current phase of the project is investigating the rise and development of the Aksumite state (late 1st millennium BC- early 1st millennium AD). Particular emphasis has also been given to the study of the involvement of Aksum in the trade network from the Mediterranean to the Indian Ocean. Excavations at Bieta Giyorgis were aimed at testing the hypothesis, based on traditional Ethiopian sources, that the hill was an area of early development at Aksum. An important goal of this project is to investigate the origins and urban development of Aksum within its environmental setting. The project includes research in archaeology, paleo-ethnobotany, archaeo-zoology, ethno-archaeology, history, geology and geomorphology, digital technologies as well as systematic mapping and conservation. This paper will present the preliminary results of GIS and Remote Sensing applications concerning the area of Aksum, comparing 2D and in 3D digital data including aerial photos (1:60,000), satellite images (Landsat TM, SPOT XS), cartographic data and landscape documentation. Finally, some important archaeological questions will be raised: how and what can we perceive from the archaeological landscape? What are the relationships between aerial photo-interpretation, satellite imagery, and the landscape? How can 3D reconstruction help us to reconstruct and to interpret archaeological landscapes? The importance and problems of this project have been exacerbated by the "war situation" between the countries of Ethiopia and Eritrea.

Key words: Aksum, archaeological landscape, remote sensing, 3D GIS, virtual reality

1. Introduction

Field archaeology, research and work in progress in the area of Aksum in Ethiopia (figure 1), involved articulated problems that require a multidisciplinary approach. In this paper, we discuss the complex issues of the virtual reconstruction of the archaeological and diachronic landscape, starting with GIS and remote sensing applications.

The project involves the application of advanced digital technologies in order to provide a detailed reconstruction of the archaeological landscape: analysis and classification by GIS and remote sensing, as well as interpretation and communication through virtual reality and visual information systems.

The innovative component of the project is its multidisciplinary approach, starting with the acquisition of the data on the ground and, then, the creation of predictive maps of the geomorphologic and anthropic landscape, including all paleoenvironmental factors (this work is ongoing).

The possibility of interacting with the reconstructed landscape in real time produces new cognitive scenarios in visual anthropology for the scientific communication and the collective knowledge (mindscape). We have created a virtual and diachronic museum of the entire territory, on the basis of visual cultural information and of the complexity of the eco-cognitive systems.

1.1. The Aksum project (Ethiopia)

Since 1993 the Istituto Universitario Orientale, Naples (Italy), and Boston University, Boston (USA) archaeological expedition at Aksum (Ethiopia) have been conducting investigations at Bieta

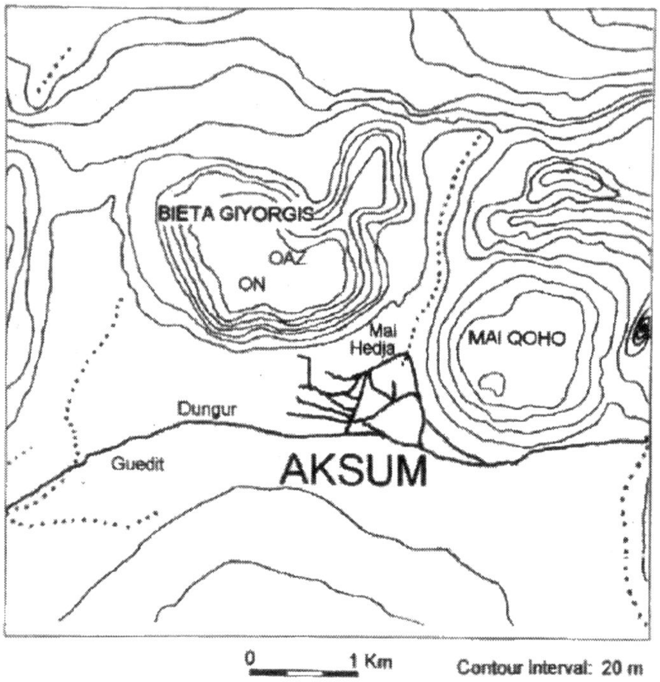

Figure 1: Map of the excavation area of Bieta Giyorgis.

Giyorgis (figure 1), a hill to the northwest of Aksum. Six field seasons have been directed by Rodolfo Fattovich (IUO) and Kathryn A. Bard (BU) (in 1993, 1994, 1995, 1996, 1997 and 1998, see: Bard et al. 2000).

The general goal of the IUO and BU Expedition is to study the rise and development of the Aksumite state (late 1st millennium BC- early 1st millennium AD). Excavations at Bieta Giyorgis have been concerned with testing the hypothesis, based on traditional Ethiopian sources, that the hill was an area of early development at Aksum, the capital city of the kingdom.

An important goal of this project is to investigate the origins and development of Aksum within its environmental setting. The project includes research in archaeology, palaeo-ethnobotany, archaeo-zoology, ethno-archaeology, history, geology and geomorphology, as well as systematic mapping and conservation.

The Expedition has been funded by the Italian Ministry of Foreign Affairs (Rome), CNR (Rome), the Ministry of University, Scientific and Technological Research (Rome), the National Geographic Society (Washington DC). Minor contributions were also provided by the African Studies Center, BU (Boston), Michigan State University (for ethnohistorical and ethnoarchaeological research), and NASA, Langley Research Center, Atmospheric Sciences Division, Hampton, VA (for geological investigation).

1.2. The Kingdom of Aksum: historical outline

Aksum is the site of the capital city of the ancient Aksumite kingdom that dominated the southern Red Sea in the early through mid-1st millennium AD. At its height, Aksum was one of the great kingdoms of the ancient world, involved in a trading network that stretched from the Mediterranean to India. At that time, the Aksumite kingdom was a partner of the Roman/Byzantine Empire onhe Red Sea trade route to the Indian Ocean. Recent excavations at Bieta Giyorgis hill indicate that the kingdom emerged from a local complex society of the late 1st millennium BC (the Proto-Aksumite Phase, approx. 400/300-100 BC).

The Aksumite kingdom began to decline in the 7th century AD, largely due to Islamic expansion along the African coast of the Red Sea. The Aksumite kingdom eventually disappeared by the 10th century AD (Late Aksumite Phase, ca. AD 700-900).

1.3. The Capital city

Aksum is located at the edge of a broad plain, and is dominated by the hills of Bieta Giyorgis and Mai Qoho, which are separated by a seasonal stream, Mai Hedja. At present, the ancient settlement is partially covered by the modern town. Aksum is one of the major archaeological areas in Ethiopia, and one of the most impressive sites in sub-Saharan Africa. The town is the main religious centre of the Ethiopian Orthodox Church, and is included in the UNESCO World Heritage List.

The archaeological area of Aksum includes (figure 1):

1. The settlement area of the ancient capital city, at the base of Bieta Giyorgis hill, and in the valley between Mai Qoho and Bieta Giyorgis (figures 1, 2).
2. Cemetery areas with stelae fields, along the Mai Hedja, at the base of Mai Qoho hill, and to the south-west and west of the settlement area.
3. Bieta Giyorgis hill, with a large settlement (Ona Nagast), a cemetery with stelae (Ona Enda Aboi Zewgè), and the remains of two Aksumite churches.
4. Two monumental subterranean tombs (Enda Kaleb and Gabra Masqal), and some pit-tombs on the top of Mai Qoho hill.

Many smaller sites have also been recorded around the main archaeological area. They form the rural hinterland of Aksum. The occurrence of archaeological sites in the area is usually indicated on the surface by concentrations of stone rubble left behind from collapsed structures.

1.4. The excavated sites

The IUO/BU expedition has conducted excavations at the sites of Ona Enda Aboi Zewgè, a large stelae field and cemetery, and Ona Nagast, the settlement associated with the cemetery.

The site of Ona Enda Aboi Zewgè (OAZ) extends over an area of about 17 hectares, and is partly covered by a modern village. Over 100 stelae (possibly up to 300) are now visible on the surface, including both roughly hewn monoliths and ones which were more carefully carved, with rounded tops, and are up to 10 m high. The largest stelae are mainly concentrated on the southern side of the site.

The site of Ona Nagast (ON) is located on a slope of Bieta Giyorgis hill about 500 m to the west of OAZ. The site covers an area exceeding 12 hectares. Part of the ancient settlement is covered by a modern village, and in the centre of the site stands a large rock-cut cistern, it is of unknown age but most likely ancient.

The general topography of Ona Nagast, and particularly the three terraces to the east sloping down to the Mai Lahlah, suggests a carefully planned complex. Results of excavations, which have unearthed substantial monumental structures, specialised functional areas, and non-elite domestic areas, suggest that the site was occupied for at least 1000 years.

1.5. Environmental setting

Aksum is located on the Tigrean Plateau in northern Tigray about 22 km to the west of Adwa at an average elevation of 2200 m. The geographical co-ordinates are 14° 7' 8" N, 38° 43' 46" E. The hills of Bieta Giyorgis and Mai Qoho dominate the present town to the north-west and east, respectively. Some hills, with an average elevation ranging between 2406 m and 2289 m, surround Aksum, delimiting a roughly circular plain approximately 10 km in diameter, which gently slopes from north to south and south-east.

Two main eco zones can be distinguished within the region;

1. The plain around Aksum, with a low gradient and highly fertile land, which is optimal for plough cultivation;
2. Hills with a steep gradient or naturally terraced land, which are less favourable for ox-plough cultivation.

The rock sequence exposed in the Aksum area includes;

1. Variegated sandstone in bedding sets 2 m thick (Takkaza Sandstone), with a total thickness of at least 320 m.
2. Flood basalts (Emba Ayba Basalts), with a maximum thickness of about 600 m, and rare tuffs that form extensive outcrops.
3. Blocky trachtyc lavas, and large platforms of basalts elevated by subsurface intrusions and plug domes (Hyperalkaline Silicic Lavas).
4. Unconsolidated or semi-consolidated clays, silts, sands, and gravels deposited in fluvial or colluvial environmental settings at the top of the rock sequence in the area (Quaternary Sediments).

1.5.1. Bieta Giyorgis

Bieta Giyorgis rises 200 to 250 m above the surrounding terrain (figure 7). The hill itself is comprised of three distinct physiographic sections: the steep flanks; the horse-shoe shaped, gently sloping Ona Nagast plain; and the central uplands of Beta Giyorgis. The drainage pattern on the hill is radial and streams are intermittent.

The flanks of Bieta Giyorgis include the very steep slopes and cliffs that form the perimeter of the mountain. These slopes are littered with a discontinuous cover of large boulder colluvium.

The Ona Nagast plain is of a horseshoe-shape, a gently undulating plain that surrounds the central uplands on three sides. The plain, which is about 175 m above the Aksumite plain, is bounded on its outer edge by a knobby rim. The knobs are underlain by bare to thinly veneered fine-grained igneous rock. The south-eastern part of the plain is underlain by stratified sedimentary rock. Elsewhere the plain appears to be formed directly on the igneous rock. The streams generally follow the joint pattern in the igneous rock or zones of weaker metamorphic rock. The central uplands consist of a rolling terrain underlain by relatively homogeneous, but jointed, igneous rock.

1.5.3. Environmental history of the Aksum region

Paleo-environmental evidence for the Aksum region is very scarce. Four major aggradation episodes occurred during the last millennium and may represent periods with increasing soil erosion due to heavier rains and/or more intense human disturbance. These episodes are tentatively dated to;

1. AD 100-350 (heavy rainfall with strong periodic floods, on a landscape already partially deforested and degraded because of intense land-use).
2. AD 650-800 (a period of deep soil erosion due to very intense land-use and very heavy periodic rains, combined with a progressive abandonment of settlement).
3. Late 1st millennium AD (more soil erosion due to a late abandonment or eventual destruction of settlement).
4. 19th-20th centuries (recent soil erosion due to more intense land use on a greatly denuded landscape).

Preliminary studies of pollen samples taken from archaeological sediments at Ona Enda Aboy Zewgè and Ona Nagast indicate that:

- In the middle to late 1st millennium BC the vegetation cover on Bieta Giyorgis hill was dominated by shrubs and herbaceous plants characteristic of open vegetation and areas of human settlement. The very low arboreal frequency in the samples (and virtual absence of tree pollen normally transported long distances by wind) suggests that trees were not common components of the vegetation pattern in the general area of Aksum.
- In the second half of the first millennium AD the general vegetation pattern on Bieta Giyorgis was very similar to the pattern of today: a predominance of grasses and shrubs with only a few isolated trees near structures, in gorges along seasonal streams, and on rocky slopes (13: BG).

At present we do not have pollen data for the first half of the first millennium AD.

2. Remote sensing and GIS applications

2.1. The digital Project

The Aksum Project is a work in progress but, in order to obtain good results from the outset, we have constructed a "conceptual model" of the project (figure 2) on the basis of the available data which includes;

- Vector and raster digitalisation (maps, aerial photos, morphological features, etc.)
- DEMs creation (micro-macro scale).
- Satellite image classification (multi-spectral).
- Integration and overlay of raster and vector data: aerial photos, satellite images, DEM, multi-layered surfaces.
- 3D processing and VR navigation through the archaeological landscape.
- Multi-spectral classification of Landsat TM and SPOT XS images.
- Predictive archaeological maps on the basis of satellite training regions.
- GIS applications (spatial analyses).
- Creation of landscape virtual models in OpenGL (format Openflight, .flt) for the visualisation and navigation on the

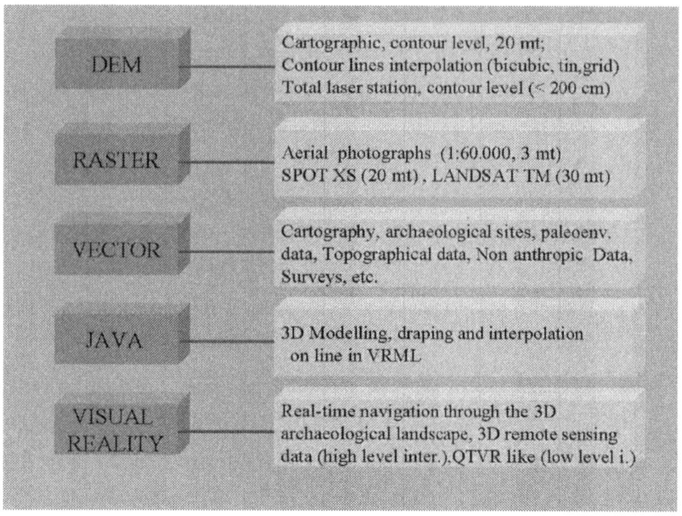

Figure 2: The scheme of the project.

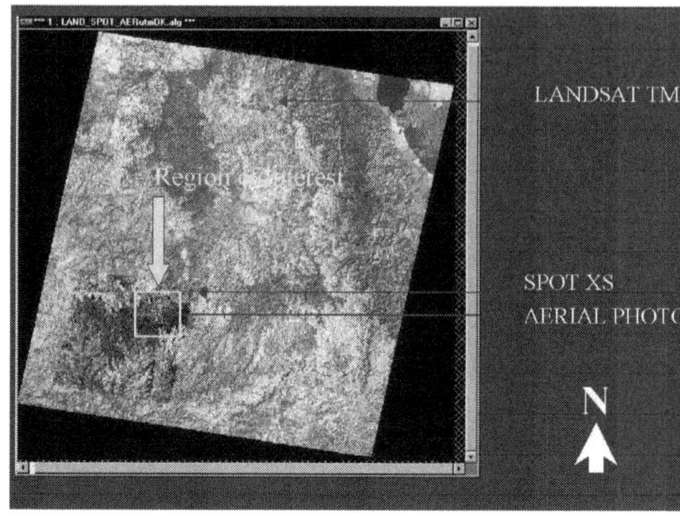

Figure 3: Landsat TM, SPOT XS and aerial photos (1964) geo referenced and mosaiced.

workstation in real time (by the Audition freeware viewer, www.cg2.com) and for the virtual theatre (using the Multigen software on SGI Onyx 2).

- 2D-3D spatial analysis with GIS applications (Arcview and Idrisi).
- Creation of digital mosaics of aerial photos and satellite images (with histogram matching of the images for obtaining homogeneous results in grey levels and in RGB components)
- Virtual reconstruction of cognitive archaeological landscapes.
- VR applications through remote sensing data and archaeological landscapes reconstructed by OpenGl applications.
- VR reconstruction and visualisation of the archaeological landscape through the virtual theatre.

In a few words remote sensing techniques and GIS applications are used for the virtual reconstruction of the archaeological landscape at the micro and macro-scale. Indeed, the 3D exploration and navigation of the virtual landscape allows us to suggest new and unexplored hypotheses and interpretations.

2.2. Data-entry

The choice of the sources has been very important because of the difficulties to obtain, first of all, detailed and useful cartographic documentation (these difficulties increased following the war between Ethiopia and Eritrea). Furthermore, it has been fundamental to integrate different data sources, such as printed documentation and digital data. Therefore, we have used the following principal data-sets:

- Landsat TM 7 bands (30 m resolution, multi-spectral, figure 3).
- SPOT XS, 3 bands (20 m resolution, multi-spectral, figures 3, 5).
- Aerial photos (1951, 1964, 3 m resolution, 1:60,000, figure 4).
- DEMs (by total laser station, 1:500 and by cartographic contour lines, 1:25,000).

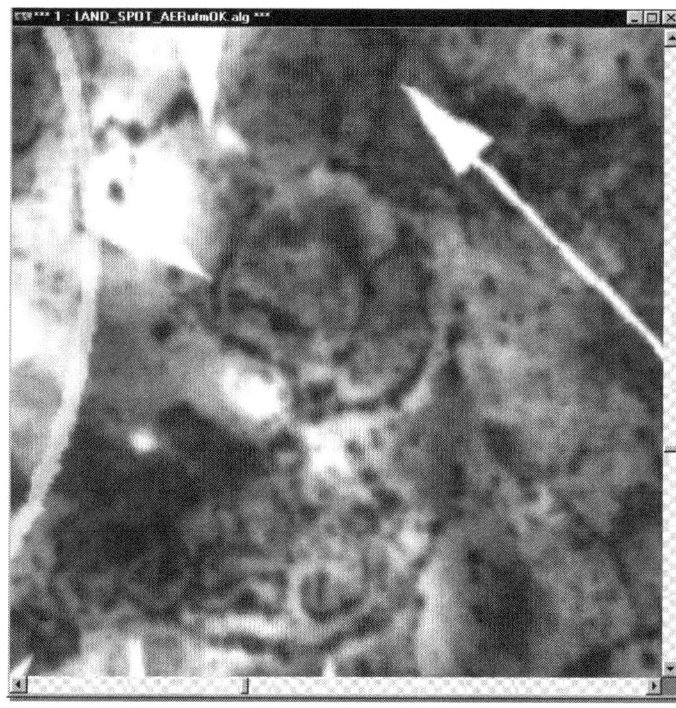

Figure 4: Identification of an archaeological site by an aerial photo.

- Aerial photo-interpretation (feature identification, cropmarks, vector thematic layers).
- Archaeological excavation maps (vector thematic layers, figure 1).
- Archaeological survey maps (vector thematic layers).
- Territorial databases.
- Geological and paleoenvironmental data (vector thematic layers, soil maps, digital classifications).

2.3. Analysis and classification

Having such a complex archaeological data-set, digital applications have been oriented towards a construction of multi-layered models, i.e. a 2D and 3D representation through GIS overlay and

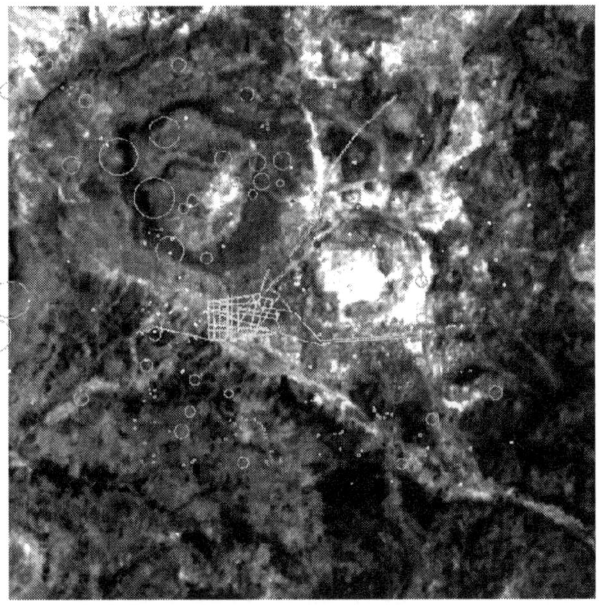

Figure 5: SPOT XS with vector themes corresponding to archaeological sites, roads and urban areas.

remote sensing applications at a micro- and macro-scale. In this first phase, research mainly involved the following methodological applications: multi-spectral satellite classification, panchromatic classification, DEM construction, overlay of raster and vector data, and 3D multi-layered visualisations.

Multi-spectral classification is aimed primarily at classifying the landscapes into macro-classes, and, in particular, at suggesting predictive models for the identification of the archaeological sites and cultural features. The combination of all these information layers is providing new interpretations of archaeological evidence on the basis of the diachronic visibility of the territory and the ancient settlements. Then, the integration of 2D with 3D data and real time visualisation allows the construction of cognitive models of the landscape, with analysis of spatial features and an overlay of all the information acquired during fieldwork, as well as by remote sensing data. In fact, dynamic perception, acquired through 3D virtual models, can open different directions to research on the basis of draping and DEM resolution. This work is only at the initial stage but, by combining all paleo-environmental data, we have obtained a diachronic reconstruction of the archaeological landscape. Problems concern the availability of data. Ongoing research must provide more detailed description of the territory, i.e., through soil maps, geomorphologic and pedological analyses, palynological data etc. Thus, knowledge of the anthropic landscape is at an early stage and new archaeological and geo-archaeological surveys must be included in future fieldwork.

2.3.1. Multi-spectral classification and image processing

In remote sensing applications thematic information can be extracted from imagery through multi-spectral classification; the results of the classifications depend mainly on the image resolution and on the number of spectral bands. Classification is a statistical process which groups homogeneous pixels into areas of interest based upon a concept referred to as spectral pattern recognition. It can be categorised into two methods: supervised (human assisted), or unsupervised (clustering) techniques. Each classification serves a particular purpose, and the two methods are often used in conjunction. Supervised classification uses 'training' regions drawn on the image (typically with vector polygons), then finds all other pixels with similar spectral characteristics. Unsupervised classification works with a nominal class and categorises all pixels into classes with similar spectral signatures, creating classes according to semi-automatic parameters. Results from the process are typically in the form of a thematic map that can be used to solve a particular problem or to provide important data unobtainable from other sources. Several classification algorithms are used in image processing, such as parallelepiped, minimum distance to means, and maximum likelihood. In our project the processed satellite data is taken from a Landsat TM (7 bands, for first processing see Bard, Fattovich, Petrassi, and Pisano 2000, figure 5, 6), and from a SPOT XS (3 bands). Landsat TM is a second generation earth resources satellite, and combines reasonable spatial resolution (cell size of 30 meters by 30 meters) with a range of spectral bands (7 bands in visible and near, short and mid infrared wavelengths). The first band covers the range from 450 nm to 520 nm, which roughly corresponds to blue light in the visible spectrum. The abbreviation "TM" refers to the Landsat Thematic Mapper which has six bands of interest:

- TM1 covers 450-520 nm
- TM2 covers 520-600 nm
- TM3 covers 630-690 nm
- TM4 covers 760-900 nm
- TM5 covers 1550-1750 nm
- TM7 covers 2080-2350 nm

Detailed information of the SPOT image (SPOT5 covers 1580-1750) is as follows:

- Scene ID 2 135-322 93/03/23 08:05:32 1X
- K-J identification 135-322
- Date 93/03/23
- Processing level 1B
- Spectral mode XS
- N134756/E0381029 N134257/E0384306
- Azim.: 120.0 Elev.: 64.5 Gains: 454
- N. of lines 3003
- N. of pixels per line 3182
- 14°8' N 38°43' E (Aksum)

The first step of processing multi-spectral data is to combine the sequences of the spectral bands in order to provide a good visualisation of the territory with respect to research goals (i.e. visibility of soils, vegetation, geomorphology, etc.). For the Landsat TM, the following sequences of bands were chosen: 7, 4, 1 and 5, 3, 1; while for the SPOT XS, we have used the following different combinations: 3, 2, 1 and 1, 2, 3. The combination of spectral bands in pseudo colour is very important because it greatly influences interpretation of the imagery.

Classification and analysis of the raster data have been directed toward the integrated digital processing of the remote sensed (panchromatic and multi-spectral) data, and considering the significant potential of integration of multi-spectral Landsat TM (30 m)

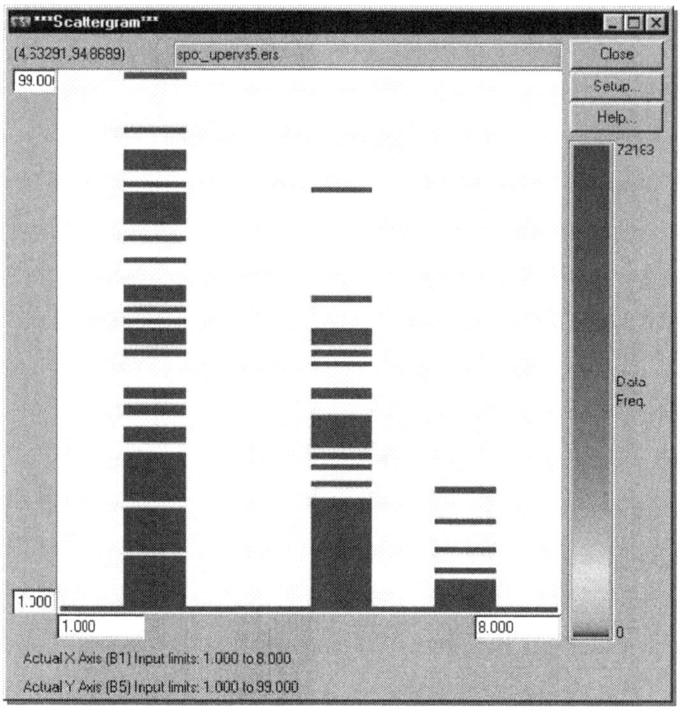

Figure 6: Digital histograms (multi-spectral signatures) corresponding to possible archaeological sites.

data with SPOT XS imagery (20 m), and, finally, with panchromatic data (aerial photos, 3 m, 1:60,000). Whereas the progress of the project will permit a detailed multi-layered classification, the following final data should be produced:

- Classification of soils, vertisoils and erosion areas;
- Classification of damp areas;
- Classification of vegetation, cultivation and crop-marks;
- Classification of archaeological sites on the basis of visible anthropic structures;
- Geomorphologic classifications;
- Classification of anthropic features by aerial photo-interpretation;
- Predictive maps derived from digital sampling of known surface structures;
- Image draping (satellite, aerial photos) over the DEMs in micro and macro-scale for analysis of connections between spectral signatures and digital elevation models;
- Multi-layer visualisation of all geo referenced raster data;
- Data-entry of all digital data in GIS.

At this stage of our work the potential of a multi-spectral classification is reduced by the lack of information on the Aksum regional landscape with respect to soils, rocks, vegetation and cultivation. However, preliminary results of the supervised classification of the Landsat TM imagery have been obtained from two different processes, with samples of 7 and 10 training regions.

In the first instance, seven classes have been distinguished on the basis of the spectral signature: vegetation (unspecified), forests, damp areas, type 1 sites (circular anthropic features), type 2 sites (terraced sites), urban areas, unclassified. In the second case, ten regions have been classified: 4 types of vegetation, forests, damp areas, urban areas, 3 types of anthropic features (2 types of features corresponding to probable archaeological sites, figure 6). In the case of archaeological sites or of anthropic features used to identify probable ancient settlements, several known areas have been sampled (figure 6). For these supervised classifications maximum likelihood algorithms have been used, but it is impossible to categorise these results as definitive, because the training regions have been chosen mainly through cartographic processes and without detailed ground truthing (this activity is planned for 2001). At a preliminary level, the identification of some anthropic features and probable archaeological sites (figure 5) is encouraging. The use of the maximum likelihood algorithm has allowed the identification of two different anthropic features (figure 5), specifically identification of features with circular shapes. In fact, with the superimposition of thematic maps created from supervised classification over the aerial photos (3 m resolution) it is possible to calculate that over 60% of the pixels belong to circular anthropic shapes (features 1), although the final analyses are still in progress.

Other important results have been obtained by data fusion, probably the best method for combining multiple types of data into a single view (for example, sharpening a Landsat TM image with a SPOT XS image). Thus, in our case, data fusion allows the integration of spectral properties (not available for panchromatic data, such as aerial photos) with panchromatic high resolution aerial photos of Aksum: the final result corresponds to "multi-spectral" imagery of 3 m of resolution. Furthermore, using this process it is also possible to superimpose thematic layers classified by supervised algorithms over the aerial grey scale photos. It is then possible to verify in overlay and in detail the correspondence between classified pixels (spectral information) with traces, crop-marks and anomalies visible from on aerial photos (analogue information).

Other interesting results have been obtained using the Brovey transform, a method to fuse different data together using one image (for example Landsat TM) for spectral or colour information, and another image for spatial or sharpness. During processing the Brovey transform has been applied with a panchromatic aerial photo and Landsat TM 7 band imagery and with a panchromatic aerial photo (1964) and SPOT XS (figure 3).

Despite the difficulties, this first attempt to classify regions of interest through a multi-layered approach provides interesting results through its potential to overlay all available information layers including:

- Overlay of vector and raster data (topography, contour levels, archaeological sites, geo-archaeological and paleo-environmental data).
- Diachronic analysis of raster data in transparency and on different layers.
- Overlay and multidimensional visualisation of all classified data.
- Creation of thematic maps and predictive models of ancient settlements systems.
- Aerial photos of the top of Beta Giyorgis hill with possible features identified in the aerial photos.
- Data fusion of panchromatic and multi-spectral data.

2.3.2. DEMs processing and 3D remote sensed data visualisation

DEM processing is currently still in progress because new topographic survey is still planned for Beta Giyorgis hill. A micro-scale and a macro-scale DEMs have been processed: the micro-scale DEM has been created from mapping at a 1:500 scale; while the macro-scale DEM has been created digitising contour levels at a 1:25,000 scale. One of the principal activities of the project involves the draping of the raster and vector data over the DEMs in order to construct 3D and multi-layered models of the landscape: each draped layer can represent different types of data (real or interpreted) and only in 3D is its level of information visible (virtual dynamic perception). Furthermore, by integrating all 2D data within 3D visualisations it has been possible to plan the following applications:

- Different interpolation methods for DEM creation: kriging, cubic-spline, nearest neighbour, grid, TIN.
- Structural filters for micro-topographic visualisation of the DEMs.
- Digital segmentation techniques for identification of archaeological and geomorphologic structures.
- Digital enhancement filtering (colour drape algorithm).
- 3D visualisation of the SPOT and Landsat images draped over the DEM.
- 3D visualisation of the DEM created by total station survey (excavation area, figure 7)
- 3D visualisation of excavation areas (micro-DEM) and aerial photos draped over the micro-DEM.

3.1. Cognitive archaeological landscapes: toward the virtual mindscape

"The map is not the territory…" (Bateson 1973, Bateson 1980), and, in digital language, a GIS is not the landscape, because it cannot represent such a complex data set as a landscape. In recent years the relationships between landscape and cognitive archaeology has been studied with two significant contributions: *The ancient mind* (Renfrew and Zubrow 1994) and *Semiotics of Landscape: Archaeology of Mind* (Nash 1997). Otherwise, post-processualist archaeologist are now studying GIS spatial applications (Forte and Cremaschi 1999, Gaffney and Van Leusen 1995, Gaffney et al. 1996, Johnston 1995, Van Leusen 1999) in order to reconstruct past landscapes (mainly "resource landscapes") on the basis of viewshed and cost surface analyses (see Van Leusen 1999). Since 1987 a significant number of these spatial applications have been undertaken with cognitive aims (Gaffney and Van Leusen 1995). The problem is very complex because it involves methodological, epistemological and technological aspects: how can we represent archaeological landscapes? What is the perception of the landscape? What is the difference between the real and imaginary landscape? How is it possible to describe an archaeological landscape? The complexity of the problem is evident when analysing, describing and representing an ancient landscape, and spatial techniques as well as epistemological and anthropological approaches are needed.

We are aware that these interesting problems need much discussion and in this paper, we would like to present a brief contribution relating to the problems of virtual digital technologies and their use for restoration of the archaeological landscape. Moreover, we believe that a cognitive landscape reconstruction needs an integrated methodology consisting of spatial digital technologies (GIS, remote sensing, virtual reality) and epistemological discussion.

Figure 7: 3D visualisation (TIN model) of the archaeological region with data interpolated by a total laser station.

The conditions for interpreting an archaeological landscape depend on many factors of visual representation. Visibility, perception, cognition and communication are the most important factors in the knowledge of landscape, especially ancient landscapes. A distinction between perception and cognition has been suggested by Van Leusen (1999) where perception is "the simple act of being aware of the landscape…" and cognitive archaeology, "the study of past ways of thought as inferred from material remains" (Renfrew and Zubrow 1994). According to Nash (1997) the "landscape, comprising natural features, is a series of stories that are constructed through time and space and formulate a series of histories; not history". Therefore, with respect to the reconstruction of cognitive archaeological landscapes these questions should be raised:

- What is the present landscape? (today).
- What is the landscape we perceive? (today and through our mind).
- What was the ancient landscape? (in the past).
- What was the landscape to the ancient people? (in the past, through the ancient mind).
- *What is the ancient landscape we perceive?* (at present, but also through ancient minds and mental maps).

A first attempt to classify the archaeological landscapes on the basis of a cognitive approach is given below (obviously all terms interact each other).

Cognitive qualities of the archaeological landscape / Description - Significance

Social-symbolic:

Socio-ritual knowledge and the consciousness and manner in which people perceive objects within their surroundings is fundamental to the formulation of a sense of belonging.

Geometrical:

The geometry of the landscape is the property used to describe (intervisibility, DEMs, viewshed, panoramas) surfaces and volumes.

Power-hierarchical:

In many cases the ancient construction of archaeological landscapes follows the rules of power, thus the perception is through hierarchical passes (layers of importance).

Visual:

The visibility is the condition to view territories (own and other peoples) without mental maps (in theory).

Artefact:

"Landscape employs a language of artefact, an object which is created by people and embodies the principles of organisation and categorisation within the human experience" (Nash 1997).

Mental:

Perception of the landscape through mental maps constitutes the grammar of the visual anthropology.

Narrative:

The preservation of a memory of the landscape means the preservation of a cultural identity. A narrative landscape describes the quality of a collective memory (i.e., through myths, imaginary places, places of memory).

Semiotic:

Any landscape is understandable by different languages and grammars. The construction of anthropic landscape involves the use of a common language for expressing the mind of the community.

Connected-inter-connected:

Landscape is a context where the whole is more significant than the sum of each component. Connections belong to the context.

Natural:

The quality of the natural in the perception of landscape is purely theoretical, including a neutral knowledge without giving hierarchical values.

Spatial-temporal:

"Landscape is a series of spaces which become places, thus establishing territory... although we perceive ideas and place them within our minds, our body too becomes a part of the landscape - it interacts with and becomes a component of the landscape, forming a sense of space" (Nash 1997:1-2).

Simulated:

Within a virtual reality reconstruction there is a passive interaction (the user navigates and visualises the model) and an active interaction (with behavioural exchange between the environment and the events).

Virtual-dynamic:

Our perception of landscape can be through the dynamic perception/interpretation of virtual models (virtual walkthrough, flythrough), requiring 3D navigation.

Mindscape:

Mind of landscape: the way to read the landscape through cognitive representations.

In our project virtual reality applications are important for the presentation of data because through virtual models the landscape becomes a cognitive landscape. We are aware that we are only at the beginning of this experiment, but we think that cognitive interaction increases the level of knowledge and communication (Forte and Beltrami in press, Barceló et al. 2000, Forte 2000).

Following this approach, three levels of interaction have been chosen for the project:

- Level 1 (2D) - QTVR applications used in order to generate virtual panoramas within the 3D models of reconstructed landscape (with different textures);
- Level 2 (3D) - VR navigation through the remote sensing data (using the OpenGl flythrough of Er Mapper);
- Level 3 - Virtual navigation through the reconstructed landscape with billboards, vegetation, paleoenvironmental data, architectonic and extruded models (made with Terravista, see below, figure10).

4. Terravista

The last stage of the project is to complete the reconstruction of the archaeological landscape, converting the GIS data into a new environment for virtual reality applications; this is possible with specific software (Terravista 2.1).

In January 2000, a program for scientific co-operation was established between ITABC (Institute of Technologies Applied to Cultural Heritage) and the American software company Terrex (www.terrex.com) for the application of the Terravista 2.1 software (ITABC is the only reliable user for cultural heritage in Europe for this kind of license). Terravista is a high level and high performance software for virtual 3D reconstruction of geo referenced landscapes starting from GIS data, and is capable of generating models by OpenGL libraries (www.opengl.org). The OpenGL working environment provides high quality graphics on inexpensive PCs, because accelerated graphic cards are used. OpenGL is the premier environment for developing portable, interactive 2D and 3D graphics applications. Since its introduction in 1992, OpenGL has become the industry's most widely used and supported 2D and 3D graphics application programming interface (API), bringing thousands of applications to a wide variety of computers. OpenGL fosters innovation and speeds application development by incorporating a broad set of rendering, texture mappings, special effects, and other powerful visualisation functions.

Encouraging experiments performed with Terravista have demonstrated a high level of interactivity and practicality for OpenGL models (combined with accelerated graphic cards API-OPENGL). These results open new perspectives for versatility of applications for non-immersive virtual reality, mainly in connection with virtual reconstruction of territorial contexts and with three-dimensional visualisation of geo-referenced data.

Standardisation of software procedures and increasing use of OpenGL graphic libraries for archaeological applications will direct research towards the organisation of specific multidisciplinary databases (2D-3D). Virtual reality applications will also be undertaken for visualisation of classified remote sensing data.

Terra Vista™ is a software that rapidly constructs digital landscapes representing both real and imaginary worlds. These landscapes are optimised for drawing, which makes them suitable for use in applications such as visual simulation, where a scene must be updated very quickly. The approach for constructing virtual worlds differs from other modelling systems in two ways:

1. A high degree of automation allows the creation of potentially huge worlds.
2. Existing sources of digital cartographic data is used as an input to its automatic processes (Terravista User's Guide).

The combination of these two features allows the program to quickly create virtual worlds that are highly representative of the real world. So, simulation is achieved through three processests: (1) Data Acquisition, (2) Terrain Database Generation, and (3) Visualisation of the Generated Terrain.

The three main types of data used by Terra Vista are as follows:

- Elevation data (DEMs), which constitutes the basis for starting the reconstruction;
- *Geospecific* imagery (raster data such as satellite imagery, aerial photos or textures created by the archaeological researches), which are the first "drape" of the landscape;
- Cultural data, including man-made structures such as buildings, roads, bridges, etc.; or natural features, such as lakes, oceans, rivers, forests, soils, vegetation. Vector files contain these features in the form of points, lines, and areas.

Additional data (textures and 3D Models) used by Terra Vista to create a terrain database are included in powerful graphic libraries. Generic or repeating textures can be used where geospecific imagery is not available. Moreover, Terra Vista supplies a library of textures that can be used on the terrain.

With respect to the features of these 3D models it is important to stress that they are constructed as LOD (levels-of-detail) models, i.e., each tile has multiple levels-of-detail (LOD, figure 8). Each LOD represents the tile at a different polygon density: less detailed for when the tile is far from the viewer; or more detailed when the tile is close to the viewer. Whether the terrain is displayed in a flight simulator or shown in a corporate boardroom on a notebook computer, the hardware on which it is rendered will determine some of the parameters for the terrain generation. The number of polygons that can be drawn per second along with the desired frame rate will determine how many polygons can be drawn per frame (texture memory can limit the amount of geospecific imagery used).

Finally, the *fly-through* visualisation is optimised for OpenGl, thus real time movement is very fast. It is also possible to set up various navigation parameters (altitude, fog, velocity, panorama, visibility).

5. Virtual Theatre

Thanks to scientific co-operation with the CINECA Visual Information Lab (Interuniversity Supercomputing Consortium, http://www.cineca.it/visit/virtualtheatre.html) the first virtual models of archaeological landscapes (created with Terravista in Openflight format) have been implemented for the Virtual Theatre (figure 9).

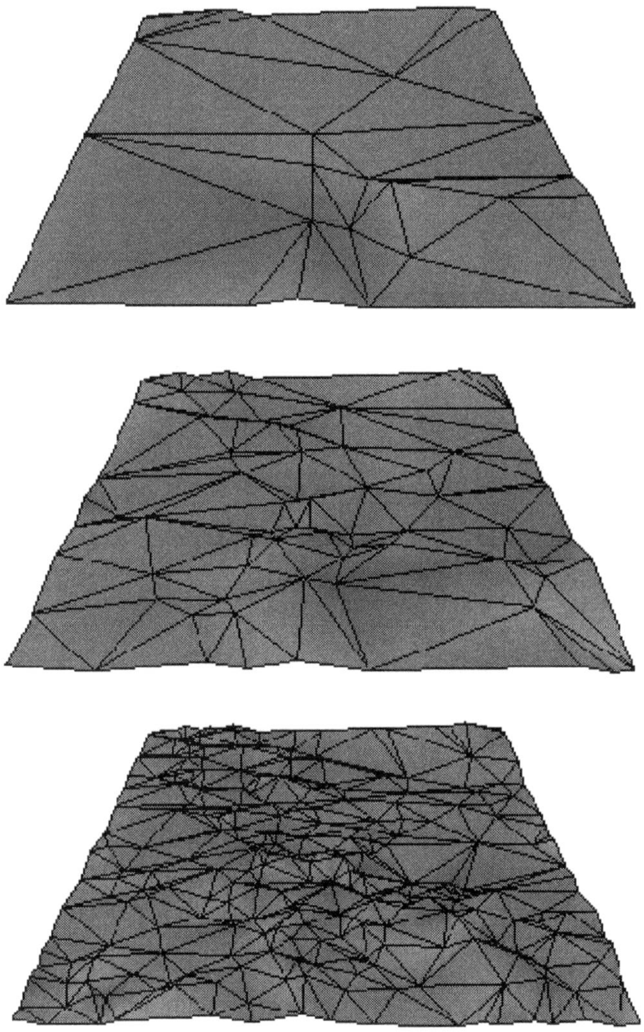

Figure 8: Terravista software: representation of 3 LOD (level of details) for different types of visualisation.

These simulations may benefit especially from viewing on a large scale in the CINECA Virtual Theatre, which aids the visualisation of scenarios through its size and "immersiveness". The Virtual Theatre is a structure for immersive three-dimensional computerised visualisation with a surround sound system and a semicircular screen that allows the spectator to experience the illusion of three-dimensional vision. The CINECA Virtual Theatre hardware platform consists of a SGI Onyx2 system with 8 R10000 processors, 4 Gigabytes of RAM, 3 "Infinite Reality2" graphic pipelines with 8 raster managers, 3 Barco projectors and a switching system integrated directly into the graphic computer system (figure 9).

Applied to research, the Virtual Theatre therefore becomes an instrument that enhances understanding. The availability of these technologies opens up new perspectives and a new way of interpreting well known data, re-appropriating he original sense of the word "theatre", in which the Greek intended a space for observing in order to understand.

Visualisation obtainable through VR crystal eyes (with stereoscopic view) enhances the level of the viewers' perception as generated virtual models can be experienced as holograms.

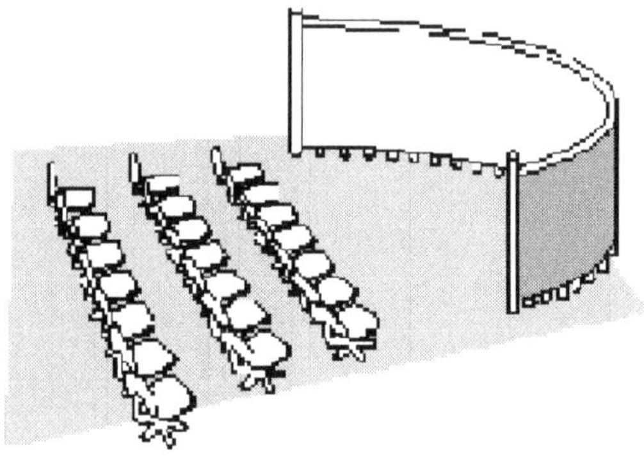

Figure 9: The virtual theatre.

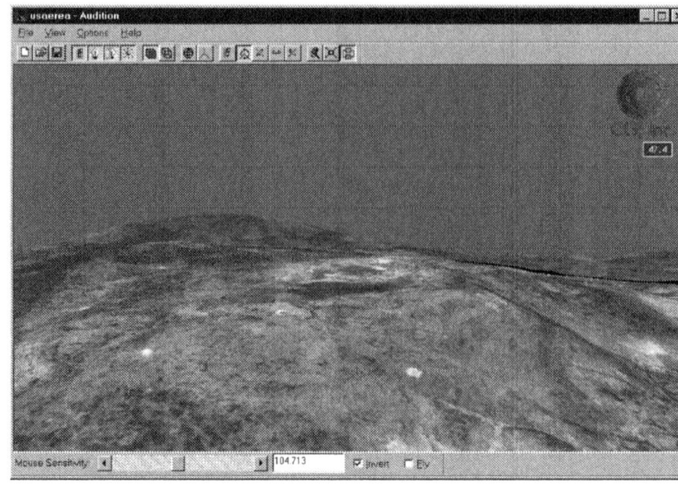

Figure 10: Virtual reconstruction of the Aksum archaeological landscape (real time navigation).

6. Perspectives and conclusions

In the above-discussion the importance of reconstructing the archaeological landscape (figure 10) through remote sensing applications has been highlighted. However, for a cognitive landscape we need to add other facets of simulation. Therefore, in the Aksum project, the archaeological landscape data will be supplied to CINECA in Multigen OpenFlight format (.flt, the same format used by the Terravista models). Using IRIS Performer libraries and the developing environment software of the MultiGen-Paradigm Vega, archaeological landscape models will be integrated with real-time dynamic behaviour and multimedia data, producing visual simulations such as:

- geomorphological simulation (e.g.,. landslides, terrain morphology variation, terraces),
- paleo-environmental events (former water courses),
- settlement simulation (population expansion, settlement planning),
- other dynamics (deforestation processes, cultivation, irrigation, etc).

In conclusion, the 3D reconstruction of archaeological landscapes through virtual reality experience (with the addition of remote sensing classification) can provide a fundamental contribution to the development of the project and to the construction of cognitive models. In fact landscape patterns are the results of economic experience, mental maps, cultural ideas and topographic reality (Renfrew and Zubrow 1994, Forte 2000). Furthermore, it is important to acquire new tools for understanding the past, including spatial techniques (GIS) and visual information dynamic systems (VR). Finally, the multi-layered perception of 3D archaeological models through advanced virtual environments (virtual theatres and mindscapes) opens new perspectives for cognitive landscape archaeology, perhaps as a first step for restoring what "ancient collective memories" represented.

Acknowledgements

Special thanks are due to all members of the IUO/BU joint archaeological expedition to Aksum, and in particular to Dr Livio Crescenzi, (Soprintendenza Archeologica. per il Lazio, Roma.), Mr. Antonio Sollazzi (Soprintendenza Archeologica per il Lazio, Roma), and Mr. Franco Cioffi, (Soprintendenza Archeologica per il Lazio, Cassino). This project has been supported by funds from the Ministero Affari Esteri, CNR, Ministero per l'Università e la Ricerca Scientifica, Rome (Italy), and the National Geographic Society, Washington DC (USA). Other thanks are also due to the Visual Information Lab of the CINECA and particularly to Eng. Antonella Guidazzoli.

References

BARCELÓ J., FORTE, M. and SANDERS, D. (eds.), 2000. *Virtual reality in archaeology*, Oxford, ArcheoPress (BAR International Series S 843).

BARD, K.A., FATTOVICH, R., PETRASSI, L. and PISANO, V., 2000. *The Aksum archaeological area: a preliminary assessment*, Istituto Universitario Orientale, Napoli.

BATESON, G., 1973. *Steps to an Ecology of Mind*. Paladin Books.

BATESON, G. 1980. *Mind and Nature - A Necessary Unity*. Bantam Books.

FORTE, M., 2000. About virtual archaeology: disorders, cognitive interactions and virtuality. In Barceló, J., Forte, M. and Sanders, D. (eds.), *Virtual reality in archaeology*, Oxford, ArcheoPress (BAR International Series S 843).

FORTE, M., 2000. Archaeology and virtual microtopography: the creation of DEMs for reconstructing fossil landscapes by Remote Sensing and GIS applications. In Lock, G. (ed.), *Beyond the Map: Archaeology and Spatial Technologies*. NATO Science Series, IOS Press Amsterdam: 199-214.

FORTE, M. and BELTRAMI, R., in press. A proposito di virtual archaeology: disordini, interazioni cognitive e virtualità, *Archeologia e Calcolatori*, 11.

FORTE, M. and CREMASCHI, M., 1999. Reconstructing a fossil landscape by remote sensing and GIS applications: sites, virtual models and territory during the Middle Bronze Age in the Po Plain (Northern Italy), *Archeologia e Calcolatori*, 10.

GAFFNEY, V. and Van LEUSEN, P.M., 1995. Postscript – GIS, Environmental Determinism and Archaeology. In Lock, G. and Stančič, Z. (eds.), *Archaeology and Geographical information Systems: A European perspective*, Taylor and Francis: 367-382, London.

GAFFNEY, V., OŠTIR, K., PODOBNIKAR, T. and STANČIČ, Z., 1996. Spatial analyses, field survey, territories and mental maps on the Island of Brac, *Archeologia e Calcolatori*, 7: 27-41.

NASH, G. (ed.), 1997, *Semiotics of Landscape: Archaeology of Mind*, Oxford.

MALINA, J. and VASICEK, Z., 1997. *Archeologia. Storia, problemi, metodi*, Electa, Milano.

RENFREW, C., ZUBROW, E. (eds.), 1994. *The Ancient Mind: elements of cognitive archaeology*, Cambridge UP.

TAYLOR, P.J. and JOHNSTON, R.J., 1995. GIS and Geography. In Pickles, J. (ed.), *Ground Truth. The social implications of Geographic Information Systems*, 51-63, New York, London.

Van LEUSEN, M., 1999. Viewshed and Cost Surface Analysis Using GIS (Cartographic Modelling in a Cell-Based GIS II). In *New Techniques for Old Times*, Computer Applications and Quantitative Methods in Archaeology, Proceedings of the 26th Conference, Barcelona, 1998, BAR 757.

Archaeological Data Spaces: Spatial Aggregation and Large-Scale Knowledge Environments

Cornelius Steckner

FOCAS, D-50996 Cologne, Etruskerstr. 22, Germany

e-mail: csteckner@hotmail.com

Abstract

Does archaeology only apply computer methodologies? In the field of spatial exploration archaeological practice offers skilled methods that computer sciences should adapt. Among the adapted is the archaeological shape of the description vocabulary. This is a text based method used to describe and classify shape amalgamation as generalised cylinders. This classification allows the aggregation of the object space to proper words and for computer-code to improve AI based robot vision. Another specific archaeological spatial aggregation describes the process of civilisation. In this view exhibitions and museums are multidimensional and multi-user spatial aggregations designed to meet socially accepted knowledge and information metaphors. Such archaeological conceptualisation is also increasingly meeting the demands of the forthcoming distributed data space protocol (DTSP) and related large scale virtual environments (X3D).

Key words: spatial aggregation, distributed space protocol, large-scale virtual environment, multi-user environment, multi-perspective environment

"We believe part of this ability to visualize and imagine must consist of skills to generate images, discover structures and relations in the images, transform the structures, and predict how the structures respond to internal dynamics or external forcing."

Yip, K. and Zhao, F. 1996.

1. Data space

By definition data space is a storage area for data, locally stored or distributed over the net. With the appearance of new protocols such distributed data spaces are increasingly becoming seen as environment-like navigable multi-user virtual information spaces. They are designed to address the human skill to discover structures and relations in the environment as well as related spatial aggregations of different density and dimensions. The most complex aggregations are architectural knowledge spaces (libraries, museums, exhibitions) which combine different space aggregation density and different media; thus the word "Europe" is of higher density than the map or the movie. The computer monitor, as the most flexible media for the mind, gives access to verbal, visual and virtual knowledge spaces. However, it has its communicative limitations as a human computer interface. Any screen addresses a specific mode of visual cognition. The screen forces the user to keep up a specific embodied access of the time-sequenced information. The screen communicating text, space and animation uses only 2D-planes. But the user's text and image reading competence makes the 2D-front-end become a universal display of multidimensional data spaces. In this event visual cognition is a mix of natural environmental and trained cultural information access. In this respect the hidden communicative basis is a wide range of common spatial metaphors used to aggregate space. The map metaphors and schemes use generalisations, and intuitively the user reacts on the level of detail in order to obtain an idea of the scale. As a consequence, metaphors such as globe, map and perspective tell us about the steps in the representation of geography related numeric data, dependent on the knowledge of the user and his communicative skill.

1.1. Spatial metaphors of communication

Any chosen metaphor of communication will help the user to find the right motivational arousal to discover structures in the aggregation by appropriate modes of access behaviour. The map is just one of these metaphors (Chalmers 1993) that offers access to large-scale environments. The method is generalised and the pattern is the page in front, ready to read. Accordingly, the text is scaled and oriented. Information technologies apply the page and illustration metaphor to the screen. Other metaphors, taken from the civilisation environment allow for a dynamic spatial access. Further architecture-like metaphors are more static but improve the information access by active flyover and fly-through modes of knowledge access. The communicative architectures of virtual information spaces (Daessler 1998) are so far classified as:

- panorama screen presentation,
- regular matrix object representation,
- virtual and abstract landscapes for environment like exploration,
- shaped objects scattered into gravity spaces,
- museum like multidimensional and multi-perspective knowledge spaces.

1.2. Virtual information space generation software

Most of the related information technologies are freely available as research programs for the UNIX/Linux environment, covering several communicative architectures and numeric information related data types (Daessler 1998). VisDB and Xgobi/XGvis import supplemental GIS data; other information explorers produce maps of their own (WebSom). Vis5d and VTK focus on environmental streaming processes, while several image understanding environments (IUE, Khoros, VTK, SNSS) are designed to analyse image records of spatial situations, or use agents (http://agents.umbc.edu/introduction/) for similarity research in stored CAD vector data and bit mapped images (SimilaryResearchSystem

S3). A good tool to study bookmark-based spatial exploration is the VRML document search engine (http://fabdo.fh-potsdam.de/infoviz for UNIX and Windows) and "The brain" (http://www.thebrain.com). These produce private information spaces that do not socially communicate as agent based systems (Thomas, 1996). Surveys on spatial information and knowledge visualisation can be found at the following homepages: http://fabdo.fh-potsdam.de/infoviz; http://www.dbs.informatik.uni-muenchen.de; http://www.hitl.washington.edu/projects/knowledge_base/research.html). The museum-style X3D knowledge representation of information spaces (http://snm.hgkz.ch/%7Emaja/begin.html) is forthcoming with the newly available data access and retrieval technologies (http://www.pitt.edu/~korfhage/viri_bib.html).

1.3. Forthcoming protocols generating spatial aggregation

The information technologies applied to spatial reasoning are referred to as "drill-down data mining technologies". The distributed data spaces and the distributed users need defined spaces of aggregation in which to meet. Consequently the spatial aggregation may be a multi-user world where agents meet. This is the tendency in the development of information technology, where the new Data Space Transfer Protocol (DTSP) meets the forthcoming development of the X3D modelling and Virtual Reality Transfer Protocol (VRTP) in support of internet 3D graphics and large-scale virtual environments (LSVE). This tendency toward LSVE meets the space-related methodology of archaeology in several aspects. Hollerbach has already focused the archaeologist's skill to discover and to describe time related spatial structures of pottery in order to improve robot visual environment cognition (Hollerbach 1975). He can show that the description and identification is based implicitly on the concept of generalised cylinders, reducing the closed surfaces of the ceramics according to cylindrical co-ordinates. In this space any object is defined by a rotating distance measure running along the axis. If the axis is straight and the distance constant, a 2D outline is sufficient to describe the entire surface. The related verbal description of these shapes uses a hierarchy according to body, neck, lip, foot and handle. A mere 105 terms were sufficient to describe and to identify 42 types of pottery. From this approach a concept of amalgamation emerged, which could describe and identify more complex structures by supplemental indentations and protrusions. Accordingly, the scientific skills used to detect structures and relations in all sorts of 3D structures and their images are a fascinating resource for applied spatial reasoning of different aggregations and scales (http://www.cs.albany.edu/~amit/spatsites.html). As a consequence archaeology is not just a field that adapts computer applications from other fields of research.

2. The spatial aggregation of archaeological space

Excavation documentation techniques are techniques that are designed to aggregate spatial information to fit the demands of printed communication. The page is the format used to combine text and numerical information with orthogonal drawings in scale, with a variety of scaled vector representations and bit-mapped perspective information.

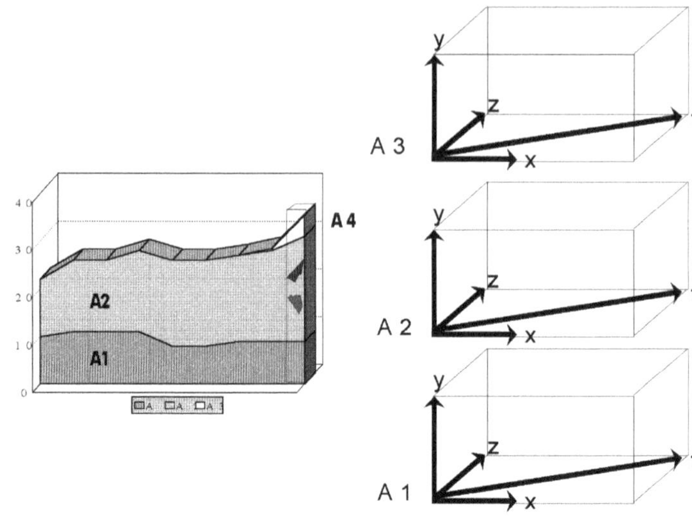

Figure 1: Strata and Sediments (left) transformed to a sequence of architectural spaces (right).

Spatial aggregation maps the strata of a site. The slicing method is illustrative for it documents every five or 10 cm on a transparent material. The result is a scaled-down 3D model of the site. Other techniques follow the strata or relate vertical and horizontal views. Slices and strata represent time, while the horizontal distribution represents the space related to a time-segment. Each stratum binds a class of relics and represents its own epoch, so each stratum has its own four world dimensions (3D + t). Accordingly, the recorded strata taken together represent a multidimensional space. By their relics the original 4D layers or epochs are conventionally restored and collected in archaeological storehouses and museums. This is where the original relics can be combined with all sorts of aggregation and site-related media in order to inform the public. But how do we express a multidimensional space?

2.1. The three age system as aggregated knowledge space

Between 1818 and 1824 Christian Thomsen (1788-1865) reorganised the Copenhagen Museum of Northern Antiquities according to the technological process of civilisation. Thus, a structure to built multi-user knowledge and information space became available. According to scientific knowledge gathered from excavations, he started his suggested tour with the Stone Age room. Then he presented the assembled Bronze Age relics, followed by the rooms with the Iron Age artefacts, each room or floor defining a knowledge space of its own. The aggregated process of civilisation became an environmental walk between ages. The walk takes time. The time consumed reflects the sedimentation of time, and the border of the room represents the border of the ages (Hotta 1992, Steckner 1995). Recent displays present spatial aggregation by additional refinements. Some displays emphasised the geographical drill-down, adapting the cardinal points of the compass with the corners of the room. Other displays emphasised morphological representations across the epochs.

2.2. Sediment and revolution

The metaphor Thomsen used (figure 1) was not the museum as such, but an architecture to display the knowledge as regards the ages of technological progress. Technological revolutions distinguish the ages by borders which separate the civilisation sediments. In relation to these sediments the idea of ages also implicitly covered the quantitative approach in archaeology. The three main strata form the structure of a 3D stacked bar composed of three sediment layers, each of them a closed surface. Transformed to an architectural structure, the arrangement of the closed surfaces makes up the knowledge space. This architectural model of the ages became adapted by all sorts of local and world-wide museums and exhibitions, each a spatial aggregation of its own. In consequence, the metaphor became the socially adapted convention to drill-down the knowledge of natural and cultural history in a multi-user environment. Media of different intensities and of different immersiveness told the story of civilisation by the multidimensional directional walk from one age to another. Optionally, the suggestive progress became emphasised by staircases and epoch related murals displaying the revolutionary environmental changes caused by technological progress. Such ideas emphasised the epoch related change in perspective. Each epoch became its own world with its own perspective. These stacked worlds are also the aim of the recent LSVE developments. Accordingly, the museum metaphor applied to large-scale VR environments is the optimal structure that can be used to express independent worlds within a unique co-ordinate system and code.

2.3. Public access feed-back

The information data types aggregated to express the knowledge space in the beginning were objects and environmental murals, associated with explanatory texts and drawings in catalogues. Spatial aggregation remained visible in the names of the sites. Then all sorts of perspective site views, panoramas and dioramas, as well as distribution maps, explained the spatial aggregation within the refined knowledge space. The information technologies were increasingly supplemented by models, movies and animation, in order to conform with the evolving public communication style. In the refined knowledge space there is not only a response to the technological progress in public communication. In addition, public access may offer feedback to spatial aggregation. Consequently, there are remote control systems, installed to change the distribution of showcases according to the demands of the social acceptance and understanding. These systems are already in use. The Disney EPCOT, Florida, which is designed as a globe map-style interface to display future developments, continuously checks the flow of people in order to respond by immediately installed attractions to avoid a bottleneck in the flow. Similarly applied Monitoring Continuous Systems (MCS: http://www.cs.utexas.edu/users/qr/papers-QR.html) in museums may look like the Yokohama Museum. A remote control and transport system allows response to the demands by immediate changes in the spatial order of the display cases (Hotta 1992). Further automation may use optimal room position of display cases and media installations to answer both, the demands of the knowledge representation and user knowledge. Of course such spatially reacting displays may be adapted for a dynamic LSVE display. Accordingly, communicative spatial reactions to the data and to the demands of the users have to be distinguished according to their freedom within their knowledge architecture constraints.

3. From text to space and back

In communication, natural speech-dependent text-based methods dominate. This covers not only the guided tour but also the labels and descriptions in catalogues. Cinema and TV-movies also have text-based structures. In these media, the written texts are cast into pictures and dialogues. In the virtual world these stories are the human readable texts of VRML-code and the scripts of povray-like photo-realistic representations and slide by slide rendered movies. Although not always, the code is usually written in human readable ASCII format. Accordingly, it is a good idea to add a further step to link spatial aggregations and related images with data spaces. The idea is to store the original data of all sorts of spatial immersiveness in ASCII based code. DTSP aims to perform this, as does the SAMOS project (Steckner 2000). The reason behind is only the issue, which highest aggregation density and also extreme reality - the manufactured object and built architecture - may be coded by words.

Accordingly there is no qualitative difference in voice, text, bit-mapped, vector and other space representing data in the data handled by archaeologists. Each site and real world object must be identified by a proper name. The image may be scanned by laser technologies, accessed from satellite bit-maps and from outlining vector-graphics, whatever is accessible. Also, any immersive VRML space script may produce photo-realistic or, alternatively, a non-photorealistic drawing-style output. The same data sets may be filtered to stereo-lithography STL and other similar computer-aided CAD building and manufacturing processes. So the drill-down techniques are in a windup loop with the 3D manufacture data processes.

3.1. Robot and agent access to archaeological data-space

In information spaces there is not just dimensional drill-down and windup by applied filters. There are also on-top techniques available. A good example of how to envelop text space is the PUEBLO software system. This software, revealed by good educational applications, runs 2D maps or images and 3D VRML representations on top of text based Multi-User dungeons (MUD). These MUD-applications show the same levels that are found in the drill-down and windup processes and data conversion and filtering. Accordingly, any of the levels may be connected in the communication system. The remote access might be in the text level, while the action is displayed by a 3D-projection, immersive or not. The projected environments are the meeting-places of remote avatars, which are agents of the distributed users. So what looks like a remote avatar is not necessarily a remote avatar. The avatar we meet in the virtual Stone Age or Bronze Age environment may be guided by remote access. This is not certain in the long run, because the avatar may also be an autonomous virtual robot merging in real-time into the virtual environment-generating program. At which level the robot generates his own environmental map will not be certain. The orientation might derive from the MUD text code or from another level up to the video-screen the human user has access to. Equivalently, the meeting place might be an aggregation of a distributed data space.

This kind of level-incursion is the idea of X3D. The idea is derived from a virtual submarine navigation by virtual sonar in virtual waters. Accordingly there are all sorts of virtual robots exploring all sorts of data aggregation by remote and autonomous processes (http://www.robotic.dlr.de). Such robots also read the human readable codes, or they have access to bit-maps or even to the visual environment, which then becomes their data space. According to the concept of data space the natural environment is also a data space.

3.2. The real and the virtual

What would happen if we could explore information like we explore the archaeological museum? The question was posed in the thesis of Mark A. Foltz "Designing Navigable Information Spaces" (1998): Like Hollerbach (1975) Foltz also first studied and examined the human skill to orient by exhibition display, while not distinguishing information and knowledge representation. In the meanwhile the first X3D environments and related aggregated information spaces appeared (http://www.infoarch.ai.mit.edu/spaces.html; http://www.infoarch.ai.mit.edu/jair/jair-space.html; http://snm.hgkz.ch/%7Emaja/museums.html). These show the transformation of the multidimensional museum aggregation into multi-user virtual space. However, even, cyber-suits and exo-skeletons with touch force feedback will not reproduce the original gravity walk in the museum.

4. The gap in the media

There is a gap in the mixed-media presentation in real museum environments and the immersive virtual information environments. In the real museum architecture visual cognition is controlled by the vestibular system and by environmental feed-back. The key and behavioural anchors within this environment are the original physical objects. But text display read from left to right will also be an anchor to assign information and orientation (Steckner 1993). These keys allow the user to relate additional media of all sorts within the architectural framework and to switch to the new perspective of each age. The user walks through the exhibition to get an impression of the environment, or to reach a goal. He stands or sits reading a map, a picture, a text, trying not to move. The visitor perceives size and material quality of the object, i.e. its size and weight. These senses confirm and map the knowledge space. Embodied intelligence guides the user in real environments. To enhance the modes of access the museum uses space segments. The knowledge space used to display the ages of technology and style is accessed by walking. The library is instead designed to preserve books and to guide users to their content, which is read in the place. There is the cinema were people watch movies on the screen and listen to lectures with slides. Each of the architectural frameworks has its own embodied style of access. The media related differences in display architecture and behaviour settings were not realised in the early pure mental concept of AI and robotics. In consequence a change in the robot programming methods is coming, a change which will have further consequences throughout the IT field. Embodied technologies are increasingly adapted for autonomous robots, and physical modelling is increasingly applied (Fishwick 1996), even such as reading weight and measure standards from bit-mapped shape information to generate information spaces (Steckner 2000).

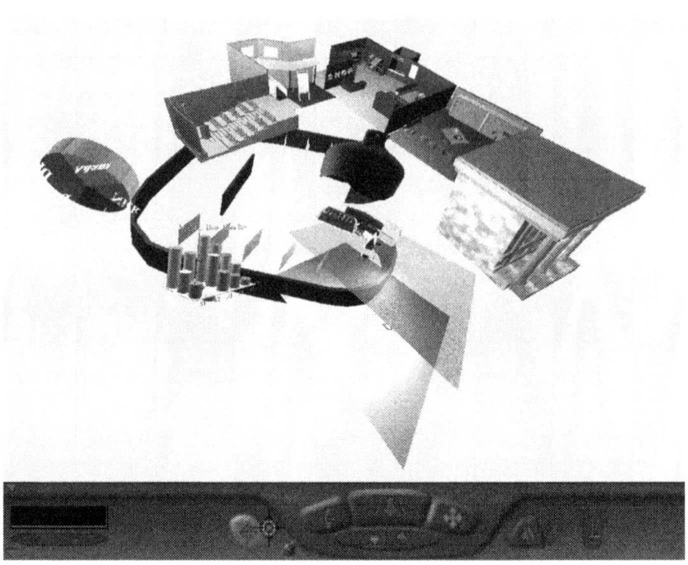

Figure 2: Layout of the spatial media and display test-bed.

5. The application

The application project designed to study the interrelations in architectural space and media access started several years ago, when I designed a media museum. There the architecture clearly showed synergy effects with the specific media, whenever the architecture was binding and separating the specific mode of access. Further research in human computer interfaces, motivational arousal, and anticipation led to an effort to analyse and to model human environmental interrelations (Steckner 1999). The application is a VRML test-bed made to interrelate specific media access styles, display styles, and spatial metaphors (figure 2).

In the present version VRML offers several features that offer a taste of X3D / DTSP LSVE-like multi-perspective virtual environments. A suggestion is to emulate the environmental switches, using a component-based virtual environment, in which components each ask for a different mode of media access, and the design emphasises the modal access. The generated world is a museum-like knowledge space displaying different information styles by an architecture compound prototype. The joined environment styles come with distinguished architectural spaces, each devoted to a dominating media. Any entrance convention is much like the opening sequence in conventional multimedia encyclopaedias, with an access-style map of the whole structure. The VRML model so far has a public entrance hall to guide the users according to their needs, a library environment to display texts, joined by a place to navigate in text-based information structures, which here are bookmarks. The tools to generate the information spaces directly from information by a similarity matrix are found in the free mSQL-based document finder (http://fabdo.fh-potsdam.de/infoviz). The text based information emphasises high contrast black-and-white text presentation, while the bookmark-converter displays navigable abstract information landscapes, where colour is also information. Then there is a cinema with seats as a hint, specialised to display movie-style information by means of an introductory movie tour through the whole environment by photo-realistic rendered presentation. The user is then taken around following the camera view and perspective. So, this place is not devoted to video-on-demand, but to the study of the differences in the information access in camera-based forced move and free navigation, all in respect to the whole architectural structure. An alternative photo-realistic slide display allows another sort of GIF-map like spatial navigation (O'Neill 1998). A show so far is outlined as complete architecture, followed by an exhibition hall with a collection of 3D site- and object-models, related photographs, and exemplary numerical and map-based information in object-style. The focus is to understand the acceptance and readability of embedded virtual real site models and object-display in relation to map and statistical display prototypes. The architectural segments are designed to answer how far real objects and abstract objects, text and images relate to each other in virtual spaces according to the access modes. The architecture of each of the rooms, according to the implemented information and display types, becomes its own environment to force the appropriate behaviour setting of the user.

References

BLOCK, A. (ed.),1990. *Cognitive models of psychological time.* Hillsdale, NJ: Lawrence Erlbaum.

CHALMERS, M., 1993. Using a Landscape Metaphor to Represent a Corpus of Documents. In Frank, A.U. and Canorri, C.I. (eds.), *Spatial Information Theory: A Theoretical Basis for GIS.* Berlin: Springer Verlag: 377-390.

DAESSLER, R. and PALM, H., 1998. *Virtuelle Informationsräume mit VRML*, Heiderlberg: dpunkt verlag.

ELIAS, N., 1969. Die höfische Gesellschaft. Untersuchungen zur Soziologie des Königtums und der höfischen Aristokratie, Neuwied und Berlin: Luchterhand.

ELIAS, N., 1987. *Über die Zeit. Arbeiten zur Wissensoziologie II*, Frankfurt am Main: Suhrkamp.

FISHWICK, P.A., 1996. *Extending Object-Oriented Design for Physical Modelling*, University of Florida (ftp://ftp.cis.ufl.edu/cis/tech-reports/tr96/tr96- 026.ps).

FOLTZ, M.A., 1998. *Designing Navigable Information Spaces*, St. Louis: Washington University (http://www.infoarch.ai.mit.edu/publications/mfoltz-thesis/; http://www.infoarch.ai.mit.edu/jair/jair-rationale.html).

FRANK, A.U. and CAMPARI, I. (eds.), 1993. Spatial Information Theory, Springer-Verlag, Berlin. In Helfrich, H. (ed.), 1996, *Time and mind.* Seattle: Hogrefe & Huber Publishers.

HOLLERBACH, J.M., 1975. *Hierarchical Shape Description of Objects by Selection and Modification of Prototypes*, MIT Artificial Intelligence Laboratory Technical Report 346.

HOTTA, K., 1992. Systematization of Displays. In Museum and Amusement Parks, *Display Designs in Japan*, 4: 16-35. Tokyo: Rikuyo-sha Publishing.

JENSEN, J., 1993. Christian Jörgensen Thomsen's Museum - a Vision of the Golden Age, *Nordisk Museologi*, 1: 17-30.

KOSSLYN, S., 1981. The medium and the message in mental imagery: A theory, *Psychological Review* 88: 46-66.

LEVINSON, S., 1998, Studying spatial conceptualizations across cultures. *Anthropology and cognitive science.* Ethos, 26: 7-24.

McNAMARA, T., 1986. Mental representation of spatial relations. *Cognitive Psychology* 18: 87-121.

O'NEILL, R. and MUIR, E., 1998. *Web developer.com guide to create 3D worlds.* New York: Wiley Computer Publishing.

PEDERSEN, B.M, 1988. *The Graphic Visualization of Quantitative Information, Procedures and Data.* Zurich: Graphis Press.

RETZ-SCHMIDT, G., 1988. Various Views on Spatial Prepositions, *AI Magazine* 6: 95-105.

SEARLE, R.J., 1958, Proper Names, *Mind* 67: 166-173.

STECKNER, C., 1993. Zum Verbund kommunikativer Lagebeziehungen. In Petzold, L., Schneider, I. and Streng, P. (eds.), *Bild und Text*, 202-209. Bratislava: Slovak Academic Press.

STECKNER, C., 1994. Museen im Zeichen der Französischen Revolution: vom evulutionären zum revolutionären Museum. In Grote A. (ed.), *Macrocosmos in Microcosmo*: 817-853, Opladen: Leske + Budrich.

STECKNER, C., 1995. SAMOS: Automatic Classification of Ancient Greek Pottery According to the Shape. In Nissan, E. and Schmidt, K. (eds.), *From information to knowledge: conceptual and content analysis by computer*, 126-150. Society of Conceptual and Content Analysis by Computer: Concepts - Content - Meaning 1, Oxford: Intellect books.

STECKNER, C.A., 1999. Anticipatorial Maps, *International Journal of Computing Anticipatory Systems* 4: 101-124.

STECKNER, C., 1999. Cross-modal visual transfer and the forming of haptic anticipatory phantom objects. In Bülthoff, H.H. and Fahle, M. (eds.), *Beiträge zur 2. Tübinger Wahrnehmungskonferenz* 100, Kirchentellinsfurt: Knirsch Verlag.

STECKNER, C., 2000. Form and Fabric. In Barceló, J.A., Forte, M., Sanders, D.H. (eds.), *Virtual Reality in Archaeology*: 121-128. Stafford: BAR International Series 843.

THOMAS, C., 1996. *To Assist the User: On the Embedding of Adaptive and Agent-based Mechanisms.* Munich: Oldenbourg-Verlag (GMD-Bericht Nr.258).

THORNDYKE, P.W. and HAYES-ROTH, B., 1982. Differences in spatial knowledge acquired from maps and navigation. *Cognitive Psychology* 14: 560-589.

TUFTE, E.R., 1990. *Envisioning Information*, Chesire, CT: Graphics Press.

WAGENER-WENDER, M., 1993. *Mentale Repräsentationen räumlicher Informationen.* Bonn: Holos.

YIP, K. and ZHAO, F., 1996. Spatial Aggregation: Theory and Applications, *Journal of Artificial Intelligence Research* 5: 1-62.

Setting Demographic Limits: The North American Case

Dean R. Snow

Department of Anthropology, The Pennsylvania State University

409 Carpenter Building, University Park, PA 16802, USA

e-mail: drs17@psu.edu

Abstract

The size of the aboriginal population of North America at the time of European discovery is a controversial issue. Estimates for the portion of the continent north of Mexico range from 1-18 million. A few scholars argue for very high numbers, usually by assertion and in the absence of empirical evidence. Many others argue that the higher estimates are preposterous, but usually atheoretically and on the basis of negative evidence. I argue that explicit theory and simple mathematical modelling allows us to establish a means to assess upper population limits at local, regional, and continental scales. This approach, along with rare empirical tests, allows us to conclude with greater confidence that the 1492 population size was near the lower end of the range, around 3.4 million.

The size of the aboriginal population of North America in AD 1500 is a controversial issue, and estimates on the size of the population have varied widely. This is an important issue because so much else in the way of subsistence, socio-political organization, settlement types, and so forth is associated with demography.

Estimates for the portion of the continent north of Mexico range from 1-18 million. Estimates have also been typically based on scanty ethnohistorical (documentary) evidence. For example, Henry Dobyns (Dobyns 1966, 1983) worked almost exclusively with documentary sources. Further, he assumed that any epidemic observed anywhere in North America in the sixteenth century must have been the local expression of a widespread pandemic. Because most of his sources dated to late in that century or to the seventeenth century, he also assumed that most or all of his scanty census data related to populations already reduced by disease. This led him to assume that most historical counts of Indians amounted to 5-10% of their earlier AD 1500 sizes. When he aggregated his inflated estimates, he came up with a total of at least ten million Indians north of central Mexico in AD 1500. Some others have proposed even larger numbers (up to 20 million) based on similar arguments.

One problem with high estimates is that archaeologists cannot provide enough addresses to accommodate ten million or more people in North America five hundred years ago. Another problem with them is that Indian enthusiasts have seized upon the inflated numbers, preferring them to lower estimates for modern political purposes. The same enthusiasts usually also attribute high ecological virtue to the same Indian populations. So we have to cope with several modern myths, some of them scholarly, some of them popular.

Douglas Ubelaker (Ubelaker 1988) attempted to provide some balance and published a paper in which he argued for a total population of less than two million people for North American (north of central Mexico) in AD 1500. He used the component chapters of the *Handbook of North American Indians* as his primary sources, although he was often working with manuscripts that were available to him at the Smithsonian Institution prior to publication, and therefore unavailable to the rest of us. He deferred to specialist authors when they offered well-reasoned estimates of aboriginal population sizes for specific populations. He came up with an estimate of 1,894,280 as shown in table 1. I have corrected two typographical errors in the table he published.

The problem with Ubelaker's work is that we have no empirical data for large parts of the Northeast and the Southeast. We also have almost no data at all for the Plains. The mounted nomadic tribes for which the Plains are famous today did not yet exist in 1500 because the domesticated horse had not yet been introduced to North America. Indeed, it is unclear to me where Ubelaker's data could have come from for this and some other regions.

My work on the Mohawk case (Snow 1995) was the first, and so far the only, successful effort to measure population size and its change over time by archaeological means. Trends in Mohawk population size from 1525 to 1776 are summarized in table 2. The results implied that Ubelaker's estimate for the continent was much closer to the mark than that of Dobyns. But for most of North America direct archaeological measurement is even less easy to do than measurement based on documents.

More recently I turned to a simpler ecological approach. I divided the continent into thirteen broad ecological zones. This was based on a simple map provided on line by the National Geographic Society. I simplified it further and remapped the ecological zones of the continent using MapInfo (figure 1). Then I looked for cases of American Indian societies for which we have both reliable pre-epidemic population estimates and reliable land area estimates. I found many, but not all, of these in various volumes of the *Handbook of North American Indians*. There are too many cases and too many sources for me to report all of them here, but I will provide them to anyone seeking to replicate my results.

Figure 2 shows tribal areas for North America in 1500. These too come largely but not entirely from volumes of the *Handbook of North American Indians*. Some come from the *Atlas of Ancient America* (Coe et al. 1986). We have population and area figures for the shaded areas shown because adequate records were made in them before epidemics and dislocations occurred. From these figures I could compute population densities. By separating them according to their ecological zones I was able to determine in a rough way the average populations densities of the zones. Table 3 summarizes the ecozones and the numbers of cases for which I found data in each of them. The aggregated cases for each ecozone constitute the sample for that ecozone.

	AD DATE					Overall
REGION	1500	1600	1700	1800	1900	Decline
Arctic	73,700	73,700	73,700	59,190	34,994	53%
Subarctic	103,400	103,400	99,750	76,350	45,535	56%
Northwest Coast	175,330	175,330	175,330	98,333	29,785	83%
California	221,000	221,000	221,000	200,000	14,825	93%
Southwest	454,200	420,000	276,260	215,950	158,283	65%
Great Basin	37,500	37,500	37,500	33,905	14,606	61%
Plateau	77,950	77,950	77,950	70,000	18,720	76%
Plains	189,100	189,100	189,100	120,330	62,656	67%
Northeast	357,700	357,700	149,360	117,260	95,457	73%
Southeast	204,400	157,400	105,125	60,370	61,701	70%
TOTALS	1,894,280	1,813,080	1,405,075	1,051,688	536,562	72%

Table 1: North American Indian population change, 1500-1900 (after Ubelaker 1988, with corrections).

Periods	Population
1755-1776	614
1712-1754	600
1693-1711	790
1689-1692	1100
1679-1688	1100
1666-1678	2000
1659-1665	2304
1657-1658	2304
1646-1656	1734
1643-1645	1760
1635-1642	2835
1626-1634	7740
1614-1625	6225
1580-1613	4575
1560-1579	2019
1545-1559	1570
1525-1544	1490

Table 2: Mohawk population change.

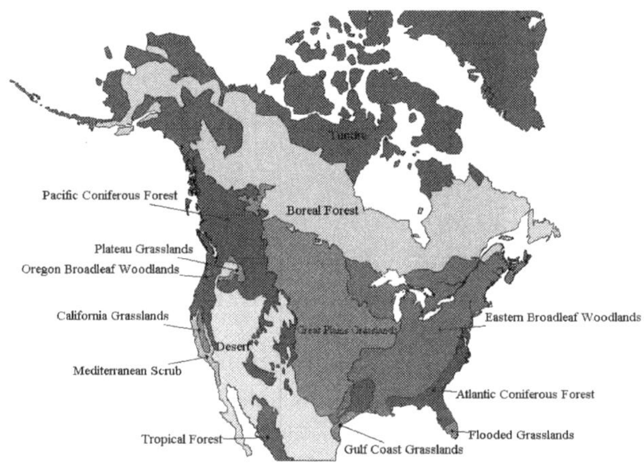

Figure 1: Ecological zones of the continent.

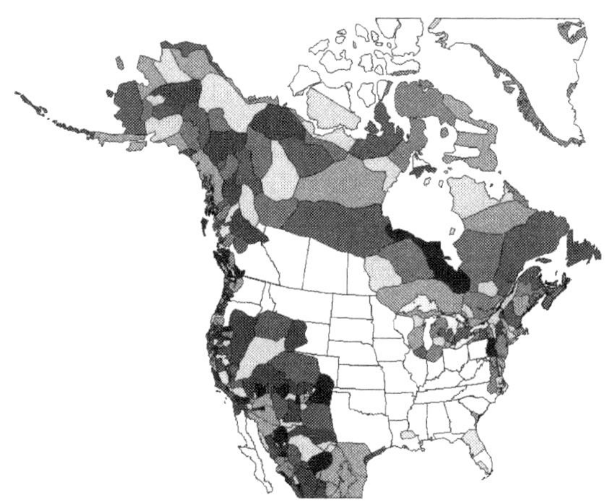

Figure 2: Tribal areas for North America in 1500.

I found 150 usable cases. In no case did I have complete data for an ecozone, but for nine of them I had enough data to compute densities. For six ecozones there were enough cases to make descriptive statistics informative. These are reported in table 4. Note that the population densities reported in table 3 are computed from aggregate sample areas and aggregate sample populations. The mean densities reported in table 4 are means of sample densities, which because of skewing are not necessarily equal to the ecozone densities reported in table 3. Unfortunately, four of the thirteen zones had no useable cases at all. These were the Great Plains Grasslands, Plateau Grasslands Oregon Broadleaf Woodlands, and Flooded Grasslands. Two of them are very small ecozones when compared to others used here. For each of these zones I had to use a rough estimate based on neighboring ecozones. My estimates were based on densities in regions having roughly similar carrying capacities. All four of the zones in question supported mainly hunter/gatherers in AD 1500, so an estimate of 0.18 persons per square kilometer is appropriate.

When I computed human population sizes by ecological zone using observed or likely densities and area totals in square kilometers, I obtained probable human population sizes for all thirteen zones. The aggregate total land area inhabited in 1500 is 23,163,390 square kilometers. The aggregate human population in 1500 was probably around 3.4 million.

This means of estimating North American population size in 1500 seems so simple and so obvious that I cannot understand why it appears not to have been used by anyone in the past. The approach has the advantage of exposing the absurdity of much higher published estimates. However, it has the persisting disadvantage of requiring extrapolations from few or no cases in seven out of thirteen ecozones. The standard deviations for the remaining six

Ecozone	Cases	Sample Area	Sample Population	Sample Density	Total Area	Total Population
Mediterranean Scrub	11	64,967	61,921	0.95	131,000	124,858
Boreal Forest	20	4,811,231	83,466	0.02	5,310,000	92,119
Atlantic Coniferous Forest	5	141,600	46,410	0.33	490,500	160,763
Pacific Coniferous Forest	41	480,767	197,745	0.41	1,976,000	812,752
Desert	15	1,284,534	226,923	0.18	2,613,000	461,607
Eastern Broadleaf Woodlands	34	910,825	324,984	0.36	2,955,000	1,054,349
Gulf Coast Grasslands	1	19,520	3,600	0.18	75,590	13,941
California Grasslands	2	40,690	46,465	1.14	52,050	59,437
Tundra	21	2,310,349	40,703	0.02	6,534,000	115,114
Great Plains Grasslands	0	0	0	0.18	2,925,000	526,500
Plateau Grasslands	0	0	0	0.18	54,550	9,819
Oregon Broadleaf Woodlands	0	0	0	0.18	14,650	2,637
Flooded Grasslands	0	0	0	0.18	32,050	5,769
TOTALS	150				23,163,390	3,439,665

Table 3: American Indian populations by ecozone.

	Valid N	Mean	Minimum	Maximum	Std.Dev.
Mediterranean Scrub	11	106.0	21.2	182.8	62.8
Boreal Forest	20	2.0	0.4	4.3	1.5
Pacific Coniferous Forest	41	117.5	4.6	665.6	123.8
Desert	15	81.5	3.2	480.0	120.5
Eastern Broadleaf Woodlands	34	60.0	8.3	320.0	81.5
Tundra	21	6.6	0.2	84.9	18.3

Table 4: Statistics for selected ecozone densities.

(table 4) are sobering. Skewing, small sample sizes, and large ranges have produced some astounding values. What we learn from the exercise is that the best course is to dampen the effects of variability within each ecozone by aggregating sample data and computing density from the sums. However, the variability within ecozones revealed by table 4 must be remembered when reasonable estimates are generated for the seven ecozones for which we have few or no cases. There is only one case available for the Gulf Coast Grasslands. Should we assume that it accurately represents the whole? Perhaps it is unrepresentatively high – or low. Thus we must rely on informed judgment, and like Ubelaker I have depended upon that of the specialists who know the individual cases best. Allowing reasonably for lacunae, the sum of the best judgment of those scholars is that the total population north of Mexico in AD 1500 was around 3.4 million people. Even if that number is 20% in error it is within a range well below the absurdly high estimates that have enjoyed popularity in recent years. Perhaps now that a more rational estimate is in hand we can get on with the solution of other important problems in Native American demography.

References

COE, M., SNOW, D.R. and BENSON, E., 1986. *Atlas of Ancient America.* Facts on File, New York.

DOBYNS, H.F., 1966. Estimating Aboriginal American Population. *Current Anthropology* 7(4):395-416.

DOBYNS, H.F., 1983. *Their Number Become Thinned: Native American Population Dynamics in Eastern North America,* University of Tennessee Press, Knoxville.

SNOW, D.R., 1995. Microchronology and Demographic Evidence Relating to the Size of Pre-Columbian North American Indian Populations. *Science* 268:1601-1604.

UBELAKER, D.H., 1988. North American Indian Population Size, A.D. 1500 to 1985. *American Journal of Physical Anthropology* 77(3):289-294.

Counting the Uncountable: A Quantitative Approach to the Religious Differences between the Roman Towns of Emona and Poetovio

Bernarda Županek

City Museum of Ljubljana

Gosposka 15, 1000 Ljubljana, Slovenia

e-mail: bernarda.zupanek@mm-lj.si

Dimitrij Mlekuž

Gradišnikova 22, 1353 Borovnica, Slovenia

e-mail: dimitrij.mlekuz@kiss.uni-lj.si

Abstract

A number of monuments dedicated to deities have been discovered in past years in the area of the Roman towns of Emona and Poetovio. We believe this corpus of monuments reflects the religious background, correlated to the dedications, of both towns. Select statistical methods were used to assess this background, and a neural network was applied for pattern-recognition tasks. A statistically significant difference was proven between the Emona and Poetovio samples, which in our opinion, is related both to the different temporal distribution of selected monuments as well as to the different religious preferences of the dedicators in Emona and Poetovio. The paper aims to point out the advantages and restrictions of a quantitative approach to the study of epigraphic monuments; an approach, we believe, which offers new possibilities for the study of epigraphic data.

Key words: quantitative methods, neural network, Roman period, epigraphic data, dedicatory inscriptions, religious background, Emona, Poetovio

"We must compare in order to make distinctions, and distinguish in order to understand."

Turcan R., 1996

1. Introduction

A considerable number of monuments dedicated to various deities have been discovered, collected and interpreted in the past years in the area of the Roman towns of Emona and Poetovio. This paper presents an application of select statistical and pattern recognition methods on the collected corpus of monuments.

Our basic premise is that the preserved corpus of monuments reflects the religious background, correlated to the dedications of Emona and Poetovio. Several questions thus arise. The paper focuses on the following: Is there a discernible difference in the religious backgrounds of both towns? And if so, what is the nature of this difference? What caused it? In an approach to these questions, statistical and pattern recognition methods have been used.

1.1. The tale of two cities

The Roman towns, Emona and Poetovio, were located in the territory of present day Slovenia (figure 1): Emona beneath the modern capital, Ljubljana, and Poetovio further eastwards, in the area of modern day Ptuj. There are some similarities between Emona and Poetovio (formerly native settlements, their decline in the fifth century, strong Early Christian communities from the fourth century onward), but the character and evolutionary course of each Roman town was rather different.

Colonia Iulia Emona was founded sometime during the course of the Octavian/Augustan period (Šašel Kos 1995). Though several authors claim that a legionary camp existed in the area before the foundation of the town, this assumption was proven to be ill-founded, supported by neither epigraphic (Šašel Kos 1995) nor archaeological evidence (Plesničar-Gec 1998).

Furthermore, it is not clear whether Emona was founded as a veteran colony. As indicated by inscriptions, it was settled with mainly civilian inhabitants from (mostly northern) Italy, and only sporadically with veterans (Šašel Kos 1995:232). Emona might have been situated in the province Pannonia initially, but it was part of Italy from the second half of the second century onwards at least. The town was quite prosperous from the first to the beginning of fifth centuries. Although it declined after the mid fifth century, sparse habitation obviously continued up to the beginning of the sixth century (Plesničar-Gec 1997).

The town of Poetovio was founded approximately one hundred years later than Emona. During the first century AD, a legionary camp was situated in this area, which was surrounded by a rapidly developing town (*canabae*). *Colonia Ulpia Traiana Poetovio* was, as its name reflects, founded under Traianus, ca. AD 103–106 (Vomer Gojkovič 1996:3), and was situated in the province of Pannonia.

Poetovio witnessed its peak during the second and third centuries (Vomer Gojkovič 1996:3), when it was the residence of the *publicum portorium Illyricum*. Immigrants from northern Italy, retired military personnel from all parts of the empire, especially from the East, and merchants of various ethnic origins all settled in this important economic and financial centre, thus giving Poetovio a rather cosmopolitan character (Šašel Kos 1993:229, 1999:153). As the onomastic study of the material indicates, the native element was pushed more into the background here than in Emona (Šašel Kos 1993:229). The period of prosperity and splendor of the largest Roman town on the territory of present day Slovenia

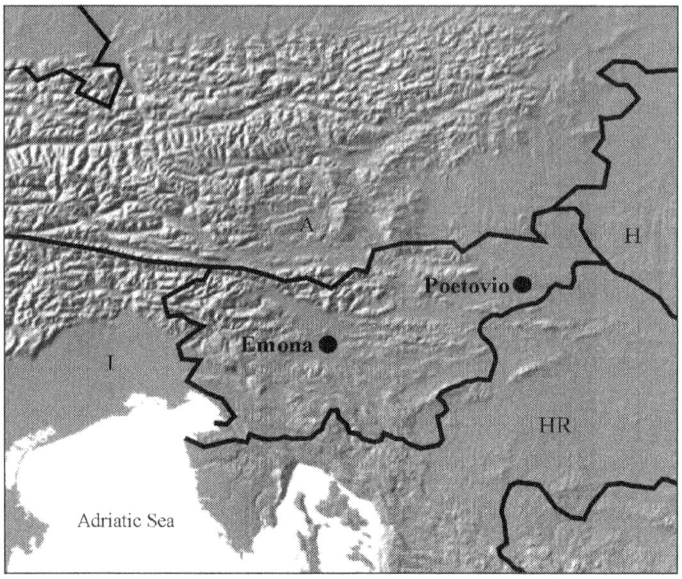

Figure 1: Location of Emona and Poetovio.

was followed by a gradual decline, and after the mid fifth century, the town was almost empty (Horvat 1999:220).

2. Counting the uncountable

2.1. Database

A relational database of dedicatory inscriptions from both towns (available on request) was created using MS Access 97. Several attributes were recorded for each monument (name of the deity, social class of the dedicator, date and original location of monument, references, etc). The database was built according to several criteria, one being at least partial preservation of the name of the deity, thus allowing for a reliable reconstruction. The dedicator's name was of equal importance for our analysis, as an establishment of her/his social class. Several problems were encountered, mainly due to the fragmented and ill-defined nature of available data. The problems that arose because of indefinite dating were confronted using a probability scheme for dates (figure 2).

Thus, two samples were obtained: one for Emona and one for Poetovio. It should be emphasized that both samples are hardly representative, due to their limited size (77 inscriptions were recorded for Poetovio, and only 23 for Emona) and several biases (depositional and post-depositional processes, collection-related biases).

2.2. Looking at patterns

Some basic statistical analyses were applied to both samples, searching for general trends concerning deities, the social class of the respective dedicators and the temporal distribution of monuments.

Among the 24 recorded deities, only five are present in both towns (Aesculapius, Diana, Jupiter, Jupiter Depulsor, and Victoria). The Poetovio sample consists of 77 monuments, of which 67% are dedicated to only three deities (Jupiter, Mithras and Nutrices), while the other deities are represented in less than 5% each. The Emona sample, with only 23 monuments, is much smaller and much more uniform (probably due to the limited size of the sample): Jupiter 21%, and 13% of Victoria and Aequorna each. Particularly noteworthy are the very similar percentages of monuments that are dedicated to Jupiter in both the Emona and Poetovio samples.

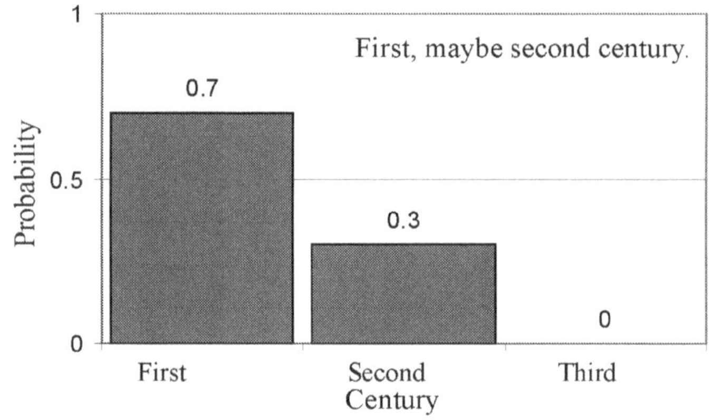

Figure 2: Example of data representated using a probability scheme.

Percentage and numerical diagrams of deities from both towns show considerable differences between the Emona and Poetovio samples (figure 3). A χ^2 test was carried out to assess this difference (Shennan 1988:65–77). The test proved that the distribution of deities through the Emona and Poetovio samples differed at a significance level of 0.001.

Some similarity is discernible in the social structures of the two towns (figure 4). Roman citizens represent the largest social group in both samples (however, we must point out that they are the most extensive group in the Emona sample anyway). In Poetovio, slaves are a strong group as well. Soldiers are the least represented group; only two were recorded for Emona and 13 for Poetovio.

Regarding the monuments by their respective dates (figure 5), the percentage of monuments in Emona is fairly equal during the first and second centuries, however it sharply declines in the third century. The situation in Poetovio is quite the opposite. The lack of monuments from the first century is followed by their peak in second, and then slight decline in third century. A temporal overlapping between the two samples can only be observed in the second and third centuries.

2.3. In search for answers

To clarify the established difference between the Emona and Poetovio samples, we conducted an analysis where the religious preferences of the defined social groups were observed and compared through several centuries. A backpropagation neural network (SNNS v4.1) was chosen (figure 6) as the primary tool in our analysis. A neural network is a computational model consisting of a number of neurodes and the connections between them (e.g. Dobnikar 1990, Kasabov 1996, Looney 1997). Neurodes respond to incoming signals by switching between states, according to the strength of the incoming signal. A neural network has similarities with the human brain: it can learn from experience and it can be thought humanlike reasoning. As such, it is often used for complex pattern-recognition tasks.

As seen above, each town witnessed a high number of various deities; only a selection of deities was represented in both towns

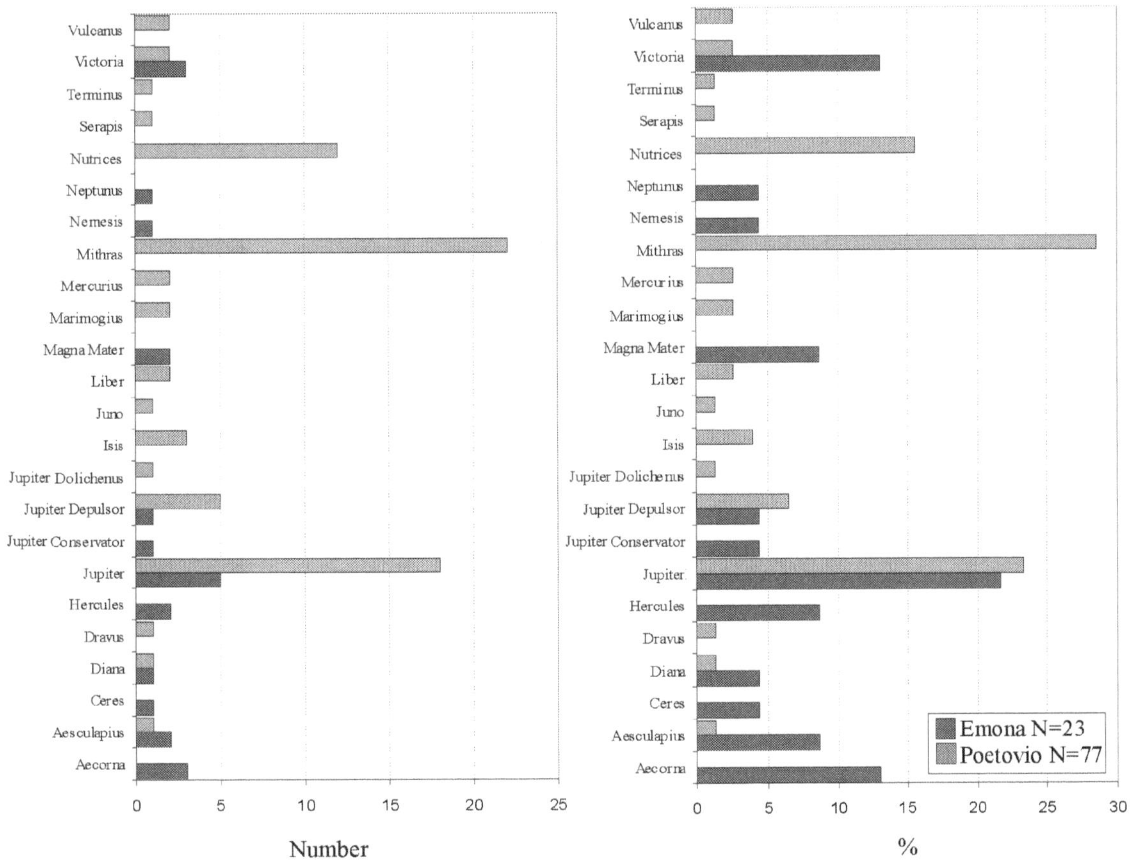

Figure 3: Percentage and numerical charts of deities from both samples.

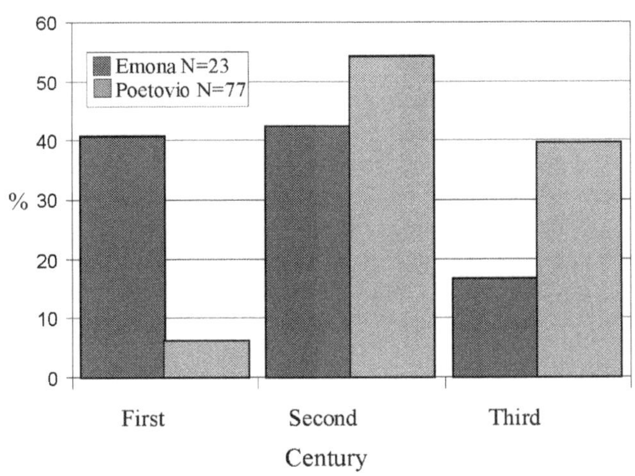

Figure 4: Percentage chart of diverse social groups from both samples.

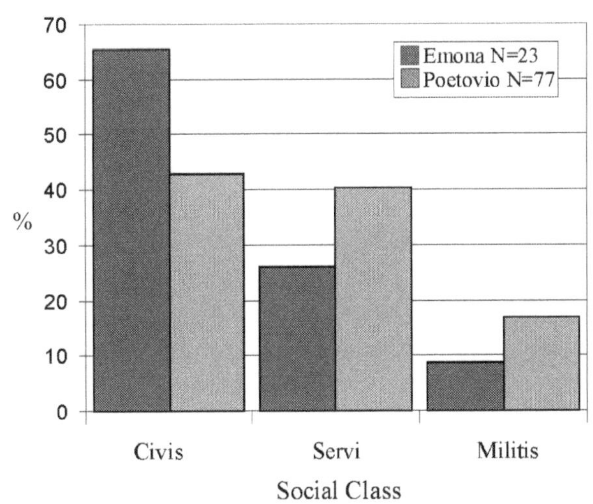

Figure 5: Percentage chart of the temporal distribution of monuments from both samples.

and thus required some classification. Deities were classified according to their respective origins and/or function into the following groups: Roman, eastern and local deities (e.g. Kolšek 1968:273, Belak 1993:233). Syncretistic deities (Jupiter Depulsor, Hercules) were represented using a probability scheme.

Dedicators were then categorized according to their respective social classes. Three categories were chosen to establish a few well-defined groups: citizen, soldier and slave. *Liberti* were included among the latter, their religious background probably not significantly changed by the transition to a different social class. Veterans were included in the soldier group.

2.3.1. Analysis

In the process of preparing the neural network, each town, social class and date were used as input variables. The class of deity was the resulting output variable. "Clear" cases (questions e.g. citizen, first century) were then presented to the neural network, and the following results were observed (figure 7).

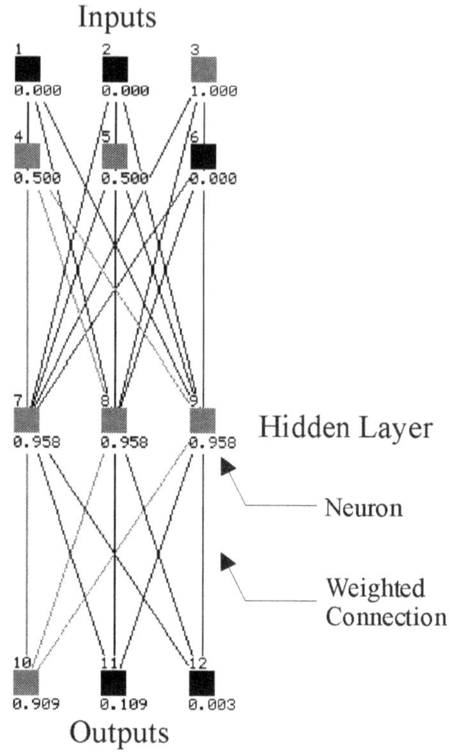

Figure 6: Neural network.

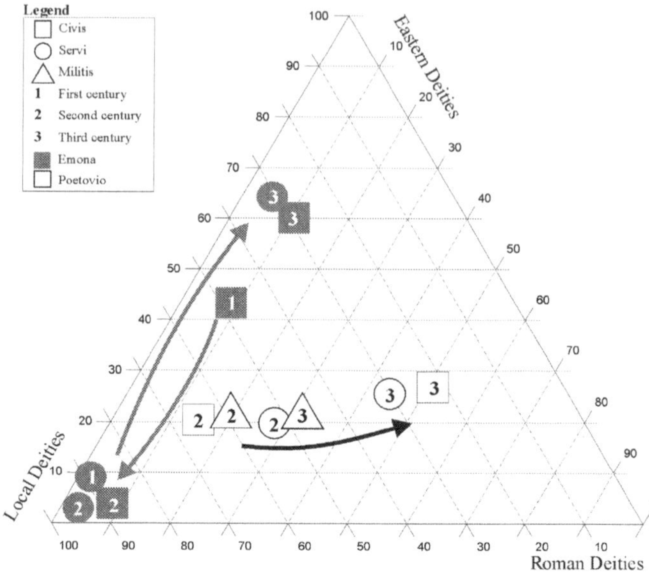

Figure 7: Analysis results.

In the Emona sample, citizens in the first century show a fairly equal preference for local and Roman deities, while in the second century, Roman deities are by far preferred. By the third century, the scarce data points towards the growing popularity of the local deities. One obvious conclusion is the relative unpopularity of eastern deities among the Roman citizens in Emona.

As for slaves, chiefly Roman deities are preferred in Emona during the first two centuries, whereas an increasing popularity of local deities is detectable in the third century. As among the citizens of Emona, eastern deities were also quite unpopular among Emona slaves.

The soldier group in the Emona sample is by far too small to allow any conclusions to be made.

The citizens of Poetovio held Roman deities in the highest esteem in the second century, with local and eastern deities coming in second place. In the third century, eastern deities are first in line, with local following and Roman far behind.

The ratio of local deities worshipped by the slave class in Poetovio is similar during both the second and third centuries. The ratio of eastern deities then rises in the third century, and the Roman one declines.

Soldiers show a similar interest in local cults in both the second and third centuries. Their interest in Roman deities considerably lessens in the third century, which is when soldiers from the Poetovio sample favour eastern deities.

The second century is marked by a high preference for Roman deities in both samples, whereas the preference for the Roman deities abruptly declines in the third century and prevails for local deities in Emona and eastern deities in Poetovio. The preference for local deities in the Poetovio sample remains at a constant level throughout the second and third century. Eastern deities were obviously never of marked importance in Emona, however they are well represented in the second century Poetovio sample and predominate almost unanimously in the third century.

2.3.2. Discussion

The religious backgrounds of both towns, as reflected in the preserved corpus of dedicatory monuments, differ significantly.

We are tempted to interpret the difference in their sample sizes as a reflection of the different populations in both towns. However, it could also be a result of varying religious habits in the local tradition, or simply an echo of diverse economic backgrounds, where stone monuments were affordable only to members of economically select groups.

The temporal distribution of monuments from Emona and Poetovio is also different. The observed temporal distribution suggests the peak of dedicative religious activity in Emona during the first century, slowly diminishing afterwards; while in Poetovio, the peak of such activity is demonstrated in the second century. We believe that the general temporal distribution of the analyzed monuments most probably reflects the different evolutionary courses of each town, as is also advocated by archaeological evidence.

A general decrease in the popularity of Roman deities in third century can be observed among slaves and citizens in both towns. However, this uniform decrease was obviously compensated diversely, as the sample from Emona shows an increase in local cults and the sample from Poetovio an adoption of eastern cults. The latter remain quite unimportant in Emona, their scant number consisting mainly of Egyptian deities. We are inclined to interpret the Poetovio sample as a reflection of the increased importance of Poetovio during the second and third centuries, as well as of the corresponding influx of new inhabitants with a predominantly eastern religious background.

Another interesting pattern is the considerable popularity of local cults among Emona citizens during the first century. It decreases severely in the second century, however, not among the Emona slaves, where local cults retain constant, though low, popularity. Again, we emphasize that all results obtained using the Emona sample are hardly representative due to the limited sample size.

3. Conclusions

The present exercise aims at pointing out the advantages and restrictions of a quantitative approach in studying epigraphic monuments; it is an approach, which we believe offers a range of new possibilities in studying epigraphic data. Select statistic analyses were applied to samples of dedicatory inscriptions from Emona and Poetovio, and a neural network was used for pattern-recognition tasks. The principle problem encountered was that of sample size. As for the results, a statistically significant difference between the samples from Emona and Poetovio was observed, which is in our opinion related both to the different temporal distribution of selected monuments and to the diverse religious preferences of dedicators in Emona and Poetovio.

Acknowledgement

We are indebted to Dr M. Lovenjak for all his advice and help assessing the social class of the respective dedicators. We would also like to thank R. Novšak for proofreading.

References

BELAK, M., 1993. Med Nutricami in Mitro. *Ptujski arheološki zbornik*: 233-239. Pokrajinski muzej Ptuj, Ptuj.

DOBNIKAR, A., 1990. *Nevronske mreže. Teorija in aplikacije.* Didakta, Radovljica.

ELIADE, M., 1996. *Zgodovina religioznih verovanj in idej II.* DZS, Ljubljana.

HORVAT, J., 1999, Petoviona – velemesto na bregovih Dravusa. In Aubelj, B. (ed.), *Zakladi tisočletij*: 215-220. Založba Modrijan, Ljubljana.

KASABOV, N.K., 1996. *Foundations of Neural Networks, Fuzzy Systems, and Knowledge Engineering.* MIT Press, Massachusetts.

KOLŠEK, V., 1968. Pregled antičnih kultov na slovenskem ozemlju. *Arheološki vestnik* 19: 273-286.

LOONEY, C.G., 1997. *Pattern Recognition Using Neural Networks.* Oxford University Press, Oxford.

PLESNIČAR GEC, L., 1997. Emona v pozni antiki v luči arhitekture. *Arheološki vestnik* 48: 359-370.

PLESNIČAR GEC, L., 1998. Emona, vojaški tabor? *Zgodovinski časopis* 52/3:331-336.

SHENNAN, S., 1988, *Quantifying Archaeology*. Edinburgh University Press, Edinburgh.

ŠAŠEL KOS, M., 1993. Petovionska vladajoča aristokracija. *Ptujski arheološki zbornik*: 219-232. Pokrajinski muzej Ptuj, Ptuj.

ŠAŠEL KOS, M., 1995. The 15th Legion at Emona - Some Thoughts. *Zeitschrift für Papyrologie und Epigraphik*, Band 109:227-244.

ŠAŠEL KOS, M., 1999. Pre-Roman divinities of the Eastern Alps and the Adriatic. *Situla* 38.

TURCAN, R., 1996, *The Cults of the Roman Empire.* Blackwell, Oxford.

VOMER GOJKOVIČ, M., 1996. Ptuj. Arheologija. In *Enciklopedija Slovenije* 10:3-5. Mladinska knjiga, Ljubljana.

Design and Performance of the Varatioscope

John W.M. Peterson

School of Information Systems, University of East Anglia

Norwich NR4 7TJ, United Kingdom

e-mail: jwmp@sys.uea.ac.uk

Abstract

Some Roman land surveys (centuriations) appear to have been used to plan features of the same period which are oblique to the survey grid. This method of oblique construction, documented by the Roman land surveyors and known in Latin as uaratio, *restricts the orientations of the features, relative to the grid, to a small set of "significant" angles with rational tangents. Useful information may therefore be gained by studying a set of linear features, such as segments of Roman roads or boundary lines, from various angles to see how well they conform to this model.*

The varatioscope is designed to give a score for each angle of observation, high scores corresponding to an angle to which the features, as a whole, have a large number of "significant" relationships. It can be used in two ways, either to test a hypothesis of centuriation or to suggest such a hypothesis, corresponding to a peak in the varatiogram. The paper describes the construction of the varatioscope as an Excel spreadsheet, results obtained, and the sensitivity of the tool to the choice of "significant" angles.

Key words: Roman, cadastre, centuriation, oblique planning, varatio

1. Varatio

In the writings of the Roman Land Surveyors the term *uaratio* refers to a process of oblique surveying. Anne Roth-Congès (1996) has shown, by careful study of the texts, that at least two approaches were employed. In *uaratio fluminis* the distance of an inaccessible point could be calculated by the construction of right angled triangles. By *uaratio in agris diuisis* the surveyors could construct, or reconstruct, lines along the hypotenuse of right angled triangles defined in a rectangular coordinate system; this sort of *uaratio* interests us here.

Even before we had this improved understanding of the texts it had become apparent to several independent workers in the last quarter of the twentieth century that linear features frequently coincide with centuriation grid intersections (Chouquer et al. 1983, Clavel-Lévêque 1991, Morra and Nelva 1977, Romano and Tolba 1996). The authors all ascribe this to design, rather than accident. If so, it a result of the use of *uaratio*.

It is even possible that one effect of *uaratio* was seen and understood much earlier. Since the variety of observed oblique relationships (1:1, 1:2, 2:3, 3:4 ...) is limited, there is a good chance that some linear features defined by a particular coordinate system, for example road segments, will be parallel. Parallel road segments have been observed in British contexts. In one case Margary (1973:60) thought it "likely" that it was the result of sighting by stars - an approach which does not seem very feasible, since it would have obliged surveyors to work at night. In another case Haverfield (1921) seems to have favoured a quite different mechanism, which may have been *uaratio*. In a study on two parallel road segments whose spacing would not allow them both to be the axes of a centuriation, he hinted at a connection between British centuriation and observable coincidences in orientation of Roman road segments. Unfortunately he did not live to enlarge on this.

2. The varatioscope

If *uaratio* was used in this way we may gain useful information by studying a set of Roman linear features - such as segments of roads or boundaries - from various angles to see how well they conform to this model.

The varatioscope is designed to give a score for each angle of observation, high scores corresponding to an angle to which the features, as a whole, have a large number of "significant" relationships.

Given an angle for the varatioscope reading, the orientation of any given feature defines a relative angle. We can put the tangent of this angle in the range 0 to 1 by taking reciprocals and changing the sign, if necessary. This defines the smallest angle between the feature and a centuriation with the current angle of the varatioscope. A score for this tangent value can then be derived from a scoring function such as that shown in figure 1. This has a value of one for "significant" tangent values and descends linearly to zero midway between them. In this particular case the "significant" values are those which, according to the author's knowledge, have been observed most often in reality.

In order to obtain a varatioscope reading for a given angle of observation, the procedure is as follows. For a segment of length l_i the tangent of its orientation is calculated from the coordinates of its endpoints and the corresponding score, s_i is derived from the scoring function. The score, S, for the population of n segments is then the normalised weighted sum of scores

$$S = \frac{\sum_{i=1}^{n} s_i l_i}{\sum_{i=1}^{n} l_i}$$

The current version of the varatioscope is implemented with an Excel spreadsheet, using a macro command to generate values of S for angles of observation between - 45° and 45° at intervals of

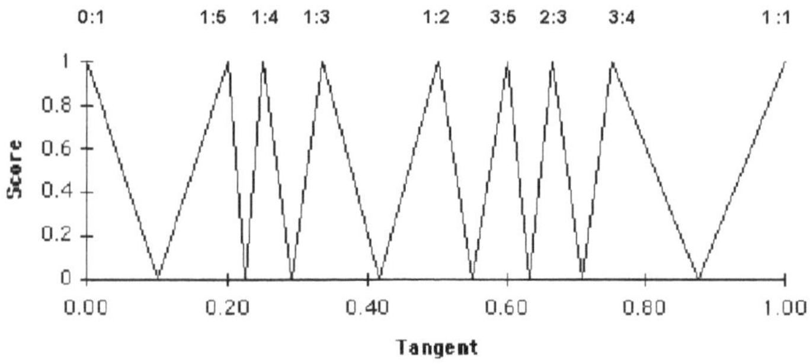

Figure 1: A scoring function for tangents of angles.

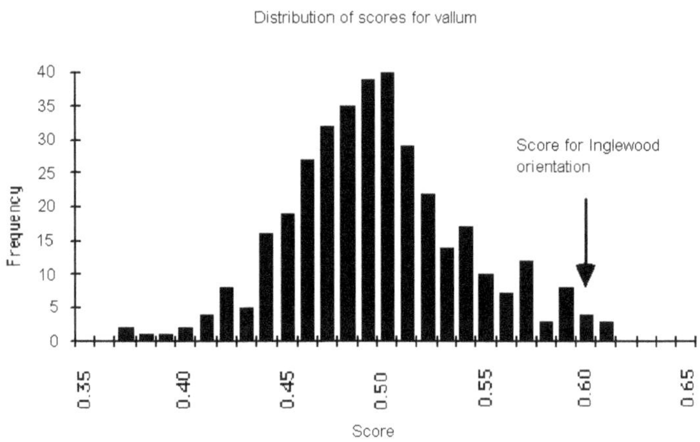

Figure 2: Score for orientation of Inglewood centuriation compared to all scores for the vallum of Hadrian's Wall.

quarter of a degree. This covers the whole range of possible centuriation angles. Because of the linearity of the scoring function between 0 and 1 the distribution of these scores has a mean of 0.5. Results suggest that it is normal, as you would expect from the central limit theorem and the arbitrary definition of observation angles.

3. Use of the results

3.1. Hypothesis testing

We may already have a hypothetical centuriation, proposed perhaps from the layout of existing surface features. The score for its orientation, with respect to a well defined set of Roman linear features in an area, can be compared with the general distribution of scores.

An example is the Inglewood centuriation in Cumbria, Britain and the vallum of Hadrian's wall. The orientation of the hypothetical centuriation, as defined by Alan Richardson (1999), is N 30.96° W (-0.5404 radians). The vallum was defined by a set of coordinates for the endpoints of 104 segment taken from the OS map of the Wall (Ordnance Survey 1964). These segments have a total length of 98.3 km. Their score relative to the angle of the proposed Inglewood centuriation is 0.600.

The general distribution of scores (figure 2) has the following parameters:

mean	0.500
standard deviation	0.044

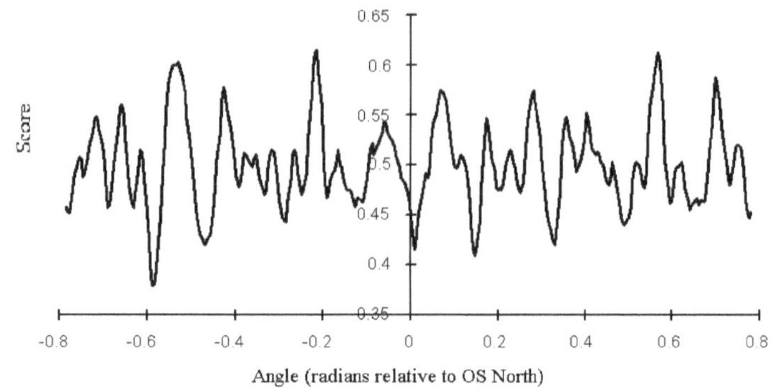

Figure 3: Varatiogram for the vallum of Hadrian's Wall.

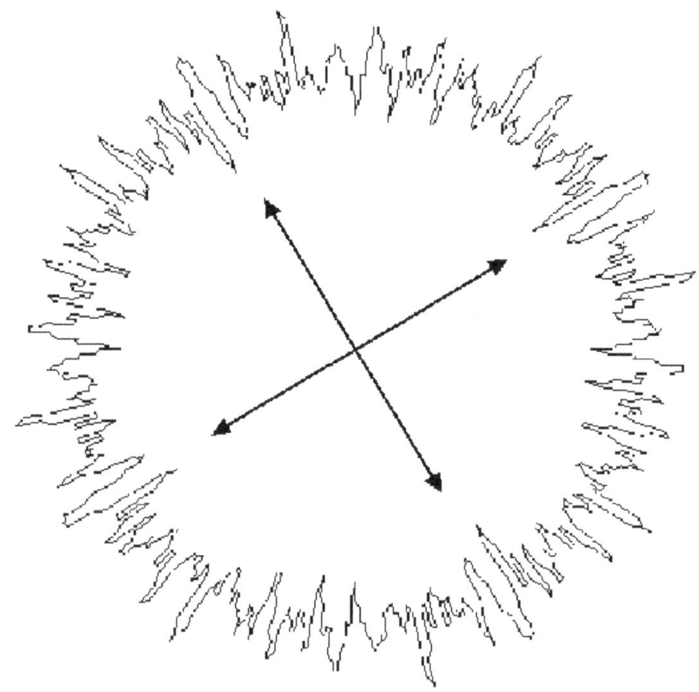

Figure 4: Varatiogram for the vallum of Hadrian's Wall in circular form. The arrows indicate the orientation of the axes of the Inglewood centuriation.

Only a small part of a normal distribution with this standard deviation exceeds the observed score of 0.600. The chance of seeing such a high value is 1.2%. The idea that we would have seen such a high score if we had been interested in some orientation other than that proposed by Richardson can be rejected at the 2% level.

3.2. Generation of hypothesis

The varatioscope output can be displayed as a varatiogram, a graph of scores for the complete range of possible centuriation angles. In the case of that for the vallum (figure 3) there are three high peaks, and four lesser ones. If the presence of a centuriation were suspected in the area, for example because landscape features had consistent orientation or Roman road segments were parallel, each peak could be the basis of an individual hypothesis to be tested against these other indicators.

So that the results of this approach might be seen more clearly, the varatiogram can also be displayed in circular form (figure 4).

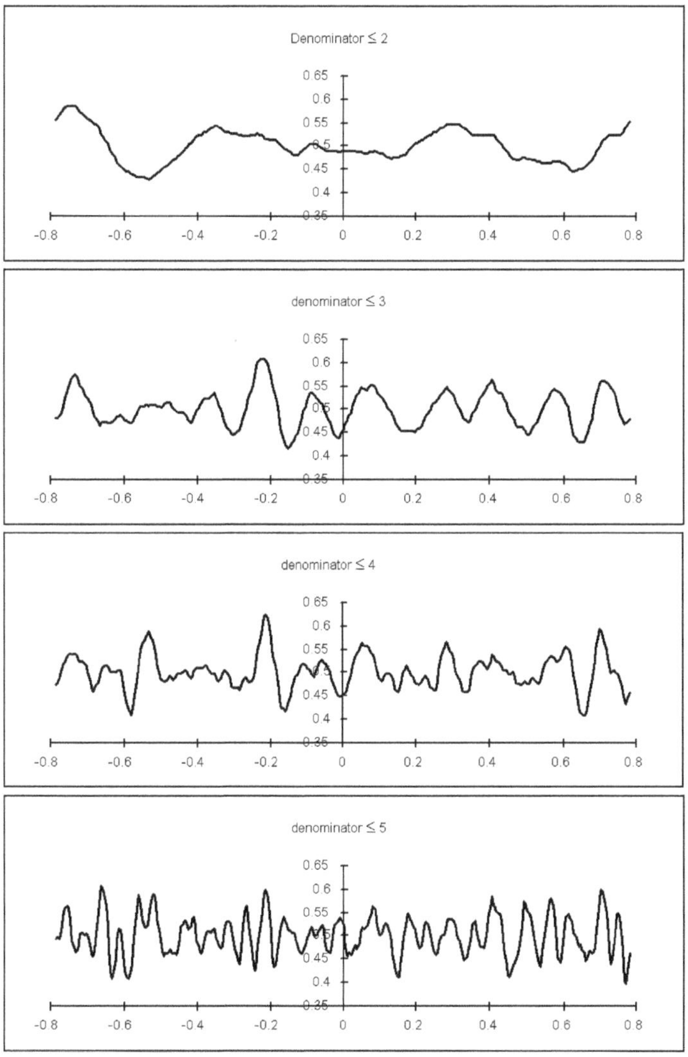

Figure 5: Varatiograms for the vallum of Hadrian's Wall, using scoring different scoring functions.

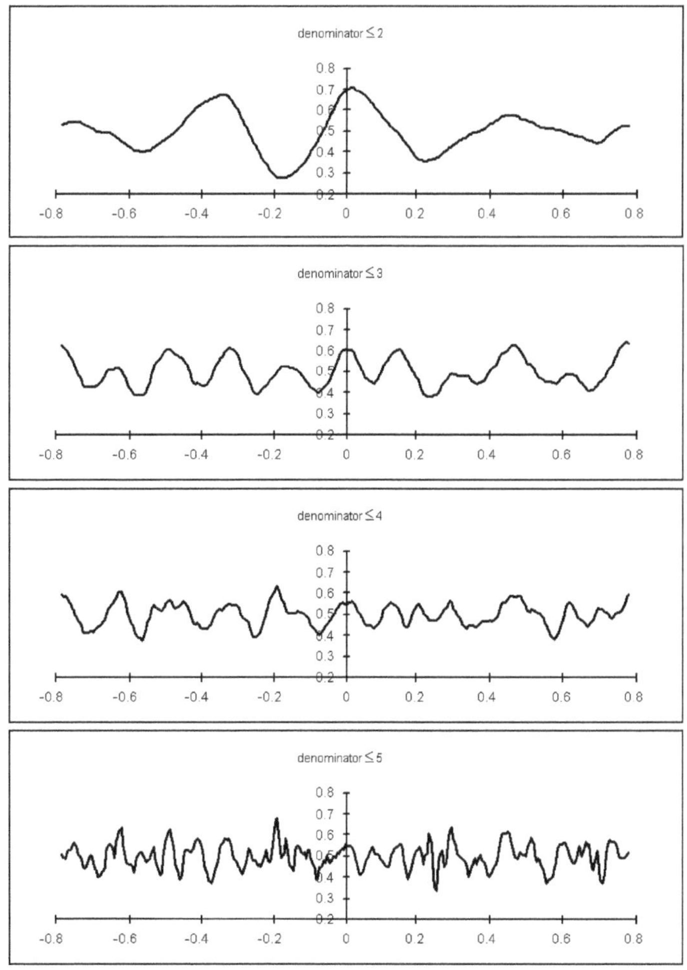

Figure 6: Varatiograms for the Wetterau limes, using different scoring functions.

4. Sensitivity

An important question about the usefulness of this approach is to how the choice of "significant" angles in the scoring function affects the varatiogram. The angles chosen as "significant" need to be matched to some well defined reality. Those chosen in the scoring function shown in figure 1 attempted to meet this criterion, but it is one of many choices which could have been made. How can we distinguish between these choices?

One approach is to use sets of angles of different sizes, each defined to include angles with tangents which are: rational, less than one and with a denominator less than a given value. The first set (denominator ≤ 2) includes 0:1, 1:1 and 1:2. The next set includes these three plus 1:3 and 2:3, and so on.

4.1. Tests of sensitivity using the vallum data

For the vallum data the results are shown in figure 5.

4.2. Trials using Wetterau limes data

The Wetterau *limes*, on the German frontier, provides another set of data. The maps in the *limes* guide (Klee 1989) show that a part of it takes advantage of the terrain, changing orientation frequently, whereas another part ignores the smaller natural features and includes several relatively long straight sections. In its relationship to the ground, it appears very similar to the vallum of Hadrian's Wall. For the purposes of this investigation this part is used, starting at Kleinkastell 21 of Strecke 4, at Langenhain, and ending at the end of Strecke 5, at Groß-Krotzenburg.

The results for different scoring functions are shown in figure 6.

5. Discussion and conclusions

From these graphs (figures 5 and 6) we can see a clear result of increasing the number of "significant" angles from 3 (denominator ≤ 2) to 11 (denominator ≤ 5). The number of peaks in the varatiograms increase in both cases. There is a greater range of options for the construction of centuriation hypotheses.

It is also clear that the position of peaks, and their height, is sensitive to this choice. This is most true for the change from denominator ≤ 2 to denominator ≤ 3, but it is also true for other sets. In the case of the Vallum, neither of the two largest sets defined in this way produces as significant a score for an input angle of -0.5404 as that produced by the set used in figure 1. This set was a particularly fortunate choice, from the point of view of supporting the Inglewood centuriation hypothesis.

The results therefore need to be handled with care, using information principally from the denominator ≤ 4 and ≤ 5 varatiograms in order to allow for the detection of the largest possible number of oblique relationships, if they exist.

References

CHOUQUER, G., CLAVEL-LÉVÊQUE, M., DODINET, M., FAVORY, F., and FICHES, J.L., 1983. Cadastres et Voie Domitienne. *Dialogues d'histoire ancienne* 9: 87-122.

CLAVEL-LÉVÊQUE, M., 1991. *Science alexandrine et jeux de construction en Biterrois*. Mélanges Étienne Bernard: 155-164.

HAVERFIELD, F., 1921. *Centuriation in Roman Essex*. Trans. Essex Arch. Soc. 15: 115-125.

KLEE, M., 1989. *Der Limes zwischen Rhein und Main: vom Beginn des obergermanischen Limes bei Rheinbrohl bis zum Main bei Grosskrotzenburg*. Stuttgart: Theiss.

MARGARY, I.D., 1973. *Roman Roads in Britain (Revised Edition)*. London: John Baker.

MORRA, L. and NELVA, R., 1977. Reciproca rotazione di tracciata delle "Centuriato" romane. *L'Universo* 57: 249-269.

ORDNANCE SURVEY, 1964. *Map of Hadrian's Wall*. Chessington.

RICHARDSON, A., 1999. *Centuriation South of Hadrian's Wall in Cumberland*. Matterdale Historical and Archaeological Society Year Book and Transactions 5: 7-14.

ROMANO, D.G., and TOLBA, O., 1996. Remote Sensing and GIS in the Study of Roman Centuriation in the Corinthia, Greece. In Kamermans, H. and Fennema, K. (eds.), *Interfacing the Past: Computer Applications and quantitative Methods in Archaeology CAA95*: 457-463. Leiden, University of Leiden. (Analecta Praehistorica Leidensia 28 II).

ROTH CONGÉS, A., 1996. *Modalités practiques d'implantation des cadastres romains: quelques aspects*. Mélanges de l'école française de Rome 108: 299-422.

Complexity in Action: "The Emergence of Agro-pastoral Societies"

Alexandra Águeda de Figueiredo Leite Velho

Instituto Politécnico de Tomar

Rua Alfredo Maia Pereira 6C, 3°Esq. 2300 Tomar, Portugal

e-mail: alexfiga@gemini.ci.uc.pt

Gonçalo Cardoso Leite Velho

Faculdade de Letras da Universidade de Coimbra

R. Dr. António José de Almeida 206 4° esq., 3000 Coimbra, Portugal

e-mail: gonvelho@ci.uc.pt

http://www.terravista.pt/AguaAlto/4003/

Abstract

This paper presents a simulation in which a fictitious society of hunter-gatherers evolves into a society based on farming. The purpose of this simulation is to show that even a simple system of rules can produce complex behaviour over time, providing a model of change with non-deterministic aspects. The model is very simple, based on agents with a small set of rules. The experiment does not try to reproduce a realistic transition to agro-pastoral societies but instead proposes to illustrate certain theoretical issues of the understanding of such a transition.

Key words: simulation, agents, emergence, hunter-gatherers, agriculture, complexity

1. Introduction

We propose a minimal model for the transition from a hunter-gatherer society to an agro-pastoral one. The application of such of a model to archaeology is relatively recent, and so far only a few examples are available.

David Thomas (1972) used this technique for his study of the Great Basin prehistory and at the same time Linda Cordell used a computer simulation to study the settlement changes within the Southwest borders of the Mesa Verde region (1972). Later, in 1978, the School of American Research hosted a seminar on archaeological simulations, later published by Sabloff (1981). In the eighties we saw important works on the American Southwest by Dove (1984) and Kohler et al. (1986), and on Mesoamérica, by Reynolds (1986).

These latter works adopted a new perspective in relation to the previous studies. The main deficiency of the previous simulations, as pointed out by George J. Gumerman and Timothy A. Kohler (1995) has to do with the main perspective used in the construction of the model. Early works modelled top to down, that is to say, first they built a global or systemic model and from there deduced the individual level.

Epistemological development in sciences such as anthropology brought forth a new point of view. The individual was enhanced. Authors like Bourdieu, Sahlins and Giddens, have focused on concrete human actions, which should be understood as the mediation of social relationships and cultural meanings. Jim Doran (1993, 1994) follows this approach and intends to provide (with a simulation) a theoretical model for the evolution of human societies.

The understanding of how these human individual practices shape tradition turns into the primary objective of an investigation.

The models based on agents are built on a bottom up perspective. The emphasis is placed on the agent as the origin of global behaviour. It is through the agents' interaction that the structure is built (Gumerman and Kohler 1995:6).

This theoretical trend was reinforced by new ideas introduced by the concept of complex adaptive systems (CAS). In complex adaptive systems agents interact and evolve by reacting to the environment and to each other. Even sets of agents animated by simple rules can interact in complex patterns that tend to get more elaborate over time. Complexity arises from simplicity.

This approach centres the analysis on the mechanisms operating at the individual level, avoiding the traditional ecological determinist approach. Global changes at the system level do not require global changes in the environment. The complex patterns that arise from individual interaction can change due to small scale fluctuations and give rise to new global arrangements.

In our simulation we intend to demonstrate this last principle. We begin by modelling bottom to up, starting with a simple model of agent interaction. We tried to keep our agents as simple as possible, in order not to globally pre-define the behaviour of the system.

Our model tests a simple hypothesis: can the change from a hunter-gatherer society to an agricultural one occur by chance, without postulating any dramatic change in external or internal conditions? In other words: can we build a model of a society based on a system of relations with the environment, and watch that system change suddenly after a long period, without any external change, only its own internal processes?

What we have constructed is completely theoretical. It does not emulate any known situation neither does it try to incorporate any type of empirical data. The model is centred on the demonstration of the evolution characteristics of a system that includes two competing potential equilibrium states, by showing that the shift from

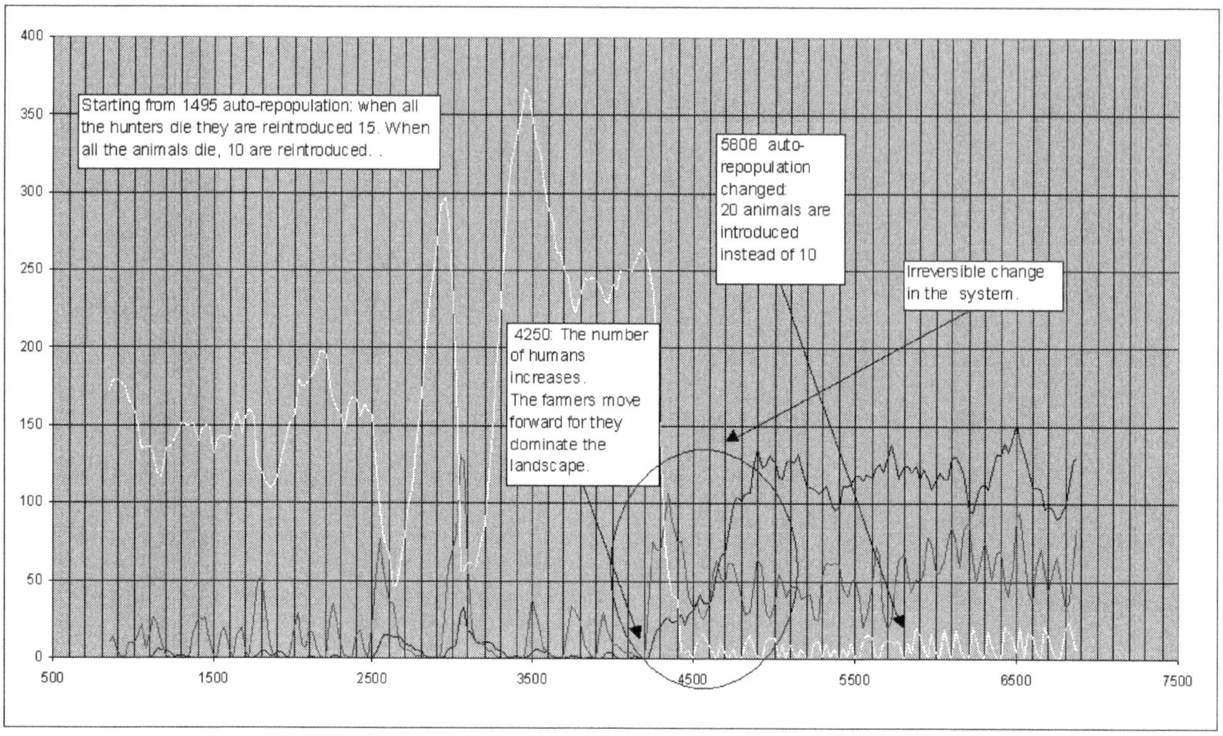

Figure 1: A long simulation: a total of 7000 cycles for a life-span of 50 cycles. Around cycle 4500 the systems shifts from predominantly hunting to farming, although no new causality was introduced.

one state to the other can be the result of very small causes. In our view, this raises the question of searching for global causes to global phenomena.

2. Overview of the Model

The model is composed of a landscape and three type of agents: cattle, hunters and farmers. The tool used for the implementation was AgentSheets (http://www.agentsheets.com).

Our aim is to create a system where all agents compete for natural resources, in our case represented by plants. The success of each type of agent is determined not only by the direct availability of the natural resources but also by the capability of other agents to gather those resources for themselves.

The landscape consists of a grid, where each square corresponds to either a water cell or a soil cell. Using AgentSheets drawing tools a specific landscape can be drawn as needed, by "painting" water and soil. There are no different types of soil, nor any representation of the relief.

The vicinity of water determines the evolution of landscape, in particular the growth of plants, which in turn determines how cattle, hunters and farmers behave. So the most determinant parameter that influences a specific run of the model is the actual drawing of lakes and rivers in the landscape. The availability of plants determines the behaviour of cattle. The behaviour of cattle in turn determines how the hunters move around. And farmers try to get close to the water to build their farms.

The interaction of all the factors can be described as follows:

- *Soil* produces two types of plants: X and Y. X (rendered light gray in our screen shots) requires close proximity to water, and corresponds to what would be grass or cereal. It grows fast and provides food for cattle and humans. Plant Y (gray areas) requires less water, grows slower, and does not provide food for cattle - it corresponds to bushes or trees. As plants are eaten they disappear from the soil (the software uses paler colours to render short plants), and need time to grow again.
- *Cattle* eat type X plants. A cattle agent has the ability to "smell" concentrations of plant X and move in their direction. As the plants are eaten, cattle move to other concentration spots in the landscape. With time the first plants to be eaten grow again and cattle eventually return to the same places. People also eat type X plants and so compete with the cattle for the same food.
- Both *hunters* and *farmers* have the same eating patterns. They can absorb plants X and Y and kill and eat cattle if they can catch them (there is a "killing" rule – if an animal is in a cell with three or more humans around it gets killed and eaten). The difference between hunters and farmers is in their moving rule. Hunters "smell" cattle concentrations, and so follow the herds around the landscape. Farmers follow water concentrations. When they find a spot close to water they "camp", and move no more. Farmers also camp when they find more than two farmers already camped. This tends to generate "villages". Plants X grow faster in the vicinity of a sedentary farmer – this corresponds to our "agriculture" effect. So the farmers' diet consists essentially of plants X, with an occasional cow that walks into the village.
- Cattle and humans reproduce themselves when a certain age/energy combination is achieved. For cattle there is the further constraint that a female and a male have an "encounter" – this allows for the future study of gender biased hunting practices, an aspect that we do not cover in this paper. With regard to humans, this rule means that the rate of reproduction of each type is connected to their ability to

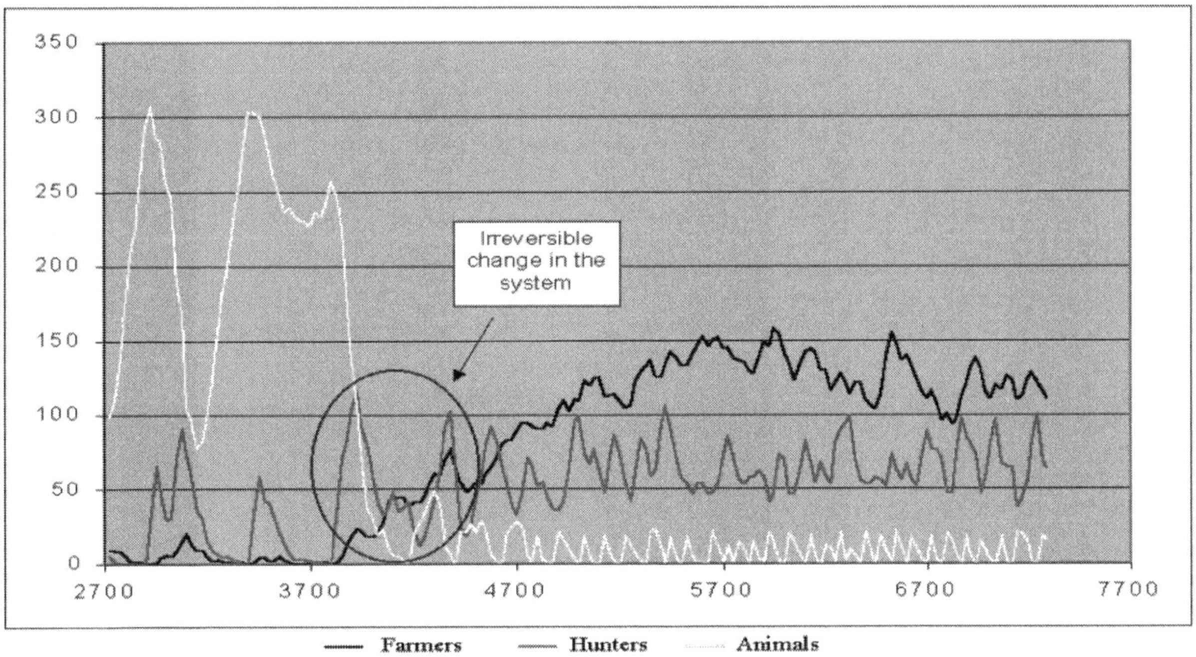

Figure 2: Another look at an irreversible transition. Note that there are no clear macro-level conditions associated to the shift.

gather resources and so obtain the necessary "energy level" for reproduction.

- When a human reproduces there is a pre-set probability that the descendant chooses a different type of life. In other words, a portion of the farmers' sons are hunters and vice-versa. This rule is very important in the understanding of our model because it provides the mechanism by which the society can choose, at any given moment, to shift from hunting to farming or from farming to hunting. There is always a mixture of hunters and farmers present. The more successful a certain type is the quicker they reproduce and become dominant. There is however always a stock of alternative types available in case the conditions of dominance change.
- Human and cattle have a pre-set maximum life-span.

The model is built in a way that concentrates the analysis on the efficacy of the alternative strategies of hunters and farmers, here reduced to different moving patterns. Instead of modelling some disruptive event that would make our society change from a predominant model to some other type, we investigate how a system of interconnected dependencies can change spontaneously, in response to fluctuations in the balance of resources.

3. Observed behaviour

Running the model consists of creating a landscape and introducing initial populations of animals and hunters. Due to the small dimension of the modelling space available the simulations frequently produce short-lived populations. A mechanism of automatic "immigration" was produced, so that when either the cattle or the humans completely disappear, new individuals are introduced. The model can be thought of as representing a portion of a landscape subject to successive migration waves of humans and animals.

The typical evolution of the simulation can be described as follows. The initial group of hunters follows the cattle around killing them whenever possible. The killing rule relates the energy of the animal to the number of humans in the cells around it. So the kills are determined by the patterns of movement of the animals and the hunters (although animals can, by the same rule, be killed if they enter an area densely populated by sedentary farmers). As the animals follow the concentration of plants, and hunters the concentration of animals, the two groups move close together but there is enough fluctuation in the moving patterns to allow for a certain randomness in the hunters' achievements.

When hunters succeed they accumulate energy and reproduce. There is a 10 % chance that a hunter's son will become a farmer. So eventually, if the hunters are successful, farmers will appear and settle down by the rivers and lakes.

The appearance of farmers creates a more complex interaction between the various elements. Without farmers the relationship between cattle and vegetables, and between hunters and cattle, is a predator-prey relationship, with the typical pattern of oscillating populations. Farmers are located in the same spots where cattle look for food. Since the presence of farming makes the plants grow faster, the herds have more food available – farming disturbs the "natural" availability of resources.

Where there is a large cattle population, the farmers have little chance of survival, since the big herds will eat their food. The existence of a group of hunters increases the survival chances of the farmers. If hunters disappear the herds will increase and eventually destroy the farming lands. However if farmers manage to hold and create a sufficiently large population they will be able to hold the cattle away, since cattle entering dense "villages" are killed by surrounding humans. And a big enough population of farmers produces hunters in quantity by the 10 % rule. So the interaction of farming, hunting and grazing is more complex than the simple predator-prey relationship. The increase of complexity has the counter-intuitive consequence of allowing the system to become more stable in demographic terms, as we will see.

The model can run for very long periods of time with the populations of hunters, farmers and cattle oscillating consider-

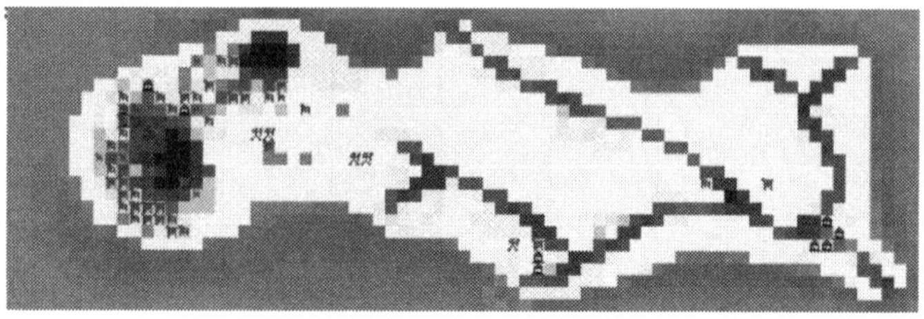

Figure 3: A compact herd, a small group of hunters and a few farmers. In many of these type of situations the humans disappear. The system reintroduces then a fresh small tribe of 15 hunters. Corresponds to cycle 2.707 of figure 1.

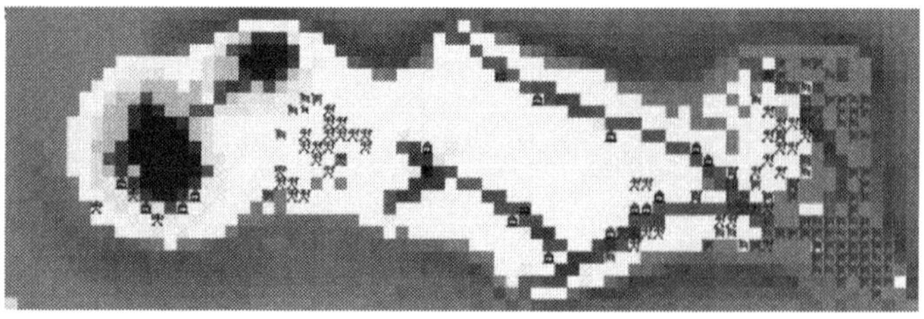

Figure 4: Another view of the first phase. The cattle is split in two herds, each with a group of hunters in persecution. The farming experiments will not survive the movements of the large herd as it latter migrates to the west to graze. By comparing with figure 4 it is possible to see the effect of the herds in the destruction of the farming experiments (lower right corner). This view corresponds to cycle 4.306 of figure 1, with the shift to a farming dominated landscape underway, although no particular aspect of the situation allows us to predict it.

ably, with large amounts of animals, a medium sized group of hunters and a minority of farmers. The farmers sometimes manage to create areas of agriculture, but their survival is closely linked to the oscillating pattern of the cattle population, produced by the predator-prey relation with hunters. In the simulation illustrated in figure 1, this period extends for 4500 cycles, with the maximum life-span of a human set at 50 cycles. So for many generations cattle rules the landscape with a hunter-group following them around, and short-lived but constant farming experiments also take place (see figures 1 and 2 for population graphs and figures 3 and 4 for landscape views of this phase).

If the simulation runs long enough there is a non-quantified chance of a significant mutation occurring in the normal pattern. This happens if farming becomes resistant to the destructive effect of the herds. This, in turn, is due to a combination of the size of the farming population, the episodic decrease of the cattle population and hard-to-quantify aspects of the spatial distribution of agents at a given moment. No simple combination of these factors could be found to be deterministic in the shift (see figure 1 for various "failures" of farming take-off). What should be made clear is this: although the shift occurs normally after a long period of time there is no new factor introduced at that moment, no "evolution" or "adaptation" in the behaviour of agents and no accumulated factor that would have reached a critical point. We believe that the same circumstances that produced long oscillating behaviour also produce the shift to a stable agriculture-based system with the qualification that the latter case has a very low probability of occurrence.

When the shift does take place the farmers become the predominant type of agents in the landscape. Most of the energy accumulated in Plant X growth ends up in the farmers, and not in cattle, as before. Hunters still exist and continue to be produced by the 10 % rule that changes the type of new humans. Cattle population is maintained at low levels by the automatic "immigration" rule and never increases too much because farmers now densely populate the grazing areas and there is always a stock of hunters around to kill the animals. Since most of the energy is gathered locally by farmers, the system becomes much more stable, and the violent population changes of the first phase disappear (check figures 1 and 2 and the landscape views of figures 5 and 6).

4. Relevant aspects of the simulation

The model does not use realistic parameters or rules, and the constraints of the tool used force considerable scale distortions. However, we consider the model relevant as it shows that even a simple interaction system that goes beyond predator-prey relationships can develop complex historical causality.

In our model the shift from a predominantly cattle and hunter system to one dominated by farmers is a macroscopic event of major proportions. The change occurs after long periods of time and consists of a reorganisation of space and the flow of resources, with a corresponding new balance among the populations of the three types of agents present. This major convolution is not, however, caused by any significant change in the causal mechanisms at work in the system, nor by any external factor. It is more appropriate to state that the system contains the possibility of undergoing a major change and that this possibility, having a low probability of happening, occurs only after extended periods of time, or does not occur at all. One interesting consequence of this is that although we have never observed a regression from predominant farming to hunting again, the possibility of that change occurring cannot be completely ruled out. In causal terms nothing really prevents it from happening, except a very low probability that we are not able to quantify.

On close inspection of the process at work in the shift period we feel that the transition is the result of a series of very small scale events occurring in a short period of time. Farmers prevail not only because the relative population of the three types of agent reached some critical proportions but mainly because the spatial distribution of agents was such that a resilient configuration was produced. So the critical factors for the major historical revolution lie in the microscopic details of the movements of agents over a short period of time. In this situation the fact that agent A turned right instead of left at point T in time may have huge consequences later on.

It is this huge difference between the scale of the "effects" and the scale of the "causes" that constitutes the most relevant aspect of the simulation.

5. Conclusions of the system analysis

Our model is consistent with the theories close to the concepts of complexity, chaos and emergence.

George J. Gumerman and Timothy A. Kohler (1995) studied the application of intelligent agents in archaeology. For these authors, society should be seen as a unit of analysis - family or/and indi-

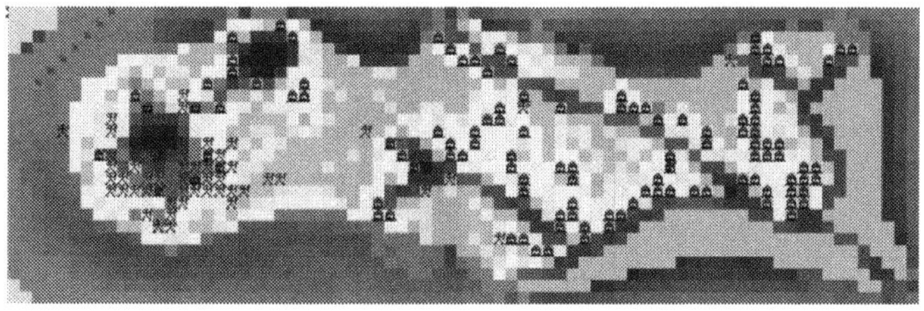

Figure 5: Farmers dominate the landscape, and the cattle has disappeared. The model reintroduces more animals (upper left). Corresponds to cycle 5.808 of figure 1.

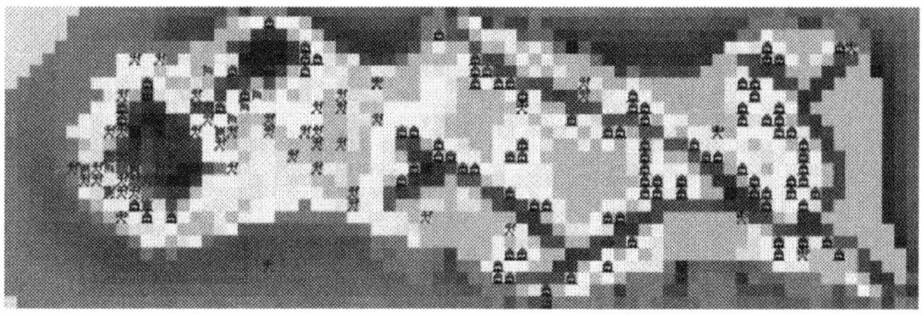

Figure 6: The new system is in place. Farms are created and disappear by the death of the farmer. A small population of animals and hunters co-exist. Corresponds to cycle 6.328 of figure 1.

vidual and space as a condition of the type of the agents' interaction and its density.

Thus culture should not be seen as homogenous and in balance with the environment aiming at a maximisation of adaptation of its members, presuppositions suggested by the processualist approach, but rather as an inherited system, a fundamental element for the understanding of the change through time.

The central aspect of our simulation is that a major global change depends on causes like a certain agent changing direction at a point in time. This is a characteristic of "chaotic" systems – the so called "butterfly effect". The "causes" can be of a very different scale in comparision to the consequences. When causes are so small that they remain undetected we call them "chance".

We can find this vision in a text by Henri Poincaré from the beginning of the 20th century:

> "A very small cause that fails to attract our attention provokes a considerable effect that we can observe, and we say then that the effect is owed to chance. If we knew exactly the laws of nature and the situation of the universe at the initial moment, we could foresee exactly what would be the situation of that same universe at a later date. But even if the natural laws held no secrets from us, we could, even then, know the future situation only approximately. If this would let us predict the next situation with the same proximity, that is all we would need, we would say that the phenomenon was predicted, that is, controlled by known laws. But that doesn't always occur: It can happen that small differences in the initial conditions cause very different final phenomena. A small error in the past will provoke an enormous error in the future. Prediction becomes impossible..." (Poincaré 1914).

These theories break the image of Nature as an unalterable order and expressed by laws that generate an eternal regularity, where time is a reversible variable.

Change can come from an agreement of non-classic factors, not explained in their totality by Marxists or Processual models. The only way to understand change is to observe the emergence of new actions and dig deep in the dirt to find as much data as possible to understand the overall picture

To find a "cause" we will have to look at a scale much smaller than that of the phenomena that we intend to explain. There is not a specific factor for change. The explanations for change lie in a multiplicity of factors, some of them microscopic. So it is necessary to have a total knowledge of the global structure at a given point in time. But we cannot grasp the whole structure, only a part of it. That is why modelling is so important. Geographers have to build maps in order to understand reality, because they cannot work with space itself in its totality, and we also have to build models to understand changes in the past.

This does not mean that the same causes do not have the same effects. What we want to point out is that the causes can be of microscopic scale, and interfere significantly with the macro scale. The possibility of reproduction of the same causes is small, because the probability of the same conjugation of tiny factors occurring is very remote. This is why a non-linear system can become unpredictable. The same effects do not always have the same causes, this happens only when we look at causes and effects at the same scale level.

We find an example of this in our system: if a certain animal changes its direction, this can allow two hunters to reach it and consequently kill it, which results in the higher probability to reproduce themselves, creating a possibility of growth in their population, which in turn can provoke a disruption in the animal population, which in turns creates the conditions for agriculture to flourish if other tiny factors related to the farmers' distribution contribute positively.

The processualist nomothetic paradigm of the search for rules for a system and determinism in human history is, according to these results, subject to revision.

Starting from one moment in which there are several alternatives of similar effectiveness from the social point of view, it is sufficient that a small scale factor privileges one of them so that it becomes dominant, and constitutes the preferential option making the other ones no longer viable. Carvalho J. Ramos (1999), showed a model where the individually created structures perpetuate in time in spite of the disappearance of the individuals that have created them. He concludes that structures appear in time, and condition subsequent individual actions and consequently perpetuate themselves.

Human behaviour is more complex than our agents'. However the conclusions that we took from the study of our "individuals" can be relevant for the understanding of certain phenomena.

Our model is not perfect, it lacks specific case data or application to a more realistic situation. We will try to work in this direction in further studies.

References

CARVALHO, J.R., 1999. O social emergente – sistemas adaptáveis complexos e interpretação histórica. *Revista Teoria das Ideias,* N° 20.

CORDELL, L., 1972. *Settlement pattern changes at Wetherill Mesa, Colorado: A test case for computer simulations in archaeology*. Ph.D. Dissertation, University of California, Santa Barbara. University Microfilms International. Ann Arbor.

DORAN, J., 1993. Simulating Prehistoric Societies: Why and How? In *Aplicaciones Informáticas en Arqueologia: Teorias y Sistemas*. Bilbao. Denboraren Argia.

DORAN, J., PALMER, M., GILBERT, G. and MELLARS, P., 1994. The EOS Project: Modelling upper Palaeolithic Social Changes. In Gilbert, N. and Doran, J. (ed.), *Simulating Societies*. London. University College of London Press.

DOVE, D.E., 1984. Prehistoric Subsistence and population change along the Lower Agua Fria River, Arizona: A model simulation. *Anthropological Research Papers* 32, Arizona State University, Tempe.

GUMERMAN, G.J. and KOLLER, T.Q., 1995. Creating Alternative Cultural Histories in the Prehistoric Southwest: Agent-based Modelling in Archaeology. *Durango Conference on Southwest Archaeology in Durago*, Colorado.

KOHLER, T.A., ORCUTT, J.D., PETERSEN, K.L. and BLINMAN, E., 1986. Anasazi Spreadsheets: The Cost of Doing Agricultural Business in Prehistoric Dolores. In *Dolores Archaeological Program: Final Synthetic Report*, compiled by D.A. Breternitz, C.K. Robinson and G.T. Gross, 525-538. Bureau of Reclamation Engineering and Research Center, Denver.

KUHN, T., 1970. *The Structure of Scientific Revolutions*. 2nd Edition. University of Chicago Press, Chicago.

POINCARÉ, H., 1914. *Science et Méthode*. Flammarion. Paris.

REYNOLDS, R.G., 1986. An adaptive computer model for the evolution of plant collecting and early agriculture in the eastern Valley of Oaxaca. In Flannery, K.V. (ed.), *Guila Naquitz: Archaic Foraging and Early Agriculture in the Oaxaca*, Mexico: 439-500. Academic Press, New York.

RUELLE, D., 1994. *O Acaso e o Caos*. Relógio d' Água Editores. Tradução de Manuel Alberto. Lisboa.

SABLOFF, J.A., 1981. *Simulations in Archaeology*. A school of American Research Book, University of New Mexico Press, Albuquerque.

THOMAS, D.H., 1972. A computer simulation model of great basin shoshonean settlement patterns. In Clarke, D. (ed.), *Models in Archaeology*: 79-102. Methuen, London.

Setting up a "Human Calibrated" Anisotropic Cost Surface for Archaeological Landscape Investigation

Michele De Silva and Giovanna Pizziolo

Dipartimento di Studi Storici e Geografici, Univesità degli studi di Firenze

Via S. Gallo 10, 50129 Firenze, Italy

e-mails: mdesilva@unifi.it, gpizziolo@unifi.it

Abstract

This paper presents an application of GIS tools for investigating one aspect of the archaeological landscape: movement through it. The authors would like to provide an experimental procedure to obtain cost surfaces related to human movement in a given geographical area, taking into consideration a settlement perspective. This research mainly focuses on the insertion of the concept of round trip movement from a given location in a calculation of cost surface through the adaptation of one function - available in the literature - related to human energy cost. After the explication of the procedure, the results and possible implications of this application are discussed using the Biferno Valley (Molise, Italy) as a case study.

Key words: GIS, cost distance, landscape archaeology, Prehistoric settlement, Neolithic, Molise

1. Introduction, research background and context

The context of this research lies within the landscape archaeology framework and it aims at investigating relationships between past human communities, their places and landscape in which they lived. Since spatial analysis was applied in archaeology, a number of different questions have arisen as regards a more in depth understanding of the relationship between archaeological sites and their contexts. Attention has been paid to methodologies which may evaluate this relationship. Within this framework, GIS applications have strongly participated in the development of Landscape Archaeology analysis, building up solutions on a cell based model, focusing on exploring different issues. These have been aimed at analysing natural resources related to site location and settlement pattern (e.g. Gaffney et al. 1995, Saile and Zimmerman 1996, Verhagen 1995), accessibility and mobility through the landscape and land subdivisions (Lock and Stančič 1995) or the influence of what is visible or perceivable in the landscape (Wheatley 1993, 1995, Llobera 1996). Based on various theories, different investigations have been undertaken, leading to a processual and post-processual debate (still currently underway) in which environmental or cognitive determinism is highlighted as an ever present risk (Van Leusen 1999). The interaction between several features of archaeological sites and the landscape which contain them has been explored in an attempt to understand the role, extension, position and other characteristics of areas surrounding -or (somehow) related to- past settlements. A wide range of elements play a role in the definitions of these portions of landscape. They may be either cultural and social elements -such as taboos, social relations at an inter and intra site level, sacred landscapes etc.- or natural elements, such as natural obstacles (slopes, streams and rivers, deserted areas), vegetation types, soil types, landscape morphology which constitute a process of complex interaction (Ingold 1986, Tilley 1994).

Despite this complexity, we can attempt to carry on landscape investigations through the use of GIS tools. In fact the GIS structure enables the correlation (and overlay) of data leading to possible comparison between different approaches, variables of a different nature and results (Stančič and Kvamme 1999). Indeed, through analyses which appreciate multiple interactions between different variables, ranging in qualitative and spatial aspects, it is possible to attempt a reconstruction of the landscape through a global approach. Consequently, if we desire to investigate a landscape archaeological theme e.g. the accessibility and use of areas related to archaeological sites, we have to confront more than one variable which are likely to be of different natures. Bearing this complexity in mind, investigations regarding what can be called an area of pertinence or interest on a site at a local scale could move from a "settlement" point of view to a territorial and landscape perspective. From among the range of variables participating in the process of interaction between communities and their territories, this work only focuses upon only one.

A particular emphasis is given to the role that movement can play in the interaction between settlements and the landscape. One may analyse movement according to different approaches but in this study our aims focus upon exploring human mobility through the landscape taking into consideration its topography in order to build up a human calibrated cost distance in relation to the slope. Thus we would like to focus upon only one of the themes related to the wide range of elements which may influence a complex human behaviour such as human mobility through the landscape.

At this experimental stage, our attention has been particularly oriented towards the methodological aspect of this issue. We then subsequently tested the methodology in different archaeological and geographical contexts.

We do not intend to propose an outright solution to the problem but simply to contribute to the methodological debate with regard to cost distance providing the setting up of a variable which can be taken into consideration during the investigation of the formation processes of settlement/landscape interactions.

2. Moving across the landscape

2.1. Setting up a framework

As we have briefly mentioned above, a range of variables should be evaluated if we are to deal with the role of movement on the perception of ancient human landscapes. This current work analyses movement from an energy consuming physical point of view, without evaluating other cultural, social or perceptive aspects which, as we know, may characterise or influence movement through space or any eventual choice of itinerary. In fact, what is considered nearby and/or easily reachable from the point of view of physical effort is not necessarily so from a social perspective. Nevertheless we believe that an evaluation of the cost of moving may provide information and additional data sets that may be subsequently compared with other kinds of variables thus constituting further discussion, and whether positive or negative key, for an interpretation of the archaeological landscape sets. In any case, within this physical framework we have introduced a cultural perspective in order to build up a model which takes into consideration one of the human needs within a settlement context. In particular we have chosen to modify the cost distance module provided by GIS software in order to be able to evaluate a variable which could be significant in a landscape archaeology investigation.

2.2. Problems of friction

Movement across space is hindered by friction. In the initial stages we can assess that from a physical point of view the natural features which may influence human movement in terms of friction can be divided into two categories: isotropic friction (i.e. independent of the direction of movement) such as vegetation density, soil types, wet or snowy areas etc. and anisotropic friction (i.e. dependent on the direction of movement) such as slope or streams. In this paper, we focus on the latter and we particularly explore how the topography of the landscape acts as anisotropic friction to movement. This work has been developed following the procedural structure provided by the software Idrisi for Windows (Eastman 1997).

It is possible to assess that the direction of movement crossing a slope does influence the intensity of friction with which one is faced. In fact, if we consider a person walking up a hill, along the line of maximum slope, they will be subject to the maximum friction, in other words friction acting at its fullest; on the other hand, if we consider a movement across the hillside the slope will, to some extent, not influence the friction which could be evaluated in terms of movement along a plain ground (figure 1).

2.3. One way movement and round trip movement

Generally, in cost distance analyses, movement is considered a one way direction, but if we are dealing with settlements, inhabited by a community, we must presume that some return trips will have to take place, perhaps in the context of daily movements there and back. Therefore, we accordingly, have to calculate the cost of a round trip movement instead of just a one way journey on which the calculation model is based.

Thus, in a one way movement from a village located in the middle of a slope (figure 2), it costs less to reach the place located at a lower altitude in relation to the village (B in figure 2) then the places located at a higher altitude (A in figure 2). But in a round trip movement through the landscape we have to take into account that people have to return to their village. In other words, we can say that an upward movement implies in itself a downward movement and *vice versa*. In this case, the cost to reach and

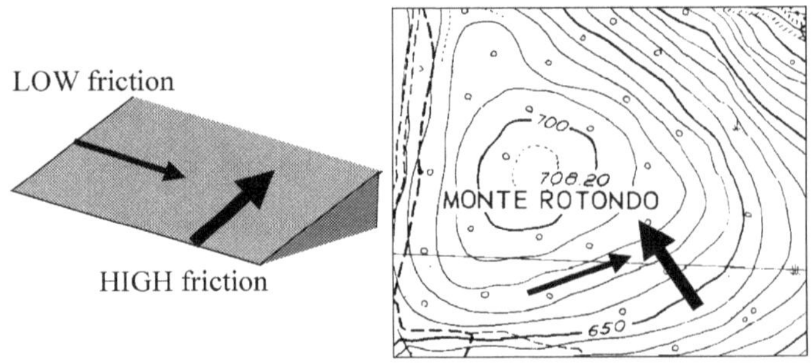

Figure 1: Topography acts as an anisotropic friction to movement, in fact friction intensity depends on the direction of movement.

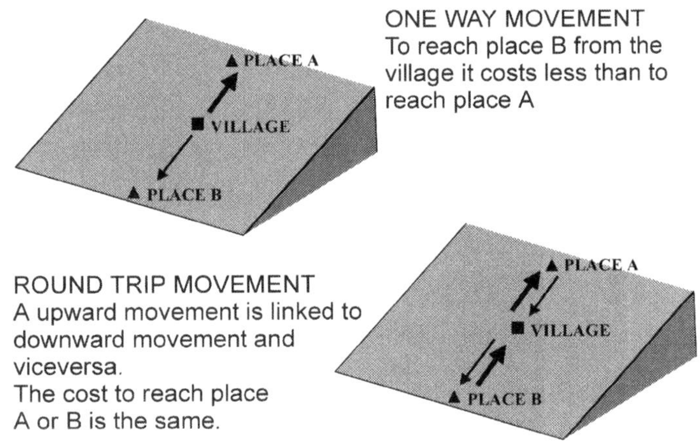

Figure 2: In the evaluation of cost to movement from a settlement we have to take into consideration a round trip movement.

come back from places at a higher or lower altitude is the same. From a settlement perspective it is clear that places at a lower altitude are not necessarily in a favourable position in relation to the village. In fact upon returning to the settlement the friction due to movement is no longer advantageous especially if one is laden down with goods to bring home.

3. Building up a cost surface related to round trip movement

3.1. Basic steps

In order to build up a cost surface related to topography, using modules provided by Idrisi for Windows, we have to follow certain steps listed below:

1. build up a Friction Surface, in other words, an image which expresses the maximum friction exerted by each portion of landscape,
2. build up a Friction Direction Surface which is a layer that expresses in which direction the movement is subject to maximum friction,
3. formulate an algorithm to describe how friction changes in relation to the direction of movement,
4. define a set of one or more sites, for instance prehistoric settlements, from which to calculate the cost surface.

3.2. Step one: The friction surface

A Friction surface is a grid in which each pixel expresses the maximum friction factor determined by the slope in relation to a friction unit measure.

Friction Surface is derived from a slope surface although it is not exactly the same. Let us give an example. A pixel that expresses a slope value equal to 20% does not have a friction value increased by 20% as opposed to a pixel on the plain ground. It is, therefore, necessary to find the relation between slope and friction. Let us assume that the unit measure (the base cost) is the cost necessary to cross a pixel on the plain ground and it is equal to one. In our cost model, friction is expressed in terms of relative friction, using the base cost as a reference. For example, if it takes 100 calories to cover a given distance on a plain ground, and 200 calories to cover the same horizontal distance on a given slope, we may indicate that the slope has a friction of 2 (Eastman 1997:15-2).

We, therefore, have to find a function to obtain a friction surface derived from a slope surface.

3.2.1. Choosing a function to represent human effort in movement

How does the effort in movement in relation to slope change when we consider people walking through the landscape?

We have to find a function which expresses the relationship between slope and effort or "cost" in terms of energy consumption. A variety of functions are available in literature (Van Leusen 1999) although attempts are currently underway to find a more appropriate and all purpose definition. Different studies to model and measure the human cost of movement are undertaken in relation to time and energy consumption.

It is not the aim of this paper to discuss either the reliability of these formulas or which ones may provide more accurate and likely models. We are, however, more interested in evaluating the possible development of a procedure which takes advantage of using one of these functions. In particular, the authors would like to test the forming of an anisotropic cost surface in relation to topography according to human behaviour in a settlement perspective (i.e. going away and coming back) realised through the adoption of a formula based on the physical effort in movement. Due to the experimental phase of this work, attention has been focused on the mathematical aspect of the formula, choosing the one that fits better into our working framework. At a future stage of research, a specific study on the best function to be adopted will be undertaken.

Thus, we can initially try to adopt the function suggested by Ericson and Goldstein (1980:23-24) to calculate the cost distance along a route in relation to a linear path, in which:

Work (i.e. moving in a particular direction) is equal to:

horizontal distance + (3.168 * distance up) +

+ (1.2 * vertical distance down)

We would like to stress here that this function has not been adopted on the basis that we believe it to be the best or the most realistic, but simply because it fits best into our working framework in this experimental phase.

3.2.2. Adapting the Ericson and Goldstein's function to a raster grid related to round trip movement

The function presented above has been developed in relation to the movement of walking along a line, but since we are dealing with landscapes and according to our framework, we have to take into consideration surfaces instead of lines expressing linear movement. We therefore have to adapt this function to a raster grid. The first stage regards the difference in height calculated for a surface. Assuming 1 pixel to be a unit measure, we can directly work out the difference in height for every pixel from the slope surface expressed as a percentage.

For instance, a pixel with a slope of 20% represents a difference in height equal to 0.2.

For every pixel along the line of maximum slope we have:

Friction Factor going upwards is equal to

1 + (3.168 * the slope as percentage)

and

Friction Factor going downwards is equal to:

1 + (1.2 * the slope as percentage)

Now we have obtained two maximum friction surfaces: one to cross every pixel upwards and one to cross every pixel in a downward direction.

When we are dealing with our cost determination we have to take into consideration that in a return journey a man in movement crosses plain and sloping areas. Therefore, each individual segment, which is not in a plain position, will be alternately crossed once in the upward slope and once in the downward slope. We can, subsequently, evaluate their cost (and thus their friction) as if the surface were crossed once in the upward slope and once in the downwards slope along the maximum slope lines.

Friction factor for a round trip is equal to:

1/2 [1+(3.168*the slope as percentage)+

+1+(1.2*the slope as percentage)]

As we can see above the average of the friction factors of these two surfaces constitute the maximum friction surface for a round trip.

3.3. Step two: The friction direction surface

Now we have to evaluate the direction of frictions.

The friction direction surface expresses the direction (in terms of azimuth) of movement which would incur the greatest cost to movement, in other words, the opposite of the maximum slope direction (figure 3). As a consequence, we can derive the friction direction surface from the slope direction surface.

3.4. Step three: Setting up the algorithm

How does the effective friction and its cost to movement change in relation to the direction of movement?

The "effective friction" to movement depends on the difference between the direction expressed by the friction direction surface and the direction of movement. The angle between the direction of movement which incurs the maximum friction and the direction under consideration will be called $\Delta \alpha$.

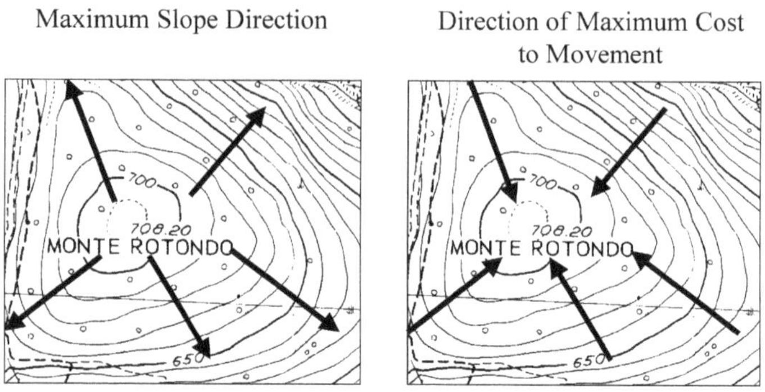

Figure 3: The direction of the maximum cost to movement is the opposite of the maximum slope direction.

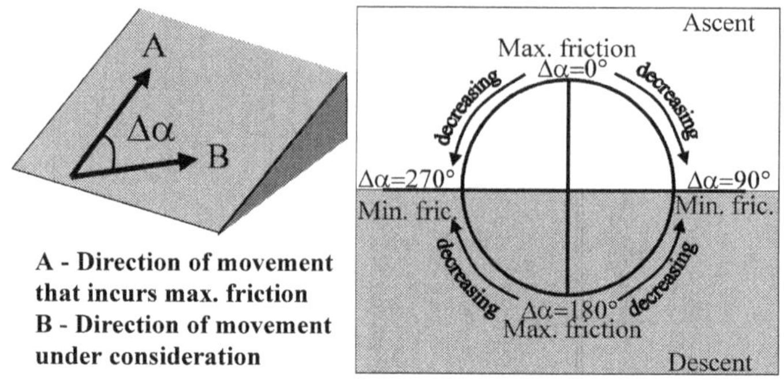

Figure 4: Assessment of how the effective friction changes in relation to the direction of movement.

In a one way trip we will have friction related to ascent that acts fully for $\Delta\alpha$ equal to zero progressively decreasing until reaching the minimum friction (i.e. horizontal plane) for $\Delta\alpha$ equal to 90° and 270°. Instead, the friction related to descent acts fully for $\Delta\alpha$ equal to 180° progressively decreasing till the minimum friction for $\Delta\alpha$ equal to 90° and 270° (figure 4).

But our "stated friction" i.e. the friction factor expressed by our friction surface, is calculated as the mean friction for a round trip.

In other words, the friction surface expresses the mean friction of each pixel along the line of maximum slope for a round trip (stated friction).

As a consequence, the effective friction has to be the same for the ascent and for the descent.

In order to satisfy our assumption, we have modified the default anisotropic function used by the module VARCOST provided by Idrisi. In fact, this algorithm evaluates friction for descent to be less than plane ground friction, that is to say as a force, and this is opposed to our above described assumptions.

The anisotropic function that we have used is the following:

$$\text{effective friction} = \text{stated friction}^{|f|}$$

where $f = \cos^k \Delta\alpha$

k = user defined coefficient (for a movement on foot we have chosen $k = 2$)

$\Delta\alpha$ = difference angle between the direction of movement that incurs the maximum friction and the direction of movement being considered.

3.5. Step four: Defining a set of points

As we have mentioned above, in order to calculate the cost surface we need a set of points from which we presume that the movement starts. Particular attention has to be paid to the choice of one or more sites from which to calculate the cost distance. If we choose more than one site it is necessary to operate a classification in order to compare homogeneous sites, especially if we are interested in defining a sort of accessibility catchment. In other words, it does not make sense to globally consider all archaeological evidence such as settlements, camp-site and off-site, artefact scatter etc. and then to investigate their reciprocal area of influence as if they played the same role in the landscape. Instead, it is fundamental to first of all set up the research goals and only then undertake the analyses. In order to obtain consistent results it is therefore necessary to stream the archaeological evidence through a classification process according to context specific criteria.

Once we have undertaken the above mentioned four steps, we can finally use an anisotropic cost distance module (in this case VARCOST provided by Idrisi for Windows) and build up a "human calibrated" cost surface related to orography.

4. Application to case studies

Once the procedure is set, we can then apply it to specific archaeological landscapes. However, it is important to remember that this cost distance due to topography may provide just one - and probably not the most important - of the variables which participate in the shaping of the human behaviour in and through the landscape. It is, therefore, necessary to integrate these results with the ones obtained from different analyses.

So far our applications have been oriented to assessing the reliability of the procedure and in evaluating its routine problems. After this phase some variation of the procedure will be attempted in order to further calibrate it.

This analytical methodology has been tested in three different archaeological contexts: Neolithic sites in the Biferno Valley (Central Italy), Calcolithic sites in Wadi Ram area (Southern Jordan) and Neolithic sites in the Altoribatejo area (Central Portugal). We have chosen contexts which are mainly related to the Neolithic culture as it is very stimulating for a settlement perspective implication. In fact, we may explore the interaction with landscape in a phase in which the concept of territory may be modified by new needs of the established communities. Needs related to round trip movements may likely become substantial when permanent activities are undertaken somewhere. In other words, in a "local to territory" perspective, movement linked to every day activities may be influenced by the "round trip" idea.

4.1. A case study in the Biferno Valley

4.1.1. Geographical context

The procedure developed so far has been applied to the Biferno Valley archaeological context, and in particular to the area surrounding the territory of Larino (Campobasso, Molise). The area selected is one of the ending structures of the Biferno lower basin and it constitutes a sort of geographical unit located on the right hand side of the valley. This area offers a range of morphological variations of hills and floodplains varying from heights of 575 m down to 25 m above sea level. However, this study area is characterised by a hilly landscape devoid of any dramatic breaks. Here the effects of past modifications are not significantly extensive

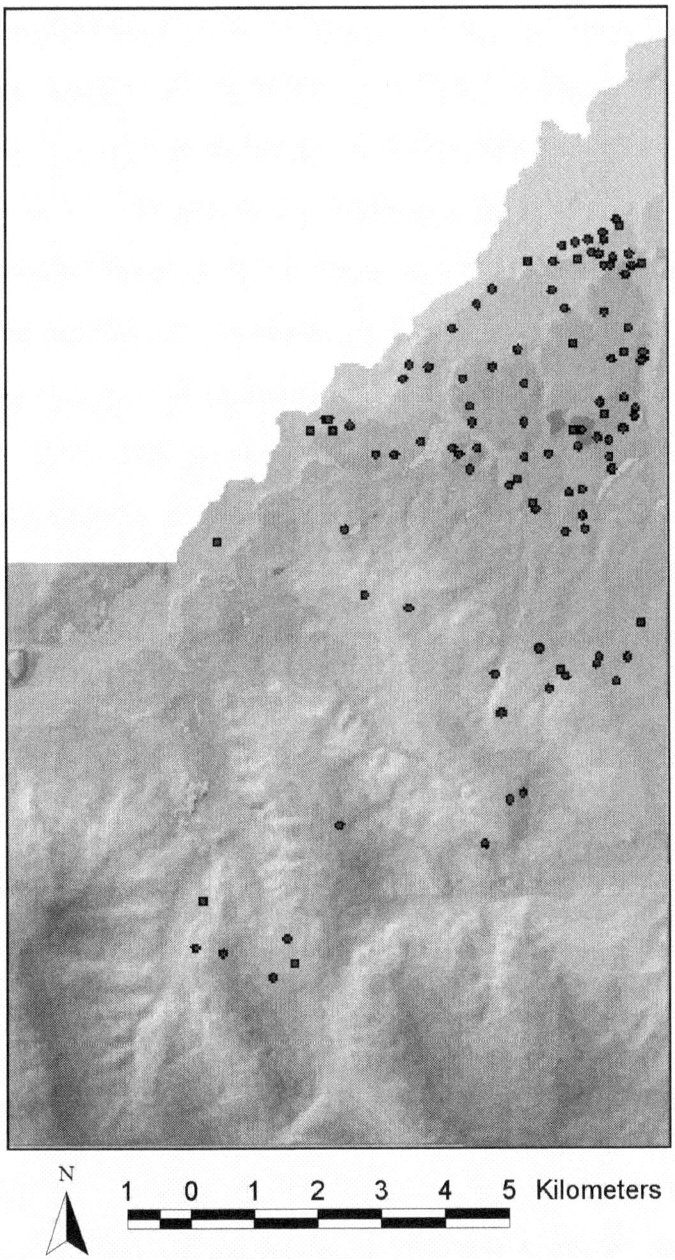

Figure 5: The geographical and archaeological context: the DEM and the distribution of Neolithic evidence.

(Barker 1995b:59) thus we can consider the morphology of the ancient landscape very similar to the actual one. By taking into consideration the study area we can observe an interesting variety of geomorphologic conditions. A DEM (figure 5) has been built originated from contour lines and spot heights derived from 1:25,000 scale topographic maps.

4.1.2. Archaeological context

As regards the archaeological context information has been derived from different sources (Barker 1995a, 1995b, De Felice 1994) resulting from systematic and non-systematic surveys conducted in the area. In the first stage a process of standardisation of archaeological data has been undertaken followed by a selection of prehistoric evidence. Then a classification of Neolithic sites has been carried out in terms of their rank, typology and chronology, taking into account the quantity, quality, type and density of archaeological evidence found in each site. As a consequence, we

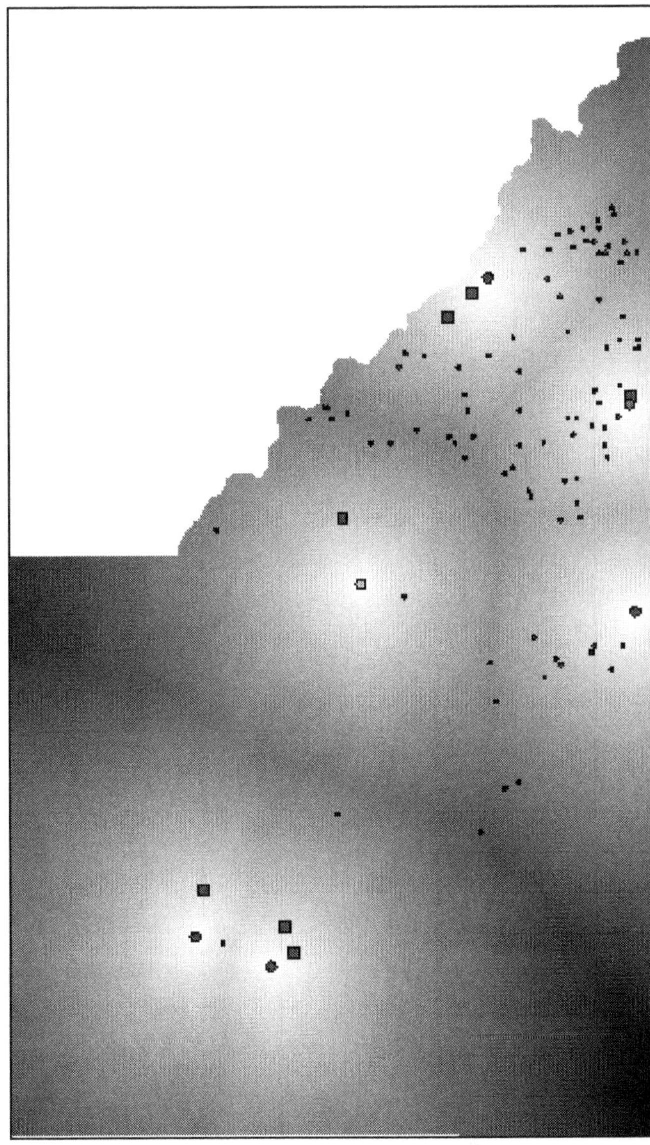

Figure 6: The cost distance surface calculated from the main Neolithic settlements (big dots in the map). Here it is visualised in a greyscale palette where the darkest tone represents the highest cost to movement.

have streamed the sites into three categories of main settlements, associated settlements - in which evidence of domestic activities are less significant or less prosperous than the one in the main settlements - and off-sites (respectively represented as big dots, squares and small dots in the figures).

4.1.3. Building up a cost surface

Adopting the procedure described in the previous paragraphs we have produced, using Idrisi for Windows, the friction surfaces and consequently the cost distance surface. The latter has been built up from the set of the main Neolithic settlements (figure 6).

In this way the relationship between the different categories of sites has been explored in terms of cost distance. A membership subdivision has been carried out among the main sites, for each pixel establishing which is the nearest settlement in terms of cost. Furthermore, the maximum "cost consuming distance" between main settlements has been calculated. This test enables us to evalu-

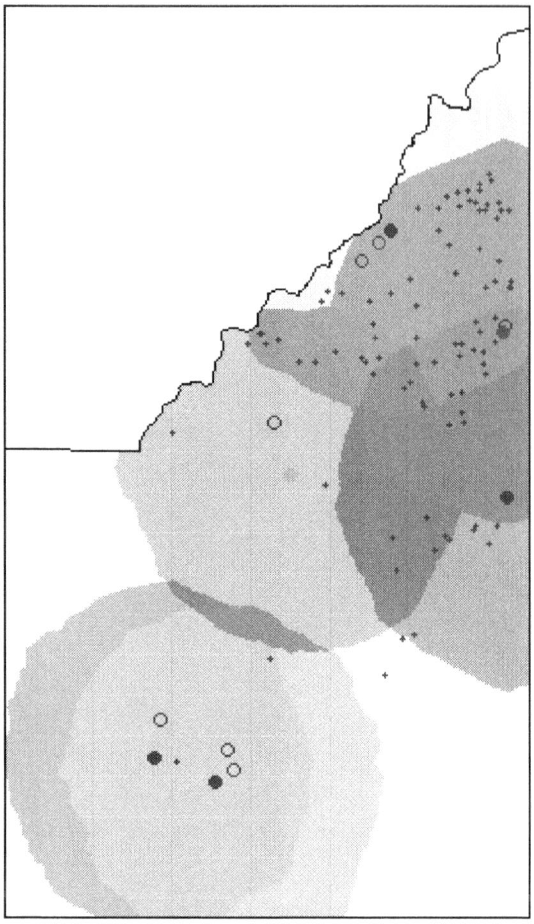

Figure 7: Cost distance buffers from the main settlements (filled dots) indicating areas reachable within a cost equivalent to the work necessary for a travel of 8 km (including a there and back journey) on plain ground.

ate a cost distance buffer for each of them (figure 7). The boundaries of buffers indicate the cost of movement equivalent to the work necessary for a travel of 8 km (including a there and back journey) on plain ground. The potential of their overlapping areas has then been explored. Subsequently, we have observed which parts of the landscape have an exclusive relationship with the settlements and which are shared by more than one. From the interpretation of this data compared with other information, we can imagine different possible scenarios. If all the settlements were of a similar rank and performed a similar territorial control, the hypothetical sharing of common areas would highlight the possible reciprocal negotiation of these small "tribes" (Alling Gregg 1987) over access to land, and/or the fact that their landscape might focus on different characteristics. On the contrary, in an alternative scenario the main settlements would play a different role over the landscape organisation and perception. The sharing of possible common territories would be planned by the main settlements and particularly exerted by the one located on the river terrace (i.e. the upper part of figure 7) in which occurred the longest and most substantial occupation. Other variables have been taken into consideration with regards to this hypothesis evaluating the presence of specific structures in the settlements, their intervisibility, the evidence of subsistence activities undertaken in the sites, the presence of artefact scatters in the hinterland of the settlements and the presence of specific geographical and morphological elements in the landscape. Regarding the latter elements we could also evaluate that despite their influence in the cost to movement they may also play a role in marking and perceiving land rights or spatial subdivisions of territories. As far as accessibility to natural resources it is quite possible that the inhabitants of the major settlement exploited the easily reachable area, represented by the soft and light soil of the river terraces, for agricultural purposes. However, it is not the aim of this paper to present a detailed report on all the different interpretations. We would simply like to present some of the problems and hypothetical interpretations connected to this application.

4.1.4. Least cost pathway

In this work the general agreement on the absence of stratified hierarchies in these first Neolithic communities has been taken into consideration. In fact our hypothetical scenarios do not deal with complex hierarchies but have only singled out a settlement located on the river terraces as the major one. Questions, therefore, arise pertinent to the investigation of the interaction between all types of archaeological evidence found in the area through the interpretative framework described in the above paragraph. Subsequently, we have attempted to analyse this issue from a movement perspective, highlighting the possible presence of itineraries or networks connecting the main sites with the major settlement. In order to investigate the reciprocal relationships between the latter and other settlements, a least cost pathway based on our cost surface has finally been calculated. The images, obtained using the module provided by Idrisi for Windows, have been performed for every main site connected to the major one and, subsequently, a sum of all the pathways has been obtained in one single image (figure 8). This highlights an interesting relationship between the position of the secondary settlements and their proximity to the least cost pathway. In other words, we can say that the pathways pass very close to the sites, or that the secondary sites are located along the least cost pathways that link the major Neolithic settlement to the main ones within the same area. Moreover, there is an interesting coincidence between two pathways connecting two main settlements to the major one. However, the interpretation of a settlement pattern or network cannot be based only on this kind of analysis notwithstanding this the above mentioned least cost pathway may constitute a source of information and provide stimulating ideas. The observations presented so far may help us to undertake further analyses and to further explore any potential in depth interpretations of the Neolithic Landscape.

5. Conclusions

We hope that this paper has succeeded in presenting some aspects related to our experimental work in its initial stages, aiming to offer further tools relevant to the exploring of the Prehistoric Landscape and in particular of the relationship between settlements and their area of interest.

In a research field as complex as this, we believe in the importance of a multivariate approach with the hope that it will lead to further investigation and potentially stimulating results.

Although we are still far from any global reconstruction of the past human landscape we can attempt to improve our understanding through a comparison and correlation of different approaches, but work is currently still underway.

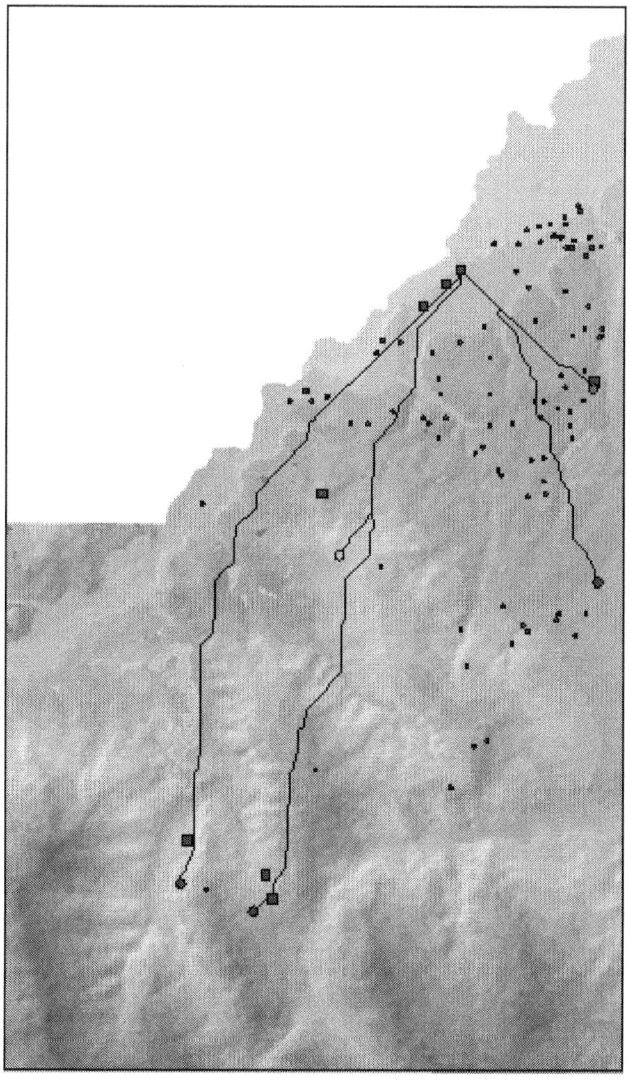

Figure 8: Least Cost Pathway (black line) connecting the major settlement located on the river terraces and the other main settlements (big dots). It is noticeable that the pathway passes by some associated settlements (squares).

Acknowledgments

Giovanna Pizziolo would like to thank Prof. G. Barker and the School of Archaeological Studies (Leicester University - UK) where she has undertaken part of this work and she is particularly grateful to Dr M.Gillings for his precious suggestions. The authors would also like to thank the Laboratorio di Geografia (Universita degli Studi di Firenze - I) for allowing access to software and hardware facilities.

References

ALLING GREGG, S., 1988. *Foragers and Farmers*, Chicago 1988.

BARKER, G., 1995a. *A Mediterranean Valley. Landscape Archaeology and Annales History in the Biferno Valley*, I, London/New York, Leicester University Press.

BARKER, G., 1995b. *The Biferno Valley Survey. The Archaeological and Geomorphological Record*, II, London/New York, Leicester University Press, 1995.

BOAZ, J.S., ULEBERG, E., 1995. The potential of GIS-based studies of Iron Age cultural landscapes in Eastern Norway. In Lock, G., Stančič, Z. (eds.), *Archaeology and Geographical Information Systems: a European Perspective*, London, Taylor & Francis.

DE FELICE, E., 1994. *Larinum*, Firenze, Olschki.

EASTMAN, R., 1989. Pushbroom Algorithms for Calculating Distances in Raster Grids, *Proceedings AUTOCARTO* 9, Baltimore: 288-297.

EASTMAN, R., 1997. *Idrisi for Windows User's Guide Version 2.0*, Clark University, Worcester MA.

ERICSON, J.F. and GOLDSTEIN, R., 1980. Work space: a new approach to the analysis of energy expenditure within site catchments. In Findlow, F.J. and Ericson, J.E. (eds.), Catchment Analysis: essays on prehistoric resource space, *UCLA: Anthropology vol.* 10: 21-30.

GAFFNEY, V., STANČIČ, Z., WATSON, H., 1995. Moving from catchments to cognition: tentative steps towards a larger archaeological context for GIS, *Scottish Archaeological Review*: 9-10, Glasgow: 41-64.

INGOLD, T., 1986. *The appropriation of nature prospective on place and space*, Manchester, Manchester University Press.

LLOBERA, M., 1996. Exploring the topography of mind: GIS, social space and archaeology, *Antiquity* 70: 612-622.

LOCK, G. and STANČIČ, Z. (eds.), 1995. *Archaeology and Geographical Information Systems: a European Perspective*, London, Taylor and Francis.

MASSAGRANDE, F., 1994. A GIS approach to the study of not systematically collected data: a case study from the Mediterranean. In Huggett, J., Ryan, N. (ed.), Computer Applications and Quantitative Methods in Archaeology 1994: 147-156.

MASSAGRANDE, F., 1995. Massagrande Federica, Using GIS with non systematic survey data: Mediterranean evidence. In Lock, G., Stančič, Z. (eds.), *Archaeology and Geographical Information Systems*, London, Taylor & Francis: 55-66.

SAILE, T. and ZIMMERMAN, A., 1996. Cattle or Crops. Applications of GIS in Central Germany. In Bietti, A., Cazzella, A., Johnson, I., Voorrips, A. (eds.), *Theoretical and Methodological Problems*, Colloquium I-II, XIII International Congress of Prehistoric and Protohistoric Sciences, Forli,: 149-158.

STANČIČ, Z. and KVAMME, K.L., 1999. Settlement Pattern Modelling Through Boolean Overlays of Social and Environmental Variables. In Barceló, J.A., Briz, I. and Vila, A. (eds.), *New Techniques for Old Times CAA98*, BAR (International Series 757): 231-237.

TILLEY, C., 1994. *A phenomenology of Landscape*, Oxford, Berg.

Van LEUSEN, M., 1999. Viewshed and Cost Surface Analysis Using GIS (Cartographic Modelling in a Cell-Based GIS II). In Barceló, J.A., Briz, I. and Vila, A. (eds.), *New Techniques for Old Times CAA98*, BAR (International Series 757): 215-223.

VERHAGEN, P., 1995. Modelling socio-economic activities in the Bronze Age of south-east Spain. In Lock, G. and Stančič, Z. (ed.), *Archaeology and GIS: a European Perspective*, London, Taylor & Francis: 187-209.

WHEATLEY, D., 1993. Going over old ground: GIS, archaeological theory and the act of perception. In Andersen, J. et al., *Computing the Past, Computer Applications and Quantitative Methods in Archaeology 1992*, Oxford: 133-138.

WHEATLEY, D., 1995. Cumulative viewshed analysis: a GIS-based method for investigating intervisibility and its archaeological application. In Lock, G. and Stančič, Z. (ed.), *Archaeology and GIS: a European Perspective*, London, Taylor and Francis: 171-185.

Note

This work has been carried out in full collaboration by the authors. In particular paragraph 1and 4 have to be attributed to Giovanna Pizziolo and paragraph 2 and 3 have to be attributed to Michele De Silva.

7. Presentation of Archaeological Data

A Digital Future for our Excavated Past

Tony Austin, Damian Robinson and Keith Westcott

Archaeology Data Service, Department of Archaeology, University of York

King's Manor, York, YO1 7EP, United Kingdom

e-mails: afa2@york.ac.uk, djr12@york.ac.uk, kaw4@york.ac.uk

1. Digital archives in archaeology

Computers have been used for archaeological research since early in the 1950s. Initial use was largely confined to specialists such as mathematicians and statisticians devoting some of their time and the processing time of a "mainframe" computer to archaeological problems (e.g. Wilcock 1999:35). In general archaeologists only started to use computers with the advent of the Personal Computer (PC) in the late 1970s. From this time on an irregular series of publications chart the rise of the PC as an increasingly necessary research tool for the archaeological community in Britain. In 1980 about 10 "micro-computers" were identified by the Museum Documentation Association (MDA) assessment of computer use in Archaeology (Stewart 1980). By 1986 the number of computers in use had risen to over 200 although some of these were being used as "intelligent terminals" to mainframe computers in university environments (Grant 1986:20). A further survey in 1989 suggests over 700 computers in use (Booth et al. 1989:4). In 2000 extrapolation from a recent survey suggests that between 2,200 and 3,000 professional archaeologists in Britain have computer access and that by the year 2004 computer access will be ubiquitous within the profession (Condron et al. 1999:24-7).

Using computers leads naturally to the creation of digital data and unsurprisingly as usage has grown the amount of data has grown and grown. Trying to quantify the amount of data produced by the archaeological community in Britain is something of a problem. Within the academic sector figures produced in 1998 suggested that as many as 1,800 of the 3,000 or so archaeological projects funded by the British Academy over a 30 year period had a digital content (Austin 1998). A broader survey by the Archaeology Data Service (ADS) in 1998 estimated the size of the digital resource produced by archaeologists in Britain as between 140 and 175 gigabytes (Condron et al. 1999:41). With the phenomenal growth in computer usage already noted the data mountain is undoubtedly much bigger today.

Modern excavations create huge amounts of digital information. Whether it is the on-site recording of the archaeology, specialist databases created during post-excavation or publication standard interpretative maps and plans, digital information has the potential to be created at every stage from assessment to publication. The existence of so much data raises problems in that it often represents primary data and sometimes is the only record of a necessarily destructive archaeological process. Synthetic accounts may also exist but here much of the functionality and complexity is lost in the move from dynamic data to a static account. An example might be the flexible query and retrieval of data from a database compared to the fixed table generated from it for inclusion in a report.

The same is true of many types of digital data; GIS, topographic and geophysical surveys, virtual reality models, CAD drawings and many others. In short, digital archives are important both in terms of re-use and of re-interpretation. The fragility of digital data should also be noted. While all storage media including paper have a finite life-span there are additional problems associated with the magnetic and optical media used to store digital data in the form of redundancy of both hardware and software technologies.

Within the discipline there has been an increasing awareness that this vulnerable data is as much a part of the primary site archive as the artefacts and paper records that have traditionally found their way into museum stores. Yet until comparatively recently there has been no secure home for such information and digital data simply languished in the virtual "desk drawers" of their creators. Clearly there was a need to create the structures; technological, managerial and administrative for the long-term preservation of and the provision of access to digital archives (e.g. Beagrie and Greenstein 1998).

The ADS was consequently founded in 1996 to provide just this kind of digital archive for the UK archaeological community. This paper looks at two specific but very different examples of digital archiving of archaeological projects undertaken by the ADS. The first involves the archives of several archaeological units based in museums and known by their penultimate resting place as the Newham Archive. When the Newham Museum Archaeological Service was closed down the digital archive eventually arrived at the ADS in the form of a large cardboard box containing dozens of floppy disks. The ADS was effectively thrown into a "rescue" situation in trying to save this data in battling against both deterioration and technological redundancy (Austin 1999). The second project was specifically designed to examine the problems associated with curating and providing access to digital archives. In 1999, the ADS, with English Heritage, the Oxford Archaeological Unit (OAU) and the Museum of London Archaeology Service (MoLAS) undertook a study into the needs and practicalities of the long-term preservation and dissemination of digital archives resultant from excavation. The Digital Archive Pilot Project for Excavation Records (DAPPER) set out to establish best practice for digital archiving in Archaeology.

1.2. The Newham Museum Archaeology Section digital archive

The Archaeology Section of Newham Museum, Greater London, was closed in late 1997 and although the closure was not entirely unexpected little thought was given to the fate of the material held on the Archaeology Service's computer system. It was not until there was a real threat of the archive disappearing that determined moves were made to save the archive, which was transferred to floppy disks and delivered to the Archaeology Data Service.

1.2.1. Accessioning the archive

The archive arrived on a total of 239 floppy discs from which a total of 6350 files were recovered. The floppy disks probably represented an image of the contents of the file server and individual hard disks on the network of the Archaeology Section of the museum. Thus it was possible to reconstruct the archive on the ADS server giving us a snapshot of the digital archive of a typical contracting archaeology unit. Initially the files were copied from the floppy discs onto the ADS server. A metadata record for each file was created as part of the accessioning process. This recorded the file name, path, format, size, last modification date and time, an abstract of the contents and comments such as problems encountered during transfer. This form of file management metadata is essential to help plan both the short term rescue of the data and also the long term maintenance of the resource. At this stage the fragility of the data was initially demonstrated when it was found that there were 25 corrupt files. The contents of around half of these files could be recovered after running scandisk on the discs although there was some loss of content. The remaining files could not be rescued and their contents were irrevocably lost.

1.2.2. The contents of the archive

The archive contains varying amounts of information for about 180 sites in northeast London that had been investigated over a period of over 15 years. Not all of the files include strictly archaeological data and there are also managerial files, project management files, correspondence and personal matter - in short all the sort of things one would expect to find on the computer system of a contracting archaeological unit. Table 1 breaks down the contents of the archive into functional categories.

The largest single category is survey data, which accounts for over a quarter of the files in the archive. This represents land surveys, borehole surveys and resistivity surveys. The second largest category is reports - Archive Reports, Level III reports, Evaluation Reports and Desktop Studies. Many of these files are draft copies created during the course of production of a printed report and consequently a substantial proportion of these files are unfinished or are duplicated in various stages of completion throughout the archive. Only the most current version of any report will be disseminated as part of the online archive, although the unfinished reports represent a rare opportunity to study the developing interpretations of a site throughout its post-excavation analysis.

Worryingly the third largest category is "unknown/no-value" which accounts for 14% of the archive. There are a number of files for which it proved impossible to determine what software was used to create them. Even worse, there are data files that cannot be associated with a site because they contain no information to specify what site they form part of the archive for and they were not in a directory associated with a recognisable site. These include over three hundred resistivity survey data files, a number of database files, and fragments of reports. Unfortunately without such contextual documentation about the site and the files associated with it the information contained on these files are to all intents and purposes lost to the scholarly community.

File Category	Percent
Survey Data	26.2
Reports and Report Drafts	22.3
Unknown format / no value	14.2
Administration / Project Management	11.0
Correspondence	10.8
Data and Databases	10.8
Project Designs and Tenders	1.7
Publication Articles	0.7
Manuals and recording templates	0.5

Table 1: Proportions of files within functional categories.

1.2.3. File formats

The Newham Museum Service began the process of computerising its post excavation practices relatively early on in the uptake of IT in the archaeological profession. Consequently a wide range of software has been used to create the files contained within its archive, which operate in both MSDOS and Windows environments. In order to archive and disseminate these files they have been migrated with no loss of information to formats that can be read by modern software. Table 2 shows the proportions of files created by the various package types.

Word processors were responsible for more than a third of the files in the archive. Five different word-processing package formats are present in the archive; the earliest files are in WordStar format progressing through Word 2 to Word 6 with occasional files in other formats. As modern word processors come with a variety of filters enabling a wide range of formats to be read and written migrating these files was a relatively straightforward task. The files created by the DOS version of WordStar, however, presented some problems as it only needed to support character printers for which 7 bit ASCII was sufficient. "Fancy" effects (emboldening, underlining) were coded into the non-printing ASCII characters. This left the eighth bit free which was used to flag the last letter in a word, presumably to help in wrapping text on the screen or page. Both Word and WordPerfect have filters to read WordStar files but they frequently cannot distinguish them from plain text files and use characters from the extended ASCII character set in place of the last character of every word. Unfortunately this makes reading these DOS WordStar files very difficult.

Table 2 appears to show that databases appear to be a significant proportion of the archive. This is misleading, however, as the majority of database files were created using dBase and consequently a proportion of the files are form definitions, report definitions and stored queries. The migration of old database files is also a straightforward task as modern database programs can also read a range of formats. Database files in the Newham archive were mainly in dBase format with a few Paradox and MS Access files, all of which could be read and thus migrated to ADS dissemination and preservation formats.

File type	Percent
Word Processor	35.1
Survey Data	26.3
Database files	17.2
Spreadsheets	10.2
Graphics	7.3
Project Management	2.9
Unknown	1.0

Table 2: Proportions of files created by software packages.

A much greater problem results from files, mainly databases, for which there is no site information. This left the data orphaned and therefore of very little value. In total about 250 files in the archive are orphaned.

Another problem arises from databases that contain coded information. In some cases the coding may be standard or easily recognisable by other specialists in the same field but there is one large database of inhumations from a monastic site in which the bones, all except for a lone patella, are coded as numbers with no explanation of the coding system. There are similar examples throughout the Newham archive, including pottery, building materials and environmental databases. This again clearly emphasised the need for effective documentation of the archive and all of its component parts before deposit. Unfortunately this could not be done with the Newham archive and in essence the data contained in these files has been lost to the archaeological community.

Spreadsheets in the archive mainly contain archaeological data tables, although they were also used in project management. DOS versions of Supercalc and QuatroPro were used before the Newham Museum Service adopted Windows and MS Excel. Unfortunately the two DOS programs are not widely supported by modern spreadsheets. A package from JASC software called Quick View Plus proved useful here. This program can view an impressive range of file formats including word-processor, spreadsheet, database, and graphics. The contents of the spreadsheets can be viewed and transferred to modern programs using copy and paste.

Many of the graphics files were created from survey data using Golden Software's Surfer and MapViewer programs. Unfortunately these packages only create files in their own proprietary formats which are not supported by other software packages. A modern version of Surfer is required to migrate these files to ADS archive standard DXF formats. This migration problem is even more acute with the CAD files created using TurboCAD 1.9. Again this is a format which is not supported by other packages nor, indeed, by modern versions of TurboCAD and thus presents major problems for migration. Effectively the functionality of the software itself will have to be preserved to enable the reuse of these files in the future. This is not a recommended archiving procedure for archaeological data (Richards and Robinson 2000) but is necessary in this case.

The Newham digital archive is probably a typical example of digital information resources of archaeological units. There are probably many archaeology units with archives of files in redundant formats, without explicit site relevance information, containing unexplained coding, and in unknown states of completion. The files may also be stored on unsuitable media in poor storage conditions. In short there may be large amounts of archived archaeological information, which can never be accessed again.

2. The Digital Archiving Pilot Project (DAPPER)

The Newham Museum Service digital archive is a depressing and salutary tale. The archive was developed as a working tool to help the Service write up and managed its archaeological projects, in this respect the archive was fit for the purpose for which it was developed. Equally the concept of digital project archiving was still in its infancy when the archive was developed and as there were no published strategies or methodologies in place to ensure the effective preservation of the data, poor condition of the archive is understandable. At a national level the ADS together with English Heritage, the Association for Local Government Archaeologists and the former Royal Commission for the Historic Monuments of England had begun a series of discussions out of which emerged the English Heritage funded Digital Archiving Pilot Project for Excavation Records (DAPPER). DAPPER aimed to demonstrate the concept of digital excavation archives in order to inform the development of best practice in this emerging field. The pilot project also aimed to encourage the re-use of these digital repositories and to explore the use of the Internet to enhance access to archaeological data. Consequently DAPPER also investigated ways to deliver the data to a potential re-user community via the Internet and in the process set up the world's first online digital excavation archive.

2.1. The excavations and their archives

In order to demonstrate the concept and potential of this new form of information provision it was necessary to choose the excavations to archive with care. After a lengthy consultation period two reasonably large and "high profile" sites were identified as being most suitable: the Royal Opera House excavation by the Museum of London Archaeology Service and Eynsham Abbey, excavated by the Oxford Archaeological Unit.

2.1.1. Eynsham Abbey

Between 1989 and 1992 the Oxford Archaeological Unit (OAU) undertook a full excavation of part of the precinct of the medieval abbey, revealing part of the cloister and south ranges, kitchens and domestic buildings. Evidence for earlier activity included the 11th century Anglo-Saxon abbey and earlier Minster, and a Bronze Age enclosure. The project was never designed as an exercise in digital data collection or management, but rapid changes in computer technology during post-excavation have meant that the project developed in this direction.

The downloadable archive contains text files, a database made available as comma delimited text, JPEG images, a 3D reconstruction of the medieval abbey available in DWG and DXF formats (i.e. AutoCAD) and numbers of digitised drawings also available as DWG, DXF and DWF files. All of the files in the Eynsham archive are also archived in ADS preservation file formats, for example, the preservation images are stored as high resolution uncompressed TIFF's.

The contents of the Eynsham digital archive can be summarised as follows:

Dataset documentation	Project background
	Excavation – aims, methodology
	Post-excavation – phasing, structures
	Specialist report summaries with database documentation and specialist bibliographies
Text files	Project documentation
Delimited text files	Architectural stone database
	Context database
	Context database keywords
	Glass lead cames database
	Vessel glass database
	Window glass database

	Lead objects database
	Post-Roman pottery database
	Environmental sample register
	Environmental sampling results register
	Small finds database
	Parallels for small finds database
	Structures database
	Tile database
Drawing files	Sections
	Phase plans
	Structure plans
	Site plans
	3-D reconstruction
Images	

2.1.2. The Royal Opera House

Following exploratory work in 1995 full-scale excavations were undertaken in 1996 by MoLAS at the site of the Royal Opera House in Covent Garden, Greater London. The excavation examined the largest area of the Saxon trading port of *Lundenwic* so far to become available and has provided a wealth of information on the form and economy of this settlement. All data from the Royal Opera House project was collected in computer based formats. This excavation was the first for which MoLAS used its state-of-the-art integrated database and GIS recording system. The online archive is currently incomplete but contains Geographical Information System (GIS) files, interpretive groupings and data files consisting of context, artefact and ecofact attribute sets. The rest of the archive will become available following publication of the site, which was projected for March 2000, and will be deposited with the ADS as part of an extension to the current project.

The contents of the Royal Opera House digital archive can be summarised as follows:

Text files	Group descriptions
Delimited text files	Field records by context
	Basic interpretation
	Roman pottery
	Post-Roman pottery
	Saxon pottery fabrics
	Registered finds
	Coins
	Loom weights
	Expansion codes for registered finds
	Assessment level animal bone
	Post assessment animal bone
	Animal Bone codes
	Animal species expansion codes
	Processed environmental samples
	Botanical remains – analysis
	Botanical expansion codes
ArcView themes	Context groups
	Trench edges
	Unreal edges

It is important to note that in neither case was the digital archive planned as part of the dissemination phase of the project and that both Eynsham Abbey and the Royal Opera House will be published as traditional monographs. The resultant digital archives are, however, remarkably similar in their content. This is due to their origin as the residues of large-scale, well-funded post excavation projects. Both archives are comprised of essentially the digital residues of the post-excavation process and represent a digital version of the "research archive" (cf. Richards and Robinson 2000).

2.2. The time taken to create the archives

The Newham Museum archive was hurriedly copied from the data server or individual hard drives before it was transferred to the ADS. Consequently the DAPPER project strove to avoid this problem through allowing the depositing units the time to adequately document their archive and to normalise their data.

The data selection and normalisation of the Eynsham Abbey excavation archive took almost twice as long as the Royal Opera House archive. This discrepancy in time is undoubtedly due to the fact that the Eynsham Abbey project was never designed as an exercise in digital data collection and management but had evolved into a digital format. Consequently much of the time allotted to the OAU was taken up in converting and normalising the numerous disparate databases into a single database structure and in converting the CAD files from the proprietary FastCAD to the DXF, CAD exchange file format. The Royal Opera House, however, was always envisaged as a "digital" project and standard file formats and software was used for both data capture and file management. Cost-effective archiving consequently can be encouraged through the use of non-proprietary software and exchangeable data formats. This is outlined in the ADS Guidelines for Depositors (ADS 1999), which is available online: http://ads.ahds.ac.uk/project/userinfo/deposit.html.

2.3. Documenting the archives

One of the aims of the DAPPER project was to inform the development of digital archiving practice. In this respect the project has been a success as it has enabled all of the participants to consider the types of documentation an excavation archive needed to facilitate its effective re-use. These recommendations will be published in the second edition of the Excavation Archiving Guide to Good Practice (Richards and Robinson 2000).

The experiences gained from working in partnership with MoLAS and the OAU have enabled the ADS to specifically identify:

- what documentation is necessary,
- the documentation already produced as part of the post-excavation process that should be included in the project level documentation, such as the site report, specialist reports and so on,
- the new forms of documentation that units need to write to accompany the archive; here the documentation of the digital archive is particularly important, if the codes used in the Newham databases, or if all the files for a particular site had been documented, a substantial proportion of the archive would today be available for reuse.

Much of this documentation can be adequately provided through the collation and editing of reports that would be written during the post-excavation or publication stages of a project. Although some additional writing is necessary and the preparation of project documentation would not represent a substantial additional burden for the units.

2.4. The cost of digital archiving

Digital archiving entails expenditure, both in the setting up and running of a digital archive and in preparing, depositing, accessioning and curating the information resources. If archiving of the digital component of fieldwork is to become a part of the everyday practice of archaeology, DAPPER also had to demonstrate the commercial case for such archives. The pilot has allowed us to quantify the effort and costs to both units and archives in the preservation and dissemination of computer-based data.

The experiences gained from DAPPER have proved invaluable in the development of the ADS charging policy (ADS 2000). More information on the charging policy, including a detailed breakdown of the costs and charging categories is available online: http://ads.ahds.ac.uk/project/userinfo/charging2.html

The central tenets of this policy are that:

- the use of ADS resources will be supported free of charge,
- archiving costs should be recovered from the body funding the archaeological investigation, or research, wherever possible,
- this would be a one-off payment collected at the point of deposit which should safeguard the long term future of the digital information.

Eynsham Abbey

Accessioning and migration

CAD	882 files	£4810
Images	24 files	£510
Database	13 data tables	£325
Spreadsheet	2 data tables	£50
Texts	2 documents	£50

Storage

CAD	477 Mb	
Images	998 Mb	
Database	2.8 Mb	
Spreadsheet	0.2 Mb	
Texts	0.1 Mb	
Total	1478.1 Mb	£739.05

Digital Archiving cost of Eynsham Abbey	£6484.05
Cost of Eynsham Abbey Project (total)	£540,000
Proportion of budget spent on digital archiving	1.2%

Royal Opera House

Accessioning and migration

GIS	3 files	£75
Database	3 data tables	£75
Spreadsheet	17 data tables	£455
Texts	2 documents	£50

Storage

CAD	4 Mb	
Database	0.1 Mb	
Spreadsheet	1.4 Mb	
Texts	0.8 Mb	
Total	6.3 Mb	£0

Digital Archiving cost of ROH Phase 1	£655
Cost of Royal Opera House Project (total)	£650,000
Proportion of budget spent on digital archiving	0.1%

The Eynsham Abbey and Royal Opera House digital archives are both essentially research archives. Yet they differ quite markedly in their cost. This is clearly due to the ways in which both sites were analysed during post-excavation. Although the stratigraphic drawings of both sites are available digitally, the use of GIS by MoLAS ensured a deposit of 3 files, as opposed to the 404 Eynsham Abbey site plan CAD files. The 80 structure plans, 15 phase plans, 2 sections, a trench plan, a composite plan and a 3-D reconstruction of the Abbey also swelled the number of Eynsham Abbey CAD files. Consequently the cost of archiving is closely related to the presentation of the drawing information. It should not be assumed, however, that GIS is preferable to CAD because it is more cost effective to archive. Both file formats have their own distinct advantages and disadvantages. The GIS structure of the Royal Opera House archive, for example, would require specialist software and training to facilitate effective understanding and reuse of the data. It is consequently difficult for a user with little knowledge of GIS to use the archive. On the other hand the Eynsham Abbey archive is structured like a traditional archaeological publication, with its separate structure and phase plans. With appropriate viewers, the CAD files can be accessed via the Internet. Consequently the CAD presentation of the data can open up access to the information for non-expert users. Yet for highly experienced users the MoLAS GIS offers a very powerful analytical tool, with interpretative functionality that is beyond CAD. It is recommended that units continue to undertake their post-excavation in the ways most appropriate to them and that the production of a comprehensive digital archive should come before matters of cost.

In its Charging Policy the ADS state that "it is usual for digital archiving costs to add an overhead of less than 5% to the total project budget (ADS 2000). DAPPER has illustrated that for projects of the scale of Eynsham Abbey and the Royal Opera House, the actual cost of digital archiving can be as little as 1% of the total project budget. It must be recognised, however, that this 1% figure relates only to large-scale excavation projects. For medium to smaller-scale projects with a variety of files, the actual cost of archiving will be represent a greater proportion of the total budget and be in the region of 3-5% of the total project budget.

2.5. Resource delivery

Both the DAPPER and Newham Museum Service archives are disseminated over the Internet via ArchSearch, the online catalogue of the ADS. ArchSearch now contains some 350,000 site index records based on the National Monuments Records of England, Scotland and Northern Ireland, regional sites and monuments records and localised gazetteers. This site-based index is increasingly being fleshed out through the deposition of resources such as the DAPPER and Newham Museum Service archives. An objects based model has been adopted to conceptualise and implement access to these resources. Active hyperlinks within the database holding the catalogue provide access to resources whose object types or classes are defined in terms of generic structure and functionality. Effectively, an objects based data structure has been bolted onto an underlying relational database (cf. Austin and Clarke 2000 for a fuller description).

One of the major data delivery issues concerned the user interface and revolved around the issue of whether a complex, visually pleasing, heavily packaged interface should be developed, or whether the raw data should be simply presented? The heavily packaged interface would be user friendly and consequently may open up access to the data. This approach, however, was rejected as it would have been prohibitively expensive to produce, would have set a precedent for all subsequent deposits and the heavily packaged nature of the data would probably present migration problems. More importantly perhaps, the packaging of data may restrict its potential for reinterpretation. Consequently it was decided to present the data in standard formats with sufficient on-line support documentation to enable the re-use of the data. This was regarded as the ideal, sustainable cost-effective solution for Research Level Archives and below.

The resource delivery issues brought the idea of who the data is aimed at into focus. The ADS receives its core funding from the Higher Education sector and consequently it is the scholarly community at which its data resources are aimed. Equally the ADS user community, as reflected in Strategies for Digital Data (Condron et al. 1999), clearly wanted raw, rather than packaged data. The simple, unpackaged DAPPER interface consequently reflects the professional archaeological community's desire for raw data.

2.6. DAPPER conclusions

The Digital Archiving Pilot Project was a necessary and timely project that enabled the Archaeology Data Service to develop and implement the world's first online digital project archive. The project also enabled its partners to think about exactly what forms of data should be incorporated into a digital project archive and how this should be documented and then delivered online. DAPPER provided the necessary funding and time for the Museum of London Archaeological Service and the Oxford Archaeological Unit to prepare their data for deposit and to reassess and develop their post-excavation procedures and documentation. Finally a critical review of DAPPER has allowed the ADS and English Heritage to consider how to ensure how other quality digital archives will be delivered to the archaeological community in the future.

DAPPER has demonstrated that an effective digital archive can be delivered for a fraction of the total project cost. For substantial fieldwork projects the DAPPER project suggests that the estimated archive deposition cost (approximately 5% of the total project budget) as reported in the ADS Charging Policy (ADS 2000) can be modified downwards, to less than 1% of the total project budget.

The results so far from DAPPER have been very encouraging. The Royal Opera House archive was released in August, followed by Eynsham Abbey in October 1999. Following the launch of the excavation archives the ADS website has received an increased number of visitors and has experienced successive months of record-breaking user figures. Indeed within the first four months of their release over 8,000 people have visited the project archives and either accessed or downloaded approximately 10,000 separate files within the archives.

When thinking about digital archiving, a more philosophical point came to mind; what is a digital archive for?

- a repository for the digital remains of project data?
- part of the project's dissemination strategy?

If a digital archive is simply a store of the digital products of archaeological projects, DAPPER has effectively delivered this form of archive at the research level and has also illustrated that it could also be implemented for lesser archives. The digital archive has the potential to develop into something more than this and an integrated archive could be part of a project's dissemination strategy. The integrated digital archive provides the opportunity to take unwieldy technical data out of the excavation report and locate it in a universally accessible environment. Modern archaeological datasets often do not translate well into the written reports, especially if much of the information is in the form of CAD drawings, GIS or relational databases, which loose their functionality when removed from their digital format. Consequently the digital archive offers the opportunity to place data in a facility that actively encourages its reuse and reinterpretation. If project data were located in a digital archive this would have profound implications for the ways in which archaeological research is published. If the correct linkages can be developed between the traditional paper or Internet publication and the digital archive this would enable the final project report to be slimmer and more synthetic and less bogged down in detailed description.

3. Conclusions

This paper has effectively illustrated through the use of DAPPER that a digital archive can be easily created from the digital residues of the excavation and post-excavation process where there is sufficient political will and funding. On the other hand Newham has demonstrated the opposite, that where the digital archive has never been considered, when the digital residues of the archaeological process are simply left to degrade, the reverse engineering of excavation archives can be an expensive and frustrating business, especially where documentation is lacking and file formats are obscure. It is becoming widely accepted amongst the archaeological community that the digital component of a project is a necessary and vital component of the primary record of a site and needs to be preserved into perpetuity. What DAPPER and Newham have illustrated is that the road to a successful deposit should ideally begin at the point of data creation and continue throughout the life cycle of the project until the point of its ultimate deposition with the digital archive. Only in this way can we truly ensure that through documentation, the use of appropriate file formats and by following adequate backing up and migration regimes that we can begin to create a digital future for our excavated past.

Acknowledgements

The authors wish to thank our colleagues at the ADS, Jo Clarke, William Kilbride, Maureen Poulton and Julian Richards for their efforts on behalf of this paper and on the DAPPER and Newham Museum projects.

Damian Robinson would like to thank Keith May and Tim Williams of English Heritage Archaeology Commissions, Ann Dodd and Alan Hardy of the Oxford Archaeology Unit and Gordon Malcolm of the Museum of London Archaeology Service for their contributions to the DAPPER project.

Keith Westcott would like to thank Sarah Richards and David Thorne, students on the MSc in Archaeological Information Systems at the University of York for their help with the Newham Museum archive.

References

ADS 1999. Guidelines for Depositors Version 1.1, http://ads.ahds.ac.uk/project/userinfo/deposit.html

ADS 2000. Charging Policy, http://ads.ahds.ac.uk/project/userinfo/charging2.html

AUSTIN, T., 1998. Tip of the iceberg in a digital sea: scale of the academic digital resource, *ADS Newsletter* 4: 4-6. Also on-line at http://ads.ahds.ac.uk/newsletter/issue4.html (18 June 2000).

AUSTIN, T., 1999. Rescue archaeology at the ADS, *ADS Newsletter* 6. Also on-line at http://ads.ahds.ac.uk/newsletter/issue6.html.

AUSTIN, T. and CLARKE, J., 2000. ArchSearch: under the bonnet, *Archaeological Computing Newsletter* 55: 15-20.

BEAGRIE, N. and GREENSTEIN, D., 1998. Managing Digital Collections: AHDS Policies, Standards and Practices: Version 1, on-line at http://ahds.ac.uk/public/srg.html.

BOOTH, B., GRANT, S. and RICHARDS, J., 1989. *Computer Usage in British Archaeology second edition 1989*, IFA Occasional Paper 3 (Birmingham).

CONDRON, F., RICHARDS, J., ROBINSON, D. and WISE, A., 1999 Strategies for Digital Data – Findings and Recommendations from Digital Data in Archaeology: A Survey of User Needs, The Archaeology Data Service. York.

DINGWALL, L., EXON, S., GAFFNEY, V., LAFLIN, S., van LEUSEN, M. (eds.), 1999. *Archaeology in the Age of the Internet*, Computer Applications and Quantitative Methods in Archaeology CAA97, BAR International Series 750 (Oxford).

GRANT, S., 1986. "Summary and recommendations". In Richards 1986: 13-32.

RICHARDS, J. and ROBINSON, D., 2000. *Digital Archives from Excavation and Fieldwork: A Guide to Good Practice Second Edition*, Oxbow Books. Oxford.

RICHARDS, J. (ed.), 1986. *Computer usage in British Archaeology*, IFA Occasional Paper 1, Birmingham.

STEWART, J., 1980. Integrated excavation and museum recording systems: methods, theories and problems, *Museum Archaeologist* 5: 11-27.

WILCOCK, J., 1999. Getting the best fit? 25 years of statistical techniques in Archaeology. In Dingwall et al. 1999: 35-51.

Virtual 3D Reconstruction of the Kiafar Site, North Caucasus, Russia

Mikhail Zhukovsky

Moscow State University, Department of Archaeology

P.O. Box 49, 105215 Moscow, Russian Federation

e-mail: mixail@dol.ru

Abstract

The VR application in archaeology is one of the most rapidly developing sections of this discipline. This paper argues current conceptualisations of VR implementation in archaeology. While the mainstream efforts in the field are dominated by the notion of increasing realism of VR models, this approach regarding VR 3D models as thoroughly assembled replicas of reality is considered to be misleading and eventually dead-ended. Alternative conceptualisation of VR as a two-step interpretation (at the data recording and data processing stages), rather then representation of initial archaeoreality is introduced. VR is proposed to be regarded as a duo system, involving a perceiving subject and an object of knowledge. In order to exist and function, this system engages an active and mobile researcher freely negotiating with and interpreting archaeological data through a medium – a VR model. The concept of realism, as well as authenticity and faithfulness of generated models, is suggested to be understood as integrity and homogeneity of initial data organized in a virtual environment.

These theoretical underpinnings are illustrated by the Kiafar project, a pioneering case-study in Russian archaeology. Its significance is confirmed by the role Kiafar plays for research, preservation and popularisation of Russian historical heritage.

Key words: VR, 3D model, conceptualisation, archaeoreality, virtual environment, non digital data

1. Introduction

The last decade has been marked by a significant rise in interest in virtual 3D modelling among the archaeological community. Some of the potential of the impressive development of virtual technology has been explored in a constantly widening range of its application in archaeological projects and case studies and presented in a number of weighty publications (Forte 1996, Earl 1999, McCullagh et al. 1999). These enthusiastic efforts were mainly concentrated on the realism of modelling, or model authenticity. This resulted in the fact, that while the presentational power of 3D models is indisputable, their contribution to archaeological interpretation is still problematic. The lack of theoretical underpinnings of existing VR models leads to a contradiction between what *we think* a model is, and what does it *truly represent*. In this paper, I intend to focus briefly on the conceptual shortcomings of VR modelling of archaeological sites. The alternative conceptualisations of VR and approaches to modelled data are to be discussed.

Another issue of enormous significance to be taken into consideration at this point is the demand for initial data from which models are produced. The currently existing digital survey data archives containing satellite images, GIS databases, digital terrain models (DTM) and other types of spatial data are fragmented. This makes the utilisation of paper copy plans, drawings, hand-managed measurements and photos that have been collected over numerous decades, inevitable. In Russia, where there are no digital data archives at all, and the survey data is stored in paper form, this problem is especially vital.

A brief outline of the Kiafar project will be given in the presented theoretical framework, and the potential of data that was initially not digital, nor was it ever supposed to be digitally processed and analysed for the creation of 3D virtual models will be demonstrated.

2. VR – seeking new approaches

The concept of VR has been recently discussed to a great extent in a number of publications (see Gillings 1999). Instead of returning to these debates here (which would take a lot of time), I intend to focus your attention on the nature of the interrelationship between *reality* and *virtual reality*. The mentioned trend, manifested by VR case studies to make produced models as realistic as possible, originates from a misleading concept of VR being a kind of reality replica (Gillings 1999, Burton et al. 1999). According to this notion, the more realistic, or authentic, the replica seems to be, the less it is considered to be a fake. Thus, the ideal VR model is supposed to be indistinguishable from what it represents. The paradox is, that the most detailed and carefully constructed model will never match the original. At first glance, the explanation of such a phenomenon lies in the specifics of an object of archaeology as a discipline. An archaeologist deals with *traces* of reality that have been transformed by time, and can be described as *archaeoreality*. This implies a notion of multiple interpretations caused by individual characters of perception. It brings us to the thesis, that any VR model is an interpretation, rather than a representation of reality.

Further on, it poses a question: is what we interpret, or better say - model, reality? The answer is negative. In fact we interpret (model) an interpretation itself. A VR model is nothing more than data processed with a given algorithm. The data we extract from reality through perception, recognition and, finally, description can be expressed with the term meta-reality, or hyper-reality. The process of modelling this data (hyper-reality) introduces also a second level of interpretation. The resulting model reflects an in-

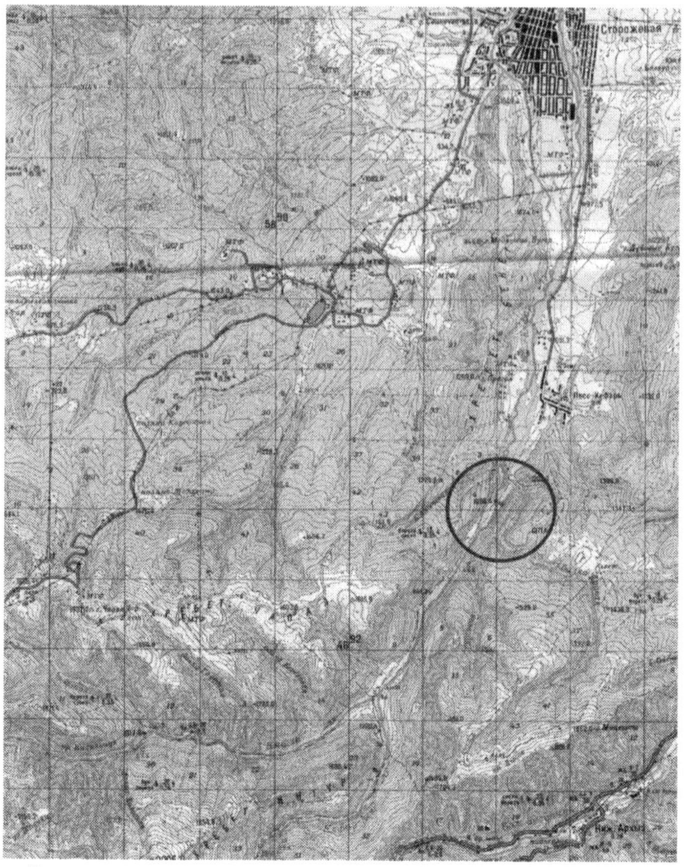

O Kjafar site area.

Figure 1: Kiafar site region map in 1:100,000 scale.

dividual (unique) interpretation of hyper-reality that was carried out by computer processing algorithms.

The nature of VR as an interpretation model involves two main elements: a perceiving subject (e.g. researcher) and an object of knowledge (e.g. modelled archaeoreality). At the production stage the link creator – data, sustains this dual system. When an external viewer is engaged, one substitutes a model creator, while a model becomes an object of knowledge itself. From the other side, a model is a medium of interaction between a perceiving researcher and recorded archaeological data. The productive interaction can be executed through two main channels: analytical functions of a model and/or visual perception. To support the first channel the model is to be an integrated part of a GIS analytical environment, as for the second, it is based on the notion of a researcher being actively involved within a model. The dynamic model generated by the VRML technique, with a mobile viewer (engaged person) navigating freely through a modelled archaeoreality environment, seems to suit the requirements. Any pre-programmed set patterns of viewers dislocation and movements are to dramatically decrease the potential of a model.

The last theoretical issue I would like to examine is the mentioned concept of realism of a model. We must realise, that ever increasing visual sophistication does not bring a model closer to reality. Realistic, in this case, does not mean real. Rather then spend time and costly efforts on stunning but static re-constructions, featuring ever unparalleled visual details, I would propose to concentrate on an alternative notion of realism. In the mainstream of formulated conceptual underpinnings the term realistic is referred to the initial data. Its integrity and homogeneity, along with volume and quality, provide true realism. Thus, the demands for authenticity and faithfulness applied to a model are incorrect, and to be redirected to the initial raw data.

Figure 2: An extract from Kiafar's topographical plan.

While working on the (for Russia, pioneering) project to virtually model one of the pilot sites of medieval time in the North Caucasus, these theoretical issues have been evoked. The low project budget made digital surveying equipment unaffordable. In this case, the initial spatial data was carefully manually obtained with simple optical theodolites and recorded and stored in a paper form. With the integrity analysis carried out through the data, it was considered sufficient for creating a preliminary virtual model.

3. The Kiafar project

3.1. Background

The Kiafar site is situated in the Karachai-Circassian Republic in the Russian North Caucasus. In the first and early second millenia AD the Iranian-speaking Alans inhabited the region. By the 10th century AD several dozens of strongholds had emerged there. Nearly all of them were equipped with complicated fortifications, well thought-out and adapted to the environment. The well-organised Kiafar township is the largest of the fortress towns of the period between the 10th and 12th century AD. The site with all its 1.8 km length is stretched along the narrow Kiafar-Argun river valley, and lies on the mountain ridge. The highest point of Kiafar, the peak, is some 180 m above the bottom of the valley, totalling approx. 1200 m above the Baltic sea level (figure 1).

Kiafar was chosen as a pilot site for 3D modelling and subsequent visualisation for the following reasons:

1. This site can boast a comprehensive survey database indispensable for adequate modelling. The archaeological expedition headed by Irina Arzhantseva has succeeded in producing a detailed 1:500-scale topographical plan with

Figure 3: Fortification wall.

Figure 4: Panorama of Kiafar site: mountain ridge completely covered with forest.

a 1m contour line interval. Providing detailed relief data, the plan also shows some 200 surviving structures as well as numerous landscape peculiarities (figure 2).

2. The remains of architectural structures at the site are very well preserved, which allows for a minimum of conjectures in the course of modelling (figure 3).
3. Kiafar, allegedly the capital of the Western Alans or a residence of a powerful prince, included all sorts of household structures, sanctuaries and fortifications. A 3D model of such a complex site provides rich opportunities for an analysis of both the juxtaposition of architectural structures and their interplay with the micro-regional landscape.
4. The mountain ridge, where the site is located, is covered with woodlands (figure 4). This fact hinders the visual estimation of the site and the degree of man-inflicted changes to the landscape. The interactions of the ancient town population with the environment, their use of the landscape and their adoption patterns can only be analysed virtually. Also, and this is also a vital issue, the same is true of the aerial and space survey. The decoded satellite and aerial data are hardly informative at all.
5. Last but not least, the Kiafar site was selected for 3D modelling to prevent its ultimate loss.

The 3D model enables us to achieve a virtual conservation of the site, and estimate and monitor the great damage done to the site with the passage of years. The information extracted from the model would help to elaborate a set of methods for the preservation of Kiafar and other archaeological sites in Russia. An adequate model of Kiafar and the adjacent territories could contribute to the creation of an archaeological national park in the North Caucasus.

2.2. Software requirements

At present a wide range of commercial GIS/CAD software products are available at constantly decreasing prices. It makes the final choice a matter of budget opportunities and particular preferences. Along with our notion to integrate the VR model into the analytical GIS environment, we were seeking a software tool with an extensive set of both GIS and CAD functions, and failed to find one. Most GIS products lack a powerful COGO engine indispensable for the purposes of sophisticated detailed 3D wireframe modelling. CAD software offers much more flexible modelling instruments and opportunities, however its analytical potential is far from being sufficient.

In the end we decided to use Autodesk AutoCAD 14 as a 3D modelling tool. The digital terrain model was created in its module EaglePoint Software LANDCADD 14, which is specially designed

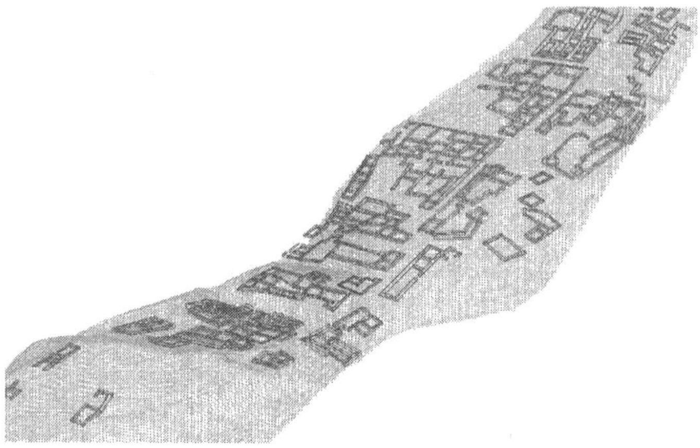

Figure 5: The remains of the buildings (3D) on the grid. Scaled to region 230x125 m.

for shaping 3D terrain surfaces, while the Autodesk 3D Studio MAX and Adobe Photoshop 5.0 were proposed for rendering the wireframe model. Budget limitations have predetermined the choice for ArcView 3.1 as a GIS tool. According to the "GIS and archaeology" survey conducted a few years ago (Moscati 1999) these software solutions seem to be the most popular for projects carried out within the framework of research institutions.

3.2. Model construction

The Kiafar project consists of a number of elements:

1. the DTM of the site and the adjacent area,
2. the models of the surviving architectural structures of the site,
3. the vegetation map of the Kiafar area,
4. the hydrological map,
5. the geological map,
6. the soil map and
7. the relatively extensive database of photos, video clips, textual descriptions.

All of these elements are to be integrated in an interactive GIS environment with VRML and HTML support, providing a viewer instant access to all types of data simultaneously. The viewer is encouraged to freely seek his position, and navigate anywhere within Kiafar's virtual environment (or archaeoreality), and explore the relative database organised with full hyperlink, cgi-script support. One of the key concepts of the proposed virtual reality is to eliminate any pre-programmed patterns of the viewer – constructed environment interactions and his engagement in the process of negotiation and interpretation of the modelled archaeoreality.

To date, we have completed only two stages of the Kiafar project, i.e. we have generated the digital terrain and architectural structures wireframe models. The 1:500-scale hardcopy of the topographical plan with a 1m contour line interval became the basis for the DTM. The contours, after being digitised, resulted in a set of polylines, that were used to create a triangular irregulated net (TIN). The TIN in turn served as the base for an elevation grid with 1x1 meter cells. We consider such a grid resolution the most appropriate for the needs of the relief model, because a less detailed elevation grid would not reflect all the peculiarities of the

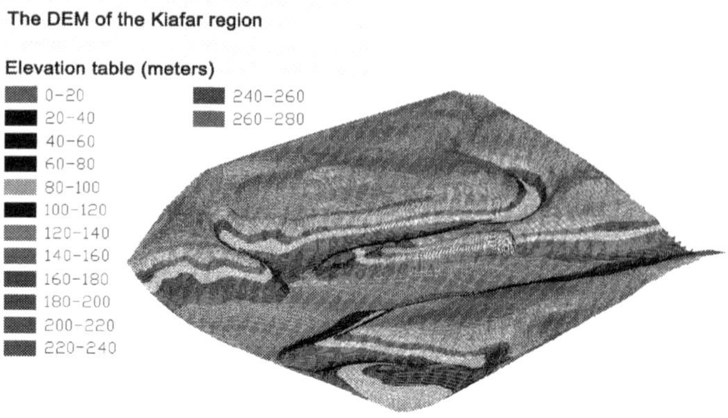

Figure 6: The DEM of the Kiafar region, 4x4 km.

micro-relief, while a denser one would offer no substantial additional information. The 1x1 meter celled grid coincides with the original topographic plan of 1m contour line intervals.

Unlike the TIN, such an elevation grid is a regular one. Hence, the architectural structures under consideration can be adequately placed in the DTM (figure 5).

After the final DTM was completed, the problem of possible errors in the course of modelling arose. The possible voids in the initial contour lines data could lead to artificial non-existent slopes/elevations, and various holes may appear on the DTM. In order to check this, we generated contour lines on the basis of the elevation grid and compared them with those of the original plan. Beyond expectation, all the virtual and original contours coincided.

At the next stage 3D wireframe models of all the architectural structures were created and referred to the DTM.

The resulting wireframe model of the Kiafar township was merged with that of the local micro-region based on a 1:100,000-scale map covering 16 square km (figure 6).

Though the project is far from being completed, even this preliminary model contributed significantly to the intra-site spatial analysis, enabling us to re-estimate the vertical juxtaposition of structures, examine the zone structure of the site, single out compounds and types of buildings within them and discriminate between residential, administrative and economic areas of Kiafar. The true potential of the virtual environment is to be exploited with a final model.

4. Conclusions

The VR application in archaeology is one of the most rapidly developing sections of this discipline. At the same time the mainstream efforts are dominated by the notion of increasing realism of details. I consider this approach to VR as thoroughly assembled replicas of reality to be misleading and eventually leading to a dead-end. Alternative conceptualisation of VR as a two-step interpretation (at the data recording and data processing stages), rather than representation of initial archaeoreality is introduced. VR is proposed to be regarded as a dual system, involving a perceiving subject and an object of knowledge. In order to exist and function, this system engages an active and mobile researcher freely negotiating with and interpreting archaeological data through a medium – a VR model. The concept of realism, as well as authenticity and faithfulness of generated models, is suggested to be understood as the integrity and homogeneity of initial data organised in a virtual environment.

These theoretical underpinnings are to be implemented in the Kiafar project, a pioneering case study in Russian archaeology. Its significance is confirmed by the role Kiafar plays for research, preservation and popularisation of Russian historical heritage.

References

BURTON, N.R., HITCHEN, M.E., BRYAN, P.G., 1999. Virtual Stonehenge: a Fall from Disgrace? *Computer Applications and Quantitative methods in Archaeology*, BAR International Series 750.

EARL, G.P., 1999. Visualising Danebury: Modelled Approaches to Spatial Theory, *Computer Applications and Quantitative methods in Archaeology*, BAR International Series 750.

FORTE, M. (ed.), 1996. *Virtual Archaeology: Great Discoveries Brought to Life Through Virtual Reality*, Thames and Hudson, London.

GILLINGS, M., 1999. Engaging Place: a Framework for the Integration and Realisation of Virtual-Reality Approaches in Archaeology, *Computer Applications and Quantitative methods in Archaeology*, BAR International Series 750.

McCULLUCH, M., MAGGI, R., PEARCE, M., and RATCLIFFE, J., 1999. From the Ground Up: Visualising Ligurian Archaeological Sites, *Computer Applications and Quantitative methods in Archaeology*, BAR International Series 750.

MOSCATI, P., 1999. GIS and Archaeology: the "Caere" survey, *Computer Applications and Quantitative methods in Archaeology*, BAR International Series 757.

Indexing and Retrieving Archaeological Resources on the Internet - A prototype Multilingual Thesaurus Application

Martijn van Leusen

Groningen Institute of Archaeology

Poststraat 6 9712 ER Groningen, The Netherlands

e-mail: p.m.van.leusen@let.rug.nl

Abstract

This paper introduces the concept of using a thesaurus for structured indexing and retrieval of archaeological information on the Internet, defines parameters for the implementation of a multilingual web-based thesaurus application, and describes the prototype under development by the ArchTerra Consortium.

Keywords: thesaurus, Internet

1. Introduction

The field of Archaeology shares in the general exponential growth of the amount of information that is being published on the Internet. Whereas the first such publications in Europe date to "gopher" sites of the early 1990s, the major search engines currently report anything up to 100,000 pages relevant to European archaeology. Over the intervening years many have lamented the resulting chaos and lack of quality control, although knowledgeable users could still reduce the chaos by using the various information filtering mechanisms offered by web index sites such as AltaVista. Champion (1997) and Van Leusen et al. (1996) provide a good overview of the situation in 1995/6, and over the past five years the general public too has been educated in the use of such mechanisms.

As the Internet became a potential source of information for the professional archaeologist too, the need for more sophisticated retrieval mechanisms (and their obverse, indexing mechanisms) becomes ever more pressing. Among the early solutions were Internet "guides" such as ARGE,[1] which provide access to a manually indexed subset of web resources, and metadata schemes such as that of the Dublin Core,[2] which allow the author of a web page to include descriptive information which can then be retrieved by web indexing robots. While these solutions provide short-term relief, the need for long-term, cross-national, and effective solutions to the problem of finding information on the Internet has been well recognised. The AQUARELLE project "Sharing cultural heritage through multimedia technologies"[3] represents an industry-driven approach to the problem, while professional archaeologists themselves have opted for the distributed, co-operative approach represented by the ArchTerra Consortium.

The cultural and "information society" programs sponsored by the European Commission as part of the 4th and, recently, 5th frameworks provide important means for experimenting with the development of solutions. Straddling the two frameworks is the ArchTerra project[4] (1999-2000), which aims to extend the invisible web of European archaeology (which is, after all, a largely western European and English-language web) to eastern Europe. In addition to putting eastern European archaeology online, the ArchTerra consortium develops distributed and multilingual tools to facilitate the publication and retrieval of archaeological information on the Internet. One of these is the online multilingual thesaurus system of which a prototype is presented in this paper.[5]

A thesaurus is a dictionary of words and phrases, grouped together according to similarities in their meaning. It contains "terms" and explanations (or "scope notes") of how those terms should be applied, and it defines relations between the terms. *Which* of these terms, explanations, and relations one selects depends on the intended use(s) and audience(s) of the thesaurus. For detailed retrieval of information from the scientific literature a large and complex thesaurus will be needed,[6] but for our goal of unlocking web resources for a very diverse audience, a very much simpler approach (targeting the proverbial intelligent 12-year old) may be taken.

1.1. Towards a prototype multilingual thesaurus application

The work of the ArchTerra Consortium is aimed at reaching the following interrelated goals:

- To establish a short list of terms adequately covering the presently available internet resources for European archaeology, with appropriate scope notes (a glossary) and translations;
- To structure these terms in a thesaurus;
- To implement a browseable user interface to the thesaurus, for submission / retrieval;
- To implement a browseable maintenance interface to the thesaurus, for submitting / discussing / activating new terms, definitions, relations, and translations.

Progress toward these goals is discussed in more detail below. Following a discussion of general design considerations in section 2, the remainder of this note is used to describe the user and maintenance interfaces of the prototype thesaurus application.

2. Design considerations

The goals specified above have been used to provide further parameters for the design of the prototype application. Parameters relating to the content of the thesaurus, the types of relations that will be recognised, the implementation of multiliguicity, and the use of standards are discussed each in turn below. It should be noted in advance that many desiderata have not been implemented in the prototype – rather, it is hoped that they may feed further and wider discussion and development at a later stage.

2.1. Multilinguicity

The requirement of multilinguicity (or, more properly, language-independence) is relevant to the design of the thesaurus application in two ways. How do we build a language independent system that will allow us to add languages as needed? And how do we ensure that submissions and retrievals will be handled correctly irrespective of the language or script being used? The latter question is a technical one, to do with the problematic representation of various European character sets by web browsing software, which we will not explore any further here. The former question leads off to the deeper issue of what constitutes a translation. True one-to-one mapping of terms in two different languages does not occur very often, and a successful thesaurus application must therefore be able to deal with the existence of multiple – multicultural – and conflicting meanings of the same term. Among the practical approaches to consider are:

a) taking an arbitrary decision, e.g. to abide by the definitions given by some authority, and
b) ensuring professionalism in the translators (avoiding debates about the quality of a translation).

Although the ArchTerra Consortium's original plan was to implement the prototype in four languages versions (English, Polish, Romanian and Bulgarian), restrictions in the available time, funding, and expertise meant that only the English version was in fact developed. However, as the prototype is language-independent, the addition of languages should be a relatively simple task.

In order to preserve language independent storage of thesaurus information, three linked database tables are used in the ARGE database structure to store information about relations, terms, and relations between terms. Another three linked tables store information about terms assigned to URLs (see figure 1). Terms, Classes, Relations and Languages are all numerically coded, and "lookup" tables are used to translate these codes into words in the user's chosen language.

2.2. Standards

As a matter of principle, the thesaurus should, where-ever possible, make use of existing international standards. For example, ISO standards or draft standards for the recording of character sets, country names, and language names can be applied without difficulty. But other standards may not be applicable for a variety of reasons; for example, the ISO date standard does not cope well with the needs of archaeological dating, because it uses hyphens as a wildcard character (so –999 expands to the year 0999 or 1999 or 2999) and does not allow the BC/AD system or negative dates. The best system for numeric representation of dates is the astronomical system, where 0 = 1 BC, -1 = 2 BC, etc (Millard, pers. comm.). It goes without saying that, where such standards differ from generally accepted usage, the standard should be used internally while the generally accepted version is presented to the user.

Existing national standards and draft standards for professional terminology, such as the English MIDAS and INSCRIPTION

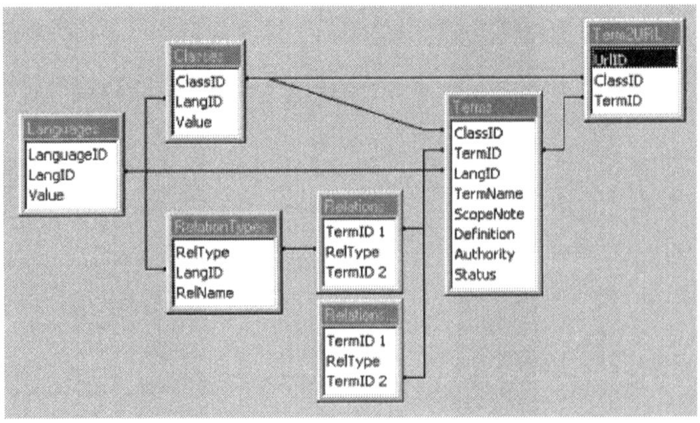

Figure 1: Term specifications are stored in table Terms, their relations in table Relations (which occurs twice because it creates "internal" links), and their assignment to URLs in table Term2URL; Tables Classes, Languages, and RelationTypes are "lookup" tables.

standards, may also be relevant to the thesaurus. MIDAS, the Monument Inventory Data Standard, is a "content" standard for recording architectural or archaeological monuments; INSCRIPTION is a set of standard "word lists" covering things like the Type of a Monument, or terms to describe archaeological periods.

It should be recognised, however, that no effective standards exist in many areas of archaeological terminology, and the creation of such standards is far outside the scope of the ArchTerra Consortium. International professional organisations such as the European Association of Archaeologists (EAA) and the Union Internationale des Sciences Préhistoriques et Protohistoriques (UISPP) should provide the forum for such work.

Worldwide standards also apply to the coding of the interfaces to the thesaurus (see sections 3 to 5), which should operate well across a range of platforms and browsers. In addition to adhering to the HTML standard, the UTF-8 (UNICODE) character set may be used by the interfaces so that western and central European, Baltic, Greek and Cyrillic scripts are all displayed correctly. It should be noted that, while the implementation of multilinguicity *per se* is straightforward, dealing with multiple character sets is not. Although the major browsers now recognise UNICODE encoding of web content, appropriate fonts may not be available to the client software to display all character sets in use across Europe correctly. We can only hope that ongoing standardisation will remove this obstacle before long.

2.3. Content

Despite the relatively restricted goals that we have set for the content of the thesaurus, this is probably the most difficult issue to settle. Relying on the experience gained earlier with ARGE, the ArchTerra Consortium settled on a guideline of including some 100 subject terms, some 30 period terms, and another 100 geographic terms in the thesaurus.

The biggest problem doubtlessly lies in how to avoid/resolve complexities or lack of clarity in the *meaning of* and *relations between* the terms. For example, archaeological periods are defined by a mixture of chronological, cultural, technological, political, and architectural criteria; it is not always clear to which area they are applicable and what are their absolute start and end dates (if any). Does "eneolithic" indicate the last phase of the Neolithic period, coming after the Late Neolithic phase (as the name implies), or is it a transitional period in its own right (in between the Neolithic and Bronze Age)? The "Roman"period begins and ends at different times in different parts of Europe, and does not exist in others. Can this be dealt with in the Scope note or is a more complex solution called for?

A second problem can be summarised in the question of how to ensure that the thesaurus provides truly equal "coverage" of all subjects, periods, and regions of Europe. For example, terminology for post-Roman periods is generally less precise than that for later prehistory; "non-preferred" terms such as "cave men" may have to be included because these are likely to be used by the public or to occur in older literature; and *local terms* (such as "sub-Apennine" for the Early Iron Age in Italy) may be needed to allow the effective inclusion of many parts of Europe.

Two approaches are available to avoid or alleviate endless disagreements about the content of the thesaurus. Firstly, definitions and scope notes describing the meaning and clarifying the applicability of terms in language appropriate to the audience will, whenever practicable, be taken verbatim from an authoritative source. Scope notes should also specify the amount of "fuzziness" in the meaning of the term. Secondly, the thesaurus will be "coarse-grained" in its description of archaeological internet resources, i.e., many potential conflicts of interpretation will be avoided because specific terms such as "Flavian" will be mapped to more general terms such as "Early Roman Empire". As a further antidote to the potential confusion, at least in the area of chronological terminology, we can include absolute chronology in our thesaurus by offering *millennia* and *centuries BC/AD* as indexing options (the BP and Cal systems of dating are not familiar to the general user, but could also be added if the need arose).

2.4. Relation types

The following standard thesaurus relations will be recognised initially by the system:

- Class (CL; top-level descriptive category to which the term belongs, e.g. "Chronological Period")
- Scope Note (SN; a brief statement about the scope of the term which could include a named authority for the definition used, geographic and temporal ranges, etc.)
- Broad Term (BT; indicates the generic term of which the current term is a specific; a.k.a. "is a kind of")
- Narrow Term (NT; indicates the specific terms of which the current term is the generic; a.k.a. "contains")
- Related term (RT; indicates all relation types not covered by BT and NT)
- Used for (UF; indicates any alternative but non-preferred terms with the same meaning; a.k.a. "is a preferred term for")
- Use (USE; indicates which other term of the same meaning is preferred over the current term; a.k.a. "is a non-preferred term for")

This can be extended at a later stage with relation types that are more specific to archaeology, e.g., Precedes (P; i.e. chronologically) and Succeeds (S; i.e. chronologically).

The importance of defining the relation types themselves may be illustrated by an example. The ISO standard for monolingual thesauri recommends that the use of BT and NT for part/whole relationships should be restricted to a few specific types of term: (a) systems and organs of the body; (b) geographical locations; (c) disciplines or fields of discourse; (d) hierarchical social structures. This would allow a BT/NT relation to exist between (a) skull and lower jaw, (b) Scandinavia and Norway, (c) Geophysics and GPR, and (d) tribe and moiety, but it would not allow (e) arms and sword.

One useful feature is the fact that terms need not be fully "linked up" to other terms in the thesaurus. For many colloquial search terms (that is, terms that non-specialist users would like to use) a simple mapping to one or more of the existing "preferred" terms in the thesaurus should be sufficient for effective retrieval. The thesaurus application would contain only the non-preferred term, its scope note, and it USE relation (to a preferred term), and any searches would take place "behind the scenes" using the preferred term. The same solution could be applied to the use of *local* terms as mentioned above, although here the relation type would be RT (related term) rather than USE.

3. The user interface

End users of the thesaurus will want to do two things with it – retrieval and submission of web resources. Logically, retrieval with the help of a thesaurus is just one among several search mechanisms, and the thesaurus option should therefore be offered in the "Search Page" of a web site. To keep the user's retrieval interface efficient and effective, the thesaurus must be activated in two stages. Primary retrieval of archaeological records should be achieved through a limited and hierarchical (i.e. using only the BT/NT relations in the thesaurus) interface, implemented either as a series of foldout menus or as a collapsible, Windows Explorer-like menu such as that used by the NAVIS web site. The results of the primary search will be accompanied by dynamically generated options for widening or narrowing searches which constitute the secondary retrieval system.

The code for the primary retrieval menu, presenting the BT/NT part of the thesaurus, may have to be periodically generated offline in order prevent the application from becoming unacceptably slow and complex (many menus with many options in the user interface). The "root" menu would contain CLASSes such as *period, region, organisation type* and *data provider*, each with no more than 10 options. The peculiar nature of the user community – a mixture of old and young, lay and professional – puts specific requirements on the design of the user interface. While users should be made aware that their actions are "filtered" by the thesaurus application, the presence of the thesaurus should be non-obtrusive and unnecessary use of jargon should be avoided. Since archaeological terminology is neither consistent nor logical, a heavy burden rests on the *definitions* and/or *scope notes* attached to the terms. These should therefore be accessible to the user at all times. Contextual information about each term can be presented to the user by employing an "onMouseOver" method (JavaScript) to show the scope note for each term; the terms selected by the user should also be echoed to the screen so that the user keeps track of his/her actions. A thorough study of interface design will be needed to upgrade the current prototype to a working "alpha" version.

The secondary retrieval menu (part of the output page resulting from a particular thesaurus query) will allow users to refine or widen their queries by hyperlinking each key term of the currently selected web resources to its thesaurus relations (see figure 2). Whilst a direct user search using the thesaurus has not been implemented at this date, a demonstration of the secondary retrieval system was implemented, and is shown below. A preliminary series of hierarchical relations (NT, BT) have been defined between the terms on the shortlist for testing purposes.

All of the above assumes that the user will be happy to use only the terms offered by the thesaurus for searching, but this is an unlikely scenario. Could the thesaurus application be taught to understand free-form keywords submitted by the user? Are there others ways in which the user can be shielded from the very structured search environment offered by a thesaurus? Among the potentially useful tools and techniques that can be explored are "mapping" terms or sets of terms to others, parsing the user's search strings to allow orthographic and syntactical variation, and searching by example:

- One of the most useful concepts in the enormously complex issue of creating an acceptable thesaurus for European archaeology is that of "mapping". A multiplicity of period terms deriving from cultural, political, chronological, and architectural criteria, for example, can perhaps be mapped to the single "yardstick" of absolute dates. "Flavian" would map to AD 68 <> AD 95 (or whatever), which would map to "Roman" in France but to "Late Iron Age" in Denmark. The implications of such mapping must be investigated. What happens if such dates change? How should overlapping periods be handled? Mapping can also be used to link two or more thesauri together, so that, for example, users can use the familiar terms from a local archaeological thesaurus for constructing queries which will be "translated" to equivalent terms in the European thesaurus before execution;
- A parser, to resolve orthographic and syntactical variation, including errors, in search strings. Many search engines already implement this as "near" searching. One simple approach would be to search case-insensitive and to ignore everything that isn't a letter, e.g., /post[^a-z]* media?eval/i is an expression that will match most of the possible spellings of that term, including the American ones;
- Searching "by example". While a lay public may not be able to handle archaeological terms well (e.g., being unaware that they should ask for "Palaeolithic" to get information about Neanderthals), this would not be a problem if they could "search by example". The thesaurus could include a list of likely "examples" (the Iceman, Stonehenge) for this purpose, and "map" these to one or more proper archaeological terms.

4. The maintenance interface

The maintenance of our prototype thesaurus application involves two separate issues: the maintenance of the thesaurus itself (adding, editing, or deleting terms and relations), and the indexing of web resources using that thesaurus. The following sections review both these issues, using examples from ARGE's prototype thesaurus maintenance interface.

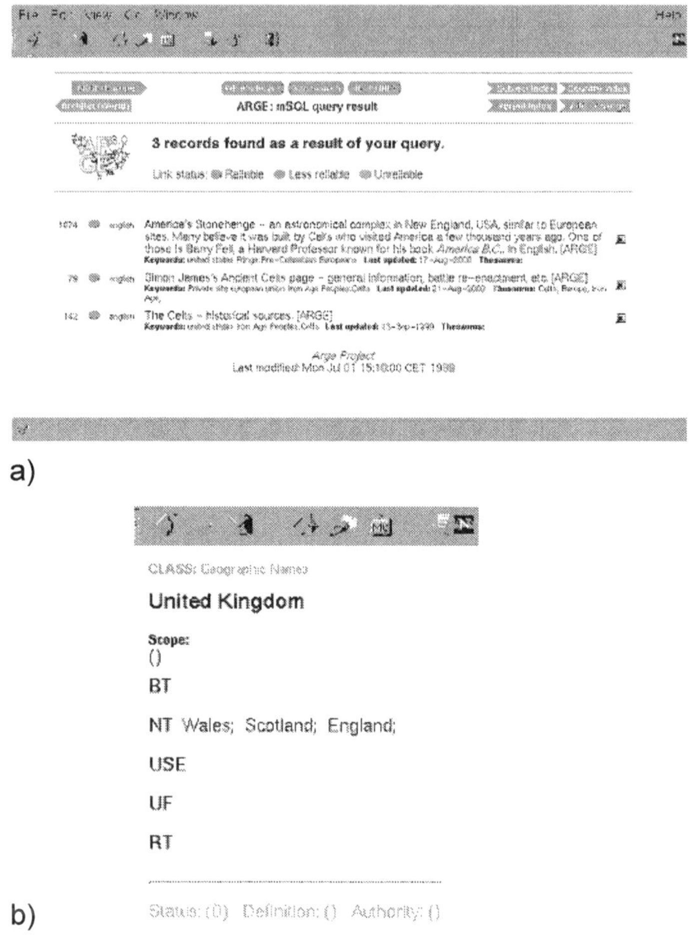

Figure 2: (a) A user query on the string "celts" results in 3 URL records being displayed by ARGE. URL 79 (Simon James's Ancient Celts page) has attached thesaurus terms " ". (b) When the user clicks on any thesaurus term, a helper window displays the current thesaurus relations of that term; when the user selects any of these, a new query is executed and new results are written to the main window. This cycle can be repeated at will.

4.1. Indexing web resources

Once the thesaurus itself is complete and available, the main task awaiting the maintainer of a web guide is to index new URLs, and to check/edit indexed URLs submitted by users. The indexing interface will present all currently available information on any URL, with options to add/change/delete key words and to visit the URL itself in a separate window. The regular maintenance interface of ARGE, which already allows key words to be added to a URL record, was extended with options to add/change/remove any number of terms from any number of CLASSes (figure 3). An option to call up the helper window containing the current thesaurus terms and descriptions was also added.

4.2. Building and maintaining the thesaurus

Building and maintaining the thesaurus itself is a task that should be restricted to an editorial board affiliated to an appropriate pro-

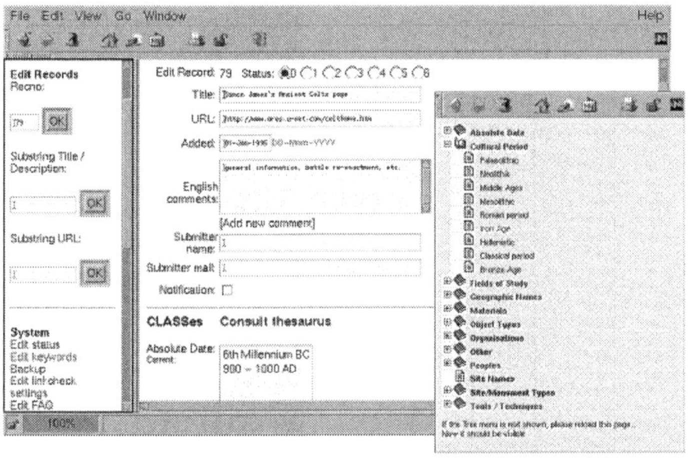

Figure 3: The URL editing page of ARGE's maintenance interface. URL 79 (Simon James's Ancient Celts page) has been selected for editing in the main frame. Any number of terms may now be selected from dynamically generated option lists for each currently defined thesaurus CLASS. Clicking on the update button at the bottom of this form updates the record across all database tables.

fessional organisation. The maintenance interface should have the following functionality:

a) Authority to performs actions should be stored in user profiles;

b) An Editor must be able to view current relations between terms, to add and edit terms, scope notes, relation types, and relations between terms; the use of definitions in the scope note requires the addition of an "authority" field;

c) A Translator should be able to add and edit translations.

The prototype thesaurus maintenance interface has been implemented as a "page" in the regular maintenance interface of ARGE (see figure 4). While the right hand frame provides links to the forms needed for adding terms to, and removing them from, the thesaurus, and for adding/editing/deleting a thesaurus relation or relation type, the left hand frame presents the hierarchical part of the current thesaurus. When clicked on an icon, the tree expands to its next level; when clicked on a term its relations are displayed in the right hand frame for editing.

5. Conclusions and further work

The exponential growth in the amount of archaeological data being published on the Internet has in recent years brought to the fore a more professional and structured approach to retrieving and indexing web resources. The use of a thesaurus to structure a body of information, as proposed and developed in prototype by the ArchTerra Consortium, is just one example of the way in which "metadata" are likely to transform our personal and professional use of the web.

The discussion, design and development of the prototype, of which this paper is intended to be a "snapshot", have brought to light a large number of issues which will need a follow-up. In no particular order these are:

- *Speed and Stability*. The prototype was developed on an HP 9000/180 workstation under a CERN HTTP server and Hughes mSQL 2.09; a more stable and fast platform will be needed for the development of an alpha version.

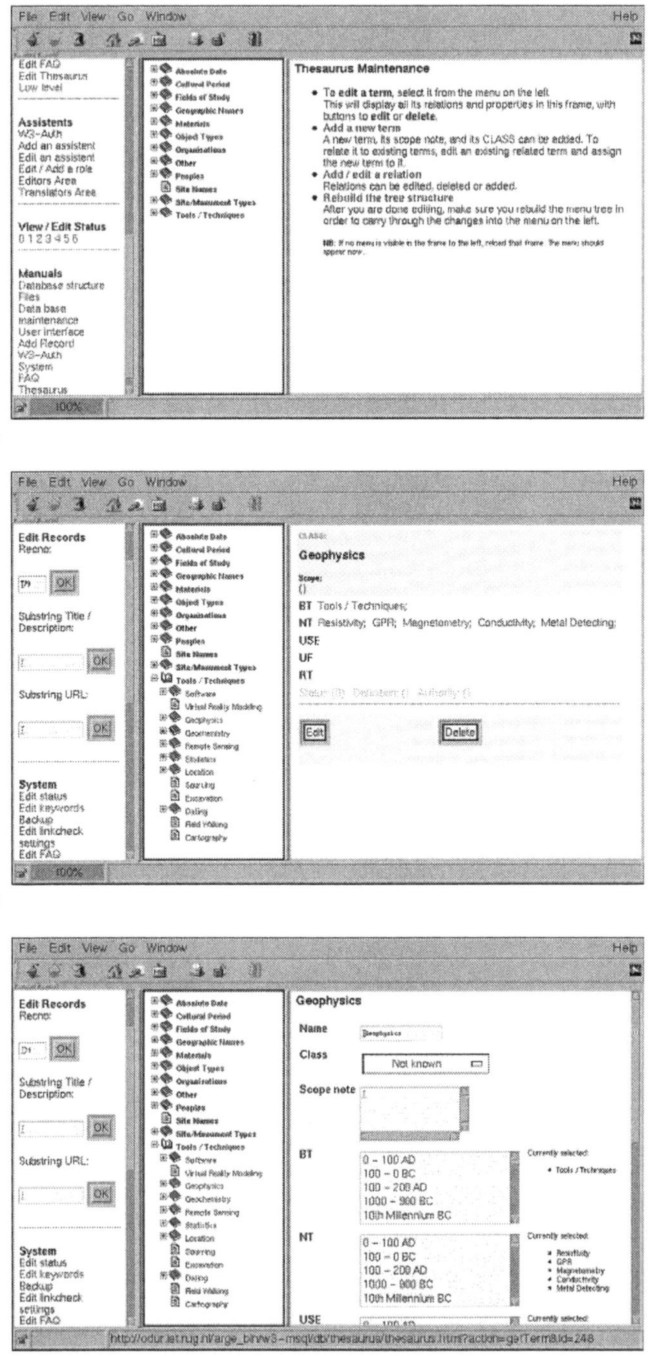

a)

b)

c)

Figure 4: Thesaurus maintenance page of ARGE (a), leading off to forms for adding or editing terms (not depicted) and for viewing (b) and editing (c) or deleting relations between terms.

- *Content*. Although the database structure underlying the thesaurus application is now in place, its current content is temporary and incomplete. In particular, relations between terms have only been inserted for testing purposes so far, and all content is in English. A priority for further work is therefore to complete the short (~250 term) thesaurus envisaged originally by the ArchTerra Consortium, and to verify that the prototype works in at least one other language. The medium of online discussion lists has been found to work well for this sort of work, which should preferably take place under the auspices of a professional organisation such as the Association of European Archaeologists. If an editorial board is established, it should have a forum and archive for discussing thesaurus maintenance, and for receiving proposals for new terms and relations from users. This implies that some sort of status tracking should also be implemented.

- *Interface design.* Little effort has gone into the on-screen layout of the prototype, and further work will have to include serious investment in studying the requirements of effective interfaces. The thesaurus environment is unlikely to be familiar to the general public, so extra care must be taken to remove unnecessary jargon, complex layout, etc.
- *Natural searching.* Among the potentially useful tools and techniques that can be explored to create a more natural user interface are "mapping" terms or sets of terms to others, parsing the user's search strings to allow orthographic and syntactical variation, and searching by example.
- *Trust and Authority.* Many archaeologists have expressed concern with the "power" wielded by an editor of a web guide and, by implication, of the editor of a thesaurus. The gist of my reply to such concerns (Van Leusen et al. 1996) has been to note that it is not a matter of power but one of trust or authority; any editor (whether traditional or online) has to earn, and can lose, the user's trust by the quality of his or her work. ARGE has piloted a system of remote online editors whose contributions are tagged with their identity, so that users can evaluate their authority if they wish to do so. Optionally, the quality of the editors could be further safeguarded by a professional accreditation system, for example with the Association of European Archaeologists (EAA).

Acknowledgements

In writing this paper, and especially the "design considerations" section, I have profited hugely from the discussions in the FISHEN e-conference on building a period thesaurus for the UK (January 2000); I have attempted to acknowledge all significant re-use of the ideas and formulations of the participants in footnotes but here would like to acknowledge their contribution as a group too. Previous versions of this paper were reviewed during another online workshop, this one organised by the ArchTerra Consortium (April 2000), and a workshop during the CAA2000 conference (Ljubljana, April 2000). Among the participants at the latter workshop, I would like to thank Martin Doerr especially for his useful presentation and comments. Further thanks go to Allard Mees for his regular e-mailed questions and criticism.

References

CHAMPION, S., 1997. Archaeology on the World-Wide Web: a user's field-guide, *Antiquity* 71: 1027-38.

INTERREG II C project "Co-operation in spatial planning" (2000-2001, no web site available yet)

TERM-IT project "Multilingual support for multimedia services" (1998-1999, http://www.mda.org.uk/term-it/welcome.htm)

Van LEUSEN, M., CHAMPION, S., LIZEE, J. and PLUNKETT, T., 1996. Toward a European Archaeological Heritage Web. In Kamermans, H. and Fennema, K. (eds.), *Interfacing the Past. Computer Applications and Quantitative Methods in Archaeology CAA95* (Analecta Praehistorica Leidensia 28), pp. 507-15. Leiden: University of Leiden.

WILL, L., 1999. Software for building and editing thesauri (http://www.willpower.demon.co.uk/thessoft.htm)

Notes

1 The World Wide Web Virtual Library for European Archaeology, at http://www.let.rug.nl/arge/ (since 1995).

2 For more information about this, go to http://www.purl.org.

3 996 - 1998, EC Telematics 2c program, at http://www.prosoma.lu/cgi-bin/show_new.py?id=4482& page=description and http://www.inria.fr.

4 For more information about this, go to http://archterra.cilea.it.

5 For further information about the prototype, including updates, go to http://www.let.rug.nl/arge/Work/thesaurus.html. The thesaurus prototype was developed at the University of Groningen by Robert Bouma, based on the mSQL database (version 2.09) and Perl (version 5).

6 See, for example, the RCHME thesaurus of Monument Types and the MDA thesaurus of archaeological objects (both at http://www.rchme.gov.uk/thesaurus/thes_splash.htm), the Bronze Age glossary produced by the Bronze Age Campaign project (Council of Europe, at http://www.cimec.ro/arh/bronz/), and the January/February 2000 archives of the FISHEN e-conference (at http://www.mailbase.ac.uk/lists/fishen/archive.html. For the "position paper" by Jeremy Oetgen which started off the FISHEN debate, see Appendix 1.

Appendix 1: Chronological terminology in the UK and Ireland

Appendices were written by Jeremy Oetgen, British and Irish Archaeological Bibliography, c/o The British Academy, 10 Carlton House Terrace, London SW1Y 5AH, United Kingdom, e-mail: j.oetgen@britac.ac.uk

Reviewed with reference to the British and Irish Archaeological Bibliography and the MIDAS standard.

Adapted from a deposition sent to the FISHEN meeting on 22 November 1999.

This contribution illustrates the diversity of chronological frameworks in use in archaeology in the British Isles and emphasises the need for an integrated scheme for the main periods that is explicit about the assumptions and evidence on which the chronology is based. It is suggested that a standard reference chronology should be set up with the participation of the whole archaeological community, along with those in related disciplines.

1. A conundrum

Did you know that both Julius Caesar and William the Conqueror landed in the south coast of England in the same period? Would it surprise you that period was the Iron Age?

2. An explanation

The RCHME periods list defines the start of the Romano-British period as AD43 - the Claudian invasion. Strict application of this "terminus post quem" would therefore classify Caesar's expeditions in 55BC and 54BC as Iron Age.

In England it is usual to accept that the Middle Ages proper began in 1066, when William defeated King Harold at Hastings. In Ireland, however, it is commonly considered that the Middle Ages did not begin until the late twelfth century AD, when Normans from England established settlements in Ireland, - furthermore, many Irish archaeologists talk of the period up to the Anglo-Norman invasion as Late Iron Age (see for example Laurence Flanagan's "A dictionary of Irish archaeology", published in 1992 by Gill & Macmillan). Thus the Norman Conquest of England could be referred to as an Iron Age event.

3. So what?

Traditionally, parallel, relative chronologies have been used in different regions to deal with the archaeological evidence as it exists in different geographical/cultural areas. We have learned to make allowances for this and we all know what we really mean by the terms Neolithic, Bronze Age, Iron Age etc, don't we?

But this relativism seems increasingly less rational as modern methods now enable archaeological evidence to be dated in much more absolute terms. Not only are scientific methods - such as radiocarbon determination and tree-ring analysis - providing ever more calendar dates with which to peg down chronologies, but increasingly sophisticated studies of artefacts are refining our ability to date occupation from cultural evidence. Integration of regional chronologies provides better opportunities for identifying cultural parallels and facilitates the study of interregional comparisons.

And, from an information specialist's perspective (as the example above shows) it cannot be assumed that we all know what we mean by the names we ascribe to archaeological periods - computers certainly don't!

At the British and Irish Archaeological Bibliography (BIAB) we analyse the content of publications relating to all aspects of archaeology from across the UK and Republic of Ireland; these publications emanate from a wide range of institutions of varying status - both professional and amateur - and many originate from abroad. We classify references according to topic, with the period being of primary importance - because experience tells us that is what people most often look for when searching for archaeological information. At present, when we ascribe a period classification to an article we need to take into account the different chronologies in use in the British Isles and consider carefully how researchers are going to expect to be able to retrieve information. If an Irish archaeologist describes a site as Iron Age, are they saying it dates within a range of c250BC to cAD1170? If so, is an English archaeologist searching for Iron Age references going to be interested in this site? Should a site in Orkney dated to AD200 be classed as "Roman", "Iron Age", or something else?

What if new evidence emerges to redefine the dating of a period? The discoverer of that evidence may consider it as incontrovertible and amend their own use of the period name immediately. It may be some time before the rest of the archaeological community learns of the new evidence or realises its significance and many will continue to use an old designation. Personal experience provides me with an example of this: at BIAB we had been using a scheme that gave the start of the Palaeolithic as 350,000BC; however, while reviewing the scheme as a prelude to this FISHEN discussion we realised that the recent hominid find from Boxgrove in Sussex had effectively pushed back to 500,000BP the earliest evidence of human occupation (and thus the start of the Palaeolithic) in Britain.

The Boxgrove example also illustrates how modern dating methods have generally transposed the prehistoric periods back in time. Early archaeologists and antiquaries conceived of the prehistoric past as being much more recent than we now know it was - and that means we need also to consider how our intellectual predecessors constructed their chronologies.

4. Proposal for the creation of a standard reference chronology defining archaeological periods

I hope that what I have just said highlights the need for a standard reference resource that provides authoritative definitions of the key periods (respecting regional and international needs). This resource should also be informative, identifying and explaining the evidence on which the dating is based. To be authoritative the resource must be inclusive of all main chronological systems used in the UK and, preferably, the whole of Ireland.

I would like the reference chronology should include dates for:

- Documented events, e.g. battles/invasions; biographical dates; kingships; documented inventions.
- Periods of currency of monument types and artefact types, certainly for those most representative of their respective periods.

- Cultural and technological innovations, e.g. introduction of metallurgy.

5. Who would use the reference chronology?

A standard archaeological reference chronology would be a valuable tool for a wide range of users (professionals, amateurs, and school and university students) and it ought potentially be of interest to researchers in related disciplines and have a variety of applications. The reference chronology should therefore be made available in a form and medium that would facilitate its use by the widest range of users.

6. Practical procedures for developing the reference chronology

I envisage that the reference chronology would be developed in the following five stages:

1. Collect information on the chronological schemes currently in use by archaeologists in the British Isles. (Out-of-date schemes might also be included for the benefit of anyone working on retrospective data.)
2. Collect any additional chronological data that are relevant (kingship lists and dates for key historical events etc)
3. Collate the received schemes
4. Establish and agree the defining characteristics of each period, based on collation of the received schemes.
5. Define the date ranges for each main period by encapsulating the accepted dates for all the component characteristics.

To maximise the cross-disciplinary potential of the reference chronology the participation should be sought of historians, palaeoclimatologists, environmental historians and experts in other related fields. Data on various chronological schemes could be collected and entered onto a simple database, a possible field structure for which is set out below.

7. Suggested field structure for recording chronological data

Fields are set out below. TPQ = terminus post quem [point after which - i.e. earliest possible date]; TAQ = terminus ante quem [point before which - i.e. latest possible date]. The dates are given in ISO format. Day and month should be included to allow for the possibility of these being known (however remote this may be!) Zeros could signify blank in these fields - although too many noughts could encourage typos! The note fields allow for source to be cited, but it might be worth setting up a separate table of sources. Both TPQ and TAQ are required for Start and End dates

Term	Code Date	Max Date	Min Order	Chronol. Uid	Term Uid	Parent Term
LOWER PALEOLITHIC	LPA	0	0	1	54	5
EARLY PREHISTORIC	EPR	0	0	1	95	0
LATER PREHISTORIC	LPR	0	0	1	94	0
PALAEOLITHIC	PA	0	0	1	5	0
PREHISTORIC OR ROMAN	UP	0	0	1	26	0
PREHISTORIC	PR	0	0	1	52	0
MIDDLE PALAEOLITHIC	MPA	0	0	2	55	5
UPPER PALAEOLITHIC	UPA	0	0	3	56	5
EARLY MESOLITHIC	EME	0	0	4	57	6
MESOLITHIC	ME	0	0	4	6	0
LATE MESOLITHIC	LME	-4000	-7000	5	58	6
NEOLITHIC	NE	-2200	-4000	6	7	
EARLY NEOLITHC	ENE	-3000	-4000	6	59	7
MIDDLE NEOLITHIC	MNE	-2700	-3500	7	60	7
LATE NEOLITHIC	LNE	-2200	-3000	8	61	7
EARLY BRONZE AGE	EBA	-1500	-2500	9	62	8
BRONZE AGE	BA	-700	-2500	9	8	
MIDDLE BRONZE AGE	MBA	-1000	-1600	10	63	8
LATE BRONZE AGE	LBA	-700	-1000	11	64	8
IRON AGE	IA	43	-800	12	9	0
EARLY IRON AGE	EIA	-400	-800	12	65	9
MIDDLE IRON AGE	MIA	-100	-400	13	66	9
LATE IRON AGE	LIA	-43	-100	14	67	9
ROMAN	RO	410	43	15	10	0
EARLY MEDIEVAL	EM	1066	410	16	11	0
EARLY MED. OR LATER	UM	0	0	16	53	0
MEDIEVAL	MD	1540	1066	17	28	0
POST MEDIEVAL	PM	1901	1540	18	16	0
MODERN	MO	0	1901	19	24	0
UNCERTAIN	UN	0	0	20	25	0

respectively in case of, say, the end of the Roman period where there is debate about how long the Roman way of life persisted.

Record fields

Name (of period/historical event):
Note on name:
Start TPQ: (ISO standard format YYYYMMDD)
Note on Start TPQ:
Start TAQ: (ISO standard format YYYYMMDD)
Note on Start TAQ:
End TPQ: (ISO standard format YYYYMMDD)
Note on End TPQ:
End TAQ: (Iso standard format YYYYMMDD)
Note on End TAQ:

Illustrative examples of completed records

Name: Bronze Age
Note on name:
Start TPQ: -23000000
Note on Start TPQ: established proprietary convention
Start TAQ:
Note on Start TAQ: none
End TPQ:
Note on End TPQ: none End TAQ: -7000000
Note on End TAQ: established proprietary convention

Name: Romano-British
Note on name: period of Roman occupation of Britain
Start TPQ: 00430000
Note on Start TPQ: literary source refs...
Start TAQ:
Note on Start TAQ:
End TPQ: 04100000
Note on End TPQ: literary source refs...
End TAQ: 04500000
Note on End TAQ: assumes persistence of local Roman administration

Appendix 2: Building a European thesaurus for archaeological periods

Many organisations have their own practice and "standard" when it comes to archaeological dating. The Royal Commission on the Historical Monuments of England provides the following list (in use in the English Heritage National Monuments Record database), which dates to 1993 and includes absolute dates and BT/NT relations (© RCHME 1993).

These and other existing documents were discussed in the FISHEN e-conference, following which a draft thesaurus of period terms for the UK was created, of which the following is a small extract (© English Heritage 2000). After review this thesaurus will become part of the UK "Inscription" standard.

Deverel Rimbury Culture
USE MIDDLE BRONZE AGE

EARLY BRONZE AGE
UF Beaker Culture
Wessex Culture
SN The first phase of the Bronze Age, associated with the more widespread use of round barrows, and the first manufacture and use of copper and bronze tools, weapons and ornaments.
CL CULTURAL PERIOD
BT BRONZE AGE

EARLY IRON AGE
UF Hallstatt Culture
CL CULTURAL PERIOD
BT IRON AGE

EARLY MEDIEVAL
CL CULTURAL PERIOD
NT ANGLIAN
ANGLO SAXON
NORSE
SAXON
RT 10TH CENTURY AD
11TH CENTURY AD
5TH CENTURY AD
6TH CENTURY AD
7TH CENTURY AD
8TH CENTURY AD
9TH CENTURY AD

EARLY MESOLITHIC
SN The earlier, post-glacial phase of the Mesolithic, characterized by the use of broad blade industries, and showing clear continuity from the preceding final stages of the Upper Palaeolithic.
CL CULTURAL PERIOD

LATE MESOLITHIC
SN The later phase of the Mesolithic, characterized by the use of narrow blade industries, with smaller and more varied microlith forms. It was during this phase that Britain was physically separated from continental Western Europe.
CL CULTURAL PERIOD
BT MESOLITHIC

LATE NEOLITHIC
UF Beaker Culture
SN The final phase of the Neolithic, associated with the construction and use of henges, and increasingly diverse (and regional) styles of pottery, lithics, etc.
CL CULTURAL PERIOD
BT NEOLITHIC

LOWER PALAEOLITHIC
UF Acheulian
SN The earliest phase of the Palaeolithic, characterized by the use of bifacial tools, such as handaxes.
CL CULTURAL PERIOD
BT PALAEOLITHIC

Maglemosian
USE MESOLITHIC

MEDIEVAL
UF Middle Ages
CL CULTURAL PERIOD
RT 11TH CENTURY AD
12TH CENTURY AD
13TH CENTURY AD
14TH CENTURY AD
15TH CENTURY AD

Even if all countries of Europe were to agree on such nation-wide standards (and not just for period terminology but also for other classes of archaeological terms), the next step of merging such professional thesauri across national, cultural, and language borders will take a huge effort by specialists. The ArchTerra Consortium has undertaken to explore these issues using the following restricted set of period terms for eastern Europe.

ID	Term
1	Ariușd
2	Aurignacian
3	Basarabi
4	Biskupin
5	Bodrogkeresztur
6	Boian
7	Bronze Age
8	Cucuteni
9	Culture
10	Dacian-Getian
11	Gumelnița
12	Hallstatt
13	Hamangia
14	Iron Age
15	Karanovo
16	La Tene
17	Mesolithic
18	Monteoru
19	Neolithic
20	Otomani
21	Palaeolithic
22	Tiszapolgar
23	Urnfield Culture
24	Vinca
25	Vucedol

ARCHAVE: A Virtual Environment for Archaeological Research

Eileen L. Vote

The Shape Lab, Engineering, Brown University

182-184 Hope Street, Box D, Providence, RI, 02912 USA

e-mail: vote@lems.brown.edu

Daniel Acevedo and David Laidlaw

Computer Science Department, Brown University

115 Waterman St. 4th floor, Box 1910, Providence, RI, 02906 USA

e-mails: daf@cs.brown.edu, dhl@cs.brown.edu

Martha Sharp Joukowsky

Center for Old World Archaeology and Art and Department of Anthropology, Brown University

Box 1921, Providence, RI, 02912 USA

e-mail: martha_joukowsky@brown.edu

Abstract

We will present an interactive system to perform archaeological analysis with site features, topography, architecture, artefacts and special finds from the Brown University Excavations at the Great Temple site in Petra, Jordan. The system is significant because it allows a user to interact with three dimensionally referenced excavation data in a CAVE - CAVE Automatic Virtual Environment (Cruz-Neira 1993) (a 3m x 3m room where users are immersed in a virtual environment through stereoscopic projection on three walls and the floor). Through user studies, we will investigate the types of analysis archaeologists can perform in the system and compare it to standard analysis methods using a database and maps of the site and excavation trenches.

Key words: virtual reality, CAVE, 3D database, spatial analysis, GIS, reconstruction

1. Introduction

The SHAPE Lab was created with a grant from the United States National Science Foundation in 1999 in an effort to develop scientific tools useful in archaeological site and artefact reconstruction and analysis using data from the Brown University Great Temple excavations in Petra, Jordan[1] (figures 1, 2). It is a multidisciplinary endeavour and involves the departments of Computer Science, Engineering, Applied Math, Old World Archaeology and Art and Anthropology at Brown University.

One of the key projects that the lab has been collaborating on is ARCHAVE, a software system that utilizes a virtual environment as an interface for archaeological research and analysis. We will present the system we are currently developing and user studies we plan on conducting to test the functionality of the system in several virtual reality environments.

2. Archaeology

The basic goal of the excavation process is to collect and analyse as much physical evidence as possible to answer questions about the site and related culture. However, one of the biggest problems the archaeologist faces in completing this task is that of cataloguing and storing the plethora of artefacts removed during the excavation process. Later, during analysis proceedings, modelling aspects of the data and establishing necessary relationships between architectural finds, stratigraphy, site features and different artefact types is also problematic.

The database of finds for the Great Temple excavation contains more than 200,000 entries, recovered during excavations that commenced in 1993. On site, architectural finds are surveyed in their in situ positions. Artefacts are located and recorded in the site database in their relative positions by loci/trench with a number of feature attributes such as object type (bone, pottery, coin, metal, sculpture, etc.), use, colour, size, key features, date given, etc. In some cases, architectural finds, sculpture and other objects that are particularly notable are photographed and drawn for further analysis and comparison with other objects.

3. Current analysis method

As the excavation proceeds there is a strong need to correlate all the objects to observe patterns within the data set and perform standard analysis. Methods for this type of analysis vary widely depending on site features, excavation strategy and data.

A quantitative analysis of all materials grouped and sorted in various ways was employed in The Great Temple five-year report published in 1998. The generated reports show statistics about the percentages of different artefacts and their find locations i.e., "pottery by phase, pottery by area, frequency of occurrence of pottery by area," etc (Joukowsky 1998). This type of analysis is informative because it can help archaeologists do a variety of statistical analyses using fairly comprehensive information from the database. It can also allow the archaeologist to quantify obvious patterns within the data set. Unfortunately, there are many factors that cannot be represented well in a traditional database and in reports generated from it. Specific attribute data, location and re-

Figure 1: Aerial of the Great Temple site in Petra, Jordan.

Figure 2: Aerial showing the Temple Proper area of the Great Temple precinct after the excavation season, 1999.

lational data between artefacts and site features cannot be represented in this method. For example, architectural features that have physical clues (stucco finishes, construction irregularities, evidence of building additions etc.) may be useful only when viewed in a context with associated finds and site features.

Besides obvious conclusions that can be made when objects are correlated spatially, combinations of artefacts when viewed by a trained eye, together in their original spatial configurations, can yield important and unlikely discoveries. Lock and Harris suggest:

> "Vast quantities of locational and thematic information can be stored in a single map and yet, because the eye is a very effective image processor, visual analysis and information retrieval can be rapid (Lock and Harris 1992)."

Although the idea of processing information visually would seem to be a more intuitive and thus effective way of processing 3D data, it has not been proven. More graphical methods of analysis have been explored in GIS systems that overlay multiple types of 2D graphic representations of data such as maps, plans and raster images together with associated attribute data in an attempt to present relationships between spatial data. However, it is strongly suggested that GIS systems are not sophisticated enough to provide a thorough description of z dimension (height) relationships.

> "The spatial relationships between the artefacts, other artefacts, site features, other sites, landscape elements and environmental aspects present a formidable matrix of alternative individual categorizations and cross-combinations to be searched for information (Lock and Harris 1972)."

A system or method that can allow formal analysis that uses the data in all three dimensions and can link important attribute data may resolve some of the analysis issues noted.

4. What we propose

The Great Temple artefact and site recording methodology will allow analysis of the findings in three dimensions because it maintains the x, y, and z coordinates for its objects. Therefore, in the system we are developing, we will assess the usefulness of ana-

a) b) c)

Figure 3: (a) Users in the CAVE virtual environment navigating through the temple reconstruction. (b) User in the system showing the excavation trenches from all years with the architectural reconstruction of the temple. (c) Immersion in two excavation trenches with multivariate visualization of pottery concentrations and bone.

lysing three-dimensional characteristics and associated attributes of the objects and features of the Great Temple site. Also, following what Forte proposes (Forte 2000), we believe that a virtual environment will be particularly useful in helping the Great Temple researchers understand their data to develop new conclusions and hypotheses about the history and evolution of Nabataean culture.

5. The system

The ARCHAVE system displays all the components of the excavation in the context of an architectural reconstruction of the Great Temple or, if the user prefers, an in situ model of the architectural remains (figure 3a). Like the excavation site, the virtual site is divided into the grid of excavation trenches representing the different areas that were excavated over the last seven years (figure 3b). Each trench is modelled so that the user can look at the relative layers or loci the excavator established during the removal of debris in that trench. As the user dictates, information about artefacts can be viewed in the locations where they were found.

For example, a user can see bulk pottery finds in a trench or group of trenches. Bulk pottery finds will be indicated by a colour range where the darkest colour represents the highest concentrations throughout the site (figure 3c). In addition to viewing bulk pottery finds, it may be helpful to see relevant coin finds, bone, glass, stone, metal, architectural fragments or special finds also. Additional object types such as bone shown with pottery are defined by range of texture along with the existing colour range. As new information is added, the user can start to form hypotheses or make conclusions about the relationships between the artefacts and the architecture or site that could not be made otherwise.

A task we envision for the system is to establish the building chronology or phasing of the Great Temple by using artefact attributes such as relative dates and grouping them with architectural elements and other site features. We believe that such a task can be achieved better in the system because it is easier to associate objects in all three dimensions. Therefore, objects that cannot be related in 2D map-based GIS systems can be accommodated here. We will test the user's ability to do such a task by an interactive method that combines queries of the artefact and architectural fragments database.

As the system is developed, we hope to integrate functions that will allow the user to mark areas and elements that have been analysed for a cumulative effect. When the user comes to conclusions in the virtual site, it can be annotated to aid the user later or to key other users.

In a fully developed environment, a researcher will be able to study the site by navigating through different stages of the excavation or important historical phases of the building. He or she will also query the database of artefacts using speech recognition, gesture-based commands, or automatic query generation depending upon the state of the user in the virtual site.

6. User studies

As Doug Bowman proposes, In an effort to compare results from the system and analyse the usefulness of various VR interfaces, we are designing user studies that utilize the following virtual environments: a Cave, Barco Baron table, head-mounted display and the desktop (figures 4-6) (Bowman 1999).

In testing this application we hope to answer the following specific questions: What virtual reality environment, if any, performs better in giving archaeologists the adequate interface and contextual information they need for analysis? What context is necessary for performing archaeological tasks? Which interaction techniques allow the user to navigate through an archaeological site and access a database of artefact information? How do we display the results of those queries in a way that he or she can gain maximum insight about the data?

We will test the system by allowing users to view and interact with the find information in the VR environments. As they do

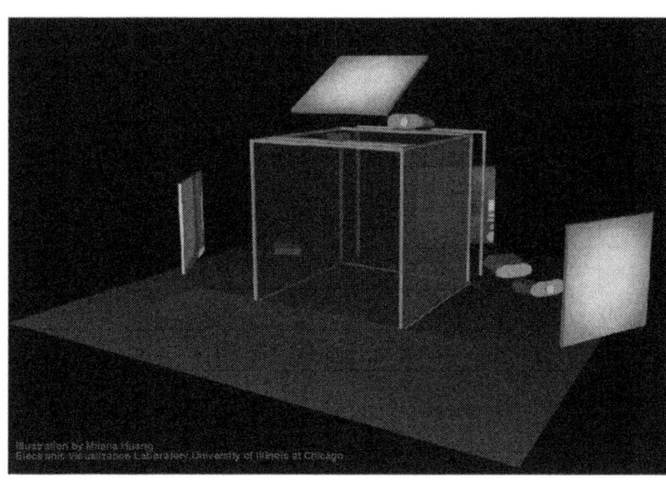

Figure 4: CAVE virtual reality environment.

a) b)

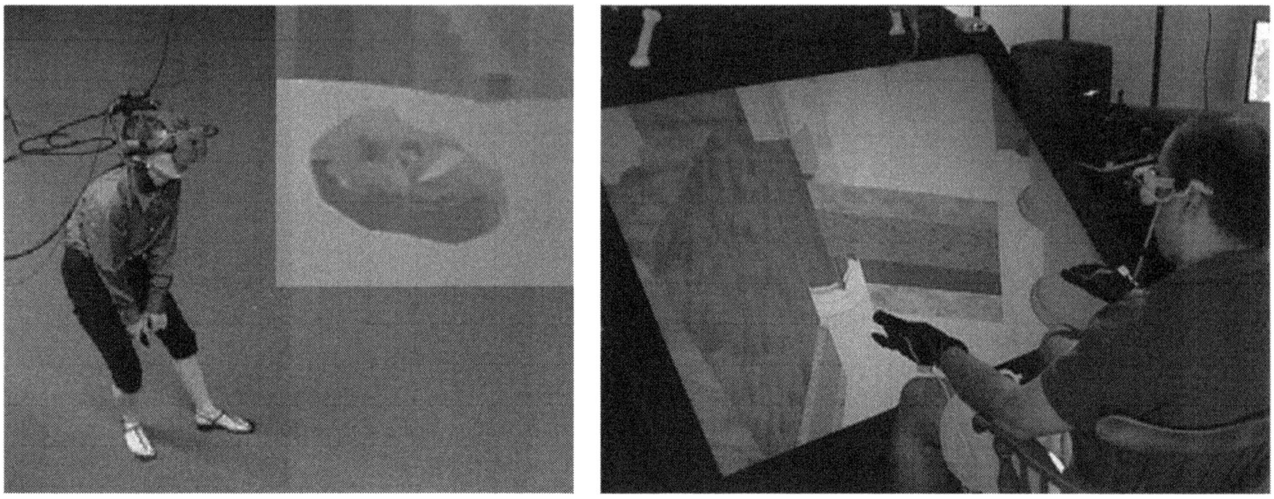

Figure 5: (a) User using a head mounted display to investigate artifacts in the system. (b) User navigating through trenches using the Barco Baron workbench system.

this, they will be asked to answer basic questions about their findings in the system. Users will also be asked to attempt to answer similar questions using the traditional site database and maps available in Filemaker Pro on the desktop.

From a number of informal tests and demonstrations, we have observed that users get a good sense of immersion in the system and those who have visited Petra report that using the system is similar to being at the actual site. However, we have also observed that, because the Great Temple is large and they are navigating inside the system at full scale, users want to look up to see important parts of the edifice. In the CAVE this is problematic because the ceiling is not a display surface. Another problem we have detected is that scale is hard to convey accurately. We are currently developing a pilot user study to compare how important this factor is in the different platforms and how it affects the archaeologists studying the site.

Figure 6: The desktop VR system.

7. Conclusions

The system we have described is significant because it allows archaeologists to preserve and visualize the data they collect in three dimensions. Implementing the system in a VR environment allows archaeologists to better understand the context of the excavation data and correlate the large quantity of artefacts with features of the site and architecture.

Through user studies we will gain insights into the way users interact with ARCHAVE and the differences between analysis performed in the system and more standard analysis using available databases, maps, photos and drawings.

References

BOWMAN, D.A., 1999. *Interaction Techniques for Common Tasks in Immersive Virtual Environments - Design, Evaluation and Application*, PhD Dissertation, Georgia Tech.

FORTE, M., 2000. *About Virtual Archaeology: Disorders, Cognitive Interactions and Virtuality*, BAR International Series 843, Archaeopress, England: 247-259.

JOUKOWSKY, M.S., 1998. *Petra Great Temple: Volume I: Brown University Excavations 1993-1997*, E.A. Johnson Company, USA: 245.

LOCK, G. and HARRIS, T., 1972. A Provisional model of an Iron Age society and its settlement system. In Clarke, D.L. (ed.), *Models in Archaeology*, London: Methuen: 801-869.

LOCK, G. and HARRIS, T., 1992. Visualizing spatial data: The importance of Geographic Information Systems. In Reilly, P. and Rahtz, S. (eds.), *Archaeology in the Information Age. A global perspective*, Routledge, London: 81-96.

Note

1 Any opinions, findings, and conclusions or recommendations expressed in this paper are those of the author(s) and do not necessarily reflect the views of the United States National Science Foundation.

Web Access to an Archaeological GIS

Andrea D'Andrea
CISA – Istituto Universitario Orientale
Vicoletto I, S. Maria ad Agnone 8, 80139 Napoli, Italy
e-mail: dandrea@iuo.it

Franco Niccolucci
University of Florence
Via Pisana 56, 50143 Firenze, Italy
e-mail: niccolucci@unifi.it

Marco Crescioli
Unirel SRL
Via Volturno 12, 50019 Sesto Fiorentino (FI), Italy
e-mail: marco@unirel.it

Abstract

After examining the impact of publishing data on archaeological theory and methods on the Internet, we propose a client-server solution characterised by low costs and high processing power and which is at the same time easy to use. On the client side it requires only a commercial browser with no particular computer skill of the user. On the server side it is based on the Linux operating system with some freeware packages, integrated by a module created by us. This system, developed around an archaeological case study, is accessible on the Internet and has been conceived to host additional archaeological GIS applications.

Key words: GIS, Internet, remote access

1. GIS, theoretical issues and new ways of communication

The introduction of information technology in archaeology represents an element that is changing important aspects of the archaeological theory (Limp 1997, Moscati 1998, Voorrips 1997, Crescioli et al. 2000) such as, for instance, the investigation potential of 3D landscape using viewshed and cost surface analysis. Creating a new space-time framework, virtually constructed on the computer, has had important consequences both on the methodology and epistemology, contributing to the outline of what has been called a new perceptive metaphor of space (Wheatley 1993, Gillings 1997, Bampton 1997).

The use of GIS has already oriented the research in new previously unavailable directions, determining an innovative cognitive perspective of the landscape, which traditional methods could not explore (Barceló et al. 2000); using alternative economic, social and political models for the cultural investigation of the territory contributes to the social explanation of archaeological facts more than a simple description of the geometry of observed structures (Harris and Lock 1995).

However, notwithstanding significant advances, the use of GIS is not yet uniform. This is caused by different study and research backgrounds (arising from different traditions), which sometimes cause the unawareness of the methodological value of this new tool. Probably only a new and different way of communication between computer applications may defend computerised solutions (especially the GIS system) from a tenacious and resistant academic scepticism that prefers – at least in Italy – to consider the computer role as a modern and sophisticated "household appliance", ignoring the important consequences of its use on the theoretical and methodological side.

An obstacle in this evolution is presented in the limited visibility of GIS applications, which, so far, remain a theoretical subject of conferences instead of becoming a widely diffused technology. Every team, oriented into the construction of its own specific application instead of comparing the existing solutions, starts a new project ignoring the choices, methods and systems already adopted by others. Such behaviour should not be ascribed only to the limited co-operative attitude of different teams, but also to the limited circulation of existing applications. This results in the fact that most researchers ignore the procedures and methods adopted in individual solutions. An interesting recent debate on GIS and archaeological theory has therefore been confined to an abstract level, making it impossible to compare systems, and leaving only the comparison of system abstractions.

A necessary condition for this is the availability and visibility of applications: only thus, with direct access (even remote) to different solutions will it be possible to guarantee a real comparison, and avoid a mere quarrel between different schools and tendencies.

Even if it is not effective and manageable, including paper records in archaeological publications allows one to check data and to reconstruct the scientific process of discovery. In a similar way, the use of computers will significantly contribute to the dissemination of data, and not only the final synthesis and reconstruction. However, it is inconceivable that every researcher could learn to use the software programs necessary to understand the aims and strategies of another research project. This limitation in data availability and access underlines a reflection based on technical problems as well as on methodological questions: it is perhaps the absence of user-friendly interfaces that excludes most scholars from the debate. By limiting the participation to computer-skilled archaeologists only, GIS applications become the object of the study instead of being a tool which would ease archaeological investigations and ensure more widely spread data circulation.

Our approach, based on the integration of GIS into the WWW (World Wide Web), aims at going beyond the limits of their application to archaeology: the difficulty of accessing information without the software used to create GIS, and the need to know how to operate different programs, with learning curves that always require a substantial investment of resources.

Existing GIS-Web applications (for a list see, for instance, Web GIS and Interactive Mapping Sites, or Ploewe 1997) have, in our opinion, several drawbacks. The first drawback is the cost which prevents most archaeological budgets from using GIS-Web applications. Secondly, a large number of applications require considerable computer skills for setting up: the rare freeware ones pay little attention to unskilled use and are not completely bug-free. Perhaps this is why no archaeological application of this technology has as yet been proposed, at least as far as we know.

So far Web use for archaeological research is not satisfactory even as far as the transmission of data and scientific results is concerned. Web sites have turned from embryonic presentation places, in a synthetic form and with limited illustration features, into real archaeological portals. The latter undoubtedly have the merit of favouring the circulation of data and information, but are often unable to go beyond the "search engine" logic (Guermandi 1999, Hermon and Niccolucci 2000).

Recently the Internet has begun to have a new impact on archaeological research thanks to the availability of papers, monographs and journals. Using the Internet in order to improve communication between professionals concerns also the diffusion of reports of still unedited investigations. This is particularly true in specific conditions: unequal development in different regions is typical for Internet archaeology, therefore successful anglo-american research or organisational models such as ADS (Archaeology Data Service) cannot be exported without adapting them to the local needs, culture, tradition and "prejudices".

In these cases, the Internet is still considered only as a modern (and extremely powerful) technological channel for information transmission, often substituting the traditional scientific printed publications with its high costs, but with a similar logic. From this perspective, Internet publishing will produce an improvement in the diffusion of scientific knowledge allowing a faster and wider distribution of contributions otherwise destined to remain accessible only to the chosen few. However, users and information providers are still unaware of the potential use of the Internet not only as a communication medium, but also as a tool to share and interchange processed and raw data.

Publishing alphanumeric and geographical data on the Internet, both with open remote access (Internet) and with restricted access (Intranet) will have two important consequences on scientific cooperation (Bogdanovich et al. in press, D'Andrea and Niccolucci 2000, Hermon and Niccolucci 2000). On the one hand it will induce data re-use, since other researchers will be able to use the same data to start new investigations, possibly from different perspectives. On the other hand, the network availability of data, and not only of the results, will ease the reconstruction of the scientific discovery process, allowing everybody, at least in theory, to re-examine the results starting from material sources as represented by their records. This consideration would naturally introduce the issue of metadata, which is not the central theme of the present paper and would lead us too far from its aim. However, it is necessary to be extremely cautious in taking for granted that there is an agreed – and simple – generalised metadata structure. In our opinion, many countries have still a very low metadata awareness and their attention focuses on semantic structures, thesauri and dictionaries, created by single teams. Therefore, to avoid another metadata proposal, we chose to take the easy way out, allowing a discursive description of data and leaving the responsibility of the description level to the authors, with the only constraint that an English translation is to be supplied: an issue that receives little attention by English-speaking people, being considered as obvious, what in other countries is decidedly not.

2. Publishing a GIS on the WWW: the features

In this paper we present a preliminary step in the path outlined above. Nonetheless, it has some features that deserve attention and that we wish to stress.

- The software we use is freely available, state-of-the-art and powerful.

  It often happens that the search for software to be used in an application stops at the best seller, however this often proves later on to be limited and unreliable. The little effort required to learn a new package is returned in terms of reliability, power and money saving. This is perhaps in conflict with the supporters of the "household appliance" theory, who love off-the-shelf solutions that allow the most computer unskilled archaeologist to develop a computer application by themselves: unfortunately for them, some competence is required, here as everywhere else, in order to build something effective.

- The customisation requires only a few new software codes.

  This ensures that the product will be reliable, since most of it relies on long tested packages, and will ease maintenance and updates. Moreover the source code is publicly available (and can be easily read and understood), and everybody can check and hopefully improve it. The source codes of base packages are also available, so that they may be extended or otherwise changed if needed, for instance for linking additional libraries or modules.

- The user needs only a commercial browser to access data, that is to interrogate the GIS.

  This allows a wider spread and easier use, ensures platform compatibility and will benefit of future improvements in the technology of browsers. In our opinion, browsers will be the glue to link different applications, since they already encapsulate (and this will only improve in the future) the tools to manage different protocols of information storage and transmission, as, for instance, XML (eXtensible Markup Language), (Goldfarb and Prescod

1998) and LDAP (Lightweight Directory Access Protocol), (Loshin 2000).

- Data formats are de-facto industry standards.

  This reduces the cost of data migration from other applications and guarantees future conversion to new format standards.

- It has been conceived having archaeological applications in mind.

3. Case study

Starting from a previous experiment of remote access to an archaeological GIS, our case study considers the interrogation and the processing of alphanumeric and geographic data concerning the Etruscan cemetery of Pontecagnano (D'Andrea and Niccolucci 2000). Data was acquired during the investigation of a vast funerary area discovered by the expansion of the modern centre, and more than 8,000 burials were discovered, dating between the ninth and the third century BC. This vast amount of data has been stored and managed with a GIS, which aims at supporting scientific research, at improving administrative preservation activity and at the integration with the local Master Plan (D'Andrea 1999). This system is based on Mapinfo and consists of some thematic layers: modern cartography (at different scales: 1:10,000, 1:2,000); the cadastre (1:4,000); aerial photos; the limits of explored areas; and finally the tombs. Alphanumeric data concerning each tomb, such as its description, the osteologic remains and the goods found in the grave, are related to the graphical objects by means of a geo-relational model.

The first experiment was based on MapExtreme, a map server by Mapinfo based on Java classes. The prototype used a trial, limited time version with full functionality. This choice was immediately abandoned for the high cost (over 20,000 US$ at the time) and a development framework with low customisation possibility. However, it was an useful test to verify the potential of such an approach.

4. Publishing a GIS on the WWW: the application

The system we present uses the following software.

4.1. Server side

Operating system: Linux (SuSe distribution)
HTTP server: Apache with PHP3 module
GIS package: GRASS version 5.0

Apache is a well known web server, the most diffused in the world, and PHP3 is a scripting language used to generate HTML pages and to interface them to several packages as database management systems. PHP3 is included in the majority of Linux distribution.

GRASS is downloadable under the GNU licence from several sites in Europe and the USA. It is very powerful, has almost any conceivable function and now is endowed with a very nice GUI (Graphical User Interface) that substantially simplifies its use and makes it comparable with most other GIS. It also has an important feature for our aims, i.e. it works silently in the background obeying the command lines sent to him and sending the results to a "virtual" monitor, from which they can be picked and sent (after format conversion) to any other program, user's browser included. This is the mechanism we adopted to display maps, as explained in greater detail below. The format conversion is performed by a set of libraries included in the Linux distribution.

From this point of view, our solution derives from Grasslinks, the well-known University of Berkeley web Grass software. However, it has been completely re-engineered and greatly simplified, in order to be prepared for future improvements that may include the use of java applets and/or XML related archives instead of a relational database.

4.2. Client side

As already stated only a browser is necessary on the user side. The user gets HTML pages with maps included in them as GIF files, so storing the downloaded data for future use locally needs only the browser "Save as" function. The session history is also based on the "Back" and "Forward" browser buttons: this allows flawless navigation through the searches with no impact on the connection, simply using the cache memory of the browser. However, creating new maps, zooming and moving in the map window require a new computation, in order to always maintain the highest level of accuracy of the displayed map, which is re-calculated and re-drawn by the server using the original data. This is particularly important when zooming in: the resolution of the original raster or the precision of the original vector map avoids enlarging the displayed image based only on the poor screen resolution of the transmitted map.

Transmitted data is very "light" in terms of storage, so the system works quickly even with a limited bandwidth: a complicate map usually requires sending and image of no more than 30Kbyte.

4.3. Data formats

Raster maps may be stored in any of the standard raster formats, which can easily be converted to a web format (GIF, JPEG or PNG). Vector maps are archived as DXF files. Practically any GIS software has filters to export data to these formats, so data transfer from any previous application is easily performed.

4.4. The database

The database is based on PostgreSQL, a relational database management system (RDBMS) that can be linked to GRASS and interfaced by PHP3. PostgreSQL is included for free in the Linux distribution. This RDBMS is very powerful and flexible and may be customised according to the needs, introducing new functions and data types. The database will store alphanumeric data related to the case study and will also link the fuzzy database presented in this proceedings (Niccolucci et al. this volume) to its geographic information.

Connection to different kind of archives, such as free-text documents, is still under study, but there are plans to develop a prototype which links XML-encoded documents to geographic information contained in them within the year 2000.

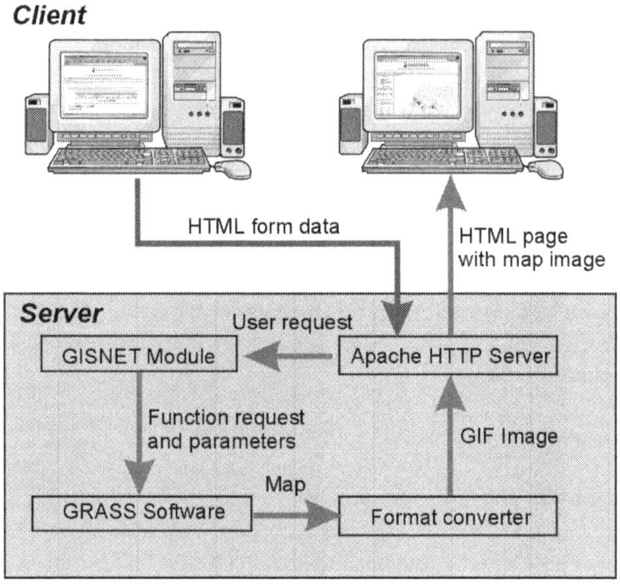

Figure 1: Functional scheme of the map server.

4.5. How it works

From a functional point of view, the system is very simple. It consists of a CGI module (a short Unix shell script) that intercepts the browser request to the server and asks GRASS to prepare the desired map. After converting the result to a web-compatible graphical format (GIF, at present), it passes the image to the HTTP server that sends it back to the browser together with the HTML page. Introducing new GIS functions thus requires only adding more CGI modules, similar to the existing one. The mechanism, shown in figure 1, relies heavily on the browser-server dialogue characteristics and is therefore very robust. Moreover, it may include restriction in a very simple way, for instance to limit access during investigation only to authorised users.

In detail, the definition of the task is performed by means of some HTML forms filled by the user, choosing options with the usual mouse clicking. This creates a string of parameters that are passed on to the server and activate the map creation

The menu page accesses information which is contained in configuration files or, as far as the map list is concerned, within the GIS itself, so the display is always automatically updated to the latest version of the GIS and variations in the GIS structure require no further maintenance.

Since the user has only to choose among the menu options, no computer skill is required.

5. Conclusions

In our opinion, this application should prove the feasibility of what we stated as the inevitable future of Internet use in archaeology, namely data sharing and co-operative research. However, we conceived it also as a service, addressing two categories of archaeologists.

The first one consists of information providers, to whom we offer a simple, inexpensive and effective way to publish their information. Neighter a science-fiction laboratory nor a computer guru is needed for that, and also a small budget can afford the cost of maintaining a web site with the above described tools. The second group is formed by the public body of archaeologists, who can thus benefit from a powerful tool for their research with a small investment of learning how to use the Internet, which nowadays they would be obliged to learn anyway.

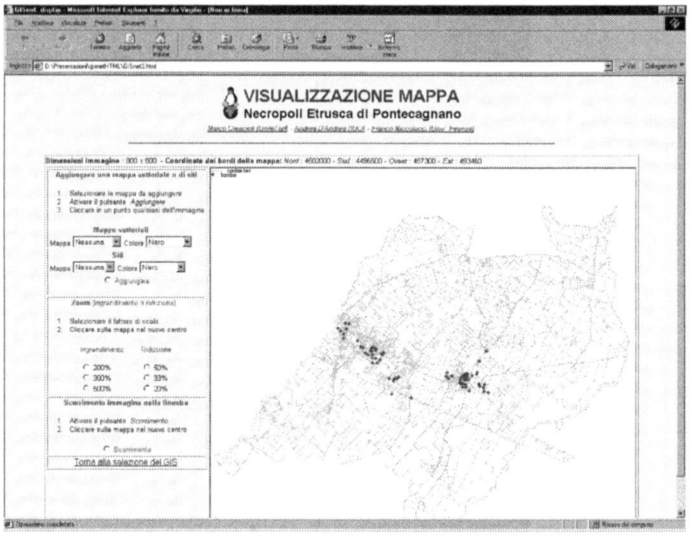

Figure 2: A snapshot of the screen showing the area of tombs superimposed on the cadastre map.

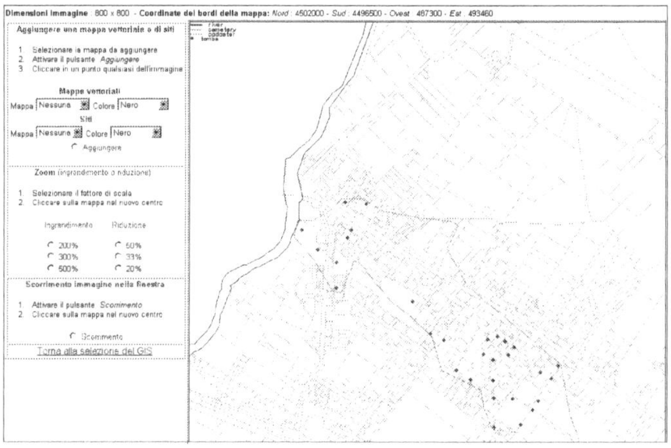

Figure 3: A zoom of the previous screen with the additions of the limits of urban cemetery.

References

BARCELÓ, J.A., SANDERS, D. and FORTE, M., (eds.), 2000. *Virtual Reality in Archaeology*, BAR International Series 843, Archaeopress, Oxford: 9-36.

BAMPTON, M., 1997. Archaeology and GIS: the view from outside. *Archeologia e Calcolatori* 8: 9-26.

BOGDANOVIC, I., VICENTE, O. and BARCELÓ, J.A., in press. A Theory of Archaeological Knowledge Building by using Internet: The Diaspora Project. *Proceedings of the CAA International Conference, Dublin 1999*.

CRESCIOLI, M., D'ANDREA, A., and NICCOLUCCI, F., 2000. A GIS-based analysis of the Etruscan cemetery of Pontecagnano using fuzzy logic. In Lock, G. (ed.), *Beyond the Map: Archaeology and Spatial Technologies*. IOS Press, Amsterdam: 157-179.

D'ANDREA, A., 1999. Il GIS nella produzione delle carte dell'impatto archeologico: l'esempio di Pontecagnano. *Archeologia e Calcolatori* 10: 227-237.

D'ANDREA, A. and NICCOLUCCI, F., 2000. *A Web based access to GIS integrating geographical databases through the WWW.* 2nd International Congress on "Science and technology for the safeguard of cultural heritage in the Mediterranean basin", 5-9 July 1999, Elsevier: Paris: 83-85.

GILLINGS, M., 1997. Not drowning but waving? – ReHumanising GIS, the Tisza floodplain revisited. In Bietti, A., Cazzella, A., Johnson, I. and Voorrips, A. (eds.), *Theoretical and Methodological Problems*, Proceedings of the XIIIth International Congress of Prehistoric and Protohistoric Sciences, vol. 1, Forlì: 69-84.

GOLDFARB, C.G. and PRESCOD, P., 1998. *The XML Handbook.* Upper Saddle River (NJ, USA): Prentice Hall PTR.

GUERMANDI, M.P., 1999. Dalle base dati alla rete: l'evoluzione del trattamento dei dati archeologici. *Archeologia e Calcolatori* 10: 89-99.

HARRIS, T.M. and LOCK, G.R., 1995. Toward an evaluation of GIS in European archaeology. The past, present and future of the theory and applications. In Lock, G. and Stančič, Z. (eds.), *Archaeological and Geographical Information Systems: a European Perspective*, 349-365. London: Taylor & Francis.

HERMON, S. and NICCOLUCCI, F., 2000. *The impact of shared information technology on archaeological scientific research.* CRIS2000 Conference, Helsinki April 2000, available at (as of 15/08/2000) ftp://ftp.cordis.lu/pub/cris2000/docs/hermon_fulltext.pdf

LIMP, W.F., 1997. Developing methodologies in the analysis of spatially referenced data and their impacts on archaeological method and theory. In Bietti, A., Cazzella, A., Johnson, I. and Voorrips, A. (eds.), *Theoretical and Methodological Problems*, Proceedings of the XIIIth International Congress of Prehistoric and Protohistoric Sciences, vol. 1. Forlì:115-125.

LOSHIN, P., 2000. *The Big Book of Lightweight Directory Access Protocol RFCs.* New York: Academic Press.

MOSCATTI, P., 1998. GIS applications in Italian archaeology. *Archeologia e Calcolatori* 9:191-236.

NICCOLUCCI, F., D'ANDREA, A. and CRESCIOLI, M., this volume. *Archaeological applications of fuzzy databases.*

PLOEWE, B., 1997. *Gis Online.* Information Retrieval, Mapping and the Internet Santa Fe, Onword Press

VOORRIPS, A., 1997. Archaeological theory and GIS, any relations? In Bietti, A., Cazzella A., Johnson, I. and Voorrips, A. (eds.), *Theoretical and Methodological Problems*, Proceedings of the XIIIth International Congress of Prehistoric and Protohistoric Sciences, vol. 1. Forlì: 209-214.

WHEATLEY, D.W., 1993. Going over old ground: GIS, archaeological theory and the act of perception. In Andersen, J., Madsen, T. and Scollar, I. (eds.), *Computing the past: Computer Applications and Quantitative Methods in Archaeology,* CAA92: 133-138. Aarhus University Press, Aarhus.

Documents available on the World Wide Web:

ADS (Archaeology Data Service), http://ads.ahds.ac.uk/ (15/10/2000)

Grasslinks, http://www.regis.berkeley.edu/grasslinks/ (15/10/2000)

Web GIS and Interactive Mapping Sites, http://sunsite.berkeley.edu/GIS/intergis.html (15/10/2000)

ArchTerra: Extending the European Archaeology Web over Bulgaria, Romania and Poland

Martijn van Leusen

Groningen Institute of Archaeology

Poststraat 6 9712 ER Groningen, The Netherlands

e-mail: P.M.van.Leusen@let.rug.nl

Andrzej Prinke

Poznan Archaeological Museum

Ul. Wodna 27, 61-781 Poznan, Poland

e-mail: aprinke@man.poznan.pl

Abstract

From the beginning of 1999 until the end of 2000, the ArchTerra[1] project (an International co-operation project under the EU Copernicus programme) has been active in the creation of an Internet infrastructure and content for Central European archaeology. This paper provides a brief overview of its goals and current results.

Keywords: European archaeology, Internet

1. Introduction

The networking and internationalisation of existing academic resources in archaeology, of the historical and cultural heritage, as well as of the research efforts of the academic community, using computer information and communication technologies is a recent trend with a rapidly growing impact on archaeological research, management and education. The reason for this is twofold: on the one hand, *the international character of the archaeological material itself*, and, on the other, *the ability of these technologies to overcome the difficulties of sharing these resources across the national borders*. One additional reason (although this development is still in its infancy) is the ability of these technologies *to facilitate group collaboration.*

Internet facilities are increasingly being used by professional archaeologists in the European Union for a variety of purposes; a review of the situation in 1995 has been published by Champion and colleagues (1996). Since then an informal but truly European virtual research network in the field of archaeology has grown up, which maintains its own virtual space of shared resources, discussion forums, international organisations, and electronic journals. The network was first established in late 1994, emulating its North American counterpart *ArchNet* which was co-ordinated at the University of Connecticut. The current state of European trends toward virtual research collaboration can be listed on the basis of the information technologies used:

- The Archaeological Resource Guide for Europe (ARGE, maintained at the University of Groningen) is the WWW "Virtual Library" for European Archaeology, providing one-stop access to over 1,500 sources of information.[2] It connects to other European servers acting as national hosts of the European Archaeology Web (known as ArchWebs), and allows the sharing of any type of resource over the Internet: public domain software, public presentations of information materials of common interest, educational and research materials in Archaeology, notice boards for public announcements, etc.
- Professional archaeologists, including the European Association of Archaeologists, have their own international mailing lists, which allow networking of researchers on a personal basis.[3] Many of these are currently hosted by the UK-based mail server *mailbase* at Newcastle (UK).[4] One specific list dedicated to the build-up of Europe-wide information services for archaeologists, is supervised by one of us (Van Leusen) at the University of Groningen.[5]
- International availability of research results is promoted by the publication of several electronic journals over the Internet, such as *Internet Archaeology* (based at York, UK), and the *Archaeological Research Repository* (at the University of Sheffield, UK).[6] These are based on maintenance of dynamic access to databases and files over the WWW.
- Public and professional access to archaeological objects and databases is promoted through virtual museums of archaeology, such as *the Virtual Magna Graecia* and *Virtual Pompeii* museums,[7] and by libraries such as the *AmberWeb* pages (Milano).[8] These employ contemporary technologies for 3D modelling, multimedia and interactive programming (HTML, CGI, VRML, and Java).

The informal nature of these networking activities has meant that their efficiency and effectiveness depend on, and are limited to, the services which happen to be made available by external sources. Unless more concerted efforts can be developed, activity remains limited to the relatively few archaeologists who can spare the time and have the interest. Crucial elements of a transnational archaeological information infrastructure, such as multilingual search engines and managed discussion lists, thus cannot be realised, and inequalities between countries are emphasised rather than removed. In addition, ignorance of the possibilities (potential or existing) of the WWW is still pervasive among many professionals, and this is compounded by the virtual absence of basic network access for archaeologists from Central Europe.[9] Although the Central European countries have an important place in the historical and cultural development of Europe, they therefore do not currently have a substantial participation in the exchange of information and expertise regarding European archaeology.

To help redress these failings, the ArchTerra project (an international co-operation project under the EU "Copernicus" programme), was formulated and successfully submitted for EU funding in 1998.[10]

2. Aims of the ArchTerra project

The ArchTerra project aims were to help redress the current imbalances in access to European networking facilities for these academics, and to provide the impetus for an active expansion of archaeological communication and information services both within Central Europe and between the EU and Central Europe through the Internet. Its objectives were:

- To establish the technical infrastructure and software tools needed to allow all researchers in the field of Archaeology to join a Europe-wide Archaeology Web, in the form of national WWW hosts of the ArchWeb network in the three participating countries of Central Europe (Bulgaria, Romania, and Poland).[11] These gateways are located at the major research organizations responsible for archiving, maintenance and supply of information in these countries. The project's aim was to make direct access to these facilities a reality for all researchers employed by the participating organisations;
- To provide practical demonstration of the trans-national nature and urgency of archaeological research and management, and the benefits and efficiencies of Internet use, to professional and general users alike. End users were to be able to access both the presently available on-line electronic resources and a core set of demonstration resources from Central European archaeology (including web pages, museum databases, live presentations and virtual exhibitions), and the creation of the latter was a core objective of the project.[12] In addition, a unified generic information system for Internet access to relational databases was to be developed jointly by programmers and museum workers.[13]
- To strengthen existing scientific relations between the EU and the countries of Central Europe and to foster long-term joint initiatives for collaboration, demonstrating the richness and fragility of the European archaeological heritage by bringing together partners and collections from across Europe. To this end, specific hurdles to international collaboration (dealing with different alphabets, creating multilanguage and multicultural thesauri of terms, compiling international heritage legislation) were to be explored.[14]

3. Project impact

International collaboration in the study and preservation of our common archaeological heritage, the goal of the ArchTerra project, has not been operating very well across Europe and hardly at all between European Union and Central European Countries. This is partly due to lack of (technical and financial) means, but of greater importance has been the lack of a focused drive to exploit modern information technology. In addition, even though it may be aware of the benefits IT could bring, archaeology as a profession does not have either the means or the technical know-how to bring this about on its own.

Figure 1: Project and INCO logos; caption to include project contract number 977054.

One important advantage of the relatively small size of the European archaeological research community is that it is possible for the proposed project to have a major impact on working practices. In effect, the project can diminish the current problems of lack of infrastructure and IT experience in Central Europe; the availability of a range of freeware WWW tools has already removed the problem of the lack of user friendliness in client systems.

The ArchTerra project activities may be grouped under the four headings of technical installation (hardware, software, networks), transfer of expertise (through workshops, visits, and discussion lists), creation of new content (data mining, software development, WWW publishing), and dissemination (conference papers, guides, web hosts). These activities already have allowed researchers to work more effectively from their own countries, contacting their national and international colleagues without cost or delay. In this, the project has closely followed the ideas and recommendations of the second CEC/EU Forum in Prague, thus in effect preparing the CEC for the *Global Information Society* in the field of Archaeology.[15]

The ArchTerra project has yet another valuable European dimension, in that it has opened for the rest of Europe some of the less well known achievements of the academic community in Central European Countries. Due to the present economic difficulties and the existing language barriers these have so far largely remained unknown outside specialist circles.

References

NEGRONI CATACCHIO, N., GUIDETTI, L., FERRARI, R. and Van LEUSEN, M., 2000. MEDIOLANVM. European Fieldwork server, CAA Conference, April 18-21, 2000, Ljubljana, Slovenia, Bollettino CILEA, vol. 73.

Van LEUSEN, M., CHAMPION, S., LIZEE, J., and PLUNKETT, T., 1996. Toward a European Archaeological Heritage Web. In Kamermans, H. and Fennema, K. (eds.), *Proceedings of the 23rd CAA conference, Leiden, 1995*, "Analecta Praehistorica Leidensia", vol. 28.

VASSILEV, V., STOEV, I., GAYDARSKA, B., ALEXANDROV, S., NEHRIZOV, G. and VAKLINOV, M., 1999. *Museum Information Systems: CIDOC data model implementation in the ArchTerra project.* Bollettino CILEA No.69, Settembre 1999, URL, http://www.cilea.it/~bollettino/69/datamodel.pdf.

Notes

[1] The ArchTerra project home page is located at http://archterra.cilea.it.

[2] ARGE is located at http://www.let.rug.nl/arge.

[3] The EAA is located at http://www.e-a-a.org.

[4] Mailbase is located at http://www.mailbase.ac.uk.

[5] The address of this list is http://www.mailbase.ac.uk/lists/eahw/.

[6] At http://intarch.ac.uk and http://www.shef.ac.uk.

[7] The Virtual Magna Graecia (URL: http://www.abramo.it/archeol.htm) and Virtual Pompeii (URL: http://www.ion.sgi.com) museums.

[8] AmberWeb pages at URL http://amberweb.cilea.it/.

[9] The number of different definitions of "Central Europe" is staggering and it would require too much space here to discuss them all in a politically correct manner; suffice it to say that it includes the countries of the former Eastern Bloc.

[10] A project summary can be found by searching on "archterra" on the EU information server CORDIS at http://www.cordis.lu; the full text of the proposal can be found at URL: http://www.cimec.ro/arheologie/archterra/inco7054eng.html

[11] Now active at the following addresses: http://www.muzarp.poznan.pl/archweb (Poland), http://www.archweb.cimec.ro (Romania) and http://archweb.nbu.bg (Bulgaria).

[12] All these products can be found via the project home page (see note 1).

[13] A full description of this system, developed according to CIDOC standards, was published by Vassilev et al. 1999 and can be found at URL: http://www.cilea.it/~bollettino/69/datamodel.pdf.

[14] This exploration has led to the development, among other products, of a prototype multilingual thesaurus management system (Van Leusen this volume) and of a multilingual glossary. Full descriptions can be found via the project home page.

[15] As laid down in the so-called Bangemann report, URL: http://www.ispo.cec.be/infosoc/backg/bangeman.html

8. Public Access to Archaeological Heritage

Using Virtual Reality to Improve Public Access to Heritage Databases over the Internet

Mike J. Pringle

National Monuments Record Centre

Kemble Drive, Swindon SN2 2GZ, United Kingdom

e-mail: mike.pringle@rchme.co.uk

Abstract

This paper presents a project, carried out at English Heritage's National Monuments Record Centre (NMRC), that has explored the potential of a public, three-dimensional interface (3Di) to present heritage material over the Internet. The paper begins with the project impetus, and the importance of pre-implementation research, before discussing some of the issues inherent in Information System (IS) development, especially the need to target specific users. The Human Computer Interface (HCI) is discussed, and the comparative benefits, and associated problems, of presenting information with graphics and text are outlined. The paper then presents an overview of the project, describing how a novel information architecture, which prompts the user to filter monument information by location, type or time period (Where, What and When), has been applied to a prototype 3Di. It then describes how text, icons and three-dimensional graphics have been utilised to improve interface usability, within the prototype. Finally, the paper outlines the 3Di's information output, and possible future developments.

Key words: human computer interface, virtual reality, databases, Internet

1. Introduction

The continuing development of the Internet, and associated technologies, is matched by an equally rapid growth in new business approaches to its use. In the Heritage industry this growth is dominated by a desire to make material more widely available, and a corresponding necessity to find practical, efficient and usable ways to do so. In English Heritage's case this desire is driven, in part, by a more general ethos promoted in a British government white paper, Modernising Government, which accepts the need for, and recognizes the benefits in digital delivery of government transactions (UK Government 1999). The British government's targets are essentially concerned with making all dealings (with the government) available across the Internet within the next few years. One of the major areas that such digital delivery will impact upon, is in the making available of greater quantities, and more varied types of information to more diverse, and greater numbers of people. The Internet offers all organisations, large and small, an opportunity to increase the dissemination of their information.

Unfortunately, because of the technological novelty, and the volume and complexity of information available, taking advantage of the opportunities afforded by the Internet is not as straightforward as it may first appear (Barrett and Maglio 1998). Computers in general, and Internet and World Wide Web (WWW) technologies in particular, are still at an immature stage of development. Furthermore, our understanding of the delivery of information, via these new mediums, is still at an early stage in the necessary exploration process. For those involved in this process it is easy to be drawn into the implementation of bottom-up, technological solutions, but, this approach in isolation will not always lead to the resolution of an information-centred problem. To enable useful mapping between any information problem area and the diversity encompassed by the current technological solution area, extensive research is required into both.

2. Developing information systems

For the development of usable, computer-based, Information Systems (IS), it is imperative that pre-implementation research is conducted into the available *information*, as well as the available delivery *systems*. In both cases, understanding of these elements, and of the influence they have on the way an appropriate HCI is designed, is dependant on the end-purpose of the IS: the user, or intended receiver of the information.

2.1. Information

Understanding of any specific information set necessarily requires understanding of information at a slightly higher level, where it can perhaps best be described in terms of its part within a larger, user-centred process. This process is illustrated, in simplified form, in figure 1.

O'Brien and Polovina (1997) give two definitions of information: the first suggests that information results from the emphasising of salient features of data; the second, that information is the organisation of data trends into a form that can be understood in a broader context, and communicated as potential knowledge to others. In both cases it is essentially the interpretation placed on *data* that transforms, or develops it into *information*. Information, by its nature, is primarily concerned with interpretation (Davies and Leclington 1991), and not the simple listing of facts, figures, measurements or recordings associated with data. Thus, for any given data-set to have value within an IS, it is important that an appropriate interpretation is placed on it. This can only be fashioned through an understanding of the end-purpose of the particular system, and, in all IS cases, this must be to supply information to the *user*. Creating a profile of who the end-users are, and what their requirements may be, can be easily overlooked or underplayed, but is a crucial part of IS development if any interpretation, that is imposed on the underlying data, is to be appropriate.

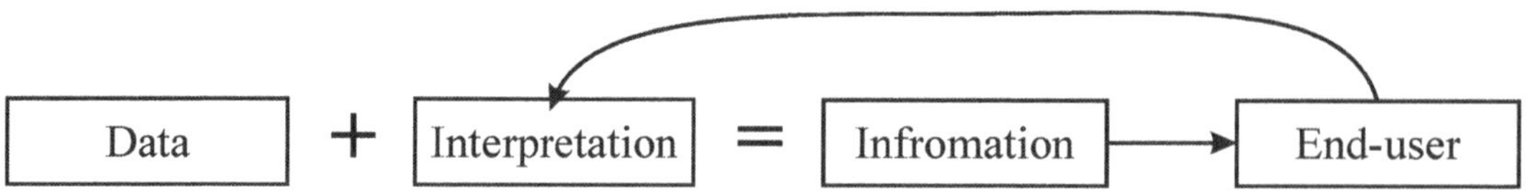

Figure 1: The simplified information process.

For small easily identifiable groups, the Internet can be used simply as a transmission device for very low-level, user-specific, information. However, as a vast IS it also offers the perhaps more attractive possibility of enabling large scale communication of information to much larger, but far less well-defined groups (Jackson 1997). Unfortunately, for these groups, user-requirements are inevitably much harder, or even impossible, to identify. Many current solutions to this problem seem to offer an "accessibility for all" policy where the information is made available on a website, with an interpretation, or interpretations, of the data intended to meet as many different user-requirements as possible. This apparent solution is restrictive in two ways:

- It relies on non-identified users "discovering" and visiting the website. This restriction is endemic across the Internet, and can probably only be resolved by concerted "advertising campaigns" to raise awareness for the site. However, because of the ill-defined nature of the target audience, directing any such campaign is obviously problematic.
- Where a large undetermined audience is assumed, only a small percentage of users (within this "audience") are likely to find the information appropriate to their needs. Furthermore, those users who believe the information to be of value to them, will almost certainly have to further interpret the material in order to fulfil their own individual requirements. This is because the interpretation placed on the data is inevitably going to be too specific for some users, and too general for others.

2.2. Systems

Both of the previously mentioned restrictions can be avoided to some degree if a genuine user-group is identified and targeted. For example, in an internal system designed for professional use, the HCI can elicit manifold, very specific, queries from a user, because a high degree of understanding, of both terminology and data structure, is assumed. Such a system can then return information very tightly aligned to the needs of the user. The interpretation of the data in this sort of case, can closely mirror the underlying structure, and utilize low-level, or narrow, terminology which enables specialist-user retrieval of information. However, as Tudhope et al. note, if the system is being designed to disseminate information to less specialized users, it cannot be assumed that they will understood either the terminology used, or the structure of the underlying data (Tudhope et al. 1995). One approach that demonstrates this problem, is the Internet search engine, which is designed to satisfy a multiplicity of users, searching a vast uncharted database, with an infinite number of different needs. Such systems are usually queried via high-level, broad keyword searches, but invariably return substantial quantities of information that bear very little relation to a user's requirements.

Whether the system is designed for a specialist or a general user, to enable successful information retrieval there is a requirement for him/her to have some understanding of the terminology and structure of the underlying data. The greater the user's understanding, the more successful his/her information retrieval is likely to be. A user wishing to discover the meaning of a collection of curvilinear earthen banks, ditches and large stones, must know what such a collection is labelled as, in any database, before he/she can access the available information. Much current Internet practice, to assist the user in finding underlying information, is dominated by lists of "suggestions" or classifications, that narrow down the query options. These can take the form of; for example, menus, buttons, icons, or hyper-text links, all relating to categories or sub-categories of the HCI designer's interpretation of the underlying data. Such categories are generally contrived in anticipation of the needs of assumed users, but it is very easy, especially when trying to disseminate greater quantities of information to ever-larger audiences, to anticipate too many options, and consequently develop an HCI that is too complex. Every button or link in a hyper-media system can be thought of as a question to the user: "Is this the category you require or are interested in?" The more questions the system asks, the more like sitting an exam the user's task becomes!

Ultimately the purpose of an IS, is to enable users to interact with, or navigate, complex information successfully. Users should, ideally, be able to communicate their requirements to the system with minimum effort (answering as few questions as possible). These requirements should then be met within a time-frame that is acceptable to the users.

2.3. Human computer interface

The HCI is effectively the "front-end" of any system designed to permit user interaction with, or control of, a single (or multiple) process, whether this be the operating of a manufacturing plant or the manipulation of numerical data (Inverso and Sokoll 1997). For Internet development, the HCI can be seen as the enabling communication facility between users and the information held in a computer (or collection of computers). The HCI provides a user with input devices for giving instructions or requests, and output devices to present responses or information (Murray 1998).

The HCI can be likened to a bridge between the user and information (Balint 1995), and the wider the bridge is, the less likely it is to become blocked or bottlenecked. In information engineering terms, the bandwidth of the communication channel between user and information needs to be as broad as possible, to permit maximum access and fluidity of movement through the information. Greater bandwidth is achieved by developing modes of communication (Sharma et al. 1998). This may mean advancing hardware capabilities, such as head-mounted displays and sensor-gloves, or improving the computer's ability to understand humans: voice, gesture, body position or even gaze (Billinghurst 1996, Brooks 1996, McKillip and Corona 1997). However, it can also mean improving the way information is presented. Brooks, whilst acknowledging the importance of sound and touch, proposes that our broadest-band channel is the sense of vision and of all the mechanisms for increasing communication between user and computer, it is graphics that relates most directly to the way we naturally interpret and communicate information (Brooks 1996, Pringle 2000).

Computer graphics may indeed increase communication between user and machine, but the use of text-based interaction is far too valuable an asset to discard. Graphics and text are both powerful modes of communication and it would seem sensible to assume that if they could be integrated in a way that utilized the strengths of both, then a highly usable HCI could be developed. However, a major challenge to the HCI designer, is to avoid the potential pitfall of clogging up the screen with an unmanageable quantity of pictures, icons, advice, buttons and links, in an attempt to give users maximum information about the underlying data. Given the range of users and the complexity of information now potentially available, the HCI is in constant danger of becoming overloaded, and therefore more difficult for users to understand or navigate. The project which this paper introduces, the English Heritage 3Di, has focused on a novel combination of graphics and text, intended to simplify the HCI in such a way as to improve the users' decision-making process.

2.3.1. Graphics

The development of computer graphics over recent years has meant new, and often dramatic ways of presenting information have evolved, including VR. In the earlier developmental stages of any new technology much of the research is dominated by technical issues, and it is only as the tools and working methods reach some sort of maturity that advantageous applications become apparent. This paper is not concerned with any technical advancement of VR, but instead suggests a novel application of the technology that is based on an acceptance of its current state of development.

VR comes in various forms but essentially, is a three-dimensional computer graphics capability which presents an interactive interface, with objects that can be viewed and manipulated in realtime. Warwick et al. (1993) define it as the "science of integrating man with information", emphasising the fact that computer-generated interactive virtual environments can be based on either real or imaginary factors, and suggesting that their purpose is the representation of information via synthetic experience. The idea of information being the core of VR is supported by Witmer et al. (1996) in their discussion of VR as a method of transfer of route knowledge, and by Machover and Tice (1994) who also add that VR facilitates the turning of information into knowledge. Putting the media-hyped image of VR (as a surreal, artificial world into which the participant is immersed via various futuristic gadgets) aside, VR has matured into a practical and powerful imaging tool for a vast number and variety of applications concerned with the transfer of information (Pringle and Moulding 1997).

As a high-quality three-dimensional computer graphics capability, VR enables pictorial representations to be created and displayed, with apparent depth, on a computer screen. The scope for designing worlds, and the objects within them, is limitless; real-world constraints, such as gravity, dimension or even common-sense, do not have to be applied. A VR world can represent the "real" conditions of a hospital operating theatre, a flight simulator or a racing car in a wind tunnel. Alternatively it can portray distant galaxies, or reconstructions of Pompeii just before Vesuvius erupts. Such boundless opportunity is available in any graphics medium, dependent only on the skill and imagination of the illustrator, but one of the core elements of VR, that separates it from other graphical mediums, is the level of interactivity that it offers to users. Theoretically, it is the ultimate in computer user-friendliness, allowing one or many people to interact with computer-generated objects and worlds in the same way that they would interact with the real-world (or other) equivalents. Within a VR "image" users can apparently fly to distant galaxies or, if they so wish, stand on the streets of Pompeii! The potential uses of such an interactive and intuitive medium for presenting information, are boundless, but there are essentially two approaches to current VR development: modelling the real-world, and abstract visualization.

Archaeology has already made good use of three-dimensional VR graphics in the reconstruction of long lost "real" buildings and artefacts, but it is not only concrete, or naturalistic elements of the real world that can be illustrated (Boyd 1998). VR illustrations, like any other graphical illustrations, are merely vehicles for elucidating, or clarifying, information to the user. It is clarification, not realism or accuracy, that is at the centre of any illustration, and it is critical to consider this when new technologies or methods for illustration are being developed. The main goal of an illustration must be to fulfil the needs of its intended user; to clarify the information in a manner, and at a level of abstraction, appropriate to the intended audience. On a map, an entire city can be abstracted down to the level of a black dot. This is sufficient for a user of a map of the world, but useless for someone requiring a town street-guide. Suitable abstraction of this kind is a necessity because, as Miller and Richards point out, it would be impractical for a map to show *all* the data (Miller and Richards 1995, Pringle 2000).

The 3Di prototype project uses three-dimensional VR graphical illustrations to assist in the dissemination of heritage database information. The overriding purpose of the graphics is to clarify users' understanding of the underlying database, and thus improve their ability to navigate the data as information. Such illustrations are not an attempt to replicate reality, nor are they records of any given site. Instead, they represent sets, or classes, of information. A user "flying" around a VR world composed of models like this is in fact moving through a multi-dimensional array of information. This approach combines aspects of both real-world modelling and abstract visualization. These can also be seen as major influences in two traditional archaeological illustration approaches: the representation of real-world form used in many reconstruction drawings and models; and, the abstract data approach inherent in maps, diagrams and other recording modes.

Representing abstract sets of information with pseudo-realistic, very familiar three-dimensional metaphors can be a powerful way to visualize large, and complex, database structures. The familiarity of real-world form is augmented by the opportunity for the user to move through the metaphors in a way that conforms, to some considerable extent, to natural human movement. Although, the user does not actually enter the metaphorical world, he/she can, with confidence, navigate a route through the information via the computer's input devices (mouse and keyboard). This enables the user to explore, or search, the available information, intuitively and productively.

2.3.2. Text

Brooks suggests that the human sense with the broadest information channel is vision, and that to exploit this fact we should be turning to computer graphics in order to improve information flow from the machine into the human brain (Brooks 1996). Three-dimensional, navigable graphics provide an extremely effective way of presenting information; a single pictorial metaphor can elucidate an enormous quantity of information. However, graphics is not the most commonly used mode for representing information across the Internet, nor is it necessarily the most efficient. Although computer games, and much contemporary science fiction, present an electronic culture of impressive graphics and virtual worlds, the Internet is, like other predominant communication mediums, dominated by language or more specifically, text (Glazier 1997).

Textual abstractions over a database are, or can be, very powerful mechanisms for enabling a user to determine specific sets or subsets of information. A series of high-level keywords, or categories, pre-determined by the interface designer, can help the user to get to the information that they require. This does have problems, as mentioned in section 2.2, in that it is very difficult to anticipate the level of understanding that the user will have in relation to the underlying data structure and categorization. Nonetheless, if a user-group is well defined and the categories are thoroughly thought out (and preferably tested with user-group representatives) they become a powerful aid in any IS. The vast majority of IS rely on this approach; for example, it has become an accepted way to enable user queries in many web-site structures. This approach is not really different to the way that text-based products have always been "queried". It is simply a development, or adapting, of the chapter headings, content pages and indexes that have long been used successfully in the printed book.

However, because textual abstractions were originally conceived for linear, book-style information, they can often be less appropriate for the multi-directional, hypertext structure of computerized "cyberspace". The huge quantity of information made available by new computer technologies means that new ways of dealing with information have had to evolve concurrently with the technology (Strawhorn 1981). The categorization approach can be further augmented, or even replaced, by an alternative text-based query mechanism that has evolved with computer technology: keyword searches. This approach allows a user to key in their own choice of words into a simple text box, and then receive a selection of the underlying data that pertains to those words. Again, as mentioned in section 2.2, this relies on the user having some knowledge of the underlying data structure or terminology (Ahlberg et al. 1992). Also, although the addition of Boolean logic (the use of AND, OR and NOT) can help to isolate very specific areas of the data-set, the user can waste much time and effort making queries that return either too many or too few matching items (Jones 1998). Nonetheless, the advantages of applying commonly accepted language (text) into the structure of an interface is indisputable. The difficulty is how to identify, and best employ, suitable terminology.

3. The English Heritage 3Di

Abstract representation of a database, whether it is through image or text-based abstractions, can only be of value if the overall interface is itself usable. Greater usability can be assured by either increased intuitiveness in the database structure and/or more intuitive information navigation methods built into the interface. The English Heritage 3Di project's approach has focussed on the latter, by developing a novel interface architecture that maximizes navigational usability whilst reducing the user's decision making process. The approach taken was based in part, on combining the strengths of both text and graphics, and as such offers the user new and potentially beneficial alternatives to information access (Ucko 1992). However, the benefits of combining graphics and text can only be fully exploited by determining a purposeful relationship between the two mediums, and by incorporating them into an interface that also includes fluid and intuitive information navigation mechanisms.

3.1. Prototype 3Di overview

The main purpose of the 3Di project was to develop a simple-to-use interface prototype, using, for experimental purposes, an established database relating to some 300,000 monuments in England. The prototype 3Di was designed an appropriate level of understanding, and therefore usability, for a conceptual "intelligent-twelve-year-old", and to present information predominantly through three-dimensional VR models and HTML text. The 3Di design also included text-labelling, iconographic navigation aids, and an automatic mechanism for presenting a contextual record of the user's activity. To utilise these various modes of presentation as fully as possible, and to increase HCI effectiveness and efficiency, the prototype 3Di project explored a novel IS approach based on clarifying, and reducing the number of, user-decisions.

3.1.1. HCI architecture

The design of the prototype 3Di architecture was based on a three-tier layering of information navigation. The top, or entry level of the prototype 3Di was designed to provide a contextual introduction to English Heritage's National Monuments Record Centre (NMRC), which supplied the underlying database for the prototype development. The "home" page of the 3Di prototype contained only an introductory paragraph and the English Heritage logo. Keeping the decision making process as straightforward as possible, at this point, a user would only have to decide whether to enter the system or not. By mouse-clicking on the logo he/she would be taken on a pre-programmed fly-through of a VR model of the NMRC, illustrated in figure 2, then into a room within the centre, and to a VR model of a computer monitor on a desk. This particular set of metaphors was chosen to reassure the user with familiar contextualisation; the user should not feel as though he/she is entering a complex computer system, rather that he/she is simply accessing a building that has useful, computer-based, information available within it.

The VR model, of the computer monitor, presented three iconographic VR models which comprised the core of the second level of information for the user. The three models were augmented with textual descriptions and labelling, and represented categories relating to monument *location (WHERE)*, *type (WHAT)* and *date (WHEN)*. These can be seen in figure 3. The principle behind such a choice of categories is radically different from the most common approaches to information categorisation currently found on the Internet. As mentioned in section 2.2, it is very common to present users with lists of classifications based on an interpretation of the underlying database, and in this respect, the three cat-

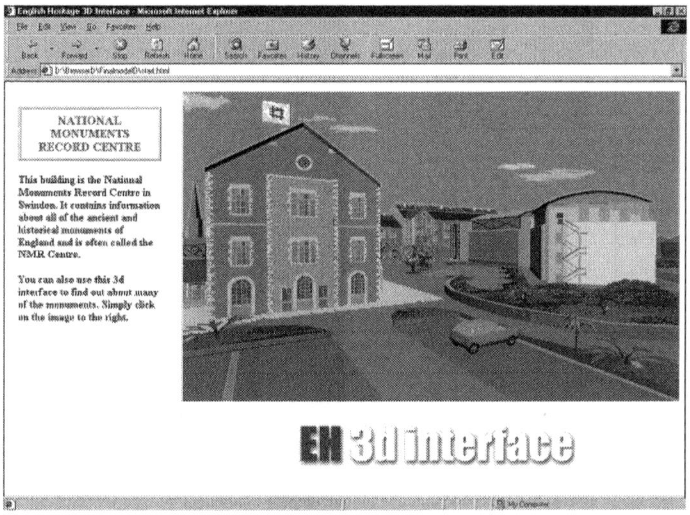

Figure 2: VR model of the National Monument Record Centre.

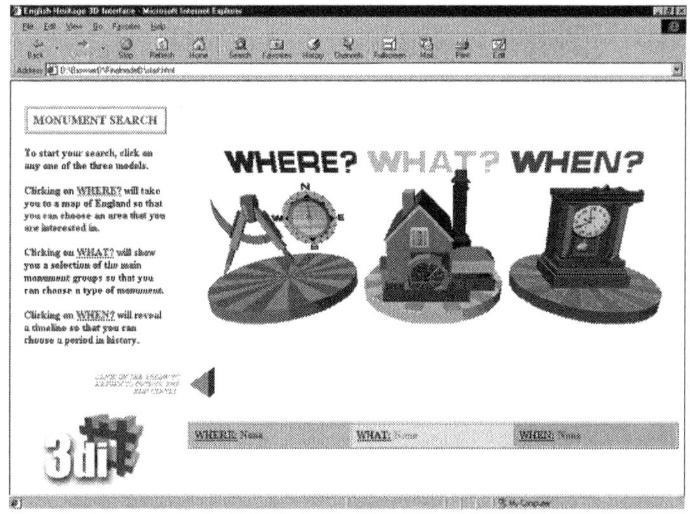

Figure 3: The three iconographic VR models representing location, type and period.

egories are no different. However, instead of determining large numbers of obvious subject divisions, of the underlying data, this paper proposes that identifying commonality across fairly low-level data entities can provide a usable method for establishing high-level categorization.

Within the 3Di project the system was designed to provide access to a database that related, exclusively, to historic and archaeological monuments. It therefore seems reasonable to assume that *all* users of such a system would be seeking information relating to the data at a monument (single entity) level. By extracting or identifying some of the essential components of *every* monument it is possible to compile a very clear interpretation, of the entire dataset, at the interface. From a designer's point of view the question is "What, in the database, are the lowest common denominators that are applicable to *every* user?"

The entry level, as mentioned above, leads the user through a model of the NMRC, and offers no alternative user-choices. This follows a basic assumption that every user would, by virtue of the fact that they were using the system, be interested in the information held in the NMRC. This is the lowest common denominator, applicable to every user, at this very high-level interpretation over the data. The second level of information relates more directly to the underlying data, but still retains a fairly high-level of abstraction, identifying location, type and date as the only categories needed for further navigation of the system's information. Every monument must: exist in a place; be a particular type or class; and belong to at least one time period. These categories do not divide the contents of the database into three separate sections, since they all apply equally to all the monuments within the database.

Within the prototype the three main categories have been represented in the alliterative style of WHERE, WHAT and WHEN, with the intention of appealing to user-group representatives in a later evaluation process. Selecting any one of these three categories would lead the user to a more detailed third level of information that relates specifically to the category chosen, and would present a user with a group of VR models representing appropriate sub-categories: WHERE would present the user with a VR map of England, divided into districts and counties; WHAT, a collection of VR models representing various classes of monument, such as Defence and Religion; and WHEN, a timeline illustrated with dates, period names, and a VR typology of houses. An example of part of one of these sub-category groups, WHAT, is illustrated in figure 4. It is highly improbable that any user would come to the system without already having enough knowledge to be able to identify one of these sub-categories as appropriate to their own requirements or interest. To increase user-confidence about which of the sub-categories would be appropriate to his/her needs, they have been conceived to be obviously different from one another; for example, a user wanting information about churches should not have any difficulty choosing one sub-category from Defence, Religion or Farming (three of twelve sub-categories). Assuming that users are interested in monuments in general, then it does not seem unreasonable to also assume that each user has a particular type of monument, geographical area, or time period, in mind.

When a selection has been made at this third level, the user would be given an option to view a list of *all* monuments corresponding to the selection; for example, if the user has selected WHERE and then the sub-category of Wiltshire, the system would offer a list of all the monuments in Wiltshire. The system would also give the user an indication of how many returns (monuments) could be included in such a list. However, because the three, second-level

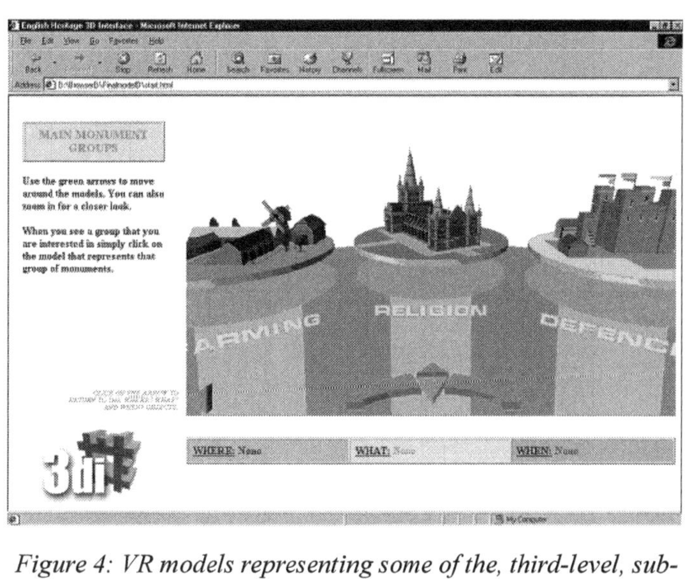

Figure 4: VR models representing some of the, third-level, sub-categories of WHAT.

categories (WHERE, WHAT and WHEN) are applicable to *all* monuments, the 3Di architecture can then offer the two remaining categories to prompt the user into further refining of the possible returns. Having chosen a county and been informed how many monuments this list contains, the user could, if they so wished, select *another* second-level category. By selecting WHEN and then Medieval, the user will have isolated the possible results to only those monuments in Wiltshire that are dated to the Medieval period. The user could then view this list, or narrow the possible results down further by selecting a sub-category from WHAT; for example, Religion.

This multi-pronged refining, or filtering mechanism, which is illustrated in figure 5, consistently urges the user to make a selection from one of the unselected fields. It also re-presents the choices from the previous level in case new information prompts a change of mind. This format enables the user to move down the structure very efficiently and purposefully. The structure, whichever path the user chooses, will result, ultimately, in the rapid identification of a small, and relevant, group of monuments. The following example demonstrates how, even with quite a large database, a non-expert user can very quickly refine the potential results down to a manageable number.

The NMRC database, over which the 3Di has been built, contained, as noted at the beginning of section 3.1, some 300,000 monument records. If a scenario was contrived where a user selected the county of Derbyshire (WHERE), the potential results set would be instantly narrowed down to 4,800 monuments: the number of monuments within Derbyshire. By further selecting Medieval (WHEN), this results set would be refined to 929 monuments. Finally, if the user selected Defence (WHAT), the system would return a list of 31 monuments. This is the total number of Medieval Defence monuments in Derbyshire. Because the user has determined the three elements common to the returned list (Derbyshire, Defence and Medieval), it is highly probable that the contents of the list, or some of them, will fulfil the user's information requirements.

This architecture, instead of offering lengthy lists of subject-divisions to the user, asks him/her to consider only *one third* of a query at a time, then offers a multiple-choice of fully illustrated answers, before attempting to elicit the next third.

3.1.2. Interface content

The prototype 3Di was designed to present information using a combination of graphics and text, with the graphical content designed around real-world metaphors centred on the NMRC building, and the three iconographic VR models representing WHERE, WHAT and WHEN. Although these second-level models are based on familiar real-world objects (compass, building and clock), they represent abstract components that are, essentially, common to all monuments, and do not correlate to straightforward subject-based divisions of the database. Although iconographic representations of abstract sets of data can potentially increase clarity for a user, the use of 3D graphics does not, in isolation, significantly increase the usability of an HCI. Effective visualization, at the HCI, is dependant on the way that the underlying data is filtered and presented to the user, and on the relationship between the 3D models (in this particular case), the data and other elements of the HCI, such as text (Steinmann and Chorofas 1996).

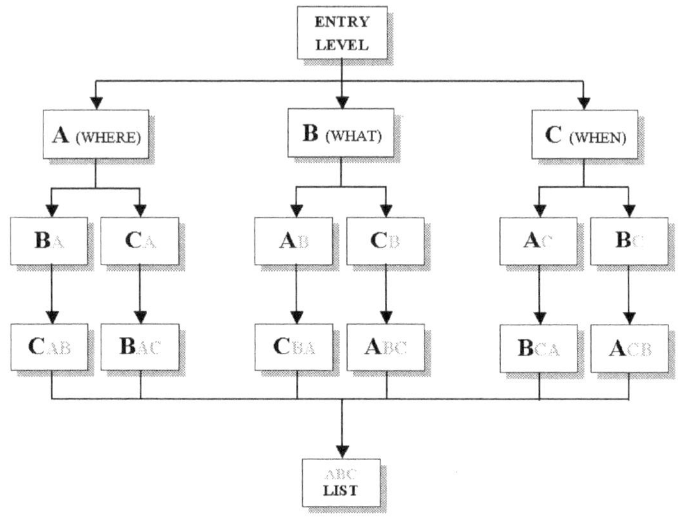

Figure 5: Three-pronged filtering mechanism. (Bold letters indicate possible choices and grey letters represent choices that have already been made.)

The overall design strategy for the prototype 3Di was to create a clean, uncluttered interface with images and language aimed at making the system comfortable to use for a member of the targeted "intelligent-twelve-year-old" user group. To this end, the emphasis within the design was placed on the three main icons (WHERE, WHAT and WHEN), which, although representing the second-level of abstraction over the data, actually form the core of the HCI's user-navigation capability. If a user selected any of the three main icons they would be able to view its appropriate sub-categories (counties, monument types or time periods), where they would be restricted to selecting one of the available sub-categories, returning to the previous level, or moving to either of the two remaining sub-category groups.

On the assumption that text and graphics form a stronger structural base when combined, the three main icons, and the models comprising the third-level sub-category groups, were clearly labelled (with text headings), and further clarified with mouse-over tips. Furthermore, a paragraph of explanatory body-text, and a record of user-history were designed into the overall framework. The body-text, which would alter according to user-selections, was included to explain available user-choices, prompt further decisions, and give an indication of how many returns could be expected from a search based on current selections. The text also included hyper-text links that corresponded to the three main icons; clicking on one of these links would have the same effect as selecting its equivalent icons. Additionally, the text provided a "written" record of user-history, relating any selections that the user had already made.

The element of user-history is a very important consideration of the design of any IS interface; such a facility can not only inform a user where he/she has already been, but the information can enable the system to offer sensible options for where the user may like to go next (Hightower et al. 1998, Rossi et al. 1997). Consequently, as well as the textual inclusion, two other elements of the prototype 3Di were designed to include, and exploit the advantages of, user-history:

- *The history table.* The purpose of the history table was to provide a constant reminder of the user's activity; keeping

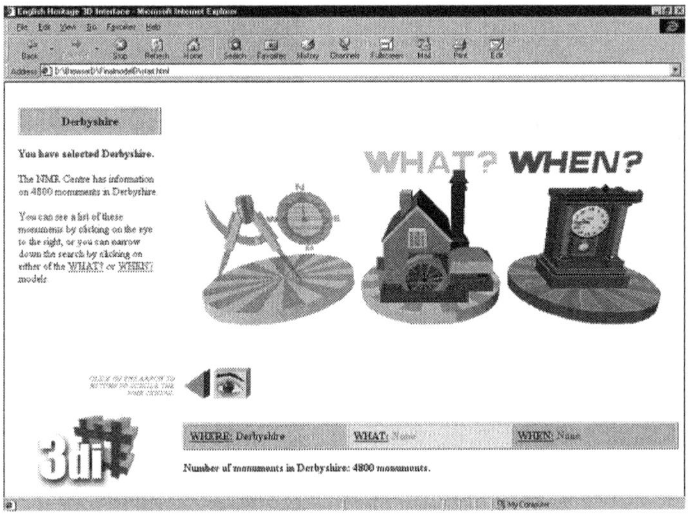

Figure 6: The second level icons with WHERE (Derbyshire) already selected.

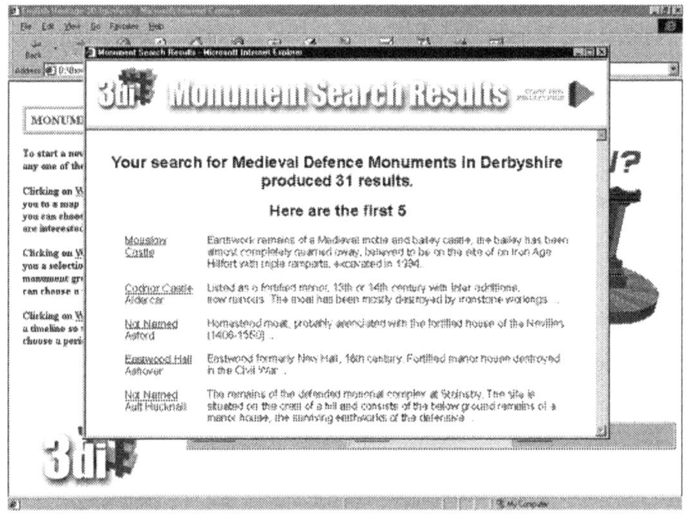

Figure 7: The returned list, with links to pages relating to each monument.

a record of any county, monument type or time period that the user had already selected and the number of possible returns that such choice, or choices, would generate. The record also contained hyper-text links and these, like those in the body-text, would if selected take the user directly to the appropriate sub-category level. This duplication of links helps to ensure that, whichever area of the screen, or whichever presentation mode the user chooses to respond to, the available options do not change.

- *The three main icons.* These also provided an element of user-history record. When a sub-category has been selected, its corresponding icon loses its text labelling and takes on a monochrome appearance, to indicate that it has already been selected. An example of this is shown if figure 6, where the County of Derbyshire (WHERE) has already been selected. This principle simply follows an idea commonly used on hyper-text links, where they change colour after having been visited. However, just as with hyper-text links this would not disable the icon; a user could select the same icon again, if so desired. Also, to further assure the user, and to help him/her avoid becoming disorientated, a colour coding scheme has been applied throughout the design. This code was conceived to give a consistency to each of the three main categories; for example, all the images, headings, and background in the history table, associated with WHERE are represented in blue.

The points outlined so far in this section, essentially comprise the content of the 3Di that is presented to the user. Two other elements were necessarily included: the back button and the view button. The back button was included to enable the user to return to a previous level by reloading all the frames within the browser window (the 3Di contains six different frames but default browser back buttons only reload the last frame to have changed. The purpose of the view button was for the user to click on when ready to view the list of available returns (monuments) that conforms to the selections that have been made. Background details of the information that would appear on this list, and the form that it would take, are discussed in the following section.

3.1.3. Information output

For the purpose of development the prototype 3Di utilized the database of an already established system. At the time of development this system was not available to the public, having been developed for the exclusive use of professionals interacting with data relating to archaeology and the built heritage. Consequently, its interface was designed on an assumption that users would understand both the complexity of the underlying data structure, and the sophisticated terminology used within it. Using the database of such a system, without amendment, in the development of the prototype 3Di, demonstrated two useful attributes of the HCI approach outlined in this paper:

- The addition of a *user-appropriate* interface can radically alter, and improve, the relationship between user and data; the level of interpretation, or abstraction, over the data-set can be manipulated to suit the needs of an identified user-group.
- Much of the information value in a database is not restricted to the delivery of low-level detail (e.g. monument unique identifiers or grid references), but instead can be related more to contextual interpretations; for example, a simple list of all the Medieval Defence monuments in Derbyshire.

However, although many users may be satisfied with this particular level of detail, the system is not restrictive. The architectural principles applied in the prototype 3Di would, as mentioned earlier, permit the user to extract a higher level of information; for example, the user may request to view a list of all the monuments in a given county, or all Religious monuments of a particular period. It could also allow the user to access lower level material from within the database; the final list of returns (an example of which is illustrated in figure 7) is composed of monument names with brief descriptions of each monument. Each name is linked, by hypertext, to an automatically-generated HTML page devoted to lower level information relating to that particular monument. This information relates to material stored in the database that has been input as part of a professional data collection policy, and as such is slightly restrictive. The project will examine the possibilities of including further very monument-specific information; for example, photographs, or even bus timetables, where available.

4. Future work

The prototype model described in this paper was developed for experimentation into novel HCI principles. To assert the validity of these principles further, the prototype 3Di will be subject to usability experiments with the inclusion of user-group representatives. When such experimentation has been conducted, and assuming that the results are positive, it is hoped that a fully operational application will begin to be developed, and then implemented. Development of a full application of the 3Di will, if it goes ahead, include research into other graphics mediums (including two-dimensional graphics), and will endeavour to take advantage of any applicable Internet, or computer, innovations that are available at the time of development. In the first instance such an application will probably be aimed at school children, and as such, is likely to be delivered as an executable file, either downloaded from the Internet or on CD-ROM. This file could load the 3Di onto the host machine, and then automatically connect to the database across the Internet. This circumvents the potential problems of slow download times, and the difficulties of obtaining and installing browser plug-ins, both of which are likely to be problems for high graphic content web-sites for some time yet.

The model described in this paper presented information at a very high-level of abstraction (the model of RCHME), and also at a low-level, such as specific monument details. It was also designed to deliver information at levels of abstraction somewhere between the two extremes; for example, a list of all Defence monuments. However, the scope for delivering information can easily be expanded to include other levels of abstraction, that could possibly fulfil different user requirements; for example, if a user selected a county he/she could be offered a subsidiary level of information that supplied general information relating to the monuments of that county. Perhaps, if a user selected the category of Religion, he/she could be offered an opportunity to "fly through", or click-on, an educational VR model based on Religious monuments. It is anticipated that future development will examine these possibilities.

5. Summary

The complexity of computing information systems, with differing user needs, complex data-structures, varying delivery mechanisms, as well as information content and graphical presentation, means that HCI design is an increasingly multi-disciplinary affair (Bulas-Cruz et al. 1999, Johannsen 1995, Santhanam and Batra 1998). The practical reality of most development scenarios is that there is not always going to be an entire development team, with experts devoted to each and every aspect of the system. Nonetheless, it is imperative, if an interface is to be of value, that the development process recognises the diversity of inherent elements, and responds to them in a positive way. This paper has presented a project that has acknowledged the complexity of designing an Internet-based interface, and has outlined some of the ways in which the project has attempted to resolve several of the pertinent issues.

The prototype 3Di has been designed to present the user with an uncluttered, and intuitive interface to complex heritage data. The HCI was constructed with familiar, real-world metaphors (three-dimensional graphic images), combined with instructive text, clear and varied labelling, and simple iconography. It presented a contextual picture of the underlying data that imparted a high level of information value at the interface. The simplicity and clarity, offered by the interface, was contrived to enhance the user's ability to decide on an appropriate course of action, and to diminish his/her chances of becoming disorientated or lost within the interface. The system user, who should always be the key consideration in any IS development, even with little or no specialist knowledge, should be able to navigate information in a quick, fluid, and most importantly, productive manner.

References

AHLBERG, C., WILLIAMSON, C. and SHNEIDERMAN, B., 1992. Dynamic Queries for Information Exploration: An Implementation and Evaluation Graphical Interfaces for Drawing, Exploring, and Organizing, *Proceedings of ACM CHI' 92 Conference on Human Factors in Computing Systems*: 619-626.

BALINT, L., 1995. Adaptive Human-computer Interfaces for Man-machine Interaction in Computer-integrated Systems, *Computer Integrated Manufacturing Systems,* Vol. 8, Part 2: 13-142.

BARRETT, R. and MAGLIO, P.P., 1998. Informative things: How to attach information to the real world, *Proceedings of the ACM Symposium on User Interface Software and Technology*: 81-88.

BILLINGHURST, M. and SAVAGE, J., 1996. Adding Intelligence to the Interface, *Proceedings of VRAIS,* IEEE.

BOYD, D.S. and ATHOUSSAKI, H., 1998. VRML: A Designer's View. In Vince, J. and Earnshaw, R. (eds.), *Virtual Worlds on the Internet*, IEEE Computer Society Press.

BROOKS, F., 1996. The Computer Scientist as Toolsmith, *Communications of the ACM,* Vol. 39, No. 3: 61-68.

BULAS-CRUZ, J., MORGADO, L., MELO-PINTO, P., ABREU, M., LOBO, H., GUEDE, M., SANTOS, A., BORGES, J., BICHO, J., BARROSO, J. and REIS, A., 1999. Beyond traditional Web page designs - a communication language between designers and web page developers. In *New Techniques for Old Times, CAA 98, Computer Applications and Quantitative Methods in Archaeology,* BAR International Series 757.

DAVIES, L.J. and LEDINGTON, P.W.J., 1991. *Information in Action,* Macmillan, England.

GLAZIER, L.P., 1997. "Our words were the form we entered": a model of world wide web hypertext, *Proceedings of the Eighth ACM Conference on Hypertext*: 24-28.

HIGHTOWER, R.R., RING, L.T., HELFMAN, J.I., BEDERSON, B.B. and HOLLAN, J.D., 1998. PadPrints: Graphical multiscale web histories, *Proceedings of the ACM Symposium on User Interface Software and Technology*: 121-122.

INVERSO, D. and SOKOLL, R., 1997. Optimum Human-Machine Interface Design, *Control Engineering,* Vol. 44, Part 12: 93-97.

JACKSON, M.H., 1997. Assessing the Structure of Communication on the World Wide Web, *Journal of Computer-Mediated Communication,* Vol. 3, Issue. 1.

JOHANNSEN, G., 1995. Knowledge-based Design of Human Machine Interfaces, *Control Engineering Practice,* Vol. 3, No. 2: 267-273, Elsevier Science Ltd.

JONES, S., 1998. Graphical Query Specification and Dynamic Result Preview for a Digital Library, *Proceedings of the ACM Symposium on User Interface Software and Technology*: 1-8.

MACHOVER, C. and TICE, S.E., 1994. Virtual Reality, *IEEE Computer Graphics & Applications,* January 1994.

McKILLIP, R.C. and CORONA, B.M., 1997. Advanced Human Computer Interface for the Army After Next, *Proceedings SPIE,* Vol. 3080: 203-211.

MILLER, P. and RICHARDS, J., 1995. The Good, the Bad, and the Downright Misleading: Archaeological Adoption of Computer Visualization. In Huggett, J. and Ryan, N. (eds.): *Computer Applications and Quantitative Methods in Archaeology,* Tempus Reparatum, BAR International Series 600.

MURRAY, N., GOBLE C. and PATON, N., 1998. A Framework for Describing Visual Interfaces to Databases, *Journal of Visual Languages and Computing,* Vol. 9: 429-456.

O'BRIEN, S. and POLOVINA, S., 1997. Web Design and Information Architecture, *Proceedings of the Sixth International World Wide Web Conference,* Santa Clara, California, USA, April 7-11, 1997.

PRINGLE, M.J. and MOULDING, M.R., 1997. Applications for Virtual Reality, and Associated Information Technology. In: *Illustration of Archaeological Material, Graphic Archaeology, AAI&S.*

PRINGLE, M.J., 2000. *The Use of Virtual Reality for the Visual Presentation of Archaeological Information,* Cranfield University, England.

ROSSI, G., SCHWABE, D. and GARRIDO, A., 1997. Design reuse in hypermedia applications development, *Proceedings of the Eighth ACM Conference on Hypertext*: 57-66.

SANTHANAM, R. and BATRA, D., 1998. The Human-Computer Interface in Information System Design, *The DATA BASE for Advances in Information Systems,* Vol. 29, No. 1: 25-30, Winter 1998.

SHARMA, R., PAVLOVIC, V.I. and HUANG, T.S., 1998. Toward Multimodal Human-Computer Interface, *Proceedings of the IEEE,* Vol. 86, No. 5.

STEINMANN, H. and CHOROFAS, D.N., 1996. *The New Wave in Information Technology,* Cassell, London.

STRAWHORN, J.M., 1981. Future Methods and Techniques. In Hills, P. (ed.), *The Future of the Printed Word,* Open University Press.

TUDHOPE, D., TAYLOR, C. and BEYNON-DAVIES, P., 1995. Semantic Navigation Tools for Museum Hypermedia, *Computers and the History of Art,* Vol. 5.2: 51-63, Harwood.

UCKO, P.J. 1992. Foreword. In Reilly, P. and Rahtz, S. (eds.), *Archaeology and the Information Age: A Global Perspective,* Routledge.

UK Government 1999. *Modernising Government, UK Government White Paper,* Her Majesty's Stationary Office, London.

WARWICK, K., GRAY, J. and ROBERTS, D., 1993. *Virtual Reality in Engineering,* IEE, London.

"Observing the Game": What can Access Statistics Really Tell Us?

William Kilbride

Archaeology Data Service, Department of Archaeology, University of York

King's Manor, York YO1 7EP, United Kingdom

e-mail: wgk1@york.ac.uk

Judith Winters

Internet Archaeology, Department of Archaeology, University of York

King's Manor, York YO1 7EP, United Kingdom

e-mail: judith@minerva.york.ac.uk

Abstract

An ubiquitous feature of the "information age" is the hype that surrounds user statistics. Web sites are becoming an increasingly important and increasingly sophisticated marketing tool, and the success or failure of commercial web sites depends on meeting important thresholds of hits per day. Archaeology, at least within education, is shielded from these pressures, though we take pleasure in noting the more obscure locations or are disappointed when there seems to be a dip in demand for our all-important web servers. User statistics can provide an enormous wealth of information about who is visiting our web sites, where they come from, what interests them and when they work. Such issues can be discussed in theory, though in practice interpretation is hampered by some basic problems.

Using access statistics gathered by the Archaeology Data Service and Internet Archaeology, this paper explores the trends represented in them. This analysis, informed by a discussion on the place of the Internet in wider social agenda, suggests that much can be gained, but only if questions are formulated with care. It will be argued that the user populations engage with the digital content of web sites, partly in response to the nature of that content, and partly on the basis of existing working practices. It is proposed that virtual heritage services are not a threat to conventional engagements with the archaeological record, but that they supplement that engagement for particular audiences.

Key words: Internet, accessibility, Access, server statistics, usage patterns, interactivity, readership, online resources, heritage services, working patterns

1. Introduction

If anyone should understand computers and the impact of the Internet, archaeologists should. The impact of technology on social reproduction, and the reproduction of technology through social practice lie at the heart of much archaeological research, though admittedly these are often written in the past tense. There are a number of readily available tools that may allow us to investigate the social impact of the Internet, and the impact of society on computers, but the most frequently quoted of these come in the form of user statistics. Web sites are important in commercial marketing, and their success or failure depends on meeting important thresholds every day.

As archaeologists, we are largely insulated from these pressures, though we take pleasure in noting the apparently insatiable public appetite, or the relatively unlikely locations of our audiences. The world, we are told is getting smaller. Numerous national and international policy initiatives are establishing mechanisms to open information sources and to distribute electronically data that otherwise would have been unavailable. These strategies are presented in context with wider policies of social inclusion, using the strengths of the cultural and heritage sectors to foster participation in the body politic. More than a decade of research shows that the Internet and computing technologies can indeed give rise to specific forms of community participation, specific forms of interaction and practice which are impossible or impractical by other means. The user statistics we generate provide a wealth of information on who is visiting our web sites, when and to an extent how they obtain (or fail to obtain) cultural heritage information. To a certain extent, these figures give us an insight into the way that computing and information technologies, in particular the Internet, inform the development of archaeological research, and the relationship that archaeology has with its public. But interpretation of these numbers is problematic, and the conclusions dependent upon any number of problems. This paper will explore the issues associated with evolving information policies and the community building exercises they seek to support. In particular, we shall highlight the strengths and weaknesses of the interpretation of figures accumulated at the Archaeology Data Service (ADS) (http://ads.ahds.ac.uk/). The interpretation of these statistics is facilitated by comparison with similar figures for the electronic journal, Internet Archaeology (http://intarch.ac.uk). As will become apparent for technical reasons, Internet Archaeology provides the only reliable and comparable statistics for this analysis.

2. ADS access statistics

The ADS, like many web servers, retains a certain amount of information about the users who visit the web site. This data can be used to generate some fairly crude statistics, giving a rapid, if perhaps unreliable insight into the use of the ADS's online catalogue, ArchSearch (http://ads.ahds.ac.uk/catalogue/), and the various documents that are associated with it. These figures reveal almost constant growth, but with enough dips and peaks to show that there are other variables underlying the growth (see figure 1). We can identify certain peaks with particular events in the history of the organisation, such as the launch of ArchSearch, the online catalogue in September 1998, and the addition of individual elements to the catalogue. The launch of the ADS Library, with back issues of the Council for British Archaeology's Research Reports and Occasional Papers in March 2000 brought an unprecedented demand on the server. These figures are calculated on a daily, weekly, monthly, and annual basis, giving slightly different insights into the demand. Thus, the year from April 1999 to March 2000 brought just over 764,000 successful requests to the server, while the very last month of that year, March was a busiest single month, with around 90,000 requests. The week between 19th and 26th March was the busiest single week on record, with over 26,000 hits. Not surprisingly, the quietest week was from 26th December, with fewer than 6,000 hits. Saturdays and Sundays are relatively quiet: the weekdays proving more popular.

In addition to identifying the peaks and dips in demand, the figures also give an insight into the users of that data. Thus, the log files for the same period identify different users and their contexts. Not surprisingly, the UK is the largest single user community, responsible for around 40% of the data downloaded. The ADS is based in the education sector, and the work we do is of particular relevance to the UK higher education sector. This is evident in our statistics also, with something like 38% of the data going to UK universities. Other countries are in evidence in a variety of guises, with the US and Europe taking up most of the rest of the demand. Many Internet service providers operate without an identifiable national domain name or with unresolved numerical addresses. Many of these are based in the United States, though they are more widely dispersed, representing almost 35% of the demand. A further 10% is served to the US education community.

In addition to giving us a broad idea of who uses the service, and when, it is, of course, possible to identify what particular resources are in demand, and how those users came to be referred to the ADS. We can identify the number of requests for individual directories and pages within these directories, showing which of the ADS's Guide to Good Practice Series are more popular, and the requests which users make on the catalogue. We can also identify the last page which the user visited, and thus get an insight into how users came to the catalogue. For example, two hundred users came to the ADS through the BBC's news server when it published an article about the ADS on their online news service.

3. Access 1: The Internet and the community (in the UK)

The increasing demand evident in these figures is hardly surprising given the rapid growth of Internet services and the numbers of people connecting to the Internet. Increasing numbers of consumer services are migrating to electronic formats, with the financial and retail sectors developing strategies for electronic commerce, encouraging, and in some cases forcing traditional customers to adopt novel practices that depend upon the Internet and allied technologies.

Access to information is one of the corner stones of political rhetoric in the UK and EU, a drive that may have particular implications for the heritage sector and for archaeology. Such policy ini-

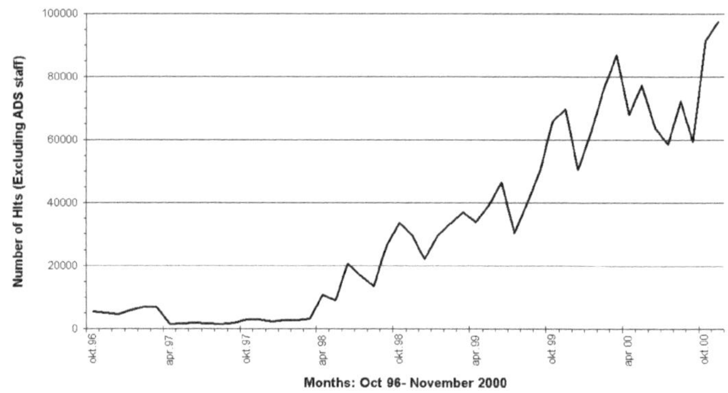

Figure 1: Hits on the ADS website Oct 96-Nov 2000.

tiatives depend on the rapid development of physical infrastructure and content, both of which are becoming priorities of government (e.g. EU 2000, Cabinet Office 2000). These initiatives are not formulated in isolation, but invariably form parts of wider social and economic policies. The United Kingdom government now has an evolving policy on information, aiming to use this investment not simply to disseminate relevant government documents, but, we are told:

> "The Government has a mission to modernise - renewing our country for the new millennium. We are modernising our schools, our hospitals, our economy and our criminal justice system. We are modernising our democratic framework" (Tony Blair, foreword to HMSO 1999).

Allied to this, ambitious policies to combat "social exclusion" have moved the accent from the conventional concerns of health and welfare, to wider issues of social and cultural regeneration, with some emphasis being placed on the cultural sector as a means of engaging actively with groups disenfranchised by conventional social and economic policies (e.g. SEU 2000). The move towards engagement for social regeneration is paralleled by a recognition of the importance of these new technologies in economic developments, represented by exercises such as the Department of Trade and Industry's *Information Society Initiative* (http://www.isi.gov.uk/isi/). Investing in communications infrastructure to facilitate social and cultural inclusion will provide the foundations for a wider economic engagement, by raising standards in universal IT skills, and providing a platform by which business can expand its use of communications technology. For many years governments have supported industry with the infrastructure of roads, railways and ports. In the "information age" the "knowledge economy" will receive a similar sort of investment. This, at least is the political theory of community building on the Internet.

On a more practical level, however, a number of important steps have been taken to realise these ambitions: at least to provide the tools that will make these theories a reality. The "open government" "learning society" and "knowledge economy" of political aspiration are supported by a number of concrete policy initiatives that will undoubtedly change the nature of archaeological practice, for better or worse. The most important of these developments, at least from an historical perspective, is the introduction of legislation on the freedom of information (inter alia House of Lords 2000). Knowledge acquired and maintained at public expense for the public good will soon be necessarily put into the public domain. The precise implications of this legislation will not be clear for some time, but plans are already afoot to exploit the potential of the Internet to distribute such information. Parallel legislation on the management of public bodies to achieve "best value" are already in place and will have an extraordinary impact on local government, the planning process and the degree to which different agencies share expertise and information (HMSO 1999a). These and related projects provide legal substance to some of the rhetoric. In practical terms, a number of organisations exist which are either already delivering these objectives or are some way down the line in implementing them. In the Higher and Further Education sector, the Joint Information Systems Committee (JISC) has a formidable track record in providing technical assistance and support to large numbers of users in numerous locations, experience gained in working with "JANET", the academic network which connects the Higher and Further Education sectors to each other and the Internet (http://www.jisc.ac.uk). JISC has developed a coherent strategy for providing data to its networked community, for the benefit of research and learning with what it describes as a "Distributed National Electronic Resource". The Archaeology Data Service, located in a university but working for the whole research community is an example of the Distributed National Electronic Resource in practice. In the secondary and primary education sectors, the British Educational Communications Technology Agency (BECTA) is leading developments for a National Grid for Learning (http://www.ngfl.gov.uk/). In the museum sector, the Museums and Galleries Commission have been working to bring data about collections together from numerous museums and present it in a single portal operated by the Cornucopia project (http://www.cornucopia.org.uk/), while the electronic "Twenty Four Hour Museum" (http://www.24hourmuseum.org.uk/) has been recognised for its efforts by being designated as a National Museum. In different ways, these different initiatives are implementing and helping to inform strategies to put communities in touch with each other and with cultural and heritage datasets.

4. Access 2: The virtual community

Plans for the creation of "virtual communities" and "information societies" are far from new, and their analysis has become a regular feature of the academic world (see Hobart and Schiffman 1998 for an impressive historical insight into what might not be such an unusual transformation after all). For many years now, various researchers have attempted to understand and anticipate the impact of computers and communications technologies on social practices, and upon our own discipline also (inter alia: Harnad 1991, Huggett 1995, Holtorf 1999 see also Internet Archaeology 6 and Exon et al. 1999). Although experiments and studies have investigated all sorts of computer-mediated communication, these have been largely undertaken from a psychological perspective, and tend to be deterministic in nature, assuming that the social outcomes described derive from the material characteristics of the technology (Markus 1994). In that respect, they investigate the technology in question without making reference to the expectations or aspirations of the people operating the systems (Walther 1992). No good archaeologist would be satisfied with such an approach. Lab based studies suggest that a reduced number of lines of communication result in aggressive, task oriented behaviour, whereas real world anthropology suggests that the register of computer mediated communication reveals the development of "speech communities" with discursive norms and an awareness of shared or fragmented histories (Cherny 1999:247-295). It is these discursive practices more than anything that have been used to justify the claim that these are real and vibrant communities. Thus, the Internet allows for a form of purposeful social action, not previously possible.

Such purposeful action and interaction occurs right across the Internet, by email and by discussion groups, but historically has been the particular provenance of "Muds" and "Moos" (Cherny 1999:1-31). Muds – multiple user dungeons – developed as gaming tools in the late 1980s as a means of putting numerous, geographically isolated fantasy role playing gamers together into one playing environment. Thus, different players encountered each other in a form of social interaction, working out and developing new storylines for games that became as variable as the characters present there. Moreover, the introduction of programmed and specified objects into these dungeons gave users a consistent content with which players could interact, providing a foundation for shared experience and knowledge, and thus a subject matter for coherent social discourse.

The technology of the MUD is not, however, restricted to fantasy battles or make-believe monsters. The multiple-object oriented (MOO) technology that supported these adventures was soon abstracted into other contexts, with different types of interaction and new content. MOO's proliferated in the early 1990s. The most renowned of these, called LamdaMoo, hit the headlines in the early 1990s with many thousands of participants taking on many different characters in a hippy-style commune, imagined as a large house in which each participant was invited to build their own space, then encounter other participants in their own unique spaces. The population of LambdaMoo demonstrated consistent learned behaviour with the terms of social engagement being developed ad hoc. The infamous LambdaMoo "cyber-rape" and the vigorous protests which it provoked are evidence that the communities thus created developed within the constraints of the technology, but in directions that are the property of their populations (Cherny 1999:31).

MOO's and MUDS are still with us. The multiple-object oriented technologies of the early 1990s were often based on technologies that now seem to be anything but interactive. Combined with an expansion of bandwidth and more flexible communications protocols, they have given rise to new forms of engagement with digital content and populations. Specifically, the almost unbridled proliferation of web "Chat Rooms" and related chat relay services on the World Wide Web have inherited and consolidated social discourse on the Internet. Chat services have a thoroughly deserved reputation for absurd triviality (Arnaut 1998), yet the technology that facilitates such inanity is far from trivial. MOO configurations have been applied in a number of contexts, most usefully in the development of "virtual learning environments" (e.g. http://www.cose.staffs.ac.uk/). These teaching tools use multiple media, and many objects to bring students and teachers together with specific materials for discussion and experimentation. Focussed on specific learning outcomes with managed interactions, or released to open ended learning, these tools are being used to put increasingly large numbers of students in touch with each other and with increasingly hard pressed resources. Used in this way, these virtual environments have the benefit that in addition to the manifest subject matter, students and staff rehearse or develop transferable skills in computing and IT, thus enhancing their own digital creativity.

Of course, there are numerous other expressions of community life and social action on the Internet. In this respect all that is being suggested here is that the Internet is not used for its own sake, but for the accomplishment of given, and often very conventional, tasks. Online consumerism, on line teaching and learning, electronic protest, hero worship and romance are all prevalent. In these contexts, and demonstrably in others, the communities formed are supplementary to, not in competition with, other forms of community interaction. In this sense, it is the communities that drive content, as well as the content that forms communities.

Consequently, the relative value of access statistics may lie in estimating the scale and configuration of the communities that use and explore this content. If the aim is to establish a local community, then large numbers of external participants may represent a failure to achieve such an objective. Where the aim is to stimulate certain practices at certain times in the day or week, then this too may be represented in the relevant statistics. Moreover, constant levels of activity through the day or week, when contrasted with relatively localised populations may demonstrate the migration of working practices from the normal working day. Analysis of these figures is only one part of a wider and much more challenging anthropology of the communities in question.

5. What access statistics really *don't* tell us

We may wish to explore our digital communities from the comfort of our own offices, but like all anthropological analyses, the data we use has its own inconsistencies, flaws and inadequacies. It is as well to be aware of these before taking that analysis forward.

First and foremost, and perhaps most critically, the number gathered by our automated processes don't represent the real numbers of hits or user sessions on any given web-site. There are numerous problems that distort the figures. On one hand, the use of large numbers of files inserted into pages can, in some circumstances lead to a gross overestimate of the real numbers involved. This is particularly the case where graphics are used. In addition the constant use of the web site by its developers can count in the number of hits, but for uses that are hardly relevant to the exercise. Both can be overcome if the data gathering tools are configured adequately: graphics and local users can be filtered out. On the other hand, web and personal browser caches, established to reduce network traffic, can lead to a gross underestimate of the popularity of a site. Thus, instead of loading images directly across the Internet, a copy of the data saved locally will suffice, meaning that a genuine hit is not registered. In large networks, a combination of external cache, proxy and mirror services means that even when the pages are loaded "as new" the web server may not register a hit in the normal way. This problem is particularly evident in the ADS access statistics where the University of York is responsible for an inordinate proportion of use compared to other higher education institutions even though office terminals are specifically excluded from the figures. The reason for this is, in part, the configuration of national and local university cache and proxy services that are configured specifically to reduce the number of hits on the ADS server. There is no obvious way round this problem, other than to recognise that the numbers produced are interval numbers, not ratio ones (Shennan 1988:11-12).

Then, there is the not inconsiderable problem of understanding. While we may be able to track down an individual user session, and know something about the pages which appear to interest them, user statistics cannot tell us whether the end user actually understood the content of the page. In this respect at least, web delivery is entirely consistent with conventional publication. Authors and publishers can know how many copies they print and sell, but not how many copies are read or understood. It is perhaps true that web delivered data will escape the ignominy of propping up tables or soaking up coffee, but in the same way, once the data is released over the Internet, the author and publisher have limited control of what people do with the data presented in this way. This problem is normally presented in concerns about copyright and the ownership of information, but from other perspectives, it is more serious than that. From the perspective of teaching and learning, the degree to which students have attained the level of expertise expected cannot be assumed simply because the students have visited the relevant web site, or completed the tasks set. To that extent, the success of web delivered teaching materials depends upon the success of a broader environment of assessment and reporting. We might have a mechanism to report what sorts of things people want from web servers, but often cannot tell what they do with that data.

Returning briefly to the cognitive perspective we abandoned at the outset, it is possible to anticipate, and thus steer users in a number of directions, predisposing them to certain forms of activity. Ideally, resources should not be more than three clicks from the home page. Established wisdom suggests that interest levels fall as the number of mouse clicks rise. Moreover, large files or obscure formats will restrict access to those competent enough to manage such resources, or with higher specification connections. It is normally supposed that interactivity enhances the users' experience of web server, encouraging their participation in data sets that are otherwise incomprehensible. Yet even here conventional wisdom is disputed. One psycho-analyst has pointed out that the state of activity so sought after in computing has its inevitable counterpart in the state of passivity (Žižek 1999). Inter-activity likewise has its counterpart in *inter-passivity*: the more we attempt to create interactive environments so the theory goes, the more we risk creating passive users.

> "More precisely, the term interactivity is currently used in two senses: (1) interacting with the medium - that is, not being just a passive consumer; (2) acting through another agent, so that my job is done while I sit back and remain passive, just observing the game." (Žižek 1999:105-106).

Though presented in scurrilous tones, this theory is not as far fetched as it might seem: resources may require some work to understand, and this may take some activity. But at least the small number of users that persist will be better informed about the limitations, complexity and efficacy of the objects presented to them. This is hardly an argument for poor design, but the point is made.

In addition to problems associated with the internal coherence of the statistics, and the use that people make of the data, there are serious problems when we attempt to map our own statistics onto other similar data sets. Server side software and the logs they draw upon are configured for different purposes. Servers themselves have different characteristics: some provide data really quickly to slow or busy networks, others run slowly on faster networks. Some forms of data – especially streaming video or sounds – require time-consuming processes; other forms of data are supplied with trivial overheads. Servers crash from time to time, or are taken offline for maintenance, enhancement and repair. These all mean that, without the most rigorous constraints, the results produced for one server are not comparable with results from others, even where the logs and statistical software are implemented in the same way.

These and other limitations mean that, with the best will in the world, the user statistics we generate are not going to answer some of the more interesting and important questions about the digital communities we develop on the Internet. Moreover, as a means of monitoring the performance of the service we provide, they are perhaps the crudest of measures.

6. What access statistics really tell us

Readers may by now be under the impression that we do not rate access statistics very highly, or that we would wish to see more carefully controlled and more comparable statistics. In part this is true, but the goal of this paper is to explore what these numbers really do tell us. As will become clear, the trick is not so much knowing what questions they can answer, so much as what questions we may reasonably ask. In fact, this wealth of data provides a powerful analytical tool to answer quite specific questions about how archaeologists and others engage with the Internet. Specifically, they reveal patterns of how archaeologists, and to a lesser extent the pubic, work in relation to different types of service. They show us how archaeologists work through the day, how they work through the week, and how they work through the year. In each of these, but especially in the last two, they may be used to interrogate the relationship between virtual heritage services, and real, conventional heritage services. In order to achieve this sleight of hand, two elements will be introduced. Firstly, for good reasons, comparison with another data set will be proposed – a comparison that for unique reasons is comparable to the statistics gathered by the Archaeology Data Service. Secondly, even these comparable figures will be massaged to make them more amenable.

The comparable data sets presented here are abstracted from the logs of the web servers of the Archaeology Data Service and the electronic journal Internet Archaeology, the first fully peer-reviewed electronic journal of archaeology. Though distinct entities, both these units operate out of the Department of Archaeology at the University of York. Thus, in a technical sense they are identical. They have identical profiles in terms of internal and external caches; they both connect to the Internet from the same network with the same number of network maintenance glitches; they both use the same server software and compile their figures in the same way. Indeed, not only do both services run across the same network, but their web servers are partitions of the *same* Unix server. Thus, server maintenance time and down time are exactly the same. For example, on the advice of security experts, the server was briefly disconnected from the Internet over the "Millennium" holiday for a five day period. Had Internet Archaeology and the ADS been working from different servers with different connections to the Internet, an interruption like these could render comparisons invalid. Even so, there is still diversity in the figures. In some respects, this diversity takes us to the core of the analysis, but there is sufficient difference in the figures to warrant a further qualification: the figures presented here are not the absolute values, but the proportion of total demand at certain times, expressed as a percentage of the total.

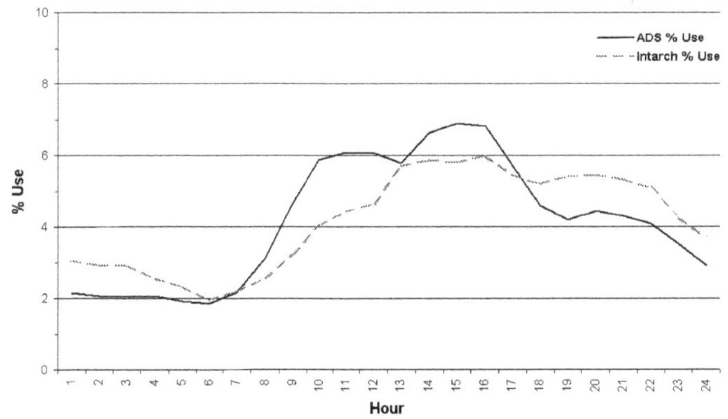

Figure 2: Working Patterns Through the Day.

This brief analysis looks at three different sets of figures based on time and the working patterns they represent. Many others forms of analysis are possible and more complex statistical modelling of these figures is certainly feasible. Moreover, an investigation of the geography underlying the use of these two web-servers might well prove revealing, and undoubtedly this informs the figures that are presented here. In this simple reconnaissance, however, we will investigate the relationship between Internet access and patterns of access to virtual heritage throughout the day, the same phenomenon through the week, and lastly, throughout the year. These figures were gathered over a twelve-month period from April 1999 to March 2000.

The levels of demand through the day are shown in figure 2. Perhaps the most striking feature of these figures is the way that the ADS profile varies so widely, while Internet Archaeology traffic fluctuates more gradually. It is clear that demand on the ADS site effectively starts around 0800 hrs, rising rapidly, and remaining constant throughout the day till around 1700hrs. The most remarkable element of this profile, however, is the clear dip in demand round lunchtime. This lunchtime dip at the ADS corresponds with a peak in demand for Internet Archaeology. Internet Archaeology remains more constant through the evening and even until quite late at night. There are, no doubt, numerous reasons for these particular profiles, but a number of explanations seem reasonable. Firstly, the content of the ADS catalogue and website is dominated by UK based resources. Though many of the documents are of general interest, it is the catalogue of over 300,000 archaeological sites and monuments in the UK that dominates demand. As we have already seen, the greatest demand for this information comes from the UK. What these figures demonstrate, however, is that users visit the site during the working day, using the catalogue for research or teaching from their desktops rather than from at home in their leisure time. It also suggests that, while the ADS catalogue may well have achieved a significant achievement in distributing resources widely across the UK, this change in the construction of space has not resulted in a change in the construction of time. Internet Archaeology, which has a more geographically diverse content and is more amenable to iterative browsing and recapitulation is clearly used in a very different way. Either, it has a more diverse audience, or an audience that is inclined to use the journal from the luxury of their own homes, and in their own spare time. A more thorough evaluation of the journal shows that it has both. (http://intarch.ac.uk/news/evaluation/index.html).

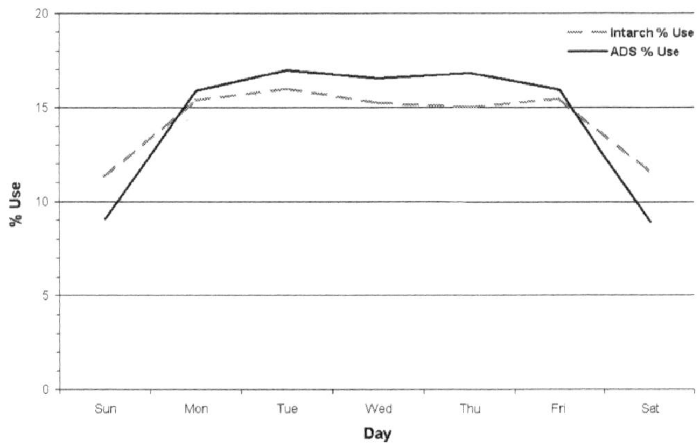

Figure 3: Working Patterns Through the Week.

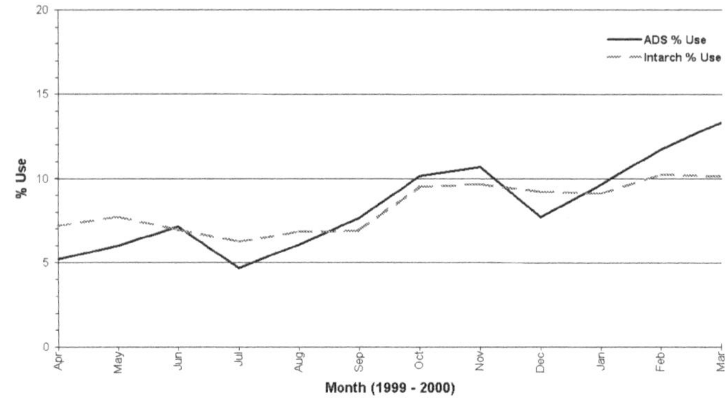

Figure 4: Working Patterns Through the Year.

The levels of demand through the week, shown in figure 3, tend to confirm this pattern. The demand on both is heavier during the week, as one might expect, but the journal tails off more gently than the catalogue at the weekends. Demand on the ADS drops to almost half of its weekday levels, whereas Internet Archaeology dips only slightly. Again, this suggests that different working practices are formed round different types of digital content. Moreover, by introducing other, conventional engagements in heritage, it would seem that the "Information Age" that we hear so many revolutionary things about, is supplementary to existing patterns of engagement. Museums and other conventional heritage sites have their largest visitor numbers at weekends: at precisely the same time as the ADS and Internet Archaeology see their least activity. Now the content of both these sites are academic in orientation, but there is sufficient public interest in both to merit a tentative conclusion about archaeology and its public. These figures suggest that virtual heritage information services really are supplementary to the public engagement in archaeology, and are not perceived by their user communities as alternatives.

This tentative conclusion is reinforced by analysis of the figures for use during the year, presented in figure 4. More than the previous sets, these figures have a clear drift, which could be eliminated by more sophisticated modelling: but this is not essential to make the points necessary here. Again, the ADS profile is more variable, the Internet Archaeology profile more constant. Again, both follow similar patterns of peaks and dips: more pronounced in one, less in the other. In both cases, the summer months of July and August are relatively flat, while the months of September, October and November, and then again January, February and March show almost constant increases. Internet Archaeology, however, remains noticeably stable throughout December, while the ADS figures drops away considerably. This pattern is strikingly different from conventional approaches to archaeological heritage, at least in the public sphere, where busiest periods are consistent with the holiday months of July and August. At least as far as the ADS figures are concerned, the opposite is the case. The peaks in demand are consistent with the peak times of the university term, so in that respect at least analysis of the figures demonstrates a congruence between the demand on the server and the ADS's principal target audience.

7. Conclusions

A number of tentative conclusions may be drawn from the preliminary analysis presented above. It is clear that moves are being made in the UK and the EU to invest more heavily in the infrastructure and content of the Internet in order to facilitate broader social goals. This is clearly consistent with existing practices on the Internet, where cultural interaction and purposeful social action are in evidence, not in isolation but as practices embedded within a wider social engagement. The means of reporting the role and impact of computing as a mechanism for facilitating broader social engagement exist in part through the analysis of access statistics, though these represent a complicated data set, which cannot be expected to supply cheap answers to expensive questions. The investment in information technology cannot therefore be audited purely by a facile appeal to access statistics as a portable indicator of performance. The figures, however, are internally consistent, and thus can be used to generate internally consistent data, where all that is required are values based on proportions of genuine numbers. Three simple analyses of two uniquely comparable sets of figures were presented here, showing that the populations engage with the digital content of web sites, partly in response to the nature of that content, and partly on the basis of existing working practices. It is clear that neither of these web servers, and by extension other similar servers, pose any immediate threat to conventional engagements with the archaeological record, but that they would appear to supplement that engagement for particular audiences.

Acknowledgements

This paper has benefited from the kind scrutiny of Tony Austin, Jo Clarke, Keith Westcott, Maureen Poulton, Julian Richards and Damian Robinson. The Archaeology Data Service is funded by The Joint Information Systems Committee and the Arts and Humanities Research Board.

References

ARNAULT, G., 1998. Internet chat rooms becoming a popular forum for business. In *New York Cyber Times*, January 26, http://www.nytimes.com/library/cyber/week/012698chat.html.

Cabinet Office 2000. *E-government: a Strategic Framework for Public Services in the Information Age*, http://www.cabinet-office.gov.uk/champions/Strategy.pdf, linked from http://www.iagchampions.gov.uk/Strategy.html.

CHERNY, L., 1999. *Conversation and Community: Chat in a Virtual World*, CSLI Publications, Stanford.

EU 2000. Communication from the Commission to the Council, the European Parliament, the Economic and Social Committee and the Committee of the Regions: *Towards a European Research Area*, http://europa.eu.int/comm/research/area/com2000-6-en.pdf, linked from http://europa.eu.int/comm/research/area.html.

EXON, S., DINGWALL, L., GAFFNEY, V., LAFLIN, S. AND Van LEUSEN, M., 1999. Archaeology in the Age of the Internet. *Computer Applications and Quantitative Methods in Archaeology* (CD-ROM).

HMSO 1999a. *Local Government Act 1999*, HMSO, London.

HMSO 1999b. *Modernising Government*. Presented to Parliament by the Prime Minister and the Minister for the Cabinet Office by Command of Her Majesty, HMSO, London.

HARNAD, S., 1991. "Post-Gutenberg Galaxy: the fourth revolution in the means of the production of knowledge", *Public Access Computer Systems Review* 2: 39-53.

HOBART, M.E. and SCHIFFMAN, Z.S., 1998. *Information Ages: Literacy, Numeracy and the Computer Revolution*, John Hopkins University Press, London and Baltimore.

HOLTORF, C., 1999. Is History going to be on my side? On the experience of writing and submitting a hypermedia. Ph.D. thesis. In *Internet Archaeology* 6, http://intarch.ac.uk/journal/issue6/holtorf_index.html.

House of Lords 2000. *Freedom of Information Bill*, House of Lords, London.

HUGGETT, J.W., 1995. Democracy, data and archaeological knowledge. In Huggett, J. and Ryan, N. (eds.), *Computer Applications and Quantitative Methods in Archaeology 1994*, BAR Int. Ser. 600: 23-26.

MARKUS, M.L., 1994. Finding a happy medium: explaining the negative effects of electronic communication on social life at work, *ACM Transactions on Information Systems* 12: 119-149.

SEU 2000. *National Strategy for Neighbourhood Renewal: a framework for consultation*, the Social Exclusion Unit, http://www.cabinet-office.gov.uk/seu/2000/Nat_Strat_Cons/default.htm, linked from http://www.cabinet-office.gov.uk/seu/index/publishe.htm.

SHENNAN, S.J., 1988. *Quantifying Archaeology*, Edinburgh University Press, Edinburgh.

WALTHER, J.B., 1992. Interpersonal effects in computer mediated interactions: a relational perspective, *Communication Research* 19: 52-90.

ŽIŽEK, S., 1999. Is it possible to traverse the fantasy in cyberspace. In Wright, E. and Wright, E. (eds.): *The Žižek Reader*, Blackwell, Oxford: 102-124.

Publishing on the Internet: The Internet as an Academic Information Source

Henriette Günther Soerensen

Soebakkevej 23, 5210 Odense NV, Denmark

e-mail: farkhes@moes.hum.aau.dk

Kaj Fredsgaard Rasmussen

Spobjergvej 72, 1, 1, 1+2, 8220 Brabrand Denmark

e-mail: farkkfr@moes.hum.aau.dk

Abstract

In this paper we will look into publishing and the Internet from various viewpoints. The focal point is the sharing and retrieving of scholarly publications. This concerns both edited and non-edited publication, as well as the need for a connection between these and structured languages.

Automated indexing of the information on the Internet is the *way of making sure that we do not get lost on the World Wide Web. This again calls for structured languages in the form of content specifying tags. There is no way to get around the use of these structuring aids, if we wish to make sure that the information thus formatted will be readable and retrievable in the future.*

Key words: sharing information, structured languages, online publishing, retrieving information, scholarly publishing

1. Introduction

Publishing on the Internet is to some extent already established in most sciences, even though the majority of scientific research never gets this far.

The printed word, in the minds of most scientists, is still the optimum media. This is due to the fact that the printed word is the more established of the two, concerning referencing and credibility. Accessibility to the Internet speaks in favour of publishing there. The only precondition is the connection.

In Denmark a large part of the archaeological material is never published because of the cost and labour requirements of traditional publishing. To overcome this, publishing on the Internet is yet again the solution, due to its lower cost.

2.1. Sharing information

In the striving towards a universal understanding of the past, information technology can provide a wider perspective and a tighter communication network through global exchange of information. This was made possible by the accelerated evolution of the processing powers of computers in combination with the growth of the Internet, its increasing speed and bandwidth (this means that more information can be transferred simultaneously). Bandwidth is one of the push-factors in generating the structured languages we will discuss later on.

Today, processing power provides a secure base for combining information with several multimedia-formats, which provides new ways of presenting data in a more efficient manner. This is an important point in archaeology, as we often need to present illustrations of parts of our publications to make these more comprehensible, and the printed media may not be the cheapest way to achieve this.

Another of Internet's positive influences on information exchange is the fact that the information presented there is accessible to anybody, anywhere, at a low or at no cost. In archaeology this is often a great help in providing a global perspective.

Sharing information via the Internet makes it possible to work in smaller or bigger groups across borders. This potential is based on the constantly increasing computer- and network capabilities. Publishing on the Internet is much more complex than described here. A large part of this complexity lies in the exact difference in the cost between traditional and other various online publishing methods. This will not be elaborated in full extent in this context.

2.2. Structured languages

Structured languages are often conceived as more complex than the normally used *word processing programs*. Transfer to various structured languages is often possible in newer versions of these programs, and there exists a range of programs available specifically for editing these types of documents.

Because of the need to preserve the original categories of formatting structured languages have become a necessity in online scholarly publishing. This provides the possibility to recreate not only the visual appearance, but also to some extent the meta-data (information *about* the contents of the data).

Structured languages are often a practical solution to publishing on the Internet, as transmission speeds on the Internet limit the size of files for most users. Structured languages refer to formatting in an efficient manner, resulting in smaller files and as a consequence faster re-assembly of the document locally.

By far the most common mark-up language is the *HyperText Markup Language* (HTML) which originally uses a predefined general *Document Type Definition* (DTD). This trait is inherited from the older SGML (see below).

In the DTD for HTML provided by the World Wide Web Consortium (http://www.W3C.org, Burnard 1995:chapter 5.1) only the formatting is defined and not the content. This means that HTML only defines the visual appearance of the text (apart from one or two meagre content defining formats like "address" (see http://hotwired.lycos.com/webmonkey, http://www.W3C.org). This visual formatting is not even interpreted in the same manner by various browsers. This is due to the fact that the formatting does not define all aspects of typography - line height, letter spacing and the like are not defined universally.

One remedy for the missing definitions of content and the lack of typographical design-tools in HTML has been, and still is, the use of Cascading Style Sheets (CSS) (Meyer 1999). These are created in order to provide the possibility of defining new formats and rules for their use. In using CSS you gain a far better control of the formatting, as well as a possibility of defining more meaningful mark-up tags, which can help the reading and indexing processes. These tags, if used, describe the content of the formatted text. An important point in this connection is that the use of meaningful tags is not obligatory. CSS are often "misused" for pure formatting reasons. Hereby HTML combined with CSS is a halfway clone towards the *eXtended Markup Language* (XML) described below.

The *Standard Generalized Mark-up Language* (defined before the advent of HTML) and the newer *eXtended Markup Language* are both defined specific and flexible (Burnard 1995). In these the user can create new DTD containing definitions of formatting, content types and rules for formatting with the defined syntax. Programs for formatting with the use of these languages are traditionally rather expensive (this applies especially to SGML tools) or difficult to obtain for other reasons.

Today we see more and more of the original HTML-editing tools extended with some XML-capabilities. The lack of programs supporting XML 1.0 as defined by the World Wide Web Consortium in 1998 (Bray et al. 1998: front page and page 1), is probably caused by the rather recent arrival of the final standard. Developers have simply not had enough time to provide full compatibility at a reasonable price.

In addition, the need for Document Type Definitions demands functionality far beyond the typical WYSIWYG structured language editor. This calls for either substantial restructuring of programs, or altogether new programs. These programs help in defining the syntax and the rules for the use of syntax: e.g. the order of statements, whether statements concern whole paragraphs or may exist inline within other paragraphs. This is slightly more complex than what the ordinary text editor can present in a comprehensible manner (Bünte 2000:200-213).

Conversion of SGML or XML formatted text is a new issue that arose from the extensive formatting possibilities in these two languages. Not all browsers, viewers, printers, WAP-phones or other viewing devices provide the same capabilities for presenting the defined formats. There are two solutions to this problem: Either the authors provide DTD's for each conceivable kind of device or they provide a conversion routine for the required adjustments. The latter are provided by the eXtensible Stylesheet Language Transformations (XSLT) currently in development by the W3C. This new language defines the rules for conversion to the *Public Domain Format* (PDF), regular HTML –code or the like (Bünte 2000:202). Such automation is important in preserving the information encapsulated in various languages, thus providing compatibility in the future.

3.1. Online publishing

Publishing on the Internet results in lower costs compared to traditional publishing, and as we often need to illustrate archaeological finds and findings, the Internet might be a preferable media to the traditionally labour intensive, slow and expensive publishing.

Today we see two different ways of publishing on the Internet. One is based on the editing of the publications, the other on publishing non-edited texts. The truth is, of course, that no publications belong strictly to one of the two.

The tradition behind edited online publications is the printed word, and the content of such documents is often easier to understand, compared to the non-edited publications. The drawback is that these are labour intensive and consequently slower due to the editing and administrative phases.

In the end it is the user who pays the eventual cost of producing these publications. The high price often connected with scholarly publishing online by publishers outside the economic safety of institutions is likely to create a barrier between those who can, and those who can not afford access to these publications.

The bulk of the savings from the use of online publishing, as compared to printed publication in scientific circles, to a great extent lies in the process of printing, translating and distributing. To give an example let us look at the periodical "Journal of Danish Archaeology": in this case the printing is prepaid by donations from foundations. Additional editing and layout is left to university employees as a part of their daily work.

As the publisher (the University) has no need of making a profit, migration of this kind of periodical to the Internet should be easily achieved, as it indeed should be the case for several university presses.

A completely different problem concerning edited publications is that the editing boards are by virtue of office in a position to control the free flow of information. They decide what is worth publishing and what is not. There exists a danger of this could to some extent lead to the monopolisation of information. On the other hand, the editing process insures a certain level of quality of the published material.

Editing on the Internet is a question of credibility and reliability. These can be used to give the reader of online documents a chance to verify the origin and the content. The editing board provides a security for the readers, that the documents are in the original form. Furthermore the editing can be expected to guarantee the quality of references and content.

In the case of non-edited publication there is no editing board to guarantee credibility and reliability. These documents can therefore be considered of a lower quality. The content is prone to contain various errors. These can occur as weak scientific argumentation, referential mistakes or bad syntax. As a result the reader must be more aware of, and critical towards these publications.

As the main problem of publications is credibility one would expect that a common consensus on formatting and content would be advantageous. For the edited publications the editing boards grant this. Since this is not the case in non-edited publications, the need for a way of providing it arises. One solution is to include a standard section in the documents. This should contain at least the name of the author, how to reach her/him, the year of publication and last update/check-up.

3.2. Layout

Consistent layout of documents enhances the content of a given text. In other words: if the text is well formed, understanding the content is much easier, due to the fact that we are used to the structured layout of the printed media.

Editing of publications may not be the most suitable solution to our specific publishing needs, as this method results in slow transfer of results within scholarly circles.

However, the use of non-edited publication on the Internet is not satisfactory either. This concept provides no way of ensuring common standardised formatting. If this is not used, reading scholarly publications *can* become a challenge.

Some kind of swift editing should therefore be the best solution providing the best of both worlds (the high speed of communication, good quality of the published material).

In this connection, the use of typographical formatting in accordance with structured languages, such as HTML, SGML and XML, can be seen as of a general interest to everybody; both to the edited and the non-edited online publications. The other advantages gained from these languages (as described in "Structured languages") underline the importance of this kind of formatting.

3.3. Retrieving information

Publishing on the Internet presents problems connected to information retrieval. Some indexing methods are more precise than others, this forces the reader and/or author to make certain choices based on their knowledge of the use of indexing methods. The need to gain knowledge of these methods in order to successfully use them may become a barrier for some readers.

Searching for information without this basic knowledge of indexing methods can be very time-consuming and render the Internet highly impractical.

The Internet *is* a more or less a chaotic place and you have to know how to use search engines as well as the important keywords in the quest for the desired information. The use of imprecise keywords leads to a multitude of results with little or no relevance. Relevance is an important factor in dealing with all kinds of information retrieval on the Internet (Lester 1997:21-22).

Basically there are two ways of retrieving information:

- One is based on automatic indexing by the software. The index search returns a million answers, but only a few of them are relevant. This originates in the way these data are collected. The indexing program has no way of telling whether a document *really* belongs to one category or the other. The solution is to count the use of specific words and/or word combinations, providing the possibility of returning not only documents containing them, but also a way of ordering the results according to their relevance.
- The other way of retrieving information is by searching lists or databases of relevant links collected manually. Such lists are maintained through the division of links into subjects. Rejecting supposedly irrelevant categories through browsing can result in the user's missing out an important information. A link list requires much care by the Webmaster in order to keep it up-to-date. Links to newer are information most likely not present in these lists. As the Internet is a vast collection of resources it is unlikely that all the categories are complete.

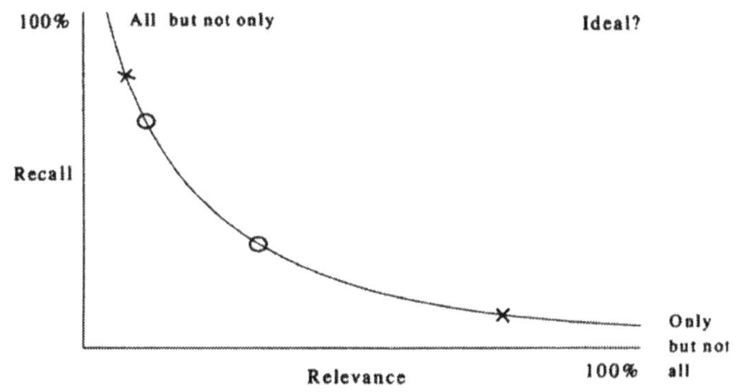

Figure 1: The recall-relevance / precision curve (Foskett 1996: fig. 2.3, p.19).

In this connection some type of stability in scholarly publishing would be advantageous, since this is the only way of ensuring that a link to the information placed here will still available after several years. This stability can be provided by the creation of online libraries maintained by public institutions, such as universities and libraries.

Databases on larger collections of links provide better possibilities in searching for cross-categories. This increases the chances of retrieving specific information on a given subject.

Index searching is most likely to retrieve all the relevant information and should be better than searching through collections of relevant links, even though it is time consuming.

From the aspect of retrieving information with the methods and problems discussed in this chapter, both the correct selection of keywords, and the two extremes of search strategies, results in the situation shown in figure 1. The recall-relevance/precision curve as described by Foskett (see Foskett 1996: figure 2.3) shows that using precise keywords and searching through link collections returns only relevant links, but not all the information available (the lower right corner of the graph). In the upper left corner the results of index searching and imprecise keywords are located. Here we retrieve all relevant information, but these are showered with irrelevant hits.

The real world is not divided as strictly into Indexing/Link collection and precise/imprecise keywords as presented here. Boundaries between these categories are loose, of course, and most search available engines can be placed somewhere between Indexing and link collection, like the keywords in most user requests can be placed in-between precise and imprecise.

4. Conclusions

Formatting text through structured languages is a condition of advanced automatic indexing of content. In scholarly publishing a shared *Document Type Definition*, when using *eXtended Markup Language* or *Standardized General Markup Language*, could be of a great additional advantage. Exchange of information between scientists would thereby be simplified due to fewer differences in document formatting. Combined with standard sections this could be a strong base for online publishing. Moreover, automated indexing is markedly enhanced, leading to improved access through search-engines.

Today one of the more popular ways of finding information is at publishers' online sites. This is because the Internet is as a whole too imprecise, unstable and not so trustworthy. The result is that the edited archaeological publications on the Internet are preferred to the non-edited ones. This could lead to some kind of economic monopoly over information, instead of leaving resources freely accessible on the Internet.

The edited and freely accessible Internet Archaeology (http://intarch.ac.uk/) is the exception that proves the rule of the impossibility of combining editing, Internet and free accessibility.

Redirecting development towards closed communities on the Internet would mean redirecting the focus from publishing for a small public in the supposedly everlasting printed media to publishing for the masses on the dangerously ever-changing Internet. If the problems concerning storage and conversion of files online should be resolved, there is a good chance that publishing here instead of publishing in the traditional way will become the long-term and permanent solution!

However, taking into consideration the problems dealt with in this article, which refer to the Internet as it is today, one can say that in general the Internet provides vast possibilities for publishing. Cheap publishing combined with good accessibility and an easy user interface (taking precautions against the problems connected with this issue) makes the Internet an open resource with a promising future as a supplement to the traditional ways of publishing.

Today the possibilities of the Internet are not well exploited regarding publishing. Many resources are wasted because the Internet is ruled by chaos and finding one's way around can cost a lot of time and patience.

Some sort of stability of scholarly publishing placed online in the form of online libraries is therefore essential. Larger libraries and universities should consider this a common goal, as these institutions are in the best position to provide it.

Acknowledgements

We are in debt to the following people for the help in preparing this text: Torsten Madsen, Lilian Ahlmann Johansen, Kristine Stub Precht and Asbjoern Romvig Thomsen.

References

FOSKETT, A.C., 1996. *The Subject Approach to Information.* Library Association Publishing, London. Fifth edition.

BRAY, T., PAOLI, J., SPERBERG-McQUEEN, C.M., 1998. Extensible Markup Language (XML) 1.0; W3C Recommendation 10-Feb-98. At: http://www.w3.org/TR/1998/REC-xml-19980210.pdf

BURNARD, L., 1995. What is SGML and How Does It Help? *Computers and The Humanities*, vol 29: 41-50. Also at: http://sable.ox.ac.uk/ota/teiedw25/

BÜNTE, O., 2000. XML auf dem Vormarsch. In *CT, Magazin für Computer Technik,* 10/2000. Verlag Heinz Heise Gmbh & Co KG: 200-213.

LESTR, R., 1997. The Need to Add Value. Towards a Worldwide Library: A Ten Year *Forecast.* In Helal, A.H. and Weiss, J.W. (eds.), *Veröffentlichungen der Universitätsbibliothek Essen* 21. Essen 1997: 13-31.

MEYER, E., 1999. CSS: If Not Now, When? At: http://webreview.com/wr/pub/1999/06/25/style/index.html *Web Review*. Miller Freeman.

Documents available on the World Wide Web:

The use of HTML and CSS: http://hotwired.lycos.com/webmonkey/

Definitions of DTD's, HTML, CSS, XML and the like: http://www.W3C.org

The online magazine *Internet Archaeology*: http://intarch.ac.uk/

Questions Raised by Electronic Publication in Archaeology

Jo Clarke

University of York, Department of Archaeology

King's Manor, York YO1 7EP, United Kingdom

e-mail: jpc105@york.ac.uk

Abstract

Electronic publication has become an accepted part of archaeology. It is seen as relatively cheap and easy to produce, giving a rapid result, as well as creating a widely accessible resource, but is it being used to its full potential?

This paper will address the issues which have been created by electronic publication in archaeology, how technological capabilities rival multimedia possibilities, and how speed and ease of publication have a bearing on the quality of presentation. Is it good enough to use the Internet for dissemination alone, or should we be focussing more upon the possibilities available through this alternative medium?

Keywords: electronic publication, Internet, non-linearity, database, images

1. Introduction

Electronic publication and the use of hypertext have been the subject of CAA papers for some years as the medium has developed (e.g. Ryan 1995). The purpose of this particular paper is to discuss the use of electronic publication in archaeology at present. In order to do this it will focus upon the transformation of a hard copy excavation report into an online hypertext document. It raises the question of whether this type of online report can be structured in such a way as to allow the reader to control and reinterpret the data presented. It covers the ability to include additional archive material from excavations, such as an online finds database, as well as clickable maps, plans and additional images. It attempts to compare the time taken to produce an online report in comparison to a traditional one and assess the longterm viability of an electronic report.

The internet strand of electronic publication has become more accepted with the appearance of journals such as *Internet Archaeology*, a peer reviewed electronic journal for archaeology which was first published in 1996 (Heyworth et al. 1996). There are now also web-sites offering information on on-going excavations (Hodder 1999), as well as guides for the public on popular tourist attractions. It seems, therefore, only a small step to start publishing excavation reports on the internet. As has been done on a few occasions so far both in *Internet Archaeology* (Wickham-Jones 1998) and on other excavation specific web-sites.

Using the Internet as a medium for publication provides new possibilities for the presentation of information. Conceptually this is not only in the way that the content of a document can differ i.e. more images, dynamic web pages etc. but it also enables the text, which can be exactly the same between a document in print and a document online, to be read in a different way. This is the difference between a linear document that is presented with a very rigid structure and a non-linear document that is basically a collection of pages that can be tied together in any number of ways.

It would appear however that the medium is not being used to its full. The amount of information presented on the Internet as if it was still a paper publication is huge. The two most usual approaches are to put the text into one large HTML page, with a linked contents list at the beginning, or not to present the document as HTML at all, but as a PDF file.

The motivation behind these two methods of electronic publication appears to be the desire to retain a traditional look to the document and also to allow it to be printed easily. It is well known that many people prefer to read text on paper as opposed to on screen, and that true hypertext documents can be seen as impractical as a publication medium because of the number of HTML pages there are to print out.

2. Aims of case study

The main aims of this case study were to create an online version of an excavation report, to expand on the information usually available in these reports and to change the format of that information. This was done in an attempt to demonstrate that the Internet is a useful medium for the publication of archaeological research, and that it is able to make information available to more people in a more user-friendly fashion.

The excavation report of the Anglian and Anglo-Scandinavian settlement at Cottam B, East Yorkshire was selected for this exercise. The excavation was carried out under the direction of Dr Julian Richards with a team from the University of York, UK. The online version of the report contains aspects that it was not possible to include in the hard copy version, i.e. a finds database, colour images, clickable maps and plans and post-excavation specialist reports in their entirety.

The excavation report was not originally intended for electronic publication and the traditional text report is published in the *Archaeological Journal* (in press). The traditional report follows a very rigid linear structure: the presentation of all the supporting data, followed by a discussion of the evidence and ending with a conclusion gathered from the facts. An online, hypertext version of such a report would not take full advantage of the medium if it did not change this structure. Hypertext allows links to be inserted in the text or within images, which if followed will expand upon the point of that text, or image. Another aim was, consequently, to reformat the report to suit the hypertext medium and to allow the reader to chose his or her own, possibly non-linear, path through the text.

Figure 1: The title page.

These aims are all to do with the presentation of the excavation report. There are, however, other aspects to this exercise: how difficult is it to create an online excavation report? How long does it take to transform information intended for archive into a form ready for presentation? Once this has been achieved, how many people will actually use this form of the report, and how does one prevent it from becoming outdated and unusable?

3. Description of the Online Excavation Report

3.1. Structure of report

The first thing the reader sees when accessing the report is the title page (figure 1). It is made to be eye-catching, the electronic equivalent of "glossy". The page contains the title, the authors, a picture of the site and four buttons giving access to the report itself. They give four different ways into the report:

- "Report" leads to the main path through the text.
- "Navigation" takes you to the "contents" page, it is a pictorial version which attempts to give a two dimensional impression of the links between the sections of the report (figure 3).
- "Finds Database" takes you to the query page of the database, from there you can pick the criteria you wish to search by and access the results.
- "Location Maps" leads to a sequence of clickable maps showing the location of the site at different scales.

The report has been structured as to avoid linearity, but the text does need to progress and contain some sense of order. To this end there is a directed path through the report, which although it is only basic, gives it a necessary backbone; something for the rest of the report to hang from. The non-linearity comes in the way that text is linked to illustrations and back to text, or to other illustrations.

The main path shown in the online version (figure 3) differs slightly from that of the hard copy report (figure 2).

Contents

Summary
Introduction
    Background and research aims
    Strategy and Methods
    Site location
Metal detector evidence
Field walking at Cottam, 1993 by B.E. Vyner
Geophysical Survey by Justin Garner-Lahire
    Methodology
    Area 1 results
    Area 2 results
    Assessment
Excavations at Burrow House Farm, Cottam, 1993 and 1995
    Introduction
    Period I: Natural and periglacial features
    Period IIA: Anglian Phase A
        COT93.1
        COT93.3
    Period IIB: Anglian Phase B
        COT93.1
        COT93.3
        COT95
    Period III: Anglo-Scandinavian
        COT95
    Period IV: Medieval and modern
The Pottery by Tony Austin
The Finds Assemblage edited by J.D. Richards
    The worked flints by John Bateman
    Miscellaneous stone items by J.D. Richards; lithology and provenance of stone objects by G.D. Gaunt
    Building material
    Non-ferrous metal objects by J.D. Richards with a contribution on a ninth-century strap end by Elaine Campbell
    Metalworking debris
    Iron objects by Patrick Ottaway
    Cottam ironwork recovered by metal detector by Patrick Ottaway
    Post-Roman coins by E.J.E. Pirie
    Bone and antler objects
Assessment of biological remains from COT93 by John Carrott, Allan Hall, Michael Issitt, Harry Kenward, Frances Large and Annie Milles
Plant macro fossils from COT95 by Allan Hall
Assessment of the bone assemblage from COT93 by Keith Dobney, Deborah Jaques and Don Brothwell
Identification of highly fragmented mammal bone from COT95 edited by J.D. Richards from report by Samantha K. Bell, Kristine Birchall and Don Brothwell
Discussion
Acknowledgements
Appendix A: Concordance of metal detector evidence
Bibliography

Figure 2: The contents page of the hard copy report.

This is because the online version is able to escape from the enforced linearity experienced by the original, and support a layered presentation. The main part of any report is the discussion and interpretation of the findings. It does, however, have to have supporting evidence. So the options available to print publication are either to include many introductory sections or many appendices, or both. This is where the ability to layer a hypertext document is useful. The main text of the report can be presented obviously, with all the detail hidden underneath. Consequently figure three illustrates the main path through the text; "Summary", "In-

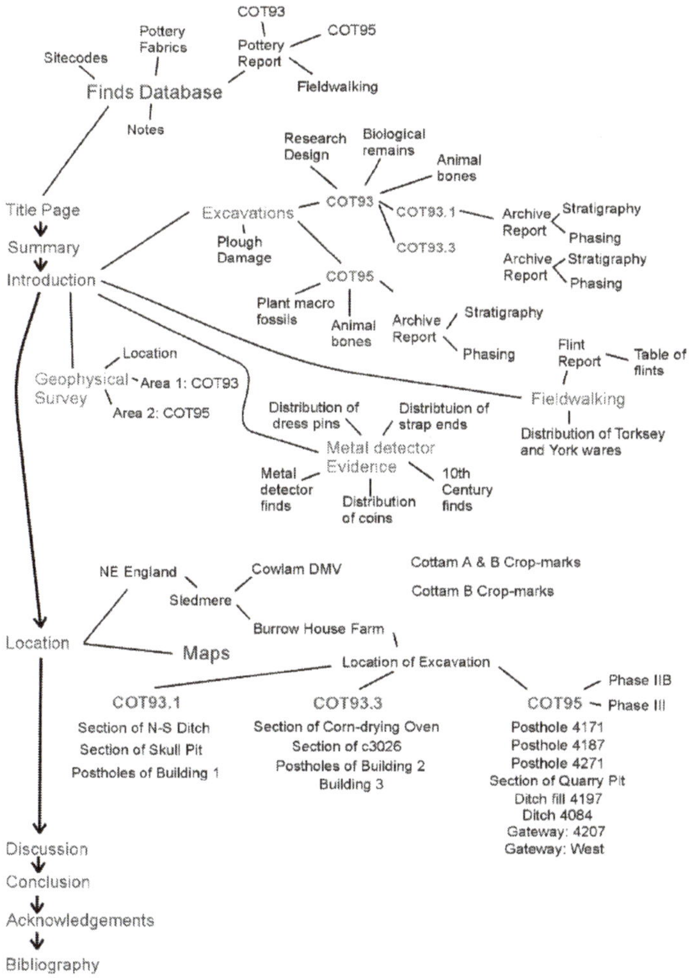

Figure 3: The contents page of the online report.

Figure 4: The navigation frame.

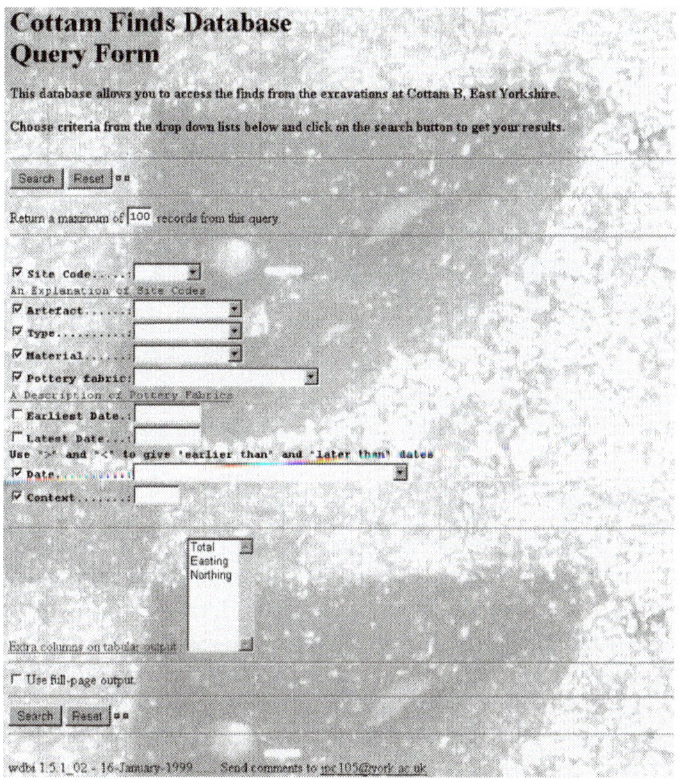

Figure 5: The query page.

troduction", "Location", "Discussion", "Conclusion", "Acknowledgements" and "Bibliography". The reader is told what the report is, what it is about, where it is, and then you get to the main interpretation of the site, without having to go through the more detailed areas of the report. The reader, however, may want to look at certain detail; hypertext makes that possible, through links to the areas of the report which are left out of the main path. This non-linearity is illustrated by the site-map or navigation page (figure 3).

The orginal idea was to make a more concise version of the traditional contents page (figure 2) into a frame at the side of the screen. This would have made navigation easy for the reader. However this did not seem feasible considering the size of contents list, and would force the report into a linear format. Instead a more pictorial form of contents page was opted for. It works by each label acting as a link to that section, and every section is represented in it. Although this form of contents page does not show every link between sections, it is able to give the reader more of an impression of the links between different areas of the report, than the traditional list would. The pictorial contents page also highlights the main path through the text.

Another aid to navigation was the inclusion of a frame at the bottom of the page containing the four buttons from the title page as well as a "home" button to allow the reader to go back to the beginning (figure 4). The combination of this small frame of navigation buttons and the "contents" or navigation page allows the reader enough freedom to roam the text as they please, and to access all pages with ease.

The database is the section of the report which differs most from the text version of the report. It allows the reader to query the small finds from the fieldwalking, metaldetecting and excavation at the site. There are three main pages that the reader sees, the query page (figure 5), the basic results page (figure 6), and the more detailed results page (figure 7). This allows the reader to drill down into the data gaining more detailed information with each "click". This is a theme which runs through the report as a whole allowing the reader to access information that matches their understanding or interest.

The results tables contain links to pictures of the artefacts and more detailed information from the specialist finds reports. It is

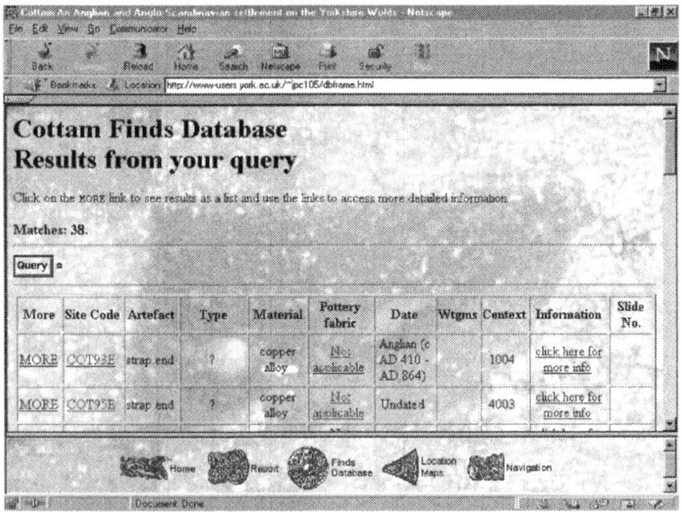

Figure 6: The basic results page.

also possible to link from sections of the text to results of particular queries, or individual database entries. This is one of the great advantages of having an online database in an excavation report. It does away with the need for lengthy descriptions and tables in the text. It also, of course, allows the reader to gain access to information not usually available.

4. Advantages and disadvantages of electronic publication

Having described the basic format of the electronic report, in this section I will discuss the methods used to create the report and any advantages or limitations I encountered using these methods.

4.1. Images

Images are a clear way in which an online report can transcend a hard copy one. It is easier to include colour images online than it is in print, due to the cost of colour printing. Also the amount of images is not limited by page limits. Images used online can also be "active" as is demonstrated by the clickable maps. The only limitation online is the amount of time it takes to download a web page containing many images. The online report does contain some large images, i.e. excavation plans, but they are not included in line with the text but are in separate pages to speed up access.

The images used are of various types, mostly scanned from photographs and drawings although some have been generated electronically i.e. by a GIS. The production of these images was time consuming and even those that were scanned from images intended for the print publication had to be cleaned and modified for use on the web.

The excavation plans were originally digitised using AutoCAD, with the intention of combining them with the geophysics, cropmarks and the excavation plans. The display of AutoCAD files was not possible however without the use of a plug-in, such as AutoDesk *WHIP!* or Voloview Express, which would reduce the accessibility of the information to many readers of the web document.

The ability to include more images in an electronic publication is an advantage but the more which are included, the more time it takes to create them, whether it is by digitisation, scanning or by

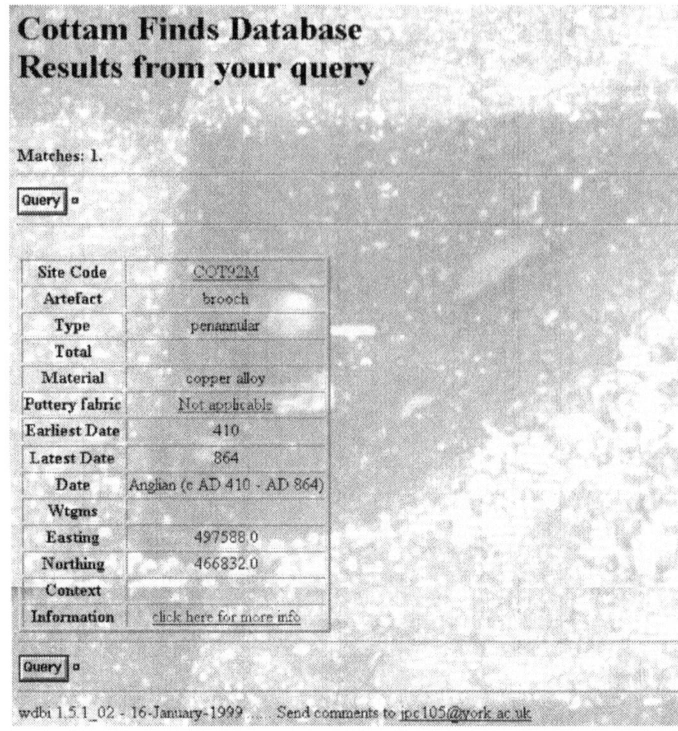

Figure 7: The detailed results page.

using a GIS. If scanning or digitising, the quality of the original also effects the time involved. Equally, it also takes time for an illustrator to draw publication pictures. Images are time consuming to produce for both Internet and print publication. Internet images are generally colour and there are more of them but they are at a lower resolution; with printed images the situation is reversed: fewer, black and white line drawings, but of better quality.

4.2. Additional material - archive/specialist reports

It is another well known advantage of online publication that disc space is not as restrictive as the number of pages in a print publication - therefore the publication of the whole of specialist reports is possible, rather than a simple summary. Although a limited number of readers may be interested in the whole report, at least if it is in electronic form the data within it can be extracted and reused.

4.3. Clickable maps

Clickable maps are a particular characteristic of electronic documents. They create a dynamic and spatial appearance to otherwise static illustrations. They are used in the sequence of location maps, which allow the reader to "zoom in" to the excavations and also to link to discussion of the images.

There are also basic distribution maps in which the reader is able to link to the database from individual finds in this case study. These distribution maps were created using the GIS software ARC/INFO and imported as images, so they do not provide the functionality of a GIS, although that can be implemented with more time.

Another use of the clickable map is in the Navigation page. The relation between pages in text was more easily portrayed using this method than it would have been with a list of text as shown in figure 3.

4.4. The database

The database allows the illustration of points made throughout the text and also lets the reader interrogate the data themselves. The report allows easy access to the database, from references to finds in the text and also, from individual database entries to descriptions in the specialist reports. It also links to illustrations of the artefacts where available.

The database runs on MySQL (http://web.mysql.com/) and is displayed on the web using a perl script called WDBI (Web DataBase Interface) (http://www.wdbi.net/). As the database is run by a cgi-script it is not client dependent, and so will run on any browser that is able to deal with HTML forms.

An online database is a feature that is only available in an electronic report and is one of the major advantages of using this type of publication. It makes a report more of an archive or research tool than a simple publication.

4.5. Archive stability

HTML provides a stable archival format; anything more proprietary is going to need more care and migration if it is to survive the years.

A last aspect of electronic publication is the need for the report to remain in a fixed place on the Internet. A most unfortunate characteristic of Internet publication at present is the transitory nature of web pages. References quoting web pages soon go out of date and all that is found is that wonderful "404 Page not found error"... This is not good practise and is where the Archaeology Data Service is essential to Archaeology.

5. Conclusions

An electronic excavation report is able to include more data and present it in a greater variety of formats than a traditional form of publication. There is a greater ability to see information from perspectives chosen by the reader, rather than being solely directed by the authors argument. An excavation report cannot be completely non-linear and reader directed but can offer more paths through the text than are available in a traditional publication.

The use of Internet technology allows a wider range of presentational characteristics, such as the finds database, the clickable maps and the use of more images, as well as GIS based techniques of displaying spatial distribution. The simple techniques are easy to use although they can be time consuming to create and maintain. The one major disadvantage in comparison to a printed volume is the need to maintain and upgrade the report as time passes and technology advances. However, it is possible to make this form of publication into both the archive and the publication if suitable file formats are used. It is also accessible to more people as an online report, as it is possible to access a web page from anywhere in the world. Both access and archival ability are very much dependent on the file formats used and the software needed to display any proprietory formats. Therefore the use of formats requiring plug-ins reduces access and should be carefully considered.

As the Internet becomes more widely used, this form of publication is certain to grow, as it is an effective way of disseminating information. Digital archives, like the ADS, are likely to become the libraries of the 21st Century, because digital information is more difficult to preserve as the growth of technology accelerates.

Acknowledgements

I would like to thank Dr Julian Richards for supplying the data used to create the online report and his subsequent feedback on the web-site as a whole.

References

HEYWORTH, M.P., ROSS, S., and RICHARDS, J.D., 1996. Internet archaeology: An international electronic journal for archaeology. In Kamermans, H. and Fennema, K. (eds.), *Interfacing the Past: Computer Applications and Quantitative Methods in Archaeology CAA95*. Analecta Praehistorica Leidensia 28: 517-523.

HODDER, I., 1999. Archaeology and Global Information Systems. *Internet Archaeology* 6 (http://intarch.ac.uk/journal/issue6/hodder/index.html).

LONGSTAFF, T.R.W., 1994. Hypertext as a medium for archaeological publication. *Archaeological Computing Newsletter* 39: 1-2.

RICHARDS, J.D., in press. An Anglian and Anglo-Scandinavian settlement on the Yorkshire Wolds. *Archaeological Journal* 156.

RYAN, N.S., 1995. The excavation archive as hypertext document? In Huggett, J. and Ryan, N. (eds.), *Computer Applications and Quantitative Methods in Archaeology 1994*, BAR International Series 600, Tempus Reparatum, Oxford: 211-219.

WICKHAM-JONES, C. and DALLAND, M., 1998. A small mesolithic site at Fife Ness, Fife, Scotland *Internet Archaeology* 5 (http://intarch.ac.uk/journal/issue5/wickham/index.html).

In Patrimonium: A Data Model for Museum and Cultural Heritage Information

Fernando Cabral

Instituto Politécnico da Universidade do Porto / Sistemas do Futuro, Lda

e-mail: fcabral@sistemasfuturo.pt

Mário Brito

Museu D. Diogo de Sousa / Instituto Português de Museus
Departamento de Ciências e Técnicas do Património da Faculdade de Letras da Universidade do Porto

e-mail: mbrito@letras.up.pt

Abstract

In this paper we presented the process that lead us to In *patrimonium, a data model for managing information in museum and cultural heritage organisations. We described the conditions of the project development, the context of Portuguese museums and the main goals for the development of the computer system.*

1. Introduction

Between 1997 and 1999 we participated in a project – GEIRA – directed by two Universities located in the north of Portugal (Universidade do Minho and Universidade de Trás-os-Montes e Alto Douro) the main goal of which was the use of the Internet and CD-ROMs to present information about Portuguese cultural heritage in the north of the country.

This project allowed us to develop a line of action dealing with the management of museums collections, paying special attention to archaeological museums. The reason for this was the importance of this kind of data in the region and also because no computer based tool for museums (in accordance to the recent international developments in museum information standards) is present on the Portuguese market.

The GEIRA project was organised in several ways according to three main themes: Science, Nature and Culture Heritage.

The Cultural Heritage theme was divided into several additional divisions. One, the museum, was defined with the following main goals:

- Establishment of a museums network
- Promotion of computer science
- Use of the Internet
- Research on computer science and museums
- Educational activity of museums with computer science.

According to an evaluation of museums' needs, we defined several activities with several levels of impact in museums' activity:

- Web pages: Development of web pages for about 50 museums. These pages present the museums, their collections and main activities.
- ISDN connections: 30 museums had an ISDN connection to provide Internet access, email and other facilities. This service included a help desk by email and phone.
- InfoMuseum: An information system for museums and museum professionals with several types of information organised in databases. An email distribution list is also being used in order to exchange information.
- Virtual Gallery: A VRML model used in order to allow museums the realisation of virtual exhibitions.
- Management of museums collections: Development of a computer tool which could be used for cataloguing and management of museum collections.

2. Management of museum collection

The development of this activity was based on the requirement of a system that could use all the international experience on standards and collections management. The increasing technological evolution with new boundaries like the Internet and public access to information, were also an important factor to this decision.

2.1. Starting points

- A system we designed in 1989, developed with 4TH DIMENSION. The cataloguing and management features included several forms of museum data and documentation: objects, library, images and conservation. However, used by 10 Portuguese museums, this system did not have the required developments, especially in the technological aspects.
- A system we designed in 1996, developed with ACCESS, designed according to the needs of certain museums and used by several museums all over the country. This system took advantage of important facilities in the integration with the Windows operating system, multimedia capabilities, Web integration and other features. However this system did still not use the international proposals from the Museum information standards.
- Experience on documentation in museums, based on a number of years of professional activity dealing with museums and other cultural institutions. For this experience the most useful were the contacts with initiatives and organisations that, especially in Europe and North America,

deal with information systems in museums and cultural heritage.

- Knowledge as regards Portuguese museums/museums professionals: their manual and computer based systems for cataloguing and managing collections, their skills in computer science and professional data training on computer science and information systems in museums.

2.2. The context

2.2.1. Evaluation of Portuguese museums

Regardless of the rising indicators on the Information Society and the Internet, museums are still not aware of the importance of their role as content providers and the multiplicity of uses that technology can have in museums: management, research, interpretation and presentation. Museums directors do not fell that the investment in computer science is a strategic one. On the other hand, the required investment, including training, is to large for their budgets.

The amount of work in a non-familiar environment such as the technological world is of a large concern: organisation of museum documentation, research, definition of standards and procedures, systems and equipment acquisition, data entry, maintenance, training, and so on.

The condition of Portuguese museums and its professionals as regards documentation systems and training in computer science is by no means excellent: computers are basically used with programs like Word and Excel; the concept of documentation still remains with the idea of card files to document the collections. When they think of using computers to deal with collections they simply reproduce the card file model they are used to.

Based on case studies, we analysed museums in relation to information: the museum as an owner and keeper of cultural information and also as an information producer. This analysis allowed us to define several patterns of museums needs and a better understanding of the complexity of museums concerning the diversity of the information they preserve and what the way they can use it.

Most museums do not have a planned documentation system nor a clear understanding of what may be the benefits of the information integration for its daily activity (collections, documentation, conservation, activities and so on).

The 1997 survey amongst the 80 museums that exists in the North of Portugal showed that very few museums[1] (to be precise only 3) used computer systems to deal with their collections and only 2 museums had their own homepages[2]. At this moment, in about the same number of museums, these numbers have radically changed: 47 museums have their own homepage[3] and 16[4] use a computer system to manage their collections.

Museums are being forced to move into the technological world. The Internet, email and other services are pressuring museums to move forward. The potential role of training and conscience in the Information Society are two very strong ideas necessary to be developed over the next years.

For these changes to take place, we think it would be useful to define special programs to do the digitalisation of the information available in museums. These programs would be important contributions to a more active role in the Information Society of museums by creating cultural contents and creating new forms of achieving their tasks.

2.2.2. State of the art: Management of museums collections, standards and terminology

This phase of the Project dealt with the study of museum documentation systems, the evaluation of several packages of software developed for the management of museum collections and the analysis of different data models and international proposals on this subject.

We based our research on several documents, namely:

- International Guidelines for Museum Object Information (http://www.cidoc.icom.org/guide/guide.htm),
- A European Core Data Standard for Archaeological Sites and Monuments (http://www.natmus.min.dk/cidoc/archsite/coredata/e_cds.htm),
- MIDAS - a manual and data standard for monument inventories (http://www.rchme.gov.uk/midas/index.html),
- Spectrum (http://www.mdocassn.demon.co.uk/spectrum.htm) and
- Normas Documentais para Museus of the Spanish Ministry of Culture.

A number of web sites from various organisations and initiatives[5] allowed us the approach to several issues like terminology developments, systems, products and metadata developments.

The complexity of museums regarding the nature of their collections, activities, organisational structures and technical know-how, increase the difficulties for the definition of standards. At that time there was no kind of co-operative work for the definition of standards in Portugal.

However, the main problem we have to consider is the different levels of approach concerning the documentation in museums. Some museums have a quite reasonable understanding about the museum documentation system. However, most of them find it hard to pull away from their daily tasks and integrate their work into a global approach.

3. Main goals

- Use of international experience as regards standards:

  A lot of work on museum documentation systems is being performed by numerous organisations and initiatives. This is taking place at a moment when globalisation is an important issue and public access to information is becoming a common goal for museums and cultural heritage organisations. Different systems, different software packages and even different languages must not be an obstacle to easy public access to cultural information.

  The use of international proposals will allow us to exchange experience and carry on with a continuous development of conceptual models.

- Modularity and scalable functionality:

  The museum information system can be rather complex, depending on the complexity of the museum itself. Systems must be designed to fulfil all of the different levels and purposes of a certain museum.

The same idea must be applied to the different modules of the programme: museums must have what they need according to their organisation, purposes and technical skills.

- Integration of several collection types of data:

  Small and local museums usually have different types of collections. We wanted to be able to integrate different collections in a very understandable data model that allows a general as well as a detailed approach to collection specifications.

- Vocabulary control and thesauri:

  As a component of museum documentation system standards, the system must provide facilities to its management, encouraging research and exchange of experiences between museums.

- Public access:

  Museums must have full control of the data they allow public access to. The data accessible over the Intranet or Internet must be useful for all different types of public and must exploit all multimedia possibilities that the museum is interested in.

- Full integration with the operating system:

  It must be possible to analyse the integration with the Windows operating system through two different approaches. The first approach is through the utilisation of the operating system capabilities, while the second should be through the creation of an environment similar to the Windows operating system.

- Full Multimedia facilities:

  Full integration of digital files: text, image, sound and movie.

4. Conclusions

Since the beginning of this project we have had almost two years of experience, which have turned out to be rich in the understanding of the complex changes that are being introduced into museum practices. Technological development forces us to constantly update our use of technology and to constantly revise what we can achieve with it.

- The positive feedback exceeded our expectations:

  One of the characteristics of our data model is its simplicity and, at the same time, its complexity. This characteristic, associated with modularity and scalability, allows several different approaches according to the needs of the organisation. We have concluded that professionals who have any kind of experience in cultural heritage information systems, will find this model to be easy and understandable.

- Training on cultural heritage information management:

  The evaluation phase of the model resulted in tests in several levels of training at a secondary school and university level. The tests were performed by pupils and students who have a special interest in cultural heritage.

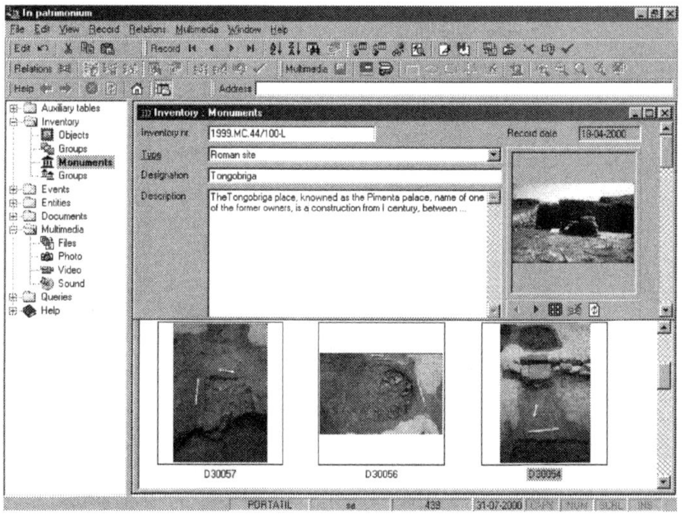

Figure 1: Page showing the organisation of information in In *patrimonium.*

5. *In* patrimonium

Prepared to work on *Windows* (98, NT or 2000) this application has been improved, either on development terms, or actualising information and interface with users. Using the most recent *Windows* programmes, this product (thanks to its module construction) allows the management of information in an integrated way. Therefore, it is possible to access information about an object, view the archaeological site where it is or was placed, the ambient where it is inserted, the related documentation, etc.

In a global management concept, *In* patrimonium integrates (in a single product) all the existent information, such as *Collections, Monuments* and *Archaeological Sites*, allowing a better management and integration of the information on cultural heritage.

5.1. Organising the information

The information is organised briefly, presiding from large forms and the specific information fields can be configured according to the user's interest.

All the information, included in the programme, can be linked. There is a module, in a certain way "revolutionary", that allows the sharing of all existing information found in the forms. All the modules can be linked between each other.

Therefore, it is possible to link Objects with Conservation and Conservation with Exhibitions; Exhibitions can be linked to files and data from the Multimedia module.

5.2. Modules

Multi-user: the configuration of different users and respective type of access. In this application it is possible to create user accounts, thus enable specific users to insert data or only consult the database. We can also create one or more database administrator accounts whose titular must have some knowledge in informatics.

Scalability: Although using a format known as Client/Server, In Patrimonium can also work with a single workstation. It is possible to upgrade it to the Client/Server format without any loss of data.

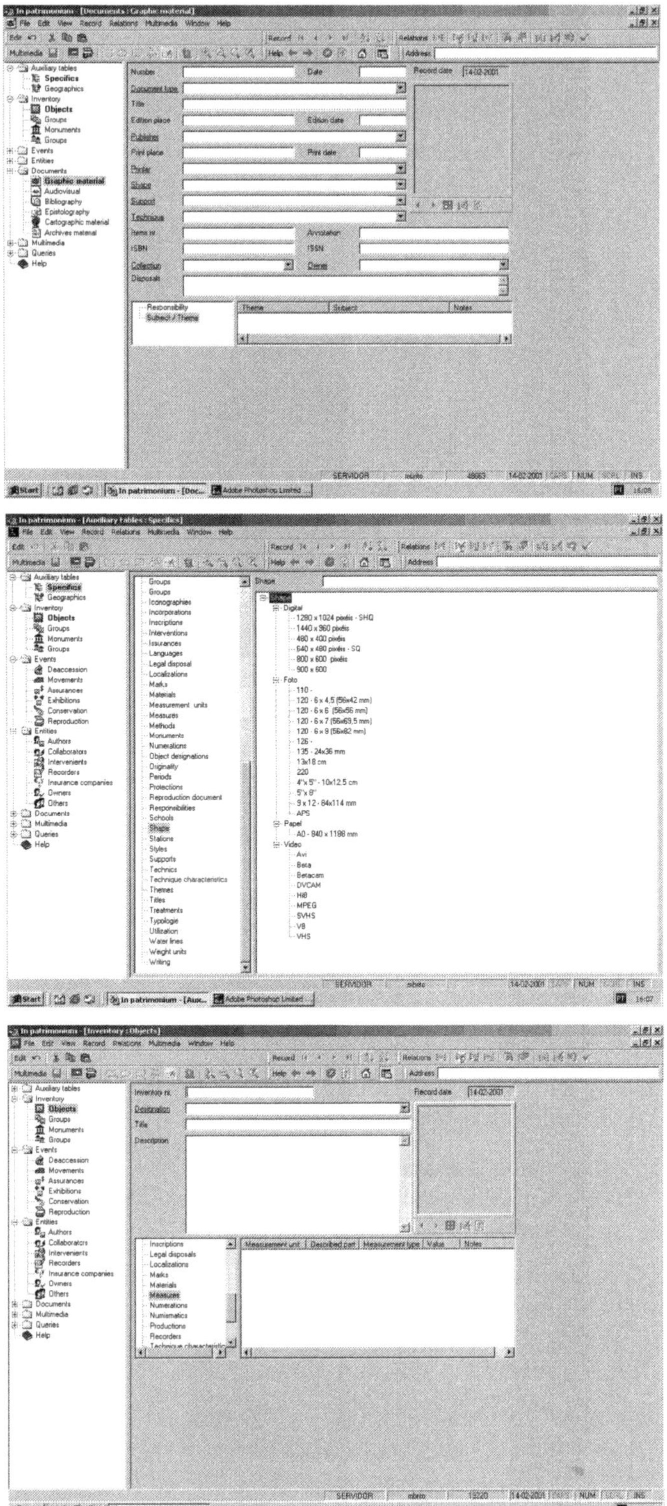

Figure 2: Three pages representing three different modules of In *patrimonium.*

Multi-language: configuration of the application to work with a linguistic interface. Besides the Portuguese version, also English, French and Spanish versions are available.

Auxiliary Tables: enable management and configuration of different *tables/terms* in order to help at the data insertion process in the remaining application modules.

Inventory: data records related to the archaeological cultural heritage and monuments, their classification; also offers the possibility of associate photos, video and sounds.

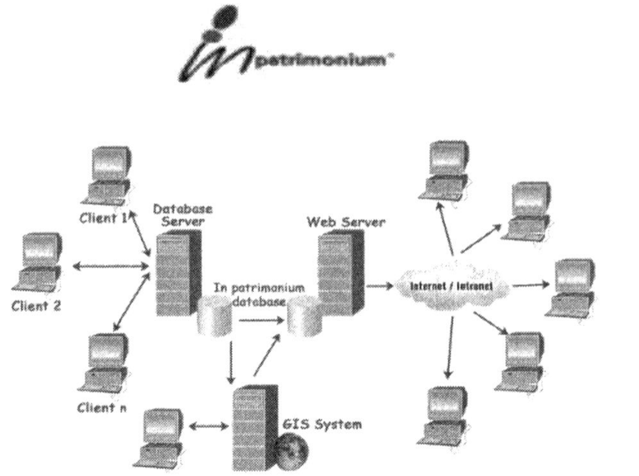

Figure 3: Physical model of In *patrimonium.*

Groups: data records related to the collections and monuments groups, with the possibility of associate photos, video and sounds.

Entities: data records related to the entities - authors, interventions, collaborators, recorders, owners and others.

Events: data records related to the events - exhibitions, conservation and restoration, movements, insurance and reproductions.

Documents: data records related to the documents - graphic material, audio-visual records, bibliography, epistemology, cartography and archives material.

Queries: query wizard – insert, change, delete and visualise queries, configured by the application users.

Multimedia: module that allows the management and treatment of photos, video and sounds.

5.3. Other functions

- Compatibility with data from other products, namely Microsoft software - Office, Word and Excel.
- Easy information access through the Internet or Intranet.
- Multimedia technology in order to use digitalisation and treat images (photographs), video and sound.
- Inter-related information with easy query access due to records self-association, as for example, of museum objects with records of other modules (documents, entities, events, multimedia, etc).
- Configuration that allows working with interfaces in several languages.
- Client/Server system (scalability), without a user limit.
- The information included in the application can be integrated with GIS.

5.4. Public access

Thanks to a special Public Access Module called *In* web, In patrimonium can safely be consulted by the general public. *In* web was developed for organisations that want to make their collections, archaeological sites and monuments available on the Internet or Intranet. This interface for public access is fully adaptable to suit the house style of a museum/institution. Instead of being a product itself, with this module, the institution staff can determinate which fields and records are shown and in what way the public can retrieve and view the information.

System requirements

Requisites	Minimum	Recommended
Operating system	Windows 95	Windows 98 / NT/ 2000
RAM	32 MB	128 MB
Disk space for *In* patrimonium	15 MB	15 MB
Disk space required	300 MB	1 GB
Processor	Pentium	Pentium II
Video card	SVGA	SVGA

Notes

1 Museu D. Diogo de Sousa, Museu Nogueira da Silva and Museu do Ferro de Moncorvo.

2 Registered in SAPO, most popular research engine in Portugal.

3 According to ICOM-Virtual Library Museums- Portugal (http://www.letras.up.pt/dctp/icom/vlmpt.html) maintained by Mário Brito and Alexandre Matos.

4 Casa Museu Abel Salazar-Porto, Casa Museu de Camilo-Vila Nova de Famalicao, Museu Abade de Pedrosa-Santo Tirso, Museu Alberto Sampaio-Guimaraes, Museu de Lamego-Lamego, Museu do Abade de Baçal-Bragança, Museu do Ferro e da Regiao de Moncorvo-Moncorvo, Museu Municipal de Penafiel-Penafiel, Museu Regional de Arqueologia D. Diogo de Sousa-Braga, Museu Terra de Miranda-Miranda do Douro, Paços da Seda-Macedo de Cavaleiros, Sala-Museu Municipal de Arqueologia de Mogadouro-Mogadouro, Museu Nogueira da Silva-Braga, Museu da Faculdade de Belas Artes da Universidade do Porto-Porto, Museu dos Biscaínhos-Braga, Museu de Serralves-Porto, Divisao de Museus da Câmara Municipal do Porto, Casa-Museu Teixeira Lopes.

5 Please refer to Cidoc web Site (http://www.cidoc.icom.org).

Can Schoolchildren Digitise Their History?

Helene Simoni and Kostas Papagiannopoulos

Landscape Archaeology Group

Kamenitsa Alissos, 25002 Vrahneika, Greece

e-mail: kamenits@otenet.gr

Abstract

Increasing numbers of Greek schools are introducing local history projects in class nowadays. Our experience shows that children can carry out their own research. Children, given the opportunity, can be instructed on how to carry it out with reliable scientific results. By developing concepts of time and space, children can record their finds more accurately. Particularly space is a very critical factor because children must learn to transfer onto paper the spatial elements of the information that exists in the landscape and in the minds of the inhabitants, such as place-names, field systems, landuse, population shifts etc. The conversion of mental maps to digital ones is facilitated by the use of modern tools like G.I.S. Close collaboration between schools and the Archaeological Service can result in the establishment of a network, through which the children's finds will reach professional archaeologists and historians for further evaluation and treatment.

Key words: education, local history, space, mental map, digital map, archaeology, GIS

1. Current situation in education

Archaeology has never been taught as an independent subject in the traditional Greek education system, whereas elements of archaeology were to be found primarily in history textbooks and secondarily in geography and language textbooks. References to archaeological finds and research are given in the syllabus with emphasis placed on their significance as examples of art and craftsmanship.

On the other hand, history and geography are the subjects, which along with language contribute more than any other subject to the promotion and consolidation of national identity (Abdela 1997:34). They are therefore, very important for the influence that they exercise upon students of all levels, despite their inferior position as ranked among subjects at Greek Schools, where mathematics or literature are held at higher esteem.

The realisation of the syllabus content, according to the curriculum, relies on the textbooks. History textbooks, for instance, are focused on Greek national history, particularly on extensive accounts of events (battles, treaties etc.), names and dates, and they do not fulfil their original educational targets. Students are given a great deal of information passively, since it is the teacher who speaks and they who listen. That makes history a rather boring subject and one that schoolchildren tend to consider light and trivial (Papastamatiou 1984:98-99, cf. Palla 1994:224-236).

The subjects of history and geography can give a very confused idea of time and space. Both concepts are defined with various definitions that all point in the same direction, to prove the continuity of the Greek culture from as far as we can reach into the past up to now and in the future, and from what nowadays constitute the Greek state to the rest of the world, where the Greek civilization has influenced the others and remains vivid thanks to the modern Greeks. The fact that Greeks appear to have had certain virtues, since time immemorial, is deprived of the particular conditions under which the inhabitants of a region in or outside the official boundaries of the modern state were forced or encouraged to act (e.g. to revolt, emigrate, fight, die etc.) (Ventoura 1997:404-5).

2. New trends in education

Even so, modern books of history and geography are undergoing changes and improvements that allow children to participate more actively in the learning process. In addition to text, there are diagrams and tables and graphs that familiarize children with a more quantitative approach to time and space. Children are not considered simple receivers of the messages that textbooks or their teachers transmit. Otherwise, detailed studies have shown that children between the ages of 12-15 have fully developed the concepts of time, space and cause, and can pose complex queries and offer solutions (Jahoda 1963).

However, most children up to this age are not familiar with scientific research. This certainly has to do not only with their stage of maturity, but with the education system as a whole, which tends to separate the real world from school activities. Therefore, the traditional method of teaching and the use of textbooks have been strongly criticised for their weakness to play the roles, which scientific work demands. Teachers and students are now encouraged to use extra-mural activities and study from the sources, while the introduction of environmental education in the 1990s has alerted them to key issues, such us the place of humans in nature, the interaction of man and the environment in the past, present and future. Another change is the shift from the "general" to the "particular", from "anywhere" to the "local".

The idea of school children writing their history is not new and has been developed elsewhere more extensively. Besides, must we be pessimistic as to the extent a pupil or a group of pupils can process their finds and proceed to interpret them, as well as their implications for the past in general? "*New History*", which emerged as a modern theory on history teaching in the 1970s and the 1980s, introduced a rather optimistic and practical perspective. According to this theory, the only alternative solution is to try to get the children, as far as possible, to carry out their own research. Rogers (1978) stresses that students may be introduced to research from the sources and obtain a permanent experience in this work. The teacher may use the appropriate way every time, depending on the level of the development of each student. Rogers quotes Bruner's theory to support his ideas. Bruner, in contrast to Piaget, claims that the three modes of representation (enactive, iconic, symbolic) coexist in children's thought process, though in the younger ones the two first dominate. Thus, schoolchildren can deal with all kinds of meanings, corresponding to the level of their abilities, of course (Worell and Stilwel 1981:168). Applying it to history teaching, it is possible to study any subject of history, without the teaching approach being limited by age limitations, as corresponds to the traditional method.

"New History" was understood as a threat to the national histories and the traditional perception of what deserves to be taught. Even scholars, like Thompson (1984:173), remark that there is a risk of developing in the students an extremely subjective image for the past and doubts for the effectiveness of their actions. Students' projects within the framework of the environmental education and local history are the best response to any and all hesitations and fears. Environmental education, as a response to the traditional school crisis, is offered on a voluntary basis at schools - and pedagogic teams work using project methods. A great deal of such mini projects set up by the children every year all over Greece and a large amount of information, occasionally fairly original, is collected and processed (Flogaiti 1993). Environmental education consists of several sections, one of which is local history. Pure local history projects may reach 20% or more out of the total every year. Students themselves can carry out a pioneering research in their region, produce new material on local history and consequently become recorders and guardians of a cultural heritage not yet recorded.

The Ministry of Education is now willing to introduce local history into the curriculum as an optional activity in the framework of general history, at the same time keeping this activity far from the traditional school practice. The fact that instead of one textbook for the children, there will be three different manuals for the teachers, written by three different teams of authors that will refer to local history theory and methodology, demonstrates the re-orientation of regional studies in the Greek education system; especially in Gymnasium, which corresponds to the very productive ages between 12-15 years. We are delighted that one book of the three, was written by the present authors plus a third colleague, S. Fragoulis. Therefore, the ideas that are put forward here will circulate throughout Greece from year 2001 onwards; and we hope that they will meet a creative response all over the country and contribute to a wider reform concerning the image of history and archaeology at Greek schools. The concept of the manual is that local history cannot be taught, but is to be discovered by the children themselves, who will carry out research in their regions. Children will explore the spatial and temporal dimension of local history and will disseminate it to the public through various activities (Papagiannopoulos et al. in press).

This is feasible if a close collaboration between educators and local researchers is established. The educators will share their pedagogic qualification, and the researchers the sort of questions they expect the children to explore. It further means that all parts involved in school education, especially teachers, have a great responsibility in setting up projects, in developing a self-critical eye and in keeping their work separate from any influences, which tend to falsify and lead to altering its meaning.

3. Understanding space

Before setting out the research project, the pupils need to know about the means that will be used. Certainly the goals have already been formulated, but they are usually general and widely accepted, such as: "preservation of local identity" and so on, which give a very positive impression, but can be accomplished in hundreds of different ways. On the contrary, the means are more specific, more concrete and give a clearer picture to what is going to take place within the following days or months of the project.

Mentioning maps as a means is not enough. Children know maps very well because they use them extensively in their school practice. They use them as static drawings on a sheet of paper. They are not taught to link them with the basic concepts of space and time, and they have not had the opportunity to think about these concepts and their representation on maps in a creative way. And yet, space and time through their tight interaction play significant roles in the course of man's life and the evolution of historical facts. To make this more self-evident, we have applied the most recent local history as our starting point and terminal. The reason is obvious. We would rather reduce time to the one, which survives in the living memories of the older inhabitants of a region and the abundant documents of the Community Archives. It leaves us with free hands to make full use of space, i.e. the environment, rural or urban, that is unfolded before the eyes of the children, and let them discover its several dimensions. Maps will provide the means to record all the new discoveries.

A map is the output of an elaborate procedure that aims at representing the earth's surface or part of it on paper. Maps allow us to calculate distances, find our way to any destination, choose a suitable location to build our house, decide where to spend our holidays etc., based on a series of geographical data (environmental, cultural etc.) that is represented on it. However, we rarely use maps in our everyday life, and yet we manage to find our way home, to plant flowers at the right side of our garden, to play "hide and seek" and not get lost. All these common habits are performed by means of a series of geographical data that is stored in our minds instead.

Children must get familiar with this kind of data that exists in their own minds and in everybody else's, and they must be able to recognize it. First they can be given examples of their daily routine that can strengthen their comprehension of the geographical dimension. Since everything they do and watch others doing is linked to a specific location on the ground and is associated with the use of space, they can think of the kinds of conventional maps that correspond to the mental maps, the ones they keep in their minds (figure 1). The significance of a conventional map is equal to the significance of a mental map because it comprises a physical and tangible document that anyone can access at any time. Moreover, its preservation depends less on its owner, since it can survive after the owner is dead; unlike the mental one.

So far, the creation of maps has been the privilege of a few, the most powerful ones, who possessed knowledge, equipment, and authorization. Officials who played a role in public administration could intervene in order to express their opinions and ideas through the creation, production and circulation of maps (Pavlou 1995:31). We do not mean that all interventions were deliberate. Many times they were guided by the subconscious and habit. We still draw Europe in the centre of the international map. Though

MENTAL and CONVENTIONAL MAPS

• Patras is within 20 mins of driving distance from Kato Achaia.	➽ Topographical map
• My parents and most of our neighbours have their own olive groves.	➽ Land use map
• Sea is contaminated near Patras, but I can swim in Tarandella.	➽ Pollution map
• My friends and I have explored the springs of the Peiros river.	➽ Hydrology map
• The old National Road is picturesque but the new one is safer.	➽ Road network map

Figure 1: The conversion of mental maps into conventional ones with examples from the children's everyday life.

we now know the shape of Earth, we have not abandoned a tradition that dates to the period of the great discoveries when the rest of the world served the interests of the European hegemonies. The "omphalos" syndrome is found in other cultural traditions as well. Chinese cartography of the imperial period show the rest of the world existed on the fringes of the Chinese empire (Smith 1998:53). The maps depicting the Holy Lands dating from the 15th to the 19th century was a means to reinforce the religious sentiment, rather than show the true location of places. This is the only way to account for the lack of accuracy and the absence of a consistent scale, projection and orientation throughout the same piece of map. Even time distortions were allowed to show together anachronistic scenes from the Old and the New Testament (Rubin 1992).

Whereas the above-mentioned examples might be considered exaggerating, there is no point in trying to identify a "true" or "false" map. Small or big, black or white, deliberate or unconscious, "lies" can be found in all maps. Truth is suppressed to help the user see what needs to be seen (Monmonier 1996:25).

It is worth examining the kind of maps that are used at school and the messages they convey. The different types show the country or separate continents or the whole of the globe. Children get a good picture of the plan of countries or provinces, but the rest is full of dots and very thick lines. Children are required to recognize the plan of the Peloponnese, Greece, Europe, Australia, the Nile etc., but no sign of their region is represented in anything more than a dot or a label. As a result, the same people, kids or adults, who can easily recognize the plans of countries and continents do not recognize the plan of their own place and cannot describe it. And yet, this sort of map is necessary because they offer the details needed to promote a good and reliable understanding of the space and the features that are included in it. However, they do not exist in commercial form because they will not sell, while the versions that the Army has produced are considered confidential and can be obtained after some bureaucracy has been involved.

The lack of more reliable alternatives has led to the heavy use of certain unsatisfactory and inadequate maps. Local history projects cannot rely on them. Fortunately the development of technology in the field of Information has been very rapid and cartography is becoming affordable by more and more individuals nowadays. The increasing number of individuals who can now own reliable and inexpensive hardware and software has given the opportunity to more people to access fields that were forbidden before. Cartography can be considered one of them. The existence of cheap and easy Geographical Information Systems provides a valuable toolkit to the children for their projects.

4. Digitizing space

The operation with the tools is another delicate stage. No matter how well certain theoretical issues have been forwarded to the children, it is practice that proves how well they have been understood. At first they can practice manually with black and white copies of existing maps. The 1:50,000 topographical maps of the Army are a good idea. Children may try to paint the outlines of rivers, roads, settlements, forests, contours, sea etc. with different colours. If this is done correctly, they should end up classifying them into three categories, thus discovering the three topological concepts - point, line and area, by which every geographical phenomenon can be represented. A fourth one might be the text on the map, that is, labels identifying what each geographical phenomenon is, and these comprise their non-spatial attributes. By drawing only one type of information (e.g. only churches or only the main rivers etc.) on tracing paper overlaid on the base-map, they create a thematic map and they realize how many themes one usual map contains.

If a map is a set of points, lines, and areas that have non-spatial attributes, their location in space with reference to a coordinate system is what makes a map complete and usable (Burrough 1992:13). The topographical maps of the Army use a standard system of coordinates, while children at this age have already learned how to measure coordinates in their mathematics class. Therefore, they can practice by finding out the coordinate pair of a few points and locating at least 2 pairs for each of the other features.

The final characteristic that needs to be explained is the scale. Students most probably won't have the luck to be able select a scale for their maps but will use the available maps as their basis. The scale shows the degree of precision and accounts for the missing details. Again, students are equipped with the knowledge required to understand how it works and are familiar with the mathematical calculations that need to be done.

Manual training with maps can conclude in the creation of their own maps of their neighbourhood or a plan of their school. As soon as the map is ready they will realize that no matter how good this is it is a finished job. If anyone wishes to correct something or move it, or erase it, or enlarge it, or re-orient etc., it will be difficult and not very neat to do it on the same map and will have to start all over again in order to draw a new updated version. Unfortunately, the creation of analogue maps is a difficult and not easily repeated task. That is why we turn to Information Technology, which offers better alternatives.

Nowadays, most students are familiar with the use of a computer and almost all schools throughout Greece have a classroom with PCs. Internet is also open to many schools, and every year more schools access it. Therefore, there are many open gates that children are allowed to enter through with the use of existing hardware and software. Fortunately, economical GIS software can be either purchased or downloaded

The difference between a digital and an analogue map should be well demonstrated. What will interest the children more, and will be of benefit for the successful realization of the project, is the emphasis on the quantitative aspect of the digital map. The analogue map is a static, qualitative document. It contains only the information that is displayed on a flat printed piece. The dynamic character of the digital map allows a quantitative approach to spatial analysis, which is what we need for our Local History projects. It is not difficult for an expert to show children the basic steps from digitising to handling their data in a Geographical Information System. If they are adequately trained in advance, certain tasks can be performed by children using GIS.

The difficulty lies on the quantitative character of every piece of information; consequently, this is where the emphasis should be placed. Behind every single pixel of the screen there is a number, which corresponds to a piece of information (we remind the children that its coordinates and its non-spatial attributes characterize every piece of information). Therefore, in order to produce a good digital map it is not only important to draw accurate lines, but also to link the drawings with the right numbers which comprise the map's database. Children will have the chance to view the matrix with the numerical data that are behind any map and not only the pictorial display of it. They will also have the chance to process these numerical data like any other numbers and see what happens when one adds, subtracts, multiplies, etc. maps.

The fact that students will create maps on the screen with great ease and speed, and will manipulate them as they wish, can be enhanced by asking them to compare their experience with hand-drawn maps and with digital ones. The answers of the students will show the degree of progress and will test how ready they are to proceed to the implementation of the Local History project.

5. Digitising history

Children with previous experience in environmental education projects know well that a good and modern historical investigation requires constant effort in groups. However, the degree of participation and interaction between the members of each group heavily depends on good information and comprehension. When a satisfactory degree of familiarity with the basic concepts has been accomplished, the central tasks of the project can be put in action. Subjects are unlimited and inexhaustible, given that the approach of any studied object can be either vertical (with regards to exhausting details and observations in depth), or horizontal (with references to relevant objects from the same or other cultural backgrounds and comparative analyses). So, the criteria of choice depend on factors such as: age, maturity and interests of pupils, qualifications of their teachers, the place where the school is situated, familiarity of the team with local history, time availability etc. The team, according to their desires and the topic they are going to study, will choose the type of techniques to access the topic. There are techniques that have been used successfully abroad and some of them have been applied in programs of museums and environmental education in Greece.

The pedagogic target is for the students to:

- cultivate imagination
- develop initiatives
- experience different roles in interpreting themes from the past (Abdela 1998:121-9).

The scientific target will be achieved depending on the amount of original evidence that will be produced. Therefore, it is necessary for the teacher to have a regular contact with the local Archaeological Service or any other research group. Children can carry out their fieldwork either by interviewing the locals or examining the Community Archives or walking in the fields and recording surface finds. Although the latter requires an additional skill in "reading" the earth's surface and distinguishing cultural material among other objects that lie on the ground, the children carry out the former very successfully.

Through interviews, the team explores unknown aspects of the local life and history and comes across with written and oral information that exists only in the narratives of the inhabitants and the documents of their private or public archives. Interviewing is a way of letting people share with you what they know, heard, experienced. When the interviewer is a child from the local school, preferably a grandchild or nephew or niece etc., the interviewee opens up more easily and more freely.

Fieldwalking is the most direct technique for accessing the landscape. Children have the chance to approach the landscape and its monuments not only as historical features but also as tangible objects, which can be described, drawn, calculated, and modelled (Dyer 1983:17). The use of a printed topographic map - as detailed as possible (e.g. scale: 1:5,000) - during the survey is a necessity since the team must record spatially all the information that they hear or they see in person.

The team is reminded that interviewees most often are not used to printed maps because they use their own mental maps to find their way. Therefore, the team must insist on getting the information that will help transfer a piece of knowledge from mental to conventional maps. Two elements are needed: location and non-spatial attributes expressed in numbers. For example, when talking about an old mill, the interviewee will not give the exact X, Y pair of coordinates, but will give a distance from distinct features, such as a river, road, or nearest farm and maybe a description of the spot etc. that will lead the interviewer to find the most possible location and record the coordinates. The non-spatial attribute can be any number that the team has agreed to use for mills. When all buildings have been recorded and given numbers, then the rest of the landscape will be Nr 0.

Some cases in which archaeological and historical research has benefited from children's groups are the following:

- Field Boundaries: The use of old aerial photos is very helpful in this case. The team draws the outlines of the land plots, records the names of the owners and identify relics of old field systems. This activity is very useful because traces of early (Roman, Venetian etc.) cadastral organization come to light. Moreover, a modern cadastral map is created, which is something the Greek state lacks at the moment, and the first official maps are not expected sooner than 2015!
- Place-names: The team collects all the place-names that the inhabitants use to identify all the localities in the community. The area that each place-name defines is drawn on the map. They are classified in macro- and micro-toponyms.

Their analysis may focus on language and meaning. In this way, hypotheses can be tested as to what was the origin of the older inhabitants (in a typical Peloponnesian village we may come across Greek, Roman, Byzantine, Venetian, Frankish, Slavic, Turkish and Albanian place-names), and how different generations perceived, and described the landscape.

- Land-use: There are plenty of old and new maps and aerial photos that display the use of land in different periods of time, such as maps dating back to 17th – 19th centuries AD, photos from 1940 and more recent ones. Problems of cartography are very well known, but sometimes we are lucky enough to identify special features (castles, forests, ruins etc.), which are not reported elsewhere. Besides, by comparing the different pictures of the same area we can find out features that existed in the older pictures but have been eroded away. The opposite may happen too. There are sites that were not visible in the past because of dense vegetation, but have been revealed due to deforestation and erosion.
- Surface survey: Surface survey is an activity that appeals to children but requires very delicate treatment. The children should proceed to the recording of surface material with care not to distort any existent archaeological patterns. The results of their effort, namely the amount of artefacts (mainly pottery) they have recorded, can be digitised and processed so that an archaeological map of the study area may be created. The Archaeological Service cannot carry out intensive surveys in all villages of its jurisdiction and such maps will be very much appreciated even if their precision is questioned.
- Settlement Mobility: Not all habitation sites of a settlement were founded contemporaneously. Children can locate different periods in which whole districts or isolated buildings were built. They will next create a set of maps, each one representing a different settlement phase of the village or town. They can remark on the expansion or relocation or shrinkage etc. of the inhabited area. The location of houses in relation to distinct features like church, graveyard, roads, town hall, towers, gardens and so on is crucial and can be investigated.

6. Conclusions

The question that emerges is whether children can really contribute to the research of their region's past. Right off we can answer "yes". Children can carry out research and they have the ability and the knowledge to compute it and convert it to computerized formats, create databases and manipulate it further provided they are properly instructed.

In fact, there are cases where children can do more than researchers do. They can discover new data with greater ease and speed than researchers. Children live in the study area, spend 365 days a year there, listen to the stories their parents and grandparents tell, travel across the landscape all the time. External researchers can spend only a few weeks every year in the region and will always remain strangers. Due to time and money constraints, professionals tend to cover research at a macro level, leaving the aspects that belong to the micro fairly untouched.

This difference should not be considered as a cause for antagonism between professionals and teenagers. On the contrary, a close collaboration can be established between schools that are interested in participating in such projects, the Archaeological Service, the local authorities and any other public or private agent with relevant activities and interests. The result can be a central database that will cover the whole province or all the provinces that are found within the jurisdiction of the Archaeological Service. Through the establishment of a network between all the contributing parts, every new piece of information accompanied with its spatial and non-spatial attributes will reach the central headquarters of the network, preferably inside the Archaeological Service and will be distributed to the relevant experts. These experts will be responsible for evaluating it and acting accordingly.

Certainly children will enjoy and take pride in the idea that they can participate in the enrichment of a Regional or even National Cultural Inventory. However, they should remember that discoveries are not achieved by accident. On the contrary, every discovery hides behind it several years of work and labour by a group of specialized scholars, testing again and again their ideas before concluding and always doubting it. Therefore, the children ought also to respect their work, feel responsible for it and have a critical eye.

Acknowledgements

We would like to thank the organizers of the CAA2000 for Helene's travel costs were partly covered by the CAA2000 and partly by the University of Ljubljana, Dept. of Archaeology in collaboration with the National Hellenic Foundation of Research.

References

ABDELA, E., 1997. Η συγκροτση της εθνικης ταυτοτητας στο ελληνικο σχολειο: «εμεις» και οι «αλλοι». In Frangoudaki, A. and Dragona, T. (eds), *« Τι ειν' πατριδα μας;»: Εθνοκεντρισμοσ στην Εκπαιδευση*: 27-45. Αθηνα: Αλεξανδρεια.

ABDELA, E., 1998. Ιστορια και σχολειο. Αθηνα: Νησος/ Τετραδια 6.

BURROUGH, P.A., 1992. *Principles of Geographical Information Systems for Land Resources Assessment.* Oxford: Clarendon Press.

DYER, J., 1983. *Teaching Archaeology in Schools.* Princes Risborough: Shire Publications no. 29.

FLOGAITI, E., 1993. *Περιβαλλιντικη Εκπαιδευση.* Αθηνα: Ελληνικες Πανεπιστηιακες Εκδοσεις.

HARLEY, J.B., 1988. Maps, Knowledge and Power. In Cosgrove, D. and Daniels, S. (eds.), *The iconography of the Landscape: Essays on the symbolic representations, design and use of past environments*: 277-312. Cambridge: Cambridge University Press.

JAHODA, G., 1963. Children's concepts of time and history. *Educational Review* 15: 87-104.

MONMONNIER, M., 1996. *How to lie with maps.* Chicago/London: The University of Chicago Press.

PALLA, M., 1994. Ο μαθητικοσ λογοσ για την Ιστορια: Οψεις ετνικης συνειδησης. *Σεμιναριο* 17: 224-236.

PAPAGIANNOPOULOS, K., SIMONI, H. and FRAGOULIS, I., in press. *Η Τοτικη Ιστορια ως πεδιο οπουδης στα πλαισυα οχολικης παιδειας.* Αθηνα: Υπουργειο Ετνικησ Παιδειας και Θρησκευματον, Παιδαγογικο Ινστιτουτο.

PAPASTAMATIOU, N., 1984. Αξιολοφηση του ενδιαφεροντοσ τον ματητον της τριτης ταξης για τα μαθηματα των Φυσικων σ' εϖα Γυμνασιο. *Νεα Παιδεια* 31: 97-104.

PAVLOU, Á., 1995. Μερικα προβληματα της Ελληνιης Χαρτογραφιας και προτασεις για πολιτιστικο χαρτογραφικο εξοπλοσμο. *Πρακτικα του 1ου Εθνικου Συνεδιου Χαρτογραφιας: Προβληματα και προοπτικες της Ελληνικες Χαρτογραφιας*: 29-32. Θεσσαλονικη: ΧΕΕΕ.

ROGERS, P.J., 1978. *The New History-Theory into practice.* London: Historical Association.

RUBIN, R., 1992. Ideology and Landscape in early printed maps of Jerusalem. In Baker, A.R.H. and Biger, G. (eds.), *Ideology and Landscape in historical perspective: Essays on the meaning of some places in the past*, 15-30. Cambridge: Cambridge University Press.

SMITH, R.J., 1998. Mapping China's World Cultural Cartography in Late Imperial Times. In Wenhsin, Y. (ed.), *Landscape, culture and power in Chinese Society*: 52-109. Berkeley: Institute of East Asian Studies, University of California.

THOMPSON, D., 1984. Understanding the Past: Procedures and Content. In Dickinson, A.K. and. Rogers, P.J. (eds.), *Learning History*: 168-186. London: Heinemann Educational Books.

VENTOURA, L., 1997. Τα βιβλια γεογραφιας: αντιφασεις στο ανθρωπιστικο και εκσυγχρονιστικο μηνυμα. In Frangoudaki, A. and Dragona, T. (eds.), *«Τι σιν η πατριαδα μας;»: Εθνοκεντρισμος στην Εκπαιδευση*: 401-441. Αθηνα: Αλεξανδρεια.

WORELL, J. and STILWEL, W., 1981. *Psychology for teachers and students*. New York: McGraw Hill.

www.ingramcontent.com/pod-product-compliance
Ingram Content Group UK Ltd.
Pitfield, Milton Keynes, MK11 3LW, UK
UKHW061844041026
15591UKWH00015B/902

9 781841 712253